Theoretical Physics: Classical and Modern Views

THEORETICAL PHYSICS

Classical and Modern Views

GEORGE H. DUFFEY

South Dakota State University

HOUGHTON MIFFLIN COMPANY
Boston

ATLANTA | DALLAS | GENEVA, ILLINOIS | HOPEWELL, NEW JERSEY | PALO ALTO

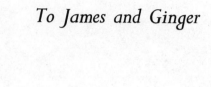

To James and Ginger

Preface

"THEORETICAL PHYSICS: Classical and Modern Views" surveys important methods for treating the behavior of particles and material continua. The text covers only a limited number of topics at the intermediate level, but each topic is discussed with enough detail so that useful results are obtained. Some of the more mathematical material may be omitted if the students are familiar with it, or are willing to accept the pertinent theorems without detailed proof.

On studying the universe and making measurements, we find that physical properties may be scalars, vectors, or polyads. A vector, or a polyad, may be treated as (a) an entity, or (b) an array of numbers that transforms in a certain way. The text develops the former view first, employing the usual boldface notation. Component notation is then introduced. Index procedures, including uses of the summation convention and the permutation symbol, are explained.

In Newtonian theory, we assume time to be the same everywhere, but space need not be considered absolute. All we have to employ is an inertial frame, with respect to which the conventional equations are valid. Systems consisting of a single particle or body, and of two or more particles or bodies, interacting with fields produced by the surroundings are analyzed in the text.

Matrices can be used to transform geometric or physical properties representable as conformable operands. Thus, a matrix may transform the coordinates of each point in space, or permute various parts or regions, effecting a rotation, a reflection, and possibly a distortion of either the reference axes or the physical system. Alternatively, a matrix may relate different physical quantities, such as the components of angular velocity and angular momentum.

Systems with constraints and/or symmetries may not be efficiently described with conventional vectors and coordinates, but certain functions of these can be effective. To exploit this fact, we write the equations of mechanics in a form involving generalized coordinates, generalized velocities, and generalized forces. Simple examples are introduced to illustrate feasible procedures.

Multidimensional vibrating systems are considered in the linear approximation. The pertinent secular equation is set up and solved for the normal frequencies. Substituting each of these back into the generating set of simultaneous equations then leads to the combination of generalized coordinates whose variation by itself corresponds to a particularly simple, or normal, mode of vibration. The effects of nonlinearities in a mode are considered.

The text then describes groups of operations that transform common symmetric systems into themselves. Useful theorems governing matrix representations and bases for such representations are established. These are particularly important because they apply to any expressions that the operations mix linearly, regardless of the physical theory that is being invoked. To illustrate their use, they are employed in simplifying the equations of motion for a vibrating system.

Generalized momentum is introduced as the derivative of the Lagrangian with respect to the temporal derivative of the corresponding coordinate. Such an expression is constant whenever the corresponding coordinate is absent from the Lagrangian. Thus, its constancy indicates a kind of symmetry within the mechanical system.

To describe the behavior of a part of nature, we may employ relationships connecting neighboring or consecutive states—differential laws. Alternatively, we may pick out global or cumulative features—integral relationships. Neither is more fundamental than the other, but the study of the alternate formulations gives us additional insight and additional computational possibilities.

Mechanics itself can be formulated as Hamilton's principle, involving the variation of an integral. As preparation for developing it, we consider variations of properties, functions, and integrals in considerable detail.

The classical forces acting on a particle or element of material can be attributed to direct interaction with similar neighboring units and to indirect interaction through gravitational and electromagnetic fields with distant particles or elements. The text includes mathematical procedures for working with such influences.

The nature of the electromagnetic field is discussed in some detail. Maxwell's equations are formulated. Systems of charged particles whose movements are dominated by electromagnetic interactions are discussed.

Experiments show that time is always bound up with space. The Newtonian bifurcation does not really make sense, so properties of Minkowski space-time are developed. The necessary generalizations of mechanics are then introduced.

The author believes that the big jump from classical tradition occurs not with this development of relativity but with that of quantum mechanics. Some of the classical basis for the latter is laid in a chapter on spinors. Finally, a nontensor presentation of some of general relativity appears.

In each chapter is a list of questions for checking comprehension of the material. The problems have been formulated to aid in mastering the methods. Additional material and alternate viewpoints can be found in the references listed at the end of each chapter, both books and journal literature.

The author wishes to thank Dr. Stanley A. Williams of Iowa State University for carefully reading the manuscript and making detailed suggestions and comments thereon. Acknowledgment is also due Dr. Joseph H. Macek of the University of Nebraska and Dr. Frances J. Anderson of the University of Minnesota for their worthwhile comments.

George H. Duffey

Contents

1 / *Elements of Vector Physics*

1.1 General considerations

A drawing in a child's magazine contains faces as parts of various objects. The drawing is so skillfully done that no face is immediately evident. Yet each is there and does appear when a viewer's mind separates it from its surroundings.

In like manner, simple *patterns* exist in nature. Most of these are interwined with other relationships, so a person may make many pertinent observations without perceiving a particular scheme. But once pointed out and described in enough detail, the pattern becomes apparent to each of us.

In physical studies, the observations are made as precisely as possible. *Numerical* results are obtained and analyzed. Patterns are formulated as equations. These are then available for correlating new data and making predictions.

Any aspect of a system that is described by a single number is called a *scalar*. As examples, we find the properties mass, charge, time, temperature, energy. The mathematics for working with such quantities include the arithmetic, algebra, and calculus with which the student is already familiar.

Other properties correspond to directed line segments. Thus, we have a particle's displacement, velocity, momentum, acceleration, force. These are called *vectors*. In working with them, we are essentially concerned with the connections between points in ordinary space. The algebra needed to manipulate such entities will be reviewed in this chapter.

Still more complicated are the properties that link different vectors. For example, the moment of inertia acts to convert the angular velocity of a body into its angular momentum. The permittivity of a dielectric acts to transform the electric intensity into the electric displacement. Such operators involve *dyads* or *polyads*. They can be represented by arrays of numbers that behave in a prescribed manner.

The aggregate of quantities describing any of these properties is called a *tensor*. Scalars, vectors, dyads, triads are said to be tensors of *rank* 0, 1, 2, 3, respectively. Polyads are tensors of rank 2 or greater.

The geometry of space was studied by the ancient Egyptians, Babylonians, Chaldeans, and Greeks. Their results were collected and systematized by Euclid of Alexandria. The boldface notation that we will employ is due to Oliver Heaviside; the shortened index notation to Curbastro Ricci, Tullio Levi-Civita, and Albert Einstein.

1.2 Defining vector processes: addition and subtraction

Any aspect of a physical system, or any abstraction, that behaves as a single number is called a scalar, while anything that behaves as a directed line segment is called a vector. In our analyses, each scalar, numerical component, and individual quantity will be represented by an italic symbol; each vector by a boldface letter. Because a person sees and understands simple geometric relationships directly, the basic vector processes will be defined as operations with lines.

We first note that any line segment can be shortened or lengthened without changing its direction. A given portion or multiple can be taken as the unit in which its length and the lengths of other vectors are expressed. Multiplying a given vector by a scalar merely multiplies its length by the scalar.

Furthermore, a line segment can be moved parallel to itself to a new location. In Euclidean space, which we will employ until we get to Einstein's theory, parallel displacement of the line segment along any closed path leads back to the original segment. So each displaced segment differs only in position from the original one and we consider each as the same vector. But a change in sign is interpreted as a reversal in direction. See figures 1.1 and 1.2.

Addition of two vectors is represented by erecting the second vector on the end point of the first. The sum is the straight line drawn from the beginning of the first vector to the end of the second one. Thus, the solid lines in figure 1.3 show **A** + **B**,

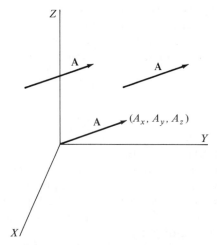

FIGURE 1.1   Representations of vector **A**.

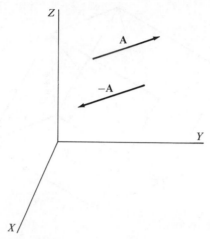

FIGURE 1.2   Effect of a change in sign.

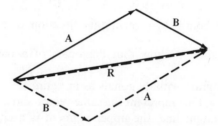

FIGURE 1.3   Sum **A** + **B** = **R** formed by solid vector **A** plus solid vector **B** and its permutation **B** + **A** = **R** formed by dashed vector **B** plus dashed vector **A**.

while the dashed lines show **B** + **A**, from a given initial point. Note that the same resultant **R** is obtained either way:

$$\mathbf{A} + \mathbf{B} = \mathbf{B} + \mathbf{A} = \mathbf{R}. \tag{1-1}$$

Since the elements combine in either order to give the same answer, vector addition obeys the *commutative law*.

In figure 1.4 vectors **A**, **B**, **C** form the edges of a parallelepiped. The definition of addition implies that the sum of any two of these gives a diagonal on a face, while the sum of a diagonal and the vector not used in forming it yields a diagonal through the solid. Diagonal **R** can be obtained in the three ways:

$$(\mathbf{A} + \mathbf{B}) + \mathbf{C} = \mathbf{A} + (\mathbf{B} + \mathbf{C}) = (\mathbf{C} + \mathbf{A}) + \mathbf{B} = \mathbf{R}. \tag{1-2}$$

How the elements are associated in performing the additions does not affect the result: the *associative law* applies and the parentheses may be omitted.

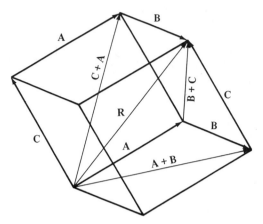

FIGURE 1.4   Adding three vectors.

To *subtract* **B** from **A**, we add negative **B** to **A**:

$$\mathbf{A} - \mathbf{B} = \mathbf{A} + (-\mathbf{B}). \qquad (1\text{-}3)$$

Negative **B** is obtained from **B** by reversing its direction, as figure 1.2 shows.

EXAMPLE 1.1   Use vectors to show that diagonals of a parallelogram bisect each other.

Construct a parallelogram with diagonals as in figure 1.5. Since opposite sides are parallel and equally long, they represent the same vector and are so labeled. Since each diagonal is a single straight line, the upper segment is a constant times the vector representing the lower segment.

Now, add vectors **C** and $d\mathbf{D}$ in the left triangle,

$$\mathbf{C} + d\mathbf{D} = \mathbf{B};$$

add vectors **D** and $c\mathbf{C}$ in the right triangle,

$$\mathbf{D} + c\mathbf{C} = \mathbf{B};$$

eliminate **B**,

$$\mathbf{C} + d\mathbf{D} = \mathbf{D} + c\mathbf{C};$$

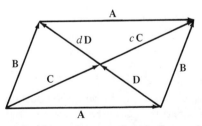

FIGURE 1.5   Parallelogram with sides and parts of each diagonal labeled.

and rearrange to make the left side a multiple of **C**, the right side a multiple of **D**:

$$(1 - c)\mathbf{C} = (1 - d)\mathbf{D}.$$

But **C** cannot be a multiple of **D** because **C** and **D** are not parallel. Therefore

$$1 - c = 0, \quad 1 - d = 0,$$

and

$$c = 1, \quad d = 1.$$

That is, the diagonals are bisected.

1.3   Multiplication

We can multiply vectors in three very useful ways, obtaining a scalar, a vector, or a polyad.

Let us consider two distinct vectors, **A** and **B**, with a given length taken as the unit of distance. Let the number of such units in **A** be A and the number in **B** be B. We then call quantity A the *magnitude* of **A**, quantity B the magnitude of **B**.

Dividing a vector by its magnitude reduces it to a vector of unit length, a *unit vector*. Thus, $\mathbf{A}/A$ is a vector of unit magnitude having the direction of **A**; $\mathbf{B}/B$ is a unit vector having the direction of **B**.

We may project any vector onto another by (a) drawing the vectors from the same initial point and (b) dropping a perpendicular from the end point of the vector to be projected onto the line through the other vector (see figure 1.6). The length subtended is called the *projection*. From the definition of cosine, the projection of **B** on **A** is

$$B \cos \theta. \tag{1-4}$$

Similarly, the projection of **A** on **B** is

$$A \cos (-\theta) = A \cos \theta. \tag{1-5}$$

The *scalar* (or *dot*) *product* of **B** with **A** is defined as the projection of **B** on **A** times the magnitude of **A**. Consequently,

$$\mathbf{B} \cdot \mathbf{A} \equiv (B \cos \theta)A = AB \cos \theta$$
$$= (A \cos \theta)B = A \cos (-\theta)B \equiv \mathbf{A} \cdot \mathbf{B}. \tag{1-6}$$

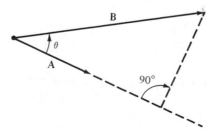

FIGURE 1.6   Projecting vector **B** onto vector **A**.

The second and third steps are true because quantities commute in an algebraic product. We see that $\mathbf{B} \cdot \mathbf{A}$ also equals the projection of $\mathbf{A}$ on $\mathbf{B}$ times the magnitude of $\mathbf{B}$, the expression $\mathbf{A} \cdot \mathbf{B}$. Scalar or dot multiplication is commutative.

Whenever $\mathbf{A}$ and $\mathbf{B}$ are parallel, θ is 0 or π and

$$\mathbf{A} \cdot \mathbf{B} = AB \cos \left(\frac{\pi}{2} \mp \frac{\pi}{2} \right) = \pm AB. \tag{1-7}$$

But when they are perpendicular, θ is $\pm \pi/2$,

$$\mathbf{A} \cdot \mathbf{B} = AB \cos \left(\pm \frac{\pi}{2} \right) = 0, \tag{1-8}$$

and the two vectors are said to be *orthogonal*.

Any two vectors $\mathbf{A}$ and $\mathbf{B}$ can be drawn from the same initial point. Then from the end point of each, the other vector can be drawn to complete a parallelogram. Whenever the two vectors do not possess the same direction, we can construct a unit vector $\mathbf{n}$ perpendicular to the plane of $\mathbf{A}$ and $\mathbf{B}$ so that $\mathbf{A}, \mathbf{B}, \mathbf{n}$ form a right-handed system, as shown in figure 1.7.

The *vector* (or *cross*) *product* of $\mathbf{A}$ with $\mathbf{B}$ is defined as the vector having the direction of $\mathbf{n}$ and a magnitude equal to the scalar area of the parallelogram. Since this area equals the base times the height, and the height is B times the sine of θ, from the definition of sine, we have

$$\mathbf{A} \times \mathbf{B} \equiv \mathbf{n}AB \sin \theta. \tag{1-9}$$

Because the planar area of the parallelogram has an orientation defined by $\mathbf{n}$, as well as the magnitude $AB \sin \theta$, it may be identified with the vector $\mathbf{A} \times \mathbf{B}$.

Applying this definition to $\mathbf{B} \times \mathbf{A}$, we obtain a vector with the direction $-\mathbf{n}$, as figure 1.7 illustrates. Hence

$$\mathbf{B} \times \mathbf{A} = -\mathbf{A} \times \mathbf{B}. \tag{1-10}$$

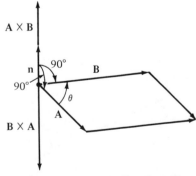

FIGURE 1.7   Parallelogram and vectors representing $\mathbf{A} \times \mathbf{B}$ and $\mathbf{B} \times \mathbf{A}$.

Whenever interchanging the order in which two elements are combined changes the sign of the result, the binary operation is said to be *anticommutative*. Formula (1-10) tells us that cross multiplication follows this anticommutative law.

Whenever **A** and **B** are parallel, angle θ is 0 or π and formula (1-9) yields zero:

$$\mathbf{A} \times \mathbf{B} = \mathbf{n}AB \sin\left(\frac{\pi}{2} \mp \frac{\pi}{2}\right) = 0. \tag{1-11}$$

Whenever **A** and **B** are perpendicular, θ is $\pm\pi/2$ and

$$\mathbf{A} \times \mathbf{B} = \mathbf{n}AB \sin\left(\pm\frac{\pi}{2}\right) = \pm\mathbf{n}AB. \tag{1-12}$$

Three noncoplanar vectors **B, C, D** can form the edges of a parallelepiped as figure 1.8 indicates. Vector **A** can also be constructed as the product **C** × **D**. The area of

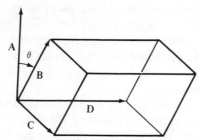

FIGURE 1.8   Parallelepiped whose volume is the dot product of **A** = **C** × **D** and **B**.

the base bordered by **C** and **D** is then magnitude A, while the distance of the opposite side (the top) from this base is $B \cos \theta$. Multiplying these quantities yields the volume

$$V = AB \cos \theta = \mathbf{A} \cdot \mathbf{B} = \mathbf{B} \cdot \mathbf{A}. \tag{1-13}$$

which the definition of **A** expands into the *triple scalar product*

$$V = \mathbf{C} \times \mathbf{D} \cdot \mathbf{B} = \mathbf{B} \cdot \mathbf{C} \times \mathbf{D}. \tag{1-14}$$

Parentheses are not needed in the last two expressions since the alternative interpretations **C** × (**D** · **B**) and (**B** · **C**) × **D** are meaningless.

From figure 1.9, we see that the projection of **B** + **C** on **A** is composed of the projection of **B** on **A** and the projection of **C** on **A**:

$$(B + C)\cos \gamma = B \cos \alpha + C \cos \beta. \tag{1-15}$$

Multiplying both sides of this equality by the magnitude of **A** yields the *distributive law* for scalar multiplication

$$(\mathbf{B} + \mathbf{C}) \cdot \mathbf{A} = \mathbf{B} \cdot \mathbf{A} + \mathbf{C} \cdot \mathbf{A}$$

or

$$\mathbf{A} \cdot (\mathbf{B} + \mathbf{C}) = \mathbf{A} \cdot \mathbf{B} + \mathbf{A} \cdot \mathbf{C}. \tag{1-16}$$

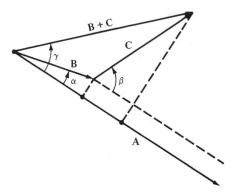

FIGURE 1.9   Angles involved in projecting **B**, **C**, and **B** + **C** on **A**.

A planar element of a *closed* surface is represented by an outward pointing normal vector of magnitude equal to its area. The element can be projected onto any given plane by projecting the vector onto the appropriate normal to the plane. If all elements of the surface are projected onto the plane, a zero vector results, since as much of the projected area is negative as positive. Therefore, the vector representing the whole closed surface is zero.

Similarly, the vector representing a closed prismatic surface is zero. So in figure 1.10, the outward pointing vector representing the area of the side bounded by **B** and **A** plus that for the side bounded by **C** and **A** must equal the inward pointing vector representing the area of the third side:

$$\mathbf{B} \times \mathbf{A} + \mathbf{C} \times \mathbf{A} = (\mathbf{B} + \mathbf{C}) \times \mathbf{A},$$

whence

$$\mathbf{A} \times \mathbf{B} + \mathbf{A} \times \mathbf{C} = \mathbf{A} \times (\mathbf{B} + \mathbf{C}). \tag{1-17}$$

We thus obtain the distributive law for vector multiplication.

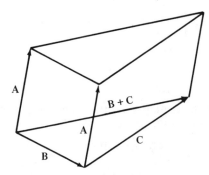

FIGURE 1.10   Prismatic surfaces bounded by **B** and **A**, by **C** and **A**, by **B** + **C** and **A**.

The *dyadic product* of **A** with **B** consists of **A** and **B** in order:

$$\mathbf{AB} = AB\,\frac{\mathbf{A}}{A}\frac{\mathbf{B}}{B}. \tag{1-18}$$

Operators containing such products will be considered in section 1.7.

EXAMPLE 1.2   Construct vector equations that identify the points lying in a plane.

Let $\mathbf{r}_0$ be the radius vector locating a fixed point in the given plane and let $\mathbf{e}_1$, $\mathbf{e}_2$ be perpendicular unit vectors drawn in this plane from the fixed point, as figure 1.11 shows. Choose some other point in the plane and let s_1 be its distance from the axis through $\mathbf{e}_2$, s_2 its distance from the axis through $\mathbf{e}_1$. If these distances are properly signed, the displacement of the point from the origin is

$$\mathbf{r} = \mathbf{r}_0 + s_1\mathbf{e}_1 + s_2\mathbf{e}_2.$$

To simplify this equation, dot multiply both sides with $\boldsymbol{\sigma}$, a vector perpendicular to the given plane:

$$\boldsymbol{\sigma} \cdot \mathbf{r} = \boldsymbol{\sigma} \cdot \mathbf{r}_0.$$

The last two terms drop out because $\boldsymbol{\sigma} \cdot \mathbf{e}_1$ and $\boldsymbol{\sigma} \cdot \mathbf{e}_2$ are zero, by equation (1-8). A particular $\mathbf{r}_0$ is the perpendicular **d** dropped on the plane from the origin. Because $\boldsymbol{\sigma}$ and **d** are parallel, we have

$$\boldsymbol{\sigma} \cdot \mathbf{r} = \boldsymbol{\sigma} \cdot \mathbf{d} = \sigma d.$$

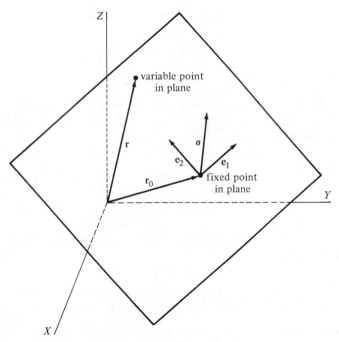

FIGURE 1.11   Vectors used in describing a plane.

Letting σ itself be **d** yields the formula

$$\mathbf{d} \cdot \mathbf{r} = d^2,$$

which identifies the points in the plane by means of the distance of the plane from the origin.

1.4   Composing and resolving vectors

One can draw vector **B** from a given reference point, then vector **C** from the end point of **B**, then a noncoplanar vector **D** from the end point of **C**. The resultant **R**

$$\mathbf{B} + \mathbf{C} + \mathbf{D} = \mathbf{R} \tag{1-19}$$

is a diagonal of the parallelepiped in figure 1.8. Now, varying the magnitudes and signs of **B**, **C**, and **D** would cause the end point of **R** to move over all space. Vector **R** itself would reproduce all possible 3-dimensional vectors in turn.

But each of the constituent vectors is a number times a vector of unit length. If the unit vectors are labeled $\mathbf{e}_1$, $\mathbf{e}_2$, and $\mathbf{e}_3$, an arbitrary vector **A** is then given by

$$A_1\mathbf{e}_1 + A_2\mathbf{e}_2 + A_3\mathbf{e}_3 = \mathbf{A}. \tag{1-20}$$

The multiplying numbers A_1, A_2, A_3 are the *components* of **A** with respect to the *base vectors* $\mathbf{e}_1$, $\mathbf{e}_2$, $\mathbf{e}_3$.

Starting with a given **A** and given $\mathbf{e}_j$'s, we can find the A_j's needed to satisfy formula (1-20) by trial and error. The resulting three numbers, an ordered triplet, define the vector **A**.

When the base vectors are mutually perpendicular, the given vector is very easily resolved. For then, the dot product of (1-20) with $\mathbf{e}_j$,

$$A_1\mathbf{e}_1 \cdot \mathbf{e}_j + A_2\mathbf{e}_2 \cdot \mathbf{e}_j + A_3\mathbf{e}_3 \cdot \mathbf{e}_j = \mathbf{A} \cdot \mathbf{e}_j, \tag{1-21}$$

reduces to

$$A_j = \mathbf{A} \cdot \mathbf{e}_j. \tag{1-22}$$

In words, the jth component of **A** equals the projection of the given vector on $\mathbf{e}_j$. Also when the base vectors are mutually perpendicular, the Pythagorean theorem tells us that the magnitude of **A** is

$$|\mathbf{A}| = A = (A_1{}^2 + A_2{}^2 + A_3{}^2)^{1/2}. \tag{1-23}$$

We associate the base vectors with a coordinate system by letting each one specify the direction along which a single coordinate changes at each point in space. Such a direction and the corresponding base vector generally vary with the point. (If one of the coordinates is the radius vector drawn from the origin, the direction in which it alone changes varies with its orientation.) In the Cartesian coordinate system, each such direction is constant.

In this system, the base vectors are often labeled **i, j, k**, as figure 1.12 shows. Then equation (1-20) takes on the form

$$\mathbf{A} = A_x\mathbf{i} + A_y\mathbf{j} + A_z\mathbf{k}, \tag{1-24}$$

in which the components A_x, A_y, A_z are projections of **A** on the x, y, z axes. The geometric significance of this equation is illustrated in figure 1.13.

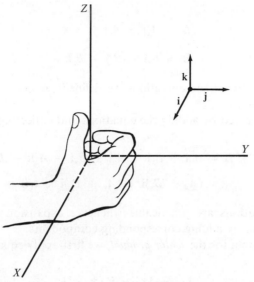

FIGURE 1.12   Base vectors for any point in a right-handed rectangular coordinate system.

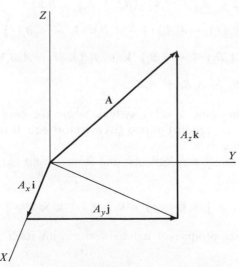

FIGURE 1.13   Summing the vectors formed from the components.

1.5   Standard operations on Cartesian components

Any two vectors **A** and **B** can be resolved along given x, y, and z axes, as we have just described.

$$\mathbf{A} = A_x\mathbf{i} + A_y\mathbf{j} + A_z\mathbf{k}, \tag{1-25}$$

$$\mathbf{B} = B_x\mathbf{i} + B_y\mathbf{j} + B_z\mathbf{k}. \tag{1-26}$$

Let us now consider what the operations we defined previously do to the resulting rectangular components.

The *sum* is constructed by adding the equations and collecting the vectors that are parallel:

$$\mathbf{A} + \mathbf{B} = A_x\mathbf{i} + B_x\mathbf{i} + A_y\mathbf{j} + B_y\mathbf{j} + A_z\mathbf{k} + B_z\mathbf{k}$$

$$= (A_x + B_x)\mathbf{i} + (A_y + B_y)\mathbf{j} + (A_z + B_z)\mathbf{k}. \tag{1-27}$$

The same manipulations are practicable whenever expansion (1-20) is employed; vectors can be added by adding corresponding components.

In deriving a formula for the *scalar product*, we first apply equations (1-7) and (1-8) to the base vectors:

$$\mathbf{i} \cdot \mathbf{i} = \mathbf{j} \cdot \mathbf{j} = \mathbf{k} \cdot \mathbf{k} = 1, \tag{1-28}$$

$$\mathbf{i} \cdot \mathbf{j} = \mathbf{j} \cdot \mathbf{k} = \mathbf{k} \cdot \mathbf{i} = 0. \tag{1-29}$$

These equations are then used to reduce the result of dot multiplying vectors **A** and **B** of (1-25) and (1-26):

$$\mathbf{A} \cdot \mathbf{B} = (A_x\mathbf{i} + A_y\mathbf{j} + A_z\mathbf{k}) \cdot (B_x\mathbf{i} + B_y\mathbf{j} + B_z\mathbf{k})$$

$$= A_xB_x\mathbf{i} \cdot \mathbf{i} + A_yB_y\mathbf{j} \cdot \mathbf{j} + A_zB_z\mathbf{k} \cdot \mathbf{k} + A_xB_y\mathbf{i} \cdot \mathbf{j} + A_xB_z\mathbf{i} \cdot \mathbf{k}$$

$$+ A_yB_x\mathbf{j} \cdot \mathbf{i} + A_yB_z\mathbf{j} \cdot \mathbf{k} + A_zB_x\mathbf{k} \cdot \mathbf{i} + A_zB_y\mathbf{k} \cdot \mathbf{j}$$

$$= A_xB_x + A_yB_y + A_zB_z. \tag{1-30}$$

In the Cartesian system (and in other systems where the base vectors are mutually perpendicular), the scalar product of two given vectors equals the sum of products of corresponding components.

Since **i**, **j**, **k** are mutually perpendicular and form a right-handed system, equation (1-12) tells us that

$$\mathbf{i} \times \mathbf{j} = \mathbf{k}, \quad \mathbf{j} \times \mathbf{k} = \mathbf{i}, \quad \mathbf{k} \times \mathbf{i} = \mathbf{j}. \tag{1-31}$$

Furthermore, the cross product of a base vector with itself is zero, according to equation (1-11):

$$\mathbf{i} \times \mathbf{i} = \mathbf{j} \times \mathbf{j} = \mathbf{k} \times \mathbf{k} = 0. \tag{1-32}$$

Consequently, *cross multiplying* (1-25) and (1-26) yields

$$\mathbf{A} \times \mathbf{B} = (A_x\mathbf{i} + A_y\mathbf{j} + A_z\mathbf{k}) \times (B_x\mathbf{i} + B_y\mathbf{j} + B_z\mathbf{k})$$

$$= A_xB_x\mathbf{i} \times \mathbf{i} + A_yB_y\mathbf{j} \times \mathbf{j} + A_zB_z\mathbf{k} \times \mathbf{k} + A_xB_y\mathbf{i} \times \mathbf{j} + A_xB_z\mathbf{i} \times \mathbf{k}$$

$$+ A_yB_x\mathbf{j} \times \mathbf{i} + A_yB_z\mathbf{j} \times \mathbf{k} + A_zB_x\mathbf{k} \times \mathbf{i} + A_zB_y\mathbf{k} \times \mathbf{j}$$

$$= (A_yB_z - A_zB_y)\mathbf{i} + (A_zB_x - A_xB_z)\mathbf{j} + (A_xB_y - A_yB_x)\mathbf{k}$$

$$= \begin{vmatrix} \mathbf{i} & \mathbf{j} & \mathbf{k} \\ A_x & A_y & A_z \\ B_x & B_y & B_z \end{vmatrix}. \tag{1-33}$$

Because of its symmetry, this determinant expression for the cross product is easily remembered.

The volume of the parallelepiped in figure 1.8 is the triple scalar product $\mathbf{B} \cdot \mathbf{C} \times \mathbf{D}$. With the formulas already obtained, this product expands as follows:

$$\mathbf{B} \cdot \mathbf{C} \times \mathbf{D} = (B_x\mathbf{i} + B_y\mathbf{j} + B_z\mathbf{k}) \cdot \begin{vmatrix} \mathbf{i} & \mathbf{j} & \mathbf{k} \\ C_x & C_y & C_z \\ D_x & D_y & D_z \end{vmatrix}$$

$$= \begin{vmatrix} B_x & B_y & B_z \\ C_x & C_y & C_z \\ D_x & D_y & D_z \end{vmatrix}. \tag{1-34}$$

1.6 Two other coordinate systems

When the pertinent physical region is symmetric about an axis or about a point, cylindrical or spherical coordinates may be used with advantage. Vectors representing certain physical aspects may then be resolved along the corresponding base vectors.

Consider a vector $\mathbf{A}$ at some point in the cylindrical coordinate system of figure 1.14. Base vectors $\mathbf{l}$, $\mathbf{n}$, $\mathbf{k}$ are constructed to show the directions in which the point would move if only r, only ϕ, or only z varied. Equation (1-20) becomes

$$\mathbf{A} = A_r\mathbf{l} + A_\phi\mathbf{n} + A_z\mathbf{k}. \tag{1-35}$$

But since $\mathbf{l}$, $\mathbf{n}$, $\mathbf{k}$ are mutually perpendicular, each of the components is a projection of $\mathbf{A}$:

$$\mathbf{A} \cdot \mathbf{l} = A_r, \qquad \mathbf{A} \cdot \mathbf{n} = A_\phi, \qquad \mathbf{A} \cdot \mathbf{k} = A_z. \tag{1-36}$$

Now, the angle between $\mathbf{l}$ and the x axis is ϕ, the angle between $\mathbf{l}$ and the y axis is $90° - \phi$, and the angle between $\mathbf{l}$ and the z axis is $90°$. So the projection of $\mathbf{l}$ on $\mathbf{i}$ is $\cos \phi$, the projection on $\mathbf{j}$ is $\cos (90° - \phi) = \sin \phi$, and the projection on $\mathbf{k}$ is 0. Letting $\mathbf{A}$ in (1-24) be $\mathbf{l}$ then yields

$$\mathbf{l} = \cos \phi\mathbf{i} + \sin \phi\mathbf{j}. \tag{1-37}$$

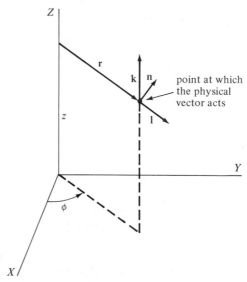

FIGURE 1.14   Base vectors for the point (r, ϕ, z) in cylindrical coordinates.

Similarly projecting **n** on the rectangular axes leads to the formula

$$\mathbf{n} = -\sin \phi \mathbf{i} + \cos \phi \mathbf{j}, \qquad (1\text{-}38)$$

while

$$\mathbf{k} = \mathbf{k}. \qquad (1\text{-}39)$$

Figure 1.15 shows how the base vectors are constructed for a given point in a spherical coordinate system. Expanding vector **A** as in equation (1-20) yields

$$\mathbf{A} = A_r\mathbf{l} + A_\theta\mathbf{m} + A_\phi\mathbf{n}. \qquad (1\text{-}40)$$

Since **l**, **m**, **n** are mutually perpendicular, the components A_r, A_θ, A_ϕ are the projections of **A** on **l**, **m**, and **n**, respectively.

Here, the angle between **l** and the z axis is θ, while the angle between **l** and the xy plane is $90° - \theta$. So its projection on **k** is $\cos \theta$, while its projection on the xy plane is $\sin \theta$. Since the angle between this second projection and the x axis is ϕ, the angle from the projection to the y axis $90° - \phi$, the net projection on the x axis is $(\sin \theta) \times (\cos \phi)$ while the net projection on the y-axis is $(\sin \theta)(\sin \phi)$. Therefore

$$\mathbf{l} = \cos \theta \mathbf{k} + \sin \theta \cos \phi \mathbf{i} + \sin \theta \sin \phi \mathbf{j}. \qquad (1\text{-}41)$$

Similarly,

$$\mathbf{m} = -\sin \theta \mathbf{k} + \cos \theta \cos \phi \mathbf{i} + \cos \theta \sin \phi \mathbf{j}. \qquad (1\text{-}42)$$

$$\mathbf{n} = -\sin \phi \mathbf{i} + \cos \phi \mathbf{j}. \qquad (1\text{-}43)$$

In discussing the kinematics of a particle, we will return to these systems.

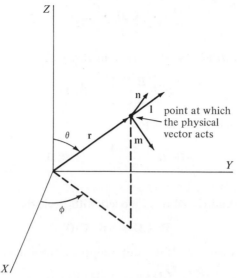

FIGURE 1.15   Base vectors for the point (r, θ, ϕ) in spherical coordinates.

EXAMPLE 1.3   Test formulas (1-41), (1-42), (1-43) by showing that they imply the orthogonality of **l**, **m**, **n**.

From (1-8) the vectors are mutually orthogonal if the dot product between each possible pair is zero. Substituting the pertinent components into (1-30) yields

$$\mathbf{l} \cdot \mathbf{m} = (\cos \theta)(-\sin \theta) + \sin \theta \cos \theta \cos^2 \phi + \sin \theta \cos \theta \sin^2 \phi$$

$$= -\sin \theta \cos \theta + \sin \theta \cos \theta (\cos^2 \phi + \sin^2 \phi)$$

$$= 0,$$

$$\mathbf{l} \cdot \mathbf{n} = 0 + \sin \theta \cos \phi (-\sin \phi) + \sin \theta \sin \phi \cos \phi$$

$$= 0,$$

$$\mathbf{m} \cdot \mathbf{n} = 0 + \cos \theta \cos \phi (-\sin \phi) + \cos \theta \sin \phi \cos \phi$$

$$= 0.$$

1.7   Polyads and polyadic polynomials

A property that links a certain vector to other vectors is described by a set of dimensional numbers. Each number is affiliated with two or more directions, or unit vectors, in sequence, and can be factored and distributed among the vectors. The number-vector complex is called a *polyad*; a sum of such elements representing the property is called a *polyadic* (or polyadic polynomial).

When the polyad contains only two vectors, as the expression

$$CD \frac{\mathbf{C}\,\mathbf{D}}{C\,D} = \mathbf{CD} \qquad (1\text{-}44)$$

does, it is said to be a *dyad*. When it contains three, as

$$BCD \frac{\mathbf{B}\,\mathbf{C}\,\mathbf{D}}{B\,C\,D} = \mathbf{BCD} \qquad (1\text{-}45)$$

does, it is a triad. When it contains four, as

$$ABCD \frac{\mathbf{A}\,\mathbf{B}\,\mathbf{C}\,\mathbf{D}}{A\,B\,C\,D} = \mathbf{ABCD}, \qquad (1\text{-}46)$$

it is a tetrad.

Dot multiplying polyad (1-44) from the left with **B** yields

$$\mathbf{B} \cdot \mathbf{CD} = (\mathbf{B} \cdot \mathbf{C})\mathbf{D} \qquad (1\text{-}47)$$

the scalar $\mathbf{B} \cdot \mathbf{C}$ times vector **D**. If it is dot multiplied from the right by **F**, the result is

$$\mathbf{CD} \cdot \mathbf{F} = \mathbf{C}(\mathbf{D} \cdot \mathbf{F}), \qquad (1\text{-}48)$$

the vector **C** times scalar $\mathbf{D} \cdot \mathbf{F}$.

Standard dyads are formulated from the base vectors:

$$\mathbf{ii}, \quad \mathbf{jj}, \quad \mathbf{kk}, \quad \mathbf{ij}, \quad \mathbf{jk}, \quad \mathbf{ki}, \quad \mathbf{ji}, \quad \mathbf{kj}, \quad \mathbf{ik}. \qquad (1\text{-}49)$$

If each of these is multiplied by a different parameter and the result summed,

$$I_{xx}\mathbf{ii} + I_{xy}\mathbf{ij} + I_{xz}\mathbf{ik} + I_{yx}\mathbf{ji} + I_{yy}\mathbf{jj} + I_{yz}\mathbf{jk} + I_{zx}\mathbf{ki} + I_{zy}\mathbf{kj} + I_{zz}\mathbf{kk},$$
$$\qquad (1\text{-}50)$$

we get the dyadic that transforms a vector

$$\boldsymbol{\omega} = \omega_x\mathbf{i} + \omega_y\mathbf{j} + \omega_z\mathbf{k} \qquad (1\text{-}51)$$

to a new vector **L** as follows:

$$
\begin{aligned}
\mathbf{L} &= (I_{xx}\mathbf{ii} + I_{xy}\mathbf{ij} + I_{xz}\mathbf{ik} + \cdots) \cdot (\omega_x\mathbf{i} + \omega_y\mathbf{j} + \omega_z\mathbf{k}) \\
&= I_{xx}\omega_x\mathbf{ii} \cdot \mathbf{i} + I_{xy}\omega_y\mathbf{ij} \cdot \mathbf{j} + I_{xz}\omega_z\mathbf{ik} \cdot \mathbf{k} + I_{yx}\omega_x\mathbf{ji} \cdot \mathbf{i} + I_{yy}\omega_y\mathbf{jj} \cdot \mathbf{j} \\
&\quad + I_{yz}\omega_z\mathbf{jk} \cdot \mathbf{k} + I_{zx}\omega_x\mathbf{ki} \cdot \mathbf{i} + I_{zy}\omega_y\mathbf{kj} \cdot \mathbf{j} + I_{zz}\omega_z\mathbf{kk} \cdot \mathbf{k} \\
&= (I_{xx}\omega_x + I_{xy}\omega_y + I_{xz}\omega_z)\mathbf{i} \\
&\quad + (I_{yx}\omega_x + I_{yy}\omega_y + I_{yz}\omega_z)\mathbf{j} \\
&\quad + (I_{zx}\omega_x + I_{zy}\omega_y + I_{zz}\omega_z)\mathbf{k}.
\end{aligned} \qquad (1\text{-}52)
$$

Note how each component of vector **L** depends linearly on each component of vector $\boldsymbol{\omega}$. Using indices, we can treat such dependences directly, as we will first see in section 1.9. Equations like (1-52), and the simpler forms to be obtained, relate (a) angular momentum to angular velocity, (b) electric displacement to electric intensity, and (c) a rotated and/or reflected vector to its initial state.

1.8 Systematizing computations and manipulations

The procedures that we have employed so far cannot cope with more complicated relationships among vectors and other nonscalars. On the other hand, the requisite combinations can be expressed through operations on components labeled by numerical indices. Standard algebraic methods and their extensions to be discussed next are then applicable.

Initially, a suitable system of base vectors e_1, e_2, e_3, is chosen. Each constituent of the expressions under study is resolved with respect to these vectors. To reduce the number of written symbols, we replace each subscript (and superscript) number with a letter, or *index*, that equals each of the pertinent integers in turn. Any index that is not summed over, that may be any of the allowed numbers without restriction, is said to be *free*.

In some discussions, a certain element, labeled by one or more indices, is split into factors, each of which is labeled by the same letters. If summation was not carried out over the indices in the original form, it is not carried out in the final form where the letters are repeated.

But in many discussions, an element labeled by a certain letter is multiplied by another element labeled by the same index and summation is carried out over the index. Or, two indices on a given element may be made equal and summation carried out over the resulting letter. These circumstances led Einstein, and will lead us, to adopt the *summation convention*: Each term in which an index appears more than once is summed over all allowed values of the index, unless the repeated index is specifically identified as a free index. The usual summation sign $\sum$ is then superfluous and is omitted. Clutter is thus reduced and readability increased.

Introduction of the index notation and employment of the summation convention simplify the left side of (1-20) in the following manner:

$$\mathbf{A} = \sum_{m=1}^{3} A_m \mathbf{e}_m \equiv \sum_{n=1}^{3} A_n \mathbf{e}_n \equiv A_m \mathbf{e}_m \equiv A_n \mathbf{e}_n. \tag{1-53}$$

The resolution of a second vector **B** can be similarly reduced:

$$\mathbf{B} = \sum_{m=1}^{3} B_m \mathbf{e}_m \equiv \sum_{n=1}^{3} B_n \mathbf{e}_n \equiv B_m \mathbf{e}_m \equiv B_n \mathbf{e}_n. \tag{1-54}$$

A repeated index can be replaced by any other letter not being used since each would run over the same integers. As a consequence, such an index is said to be *dummy* in character. Independent summations must always be indicated by different indices. Thus,

$$\mathbf{A} \cdot \mathbf{B} = (A_m \mathbf{e}_m) \cdot (B_n \mathbf{e}_n) = (A_n \mathbf{e}_n) \cdot (B_m \mathbf{e}_m)$$
$$\neq (A_m \mathbf{e}_m) \cdot (B_m \mathbf{e}_m). \tag{1-55}$$

The last form does not represent the dot product because it does not include any cross terms, such as $A_1 B_2 \mathbf{e}_1 \cdot \mathbf{e}_2$.

The *Kronecker delta* is defined by the statements

$$\delta_{mn} = 1 \qquad \text{if} \quad m = n, \tag{1-56}$$

$$\delta_{mn} = 0 \qquad \text{if} \quad m \neq n. \tag{1-57}$$

With it, the condition that the base vectors be mutually orthogonal can be written simply as

$$\mathbf{e}_m \cdot \mathbf{e}_n = \delta_{mn}. \tag{1-58}$$

We will assume that this condition holds in the following discussion.

We can then reduce the dot product as follows:

$$\mathbf{A} \cdot \mathbf{B} = A_m B_n \mathbf{e}_m \cdot \mathbf{e}_n = A_m B_n \delta_{mn} = A_m B_m = A_n B_n. \tag{1-59}$$

Incidentally, compare these manipulations with those in (1-30).

When orthogonal base vectors are used, the vector product expands into a 3×3 determinant, as (1-33) shows. But writing such a determinant in summation form requires a symbol that eliminates all combinations of elements except those whose indices differ. The symbol must also apply the proper sign to the nonzero terms.

A suitable expression is the *permutation symbol* defined by the equations

$$\varepsilon_{rmn} = 1 \qquad \text{if} \quad rmn \text{ is } 123, 231, \text{ or } 312, \tag{1-60}$$

$$\varepsilon_{rmn} = -1 \qquad \text{if} \quad rmn \text{ is } 213, 321, \text{ or } 132, \tag{1-61}$$

$$\varepsilon_{rmn} = 0 \qquad \text{if} \quad rmn \text{ has any two numbers the same.} \tag{1-62}$$

Note that interchanging any two of the indices in ε_{rmn} alters its sign:

$$\varepsilon_{rmn} = -\varepsilon_{mrn}, \tag{1-63}$$

$$\varepsilon_{rmn} = -\varepsilon_{nmr}, \tag{1-64}$$

$$\varepsilon_{rmn} = \varepsilon_{nrm}. \tag{1-65}$$

For, when we construct a vector $\mathbf{C}$ by the formula

$$\mathbf{C} = \varepsilon_{rmn} \mathbf{e}_r A_m B_n. \tag{1-66}$$

and evaluate each component, dropping all terms that are zero by (1-62) and determining the remaining ε's by (1-60) and (1-61), we obtain

$$\begin{aligned} C_1 &= \varepsilon_{1mn} A_m B_n = \varepsilon_{123} A_2 B_3 + \varepsilon_{132} A_3 B_2 \\ &= A_2 B_3 - A_3 B_2, \end{aligned} \tag{1-67}$$

$$\begin{aligned} C_2 &= \varepsilon_{2mn} A_m B_n = \varepsilon_{231} A_3 B_1 + \varepsilon_{213} A_1 B_3 \\ &= A_3 B_1 - A_1 B_3, \end{aligned} \tag{1-68}$$

$$\begin{aligned} C_3 &= \varepsilon_{3mn} A_m B_n = \varepsilon_{312} A_1 B_2 + \varepsilon_{321} A_2 B_1 \\ &= A_1 B_2 - A_2 B_1. \end{aligned} \tag{1-69}$$

The final forms are the components of $\mathbf{A} \times \mathbf{B}$; we do have

$$(\mathbf{A} \times \mathbf{B})_r = C_r = \varepsilon_{rmn} A_m B_n. \tag{1-70}$$

In various expressions, products of ε's occur. These can be reduced with the following identities:

$$\varepsilon_{rmn}\varepsilon_{rmn} = 6, \tag{1-71}$$

$$\varepsilon_{rmn}\varepsilon_{rmq} = 2\delta_{nq}, \tag{1-72}$$

$$\varepsilon_{rmn}\varepsilon_{rpq} = \delta_{mp}\delta_{nq} - \delta_{mq}\delta_{np}. \tag{1-73}$$

Since the first two involve summations of the last one, let us check only (1-73) in detail.

When all three numbers 1, 2, 3 occur among m, n, p, q, then index r is equal to one of these numbers in each term of the sum $\varepsilon_{rmn}\varepsilon_{rpq}$ and the corresponding ε is zero. The left side of (1-73) must then vanish. Furthermore, one cannot have either $m = p$ and $n = q$, or $m = q$ and $n = p$. So a delta in each term on the right side of the equation is zero and this side does equal the left side.

Secondly, when $m = n$ and/or $p = q$, two indices in one factor (at least) in each term on the left are equal and that side is zero. Moreover, the terms cancel on the right so the right side is also zero.

Thirdly, if $m = p$, $n = q$, and $m \neq n$, the only term on the left different from zero is the one for r not equal to m or n. Then both ε's are either $+1$ or -1 and the product is $+1$. On the right, the first term is $+1$ and the second 0. So both sides equal $+1$.

Fourthly, if $m = q$, $n = p$, and $m \neq n$, only one term on the left differs from zero. In it one ε contains a single interchange of indices from the other, so one equals $+1$, the other -1, and the product is -1. On the right, the first term is 0, the second -1, so both sides equal -1.

EXAMPLE 1.4   List and illustrate the properties that may be exploited in manipulating vector expressions.

From the text we obtain the following items:

(a) Expansion in terms of orthogonal components. For instance,

$$\mathbf{A} \cdot \mathbf{B} = A_m B_m \qquad \text{and} \qquad (\mathbf{A} \times \mathbf{B})_r = \varepsilon_{rmn} A_m B_n.$$

(b) Dummy nature of repeated indices. Thus

$$A_m B_m = A_n B_n.$$

(c) Commutativity of numbers in a product. Thus

$$A_m B_n C_r = A_m C_r B_n.$$

(d) Behavior of the permutation symbol when indices are shifted. Thus

$$\varepsilon_{rmn} = \varepsilon_{nrm} = -\varepsilon_{rnm}.$$

(e) Effect of the Kronecker delta. For example,

$$\delta_{nm} A_m = A_n.$$

(f) Identities for products of ε's. The pertinent one is

$$\varepsilon_{rmn}\varepsilon_{rpq} = \delta_{mp}\delta_{nq} - \delta_{mq}\delta_{np}.$$

(g) Distributive property of numbers. Thus

$$A_m(B_n + C_n) = A_mB_n + A_mC_n.$$

In the following derivation, a subscript on an equality sign indicates the property from the above list which justifies the equality.

EXAMPLE 1.5   Expand $\mathbf{A} \times (\mathbf{B} \times \mathbf{C})$ in terms of vectors $\mathbf{A}$, $\mathbf{B}$, and $\mathbf{C}$.
    We write

$$\begin{aligned}
[\mathbf{A} \times (\mathbf{B} \times \mathbf{C})]_r &=_a \varepsilon_{rmn}A_m(\mathbf{B} \times \mathbf{C})_n =_a \varepsilon_{rmn}A_m\varepsilon_{npq}B_pC_q \\
&=_c \varepsilon_{rmn}\varepsilon_{npq}A_mB_pC_q =_d \varepsilon_{nrm}\varepsilon_{npq}A_mB_pC_q \\
&=_f (\delta_{rp}\delta_{mq} - \delta_{rq}\delta_{mp})A_mB_pC_q \\
&=_{g,c} (\delta_{rp}B_p)(\delta_{mq}C_q)A_m - (\delta_{rq}C_q)(\delta_{mp}B_p)A_m \\
&=_e B_r(C_mA_m) - C_r(B_mA_m) =_a [\mathbf{B}(\mathbf{A} \cdot \mathbf{C}) - \mathbf{C}(\mathbf{A} \cdot \mathbf{B})]_r.
\end{aligned}$$

The net result,

$$\mathbf{A} \times (\mathbf{B} \times \mathbf{C}) = \mathbf{B}(\mathbf{A} \cdot \mathbf{C}) - \mathbf{C}(\mathbf{A} \cdot \mathbf{B}),$$

is called the   BAC CAB   rule for simplifying the triple vector product.

1.9   Transforming Cartesian components

A given vector is described by its components with respect to a set of base vectors. Any change in these reference vectors, associated with a reorientation of one or more of the coordinate axes and/or with a movement in the base position (the point at which the given vector acts), causes the components to change. If the frame is Cartesian, as we will here assume, a shift in base position has no effect.
    A given transformation can also be associated with a reorientation of the vector. Indeed, rotation of the reference axes alters the components in the same way as rotation of the vector in the opposite direction by the same angle.
    Let two sets of base vectors be constant and orthogonal as figure 1.16 shows. Then

$$\mathbf{e}_m \cdot \mathbf{e}_n = \delta_{mn}, \tag{1-74}$$

$$\mathbf{e}_m' \cdot \mathbf{e}_n' = \delta_{mn}. \tag{1-75}$$

Resolve an arbitrary vector $\mathbf{A}$ with respect to each set

$$\mathbf{A} = A_n\mathbf{e}_n = A_n'\mathbf{e}_n' \tag{1-76}$$

following equation (1-20).
    Then operate on both sides of the last equality with $\cdot \, \mathbf{e}_r$

$$A_n\mathbf{e}_n \cdot \mathbf{e}_r = A_n'\mathbf{e}_n' \cdot \mathbf{e}_r \tag{1-77}$$

and reduce with equation (1-74)

$$A_r = A_n'\mathbf{e}_n' \cdot \mathbf{e}_r. \tag{1-78}$$

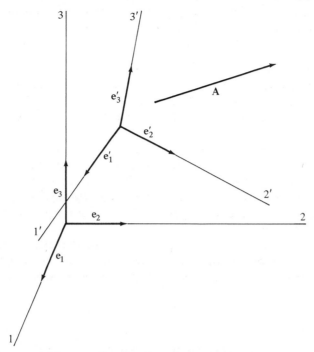

FIGURE 1.16   Two sets of base vectors for Cartesian components of **A**.

Also, operate with $\cdot\, \mathbf{e}_r'$ to get

$$A_n \mathbf{e}_n \cdot \mathbf{e}_r' = A_n' \mathbf{e}_n' \cdot \mathbf{e}_r' \tag{1-79}$$

or

$$A_n \mathbf{e}_n \cdot \mathbf{e}_r' = A_r'. \tag{1-80}$$

Let us represent the projection of $\mathbf{e}_n'$ on $\mathbf{e}_r$ as a_{nr}

$$\mathbf{e}_n' \cdot \mathbf{e}_r = a_{nr}. \tag{1-81}$$

Then,

$$\mathbf{e}_n \cdot \mathbf{e}_r' = \mathbf{e}_r' \cdot \mathbf{e}_n = a_{rn}, \tag{1-82}$$

and equation (1-80) becomes

$$A_r' = a_{rn} A_n. \tag{1-83}$$

According to (1-76) and (1-22), the coefficients of $\mathbf{e}_n$ in the expansion of $\mathbf{e}_r'$ are the projections of $\mathbf{e}_r'$ on $\mathbf{e}_n$. Since these are given by (1-82), we have

$$\mathbf{e}_r' = a_{rn} \mathbf{e}_n \tag{1-84}$$

whence

$$\mathbf{e}_r' \cdot \mathbf{e}_s' = (a_{rm}\mathbf{e}_m) \cdot (a_{sn}\mathbf{e}_n) \tag{1-85}$$

or

$$\delta_{rs} = a_{rm} a_{sn} \delta_{mn} = a_{rn} a_{sn}. \tag{1-86}$$

Since numbers can be commuted in a product, sum $a_{rn}a_{sn}$ is not essentially different from sum $a_{sn}a_{rn}$ and set (1-86) contains only six independent equations limiting the nine a_{rn}'s. A similar treatment of $\mathbf{e}_r$ leads to

$$\delta_{rs} = a_{nr}a_{ns}. \tag{1-87}$$

Because the properties of all the base vectors are used in deriving either set of *orthogonality relationships*, (1-86) or (1-87), the second set may not introduce any additional limitations beyond those introduced by the former. To see that they do not, consider the *independent* steps needed in changing the unprimed axes to primed ones.

In the first step, the reference system is rotated counterclockwise by angle ϕ about the z axis as in figure 1.17. An intermediate axis is formed where the x axis stops. The second operation consists of rotation counterclockwise by angle θ about this new axis. Thirdly, the system is rotated counterclockwise by angle ψ about the z' axis. The final operation, not shown in figure 1.17, is a translation of the origin. This has no effect on the components.

Hence, three independent angles are needed to express the orientation of the primed system with respect to the unprimed one. Since set (1-86) already imposes six conditions on the nine cosines, equations (1-87) do not really introduce any additional conditions.

1.10 Handedness

A particularly interesting reorientation occurs when each particle or point in the given system is shifted to the position its image had occupied behind (a) a mirror plane or (b) a reflecting point. The resulting rearrangement, called a *reflection*, does

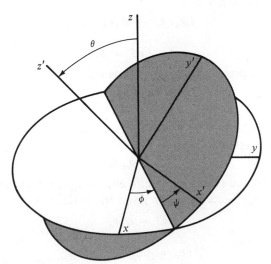

FIGURE 1.17   The Eulerian angles ϕ, θ, ψ used to express the orientation of the primed coordinate system with respect to the unprimed one.

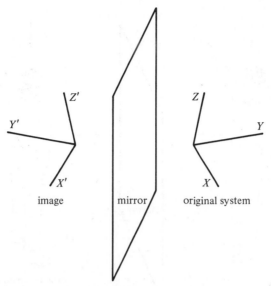

FIGURE 1.18 Change of a right-handed coordinate system to a left-handed one on reflection.

preserve the distances between points and the angles between lines. But it changes the handedness of objects and of coordinate systems (see figure 1.18). Because angles are preserved, equations (1-86) and (1-87) still apply; another criterion is needed to distinguish the reflection from a rotation.

A suitable criterion involves the triple scalar product. Suppose the transformation to be tested changes the mutually orthogonal unit vectors $\mathbf{e}_1$, $\mathbf{e}_2$, $\mathbf{e}_3$ to the set $\mathbf{e}_1{}'$, $\mathbf{e}_2{}'$, $\mathbf{e}_3{}'$. Each of the primed vectors is then projected onto the original unit vectors as in equation (1-81) and the product

$$\mathbf{e}_1{}' \cdot \mathbf{e}_2{}' \times \mathbf{e}_3{}' = \begin{vmatrix} a_{11} & a_{12} & a_{13} \\ a_{21} & a_{22} & a_{23} \\ a_{31} & a_{32} & a_{33} \end{vmatrix} \tag{1-88}$$

calculated using (1-34). Whenever the transformed vectors form a right-handed system, $\mathbf{e}_2{}' \times \mathbf{e}_3{}'$ is $\mathbf{e}_1{}'$ and the triple scalar product is $+1$. Whenever they form a left-handed system, $\mathbf{e}_2{}' \times \mathbf{e}_3{}'$ is $-\mathbf{e}_1{}'$ and this product is -1. So when the determinant of the a's equals -1, a reflection has occurred.

It is of some interest to use unit vectors in illustrating the conversion of right to left on reflection. In figure 1.19, the forward directions are represented by displacements of unit distance ahead—indeed, by unit vector $\mathbf{e}_{\text{front}}$ in the original system and by its image $\mathbf{e}'_{\text{front}}$ in the transformed system. The direction straight up is defined by a single displacement of unit distance upward—that is, by unit vector $\mathbf{e}_{\text{up}}$ and its image

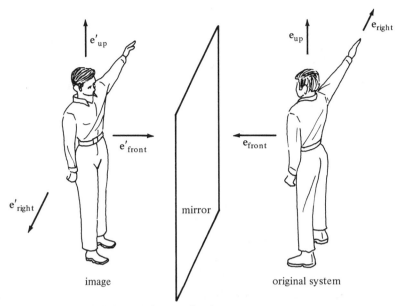

FIGURE 1.19   Change of right to left on reflection.

e_{up}'. Direction right, on the other hand, is determined by the right-hand rule from the forward and upward directions. Thus in the original system, the unit vector defining it is

$$e_{right} = e_{front} \times e_{up},$$

(1-89)

while in the image,

$$e'_{right} = e'_{front} \times e_{up}'.$$

(1-90)

We see from figure 1.19 that e'_{right} is the negative of the reflection of displacement e_{right}. While the man has his right hand extended, the image has its left one extended. However, e_{up}' is the reflection of displacement e_{up} and e'_{front} is the reflection of e_{front}.

A vector is said to be *regular* or *polar* if it reflects as a displacement. Vectors e_{up} and e_{front} are of this type. A vector is said to be *pseudo* or *axial* if its reflection is the negative of the reflection of the displacement representing it. Vector e_{right} is pseudo.

One can likewise distinguish between scalars. A scalar is *regular* if it is unchanged on reflection. A scalar is called *pseudo* if it changes sign in the reflection. The triple scalar product of regular vectors is a pseudo scalar. Thus in figure 1.20, $C \times D \cdot B$ is positive while $C' \times D' \cdot B'$ is negative.

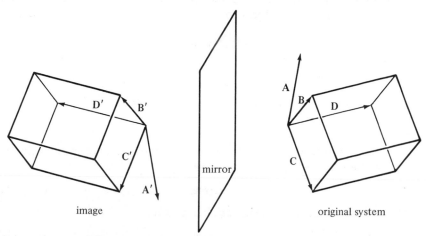

FIGURE 1.20   Reflection of a parallelepiped and the vectors used in calculating its volume.

1.11   Applying calculus to expressions containing vectors

In a physical system, measurable properties tend to vary smoothly with position and time. Infinities are absent and discontinuities appear only as idealizations. As a consequence, the pertinent mathematical forms can be differentiated and integrated.

We could derive the necessary formulas in a fundamental way, employing appropriate definitions and properties of limits. It is simpler, however, to break each standard form down into vector factors, expand these with respect to constant base vectors $\mathbf{e}_1$, $\mathbf{e}_2$, $\mathbf{e}_3$, and apply the usual formulas to the varying numerical parts. The results are then converted back to expressions free of the base vectors.

As an example, consider the vector product $\mathbf{A} \times \mathbf{B}$ in which $\mathbf{A}$ and $\mathbf{B}$ depend on time t. The expansions that have to be considered include

$$\mathbf{A} \equiv \mathbf{A}(t) = A_m(t)\mathbf{e}_m \equiv A_m\mathbf{e}_m, \tag{1-91}$$

$$\mathbf{B} \equiv \mathbf{B}(t) = B_n(t)\mathbf{e}_n \equiv B_n\mathbf{e}_n, \tag{1-92}$$

and

$$\mathbf{A} \times \mathbf{B} = \varepsilon_{rmn}\mathbf{e}_r A_m B_n. \tag{1-93}$$

Differentiating both sides of each of these equations with the base vectors constant yields

$$\frac{d\mathbf{A}}{dt} = \frac{dA_m}{dt}\,\mathbf{e}_m, \tag{1-94}$$

$$\frac{d\mathbf{B}}{dt} = \frac{dB_n}{dt}\,\mathbf{e}_n, \tag{1-95}$$

$$\frac{d}{dt}\,(\varepsilon_{rmn}\mathbf{e}_r A_m B_n) = \varepsilon_{rmn}\mathbf{e}_r\,\frac{dA_m}{dt}\,B_n + \varepsilon_{rmn}\mathbf{e}_r A_m\,\frac{dB_n}{dt}. \tag{1-96}$$

Note that dA_m/dt and dB_n/dt are the pertinent components of $d\mathbf{A}/dt$ and $d\mathbf{B}/dt$. Consequently, equation (1-96) reduces to

$$\frac{d}{dt}(\mathbf{A} \times \mathbf{B}) = \frac{d\mathbf{A}}{dt} \times \mathbf{B} + \mathbf{A} \times \frac{d\mathbf{B}}{dt}. \qquad (1\text{-}97)$$

In like manner, we obtain

$$\frac{d}{dt}(\mathbf{A} \cdot \mathbf{B}) = \frac{d\mathbf{A}}{dt} \cdot \mathbf{B} + \mathbf{A} \cdot \frac{d\mathbf{B}}{dt} \qquad (1\text{-}98)$$

and other formulas.

The *radius vector* $\mathbf{r}$ drawn from the origin to a given particle depends on the time t. If its three components are x_1, x_2, x_3, we have

$$\mathbf{r} = x_m(t)\mathbf{e}_m. \qquad (1\text{-}99)$$

Instead of t, we may employ a related variable s. Then,

$$\mathbf{r} = x_m(s)\mathbf{e}_m. \qquad (1\text{-}100)$$

As long as the base vectors are constant, the differential of (1-100) is

$$d\mathbf{r} = dx_m\mathbf{e}_m = \left(\frac{dx_m}{ds}\,ds\right)\mathbf{e}_m. \qquad (1\text{-}101)$$

Since the initial point of $\mathbf{r}$ is fixed at the origin of a reference frame, any change $\Delta\mathbf{r}$ occurring when t increases by Δt is a displacement of the particle. The ratio of $\Delta\mathbf{r}$ to Δt represents the average velocity in that frame during the given interval. The limit of $\Delta\mathbf{r}/\Delta t$ for vanishing Δt is no longer an average, but simply the *velocity* $\mathbf{v}$ at the given time t.

Formally dividing the first equality in (1-101) by dt yields

$$\mathbf{v} = \frac{d\mathbf{r}}{dt} = \frac{dx_m}{dt}\,\mathbf{e}_m. \qquad (1\text{-}102)$$

Letting a dot over a letter indicate differentiation with respect to time simplifies this result to

$$\mathbf{v} = \dot{\mathbf{r}} = \dot{x}_m\mathbf{e}_m. \qquad (1\text{-}103)$$

The limit of change in velocity of the particle over change in time as the latter vanishes is called the *acceleration* $\mathbf{a}$. Here

$$\mathbf{a} = \frac{d\mathbf{v}}{dt} = \frac{d^2\mathbf{r}}{dt^2} = \frac{d^2x_m}{dt^2}\,\mathbf{e}_m \qquad (1\text{-}104)$$

as long as the base vectors are constant. Alternatively, we write

$$\mathbf{a} = \dot{\mathbf{v}} = \ddot{\mathbf{r}} = \ddot{x}_m\mathbf{e}_m. \qquad (1\text{-}105)$$

A vector $\mathbf{A}$ acting on the particle may vary with position of the particle, that is, with $\mathbf{r}$:

$$\mathbf{A} = \mathbf{A}(\mathbf{r}) = A_m(\mathbf{r})\mathbf{e}_m. \qquad (1\text{-}106)$$

We can then set up the *line integral* $\int \mathbf{A} \cdot d\mathbf{r}$ in which each infinitesimal element is the scalar product of $\mathbf{A}$ with $d\mathbf{r}$. Introducing the components of $\mathbf{A}$ and $d\mathbf{r}$, from equations (1-106) and (1-101), leads to

$$\int \mathbf{A} \cdot d\mathbf{r} = \int A_m \frac{dx_m}{ds} \, ds. \qquad (1\text{-}107)$$

An example of such a line integral is the work done on a particle by the force $\mathbf{F}$ acting over the distance the particle moves:

$$W = \int \mathbf{F} \cdot d\mathbf{r}. \qquad (1\text{-}108)$$

EXAMPLE 1.6   Calculate the work done on a particle, as it moves along a semicircle of radius a, subject to an attracting force proportional to the distance of the particle from its final position.

Place the origin at the center of the semicircle. From it draw $\mathbf{a}$ to the attracting center and $\mathbf{r}$ to the particle. Draw $\mathbf{d}$ from the attracting center to the particle. Label the angles as indicated in figure 1.21.

Lay out the x axis along $\mathbf{a}$ and the y axis perpendicular to it as shown. Parametric equations for the curve are then

$$x = a \cos \alpha, \qquad y = a \sin \alpha,$$

while the components of force are

$$F_x = kd \cos \beta, \qquad F_y = -kd \sin \beta.$$

From the definition of sine, distance y of the particle above the x axis is also given by

$$y = d \sin \beta,$$

whence

$$d \sin \beta = a \sin \alpha.$$

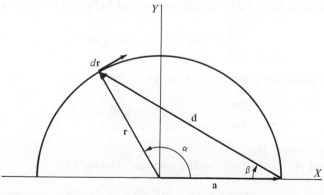

FIGURE 1.21   Pertinent vectors and angles for describing a particle moving along a semicircle to an attracting center.

On the other hand, the projection of **d** on the x axis equals a minus the projection of **r** on **a**:

$$d \cos \beta = a - r \cos \alpha = a(1 - \cos \alpha).$$

Using these relationships to eliminate β from the force components yields

$$F_x = ka(1 - \cos \alpha), \qquad F_y = -ka \sin \alpha,$$

while differentiating x and y gives

$$\frac{\partial x}{\partial \alpha} = -a \sin \alpha, \qquad \frac{\partial y}{\partial \alpha} = a \cos \alpha.$$

Therefore

$$W = \int \mathbf{F} \cdot d\mathbf{r} = \int_\pi^0 \left(F_x \frac{\partial x}{\partial \alpha} + F_y \frac{\partial y}{\partial \alpha} \right) d\alpha$$

$$= \int_\pi^0 \left[-ka^2(1 - \cos \alpha) \sin \alpha - ka^2 \sin \alpha \cos \alpha \right] d\alpha$$

$$= \int_\pi^0 \left[-ka^2 \sin \alpha \right] d\alpha = 2ka^2.$$

1.12   Common nonconstant base vectors

In analyzing some motions of a particle, we may employ variable base vectors with advantage.

As in the preceding section, we let vector **r** extend from the origin to the position of the particle at the given time t. When the particle is confined to the xy plane, we have

$$\mathbf{r} = r(t)\mathbf{l}(\phi). \tag{1-109}$$

Differentiating equation (1-37) for the base vector **l** then yields the formula

$$\frac{d\mathbf{l}}{d\phi} = -\sin \phi \mathbf{i} + \cos \phi \mathbf{j} = \mathbf{n}, \tag{1-110}$$

while differentiating equation (1-38) gives us

$$\frac{d\mathbf{n}}{d\phi} = -\cos \phi \mathbf{i} - \sin \phi \mathbf{j} = -\mathbf{l}. \tag{1-111}$$

See figures 1.22 and 1.23.

With the end point of **r** locating the particle, velocity **v** is

$$\frac{d\mathbf{r}}{dt} = \frac{dr}{dt}\mathbf{l} + r\frac{d\mathbf{l}}{d\phi}\frac{d\phi}{dt} = \dot{r}\mathbf{l} + r\dot{\phi}\mathbf{n}. \tag{1-112}$$

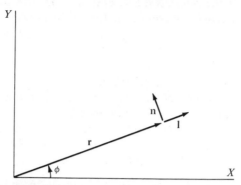

FIGURE 1.22 Plane polar coordinates.

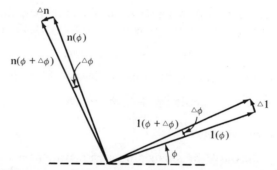

FIGURE 1.23 Magnified view of the changes in **l** and **n** caused by a change in ϕ. Note why $d\mathbf{l}$ points in the direction of **n** and why $d\mathbf{n}$ points in the direction of $-\mathbf{l}$. Also note that $|d\mathbf{l}| = l\, d\phi$, $|d\mathbf{n}| = n\, d\phi$. Therefore, $d\mathbf{l}/(l\, d\phi) = \mathbf{n}$ and $d\mathbf{n}/(n\, d\phi) = -\mathbf{l}$.

Note that a dot over a letter indicates differentiation with respect to t. The components of **v** are

$$v_r = \dot{r}, \tag{1-113}$$

$$v_\phi = r\dot{\phi}. \tag{1-114}$$

The acceleration is obtained by differentiating again:

$$\mathbf{a} = \frac{d\mathbf{v}}{dt} = \ddot{r}\mathbf{l} + \dot{r}\frac{d\mathbf{l}}{d\phi}\frac{d\phi}{dt} + \dot{r}\dot{\phi}\mathbf{n} + r\ddot{\phi}\mathbf{n} + r\dot{\phi}\frac{d\mathbf{n}}{d\phi}\frac{d\phi}{dt}$$

$$= (\ddot{r} - r\dot{\phi}^2)\mathbf{l} + (r\ddot{\phi} + 2\dot{r}\dot{\phi})\mathbf{n}. \tag{1-115}$$

Thus, the components of acceleration include

$$a_r = \ddot{r} - r\dot{\phi}^2 \tag{1-116}$$

$$a_\phi = r\ddot{\phi} + 2\dot{r}\dot{\phi} \tag{1-117}$$

The term $r\dot\phi^2$ in a_r is the *centripetal* acceleration caused by motion in the direction of increasing ϕ. For circular motion about the origin as center, $\ddot r$ is zero and

$$a_r = -r\dot\phi^2 = -\frac{v_\phi^2}{r}. \tag{1-118}$$

The term $2\dot r\dot\phi$ is called the *coriolis* acceleration.

When the particle moves out of the plane into 3-dimensional space, we may consider **r** to be a vector drawn along the shortest path from the z axis to the particle. Equations (1-37), (1-38), and (1-39) apply and we merely add the effect of varying z to the preceding equations:

$$\mathbf{v} = \dot z\mathbf{k} + \dot r\mathbf{l} + r\dot\phi\mathbf{n}, \tag{1-119}$$

$$\mathbf{a} = \ddot z\mathbf{k} + (\ddot r - r\dot\phi^2)\mathbf{l} + (r\ddot\phi + 2\dot r\dot\phi)\mathbf{n}. \tag{1-120}$$

In spherical coordinates, vector **r** is drawn from the origin to the particle and we let

$$\mathbf{r} = r(t)\mathbf{l}(\theta, \phi). \tag{1-121}$$

The velocity and acceleration then result from differentiating this expression and substituting the appropriate derivatives of equations (1-41), (1-42), (1-43). We obtain

$$\mathbf{v} = \dot r\mathbf{l} + r\dot\theta\mathbf{m} + r\sin\theta\dot\phi\mathbf{n}, \tag{1-122}$$

$$\mathbf{a} = (\ddot r - r\dot\theta^2 - r\sin^2\theta\dot\phi^2)\mathbf{l} + (r\ddot\theta + 2\dot r\dot\theta - r\dot\phi^2\sin\theta\cos\phi)\mathbf{m}$$

$$+ (r\sin\theta\ddot\phi + 2\dot r\sin\theta\dot\phi + 2r\dot\theta\dot\phi\cos\theta)\mathbf{n}. \tag{1-123}$$

1.13   Artificiality of components

A scalar property is measured by a single dimensional number that is independent of the coordinate system, but a vector property is measured by dimensional numbers that transform when the coordinate system is changed. Alterations in a Cartesian system produce the transformations described in section 1.9. The numbers that compose a polyadic (or tensor) also transform linearly.

Nevertheless, each physical entity exists independent of the coordinate system. While changes in the reference system alter the numbers describing each property, they do not alter the physical entity.

Consequently, physical laws should appear as relationships among scalars, vectors, and polyadics. Components are artificial in the sense that they reflect the nature of the coordinate system.

DISCUSSION QUESTIONS

1.1 How do physical theories originate?

1.2 What is the nature of vectors? How can they combine?

1.3 In formulating laws for combining vectors, why do we follow geometric diagrams?

1.4 How may the use of vectors facilitate certain geometric proofs?

1.5 Show that successive displacements add as vectors.

1.6 A rotation is characterized by a magnitude (the angle of rotation) and a direction (the axis). Prove that it is not represented by a vector because two given rotations of a solid body do not commute.

1.7 Set up various vector equations that define a plane.

1.8 How are the base vectors **i, j, k** used?

1.9 Show that if e_1, e_2, e_3 are not mutually perpendicular, component A_m is not the projection of **A** on e_m.

1.10 Discuss the behavior of base vectors fitting cylindrical and spherical coordinate systems.

1.11 Why are the numbers describing vectors and polyadics generally dimensional?

1.12 What operations would convert dyad **CD** to a scalar?

1.13 Represent dyad **ij** by a plane and a vector.

1.14 Why can one manipulate numbers more easily than directed lines?

1.15 How does the summation convention simplify expressions?

1.16 What is wrong with the equation

$$\mathbf{A} \cdot \mathbf{B} = (A_m e_m) \cdot (A_m e_m)?$$

1.17 Write out the determinant

$$\varepsilon_{rmn} A_r B_m C_n.$$

Why is ε_{rmn} called the permutation symbol?

1.18 Prove that

$$\varepsilon_{rmn}\varepsilon_{rmn} = 6$$

and

$$\varepsilon_{rmn}\varepsilon_{rmq} = 2\delta_{nq}.$$

1.19 Derive the transformation law

$$A_r' = a_{rn} A_n.$$

1.20 Write out the equations

$$a_{rn} a_{sn} = \delta_{rs}$$

in full.

1.21 Explain how three angular coordinates may be used to describe the orientation of a rigid body.

1.22 Show that $\mathbf{e_1}' \cdot \mathbf{e_2}' \times \mathbf{e_3}'$ equals the negative of the determinant of the a's when the initial Cartesian system is left-handed.

1.23 Why does a vertical mirror reverse left and right but leave upward and downward directions unchanged?

1.24 Show that the cross product of two pseudo vectors is a pseudo vector.

1.25 Sketch $\mathbf{A}(t)$, $\mathbf{A}(t + \Delta t)$, $\Delta \mathbf{A}$ for a vector varying (a) in length only, (b) in direction only, (c) in both length and direction. When is $d\mathbf{A}$ perpendicular to $\mathbf{A}$?

1.26 Derive

$$\frac{d}{dt}(\mathbf{A} \cdot \mathbf{B}) = \frac{d\mathbf{A}}{dt} \cdot \mathbf{B} + \mathbf{A} \cdot \frac{d\mathbf{B}}{dt}.$$

1.27 Explain what line integral represents the work done on a particle.

1.28 Carry through the derivation of the formulas for particle velocity $\mathbf{v}$ and acceleration $\mathbf{a}$ in spherical coordinates.

1.29 Discuss how physical properties exist whether a coordinate system is defined or not.

PROBLEMS

1.1 Employ vector analysis in showing that the line drawn from one corner of a parallelogram to the middle of an opposite side trisects a diagonal.

1.2 Show that (a) if a line bisects one side of a triangle and is parallel to a second side, it bisects the third side and that (b) the line segment joining the midpoints of two sides of a triangle equals one half the third side.

1.3 If l_1, m_1, n_1 are direction cosines of $\mathbf{A}$, show that

$$l_1 = \frac{A_x}{A}, \qquad m_1 = \frac{A_y}{A}, \qquad n_1 = \frac{A_z}{A}.$$

Then let l_2, m_2, n_2 be direction cosines of $\mathbf{B}$ and write similar expressions. Finally apply a vector formula to get

$$\cos \theta = l_1 l_2 + m_1 m_2 + n_1 n_2.$$

1.4 Draw the regular tetrahedron with vertices at four corners of a cube. Formulate the radius vectors that run from the center to two of the vertices and from them calculate the angle subtended.

1.5 Describe the triple products
(a) $\mathbf{A}(\mathbf{B} \cdot \mathbf{C})$,   (b) $\mathbf{A} \cdot (\mathbf{B} \cdot \mathbf{C})$,
(c) $\mathbf{A} \times (\mathbf{B} \cdot \mathbf{C})$,   (d) $\mathbf{A} \cdot (\mathbf{B} \times \mathbf{C})$.

1.6 Use the index notation in showing that

$$\mathbf{A} \cdot (\mathbf{B} \times \mathbf{C}) = \mathbf{B} \cdot (\mathbf{C} \times \mathbf{A}) = \mathbf{C} \cdot (\mathbf{A} \times \mathbf{B}).$$

1.7 Use the index notation in reducing

$$(\mathbf{A} \times \mathbf{B}) \cdot (\mathbf{B} \times \mathbf{C}) \times (\mathbf{C} \times \mathbf{A}) \quad \text{to} \quad (\mathbf{A} \cdot \mathbf{B} \times \mathbf{C})^2.$$

1.8 From the relationship between $\mathbf{r}$ and ϕ,

$$\mathbf{r} = \mathbf{A}e^{i\phi} + \mathbf{B}e^{-i\phi} + \mathbf{C},$$

eliminate the arbitrary constant vectors $\mathbf{A}$ and $\mathbf{B}$.

1.9 Find the components of $d\mathbf{A}/dt$ in spherical coordinates.

1.10 A particle is attracted toward the origin by a force inversely proportional to its distance from the x axis.
(a) What work is done by the force when the particle moves along a straight line from point $(0, a)$ to $(2a, a)$ and then along another straight line to point $(2a, 0)$?
(b) Calculate the work done when it moves along the ellipse

$$x = 2a \sin \phi, \qquad y = a \cos \phi$$

from $(0, a)$ to $(2a, 0)$.

1.11 A particle moves along the ellipse

$$x = a \cos \phi, \qquad y = b \sin \phi$$

at constant angular velocity

$$\phi = ct.$$

Calculate its speed and acceleration. Where is the acceleration directed?

1.12 Employ vector algebra to show that the line joining the midpoints of the sides of a trapezoid is parallel to the bases.

1.13 Prove that one median of a triangle trisects another.

1.14 Show that the resultant of two unit vectors issuing from a common point bisects the angle between the unit vectors.

1.15 What is the acute angle between two edges of a rhombohedron constructed inside a cube as follows? From one corner of the cube draw a vector to the middle of each adjacent face. Proceed similarly from the opposite corner. Finally add vectors parallel to those already drawn to complete the rhombohedron.

1.16 With the BAC CAB rule, show that

$$(\mathbf{A} \times \mathbf{B}) \times (\mathbf{C} \times \mathbf{D}) = \mathbf{C}(\mathbf{A} \cdot \mathbf{B} \times \mathbf{D}) - \mathbf{D}(\mathbf{A} \cdot \mathbf{B} \times \mathbf{C}).$$

1.17 Use the index notation in proving the identity

$$(\mathbf{A} \times \mathbf{B}) \cdot (\mathbf{C} \times \mathbf{D}) = (\mathbf{A} \cdot \mathbf{C})(\mathbf{B} \cdot \mathbf{D}) - (\mathbf{A} \cdot \mathbf{D})(\mathbf{B} \cdot \mathbf{C}).$$

1.18 Use the index notation in showing that

$$\mathbf{A} \times [\mathbf{B} \times (\mathbf{C} \times \mathbf{D})] = (\mathbf{B} \cdot \mathbf{D})\mathbf{A} \times \mathbf{C} - (\mathbf{B} \cdot \mathbf{C})\mathbf{A} \times \mathbf{D}.$$

1.19 From

$$\frac{x^2}{a^2} + \frac{y^2}{b^2} + \frac{z^2}{c^2} = 1,$$

derive the form

$$\mathbf{n} \cdot d\mathbf{r} = 0,$$

and so obtain vector $\mathbf{n}$, which is normal to the ellipsoid at point (x, y, z).

1.20 Find the components of $d\mathbf{A}/dt$ in cylindrical coordinates.

1.21 A particle is attracted vertically to the x axis by a force inversely proportional to its distance from the origin.
(a) What work is done by the force when the particle moves along the straight line from $(0, a)$ to $(2a, a)$ and then along the straight line to $(2a, 0)$?
(b) Calculate the work done when it moves along the ellipse

$$x = 2a \cos \phi, \qquad y = a \sin \phi$$

from $\phi = \pi/2$ to $\phi = 0$.

1.22 A particle moves along the cardioid

$$r = k(1 + \cos \phi)$$

at constant angular velocity

$$\phi = ct.$$

Calculate its speed and acceleration.

REFERENCES

BOOKS

Brand, L. *Vector Analysis*, pp. 1–45, 58–86, 110–116. John Wiley & Sons, Inc., New York, 1957.

Gibbs, J. W., and **Wilson, E. B.** *Vector Analysis*, pp. 1–136, 179–180. Dover Publications, Inc., New York, 1960.

Hanson, N. R. *Patterns of Discovery*, pp. 1–49. Cambridge University Press, London, 1958.

Heaviside, O. *Electromagnetic Theory*, pp. 34–45. Dover Publications, Inc., New York, 1950.

Hollingsworth, C. A. *Vectors, Matrices, and Group Theory for Scientists and Engineers*, pp. 1–33. McGraw-Hill Book Company, New York, 1967.

Hummel, J. A. *Vector Geometry*, pp. 1–250. Addison-Wesley Publishing Company, Inc., Reading, Mass., 1965.

Schwartz, M., Green, S., and **Rutledge, W. A.** *Vector Analysis with Applications to Geometry and Physics*, pp. 1–44, 66–68, 121–123. Harper & Brothers, New York, 1960.

ARTICLES

Bork, A. M. "Vectors Versus Quaternions"—The Letters in *Nature. Am. J. Phys.*, **34,** 202 (1966).

Dyson, F. J. Mathematics in the Physical Sciences. *Sci. American*, **211** (3), 129 (1964).

Evett, A. A. Permutation Symbol Approach to Elementary Vector Analysis. *Am. J. Phys.*, **34,** 503 (1966).

Gelman, H. Handed Products and Axial Vectors. *Am. J. Phys.*, **38,** 599 (1970).

Hestenes, D. Vectors, Spinors, and Complex Numbers in Classical and Quantum Physics. *Am. J. Phys.*, **39,** 1013 (1971).

Masket, A. V. Polar Vectors and Axial Vectors in Real 3-Space. *Am. J. Phys.*, **34,** 164 (1966).

Metzger, E. On the Use of Complex Numbers in Plane Mechanics. *Am. J. Phys.*, **40,** 924 (1972).

Stephenson, R. J. Development of Vector Analysis from Quaternions. *Am. J. Phys.*, **34,** 194 (1966).

Zaidins, C. S. An Analytic Expression for the Levi-Civita Symbols. *Am. J. Phys.*, **38,** 380 (1970).

2 / *Motion of a Newtonian Particle*

2.1 Concepts and postulates on which classical mechanics is based

The first accurate comprehensive physical theory, or pattern for the behavior of particles, was that initiated by Isaac Newton in 1687 and developed by many subsequent physicists. The scheme proceeds from the following basic assumptions.

1. Time is universal and absolute.

2. Space is Euclidean.

3. Bodies exist in space. At any given time, each of these can be treated as a *particle* located at a point, or can be broken down into elements, each of which is such a particle.

4. As time increases, every particle moves along a definite path, or *trajectory*, which can be observed and described as a smooth curve in some coordinate system.

5. This system can be transformed to one in which any particle, far enough removed from other particles to be unaffected by them, moves at constant velocity. Such a reference frame is said to be *inertial*.

6. When two particles uninfluenced by other particles interact with each other, their accelerations in inertial frames are proportional to each other. Thus, we write

$$m_1\mathbf{a}_1 = -m_2\mathbf{a}_2 \qquad (2\text{-}1)$$

where m_j is called the *mass* of particle j and $\mathbf{a}_j$ its acceleration. To obtain a unit of mass, we assign a given mass to a standard body.

7. The total *force* **F** acting on a body is given by its mass times its acceleration in an inertial system:

$$\mathbf{F} = m\mathbf{a}. \qquad (2\text{-}2)$$

8. The total force acting on a body at any instant equals the vector sum of the forces imposed by the individual agents affecting the body at that instant.

9. The force exerted by an individual agent equals the mass of the body times the acceleration it would experience if all other interacting particles were removed from influencing the body at the instant t.

36

2.2 Discussion of these postulates

Each item in section 2.1 will now be discussed, by number, from a modern point of view.

1. The human mind receives impressions *in order*, from both outside and inside its body. It recognizes periodic processes within the body and locates stimuli on the resulting time scale. Furthermore, the mind projects this scale uncritically on all other objects, regardless of where they are or how they are moving. It feels that the present instant is universal, even though signals cannot travel at infinite speed.

Common measurements support these views. We can synchronize two identical cyclic mechanisms, watches or clocks, at separated locations using light beams and check the synchronization at any later time. As long as the speed of one with respect to the other remains small, no difficulty is encountered.

So classical physicists, including Newton, did assume that the time shown by each clock pertained to all points in the universe without ambiguity. They assumed that time is continuous and representable by a single well-behaved variable t.

2. The mind also receives visual, aural, and tactile evidence of the coexistence of things. It constructs a three-dimensional space in which sources of the stimuli are mapped.

Precise and objective surveys of the space so created are possible with measuring rods and protractors. We assume that these devices are not altered by movement itself. We also suppose that simultaneity is no problem—that points simultaneous with respect to one clock are simultaneous with respect to any clock moving away or toward the given clock. In classical mechanics, we also assume that space is not grainy, that any spatial coordinate is continuous and well-behaved.

Conventional surveys do support these assumptions. Furthermore, the parallel postulate is found to hold quite accurately, as we have assumed in chapter 1. Under extreme conditions, however, a geometry in which time is bound up inseparably with space is required. This refinement was developed by Einstein and will be treated later.

3. We observe bodies and note that any visible part can be separated from the rest. But each small part may not be distinct from the field or fields associated with it. The system may not be infinitely divisible—at some stage we may reach indivisible particles.

4. In classical discussions, one assumes that every particle moves along a trajectory, even though no observer can locate a given particle with absolute certainty at any given time. By careful measurements, however, the error in position at time t can be made very small, unless the particle is submicroscopic.

When the particle is very small, no definite trajectory is found. Instead, all propagation is governed by a wave equation, and diffraction effects are observed. These will not be considered in this text.

5. The essence of *Newton's first law* is contained in the statement that inertial systems exist. Since we have no criterion for determining when a coordinate system is absolutely at rest, Newton's idea of absolute space with respect to which motion is defined has to be abandoned.

Coordinate systems that are approximate inertial frames for particular motions are readily found. In elementary laboratory work, where the bodies move over a very limited region in time and space, a frame fixed to the earth suffices. In dealing with extensive motion around the earth or between planets, a frame fixed to the center of the sun is suitable. For studying the movements of stars and cosmic rays, a nonrotating frame based on the nucleus of the galaxy is employed.

Since inertial frames are defined as systems in which an isolated particle moves at constant velocity, a second inertial frame must move at constant velocity with respect to a given inertial frame. But the velocity of the second frame with respect to the first one subtracts from the velocity of an accelerating particle in the first frame to give the particle's velocity in the second frame, as long as space is Euclidean and time universal. Because the shift in velocity is constant, the acceleration of the particle is the same in both inertial frames.

6. These statements, which constitute *Newton's third law*, do not allow for the finite rate of propagation of effects. As a consequence, equation (2-1) is accurate only when the particles are close together and the interactions are localized.

7. Note that each mass m_j is considered to be a scalar, the same in all inertial frames. With each a_j also invariant, equations (2-1) and (2-2) are not affected by any transformation from one inertial frame to another; these equations are said to be *covariant*.

Since the Newtonian equations do not distinguish between inertial systems, no system occupies a preferred position and the *relativity principle* is satisfied. Indeed, the collection of possible movements in a set of particles is the same for each system.

The force exerted by a spring balance on a body can be read from a scale. Observers find that the resulting push or pull always produces results conforming to equation (2-2). We imagine that other agents act in a similar way. Consequently, equation (2-2) is considered a law of nature; in Newton's scheme, it forms the *second law*.

In noninertial systems, subtracting the total force on a particle from its mass times acceleration yields a fictitious force:

$$\mathbf{F}_{\text{fict}} = m\mathbf{a} - \mathbf{F}. \tag{2-3}$$

Centrifugal force is an example of such a vector.

8. This is the familiar parallelogram-of-forces law. It has meaning only when individual forces are defined as in item 9 (or by some alternate rule).

9. Insofar as anyone has been able to determine, different forces act independently. None of the force that a particular agent exerts on a given body seems to depend on the presence of other agents.

2.3 Kepler's laws

Preceding Newton, Johannes Kepler analyzed the available measurements on apparent motion of the sun and planets, while Galileo Galilei studied data on falling bodies. After much anguish, both of these men discarded older geometric patterns that had long been accepted and began to assign a fundamental role to time.

Kepler studied solar and planetary data gathered by Tycho Brahe, Longomontamus, and earlier observers. Initially, he accepted the prevailing opinion that celestial bodies followed circular paths. After all, the only perfect geometric curve is the circle.

Kepler thought that since the sun was near the center of the planetary system, and was very large, it must cause the planets to move as they do. Consequently, he assumed the sun to be at rest as Copernicus had suggested. The orbit of the earth was studied, then that of Mars.

Kepler had noted that the line drawn from a central body to a planet moving at constant speed in a circle centered on the mother body swept out equal areas in equal times. He carried this idea over to the actual orbit in which the distance from the central body varied.

He found that no circular orbit about the sun could yield the observed results for Mars. Kepler then tried an egg-shaped orbit, a noncircular curve that had *one* focus, with the sun at the focus. Since this curve was not tractable, Kepler had to approximate it by a *two*-focus curve, the ellipse of proper eccentricity, even though introducing the second focus seemed unreasonable. After all, no physical entity was associated with this second unique geometric point.

But when the parameters were chosen properly, the ellipse did work. Application of the equal-area idea to this ellipse did lead to agreement with observations. (See figure 2.1.) Finally, Kepler related the period of motion to the major axis of the ellipse.

The resulting *laws of Kepler* are very accurate:
1. Each planet moves in an elliptic orbit with the sun at one focus.
2. The speed of a planet varies so that the radius vector drawn from the sun to the planet sweeps out constant area in equal time intervals.
3. The period of revolution of a planet around the sun is proportional to the three-halves power of the major axis of its elliptic orbit.

Supplementing the first law is the observation that a nonperiodic comet follows one branch of a hyperbola, with the sun at one focus.

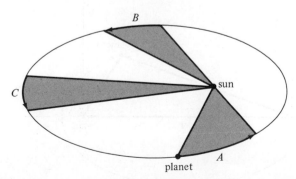

FIGURE 2.1   Orbit segments A, B, and C, of differing lengths, traversed by a planet in the same time. The shaded areas, swept out by the radius vector, are equal.

EXAMPLE 2.1   From Kepler's first law and its supplement, construct the polar equation describing the orbit of a planet or comet.

An ellipse (or hyperbola) is determined by the positions of its foci and the length of its major axis $2a$. According to the first law (and its supplement), the central body is at one of the foci.

Draw **r** from this focus to the instantaneous position of the planet, as in figure 2.2. Draw **q** from the other focus to the central body and **s** from it to the planet. Let ϕ be the angle to **r** from the line drawn from the star to the nearest point on the orbit.

Since the orbit is an ellipse (or hyperbola), the sum (or difference) of the distances to the foci is constant; and

$$s \pm r = 2a$$

or

$$s = 2a \mp r.$$

The eccentricity e is defined by the equation

$$q = 2ea.$$

The rule for vector addition yields

$$\mathbf{s} = \mathbf{q} + \mathbf{r}.$$

Dot multiply each side of this equation by itself to get

$$\mathbf{s} \cdot \mathbf{s} = (\mathbf{q} + \mathbf{r}) \cdot (\mathbf{q} + \mathbf{r})$$

or

$$s^2 = q^2 + r^2 + 2\mathbf{q} \cdot \mathbf{r}.$$

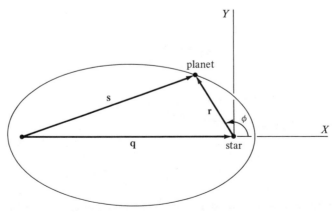

FIGURE 2.2   Vectors used in describing planetary orbits.

Introduce the relationships for s and q from above:

$$(2a \mp r)^2 = (2ea)^2 + r^2 \pm 4ear \cos \phi,$$

multiply out and rearrange:

$$\frac{1}{r} = \frac{1}{a(1 - e^2)} (1 + e \cos \phi)$$

or

$$\frac{1}{r} = \frac{1}{a(e^2 - 1)} (1 + e \cos \phi).$$

Letting $r = -r'$ and $\phi = \phi' + \pi$ converts the last equation to the form

$$\frac{1}{r'} = \frac{1}{a(1 - e^2)} (1 - e \cos \phi')$$

in which the coefficient on the right side is negative. As a coordinate, r' is related to ϕ' as r is to ϕ.

EXAMPLE 2.2   What condition does Kepler's second law impose on the polar coordinates for a planet?

Let $\mathbf{r}$ be the radius vector extending from the central body to the planet at time t, while $d\mathbf{r}$ is the change in $\mathbf{r}$ that occurs during the subsequent interval dt. If $d\phi$ is the angle between $\mathbf{r}$ and $\mathbf{r} + d\mathbf{r}$, as figure 2.3 shows, the area swept out is

$$\tfrac{1}{2}\mathbf{r} \times d\mathbf{r} = \tfrac{1}{2}r(r \, d\phi)\mathbf{n}$$

where $\mathbf{n}$ is the unit vector perpendicular to $\mathbf{r}$ and $d\mathbf{r}$ in the right-handed sense.

The rate at which this area is swept out is the corresponding derivative:

$$\tfrac{1}{2}\mathbf{r} \times \frac{d\mathbf{r}}{dt} = \tfrac{1}{2}r^2 \frac{d\phi}{dt} \mathbf{n} = \tfrac{1}{2}r^2 \dot{\phi}\mathbf{n} = \mathbf{A}.$$

As before, a dot over a variable indicates differentiation with respect to time.

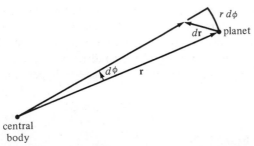

FIGURE 2.3   Successive positions of radius vector $\mathbf{r}$ and the arc that approximates the altitude of the triangle swept over.

Kepler's first law implies that the motion is in a plane and that r, ϕ are conventional polar coordinates. Kepler's second law then implies that vector **A** is constant. From the last equality, we obtain the scalar equation

$$r^2\dot{\phi} = 2A$$

with A constant.

EXAMPLE 2.3   Show that any motion governed by Kepler's second law is central if the area swept out lies in an inertial frame.

From example 2.2, the rate at which the area is swept over by the radius vector is

$$\tfrac{1}{2}\mathbf{r} \times \frac{d\mathbf{r}}{dt} = \mathbf{A}.$$

Differentiating this equation with **A** constant yields

$$\tfrac{1}{2}\mathbf{r} \times \frac{d^2\mathbf{r}}{dt^2} + \frac{1}{2}\frac{d\mathbf{r}}{dt} \times \frac{d\mathbf{r}}{dt} = 0.$$

Since the vector product of a vector with itself vanishes, the second term in this equation is zero. Multiplying the remaining term by $2m$, we have

$$\mathbf{r} \times m\frac{d^2\mathbf{r}}{dt^2} = 0.$$

Introducing (2-2) reduces this to the form

$$\mathbf{r} \times \mathbf{F} = 0.$$

Because the vector product of **r** with **F** is zero, the force is directed along the radius vector **r**:

$$\mathbf{r} \parallel \mathbf{F}.$$

Such a force is called a *central force*.

2.4   Newton's law of gravitation

Not only do the forces obtained from equation (2-2) behave as the pushes and pulls constructed intuitively, but they also obey very simple laws, except within atoms. Although motions are complicated, the classical forces are not. Indeed, Kepler's laws imply the simple inverse-square gravitational law found by Newton. Let us employ modern notation in carrying out the reduction.

We start with the equation describing an ellipse or hyperbola

$$\frac{1}{r} = \frac{1}{c}(1 + e \cos \phi) \tag{2-4}$$

and the result from example 2-2

$$r^2\dot{\phi} = 2A. \tag{2-5}$$

Let us differentiate equation (2-4) with respect to time to get

$$-\frac{1}{r^2}\dot{r} = -\frac{e}{c}\sin\phi\dot{\phi} \tag{2-6}$$

or

$$\dot{r} = \frac{e}{c}\sin\phi r^2\dot{\phi}. \tag{2-7}$$

Then eliminate $r^2\dot{\phi}$ with equation (2-5),

$$\dot{r} = \frac{e}{c}\sin\phi 2A, \tag{2-8}$$

differentiate again,

$$\ddot{r} = \frac{e}{c}\cos\phi\dot{\phi}2A, \tag{2-9}$$

and use equation (2-5) to eliminate $\dot{\phi}$:

$$\ddot{r} = \frac{e}{c}\cos\phi\frac{4A^2}{r^2}. \tag{2-10}$$

Substituting these expressions for $\dot{\phi}$ and $\ddot{r}$ into (1-116),

$$a_r = \frac{e}{c}\cos\phi\frac{4A^2}{r^2} - r\left(\frac{2A}{r^2}\right)^2 = \frac{4A^2}{r^2}\left(\frac{e}{c}\cos\phi - \frac{1}{r}\right), \tag{2-11}$$

and combining with (2-4) for ellipse or hyperbola, we derive

$$a_r = \frac{4A^2}{r^2}\left(-\frac{1}{c}\right), \tag{2-12}$$

whence

$$F_r = ma_r = -\frac{(4A^2/c)m}{r^2} = -\frac{\text{constant}}{r^2}. \tag{2-13}$$

The mass introduced here is that of the planet.

Since A is the time rate at which area is swept out by r, multiplying it by period T gives the total area of the ellipse:

$$TA = \pi ab = \pi a^2(1 - e^2)^{1/2}. \tag{2-14}$$

the expression for $c^{1/2}$ from example 2.1,

$$c^{1/2} = [a(1 - e^2)]^{1/2}, \tag{2-15}$$

converts equation (2-14) to

$$T = \frac{\pi}{A}c^{1/2}a^{3/2}. \tag{2-16}$$

Comparing this equation with the third law of Kepler, we find that

$$\frac{\pi}{A} c^{1/2} = \text{constant} \tag{2-17}$$

for a given mother body. Consequently, the coefficient of m/r^2 in (2-13) does not vary with parameters a and e; it is determined by properties of the attracting mass only.

If we assume the coefficient to be proportional to mass M of the mother body, the force law is *symmetric* in the two masses and we have

$$F_r = -\frac{GMm}{r^2} \tag{2-18}$$

where G, the *gravitational constant*, is introduced by definition. Then (2-13) yields the result

$$\frac{4A^2}{c} = GM \tag{2-19}$$

which reduces (2-16) to

$$T = \frac{2\pi}{G^{1/2}M^{1/2}} a^{3/2}. \tag{2-20}$$

These arguments can be reversed to show that the inverse-square law allows no other path but that of a conic section. But first, let us consider a more general setup in which the force is merely directed toward a point fixed in an inertial frame.

EXAMPLE 2.4   Discuss the determination of G.

The gravitational constant is one of the least accurately known fundamental constants for two reasons.

(a) Gravitational "charge" cannot be turned on and off or neutralized as electric charge can. The full inertial mass of each body enters into (2-18).

(b) Gravitational forces are very weak; G is a small number.

Nevertheless, measurements of the forces between large dense spheres, carried out with a delicate torsion balance, have yielded the value

$$G = 6.673 \times 10^{-11} \text{ N m}^2 \text{ kg}^{-2}.$$

The studies also indicate symmetry between the masses, as we have assumed.

EXAMPLE 2.5   Describe the similar establishment of Coulomb's law.

Put a charge q_1 on a small ball hung from a torsion balance, as in figure 2.4. Note that rotating the knob on top causes the charge to move through a horizontal circle. Along the tangent to this circle drawn from the point where the first charge is located, move a second ball carrying charge q_2 to distance r from the initial position of q_1. Then apply torsion by rotating the knob until q_1 is brought back to its original position, distance r from q_2. The force of interaction is obtained from the angle of rotation, the balance having been previously calibrated with known forces.

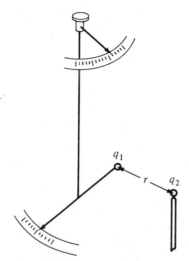

FIGURE 2.4   Essential parts of an apparatus for establishing Coulomb's law.

When we study changes in force by varying r and q's, we find that the results are represented by *Coulomb's law*,

$$\mathbf{F} = \frac{q_1 q_2}{4\pi\varepsilon_0 r^2}\, \mathbf{l},$$

where $\mathbf{F}$ is the electrostatic force acting on q_2, $\mathbf{l}$ is the unit vector drawn along the line from q_1 to q_2, r is the distance between the charges, and ε_0 is the permittivity of the space between.

2.5   Motion in a central field

Let us consider a small body, or particle, subject to a varying force $\mathbf{F}$ aimed directly toward, or away from, a given point in an inertial frame. For simplicity, let us place the origin at this point and draw a radius vector $\mathbf{r}$ from it to the particle. The velocity of the particle $\dot{\mathbf{r}}$ is designated $\mathbf{v}$ and its mass m.

The condition that $\mathbf{F}$ be directed parallel to $\mathbf{r}$ makes the cross product of $\mathbf{r}$ and $\mathbf{F}$ zero:

$$\mathbf{r} \times \mathbf{F} = 0. \tag{2-21}$$

Newton's second law, equation (2-2), converts this relationship to

$$\mathbf{r} \times m\frac{d\mathbf{v}}{dt} = 0. \tag{2-22}$$

But since

$$\frac{d\mathbf{r}}{dt} \times m\mathbf{v} = \mathbf{v} \times m\mathbf{v} = m\mathbf{v} \times \mathbf{v} = 0, \tag{2-23}$$

the two terms obtained on differentiating $\mathbf{r} \times m\mathbf{v}$ reduce to the expression on the left side of (2-22), and we have

$$\frac{d}{dt}(\mathbf{r} \times m\mathbf{v}) = 0. \tag{2-24}$$

Integrating (2-24) yields

$$\mathbf{r} \times m\mathbf{v} = \mathbf{L} = \textbf{constant}. \tag{2-25}$$

From example 2.2, the magnitude of the left side is $mr^2\dot{\phi}$; hence

$$mr^2\dot{\phi} = L = \text{constant}. \tag{2-26}$$

The vector $\mathbf{r} \times m\mathbf{v}$ is called the *angular momentum* $\mathbf{L}$.

We see that whenever the force acting on a particle is directed toward or away from a given point in an inertial frame, the particle is accelerated toward or away from this center. Consequently, it cannot move out of the plane passing through a small section of its path and the given point. Furthermore, it obeys Kepler's second law with respect to the center.

In equation (2-25), this law appears as the conservation-of-angular-momentum law. Equation (2-25) also implies that $\mathbf{r}$ is always perpendicular to $\mathbf{L}$, that is, in one plane. Consequently, the components of its acceleration are given by equations (1-116) and (1-117).

Substituting these into the radial and angular components of equation (2-2) and setting the angular component of force equal to zero lead to

$$m\ddot{r} - mr\dot{\phi}^2 = F, \tag{2-27}$$

$$mr\ddot{\phi} + 2m\dot{r}\dot{\phi} = 0. \tag{2-28}$$

Let us multiply (2-28) by r, combine the terms, and identify $mr^2\dot{\phi}$ as L:

$$\frac{d}{dt}(mr^2\dot{\phi}) = \frac{dL}{dt} = 0. \tag{2-29}$$

Integrating again yields equation (2-26).

The square of this integrated equation rearranges to

$$mr\dot{\phi}^2 = \frac{L^2}{mr^3}. \tag{2-30}$$

Formula (2-30) converts (2-27) to

$$m\ddot{r} = F + \frac{L^2}{mr^3}. \tag{2-31}$$

Equation (2-31) is simplified by the substitution

$$r = \frac{1}{u}; \tag{2-32}$$

for then, the derivatives are

$$\dot{r} = -\frac{1}{u^2}\frac{du}{d\phi}\,\dot{\phi} = -r^2\dot{\phi}\,\frac{du}{d\phi} = -\frac{L}{m}\frac{du}{d\phi}, \tag{2-33}$$

$$\ddot{r} = -\frac{L}{m}\frac{d^2u}{d\phi^2}\,\dot{\phi} = -\frac{L^2u^2}{m^2}\frac{d^2u}{d\phi^2}, \tag{2-34}$$

and the differential equation for the orbit reduces to the form

$$\frac{d^2u}{d\phi^2} = -u - \frac{m}{L^2u^2}\,F\left(\frac{1}{u}, \phi\right) \tag{2-35}$$

which is linear when $u^{-2}F$ is constant or linear. The important inverse-square law, according to which F varies as $1/r^2$, makes the final term constant.

As long as the force F is independent of angle ϕ, a second conservation law exists. Multiply equation (2-31) by the differential $(dr/dt)\,dt = dr$,

$$m\ddot{r}\,\frac{dr}{dt}\,dt = m\,\frac{d\dot{r}}{dt}\,\dot{r}\,dt = m\dot{r}\,\frac{dr}{dt}\,dt = m\dot{r}\,d\dot{r}$$

$$= F\,dr + \frac{L^2}{mr^3}\,dr, \tag{2-36}$$

rearrange,

$$m\dot{r}\,d\dot{r} - \frac{L^2}{mr^3}\,dr - F(r)\,dr = 0, \tag{2-37}$$

and integrate

$$\tfrac{1}{2}m\dot{r}^2 + \frac{L^2}{2mr^2} - \int_a^r F(r)\,dr = E. \tag{2-38}$$

The arbitrary constant E is referred to as the total *energy* while expression

$$-\int_a^r F(r)\,dr \equiv V(r) \tag{2-39}$$

is called the potential energy of the particle. The first two terms in (2-38) are said to represent the kinetic energy of the particle. Equation (2-38) itself states that the total energy of the particle is conserved.

When the particle is subject to an inverse-square force law, $F(r)$ has the form K/r^2 and

$$V(r) = -\int_a^r \frac{K\,dr}{r^2} = \frac{K}{r} - \frac{K}{a}. \tag{2-40}$$

With reference distance a infinite, equation (2-38) then becomes

$$\tfrac{1}{2}m\dot{r}^2 + \frac{L^2}{2mr^2} + \frac{K}{r} = E. \tag{2-41}$$

2.6 Consequences of linearity

There is no general method for solving a differential equation. Indeed, if a given equation is nonlinear, even the existence of a solution may be in doubt. But this question of existence can be answered by the behavior of the physical system that the equation describes.

Effective methods do exist, however, for dealing with linear equations, such as (2-35) when F is proportional to u^2. A linear differential equation is an equation in which each term is either independent of the dependent variable or contains it or one of its derivatives to the first power. The examples we are interested in have the form

$$RU = F \qquad (2\text{-}42)$$

where U is the dependent variable, F a function of the independent variables, and R the sum $R_1 + R_2 + \cdots + R_M$ with each R_j a function of the independent variables times a product of simple differentiating operators.

Any operator S is said to be *linear* if

$$S\alpha U = \alpha S U \qquad (2\text{-}43)$$

whenever α is a number (real or complex), as long as U is well-behaved. Note that R is such an operator.

Letting the inverse of R act on each side of equation (2-42) yields a solution

$$U_0 = R^{-1}F \qquad (2\text{-}44)$$

without arbitrary constants. In practice, we may factor R^{-1} and carry out the indicated operations, or decompose this inverse into partial fractions and evaluate each term by itself. The result can be checked by substitution:

$$RU_0 = F. \qquad (2\text{-}45)$$

In addition, a complete set of independent functions $U_1, U_2, \ldots, U_N$ satisfying the related homogeneous equation

$$RU_1 = 0, \qquad (2\text{-}46)$$

$$RU_2 = 0, \qquad (2\text{-}47)$$

$$\vdots$$

$$RU_N = 0 \qquad (2\text{-}48)$$

is found. Multiplying (2-45) by 1, (2-46) by constant c_1, (2-47) by constant $c_2, \ldots,$ and adding yields

$$RU_0 + c_j RU_j = F \qquad \text{where} \quad j = 1, 2, \ldots, N, \qquad (2\text{-}49)$$

whence

$$R(U_0 + c_j U_j) = F. \qquad (2\text{-}50)$$

Consequently, the linear equation is satisfied by the sum

$$U = U_0 + c_j U_j. \qquad (2\text{-}51)$$

When the differential equation is partial, the number of U_j's may be infinite. And where the U_j's form a continuum, the summation in (2-51) is replaced by an integration. When equation (2-42) is ordinary, U_0 is called the *particular integral* and $c_j U_j$ the *complementary function*.

EXAMPLE 2.6   Solve the homogeneous differential equation

$$\frac{d^n u}{dt^n} + a_1 \frac{d^{n-1} u}{dt^{n-1}} + \cdots + a_n u = 0$$

in which each coefficient is constant.

If we let

$$\frac{d}{dt} = D,$$

the given equation can be rewritten as

$$(D^n + a_1 D^{n-1} + \cdots + a_n)u = 0.$$

Since D is a linear operator, it combines with the constant coefficients $a_1, a_2, \ldots, a_n$ as an algebraic unknown would. Therefore, the expression in parentheses factors as the corresponding polynomial

$$r^n + a_1 r^{n-1} + \cdots + a_n = 0.$$

If the roots are

$$r_1, r_2, \ldots, r_n,$$

the factored form is

$$(D - r_1)(D - r_2) \cdots (D - r_n)u = 0.$$

Grouping all but one of the factors in operator R yields the equation

$$R\left(\frac{du}{dt} - r_n u \right) = 0$$

which is satisfied if

$$\frac{du}{dt} = r_n u$$

or

$$u_n = c_n e^{r_n t}.$$

Note that n is a free index.

Since any of the roots can appear in the last factor on the left of the differential equation, each of the others yields a similar solution. Adding all the exponentials so obtained yields

$$u = c_1 e^{r_1 t} + c_2 e^{r_2 t} + \cdots + c_n e^{r_n t}.$$

When two or more roots are equal, the corresponding terms are equivalent and additional ones have to be added to preserve the required number of independent arbitrary constants.

To determine how the additional expressions appear, let h be the difference between the pertinent roots when they are distinct:

$$r_2 = r_1 + h.$$

But since

$$c_1 e^{r_1 t} + c_2 e^{(r_1 + h)t} = e^{r_1 t}(c_1 + c_2 e^{ht})$$

and

$$e^{ht} \simeq 1 + ht$$

when h is small, the solution can be rewritten as

$$u \simeq e^{r_1 t}(c_1 + c_2 + c_2 ht) + c_3 e^{r_3 t} + \cdots.$$

If we let h go to zero so that

$$c_2 h = B$$

and

$$c_1 + c_2 = A$$

then we obtain

$$u = (A + Bt)e^{r_1 t} + c_3 e^{r_3 t} + \cdots.$$

For each additional root equal to r_1, an additional term is preserved in the expansion of the exponential of ht. The subsequent procedure then adds a constant times the next higher power of t to the terms in parentheses.

EXAMPLE 2.7   What kind of solution does a pair of complex roots yield?

If the algebraic equation corresponding to the homogeneous differential equation with constant coefficients has the roots

$$r_1 = a + ib \quad \text{and} \quad r_2 = a - ib,$$

then the solution to the differential equation contains the terms

$$c_1 e^{(a+ib)t} + c_2 e^{(a-ib)t} = e^{at}(c_1 e^{ibt} + c_2 e^{-ibt})$$

$$= e^{at}\left[(c_1 + c_2) \frac{e^{ibt} + e^{-ibt}}{2} + i(c_1 - c_2) \frac{e^{ibt} - e^{-ibt}}{2i} \right]$$

$$= e^{at}[(c_1 + c_2)\cos bt + i(c_1 - c_2)\sin bt].$$

2.7   The inverse-square force field

The source of the radially directed force $\mathbf{F}$ may be an approximately spherically symmetric body massive enough to be practically at rest in an inertial frame.

The origin then resides at the center of the body. And if the force varies inversely with the square of r, as it does for gravitational or electrostatic forces, we have

$$\mathbf{F} = \frac{K}{r^2} \mathbf{1} = K u^2 \mathbf{1}. \tag{2-52}$$

Substituting this force law into equation (2-35) and rearranging yields

$$\frac{d^2u}{d\phi^2} + u = -\frac{Km}{L^2}.$$ (2-53)

The particular integral of this second-order linear differential equation is

$$u_1 = -\frac{Km}{L^2}$$ (2-54)

while the complementary function is

$$u_2 = -\frac{Km}{L^2} B \cos(\phi - \phi_0).$$ (2-55)

If we choose the axes so ϕ_0 is zero, the complete solution is

$$\frac{1}{r} = u = -\frac{Km}{L^2}(1 + B \cos \phi).$$ (2-56)

This is the same as (2-4) if

$$c = -\frac{L^2}{Km}$$ (2-57)

and

$$e = B.$$ (2-58)

Thus, no solutions other than the conic-section solutions exist.

Parameter c is positive when the field is gravitational or when it is electrostatic with the charges attracting each other. When the field is electrostatic and the two charges repel, parameter c is negative and we must take

$$e = -B$$ (2-59)

for r to be positive. Then the particle moves along the negative branch of the hyperbola and

$$\frac{1}{r} = \frac{1}{c}(1 - e \cos \phi).$$ (2-60)

See figure 2.5.

For either branch, the angle α that an asymptote makes with the axis passing through the foci is obtained on setting $1/r$ equal to zero in the pertinent formula, (2-4) or (2-60):

$$\cos \alpha = \mp \frac{1}{e}.$$ (2-61)

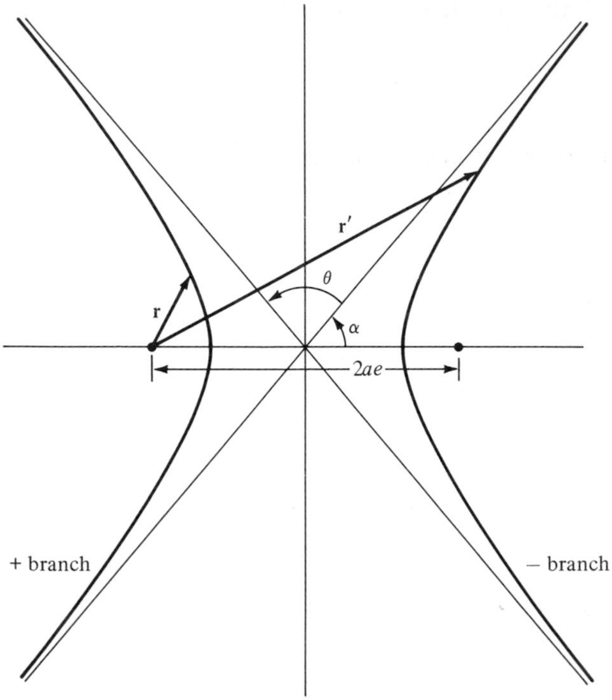

FIGURE 2.5   Hyperbolic paths for the two signs of $-K$.

Angle θ between the asymptotes is then given by

$$\frac{\theta}{2} = \alpha - 90° \quad \text{or} \quad \frac{\theta}{2} = 90° - \alpha, \tag{2-62}$$

so

$$\sin\frac{\theta}{2} = \frac{1}{e}, \quad \cos\frac{\theta}{2} = \left(1 - \frac{1}{e^2}\right)^{1/2}, \tag{2-63}$$

and

$$\cot\frac{\theta}{2} = (e^2 - 1)^{1/2}. \tag{2-64}$$

Distance r to the particle is an extremum and the time derivative of r is zero when the cosine of ϕ is $+1$. Then, (2-4) and (2-60) reduce to

$$u = -\frac{Km}{L^2}[1 \pm e(+1)]. \tag{2-65}$$

At this point the first term in equation (2-41) drops out, leaving

$$\frac{L^2}{2m}u^2 + Ku - E = 0, \tag{2-66}$$

whence

$$u = \frac{m}{L^2}\left[-K \mp \left(K^2 + \frac{2EL^2}{m}\right)^{1/2}\right] = \frac{-Km}{L^2}\left[1 \pm \left(1 + \frac{2EL^2}{K^2 m}\right)^{1/2}\right]. \quad (2\text{-}67)$$

Comparing (2-65) with (2-67), we find that the eccentricity is

$$e = \left(1 + \frac{2EL^2}{K^2 m}\right)^{1/2}. \quad (2\text{-}68)$$

Substituting this into equation (2-64) yields

$$\cot\frac{\theta}{2} = \left(\frac{2EL^2}{K^2 m}\right)^{1/2}. \quad (2\text{-}69)$$

EXAMPLE 2.8   From the definition of the number of radians in a planar angle, establish a measure for the 3-dimensional opening within the tip of a cone.

Through two lines intersecting as in figure 2.6, draw a unit circle centered on the intersection. The length of arc subtended between the lines gives the number of radians in the planar angle.

In a similar way, construct a unit sphere about the vertex of a 3-dimensional conical surface. See figure 2.7. The area subtended then yields the number of *steradians* in the solid angle.

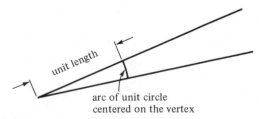

FIGURE 2.6   The arc that measures a planar angle.

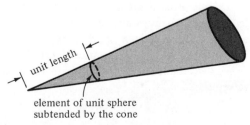

FIGURE 2.7   The area that measures a solid angle.

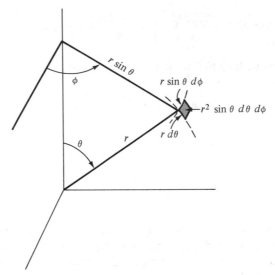

FIGURE 2.8   Small element of a sphere of radius r centered on the origin.

Figure 2.8 shows an infinitesimal element of another sphere centered on the origin. The front edge is the line $r\, d\theta$, a side edge the perpendicular line $r \sin \theta\, d\phi$. Multiplying these together yields the area

$$dS = r^2 \sin \theta\, d\theta\, d\phi.$$

A cone drawn from the origin to the edges of this element then subtends

$$d\Omega = \sin \theta\, d\theta\, d\phi$$

on a unit sphere; the opening at the vertex is $\sin \theta\, d\theta\, d\phi$ steradians large.

2.8   Rutherford scattering

The nuclear model of the atom was introduced by Ernest Rutherford to explain results from scattering experiments. In a typical setup, alpha particles impinge on a foil and are deflected through various angles by the atoms therein.

The scattering about a single atom is illustrated in figure 2.9. It is assumed that the atom repels each projectile and is heavy enough to be little affected in the process.

An element of the detector subtends a solid angle of scattered particles. This angle cuts out area $d\Omega$ from a unit sphere centered on the deflecting atom. The magnitude of this area is called the number of steradians in the angle. From example 2.8 and figure 2.10, we see that

$$d\Omega = \sin \theta\, d\theta\, d\phi \tag{2-70}$$

where θ and ϕ are spherical coordinates about the scattering center.

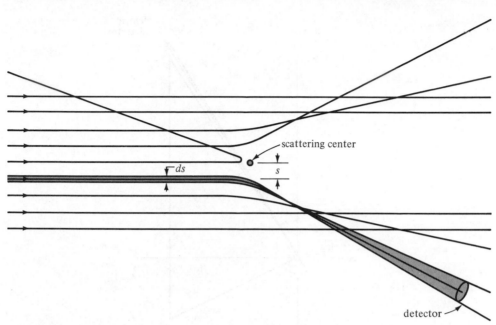

FIGURE 2.9 Paths of representative particles passing near a repelling center, with a small portion picked up by a detector.

Now, all particles arriving in a certain cylindrical shell are deflected by some angle between θ and $\theta + d\theta$. To pick up all of these would require a detector extending as a ring 360° around the axis. So if the element subtends only angle $d\phi$ about the axis, it picks up only that part of the incident shell. See figure 2.11.

Therefore, the effective area presented to the incident beam by the element is

$$d\sigma = ds(s \, d\phi) \tag{2-71}$$

where s is the distance between the incident asymptote and a parallel line through the center of the nucleus. This distance is called the *impact parameter* for the collision.

The counting rate N of a detector is proportional to the number I of pertinent particles falling on unit cross section in unit time,

$$N = IA. \tag{2-72}$$

If the efficiency is 100%, A is the area presented and the rate for area $n \, d\sigma$ is

$$dN = In \, d\sigma \tag{2-73}$$

where n is the number of exposed scattering centers in the foil.

To deduce the distribution pattern, assume that the scattering centers are very small, positively charged, heavy particles and that the interaction is electrostatic. The force acting on a projectile is then given by (2-52) with K positive. Equation (2-69) yields the angle between the ingoing and outgoing asymptotes, that is, the measured deflection angle, for given energy E and angular momentum L.

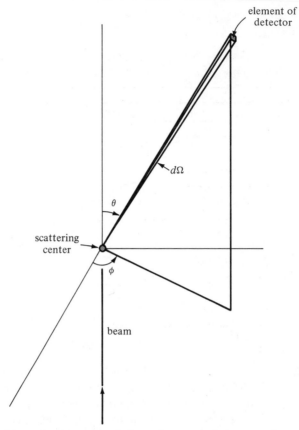

FIGURE 2.10   Solid angle $d\Omega$ subtended at the scattering center by an element of the detector.

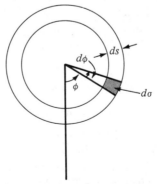

FIGURE 2.11   Cross section of the cylindrical shell containing the particles that are deflected by some angle between θ and $\theta + d\theta$. Particles from the shaded part end up at the detector.

Outside the range of the target nucleus, the second and third terms of (2-41) drop out and we have

$$E = \tfrac{1}{2}mv_0{}^2 \tag{2-74}$$

where v_0 is the initial speed. By manipulating the angular momentum (2-26) for this region, we get

$$L = mv_0 s. \tag{2-75}$$

For K we employ

$$K = \frac{q_1 q_2}{4\pi\varepsilon_0} \tag{2-76}$$

from example 2.5.

Substituting into (2-69) and inverting then gives

$$\tan\frac{\theta}{2} = \frac{q_1 q_2}{4\pi\varepsilon_0 msv_0{}^2}, \tag{2-77}$$

while differentiating and reducing yields

$$\frac{1}{2\cos\theta/2}\,d\theta = -\frac{\sin\theta/2}{s}\,ds. \tag{2-78}$$

The negative sign merely reflects the fact that θ decreases as s increases; we do not use it in finding the magnitude of $d\sigma$.

Solving for the magnitude of ds and s and substituting the results into (2-71) yields

$$d\sigma = \left(\frac{q_1 q_2}{4\pi\varepsilon_0 \tfrac{1}{2}mv_0{}^2}\right)^2 \frac{\sin\theta}{16\sin^4\theta/2}\,d\theta\,d\phi. \tag{2-79}$$

In applications, we are interested in the *differential scattering cross section* $d\sigma/d\Omega$. From (2-79) and (2-70),

$$\frac{d\sigma}{d\Omega} = \left(\frac{q_1 q_2}{4\pi\varepsilon_0 E}\right)^2 \frac{1}{(2\sin\theta/2)^4}. \tag{2-80}$$

Rutherford showed that this theory predicts how alpha particles are scattered when q_1 is the charge on an alpha particle and q_2 is the atomic number times the elementary charge, as long as the pericentron distance is greater than about 10^{-14} m. When the pericentron distance is smaller than this limit, which varies from atom to atom, the projectile reaches the nucleus and is affected by nuclear forces.

Even though the final result is valid, the theory is not strictly correct because it does not allow for quantum effects. Actually, the concept of a definite trajectory with a definite impact parameter is an idealization.

2.9   Gauss's law

Up to now, we have treated massive bodies as point masses Disregarding the spread in mass of each does not introduce so much error as one might fear, though, because a spherically symmetric body produces the same external field as a point mass. Gauss's law enables us to deduce this result.

In deriving the law, we first consider the effects of a single point mass M located at the origin 0. From (2-18), the force per unit mass acting on a test particle at distance r from 0 is

$$\mathbf{I} = -\frac{GM}{r^2}\mathbf{l}. \tag{2-81}$$

Vector $\mathbf{I}$ is called the *intensity* of the field.

Around the mass, let us construct a surface as in figure 2.12. Now, each infinitesimal part of the surface is essentially planar and is represented by a vector $d\mathbf{S}$. Projection of this on $\mathbf{I}$ (or $\mathbf{l}$) corresponds to projection of the infinitesimal area on a sphere of radius r about the origin. Such a projection equals r^2 times the size $d\Omega$ of the area that the pertinent solid angle would subtend on a sphere of unit radius based on the same center; it equals r^2 times $d\Omega$ steradians, where $d\Omega$ is positive or negative depending on whether $d\mathbf{S}$ projects positively or negatively on $\mathbf{l}$. Thus,

$$\mathbf{I} \cdot d\mathbf{S} = I \cos\theta\, dS = Ir^2\, d\Omega = -\frac{GM}{r^2} r^2\, d\Omega$$

$$= -GM\, d\Omega. \tag{2-82}$$

Integrating (2-82) over the entire closed surface yields

$$\int_S \mathbf{I} \cdot d\mathbf{S} = -\int_S GM\, d\Omega = -GM \int_S d\Omega$$

$$= -4\pi GM, \tag{2-83}$$

since the net solid angle subtended by the surface is 4π. If the mass were outside the closed surface, the integral would be zero because the net solid angle would then equal zero. (Corresponding to each positive $d\Omega$ would be a canceling negative $d\Omega$.)

Note that the conformation of the surface does not enter the final result. Why is this true? Clearly, the result depends on the canceling of r^2 in (2-82). Thus, it

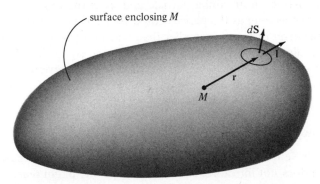

FIGURE 2.12   Surface enclosing M over which the normal component of intensity is integrated.

depends on the inverse square nature of I. Any entity decreasing with distance in the same manner must obey the same formula.

A simple example is afforded by radiation into free space from a point source. Let the strength of the source be measured by N, the number of rays emanating from it, as in figure 2.13. Then the flux intensity at distance r from it is given by

$$I = \frac{N}{4\pi r^2} = \frac{\text{constant}}{r^2}. \tag{2-84}$$

The fraction of a ray penetrating surface $d\mathbf{S}$ is equal to the flux intensity times the area perpendicular to $\mathbf{I}$, that is, to

$$\mathbf{I} \cdot d\mathbf{S} = \frac{N}{4\pi r^2} r^2 \, d\Omega = \frac{N}{4\pi} \, d\Omega. \tag{2-85}$$

So

$$\int \mathbf{I} \cdot d\mathbf{S} = \frac{N}{4\pi} \int d\Omega = N. \tag{2-86}$$

When we integrate the normal flux over the surface, we merely get the total flux from the source.

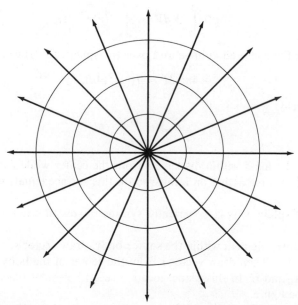

FIGURE 2.13 Spreading of rays homogeneously through successive concentric spheres from a single constant source.

If there are various gravitating particles inside a surface, each contributes additively to the field, since independent forces add. Then (2-83) is replaced by

$$\int \mathbf{I} \cdot d\mathbf{S} = -4\pi G \sum M_j. \tag{2-87}$$

When there is a continuous distribution of mass, each element is considered a particle and (2-87) becomes

$$\int \mathbf{I} \cdot d\mathbf{S} = -4\pi G \int \rho \, dV \tag{2-88}$$

where ρ is the density.

Equations (2-87) and (2-88) are mathematical statements of *Gauss's law*. They are particularly useful when the arrangement of masses exhibits planar, cylindrical, or spherical symmetry. For then, a bounding surface can be placed so all parts of it see essentially the same distribution. Since the distribution can produce only one field, I cannot vary over the surface. It can therefore be factored out of the integral.

For a spherically symmetric distribution of mass, an appropriate surface is a sphere at a given r from the center of the system. Each point on this sphere is equivalent, and the field is unique; so I is constant and perpendicular to the surface:

$$\mathbf{I} = I(r)\mathbf{l}. \tag{2-89}$$

Substituting (2-89) into (2-88),

$$-4\pi G \int_V \rho \, dV = \int_S I(r)r^2 \, d\Omega, \tag{2-90}$$

integrating over the enclosed volume and over the spherical surface,

$$-4\pi GM = I(r)r^2(4\pi), \tag{2-91}$$

and reducing yields

$$I = -\frac{GM}{r^2}. \tag{2-92}$$

Note that M is the mass within distance r from the center while I is the intensity at this distance. The force acting on a mass m on this surface equals m times $\mathbf{I}$.

EXAMPLE 2.9   Explain why a spherically symmetric distribution of mass yields a radially directed $\mathbf{I}$.

From an arbitrary element within the source body, draw vector $\mathbf{s}_1$ to the given point P, as in figure 2.14. Then draw vector $\mathbf{r}$ from the center of the body to P and pass a plane through $\mathbf{s}_1$ and $\mathbf{r}$. In this plane, locate a second element placed symmetrically below $\mathbf{r}$ as in the figure.

Now, the component of intensity $\mathbf{I}$, along $\mathbf{m}$ perpendicular to $\mathbf{l}$, caused by the first mass element is exactly opposed by that caused by the second element. To every

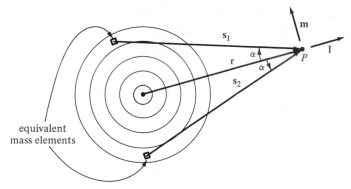

FIGURE 2.14   Corresponding elements having opposing contributions to **I** in the direction **m**.

other element above **r**, we can also find a compensating element below; the net intensity is directed along **l**. Furthermore, like elements are arranged in the same manner with respect to any other P, distance *r* from the center.

2.10   Gravitation near the earth's surface

Supplementing the analysis of Kepler on planetary motion was that of Galileo on freely falling bodies. Observations had shown that the distance fallen by such a body was proportional to the square of the falling time. Thinking in terms of the trajectory, a *geometric* concept, Leonardo, Benedetti, Varron, and, at first, Galileo supposed that the observations could be explained by assuming the body's velocity to be proportional to the space traversed. In their thinking, time was considered as separate from space and of no fundamental importance in the problem.

Later however, Galileo found that the earlier hypothesis was wrong and concluded that the velocity must be proportional to the time. Leonardo had also suggested this rule but did not realize its inconsistency with the spatial one. This corrected form of the law for freely falling bodies then led to the constant acceleration rule, which we will now consider.

As an approximation, the earth is spherically symmetric and (2-92) holds. Using (2-2) to eliminate **F** then yields

$$mg = -\frac{GMm}{r^2} \tag{2-93}$$

where *m* is the mass of a freely falling body, *g* its acceleration, *r* its distance from the center of the earth, *G* the gravitational constant, and *M* the mass of the earth. Canceling *m*, we get the result

$$g = -\frac{GM}{r^2} \tag{2-94}$$

which is constant, regardless of *m*, for small changes in *r*.

Implicit in the derivation is the assumption that coordinates erected on the surface of the earth are inertial. Acceleration due to gravity g is, however, observed on a spinning earth. In (2-93) the fictitious centrifugal force caused by this rotation should be subtracted out. Parameter g can be measured very accurately with a suitable pendulum.

EXAMPLE 2.10   What forces act on the bob of a simple pendulum?

Let m be the mass of the bob and θ the angular displacement of the supporting string from the vertical direction. Then the force of gravity $m\mathbf{g}$ acts as shown in figure 2.15. Component $mg \cos \theta$ acts along the string while component $mg \sin \theta$ acts perpendicular to it, opposing or aiding the motion.

Always opposing the motion, though, is the viscous drag. As an approximation, it is proportional to the speed of the bob $r\dot{\theta}$.

The forces acting perpendicular to the string include

$$F_\theta = -mg \sin \theta - kr\dot{\theta}$$

where k is the constant of proportionality. The forces acting radially include

$$F_r = mg \cos \theta - \tau$$

where τ is the tension in the string.

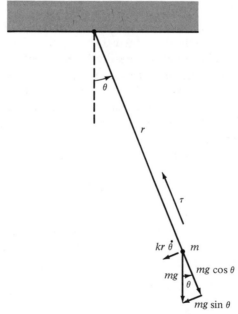

FIGURE 2.15   Forces on a pendulum bob and their resolution along string and normal to it.

EXAMPLE 2.11   Discuss the effect of viscous damping on a simple pendulum.

With r constant, (1-117) tells us that acceleration in the direction of motion equals

$$a = r\ddot{\theta}.$$

Multiplying this acceleration by the mass, setting the result equal to the corresponding force, and rearranging yields

$$mr\ddot{\theta} + kr\dot{\theta} + mg \sin \theta = 0,$$

or

$$\ddot{\theta} + 2\beta\dot{\theta} + \omega_0^2 \sin \theta = 0,$$

where

$$\beta = \frac{k}{2m} \quad \text{and} \quad \omega_0 = \sqrt{\frac{g}{r}}.$$

As an approximation valid for small angular displacements, let us replace $\sin \theta$ by θ.

$$\ddot{\theta} + 2\beta\dot{\theta} + \omega_0^2\theta = 0.$$

The auxiliary equation then yields the roots

$$-\beta \pm (\beta^2 - \omega_0^2)^{1/2},$$

and the solution

$$\theta = e^{-\beta t}\{A \cos [(\omega_0^2 - \beta^2)^{1/2}t - \theta_0]\}.$$

Since the coefficient of t in the argument is $2\pi/T$, where T is the period, we have

$$T = \frac{2\pi}{(\omega_0^2 - \beta^2)^{1/2}} = 2\pi \left(\frac{r}{g - k^2r/4m^2}\right)^{1/2}.$$

Note how the period is affected by the damping. Also note that the amplitude decreases exponentially because of the factor $e^{-\beta t}$.

2.11   Momentum and energy theorems

Newton's second law is a condition on the rate of change in momentum of a body. Integrating this condition yields a formula for the change in momentum. Transforming the condition and integrating yields a restriction on the energy of the body. Introducing the condition into the formula for rate of change in angular momentum relates this rate to a torque.

By definition, the *linear momentum* of a particle is

$$\mathbf{p} = m\mathbf{v} \tag{2-95}$$

where m is the mass and $\mathbf{v}$ the velocity of the particle in some inertial frame. Since the acceleration $\mathbf{a}$ is $d\mathbf{v}/dt$, Newton's second law (2-2) implies that

$$\frac{d\mathbf{p}}{dt} = \frac{d}{dt}(m\mathbf{v}) = m\frac{d\mathbf{v}}{dt} = \mathbf{F}. \tag{2-96}$$

The time rate of change of a particle's momentum equals the force acting thereon.
   Integrating this relationship yields

$$\Delta \mathbf{p} = \int_{t_1}^{t_2} \mathbf{F} \, dt. \tag{2-97}$$

The expression on the right is the *impulse* delivered by the force. When it is zero, the momentum does not change.
   Dot multiplying both sides of the last equality in (2-96) by the differential of distance covered $(d\mathbf{r}/dt) \, dt = d\mathbf{r}$

$$m \frac{d\mathbf{v}}{dt} \cdot \frac{d\mathbf{r}}{dt} \, dt = \mathbf{F} \cdot d\mathbf{r}, \tag{2-98}$$

reducing the differential $(d\mathbf{v}/dt) \, dt$ to $d\mathbf{v}$,
rearranging,

$$m \frac{d\mathbf{r}}{dt} \cdot d\mathbf{v} = m\mathbf{v} \cdot d\mathbf{v} = \frac{m}{2} \, d(\mathbf{v} \cdot \mathbf{v}) = \mathbf{F} \cdot d\mathbf{r}, \tag{2-99}$$

and integrating gives us

$$\Delta(\tfrac{1}{2}mv^2) = \int_{\mathbf{r}_1}^{\mathbf{r}_2} \mathbf{F} \cdot d\mathbf{r}. \tag{2-100}$$

By definition,

$$\tfrac{1}{2}mv^2 = T, \tag{2-101}$$

where T is the *kinetic energy*. So we have

$$\Delta T = \int_{\mathbf{r}_1}^{\mathbf{r}_2} \mathbf{F} \cdot d\mathbf{r}. \tag{2-102}$$

   When the integral on the right is independent of the path followed, a function of position can be introduced thus:

$$V(\mathbf{r}) = - \int_{\mathbf{r}_0}^{\mathbf{r}} \mathbf{F} \cdot d\mathbf{r}. \tag{2-103}$$

Equation (2-102) reduces to

$$\Delta T = -\Delta V \tag{2-104}$$

or

$$\Delta(T + V) = \Delta E = 0, \tag{2-105}$$

so the *total energy*

$$E = T + V \tag{2-106}$$

is then conserved.
   Function V is called the *potential energy* of the particle. In the central-force field of section 2.5, this energy V is a function of the magnitude of $\mathbf{r}$ only.
   By definition, the *angular momentum* $\mathbf{L}$ of a particle about the origin is the pseudo vector

$$\mathbf{L} = \mathbf{r} \times \mathbf{p} = \mathbf{r} \times m\mathbf{v}. \tag{2-107}$$

To see how Newton's second law governs the behavior of **L**, we differentiate (2-107) and introduce the law through (2-96) as follows:

$$\frac{d\mathbf{L}}{dt} = \mathbf{r} \times \frac{d\mathbf{p}}{dt} + \frac{d\mathbf{r}}{dt} \times \mathbf{p} = \mathbf{r} \times \mathbf{F} + \mathbf{v} \times m\mathbf{v} = \mathbf{N}. \tag{2-108}$$

Here the *torque* **N** is introduced as the pseudo vector

$$\mathbf{N} = \mathbf{r} \times \mathbf{F}. \tag{2-109}$$

When no torque acts on the particle,

$$\frac{d\mathbf{L}}{dt} = 0, \tag{2-110}$$

the angular momentum remains constant.

Corresponding theorems for many-particle systems will be developed and applied in the next chapter.

2.12 Key ideas

Until the work of Einstein, physicists thought that time could be separated from space without ambiguity and could be considered a universal variable. Space was considered to be Euclidean.

Newton and his followers also assumed that each small part of a body could be assigned to a definite part of space at any given time. Such an element then moved smoothly in space as time progressed.

This movement could be referred to an inertial frame. When no force acted on the particle, it would travel in a straight line at constant velocity. When it interacted with one other particle, the two accelerations were proportional. The masses were thought to be inversely proportional to the accelerations, and the force acting on each particle was assumed to be mass times acceleration.

Each additional particle acting on a given particle would exert the same force as if it acted alone. The different forces would add vectorially to produce the mass of the given particle times its net acceleration.

If the mass of a given particle was small with respect to the masses of all bodies interacting appreciably with it, its influence on these bodies was negligible. One might then consider the particle to move in the field of the other bodies.

DISCUSSION QUESTIONS

2.1 How can time be the same for all particles in the universe?

2.2 Why does space appear to be absolute? How is the Euclidean nature of space established?

2.3 What are particles? How does a particle seem to move?

2.4 What is wrong with the conventional statement of Newton's first law?

2.5 When does Newton's third law break down?

2.6 Why is

$$\mathbf{F} = m\mathbf{a}$$

considered to be a law of nature?

2.7 Explain the parallelogram-of-forces law.

2.8 Discuss the development of Kepler's laws.

2.9 Explain what Kepler's second law implies about the force acting on a planet.

2.10 From the geometric definition of a hyperbola, derive its equation in polar coordinates.

2.11 Explain what Kepler's first law implies when it is coupled with his second law.

2.12 What information about the force law comes from Kepler's third law?

2.13 Why is it hard to determine the gravitational constant G accurately?

2.14 Explain the remark "Although motions are complicated, forces are simple."

2.15 Derive the trajectory of a particle subject to a force varying inversely as the square of its distance from the origin of an inertial frame.

2.16 If the force $\mathbf{F}$ acting on a particle is $K\mathbf{l}/r^3$, what is the form of its trajectory?

2.17 How does one calculate the angle between the asymptotes when the path is hyperbolic?

2.18 Discuss Rutherford scattering.

2.19 How is any quantity that varies inversely with the square of the distance from its source representable as a flux?

2.20 To what approximation does the earth behave as a point mass toward a satellite?

2.21 What kind of problems can be solved with Gauss's law?

2.22 Consider Galileo's contributions to classical physics.

2.23 Discuss how yield of the support, knife edge compression, and elasticity of the string or rod affect the behavior of a pendulum. What about friction at the support?

2.24 What is the relationship between centrifugal force and gravitational force for a satellite moving along a circular path?

2.25 Cite force laws for which a potential function $V(\mathbf{r})$ does not exist.

2.26 How does Newton's second law govern the behavior of (a) the linear momentum, (b) the angular momentum, of a particle? When are **p** and **L** conserved?

PROBLEMS

2.1 Calculate the period of revolution of Mars if its major axis is 1.5237 times that of the earth.

2.2 Jupiter requires 4334 days to travel around the sun along an orbit averaging 4.833×10^8 miles from the sun, while a moon of Jupiter travels along an orbit with a mean radius of 2.62×10^5 miles and a period of 1.769 days. What is the ratio of the mass of the sun to the mass of Jupiter?

2.3 Rewrite Newton's second law in the form

$$\mathbf{F} = km\mathbf{a}$$

and evaluate k so that the gravitational constant G is numerically one when masses and distances are measured in grams and centimeters.

2.4 How far would a body accelerating at g m sec^{-2} have to fall from rest to reach the same terminal speed as one falling freely from infinite distance to the surface of the earth in the earth's gravitational field?

2.5 Obtain the formula for the distance traveled by a body falling from rest through a viscous medium in which the retarding force is proportional to the speed of the body while the gravitational field is constant.

2.6 What is the acceleration of a body falling freely down a straight hole bored through the exact center of the earth, if density of the earth were constant? How does distance r from the center vary with the time?

2.7 Given that G is 6.673×10^{-11} N m^2 kg^{-2} and that g is 9.832 m sec^{-2} at a pole 6.357×10^6 m from the center, what is the mass of the earth?

2.8 While time t is negative, the path of a particle is described in spherical coordinates by the equations

$$r = -at \qquad \text{and} \qquad \phi = -\frac{b}{t}$$

where a and b are constants. What force law does the particle follow?

2.9 Assume that a potential $V(\mathbf{r})$ exists for a particle and that the particle is attracted towards the z axis by a force proportional to the square of its distance from the xy plane and inversely proportional to its distance from the z axis. Solve for V and for the additional perpendicular force that exists.

2.10 Show which of the forces

(a) $F_x = f(x)g(t)$,   (b) $F_x = f(\dot{x})g(t)$,   (c) $F_x = f(x)g(\dot{x})$,

acting on a particle, yield an integrable equation of motion.

2.11 If the period for Venus to traverse its orbit is 224.70 days while the period for Earth is 365.26 days, what is the ratio of the major axis of Venus to that of the Earth?

2.12 Transform the equation

$$\frac{1}{r} = \frac{1}{c}(1 + e \cos \phi)$$

to rectangular coordinates and note what equation is obtained when $e = 1$.

2.13 In an inertial system a particle follows an orbit along which

$$r\phi = \text{constant}.$$

If this behavior is caused by a single attracting particle at the origin, what is the force law?

2.14 What central-force law causes a small body to move along the circle

$$r = A \cos \phi?$$

2.15 A body coasting across a fluid is subject to a retarding force proportional to its velocity. Obtain a formula for the distance traveled as a function of time t.

2.16 A body falls from rest through a viscous medium in which the retarding force is proportional to the square of the speed of the body while the gravitational field is constant. Obtain a formula for the distance fallen.

2.17 If a spherically symmetric dust cloud of uniform density ρ surrounded the sun, what additional central force would act on a planet within the cloud?

2.18 If the mass and radius of Venus are 0.8136 and 0.966, respectively, times the mass and radius of the Earth, what is the acceleration due to gravity on Venus in terms of that on Earth?

2.19 When following a circular orbit at distance a from an attracting mass, a particle moves at a certain speed. Express in terms of this speed the minimum speed that a particle must have to escape to infinity from a point on this orbit.

2.20 A particle is attracted toward the origin by the force

$$F_r = -2ar \sin \theta \cos \phi.$$

What is the simplest F_θ and F_ϕ which will allow potential $V(\mathbf{r})$ to exist?

REFERENCES

BOOKS

Arthur, W., and Fenster, S. K. *Mechanics*, pp. 85–174. Holt, Rinehart, and Winston, Inc., New York, 1969.

Greenwood, D. T. *Principles of Dynamics*, pp. 12–28, 185–228. Prentice-Hall, Inc., Englewood Cliffs, N.J., 1965.

Hanson, N. R. *Patterns of Discovery*, pp. 70–118. Cambridge University Press, London, 1958.

Hesse, M. B. *Forces and Fields*, pp. 126–156. Philosophical Library, Inc., New York, 1962.

Jammer, M. *Concepts of Force*, pp. 81–264. Harvard University Press, Cambridge, Mass., 1957.

Marion, J. B. *Classical Dynamics of Particles and Systems*, 2nd ed., pp. 45–91. Academic Press, Inc., New York, 1970.

Nelson, W. C., and Loft, E. E. *Space Mechanics*, pp. 1–226. Prentice-Hall, Inc., Englewood Cliffs, N.J., 1962.

Symon, K. R. *Mechanics*, 2nd ed., pp. 1–154. Addison-Wesley Publishing Company, Inc., Reading, Mass., 1960.

Taylor, E. F. *Introductory Mechanics*, pp. 1–177. John Wiley & Sons, Inc., New York, 1963.

ARTICLES

Arons, A. B., and Bork, A. M. Newton's Laws of Motion and the 17th Century Laws of Impact. *Am. J. Phys.*, **32**, 313 (1964).

Banerjee, K. A Note on Kepler's Third Law. *Am. J. Phys.*, **39**, 455 (1971).

Bork, A. M. Logical Structure of the First Three Sections of Newton's *Principia*. *Am. J. Phys.*, **35**, 342 (1967).

Bunge, M. Mach's Critique of Newtonian Mechanics. *Am. J. Phys.*, **34**, 585 (1966).

Chapman, S. Kepler's Laws: Demonstration and Derivation without Calculus. *Am. J. Phys.*, **37**, 1134 (1969).

Christy, R. W., and Mayhugh, M. R. Elementary Analysis of Translunar Apollo Orbit. *Am. J. Phys.*, **37**, 1103 (1969).

Collas, P. Algebraic Solution of the Kepler Problem Using the Runge-Lenz Vector. *Am. J. Phys.*, **38**, 253 (1970).

Cronin, J. L., Jr., and Jones, L. C. Velocity Hodograph of Inverse-Square Central Force Motion. *Am. J. Phys.*, **36**, 1016 (1968).

Farrell, D. E., and Tripp, J. H. Gravitational Field of a Spherical Shell. *Am. J. Phys.*, **35**, 885 (1967).

Feather, N. The Physical Basis of Newtonian Mechanics. *Contemporary Phys.*, **7**, 29, 122 (1965).

Fock, V. A. Galileo's Principles of Mechanics and Einstein's Theory. *Soviet Phys. Uspekhi*, **7**, 592 (1965).

Freeman, I. M. Why is Space Three-Dimensional? Based on W. Büchel: "Warum hat der Raum drei Dimensionen?" *Physikalische Blätter*, **19**, 12, pp. 547–549 (December 1963). *Am. J. Phys.*, **37**, 1222 (1969).

Gingerich, O. The Computer Versus Kepler. *Am. Scientist*, **52**, 218 (1964).

Halford, W. D. On the Relativistic Nature of Newtonian Mechanics. *Am. J. Phys.*, **39**, 114 (1971).

Hesse, M. Resource Letter PhM-1 on Philosophical Foundations of Classical Mechanics. *Am. J. Phys.*, **32**, 905 (1964).

Katz, A. Formulation of Newton's Second Law. *Am. J. Phys.*, **35**, 882 (1967).

Kelsey, E. J. The Effects of Gravity Decrease on Planetary Orbits. *Am. J. Phys.*, **39**, 795 (1971).

Laporte, O. On Kepler Ellipses Starting from a Point in Space. *Am. J. Phys.*, **38**, 837 (1970).

Pickering, W. H. The Grand Tour. *Am. Scientist*, **58**, 148 (1970).

Plybon, B. F. A Generalization of the Lenz Vector. *Am. J. Phys.*, **40**, 196 (1972).

Poss, H. L. Pulsar Periods and Kepler's Third Law. *Am. J. Phys.*, **38**, 109 (1970).

Rosenberg, R. M. On Newton's Law of Gravitation. *Am. J. Phys.*, **40**, 975 (1972).

Sciamanda, R. J. The Problem with Average Acceleration Problems. *Am. J. Phys.*, **39**, 650 (1971).

Siegel, S. More about Variable Mass Systems. *Am. J. Phys.*, **40**, 183 (1972).

Weinstock, R. Laws of Classical Motion: What's F? What's m? What's a? *Am. J. Phys.*, **29**, 698 (1961).

Wilson, C. How Did Kepler Discover His First Two Laws? *Sci. American*, **226** (3), 92 (1972).

3 / Simple Motions of Systems of Interacting Particles

3.1 Introduction

Newton's second law, equation (2-2), relates the second temporal derivative of the position vector for each particle in a given system to the force acting on the particle. Whenever such a force depends on the position and/or velocity of any of the other particles in the system, the corresponding differential equations are not independent. Furthermore, the interacting equations are generally not readily integrable. Nevertheless, they can be combined to form some exact and some approximate relationships that are tractable.

In chapter 2, one particle, or element of mass, was considered separate from all the rest. A system of two particles is also relatively simple, since such a system corresponds to two one-particle systems. Pertinent procedures will be discussed in this chapter.

The differential equations for three or more particles can be combined to yield (a) an equation describing translation of the system as a whole, (b) an equation describing rotation of the system about a given point, (c) equations describing vibrations that can be present, and (d) equations describing the other relative motions.

The equation for translation is similar to the equation of motion for a single particle. The first integral of this equation links the change in net linear momentum to the total impulse delivered by the net force causing the change. As long as this impulse is zero (as when no force acts), the linear momentum is conserved.

The equation describing rotation is the principal concern of this chapter. In analogy with the force law governing translation, the equation for rotation associates a torque with any change in angular momentum of the given system. Whenever the torque is not acting, the angular momentum is conserved. The angular momentum does not generally parallel the angular velocity, however.

An introduction to the theory of vibrations will appear in chapter 6. Neither directed nor random relative motions will be considered in any detail.

All movements occur subject to the energy integral. Dot multiplying each Newtonian equation by the pertinent differential distance and adding the results yields a relationship that can be integrated formally.

71

3.2 Equations governing the movements of two interacting particles

Let us start with the simplest multiparticle system, in which two relatively small bodies interact with each other and possibly with some imposed field. The two Newtonian equations for the motion will be combined to yield one equation governing movement of the center of mass and a second governing movement of one particle with respect to the other.

Suppose that m_1 is the mass of the first small body and m_2 the mass of the second small body. Let $\mathbf{r}_1$ and $\mathbf{r}_2$ be drawn from a point fixed in an inertial frame to the essential positions of m_1 and m_2 as in figure 3.1. If the force that m_1 exerts on m_2 is $\mathbf{F}_2{}^i$, while the force that m_2 exerts on m_1 is $\mathbf{F}_1{}^i$, we have

$$\mathbf{F}_1{}^i = -\mathbf{F}_2{}^i \tag{3-1}$$

from Newton's third law.

If the additional force acting on m_1 is $\mathbf{F}_1{}^e$, then

$$m_1\ddot{\mathbf{r}}_1 = \mathbf{F}_1{}^i + \mathbf{F}_1{}^e \tag{3-2}$$

from Newton's second law. And if the additional force acting on m_2 is $\mathbf{F}_2{}^e$, we similarly obtain

$$m_2\ddot{\mathbf{r}}_2 = \mathbf{F}_2{}^i + \mathbf{F}_2{}^e. \tag{3-3}$$

Superscripts i and e stand for internal and external.

Adding (3-2), (3-3), and using (3-1) to reduce the result yields

$$m_1\ddot{\mathbf{r}}_1 + m_2\ddot{\mathbf{r}}_2 = \mathbf{F}_1{}^e + \mathbf{F}_2{}^e. \tag{3-4}$$

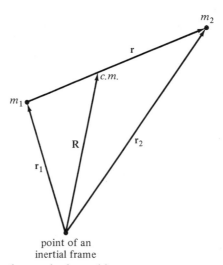

point of an
inertial frame

FIGURE 3.1   Vectors for the two-body problem.

If we let the vector $\mathbf{R}$ locating the *center of mass* be defined by the equation

$$m_1\mathbf{r}_1 + m_2\mathbf{r}_2 = (m_1 + m_2)\mathbf{R} \tag{3-5}$$

whose second derivative is

$$m_1\ddot{\mathbf{r}}_1 + m_2\ddot{\mathbf{r}}_2 = (m_1 + m_2)\ddot{\mathbf{R}}, \tag{3-6}$$

equation (3-4) can then be rewritten as

$$(m_1 + m_2)\ddot{\mathbf{R}} = \mathbf{F}_1{}^e + \mathbf{F}_2{}^e. \tag{3-7}$$

With M the total mass $m_1 + m_2$ and $\mathbf{F}$ the net force $\mathbf{F}_1{}^e + \mathbf{F}_2{}^e$, (3-7) simplifies to

$$M\ddot{\mathbf{R}} = \mathbf{F}, \tag{3-8}$$

the second-law formula for a single particle. Thus, in *translation*, the system behaves as a particle of mass M on which the total force $\mathbf{F}$ acts. The hypothetical particle is located at and moves with the center of mass.

Multiplying (3-2) by m_2 and subtracting the result from (3-3) multiplied by m_1 yields an equality that the second derivative of

$$\mathbf{r} = \mathbf{r}_2 - \mathbf{r}_1 \tag{3-9}$$

and (3-1) reduce to

$$m_1 m_2\ddot{\mathbf{r}} = (m_1 + m_2)\mathbf{F}_2{}^i + m_1 m_2\left(\frac{\mathbf{F}_2{}^e}{m_2} - \frac{\mathbf{F}_1{}^e}{m_1}\right) \tag{3-10}$$

or

$$\mu\ddot{\mathbf{r}} = \mathbf{F}_2{}^i + \mu\left(\frac{\mathbf{F}_2{}^e}{m_2} - \frac{\mathbf{F}_1{}^e}{m_1}\right). \tag{3-11}$$

In the last step, equation (3-10) has been divided by $m_1 + m_2$ and the *reduced mass* μ introduced by the formula

$$\mu = \frac{m_1 m_2}{m_1 + m_2} \quad \text{or} \quad \frac{1}{\mu} = \frac{1}{m_1} + \frac{1}{m_2}. \tag{3-12}$$

The expression inside the parentheses in (3-11) vanishes when (a) no external forces act on the masses or when (b) the external forces have the same direction and are proportional to the masses. The first condition is met by an isolated pair of interacting particles. (For example, consider an isolated diatomic molecule.) The second alternative is met by a pair of bodies in similar gravitational fields. (The earth-moon system is an example.)

When the expression inside the parentheses is zero, (3-11) simplifies to

$$\mu\ddot{\mathbf{r}} = \mathbf{F}_2{}^i, \tag{3-13}$$

the equation of motion for a hypothetical particle of mass μ in the force field established by m_1 fixed at the origin. Parameter μ is the reduced mass, $\mathbf{r}$ the distance of m_2 from m_1, and $\mathbf{F}_2{}^i$ that part of the force acting on m_2 caused by m_1.

Therefore, the theory of chapter 2 can be applied to a two-body system in which motion of the attracting center is not negligible. Variation in the direction of **r** describes *rotation* of the system about the center of mass, while variation in its magnitude describes motion of the second particle with respect to the first one. When the relative motion is cyclic, or oscillatory, it is called a *vibration*.

EXAMPLE 3.1   Can a system consisting of more than two particles ever be considered a two-particle system?

When motion of the system involves essentially (a) oscillation of one part with respect to the remaining part and (b) rotation of the system as a whole, it can be treated as a two-body system. The reduced mass for the existing mode of vibration is then given by (3-12), with m_1 and m_2 the masses of the separate units, while vector **r** represents the separation between the centers of mass of these units. When the extent of either unit from its center of mass becomes appreciable with respect to r, the distribution of mass affects the reduced mass for rotation. See sections 3.5, 3.8, and 3.9.

EXAMPLE 3.2   Describe the orbits of the components of a double star.

Assume that the components attract each other as if the mass of each were concentrated at its center. Erect a nonrotating coordinate frame on one of the centers and let radius vector **r** extend to the other center. The force that the star at the origin exerts on the other star is then given by formula (2-52) with

$$K = -GMm = -Gm_1m_2.$$

Assume that the system is in a uniform gravitational field, so (3-11) reduces to (3-13). But the r and ϕ components of (3-13) are equations (2-27) and (2-28) with $m = \mu$. Therefore, formula (2-56) derived from these equations is applicable; the end points of **r** traces out an elliptic path.

Next, draw $\mathbf{r}_1$ and $\mathbf{r}_2$ from the center of mass to the centers of the first and second stars. Then

$$m_1\mathbf{r}_1 + m_2\mathbf{r}_2 = 0$$

and

$$\mathbf{r}_1 = -\frac{m_2}{m_1}\mathbf{r}_2.$$

With **r** equal to $\mathbf{r}_2$ minus $\mathbf{r}_1$, we obtain

$$\mathbf{r} = \mathbf{r}_2 + \frac{m_2}{m_1}\mathbf{r}_2 = \frac{m_1 + m_2}{m_1}\mathbf{r}_2$$

$$= \frac{m_1 + m_2}{m_1}\left(-\frac{m_1}{m_2}\mathbf{r}_1\right) = -\frac{m_1 + m_2}{m_2}\mathbf{r}_1.$$

These equations show that the directions of $\mathbf{r}_2$ and **r** are the same while the magnitude of $\mathbf{r}_2$ is $m_1/(m_1 + m_2)$ times that of **r**. The end point of $\mathbf{r}_2$ traces out an ellipse with parameter c and semimajor axis a equal to $m_1/(m_1 + m_2)$ times the c and a for the end point of $\mathbf{r}$ in its coordinate frame. Similarly, the magnitude of $\mathbf{r}_1$ is $m_2/(m_1 + m_2)$ times that of $\mathbf{r}$, but its direction is that of $-\mathbf{r}$. Its end point traces out an ellipse with an a equal to $m_2/(m_1 + m_2)$ times that for $\mathbf{r}$.

3.3 Momentum and energy integrals for the two particles

Newton's laws impose certain conditions on the momenta and energies of a mechanical system. We can apply these conditions directly in working many problems. Particularly simple are the relationships for two interacting particles.

Let $\mathbf{r}_1$ and $\mathbf{r}_2$ be radius vectors locating the masses m_1 and m_2, respectively, in the reference *inertial* frame. Also, let M be the total mass, $\mathbf{R}$ the radius vector locating the center of mass in the inertial frame, and $\mathbf{v}_1$, $\mathbf{v}_2$, and $\mathbf{V}$ the temporal derivatives $\dot{\mathbf{r}}_1$, $\dot{\mathbf{r}}_2$, and $\dot{\mathbf{R}}$.

The *linear momentum* $\mathbf{p}_j$ of a particle is defined as its mass m_j times its velocity $\mathbf{v}_j$. For the two particles, we write

$$\mathbf{p}_1 = m_1\dot{\mathbf{r}}_1 = m_1\mathbf{v}_1 \quad \text{and} \quad \mathbf{p}_2 = m_2\dot{\mathbf{r}}_2 = m_2\mathbf{v}_2. \tag{3-14}$$

If all the mass were concentrated at the center of mass, the momentum would be

$$\mathbf{P} = (m_1 + m_2)\dot{\mathbf{R}} = M\mathbf{V}. \tag{3-15}$$

Differentiating equation (3-5), which defines vector $\mathbf{R}$ and the center of mass, yields the relationship

$$(m_1 + m_2)\dot{\mathbf{R}} = m_1\dot{\mathbf{r}}_1 + m_2\dot{\mathbf{r}}_2. \tag{3-16}$$

This combines with (3-15) and (3-14) to give

$$\mathbf{P} = \mathbf{p}_1 + \mathbf{p}_2. \tag{3-17}$$

The linear momentum of all the mass concentrated at the center of mass equals the sum of the actual linear momenta of the parts.

Newton's laws, applied to the system of two interacting particles, yield (3-8) or

$$\frac{d\mathbf{P}}{dt} = \mathbf{F} \tag{3-18}$$

which integrates to

$$\Delta\mathbf{P} = \int_{t_1}^{t_2} \mathbf{F}\, dt. \tag{3-19}$$

Whenever no net force acts on the system, momentum $\mathbf{P}$ is constant. Then the change in momentum of the second particle is equal and opposite to any change in the first.

The *kinetic energy* T_j of a particle is defined as one-half its mass times the square of its velocity in an inertial frame. The kinetic energy T of a system is defined as the sum of the kinetic energies of the parts.

$$T = T_1 + T_2 = \tfrac{1}{2}m_1v_1{}^2 + \tfrac{1}{2}m_2v_2{}^2. \tag{3-20}$$

We can construct the differential of this expression by dot multiplying each mass-acceleration product with the corresponding differential displacement:

$$m_1 \frac{d\mathbf{v}_1}{dt} \cdot d\mathbf{r}_1 + m_2 \frac{d\mathbf{v}_2}{dt} \cdot d\mathbf{r}_2 = m_1 \frac{d\mathbf{r}_1}{dt} \cdot d\mathbf{v}_1 + m_2 \frac{d\mathbf{r}_2}{dt} \cdot d\mathbf{v}_2$$

$$= \tfrac{1}{2}m_1 \, d(\mathbf{v}_1 \cdot \mathbf{v}_1) + \tfrac{1}{2}m_2 \, d(\mathbf{v}_2 \cdot \mathbf{v}_2)$$

$$= d(T_1 + T_2) = dT. \tag{3-21}$$

Newton's second law converts (3-21) to an equation

$$\mathbf{F}_1 \cdot d\mathbf{r}_1 + \mathbf{F}_2 \cdot d\mathbf{r}_2 = dT \tag{3-22}$$

involving the forces acting on the first and second particles. When the left side of this equation is an exact differential; that is, when its integral is independent of the path followed, a potential function V can be defined

$$\mathbf{F}_1 \cdot d\mathbf{r}_1 + \mathbf{F}_2 \cdot d\mathbf{r}_2 = -dV, \tag{3-23}$$

and (3-22) becomes

$$-dV = dT \qquad \text{or} \qquad dT + dV = 0. \tag{3-24}$$

Integrating (3-24) yields the conservation law

$$T + V = E = \text{constant} \tag{3-25}$$

in which T is the kinetic energy, V the potential energy, and E (the constant of integration) the total energy.

In a collision between submicroscopic bodies, kinetic energy is converted to potential energy as the particles approach each other and is changed back as they fly apart. But in a similar collision between macroscopic bodies, there is always some frictional loss, conversion of kinetic energy to heat.

EXAMPLE 3.3   A particle of mass m_2 and momentum $\mathbf{p}_2$ strikes a particle of mass m_1 at rest producing a particle of mass m_3 with momentum $\mathbf{p}_3$ and one of mass m_4 with momentum $\mathbf{p}_4$ as in figure 3.2. What formulas govern the collision?

In general, mass is conserved:

$$m_1 + m_2 = m_3 + m_4.$$

If no external forces act on the system, the components of momentum are conserved:

$$p_2 = p_3 \cos \theta_3 + p_4 \cos \theta_4,$$

$$0 = p_3 \sin \theta_3 - p_4 \sin \theta_4.$$

And if there is no net conversion of kinetic energy to other forms, then

$$T_2 = T_3 + T_4$$

or

$$\frac{p_2{}^2}{2m_2} = \frac{p_3{}^2}{2m_3} + \frac{p_4{}^2}{2m_4}.$$

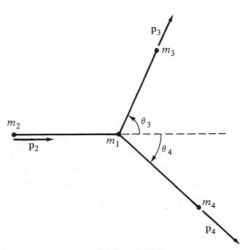

FIGURE 3.2   Conditions before and after collision of mass m_2 with mass m_1.

3.4   Relationships in a nonrotating frame erected on the center of mass

Momentum calculations can often be simplified by formulating the given problem in a reference system whose origin moves with the center of mass and whose Cartesian axes are parallel to those of an inertial system.

As an example, again consider the set of two particles described in figure 3.1. The radius vectors drawn from the center of mass to the first and second particles are the differences

$$\mathbf{r}_1{}^i = \mathbf{r}_1 - \mathbf{R}, \tag{3-26}$$

$$\mathbf{r}_2{}^i = \mathbf{r}_2 - \mathbf{R}. \tag{3-27}$$

These vectors locate the particles in the center-of-mass frame; their components are the coordinates of m_1 and m_2 in this generally noninertial system.

To relate these internal vectors to vector $\mathbf{r}$ drawn from m_1 to m_2, we employ equation (3-5) to eliminate $\mathbf{R}$, (3-9) to introduce $\mathbf{r}$, and (3-12) to simplify the mass factor. We find that

$$\mathbf{r}_1{}^i = -\frac{\mu}{m_1}\,\mathbf{r}, \tag{3-28}$$

and

$$\mathbf{r}_2{}^i = \frac{\mu}{m_2}\,\mathbf{r}, \tag{3-29}$$

whence

$$\mathbf{v}_1{}^i = \dot{\mathbf{r}}_1{}^i = -\frac{\mu}{m_1}\,\dot{\mathbf{r}} = -\frac{\mu}{m_1}\,\mathbf{v}, \tag{3-30}$$

and

$$\mathbf{v}_2{}^i = \frac{\mu}{m_2}\,\mathbf{v}. \tag{3-31}$$

The mass m_j of a particle times its velocity $\mathbf{v}_j{}^i$ in the center-of-mass system is called the momentum $\mathbf{p}_j{}^i$ of the particle in the system. Now, equations (3-30) and (3-31) allow us to convert $\mathbf{p}_1{}^i$ to $-\mathbf{p}_2{}^i$:

$$\mathbf{p}_1{}^i = m_1\left(-\frac{\mu}{m_1}\,\mathbf{v}\right) = -\mu\mathbf{v} = -m_2\left(\frac{\mu}{m_2}\,\mathbf{v}\right) = -\mathbf{p}_2{}^i. \tag{3-32}$$

The two internal momenta are equal in magnitude but opposite in direction.

The sum $\mathbf{p}_1{}^i + \mathbf{p}_2{}^i$ is zero even when external forces $\mathbf{F}_1{}^e$ and $\mathbf{F}_2{}^e$ act and the center of mass is accelerated. Momentum is always conserved in the center-of-mass frame.

Movements may be usefully referred to (a) the frame in which momenta add to zero, (b) the frame in which the particle considered to be the source of the interaction field remains at rest, and (c) the frame of the laboratory, which is an approximate inertial frame.

An elastic scattering of one particle by another is depicted in the center-of-mass frame by figure 3.3. Since the internal momenta vectors have to be directly opposed before and after interaction has occurred, angles $\theta_1{}^i$ and $\theta_2{}^i$ add up to 180°. If the process were viewed from the frame in which m_1 remains at rest at the origin, equation (3-13) would determine acceleration of the radius vector locating m_2. Observing the

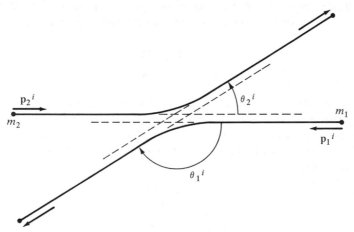

FIGURE 3.3   Trajectories of two interacting particles plotted in the center-of-mass reference frame.

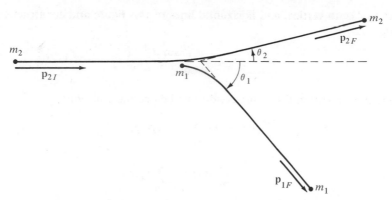

FIGURE 3.4 Two-body collision as seen from the inertial frame in which m_1 is initially at rest.

process from the inertial frame in which m_1 is initially at rest produces the results plotted in figure 3.4.

In section 2.8, the source of the field was assumed to be at rest. Now we see that the validity of (3-13) justifies this assumption even when the target nucleus acquires appreciable momentum in an inertial frame. However, the angle θ obtained for Rutherford scattering must be interpreted as the deflection of relative velocity $\dot{\mathbf{r}} = \mathbf{v}$. Since $\mathbf{v}_2{}^i$ is proportional to $\mathbf{v}$, by equation (3-31), the deflection of $\mathbf{v}_2{}^i$ is also equal to θ:

$$\theta_2{}^i = \theta. \tag{3-33}$$

EXAMPLE 3.4 Relate the deflection of the velocity of a moving particle striking a particle at rest to its deflection in the center-of-mass system.

Solve (3-27) for $\mathbf{r}_2$:

$$\mathbf{r}_2 = \mathbf{r}_2{}^i + \mathbf{R},$$

and differentiate with respect to time:

$$\mathbf{v}_2 = \mathbf{v}_2{}^i + \mathbf{V}.$$

Since the laboratory system is an approximate inertial frame, the final velocity of particle 2 in this system is related to its final velocity in the center-of-mass system as figure 3.5 shows.

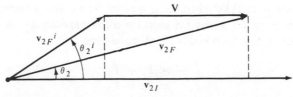

FIGURE 3.5 Pertinent velocity vectors from the center-of-mass and laboratory coordinate systems.

Project $v_{2F}{}^i$ onto vertical and horizontal lines of this figure and construct the ratio for $\tan \theta_2$:

$$\tan \theta_2 = \frac{v_{2F}{}^i \sin \theta_2{}^i}{v_{2F}{}^i \cos \theta_2{}^i + V}.$$

Then introduce equation (3-16) for conditions before the collision:

$$(m_1 + m_2)\mathbf{V} = m_2 \mathbf{v}_{2I}$$

or

$$\mathbf{V} = \frac{\mu}{m_1} \mathbf{v}_{2I} = \frac{\mu}{m_1} \mathbf{v}_I,$$

and (3-31) for conditions after the collision:

$$\mathbf{v}_{2F}{}^i = \frac{\mu}{m_2} \mathbf{v}_F.$$

Substitute the corresponding scalar expressions into the equation for $\tan \theta_2$ and reduce:

$$\tan \theta_2 = \frac{\sin \theta_2{}^i}{\cos \theta_2{}^i + (m_2 v_I / m_1 v_F)}.$$

If the kinetic energy is conserved, that is, if the collision is elastic, the initial and final relative speeds, v_I and v_F, are the same and we obtain

$$\tan \theta_2 = \frac{\sin \theta_2{}^i}{\cos \theta_2{}^i + (m_2 / m_1)}.$$

3.5   Equations of motion for N interacting particles

Let us now consider a set of N particles interacting with each other and possibly with some imposed field. The N simultaneous Newtonian equations for the motion will be combined to form one equation regulating translation of the center of mass and one regulating rotation about an appropriate point. Vibratory motions will be treated in later chapters.

Let m_j be the mass of the jth particle while $\mathbf{F}_j$ is the force acting on it at time t. A reference point is established in some inertial frame and a vector $\mathbf{r}_j$ is drawn from this point to the particle. Index j is used for all N particles in the set.

Newton's second law then yields the N equations

$$
\begin{aligned}
m_1 \ddot{\mathbf{r}}_1 &= \mathbf{F}_1, \\
m_2 \ddot{\mathbf{r}}_2 &= \mathbf{F}_2, \\
&\vdots \\
m_N \ddot{\mathbf{r}}_N &= \mathbf{F}_N.
\end{aligned}
\tag{3-34}
$$

Let us add these equations, employing the summation convention and symbol $\mathbf{F}$ for the net force. Then we obtain the equation

$$m_j \ddot{\mathbf{r}}_j = \mathbf{F} \tag{3-35}$$

in which

$$\mathbf{F} = \mathbf{F}_1 + \mathbf{F}_2 + \cdots + \mathbf{F}_N. \tag{3-36}$$

Adding the linear momenta of all the particles in the set gives us the total momentum

$$m_j \dot{\mathbf{r}}_j = \mathbf{P}. \tag{3-37}$$

Again note how repetition of the index is used to indicate the pertinent summation. Differentiating this equation,

$$m_j \ddot{\mathbf{r}}_j = \frac{d\mathbf{P}}{dt}, \tag{3-38}$$

and combining the result with (3-35) yields

$$\frac{d\mathbf{P}}{dt} = \mathbf{F}. \tag{3-39}$$

Let us define the *center of total mass M* as the point located by the radius vector $\mathbf{R}$ for which

$$M\mathbf{R} = m_j \mathbf{r}_j. \tag{3-40}$$

Then the sum in (3-37) is the derivative of $M\mathbf{R}$, and

$$M\dot{\mathbf{R}} = \mathbf{P}. \tag{3-41}$$

Equation (3-39) also becomes

$$M\ddot{\mathbf{R}} = \mathbf{F}. \tag{3-42}$$

Note that this relationship has the same form as (2-2) while (3-41) is like (2-95). As a consequence, the net force $\mathbf{F}$ accelerates the center of mass as it would a single particle of mass M at the end point of $\mathbf{R}$. We are already familiar with such motion.

To treat movement around a point in an inertial frame, we first choose the point as origin for the radius vectors. Then we add the angular momenta given by formula (2-107) to get the total angular momentum:

$$\mathbf{L} = \mathbf{r}_j \times \mathbf{p}_j = \mathbf{r}_j \times m_j \dot{\mathbf{r}}_j = m_j \mathbf{r}_j \times \dot{\mathbf{r}}_j. \tag{3-43}$$

Differentiation of (3-43) yields

$$\frac{d\mathbf{L}}{dt} = \mathbf{r}_j \times m_j \ddot{\mathbf{r}}_j + \dot{\mathbf{r}}_j \times m_j \dot{\mathbf{r}}_j. \tag{3-44}$$

The last sum vanishes because the cross product of parallel vectors is zero:

$$\frac{d\mathbf{L}}{dt} = \mathbf{r}_j \times m_j \ddot{\mathbf{r}}_j. \tag{3-45}$$

Newton's second law (3-34) converts (3-45) to

$$\frac{d\mathbf{L}}{dt} = \mathbf{r}_j \times \mathbf{F}_j. \tag{3-46}$$

Torque is defined as the cross product of lever arm with the pertinent force. The right side of (3-46) is therefore equal to the total torque $\mathbf{N}$ and

$$\frac{d\mathbf{L}}{dt} = \mathbf{N}. \tag{3-47}$$

The rate of change of angular momentum equals the torque. The angular momentum and the torque may be calculated about any point in an inertial frame.

Now, the angular momentum that the body would have if all its mass were at the center of mass is

$$\mathbf{L}^e = \mathbf{R} \times \mathbf{P} = \mathbf{R} \times M\dot{\mathbf{R}}. \tag{3-48}$$

The derivative of this is

$$\frac{d\mathbf{L}^e}{dt} = \mathbf{R} \times M\ddot{\mathbf{R}} + \dot{\mathbf{R}} \times M\dot{\mathbf{R}} = \mathbf{R} \times M\ddot{\mathbf{R}}. \tag{3-49}$$

Introducing (3-42) converts (3-49) to

$$\frac{d\mathbf{L}^e}{dt} = \mathbf{R} \times \mathbf{F} = \mathbf{N}^e \tag{3-50}$$

where $\mathbf{N}^e$ is the torque that the net force would exert if it acted at the center of mass.

EXAMPLE 3.5   Prove that in a uniform gravitational field, the center of gravity coincides with the center of mass.

The *center of gravity* is the point at which the total mass seems to be located in a given gravitational field. If all the mass were concentrated at this point, the net gravitational force and torque would not be affected.

Let the acceleration due to gravity be $\mathbf{g}_j$ at the jth particle. The total gravitational force is then

$$\mathbf{F} = m_j\mathbf{g}_j.$$

If $\mathbf{g}_j$ is constant and equal to $\mathbf{g}$, it factors out of the sum and

$$\mathbf{F} = M\mathbf{g}$$

where M is the total mass. The same formula would be obtained if all the mass M were at the center of mass.

The gravitational torque about any point chosen as the origin is

$$\mathbf{N} = \mathbf{r}_j \times \mathbf{F}_j = \mathbf{r}_j \times m_j\mathbf{g}_j = m_j\mathbf{r}_j \times \mathbf{g}_j.$$

When $\mathbf{g}_j$ equals the constant $\mathbf{g}$, it factors out of the sum. Then formula (3-40) for the center of mass can be introduced:

$$\mathbf{N} = m_j\mathbf{r}_j \times \mathbf{g} = M\mathbf{R} \times \mathbf{g} = \mathbf{R} \times M\mathbf{g}.$$

The gravitational field would exert the same torque on mass M located at the end point of $\mathbf{R}$.

Thus the mass does appear to be concentrated at the center of mass. Note how this result depends on the constancy of $\mathbf{g}_j$.

3.6 Rotation about the center of mass

When the net force acting on a physical unit is zero, its center of mass does not accelerate and the nonrotating frame based on the center is an inertial frame. Then (3-47) governs rotation about this center. But what happens when $\mathbf{F}$ in (3-42) is not zero?

Let $\mathbf{r}_j$ and $\mathbf{R}$ be vectors drawn from the origin of some inertial frame, to the jth particle and to the center of mass, respectively. The vector locating the jth particle in the pertinent frame based on the center of mass is

$$\mathbf{r}_j{}^i = \mathbf{r}_j - \mathbf{R} \tag{3-51}$$

as in section 3.4. And the transformation equation for any vector $\mathbf{r}_j$ is

$$\mathbf{r}_j = \mathbf{R} + \mathbf{r}_j{}^i. \tag{3-52}$$

Multiplying (3-51) by m_j, summing over all particles in the set, and introducing equation (3-40) yields

$$m_j\mathbf{r}_j{}^i = m_j\mathbf{r}_j - M\mathbf{R} = 0. \tag{3-53}$$

Differentiating with respect to time then leads to

$$m_j\dot{\mathbf{r}}_j{}^i = m_j\dot{\mathbf{r}}_j - M\dot{\mathbf{R}} = 0. \tag{3-54}$$

Note how the internal momenta add to zero.

Let us now transform $\mathbf{r}_j$ and $\dot{\mathbf{r}}_j$ in the angular momentum formula (3-43) with (3-52). Introduce M for the total mass, 0 for $m_j\mathbf{r}_j{}^i$ and $m_j\dot{\mathbf{r}}_j{}^i$ as in (3-53) and (3-54), and $\mathbf{p}_j{}^i$ for the internal momentum of the jth particle:

$$\begin{aligned}
\mathbf{L} &= m_j(\mathbf{R} + \mathbf{r}_j{}^i) \times (\dot{\mathbf{R}} + \dot{\mathbf{r}}_j{}^i) \\
&= M\mathbf{R} \times \dot{\mathbf{R}} + (m_j\mathbf{r}_j{}^i) \times \dot{\mathbf{R}} + \mathbf{R} \times (m_j\dot{\mathbf{r}}_j{}^i) + m_j\mathbf{r}_j{}^i \times \dot{\mathbf{r}}_j{}^i \\
&= \mathbf{R} \times M\dot{\mathbf{R}} + 0 + 0 + \mathbf{r}_j{}^i \times m_j\dot{\mathbf{r}}_j{}^i \\
&= \mathbf{R} \times \mathbf{P} + \mathbf{r}_j{}^i \times \mathbf{p}_j{}^i = \mathbf{L}^e + \mathbf{L}^i.
\end{aligned} \tag{3-55}$$

In (3-55), $\mathbf{R} \times \mathbf{P}$ is the angular momentum $\mathbf{L}^e$ that the body would have if all its mass were at the center of mass while $\mathbf{r}_j{}^i \times \mathbf{p}_j{}^i$ is the angular momentum of the body about its center of mass, $\mathbf{L}^i$.

Formula (3-52) also alters the formula for torque:

$$\mathbf{N} = \mathbf{r}_j \times \mathbf{F}_j = (\mathbf{R} + \mathbf{r}_j{}^i) \times \mathbf{F}_j$$
$$= \mathbf{R} \times \mathbf{F} + \mathbf{r}_j{}^i \times \mathbf{F}_j = \mathbf{N}^e + \mathbf{N}^i. \tag{3-56}$$

As before, $\mathbf{R} \times \mathbf{F}$ is the torque $\mathbf{N}^e$ that the sum of the forces would exert around the initial reference point if the sum acted on the center of mass, while $\mathbf{r}_j{}^i \times \mathbf{F}_j$ is the torque $\mathbf{N}^i$ that the actual forces exert around the center of mass.

Substituting (3-55) and (3-56) into equation (3-47),

$$\frac{d\mathbf{L}^e}{dt} + \frac{d\mathbf{L}^i}{dt} = \mathbf{N}^e + \mathbf{N}^i, \tag{3-57}$$

and subtracting (3-50) yields

$$\frac{d\mathbf{L}^i}{dt} = \mathbf{N}^i. \tag{3-58}$$

Thus the rate of change of angular momentum about the center of mass equals the torque acting about this center. The relationship is the same as it would be if the nonrotating frame erected on the center of mass were an inertial frame.

3.7   Why linear momentum, angular momentum, and energy vary

Equations derived from Newton's laws tell us that certain properties change only when the system interacts appropriately with the surroundings. Variation in such a property therefore indicates a certain kind of interaction.

According to equation (3-39),

$$\frac{d\mathbf{P}}{dt} = \mathbf{F}, \tag{3-59}$$

the time rate of change in net linear momentum $\mathbf{P}$ equals the net force $\mathbf{F}$ acting on the system. Integrating this equation from time t_1 to time t_2 yields

$$\Delta\mathbf{P} = \int_{t_1}^{t_2} \mathbf{F}\, dt. \tag{3-60}$$

Whenever the integral on the right is zero, as when no net force acts on the particles, the momentum associated with translation of the system as a whole is conserved:

$$\Delta\mathbf{P} = 0. \tag{3-61}$$

In a similar way, equation (3-58)

$$\frac{d\mathbf{L}^i}{dt} = \mathbf{N}^i \tag{3-62}$$

states that the time rate of change in angular momentum $\mathbf{L}^i$ equals the net torque $\mathbf{N}^i$ acting about the center of mass. Integrating this equation from time t_1 to time t_2 yields

$$\Delta \mathbf{L}^i = \int_{t_1}^{t_2} \mathbf{N}^i \, dt. \tag{3-63}$$

When the forces produce no torque around the center of mass, the angular momentum is conserved:

$$\Delta \mathbf{L}^i = 0. \tag{3-64}$$

Dot multiplying each Newtonian equation (3-34) by the distance its particle moves in time dt,

$$m_1 \frac{d\dot{\mathbf{r}}_1}{dt} \cdot d\mathbf{r}_1 = \mathbf{F}_1 \cdot d\mathbf{r}_1,$$

$$\vdots$$

$$m_N \frac{d\dot{\mathbf{r}}_N}{dt} \cdot d\mathbf{r}_N = \mathbf{F}_N \cdot d\mathbf{r}_N, \tag{3-65}$$

commuting $d\dot{\mathbf{r}}_j$ with $d\mathbf{r}_j$ on the left sides,

$$m_1 \dot{\mathbf{r}}_1 \cdot d\dot{\mathbf{r}}_1 = \mathbf{F}_1 \cdot d\mathbf{r}_1,$$

$$\vdots$$

$$m_N \dot{\mathbf{r}}_N \cdot d\dot{\mathbf{r}}_N = \mathbf{F}_N \cdot d\mathbf{r}_N, \tag{3-66}$$

and adding leads to

$$m_j \dot{\mathbf{r}}_j \cdot d\dot{\mathbf{r}}_j = \mathbf{F}_j \cdot d\mathbf{r}_j. \tag{3-67}$$

The integral of this equation is

$$\Delta(\tfrac{1}{2} m_j \dot{\mathbf{r}}_j \cdot \dot{\mathbf{r}}_j) = \int_{t=t_1}^{t=t_2} \mathbf{F}_j \cdot d\mathbf{r}_j. \tag{3-68}$$

When the expression on the right of (3-68) is independent of how the particles move between the initial and final configurations, it defines the change in a function $V(\mathbf{r}_1, \ldots, \mathbf{r}_N)$. Letting the kinetic energy $\tfrac{1}{2} m_j \dot{\mathbf{r}}_j \cdot \dot{\mathbf{r}}_j$ be T, we then have

$$\Delta T = -\Delta V, \tag{3-69}$$

whence

$$T + V = E = \text{constant.} \tag{3-70}$$

Since any increase in T comes at the expense of V, V is called the *potential energy* of the system.

The forces in $-dV$ are called *conservative forces* ($T + V$ is conserved) or *potential forces* (forces derivable from a potential V). In general, frictional forces, which dissipate mechanical energy, are also present. Furthermore, forces that act normal to the direction of motion, such as magnetic forces, may be present. These will be considered further in chapter 5.

EXAMPLE 3.6 Show that the kinetic energy associated with relative motion and rotation about the center of mass in a set of particles or mass elements is independent of how the center of mass moves.

Into formula $\frac{1}{2}m_j\dot{\mathbf{r}}_j \cdot \dot{\mathbf{r}}_j$ for the kinetic energy, introduce relationship (3-52) for the radius vector $\mathbf{r}_j$, M for the total mass, and equation (3-54) for the total internal momentum:

$$
\begin{aligned}
T &= \tfrac{1}{2}m_j(\dot{\mathbf{r}}_j \cdot \dot{\mathbf{r}}_j) = \tfrac{1}{2}m_j(\dot{\mathbf{R}} + \dot{\mathbf{r}}_j{}^i) \cdot (\dot{\mathbf{R}} + \dot{\mathbf{r}}_j{}^i) \\
&= \tfrac{1}{2}M\dot{\mathbf{R}} \cdot \dot{\mathbf{R}} + \dot{\mathbf{R}} \cdot m_j\dot{\mathbf{r}}_j{}^i + \tfrac{1}{2}m_j\dot{\mathbf{r}}_j{}^i \cdot \dot{\mathbf{r}}_j{}^i \\
&= \tfrac{1}{2}M\dot{\mathbf{R}} \cdot \dot{\mathbf{R}} + \tfrac{1}{2}m_j\dot{\mathbf{r}}_j{}^i \cdot \dot{\mathbf{r}}_j{}^i.
\end{aligned}
$$

In the final form, the first term is the kinetic energy the total mass would have if it moved at the velocity of the center of mass while the sum includes the rotational energy about the center and the kinetic energy associated with relative motion of the particles. Since the sum of cross terms is zero, from (3-54), the total kinetic energy minus the translational energy of the system as a whole is not affected by motion of the center of mass. The nonrotating frame based on this center is like an inertial frame insofar as the rotational and relative energies are concerned.

3.8 Angular momentum as a function of angular velocity

The general equations governing movement about the center of mass become quite complicated when the system contains more than a few particles. The formulas do simplify, however, when all interparticle distances can be replaced by their averages and the system is effectively *rigid*.

In principle, one can observe such a system from the nonrotating frame based on the center of mass. During any small interval of time, then, the body turns about an axis passing through this center. Transforming to an inertial frame would shift the momentary axis of rotation to a parallel line.

In either frame, a typical particle momentarily follows the arc of a circle about a certain axis of rotation. The position of the particle is located by a radius vector $\mathbf{r}$ drawn from a point that is common to successive positions of the axis. See figure 3.6.

Let ω be defined as the *angular velocity* at which the particle moves around the axis, so $\omega\, dt$ equals the angle swept out in time dt. Multiplying the radians in $\omega\, dt$ by the lever arm $r \sin \theta$ gives the distance that the particle moves:

$$ds = \omega r \sin \theta\, dt. \tag{3-71}$$

Let us construct a pseudo vector of magnitude ω lying along the axis and label it $\boldsymbol{\omega}$ as shown. Then on comparing the coefficient of dt in (3-71) with formula (1-9), we see that the coefficient equals the magnitude of the cross product of $\boldsymbol{\omega}$ with $\mathbf{r}$.

Being infinitesimal, the arc ds is practically a straight line. Furthermore, it is traversed in the direction of the product $\boldsymbol{\omega} \times \mathbf{r}$. Therefore, we can rewrite (3-71) in the vector form

$$d\mathbf{s} = \boldsymbol{\omega} \times \mathbf{r}\, dt. \tag{3-72}$$

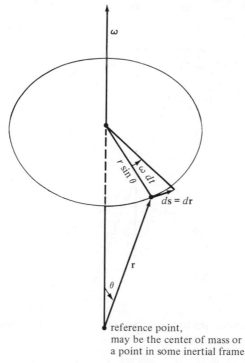

reference point,
may be the center of mass or
a point in some inertial frame

FIGURE 3.6   Vectors for describing an infinitesimal rotation. Both the magnitude and direction of ω may vary with time.

For the jth particle, $d\mathbf{s}$ is $d\mathbf{r}_j$ and $\mathbf{r}$ is $\mathbf{r}_j$. Then

$$\dot{\mathbf{r}}_j = \frac{d\mathbf{r}_j}{dt} = \boldsymbol{\omega} \times \mathbf{r}_j. \tag{3-73}$$

Equation (3-73) combines with (3-43) to produce the angular-momentum formula

$$\mathbf{L} = m_j \mathbf{r}_j \times \dot{\mathbf{r}}_j = m_j \mathbf{r}_j \times (\boldsymbol{\omega} \times \mathbf{r}_j) \tag{3-74}$$

for a rotating system that is effectively rigid. Note that the angular momentum $\mathbf{L}$ and the angular velocity $\boldsymbol{\omega}$ need not point in the same direction.

When all parts of a system move at the same angular velocity, the kinetic energy formula also becomes simple. Into the conventional kinetic energy expression, we introduce relationship (3-73) for the first $\dot{\mathbf{r}}_j$. Then we rearrange the resulting triple scalar product and use (3-74) to identify $\mathbf{L}$:

$$T = \tfrac{1}{2}m_j\dot{\mathbf{r}}_j \cdot \dot{\mathbf{r}}_j = \tfrac{1}{2}m_j\boldsymbol{\omega} \times \mathbf{r}_j \cdot \dot{\mathbf{r}}_j = \tfrac{1}{2}m_j\mathbf{r}_j \times \dot{\mathbf{r}}_j \cdot \boldsymbol{\omega}$$
$$= \tfrac{1}{2}\mathbf{L} \cdot \boldsymbol{\omega}. \tag{3-75}$$

Thus, the rotational energy of an effectively rigid body equals one-half the scalar product of its angular momentum and angular velocity vectors.

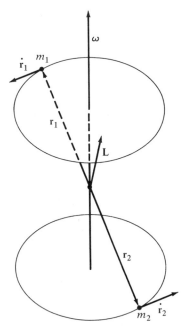

FIGURE 3.7   Particles constrained to move 180° out of phase along parallel circles.

EXAMPLE 3.7   Discuss the angular momentum of two point masses traveling 180° out of phase along parallel circles in some inertial frame.

The masses m_1 and m_2 are forced to move along the parallel circles so that m_1 is always 180° ahead of m_2. Consequently, the straight line joining the two masses intersects the axis of rotation. Let this intersection be the reference point and draw $\mathbf{r}_1$ and $\mathbf{r}_2$ to m_1 and m_2 as figure 3.7 shows.

To the angular momentum $\mathbf{L}$ about this point, the motions of m_1 and m_2 contribute $m_1\mathbf{r}_1 \times \dot{\mathbf{r}}_1$ and $m_2\mathbf{r}_2 \times \dot{\mathbf{r}}_2$, respectively. But the latter equals $m_2\mathbf{r}_1 \times \dot{\mathbf{r}}_1$ because $\mathbf{r}_2$ is $-\mathbf{r}_1$ and $\dot{\mathbf{r}}_2$ is $-\dot{\mathbf{r}}_1$. The resultant $\mathbf{L}$ is perpendicular to $\mathbf{r}_1$ and $\dot{\mathbf{r}}_1$ as shown. Whenever the two circles are distinct, ω is not perpendicular to $\mathbf{r}_1$ and $\mathbf{L}$ does not lie along ω.

The motions cause $\mathbf{r}_1$, $\dot{\mathbf{r}}_1$, and $\mathbf{L}$ to vary with time. ($\mathbf{L}$ is said to *precess* around ω.) So $d\mathbf{L}/dt$ is not zero and torque has to be applied by whatever bearings are constraining the system. The bearings may transmit the forces to an essentially massless crankshaft supporting the masses.

3.9   Moment-of-inertia matrix

Relationship (3-74) between angular momentum and angular velocity is not the simple proportionality we see in

$$\mathbf{p} = m\mathbf{v}. \tag{3-76}$$

However, the components of **L** and ω are linear functions of each other—as the components of unprimed and primed vectors are in section 1.9.

Indeed, let us expand the right side of (3-74) by the BAC CAB rule,

$$\mathbf{L} = m_j(\omega \mathbf{r}_j \cdot \mathbf{r}_j - \mathbf{r}_j \mathbf{r}_j \cdot \omega), \tag{3-77}$$

and use the result to write each component of **L**:

$$L_x = m_j[\omega_x(x_j{}^2 + y_j{}^2 + z_j{}^2) - x_jx_j\omega_x - x_jy_j\omega_y - x_jz_j\omega_z]$$
$$= m_j(y_j{}^2 + z_j{}^2)\omega_x - m_jx_jy_j\omega_y - m_jx_jz_j\omega_z, \tag{3-78}$$

$$L_y = -m_jy_jx_j\omega_x + m_j(z_j{}^2 + x_j{}^2)\omega_y - m_jy_jz_j\omega_z, \tag{3-79}$$

$$L_z = -m_jz_jx_j\omega_x - m_jz_jy_j\omega_y + m_j(x_j{}^2 + y_j{}^2)\omega_z. \tag{3-80}$$

Equations (3-78) through (3-80) transform the components of ω into those of **L**. We formally combine the equations into a single one,

$$\mathbf{L} = \mathbf{I}\omega, \tag{3-81}$$

in which the entity **I** changes vector ω to vector **L**. Or in more detail, we write

$$L_k = I_{kl}\omega_l, \tag{3-82}$$

where

$$I_{kl} = m_j[\delta_{kl}r_j{}^2 - (x_k)_j(x_l)_j] \tag{3-83}$$

and

$$(x_1)_j = x_j,$$
$$(x_2)_j = y_j,$$
$$(x_3)_j = z_j, \tag{3-84}$$
$$r_j{}^2 = x_j{}^2 + y_j{}^2 + z_j{}^2. \tag{3-85}$$

Quantity I_{kl} is called the *moment of inertia* about the *l*th axis when $k = l$, and a negative *product of inertia* when $k \neq l$. The nine elements of **I** may be arranged in a 3×3 array

$$\begin{pmatrix} I_{11} & I_{12} & I_{13} \\ I_{21} & I_{22} & I_{23} \\ I_{31} & I_{32} & I_{33} \end{pmatrix} \tag{3-86}$$

called the moment-of-inertia matrix. The multiplication in (3-81) is the matrix multiplication to be treated in the next chapter.

According to formulas (3-78), (3-79), (3-80), the moment of inertia about each axis equals the sum of the product of each particle mass times the square of its distance from the axis. To derive this relationship from (3-83) by vector algebra, proceed as follows.

Let **n** be a unit vector lying along the given axis a, as in figure 3.8. Furthermore, let the a axis be the 1 axis, so k and l are both 1 in (3-83), and $(x_1)_j$ is the projection of $\mathbf{r}_j$ on **n**, that is, $\mathbf{r}_j \cdot \mathbf{n}$. Then equation (3-83) yields

$$I_a = m_j[r_j{}^2 - (\mathbf{r}_j \cdot \mathbf{n})^2] = m_j\mathbf{r}_j \cdot [\mathbf{r}_j(\mathbf{n} \cdot \mathbf{n}) - \mathbf{n}(\mathbf{r}_j \cdot \mathbf{n})] \tag{3-87}$$

for the moment of inertia about the a axis.

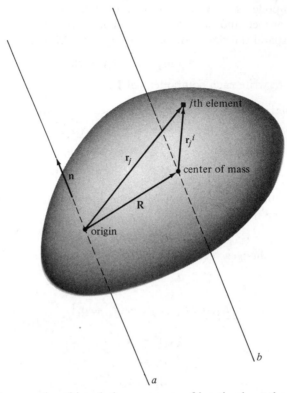

FIGURE 3.8   Vectors employed in relating moments of inertia about the *a* and *b* axes.

Applying the BAC CAB rule to the right side of (3-87),

$$I_a = m_j \mathbf{r}_j \cdot \mathbf{n} \times (\mathbf{r}_j \times \mathbf{n}), \tag{3-88}$$

and interchanging the dot and cross in the triple scalar product gives us

$$I_a = m_j (\mathbf{r}_j \times \mathbf{n}) \cdot (\mathbf{r}_j \times \mathbf{n}) = m_j (\mathbf{r}_j \times \mathbf{n})^2. \tag{3-89}$$

Let us now relate I_a to the moment about the parallel axis through the center of mass. For this purpose, we break $\mathbf{r}_j$ down into the vector $\mathbf{R}$ locating the center of mass and vector $\mathbf{r}_j^i$ locating m_j with respect to the center

$$\mathbf{r}_j = \mathbf{R} + \mathbf{r}_j^i = \mathbf{r}_j^i + \mathbf{R} \tag{3-90}$$

as in (3-52). Equation (3-89) can then be rewritten in the form

$$\begin{aligned} I_a &= m_j [(\mathbf{r}_j^i + \mathbf{R}) \times \mathbf{n}]^2 \\ &= m_j (\mathbf{r}_j^i \times \mathbf{n})^2 + M(\mathbf{R} \times \mathbf{n})^2 + 2 m_j (\mathbf{r}_j^i \times \mathbf{n}) \cdot (\mathbf{R} \times \mathbf{n}) \end{aligned} \tag{3-91}$$

where M is the total mass.

In the last expression, the sum $m_j r_j{}^i$ appears as a factor. Introducing relationship (3-53) for this sum yields

$$2(m_j r_j{}^i \times \mathbf{n}) \cdot (\mathbf{R} \times \mathbf{n}) = 0. \tag{3-92}$$

Since first sum on the right of (3-91) has the same form as the right side of (3-89) when $\mathbf{r}_j$ is $\mathbf{r}_j{}^i$, it is the moment about the b axis:

$$I_b = m_j (\mathbf{r}_j{}^i \times \mathbf{n})^2. \tag{3-93}$$

Equations (3-92) and (3-93) reduce (3-91) to

$$I_a = I_b + M(\mathbf{R} \times \mathbf{n})^2. \tag{3-94}$$

Hence the moment of inertia about any axis equals that about the parallel axis through the mass center plus the moment that would exist if all the mass of the body were concentrated at the mass center.

Moments and products of inertia may be calculated with respect to a set of axes fixed in the body. Rotation of the axes generally alters the elements of **I**. The behavior in such transformations will be considered in the next chapter.

EXAMPLE 3.8   Discuss how elements of **I** are calculated for a body.

In studying the mechanical behavior of a body, we replace the summation over particles with integration. This replacement is valid because the molecules, atoms, or ions making up the system are so small and close together. The system behaves as if the matter were continuously distributed at density ρ.

From equations (3-83), we obtain the formula

$$I_{kl} = \int \rho(\delta_{kl} r^2 - x_k x_l)\, d\tau$$

in which $d\tau$ is the element of volume.

EXAMPLE 3.9   Find the equation of motion for a compound pendulum swinging in a single plane.

Let the axis of rotation be the x axis with the origin in the vertical plane in which the center of mass moves. Let the z axis point straight down and the y axis point horizontally in the plane of motion. Choose the 1, 2, 3 axes in the pendulum so they coincide with the x, y, z axes when the pendulum is at stable equilibrium.

Since the rotation is in the yz plane, ω_y and ω_z remain zero and equation (3-82) yields

$$L_1 = I_{11}\omega = I_{11}\frac{d\theta}{dt}.$$

As long as the gravitational field is uniform over the pendulum, the center of gravity coincides with the center of mass, from example 3.5. Then the torque N_1 is

$$- Mgl \sin \theta$$

where l is the distance from the axis of rotation to the center of mass, M the total mass of the pendulum, and θ the angular displacement of the pendulum, measured from the z axis.

Setting the rate of change of angular momentum equal to the torque, as in (3-47), leads to

$$I_{11} \frac{d^2\theta}{dt^2} = - Mgl \sin \theta.$$

3.10   Quadratic relationships governing rotation

When no torque acts about the center of mass of a rigid body, the end of the angular velocity moves along the intersection of two ellipsoids fixed in the body. One ellipsoid comes from the form for the angular momentum squared; the other from the kinetic energy expression.

In deriving these ellipsoids, we first erect nonrotating x, y, z axes based on the center of mass as in figure 3.9. Then we note that the **I** array is symmetric; turning the axes to eliminate I_{12}, I_{13}, I_{23} also makes I_{21}, I_{31}, I_{32} zero. (That such a transformation is possible will be proved in chapter 4. Here we are concerned with its consequences.)

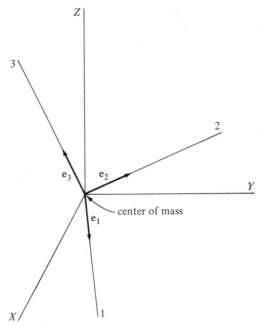

FIGURE 3.9   Instantaneous orientation of the principal axes 1, 2, and 3 with respect to nonrotating x, y, and z axes drawn from the center of mass.

At a given instant of time, the requisite axes 1, 2, and 3 may appear as in figure 3.9. About such axes, equations (3-78) through (3-80) reduce to

$$L_1 = m_j[(x_2)_j{}^2 + (x_3)_j{}^2]\omega_1 = I_1\omega_1, \tag{3-95}$$

$$L_2 = m_j[(x_3)_j{}^2 + (x_1)_j{}^2]\omega_2 = I_2\omega_2, \tag{3-96}$$

$$L_3 = m_j[(x_1)_j{}^2 + (x_2)_j{}^2]\omega_3 = I_3\omega_3. \tag{3-97}$$

The angular momentum about the mass center has the form

$$\mathbf{L}^i = L_k\mathbf{e}_k = I_k\omega_k\mathbf{e}_k \tag{3-98}$$

in which I_1, I_2, and I_3 are called the *principal* moments of inertia.

Applying the Pythagorean theorem to the components of the angular momentum yields

$$(L^i)^2 = I_1{}^2\omega_1{}^2 + I_2{}^2\omega_2{}^2 + I_3{}^2\omega_3{}^2 \tag{3-99}$$

or

$$\frac{\omega_1{}^2}{(L^i)^2/I_1{}^2} + \frac{\omega_2{}^2}{(L^i)^2/I_2{}^2} + \frac{\omega_3{}^2}{(L^i)^2/I_3{}^2} = 1. \tag{3-100}$$

According to (3-64), $\mathbf{L}^i$ is constant when no torque acts about the mass center (here the origin). Then each denominator in (3-100) is the square of a constant and ω runs from the origin to the ellipsoidal surface whose principal axes are $2L^i/I_k$ long.

From Example 3.6, the kinetic energy of a rigid body separates into a translational part and a rotational part. Employing formula (3-75) for the latter and inserting the components of $\mathbf{L}^i$ given by (3-95), (3-96), (3-97) yields

$$T^i = \tfrac{1}{2}\mathbf{L}^i \cdot \boldsymbol{\omega} = \tfrac{1}{2}L_k\omega_k = \tfrac{1}{2}(I_k\omega_k)\omega_k$$
$$= \tfrac{1}{2}I_1\omega_1{}^2 + \tfrac{1}{2}I_2\omega_2{}^2 + \tfrac{1}{2}I_3\omega_3{}^2 \tag{3-101}$$

whence

$$\frac{\omega_1{}^2}{2T^i/I_1} + \frac{\omega_2{}^2}{2T^i/I_2} + \frac{\omega_2{}^2}{2T^i/I_3} = 1. \tag{3-102}$$

Whenever no torque N^i acts, the rotational kinetic energy T^i is constant and equation (3-102) implies that point $(\omega_1, \omega_2, \omega_3)$ moves on an ellipsoidal surface with principal axes $2(2T^i/I_k)^{1/2}$ long. Since the end point of ω also moves on ellipsoid (3-100), it travels on the intersection of the two surfaces.

Angular momentum can be removed from or added to a given rigid body without altering the rotational energy, as long as both (3-100) and (3-102) are satisfied. The angular-momentum ellipsoid then contracts or expands as shown in figure 3.10.

The smallest allowed size is reached when ellipsoid (3-100) lies entirely within ellipsoid (3-102) except for two points of intersection on its longest axis, the axis for which the principal moment is smallest. Angular velocity ω then points in either direction along the axis.

Because this axis is a principal axis, the angular momentum points in the same direction. And as long as $\mathbf{L}^i$ is conserved, the axis of rotation does not turn (or

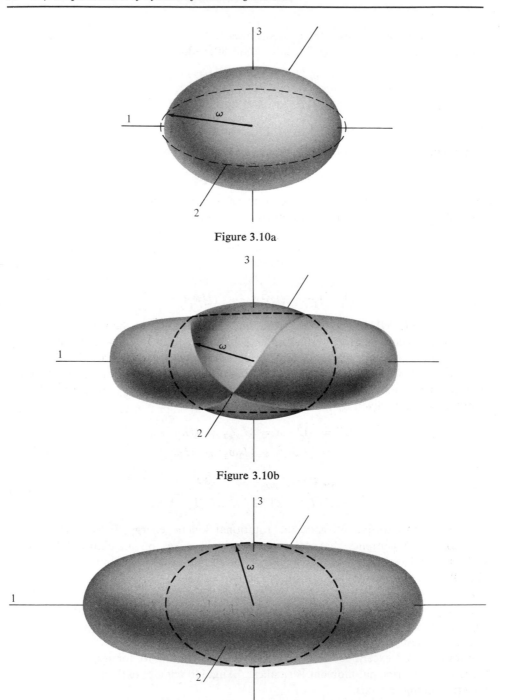

Figure 3.10a

Figure 3.10b

Figure 3.10c

FIGURE 3.10   How the angular-momentum-squared ellipsoid may intersect the rotational-energy ellipsoid.

precess). A frictional loss that is represented by a torque about the axis of rotation merely acts to decrease both L^i and T^i; it does not alter the direction of the angular momentum or angular velocity.

Only a component of torque perpendicular to the direction of ω can alter the direction of ω and cause the intersection to expand from a point to a small closed curve. Rotation is stable around the principal axis about which the moment of inertia is smallest.

Introducing such a component of torque while keeping the kinetic energy constant causes ellipsoid (3-100) to expand as long as the torque lasts. The two intersections then move along ellipsoid (3-102). No possible axis of rotation coincides with a principal axis, and ω must move about the fixed **L**, until the intersections meet at the intermediate principal axis.

A rotation about this axis is like that about the other two principal axes in that ω has the same direction as **L**, and a retarding torque about the axis merely decreases the magnitude of both L and T. But even an infinitesimal component of torque perpendicular to the axis can move ω away from this intermediate axis. Indeed if the movement makes the sign of ω_3 opposite to that of ω_1, when ω_2 is positive, the Euler equations to be derived in the next section tell us that the resulting ω continues to turn away from the intermediate axis without any torque acting. Its end point then moves along the pertinent intersection of the ellipsoids. Rotation about the intermediate principal axis is unstable.

Applying a finite component of torque that increases the magnitude of the angular momentum but does not affect the kinetic energy causes ellipsoid (3-100) to expand further. The two intersections finally contract to points on the axis along which the principal moment is greatest. Again, the axis of ω coincides with the conserved axis of **L**, and a retarding torque does not change this relationship. A small torque perpendicular to the axis acting over a limited period of time can only cause a small change in direction of ω; so rotation is also stable around the principal axis about which the moment of inertia is largest.

3.11 The Euler equations for rotation of a rigid body

From Newton's laws, we have found that a torque acting about the center of mass produces an equivalent rate of change in angular momentum. At any instant of time, this angular momentum can be resolved along the principal axes and the rates of change in the corresponding components of angular velocity calculated.

Consider an effectively rigid body rotating at time t as figure 3.6 indicates. The definitions of angular momentum and angular velocity then lead to equations (3-74) and (3-98). Note particularly that vector ω is the angular velocity of the body with respect to a nonrotating frame of reference. The components ω_1, ω_2, ω_3 in equation (3-98),

$$\mathbf{L}^i = L_k \mathbf{e}_k = I_k \omega_k \mathbf{e}_k, \tag{3-103}$$

must be measured with respect to the nonrotating frame coinciding momentarily with the principal axes of inertia.

The base vectors e_1, e_2, e_3 are anchored in the body. Differentiating (3-103),

$$\dot{\mathbf{L}}^i = I_k\dot{\omega}_k\mathbf{e}_k + I_k\omega_k\dot{\mathbf{e}}_k = I_k\dot{\omega}_k\mathbf{e}_k + L_k\dot{\mathbf{e}}_k, \tag{3-104}$$

introducing (3-73) for the rate of change in a vector like $\mathbf{r}_j$,

$$\dot{\mathbf{e}}_k = \boldsymbol{\omega} \times \mathbf{e}_k, \tag{3-105}$$

factoring out the common $\boldsymbol{\omega} \times$ and bringing in (3-103) for the last sum, yields

$$\dot{\mathbf{L}}^i = I_k\dot{\omega}_k\mathbf{e}_k + L_k\boldsymbol{\omega} \times \mathbf{e}_k = I_k\dot{\omega}_k\mathbf{e}_k + \boldsymbol{\omega} \times L_k\mathbf{e}_k$$
$$= I_k\dot{\omega}_k\mathbf{e}_k + \boldsymbol{\omega} \times \mathbf{L}^i. \tag{3-106}$$

The equation of motion (3-58), obtained on introducing Newton's second law into the derivative of the appropriate angular momentum equation, tells us that

$$\dot{\mathbf{L}}^i = \mathbf{N}^i = N_k\mathbf{e}_k \tag{3-107}$$

while the cross product in (3-106) is

$$\boldsymbol{\omega} \times \mathbf{L}^i = \begin{vmatrix} \mathbf{e}_1 & \mathbf{e}_2 & \mathbf{e}_3 \\ \omega_1 & \omega_2 & \omega_3 \\ I_1\omega_1 & I_2\omega_2 & I_3\omega_3 \end{vmatrix}$$

$$= (\omega_2 I_3\omega_3 - I_2\omega_2\omega_3)\mathbf{e}_1 + (\omega_3 I_1\omega_1 - I_3\omega_3\omega_1)\mathbf{e}_2$$
$$+ (\omega_1 I_2\omega_2 - I_1\omega_1\omega_2)\mathbf{e}_3$$
$$= (I_3 - I_2)\omega_2\omega_3\mathbf{e}_1 + (I_1 - I_3)\omega_3\omega_1\mathbf{e}_2 + (I_2 - I_1)\omega_1\omega_2\mathbf{e}_3. \tag{3-108}$$

Using (3-107) to eliminate $\dot{\mathbf{L}}^i$ and (3-108) to eliminate $\boldsymbol{\omega} \times \mathbf{L}^i$ in (3-106) produces an expanded vector equation whose components are

$$N_1 = I_1\dot{\omega}_1 + (I_3 - I_2)\omega_2\omega_3, \tag{3-109}$$

$$N_2 = I_2\dot{\omega}_2 + (I_1 - I_3)\omega_3\omega_1, \tag{3-110}$$

$$N_3 = I_3\dot{\omega}_3 + (I_2 - I_1)\omega_1\omega_2. \tag{3-111}$$

These are called *Euler's equations* for rotation of the body.

EXAMPLE 3.10   Show how free rotation about the intermediate principal axis is unstable.

In free rotation, no torque acts and equation (3-110) reduces to

$$\dot{\omega}_2 = \frac{I_3 - I_1}{I_2}\omega_3\omega_1.$$

With

$$I_1 < I_2 < I_3,$$

we find that

$$\dot{\omega}_2 < 0$$

whenever

$$\text{sign } \omega_1 = -\text{sign } \omega_3.$$

If the system is initially rotating around axis 2 with

$$\omega_2 > 0,$$

any momentary torque that moves the end point of ω along the intersection to make

$$\omega_1 \text{ negative} \quad \text{and} \quad \omega_3 \text{ positive}$$

or in the opposite direction making

$$\omega_1 \text{ positive} \quad \text{and} \quad \omega_3 \text{ negative}$$

causes $\dot{\omega}_2$ to become negative. Component ω_2 then decreases with time.

Treating the other two Euler equations similarly, we find that the magnitudes of ω_1 and ω_3 simultaneously increase. When vector ω is tipped slightly away from the intermediate axis, the perturbation tends to increase, and motion about the axis is unstable.

3.12 Rotation of a free symmetric top

When two of the principal moments are equal, Euler's equations can be readily integrated.

Consider a torque-free

$$N_1 = N_2 = N_3 = 0 \tag{3-112}$$

rigid body,

$$I_1 = \text{constant}, \tag{3-113}$$

$$I_2 = \text{constant}, \tag{3-114}$$

$$I_3 = \text{constant}, \tag{3-115}$$

for which

$$I_1 = I_2. \tag{3-116}$$

Equations (3-109), (3-110), (3-111) then reduce to

$$I_1\dot{\omega}_1 + (I_3 - I_1)\omega_2\omega_3 = 0, \tag{3-117}$$

$$I_1\dot{\omega}_2 + (I_1 - I_3)\omega_3\omega_1 = 0, \tag{3-118}$$

$$I_3\dot{\omega}_3 = 0. \tag{3-119}$$

Integrating equation (3-119) yields

$$\omega_3 = \alpha \tag{3-120}$$

with α fixed (α is merely a constant of integration). The free body thus turns at constant angular velocity about its unique axis. By (3-97), component L_3 of the angular momentum is also constant.

Now, substitute α for ω_3 in (3-117) and (3-118) and rearrange:

$$\dot{\omega}_1 = -\frac{(I_3 - I_1)\alpha}{I_1}\omega_2, \tag{3-121}$$

$$\dot{\omega}_2 = \frac{(I_3 - I_1)\alpha}{I_1}\omega_1. \tag{3-122}$$

Differentiate (3-121) and employ (3-122) to eliminate $\dot{\omega}_2$:

$$\ddot{\omega}_1 = -\left[\frac{(I_3 - I_1)\alpha}{I_1}\right]^2 \omega_1. \tag{3-123}$$

Treat (3-122) similarly:

$$\ddot{\omega}_2 = -\left[\frac{(I_3 - I_1)\alpha}{I_1}\right]^2 \omega_2. \tag{3-124}$$

Integrate equation (3-123) with the origin for the time scale at one of the points where ω_1 is zero:

$$\omega_1 = \beta \sin \frac{(I_3 - I_1)\alpha}{I_1} t. \tag{3-125}$$

Substitute this result into (3-121) and solve for ω_2:

$$\omega_2 = -\beta \cos \frac{(I_3 - I_1)\alpha}{I_1} t. \tag{3-126}$$

Note that β is a constant of integration.

Applying the Pythagorean theorem to these two components gives us

$$(\omega_1{}^2 + \omega_2{}^2)^{1/2} = \beta \tag{3-127}$$

for the projection of ω on the 1–2 plane. Because the third component is also fixed, the total angular velocity is of constant magnitude.

Solutions (3-125) and (3-126) tell us that vector ω swings around the axis of symmetry in the body with the angular velocity

$$\gamma = \frac{I_3 - I_1}{I_1}\omega_3. \tag{3-128}$$

Such a wandering of one axis around another is called *precession*. See figures 3.11 and 3.12.

Constraints can alter the precession by causing **L** to vary. Recall example 3.7.

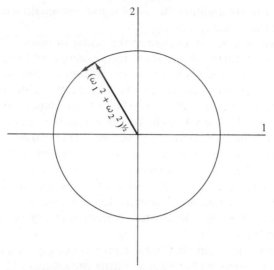

FIGURE 3.11 Precession of the angular velocity projection on the 1–2 plane around the unique axis fixed in the freely rotating body.

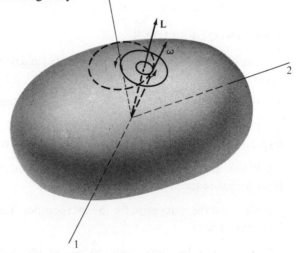

FIGURE 3.12 Precession of ω and the unique axis about the angular momentum **L**.

3.13 Highlights

A physical system is distinguished by its lasting features, expressions that change only when some influence is exerted from the outside or when some disruptive effect (such as friction) is present.

One of the most important of these attributes is the scalar energy, which does remain fixed in a conservative system. Another is the linear momentum, which is constant when no net force acts. In addition, the angular momentum does not change when no net torque acts on the system.

When a net force does act, it accelerates the center of mass of the set of particles as if all the mass were there. When a net torque acts about the center of mass, it alters the angular momentum about this center at a rate equal to the torque.

As long as all particles move around an axis through the center of mass at the same angular velocity ω, the angular velocity is related to the angular momentum through the moment of inertia $\mathbf{I}$. Then $\mathbf{I}$ might be defined as the operator that converts ω to $\mathbf{L}$. When $\mathbf{I}$ is constant, free rotation is governed by the two quadratic relationships. Accelerated rotation, on the other hand, is governed by the Euler equations.

Relative motions, when they occur among the particles, may be quite complicated. Nevertheless, they can be analyzed if their amplitudes are small or if they consist of harmonic and nearly harmonic vibrations. Certain combinations of the Cartesian coordinates then act as simple coordinates. Variation in any one of these governs a single mode of vibration.

Since the pertinent transformation of the conventional equations of motion for the vibrating system is a matrix multiplication, matrix procedures will be developed next. Since the Lagrangian language is very helpful, it will also be developed.

DISCUSSION QUESTIONS

3.1 Explain why Newton's laws commonly yield a set of simultaneous differential equations.

3.2 How does the two-body problem reduce to separate one-body problems?

3.3 Explain when the equation describing the movement of one particle with respect to the other is not affected by external forces.

3.4 When are rotation and vibration governed by the same reduced mass? By different reduced masses?

3.5 How can the mass ratio for the two components of a double star be obtained from the orbits?

3.6 Why don't frictional losses occur in the collision of one molecule with another?

3.7 With sketches, show the appearance of a two-body collision (a) in the center-of-mass frame and (b) in the frame in which one body is initially at rest.

3.8 How do center-of-mass coordinates simplify calculations?

3.9 Define (a) translation, (b) rotation, and (c) vibration for a set of particles.

3.10 Show how the kinetic energy can be split into that associated with movement of the center of mass and that associated with rotation about this center.

3.11 Show that the rate of change in angular momentum about the center of mass equals the torque acting about the center even when the center of mass accelerates in an inertial frame.

3.12 Show that the center of gravity coincides with the center of mass when a body is in a uniform gravitational field.

3.13 Why can angular velocity be represented as a vector?

3.14 When are the rotational movements of all parts of a body described by the same angular velocity vector?

3.15 Show by example how the direction of angular momentum may differ from the direction of angular velocity.

3.16 Derive the formula

$$T = \tfrac{1}{2}\mathbf{L} \cdot \boldsymbol{\omega}$$

for rotational kinetic energy. What is the analogous formula for translational kinetic energy?

3.17 Besides rotating ω, what else does $\mathbf{I}$ do when it acts on ω?

3.18 Relate the moment of inertia about one axis to the moment about a parallel axis.

3.19 How are the moments of inertia for a continuous body calculated?

3.20 Explain the quadratic relationships that govern free rotation of a rigid body. Why is rotation stable only about the largest and smallest principal axes?

3.21 Derive the Euler equations of motion. Discuss precession.

3.22 What is wrong with the following proof? A mass m is attached to a weightless hoop, which rolls without slippage on level ground. When the mass is at the top of the hoop, the kinetic energy is $\tfrac{1}{2}m(2v)^2$ and the potential energy is $mg(2r)$, where v is the forward velocity of the center of the hoop and r is its radius. But when the mass is exactly at the bottom, it is resting on the ground, and both kinetic and potential energies are zero. Therefore, energy is not conserved.

PROBLEMS

3.1 If the charge on an electron is 1.602×10^{-19} coulomb while the volt is a joule per coulomb, how many joules of kinetic energy are gained by an electron in moving freely through a potential rise of 10^6 volts? (This amount of energy is called a million electron volts or an MeV.)

3.2 Employing the MeV as the unit of energy, calculate the recoil energy when Gd^{148} emits a 3.16 MeV alpha particle. Then find how many MeV's of potential energy are converted to kinetic energy in the disintegration.

3.3 If atomic projectile B collides with the stationary free nucleus A, what fraction of the kinetic energy can be converted to potential energy? Let the mass of B be m_B and the mass of A be m_A.

3.4 What is the maximum velocity to which an electron can be accelerated when it is struck by an alpha particle moving at velocity v? Assume that the electron is initially at rest. If the alpha particle has 6.0 MeV kinetic energy, what is the final energy of the electron?

3.5 In a cloud chamber, a projectile strikes a particle at rest and is deflected through angle θ_1 while the target particle takes off at angle θ_2 from the original direction and at angle $\theta_1 + \theta_2$ from the final direction of the first particle. Express the mass ratio m_1/m_2 in terms of θ_1 and θ_2.

3.6 Dirt of mass m falls vertically on an idling horizontal conveyor belt of mass M. Calculate the kinetic energy lost. Add this to the final kinetic energy of the dirt and let m become small with respect to M. Thus get the work that is supplied to give the dirt horizontal velocity v.

3.7 Calculate the moment of inertia of a uniform rectangular parallelepiped about an axis through the center of mass parallel to edge c. What is the moment of inertia about an axis distance l from this axis through the center of mass?

3.8 A compound pendulum is swinging in a single plane. How does its period vary with the distance of the support from the center of mass? Assume that the amplitude of vibration is small.

3.9 Show that if a compound pendulum is suspended from two different positions on opposite sides of the center of mass which yield the same period, the distance between the supports equals the length of a simple pendulum having the same period.

3.10 The earth is a nearly symmetric top for which

$$\frac{I_3 - I_1}{I_1} = 0.0033.$$

What is the period of the consequent precession of the North Pole?

3.11 The kinetic energy gained by a proton or positron in falling through a potential drop of 1 volt is called an *electron volt* (eV). Now if the ratio of proton mass to positron mass is 1836, what is the ratio of proton to positron speeds when each is moving with 10.0 eV energy?

3.12 Employing one million electron volts, the MeV, as a unit of energy, calculate the recoil energy when Bi^{212} emits a 6.090 MeV alpha particle. Then find how much potential energy is converted to kinetic energy in the emission.

3.13 The reaction

$$Li^7 + p \rightarrow Be^7 + n$$

proceeds only when the protons have a kinetic energy greater than 1.88 MeV. How much potential energy has to be supplied to get the reaction to go?

3.14 If the maximum energy that an electron can receive from a single collision with a proton of given energy is 10.0 keV, what is the energy of the proton?

3.15 A deuteron is deflected elastically by a proton at rest. In a center-of-mass coordinate system resolve the components of possible deuteron velocities. Transform each to the laboratory system, set up the tangent of the corresponding angle of deflection, and maximize this angle.

3.16 A pendulum consisting of an iron ball at distance l from a fixed point is initially inclined 45° from the vertical. It is released from this position to strike a freely movable magnet at the low point in its swing. To what angle does the bob rise with its burden? Assume that the mass of the ball is m_1 and that of the magnet m_2.

3.17 Calculate the moment of inertia of a uniform sphere about any axis passing through its center.

3.18 A horizontal platform weighing 500 kg is essentially a circular disk 10.0 meters in diameter turning about a vertical axis through its center at 20.0 revolutions per minute while two objects weighing 50 kg each rest at the center. Neglect friction at the bearings and determine how the rate of rotation is affected by the objects sliding to opposite points on the rim of the disk.

3.19 The period of a given compound pendulum swinging over a small arc in a single plane is

$$2\pi \sqrt{\frac{I}{Mgl}}$$

where l is the distance from the point of support to the center of mass and M the total mass. Vary l until this period is a minimum.

3.20 Show that the time rate of change of the square of the angular velocity of a freely rotating rigid body is given by the formula

$$\frac{d\omega^2}{dt} = -2 \frac{\dot{\omega}_1 \dot{\omega}_2 \dot{\omega}_3}{\omega_1 \omega_2 \omega_3} .$$

REFERENCES

BOOKS

Bradbury, T. C. *Theoretical Mechanics*, pp. 316–484. John Wiley & Sons, Inc., New York, 1968.

Goodman, L. E., and Warner, W. H. *Dynamics*, pp. 184–465, 587–606. Wadsworth Publishing Company, Inc., Belmont, Calif., 1963.

Hauser, W. *Introduction to the Principles of Mechanics*, pp. 234–255, 282–330. Addison-Wesley Publishing Company, Inc., Reading, Mass., 1965.

Symon, K. R. *Mechanics*, 2nd ed., pp. 155–256. Addison-Wesley Publishing Company, Inc., Reading, Mass., 1960.

Taylor, E. F. *Introductory Mechanics*, pp. 178–240. John Wiley & Sons, Inc., New York, 1963.

ARTICLES

Armstrong, H. L. On the Precession and Nutation of Gyroscopes. *Am. J. Phys.*, **35**, 883 (1967).

Armstrong, H. L. Some Pitfalls in Demonstrating Conservation of Momentum. *Am. J. Phys.*, **36**, 56 (1968).

Beatty, M. F. Lagrange's Theorem on the Center of Mass of a System of Particles. *Am. J. Phys.*, **40**, 205 (1972).

Gruber, G. R. Clarification on Two Important Questions in Rigid Body Mechanics. *Am. J. Phys.*, **40**, 421 (1972).

Hood, C. G. A Reformulation of Newtonian Dynamics. *Am. J. Phys.*, **38**, 438 (1970).

Lamy, P. L., and Burns, J. A. Geometrical Approach to Torque Free Motion of a Rigid Body Having Internal Energy Dissipation. *Am. J. Phys.*, **40**, 441 (1972).

Masket, A. V. Comments on an Analytic Expression for Angular Velocity. *Am. J. Phys.*, **34**, 983 (1966).

McGuire, J. B. Binary Collisions in the Nonrelativistic Three-Particle Problem. *Phys. Rev. A*, **1**, 353 (1970).

Mott, D. L. Torque on a Rigid Body in Circular Orbit. *Am. J. Phys.*, **34**, 562 (1966).

Pearlman, N. Vector Representation of Rigid-Body Rotation. *Am. J. Phys.*, **35**, 1164 (1967).

Rutledge, H. D. Graphical Method of Teaching Elementary Collisions. *Am. J. Phys.*, **38**, 223 (1970).

Shelupsky, D. Lagrange's Three Particles. *Am. J. Phys.*, **31**, 136 (1963).

Shonle, J. I. Resource Letter CM-1 on the Teaching of Angular Momentum and Rigid Body Motion. *Am. J. Phys.*, **33**, 879 (1965).

Srivastava, S. N. A New Theorem for Moment of Inertia. *Am. J. Phys.*, **29**, 211 (1961).

4 / *Operations with Matrices*

4.1 Sets of numbers for describing a linear transformation

Each of the equations (1-52), (3-82), (1-80), and (1-83) mixes the components of a given vector linearly and homogeneously to produce the indicated component of another vector. The other components are produced on changing the free index. Now, the coefficients in the set of linear forms constitute the elements of a dyadic. Dot multiplication of this dyadic with the original vector leads to the transformed vector.

Alternatively, one can place the coefficients in a square array. The components of the initial vector can be placed in a vertical column after the square array. Combining numbers in the two factors following the rule to be formulated shortly produces the components of the final vector in a similar column. Arrays that unite in the prescribed ways are called matrices.

Some reorientations of a symmetric physical system permute parts that are equivalent. If we choose a similar point in each part, such an operation rearranges some or all of the points. When we place expressions characterizing the chosen points in a column in order, each permutation is represented by a square array of numbers acting on the column.

Any set of homogeneous linear polynomials can be considered the product of a rectangular matrix with a column matrix. Simultaneous linear equations are represented by a single matrix equation.

One can interchange the rows and columns in a matrix without changing the order of any expressions in a pertinent sequence. As a consequence, the transposed matrix represents the same physical property or entity as the original matrix. A vector can be represented by listing the components in order in a row as well as in a column. The transformation of a vector can be represented by the row matrix times a square matrix as well as by a square matrix times a column matrix.

A homogeneous quadratic expression can be considered as the product of a row matrix, a square matrix, and a column matrix. Since such expressions make up the kinetic energy and the potential energy of a linearly vibrating system, the corresponding equation of motion is a matrix equation.

4.2   The combinatorial rules that make the sets matrices

The significance of a particular set of expressions depends on how the set combines with conformable sets. Here, addition will be carried out by summing similar expressions in the contributing sets, multiplication will be defined so as to effect the transformations we have described, and combinations of transformations will be represented.

Consider something that is described by a sequence of expressions A_{jk}, with index j varying by integers from 1 to m and index k from 1 to n. Place the expressions in a rectangular array

$$\mathbf{A} = \begin{pmatrix} A_{11} & A_{12} & \cdot & \text{---row---} & \cdot \\ A_{21} & A_{22} & \cdot & \cdot & \cdot & \cdot \\ \text{column} & & \cdot & \cdot & \cdot & \cdot \\ & & \cdot & \cdot & A_{jk} & \cdot & \cdot \\ & & \cdot & \cdot & \cdot & \cdot & \cdot \\ \vdots & \cdot & \cdot & \cdot & \cdot & A_{mn} \end{pmatrix} \tag{4-1}$$

in which *element* A_{jk} occupies the intersection of the jth row and the kth column. This array is a *matrix* if and only if it combines with other such arrays according to the following rules.

To *add* or *subtract* matrix **B** from matrix **A**,

$$\mathbf{A} \pm \mathbf{B} = \mathbf{C}, \tag{4-2}$$

we add or subtract the corresponding elements:

$$A_{jk} \pm B_{jk} = C_{jk}. \tag{4-3}$$

It is necessary that **A** and **B** contain the same number of rows and the same number of columns.

To *multiply* matrix **A** by *scalar* α

$$\alpha \mathbf{A} = \mathbf{B} \tag{4-4}$$

we multiply each element of **A** by the number (real or complex):

$$\alpha A_{jk} = B_{jk}. \tag{4-5}$$

We *multiply matrices* **A** and **B**

$$\mathbf{AB} = \mathbf{C} \tag{4-6}$$

with the formula

$$A_{j1}B_{1l} + A_{j2}B_{2l} + \cdots + A_{jn}B_{nl} = A_{jk}B_{kl} = C_{jl} \tag{4-7}$$

applied to each row of **A** and column of **B** in turn. As before, repetition of an index is used to indicate the pertinent summation. When the matrices are written out, element C_{jl} is set equal to the sum of products of corresponding elements in the jth row of the first matrix and the lth column of the second one:

$$\mathbf{AB} = \begin{pmatrix} \cdot & \cdot & \cdot & \cdot & \cdot \\ A_{j1} & A_{j2} & \longrightarrow & A_{jn} \\ \cdot & \cdot & \cdot & \cdot & \cdot \\ \cdot & \cdot & \cdot & \cdot & \cdot \end{pmatrix} \begin{pmatrix} \cdot & B_{1l} & \cdot \\ \cdot & B_{2l} & \cdot \\ \cdot & \downarrow & \\ \cdot & B_{nl} & \cdot \end{pmatrix} = \begin{pmatrix} \cdot & \cdot & \cdot \\ \cdot & C_{jl} & \cdot \\ \cdot & \cdot & \cdot \end{pmatrix}. \tag{4-8}$$

Because each term in C_{jl} must have two factors, the number of columns in $\mathbf{A}$ must equal the number of rows in $\mathbf{B}$. Matrices $\mathbf{A}$ and $\mathbf{B}$ are said to be *conformable* to multiplication in the given order when this condition is met. Changing the order may destroy the conformability, but when the change is allowed, it generally leads to a different result:

$$\mathbf{AB} \neq \mathbf{BA}. \tag{4-9}$$

Matrix multiplication does not obey the commutative law.

Multiple products are built up from binary products. Thus, $\mathbf{ABC}$ may be formed as $(\mathbf{AB})\mathbf{C}$ or $\mathbf{A}(\mathbf{BC})$:

$$[(\mathbf{AB})\mathbf{C}]_{jm} = (\mathbf{AB})_{jl}C_{lm} = (A_{jk}B_{kl})C_{lm} = A_{jk}B_{kl}C_{lm}, \tag{4-10}$$

while

$$[\mathbf{A}(\mathbf{BC})]_{jm} = A_{jk}(\mathbf{BC})_{km} = A_{jk}(B_{kl}C_{lm}) = A_{jk}B_{kl}C_{lm}. \tag{4-11}$$

Whether the multiplication proceeds as $\mathbf{AB}$ times $\mathbf{C}$ or as $\mathbf{A}$ times $\mathbf{BC}$, the same result is obtained. A multiple product does not depend on how the matrices are associated as long as the order in which they appear is preserved. Matrix multiplication obeys the *associative law*.

Multiplication of any matrix by the conformable square matrix

$$\mathbf{E} = \begin{pmatrix} 1 & 0 & \cdot & \cdot & \cdot \\ 0 & 1 & \cdot & \cdot & \cdot \\ \cdot & \cdot & \cdot & \cdot & \cdot \\ \cdot & \cdot & \cdot & 1 & 0 \\ \cdot & \cdot & \cdot & 0 & 1 \end{pmatrix} \tag{4-12}$$

whose jkth element is δ_{jk} (the Kronecker delta) does not alter the given matrix. For,

$$(\mathbf{EA})_{jl} = \delta_{jk}A_{kl} = A_{jl}, \tag{4-13}$$

and

$$(\mathbf{AE})_{jl} = A_{jk}\delta_{kl} = A_{jl}. \tag{4-14}$$

Consequently, $\mathbf{E}$ is called the *unit matrix*, or the identity operator.

EXAMPLE 4.1 Show how matrix multiplication can effect the scalar multiplication of two vectors.

Resolve the given vectors along the same mutually perpendicular base vectors $\mathbf{e}_1$, $\mathbf{e}_2$, $\mathbf{e}_3$. Then place the components of the left factor in a single row, those of the right factor in a single column,

$$\tilde{\mathbf{x}} = (x_1 \quad x_2 \quad x_3), \qquad \mathbf{y} = \begin{pmatrix} y_1 \\ y_2 \\ y_3 \end{pmatrix},$$

and multiply the matrices:

$$\tilde{\mathbf{x}}\mathbf{y} = (x_1 \quad x_2 \quad x_3) \begin{pmatrix} y_1 \\ y_2 \\ y_3 \end{pmatrix} = (x_1 y_1 + x_2 y_2 + x_3 y_3) = (\mathbf{x} \cdot \mathbf{y}).$$

EXAMPLE 4.2   Formulate the cross product of two vectors as the product of matrices.

Since the result is to be a vector, it is described by a three-element row matrix or by a three-element column matrix. Consequently, the process can be represented by a row matrix times a square matrix or by a square matrix times a column matrix. Choosing the latter alternative, we let

$$\mathbf{X}\mathbf{y} = \begin{pmatrix} a & b & c \\ d & e & f \\ g & h & i \end{pmatrix} \begin{pmatrix} y_1 \\ y_2 \\ y_3 \end{pmatrix} = \begin{pmatrix} ay_1 + by_2 + cy_3 \\ dy_1 + ey_2 + fy_3 \\ gy_1 + hy_2 + iy_3 \end{pmatrix}.$$

For the elements in the final column matrix to be those given by formula (1-33),

$$(\mathbf{x} \times \mathbf{y}) = \begin{pmatrix} x_2 y_3 - x_3 y_2 \\ x_3 y_1 - x_1 y_3 \\ x_1 y_2 - x_2 y_1 \end{pmatrix},$$

we must have

$$
\begin{array}{lll}
a = 0, & b = -x_3, & c = x_2, \\
d = x_3, & e = 0, & f = -x_1, \\
g = -x_2, & h = x_1, & i = 0.
\end{array}
$$

The first factor in the matrix product is then

$$\mathbf{X} = \begin{pmatrix} 0 & -x_3 & x_2 \\ x_3 & 0 & -x_1 \\ -x_2 & x_1 & 0 \end{pmatrix}.$$

4.3   Reorientation matrices

The coordinates for any given point in a physical system can be placed in a row matrix or a column matrix, as example 4.1 shows. Each reorientation of the coordinate axes, or reorientation of the physical system about the origin, alters the matrix. Multiplying the row or column matrix with the pertinent square matrix produces the same change; it represents the reorientation. In this section, we will construct the 3×3 matrices that effect typical transformations of 3×1 column matrices.

We start with a reference frame in which the Cartesian coordinates of a typical point or particle are x, y, z. After the reorientation, these are x', y', z'. The initial and final radius vectors to the point are then

$$\mathbf{r} = \begin{pmatrix} x \\ y \\ z \end{pmatrix} \quad \text{and} \quad \mathbf{r}' = \begin{pmatrix} x' \\ y' \\ z' \end{pmatrix}. \tag{4-15}$$

Let us write the multiplier that converts $\mathbf{r}$ to $\mathbf{r}'$ as

$$\mathbf{A} = \begin{pmatrix} A_{11} & A_{12} & A_{13} \\ A_{21} & A_{22} & A_{23} \\ A_{31} & A_{32} & A_{33} \end{pmatrix}. \tag{4-16}$$

Then

$$\mathbf{r}' = \mathbf{A}\mathbf{r} \tag{4-17}$$

yields

$$\begin{aligned} x' &= A_{11}x + A_{12}y + A_{13}z, \\ y' &= A_{21}x + A_{22}y + A_{23}z, \\ z' &= A_{31}x + A_{32}y + A_{33}z. \end{aligned} \tag{4-18}$$

Equations (4-18) merely relate the two sets of coordinates. They do not tell one whether reorientation of axes or reorientation of the physical system has occurred. They do not even tell whether distortion of the physical figure has occurred; but we will assume that such distortion is absent. We will show that these equations work by actually constructing pertinent $\mathbf{A}$'s. Later we will consider the condition that $\mathbf{A}$ must satisfy if it is to introduce no distortion.

The simplest operation is the one involving no change, the *identity* operation E, for which

$$\begin{aligned} x' &= x + 0 + 0, \\ y' &= 0 + y + 0, \\ z' &= 0 + 0 + z, \end{aligned} \tag{4-19}$$

and

$$\begin{pmatrix} x' \\ y' \\ z' \end{pmatrix} = \begin{pmatrix} 1 & 0 & 0 \\ 0 & 1 & 0 \\ 0 & 0 & 1 \end{pmatrix} \begin{pmatrix} x \\ y \\ z \end{pmatrix}, \tag{4-20}$$

or

$$\mathbf{r}' = \mathbf{E}\mathbf{r}. \tag{4-21}$$

The identity operation E is represented by the unit matrix $\mathbf{E}$.

When either the axes (as in figure 4.1), or the particles, are reflected through the origin, *inversion* is said to occur. This operation causes the sign of each coordinate to change; so

$$\begin{aligned} x' &= -x + 0 + 0, \\ y' &= 0 - y + 0, \\ z' &= 0 + 0 - z, \end{aligned} \tag{4-22}$$

or

$$\begin{pmatrix} x' \\ y' \\ z' \end{pmatrix} = \begin{pmatrix} -1 & 0 & 0 \\ 0 & -1 & 0 \\ 0 & 0 & -1 \end{pmatrix} \begin{pmatrix} x \\ y \\ z \end{pmatrix}. \tag{4-23}$$

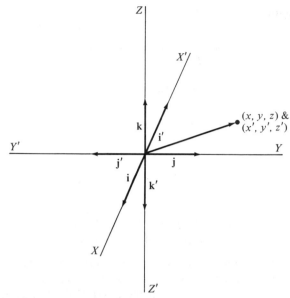

FIGURE 4.1   Inversion of coordinate axes through the origin.

Equation (4-23) is abbreviated as

$$\mathbf{r}' = \mathbf{ir} \tag{4-24}$$

while the inversion operation itself is labeled *i*.

   Reflection of either the axes or the physical system in a *plane* causes the coordinate describing distance from the plane to be replaced by its negative. Thus, reflection in the *yz* plane as in figure 4.2 changes the sign of *x*:

$$\begin{aligned}
x' &= -x + 0 + 0, \\
y' &= 0 + y + 0, \\
z' &= 0 + 0 + z.
\end{aligned} \tag{4-25}$$

These equations are all contained in the matrix equation

$$\begin{pmatrix} x' \\ y' \\ z' \end{pmatrix} = \begin{pmatrix} -1 & 0 & 0 \\ 0 & 1 & 0 \\ 0 & 0 & 1 \end{pmatrix} \begin{pmatrix} x \\ y \\ z \end{pmatrix} \tag{4-26}$$

which is summarized as

$$\mathbf{r}' = \sigma_v \mathbf{r}. \tag{4-27}$$

   The symbol σ indicates reflection in a plane, while the subscript v indicates that the plane is one of the vertical planes passing through the z axis. In (4-27), we have chosen the plane to be the *yz* plane. Operation σ_d is reflection through a vertical plane that forms part of a dihedral angle of symmetry; operation σ_h is reflection through the horizontal *xy* plane.

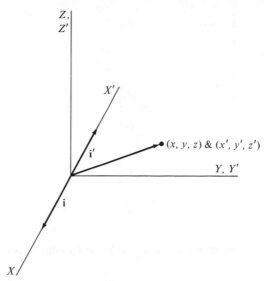

FIGURE 4.2 Reflection of coordinate axes in the *yz* plane.

Figure 4.3 illustrates *rotation* of the axes in the *xy* plane, while figure 4.4 shows the corresponding rotation of a typical point in the physical system. Applying trigonometric definitions to figure 4.3 yields

$$x' = x \cos \phi - y \sin \phi + 0,$$
$$y' = x \sin \phi + y \cos \phi + 0,$$
$$z' = \quad 0 \quad + \quad 0 \quad + z, \tag{4-28}$$

whence

$$\begin{pmatrix} x' \\ y' \\ z' \end{pmatrix} = \begin{pmatrix} \cos \phi & -\sin \phi & 0 \\ \sin \phi & \cos \phi & 0 \\ 0 & 0 & 1 \end{pmatrix} \begin{pmatrix} x \\ y \\ z \end{pmatrix}. \tag{4-29}$$

Equation (4-29) is abbreviated as

$$\mathbf{r}' = \mathbf{C}_n \mathbf{r}. \tag{4-30}$$

Operation C_n consists of either (a) clockwise rotation of the axes in the horizontal plane by $2\pi/n$ radians or (b) counterclockwise rotation of the radius vector $\mathbf{r}$, and the physical system, by the same amount. Operation C_n followed by reflection in a plane perpendicular to the axis of rotation is called S_n.

A system may be subject to rotations about various axes. If there is a unique axis about which n is greatest, it is called the *principal axis*. Twofold rotations about axes perpendicular to it are designated C_2', C_2'', . . .

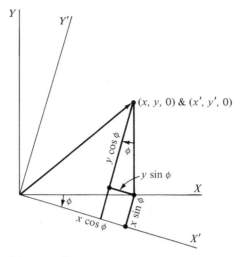

FIGURE 4.3   Rotation of the coordinate axes clockwise by angle ϕ around the z axis.

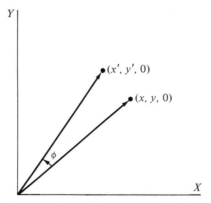

FIGURE 4.4   Counterclockwise rotation of the vector which is equivalent to the clockwise rotation of the axes in figure 4.3.

EXAMPLE 4.3   Obtain the matrices that rotate a radius vector **r** (a) 90° and (b) 120° counterclockwise around the z axis.

Setting ϕ equal to 90° in the rotation matrix of equation (4-29) yields

$$\mathbf{C}_4 = \begin{pmatrix} 0 & -1 & 0 \\ 1 & 0 & 0 \\ 0 & 0 & 1 \end{pmatrix}$$

where $\mathbf{C}_4$ represents rotation by $2\pi/4$ radians.

Setting ϕ equal to 120° in the rotation matrix gives

$$\mathbf{C}_3 = \begin{pmatrix} -\dfrac{1}{2} & -\dfrac{\sqrt{3}}{2} & 0 \\[2ex] \dfrac{\sqrt{3}}{2} & -\dfrac{1}{2} & 0 \\[2ex] 0 & 0 & 1 \end{pmatrix}.$$

Here $\mathbf{C}_3$ represents the rotation by $2\pi/3$ radians about the z axis.

EXAMPLE 4.4 What matrix represents operation S_3?
  From the definition of S_n, S_3 involves operation C_3 followed by operation σ_h:

$$S_3 = \sigma_h C_3.$$

The rotation is represented by the last matrix in example 4.3. Since reflection in the xy plane changes the sign of z,

$$\mathbf{r}' = \begin{pmatrix} x' \\ y' \\ z' \end{pmatrix} = \begin{pmatrix} 1 & 0 & 0 \\ 0 & 1 & 0 \\ 0 & 0 & -1 \end{pmatrix} \begin{pmatrix} x \\ y \\ z \end{pmatrix} = \sigma_h \mathbf{r},$$

we also have

$$\sigma_h = \begin{pmatrix} 1 & 0 & 0 \\ 0 & 1 & 0 \\ 0 & 0 & -1 \end{pmatrix}.$$

Combining the matrices for σ_h and C_3 as indicated in the first equation yields the desired result:

$$S_3 = \sigma_h C_3 = \begin{pmatrix} 1 & 0 & 0 \\ 0 & 1 & 0 \\ 0 & 0 & -1 \end{pmatrix} \begin{pmatrix} -\dfrac{1}{2} & -\dfrac{\sqrt{3}}{2} & 0 \\[2ex] \dfrac{\sqrt{3}}{2} & -\dfrac{1}{2} & 0 \\[2ex] 0 & 0 & 1 \end{pmatrix}$$

$$= \begin{pmatrix} -\dfrac{1}{2} & -\dfrac{\sqrt{3}}{2} & 0 \\[2ex] \dfrac{\sqrt{3}}{2} & -\dfrac{1}{2} & 0 \\[2ex] 0 & 0 & -1 \end{pmatrix}.$$

4.4   Permutation matrices that represent symmetry operations

Whenever a physical system possesses some spatial symmetry, there are reorientations, called *symmetry operations*, that permute similar parts. The operations do *not* alter the appearance of the system, for each transformed configuration is equivalent to the initial one.

As an example, the array in figure 4.5 is not essentially changed by any of the reorientations: E; S_4, C_2, and S_4^3 about the S_4 axis; C_2' and C_2'' about two axes perpendicular to each other and to the S_4 axis; σ_d and σ_d' bisecting the angles between the axes of C_2' and C_2''. Each of these is a symmetry operation, which leaves the spectrum of properties of the physical system unchanged.

Matrices representing a set of symmetry operations can be obtained in the following manner. Equivalent positions are located in the similar parts of the given system. A point that is equidistant from these positions and is not altered in the operations is chosen as the origin. Radius vectors are then drawn to the equivalent points.

The vector drawn to the jth part before an operation is labeled r_j; the one drawn to the same physical point after the operation is labeled r_j'. The column matrices

$$\begin{pmatrix} r_1' \\ r_2' \\ \vdots \\ r_n' \end{pmatrix} \quad \text{and} \quad \begin{pmatrix} r_1 \\ r_2 \\ \vdots \\ r_n \end{pmatrix} \qquad (4\text{-}31)$$

are constructed.

One then determines the r_k that each r_j' equals. The results are put into the first matrix of (4-31) and the consequent column matrix is factored into a square matrix

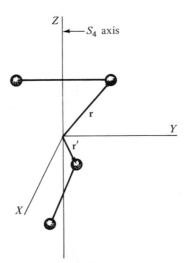

FIGURE 4.5   Four equivalent bodies arranged around a rotation-reflection axis. Vectors **r** and **r'** are related by the equation $\mathbf{r'} = S_4\mathbf{r}$.

times the second matrix. The square matrix is a permutation matrix representing the symmetry operation.

Consider the system in figure 4.6 with the first four radius vectors locating equivalent points in the first four bodies. By inspection, determine the effect of each symmetry operations on each radius vector and summarize the results as follows:

$$r_j' = Er_j = r_j,$$
$$r_j' = C_4 r_j = r_{j+1},$$
$$r_j' = C_2 r_j = r_{j+2},$$
$$r_j' = C_4{}^3 r_j = r_{j+3}. \tag{4-32}$$

Note that the indices 1, 2, 3, 4 are considered to be cyclic, with 1 following 4 in sequence.

The expressions from applying the identity operation

$$\mathbf{r}' = \mathbf{Er} \tag{4-33}$$

lead to the factorization

$$\begin{pmatrix} r_1' \\ r_2' \\ r_3' \\ r_4' \end{pmatrix} = \begin{pmatrix} Er_1 \\ Er_2 \\ Er_3 \\ Er_4 \end{pmatrix} = \begin{pmatrix} r_1 \\ r_2 \\ r_3 \\ r_4 \end{pmatrix} = \begin{pmatrix} 1 & 0 & 0 & 0 \\ 0 & 1 & 0 & 0 \\ 0 & 0 & 1 & 0 \\ 0 & 0 & 0 & 1 \end{pmatrix} \begin{pmatrix} r_1 \\ r_2 \\ r_3 \\ r_4 \end{pmatrix}. \tag{4-34}$$

The expressions from the C_4 operation

$$\mathbf{r}' = \mathbf{C_4 r} \tag{4-35}$$

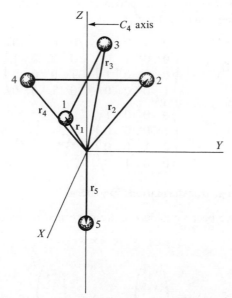

FIGURE 4.6   Configuration of five bodies about a 4-fold axis of symmetry.

yield

$$\begin{pmatrix} r_1' \\ r_2' \\ r_3' \\ r_4' \end{pmatrix} = \begin{pmatrix} C_4 r_1 \\ C_4 r_2 \\ C_4 r_3 \\ C_4 r_4 \end{pmatrix} = \begin{pmatrix} r_2 \\ r_3 \\ r_4 \\ r_1 \end{pmatrix} = \begin{pmatrix} 0 & 1 & 0 & 0 \\ 0 & 0 & 1 & 0 \\ 0 & 0 & 0 & 1 \\ 1 & 0 & 0 & 0 \end{pmatrix} \begin{pmatrix} r_1 \\ r_2 \\ r_3 \\ r_4 \end{pmatrix}. \tag{4-36}$$

For the C_2 operation

$$\mathbf{r}' = \mathbf{C}_2 \mathbf{r}, \tag{4-37}$$

we find

$$\begin{pmatrix} r_1' \\ r_2' \\ r_3' \\ r_4' \end{pmatrix} = \begin{pmatrix} C_2 r_1 \\ C_2 r_2 \\ C_2 r_3 \\ C_2 r_4 \end{pmatrix} = \begin{pmatrix} r_3 \\ r_4 \\ r_1 \\ r_2 \end{pmatrix} = \begin{pmatrix} 0 & 0 & 1 & 0 \\ 0 & 0 & 0 & 1 \\ 1 & 0 & 0 & 0 \\ 0 & 1 & 0 & 0 \end{pmatrix} \begin{pmatrix} r_1 \\ r_2 \\ r_3 \\ r_4 \end{pmatrix}. \tag{4-38}$$

Rotation by 270° about the z axis,

$$\mathbf{r}' = \mathbf{C}_4{}^3 \mathbf{r}, \tag{4-39}$$

yields

$$\begin{pmatrix} r_1' \\ r_2' \\ r_3' \\ r_4' \end{pmatrix} = \begin{pmatrix} C_4{}^3 r_1 \\ C_4{}^3 r_2 \\ C_4{}^3 r_3 \\ C_4{}^3 r_4 \end{pmatrix} = \begin{pmatrix} r_4 \\ r_1 \\ r_2 \\ r_3 \end{pmatrix} = \begin{pmatrix} 0 & 0 & 0 & 1 \\ 1 & 0 & 0 & 0 \\ 0 & 1 & 0 & 0 \\ 0 & 0 & 1 & 0 \end{pmatrix} \begin{pmatrix} r_1 \\ r_2 \\ r_3 \\ r_4 \end{pmatrix}. \tag{4-40}$$

Permutation matrices for other systems can be set up in the same way.

EXAMPLE 4.5   Use the permutation matrices just derived to show that $\mathbf{C}_4 \mathbf{C}_2$ yields $\mathbf{C}_4{}^3$.

Equations (4-36), (4-38) contain the pertinent 4×4 matrices. Multiply these and identify the result, using equation (4-40):

$$\mathbf{C}_4 \mathbf{C}_2 = \begin{pmatrix} 0 & 1 & 0 & 0 \\ 0 & 0 & 1 & 0 \\ 0 & 0 & 0 & 1 \\ 1 & 0 & 0 & 0 \end{pmatrix} \begin{pmatrix} 0 & 0 & 1 & 0 \\ 0 & 0 & 0 & 1 \\ 1 & 0 & 0 & 0 \\ 0 & 1 & 0 & 0 \end{pmatrix}$$

$$= \begin{pmatrix} 0 & 0 & 0 & 1 \\ 1 & 0 & 0 & 0 \\ 0 & 1 & 0 & 0 \\ 0 & 0 & 1 & 0 \end{pmatrix} = \mathbf{C}_4{}^3.$$

4.5   Homogeneous linear transformations in general

The operations we have been considering are governed by matrix equations

$$\mathbf{x}' = \mathbf{A}\mathbf{x} \tag{4-41}$$

in which

$$\mathbf{x}' = \begin{pmatrix} x_1' \\ x_2' \\ \vdots \\ x_n' \end{pmatrix}, \qquad \mathbf{x} = \begin{pmatrix} x_1 \\ x_2 \\ \vdots \\ x_n \end{pmatrix}, \tag{4-42}$$

and

$$A = \begin{pmatrix} A_{11} & A_{12} & \cdot & \cdot \\ A_{21} & A_{22} & \cdot & \cdot \\ \cdot & \cdot & \cdot & \cdot \\ \cdot & \cdot & \cdot & A_{nn} \end{pmatrix}, \tag{4-43}$$

or

$$x_k' = A_{kl}x_l. \tag{4-44}$$

Such transformations are said to be *homogeneous* because each term in (4-44) contains a component of **x** to the same power. They are said to be *linear* because this power is 1.

Two or more of these transformations applied in succession produce the same effect as a single transformation. Indeed, multiplying **x′** by a conformable square matrix **B**,

$$x_j'' = B_{jk}x_k', \tag{4-45}$$

and expanding x_k' according to (4-44) yields

$$x_j'' = B_{jk}A_{kl}x_l = C_{jl}x_l, \tag{4-46}$$

a homogeneous linear form with the coefficients

$$C_{jl} = B_{jk}A_{kl}. \tag{4-47}$$

These make up the transformation matrix

$$\mathbf{C} = \mathbf{BA}. \tag{4-48}$$

Equations (4-44) can yield components of **x′** that are dependent on each other, even when the components of **x** are not. Some information about the original vector is then absent from the transformed vector. No operation on such an **x′** can yield the original **x**.

When all information is preserved, though, there is a second operation, called the *inverse*, that reverses the effect of the first transformation. Then we can let $\mathbf{A}^{-1}$ be the inverse of **A**,

$$\mathbf{x} = \mathbf{A}^{-1}\mathbf{x}' = \mathbf{A}^{-1}\mathbf{Ax}, \tag{4-49}$$

and obtain

$$\mathbf{A}^{-1}\mathbf{A} = \mathbf{E}. \tag{4-50}$$

The inverse of a matrix times the matrix itself equals the identity matrix.

Multiplying the middle and left of (4-49) by **A**, then introducing (4-41), yields

$$\mathbf{AA}^{-1}\mathbf{x}' = \mathbf{Ax} = \mathbf{x}', \tag{4-51}$$

whence

$$\mathbf{AA}^{-1} = \mathbf{E}. \tag{4-52}$$

Transformation **A** cancels $\mathbf{A}^{-1}$ just as $\mathbf{A}^{-1}$ reverses the effect of **A**.

Substituting for **A** the product **AB** changes (4-50) to

$$(\mathbf{AB})^{-1}(\mathbf{AB}) = \mathbf{E}. \tag{4-53}$$

Multiplying this equation from the right, first by $\mathbf{B}^{-1}$,

$$(\mathbf{AB})^{-1}\mathbf{ABB}^{-1} = \mathbf{EB}^{-1}, \tag{4-54}$$

$$(\mathbf{AB})^{-1}\mathbf{A} = \mathbf{B}^{-1}, \tag{4-55}$$

then by $\mathbf{A}^{-1}$, yields

$$(\mathbf{AB})^{-1}\mathbf{AA}^{-1} = \mathbf{B}^{-1}\mathbf{A}^{-1}, \tag{4-56}$$

or

$$(\mathbf{AB})^{-1} = \mathbf{B}^{-1}\mathbf{A}^{-1}. \tag{4-57}$$

Thus the inverse of a product equals the product of the inverses in the opposite order.

EXAMPLE 4.6   Rewrite the set
$$5x + 3y = 2,$$
$$2x + 3y = 1$$
as a matrix equation.

Let the matrix of the coefficients act on a column matrix of the unknowns. Since each row of the result equals the left side of one of the given equations, equate it to a column matrix of the numerical terms:

$$\begin{pmatrix} 5 & 3 \\ 2 & 3 \end{pmatrix} \begin{pmatrix} x \\ y \end{pmatrix} = \begin{pmatrix} 2 \\ 1 \end{pmatrix}.$$

EXAMPLE 4.7   Show that $\mathbf{i}$ is its own inverse.

Equation (4-23) tells us that the reorientation matrix for i is

$$\mathbf{i} = \begin{pmatrix} -1 & 0 & 0 \\ 0 & -1 & 0 \\ 0 & 0 & -1 \end{pmatrix}.$$

But

$$\mathbf{ii} = \begin{pmatrix} -1 & 0 & 0 \\ 0 & -1 & 0 \\ 0 & 0 & -1 \end{pmatrix} \begin{pmatrix} -1 & 0 & 0 \\ 0 & -1 & 0 \\ 0 & 0 & -1 \end{pmatrix} = \begin{pmatrix} 1 & 0 & 0 \\ 0 & 1 & 0 \\ 0 & 0 & 1 \end{pmatrix}.$$

Matrix multiplication of $\mathbf{i}$ by itself does yield the unit matrix.

4.6   Inverting matrices

We have seen how matrices can be written for all pertinent reorientations and permutations of a physical system. Now, each operation of the set can be reversed so that the system returns to its initial state. But a matrix can be constructed for each of these inverse processes, just as for the initial operation. It is the inverse of the matrix for the initial operation.

The inverse of a square matrix that does not represent such a reversible operation can be calculated from (4-50) or (4-52). The unknown matrix is constructed with unknown elements, the indicated multiplications are carried out, and the simultaneous equations are solved.

Alternatively, we can make use of an expression for the determinant of the matrix. This determinant

$$D = \begin{vmatrix} A_{11} & A_{12} & \cdot & \cdot \\ A_{21} & A_{22} & \cdot & \cdot \\ \cdot & \cdot & \cdot & \cdot \\ \cdot & \cdot & \cdot & A_{nn} \end{vmatrix} \tag{4-58}$$

is the sum of

$$A_{11}A_{22} \cdots A_{nn} \tag{4-59}$$

and the products obtained on permuting the second indices in the factors, if a negative sign is assigned each term constructed by an odd permutation, a positive sign each from an even permutation.

The sum D is reproduced by the formula

$$\delta_{jk}D = (-1)^m A_{jl}M_{kl} \tag{4-60}$$

where m is $k + l$ and M_{kl} is the *minor* determinant left when the kth row and lth column are deleted from D. (The δ_{jk} arises because letting $j \neq k$ makes the right side correspond to a determinant in which two rows are equal.)

We get the elements of $\mathbf{E}$ on dividing $\delta_{jk}D$ by D. These elements are also obtained on multiplying out $\mathbf{AA}^{-1}$; so

$$A_{jl}A_{lk}^{-1} = \frac{(-1)^m A_{jl}M_{kl}}{D}, \tag{4-61}$$

and

$$A_{lk}^{-1} = \frac{(-1)^m M_{kl}}{D}. \tag{4-62}$$

Consequently

$$\mathbf{A}^{-1} = \begin{pmatrix} \dfrac{M_{11}}{D} & -\dfrac{M_{21}}{D} & \dfrac{M_{31}}{D} & \cdot \\ -\dfrac{M_{12}}{D} & \dfrac{M_{22}}{D} & -\dfrac{M_{32}}{D} & \cdot \\ \cdot & \cdot & \cdot & \cdot \end{pmatrix} \tag{4-63}$$

EXAMPLE 4.8 Construct $\mathbf{C}_n^{-1}$.

The inverse of C_n consists of *clockwise* rotation of the physical body by $2\pi/n$ radians about the z axis. Letting ϕ be this clockwise $2\pi/n$ changes the square matrix in (4-29) to

$$\mathbf{C}_n^{-1} = \begin{pmatrix} \cos\phi & \sin\phi & 0 \\ -\sin\phi & \cos\phi & 0 \\ 0 & 0 & 1 \end{pmatrix}.$$

EXAMPLE 4.9   Derive the Cramer rule for solving simultaneous equations.

Consider n linear equations in n unknowns

$$
\begin{aligned}
A_{11}x_1 + A_{12}x_2 + \cdots &= b_1, \\
A_{21}x_1 + A_{22}x_2 + \cdots &= b_2, \\
&\ \ \vdots \\
A_{n1}x_1 + A_{n2}x_2 + \cdots &= b_n
\end{aligned}
$$

in the form

$$\mathbf{Ax} = \mathbf{b}.$$

Multiply this by the inverse of $\mathbf{A}$,

$$\mathbf{x} = \mathbf{A}^{-1}\mathbf{b},$$

and write out the result

$$
\begin{pmatrix} x_1 \\ x_2 \\ \vdots \end{pmatrix} = \begin{pmatrix} A_{1k}^{-1}b_k \\ A_{2k}^{-1}b_k \\ \vdots \end{pmatrix}.
$$

Into the jth row, introduce formula (4-62) and obtain

$$x_j = A_{jk}^{-1}b_k = \frac{(-1)^m M_{kj}b_k}{D}, \qquad \text{with}\quad m = k + j.$$

The sum in the numerator reproduces the determinant obtained from D on replacing the jth column with the column of b_k's.

4.7   Reorienting dyadics

In section 4.3, we saw how reorienting a vector changes its components, and the column matrix representing it. Let us now consider how such an operation affects the elements of a dyadic.

We assume that the dyadic $\mathbf{I}$ links physical vector $\boldsymbol{\omega}$ to $\mathbf{L}$

$$\mathbf{I}\boldsymbol{\omega} = \mathbf{L} \tag{4-64}$$

as in (1-52) and (3-81). A reorientation of the physical system with respect to the reference axes does not destroy this relationship, but it does convert $\boldsymbol{\omega}$ to $\boldsymbol{\omega}'$, $\mathbf{L}$ to $\mathbf{L}'$, and $\mathbf{I}$ to the unknown $\mathbf{I}'$:

$$\mathbf{I}'\boldsymbol{\omega}' = \mathbf{L}'. \tag{4-65}$$

Since the physical vectors are represented by directed line segments like $\mathbf{r}$, they must transform according to (4-17):

$$\mathbf{A}\boldsymbol{\omega} = \boldsymbol{\omega}', \tag{4-66}$$

$$\mathbf{AL} = \mathbf{L}'. \tag{4-67}$$

These relationships determine how $\mathbf{I}$ must transform.

Using them to eliminate $\boldsymbol{\omega}'$ and $\mathbf{L}'$ from (4-65),

$$\mathbf{I}'\mathbf{A}\boldsymbol{\omega} = \mathbf{AL}, \tag{4-68}$$

and multiplying through by $\mathbf{A}^{-1}$ yields

$$\mathbf{A}^{-1}\mathbf{I}'\mathbf{A}\omega = \mathbf{L}. \tag{4-69}$$

No difficulty arises because each reorientation matrix $\mathbf{A}$ has an inverse.

Comparing (4-69) with (4-64) now demonstrates that

$$\mathbf{A}^{-1}\mathbf{I}'\mathbf{A} = \mathbf{I}, \tag{4-70}$$

whence

$$\mathbf{I}' = \mathbf{A}\mathbf{I}\mathbf{A}^{-1} \tag{4-71}$$

or

$$\mathbf{I}' = \mathbf{B}^{-1}\mathbf{I}\mathbf{B} \tag{4-72}$$

if $\mathbf{B}$ is the inverse of $\mathbf{A}$.

A rotation and/or reflection that transforms a vector by simple law (4-67) transforms a dyadic for the same system by law (4-71) and its equivalent (4-72). Because the physical dyadic is merely reoriented in the process, transformation (4-72) is called a *similarity transformation*. A particular similarity transformation to be considered later diagonalizes the matrix representing the dyadic.

4.8 Similarity transformations of other square matrices

Since reorientation and permutation matrices link vectors as dyadics do, they transform in the same manner. Details will be worked out here. We will also find that similarity transformations leave the trace invariant and preserve multiplicative properties.

Let radius vectors $\mathbf{x}$ and $\mathbf{y}$ be drawn from a reference origin to points a given distance from this origin in a physical body or system. Let $\mathbf{A}$ be the reorientation matrix that converts $\mathbf{x}$ to $\mathbf{y}$,

$$\mathbf{A}\mathbf{x} = \mathbf{y}. \tag{4-73}$$

Furthermore, introduce the coordinate transformation $\mathbf{P}$,

$$\mathbf{P}\mathbf{x} = \mathbf{x}', \qquad \mathbf{P}\mathbf{y} = \mathbf{y}'. \tag{4-74}$$

Because this transformation does not alter the system, the transformed vectors $\mathbf{x}'$ and $\mathbf{y}'$ are linked by a matrix $\mathbf{B}$,

$$\mathbf{B}\mathbf{x}' = \mathbf{y}', \tag{4-75}$$

that represents the same operation as $\mathbf{A}$. Using (4-74) to eliminate the primed radius vectors,

$$\mathbf{B}\mathbf{P}\mathbf{x} = \mathbf{P}\mathbf{y}, \tag{4-76}$$

and multiplying from the left by $\mathbf{P}^{-1}$ yields

$$\mathbf{P}^{-1}\mathbf{B}\mathbf{P}\mathbf{x} = \mathbf{y}. \tag{4-77}$$

Since (4-77) is a form of (4-73), we have

$$\mathbf{P^{-1}BP = A},\tag{4-78}$$

or

$$\mathbf{B = PAP^{-1}}.\tag{4-79}$$

Letting $\mathbf{Q}$ be $\mathbf{P^{-1}}$ converts this result to

$$\mathbf{B = Q^{-1}AQ}.\tag{4-80}$$

The same equations apply when $\mathbf{P}$ represents a reorientation. Matrix $\mathbf{Q}$ then represents the inverse reorientation and B is a reorientation that is merely equivalent to the initial one, A. Compare figures 4.7 and 4.8.

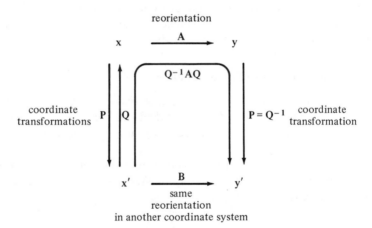

FIGURE 4.7   One interpretation of the equations relating radius vectors $\mathbf{x}$, $\mathbf{y}$, $\mathbf{x'}$, $\mathbf{y'}$.

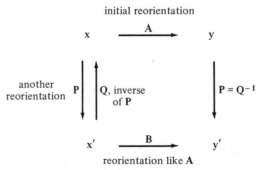

FIGURE 4.8   Another interpretation of the same equations.

When operation Q is set equal to each member of a closed set containing operation A, in turn, the B operations satisfying

$$B = Q^{-1}AQ \tag{4-81}$$

are said to belong to the same *class* as A. The preceding discussion shows that members of a class are equivalent operations.

Although elements of **A** are altered by a similarity transformation, certain functions of the elements are not. One such function is the *trace*, the sum of elements along the principal diagonal. If Tr **A** and Tr **B** designate the trace of **A** and the trace of **B**, respectively, we have

$$\text{Tr } \mathbf{A} = A_{kk} \quad \text{and} \quad \text{Tr } \mathbf{B} = B_{jj} \tag{4-82}$$

by definition.

Setting **B** equal to $\mathbf{Q}^{-1}\mathbf{AQ}$ yields

$$\begin{aligned}
\text{Tr } \mathbf{B} &= (\mathbf{Q}^{-1}\mathbf{AQ})_{jj} = Q_{jk}{}^{-1}A_{kl}Q_{lj} = Q_{lj}Q_{jk}{}^{-1}A_{kl} \\
&= \delta_{lk}A_{kl} = A_{kk} = \text{Tr } \mathbf{A}.
\end{aligned} \tag{4-83}$$

In the fourth step, Kronecker delta δ_{lk} has been introduced to describe the element of identity $Q_{lj}Q_{jk}{}^{-1}$.

The matrices representing a set of reorientations or permutations multiply as the operations. A similarity transformation of one by any other member from the set converts it to some member of the same class. It also leaves the trace invariant, according to the argument above. Since the trace characterizes both the class and the representation, it is called the *character* χ. From (4-83),

$$\chi(B) = \chi(A). \tag{4-84}$$

Let us next consider how a general similarity transformation affects multiplicative properties within a closed set of matrices. The set contains, say, **A**, **B**, and the product

$$\mathbf{AB} = \mathbf{C} \tag{4-85}$$

among other matrices.

We construct an additional matrix **U**, of the same size as **A**, with an inverse $\mathbf{U}^{-1}$. Equation (4-85) is then multiplied by **U** from the right, and by $\mathbf{U}^{-1}$ from the left:

$$\mathbf{U}^{-1}\mathbf{ABU} = \mathbf{U}^{-1}\mathbf{CU}. \tag{4-86}$$

Since

$$\mathbf{UU}^{-1} = \mathbf{E}, \tag{4-87}$$

and since **E** can be inserted anywhere in a product, we have

$$(\mathbf{U}^{-1}\mathbf{AU})(\mathbf{U}^{-1}\mathbf{BU}) = \mathbf{U}^{-1}\mathbf{CU}. \tag{4-88}$$

The transformed matrices $\mathbf{U}^{-1}\mathbf{AU}$, $\mathbf{U}^{-1}\mathbf{BU}$, $\mathbf{U}^{-1}\mathbf{CU}, \ldots$ multiply in the same way as the original ones **A**, **B**, **C**, $\ldots$; there is a one-to-one correspondence. Note that **U** may contain complex elements, so some or all of the transformed matrices may be complex.

EXAMPLE 4.10   If the symmetry operations for a given molecule are E, C_4, C_2, $C_4{}^3$, what are the classes?

Since a rotation by angle ϕ_1 plus a rotation by angle ϕ_2 about the *same* axis produces the same result as a rotation by ϕ_2 plus one by ϕ_1, the rotations commute with each other. Here

$$C_2{}^{-1}C_4C_2 = C_4C_2{}^{-1}C_2 = C_4$$

and so on. Hence no test similarity transformation of *any* of the given operations alters the operation; each symmetry operation forms a class by itself.

EXAMPLE 4.11   Show that C_3 and $C_3{}^{-1}$ belong to the same class when σ_v is also present.

Arrange three similar particles around a threefold axis as in figure 4.9. Then apply σ_v, C_3, and $\sigma_v{}^{-1}$ in succession. Note that the resulting configuration is the same as that obtained from $C_3{}^{-1}$ alone. Therefore

$$C_3{}^{-1} = \sigma_v{}^{-1}C_3\sigma_v.$$

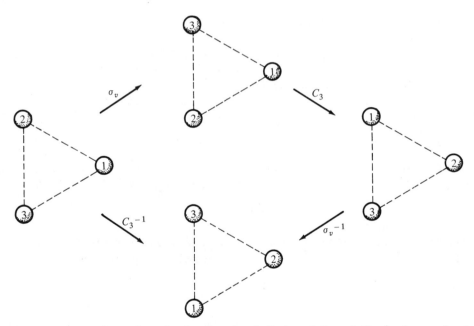

FIGURE 4.9   Transformations for showing the similarity of C_3 and $C_3{}^{-1}$ when σ_v is also a symmetry operation.

4.9   Multiplying partitioned and reduced matrices

A suitable similarity transformation may convert two or more matrices of a set representing reorientations to a form in which nonzero square submatrices lie on the principal diagonals and zeros appear everywhere else. Such transformed matrices correspond to the same operations as the original matrices (recall section 4.8.) But when two are reduced to this form, is their product, also in the set, reduced to the same form? Before answering this question, let us develop a rule for multiplying partitioned matrices.

The given conformable matrices **A** and **B** are first broken down into submatrices by partitions between

(a) the $(s\text{-}1)$th and the sth column of **A**,
(b) the $(s\text{-}1)$th and the sth row of **B**,
(c) the $(m\text{-}1)$th and the mth row of **A**,
(d) the $(m\text{-}1)$th and the mth column of **B**,

and so on. Each submatrix is designated by a small boldface letter with subscripts from the element in its upper left corner:

$$\mathbf{A} = \begin{pmatrix} \mathbf{a}_{11} & \mathbf{a}_{1s} & \cdot & \cdot \\ \mathbf{a}_{m1} & \mathbf{a}_{ms} & \cdot & \cdot \\ \cdot & \cdot & \cdot & \end{pmatrix} \quad \text{and} \quad \mathbf{B} = \begin{pmatrix} \mathbf{b}_{11} & \mathbf{b}_{1m} & \cdot & \cdot \\ \mathbf{b}_{s1} & \mathbf{b}_{sm} & \cdot & \cdot \\ \cdot & \cdot & \cdot & \end{pmatrix}. \tag{4-89}$$

Treating the submatrices as the matrix elements in formula (4-7) then yields the *block product*

$$\mathbf{A}(\text{block})\mathbf{B} \equiv \begin{pmatrix} \mathbf{a}_{11}\mathbf{b}_{11} + \mathbf{a}_{1s}\mathbf{b}_{s1} + \cdots & \mathbf{a}_{11}\mathbf{b}_{1m} + \mathbf{a}_{1s}\mathbf{b}_{sm} + \cdots & \cdot & \cdot \\ \mathbf{a}_{m1}\mathbf{b}_{11} + \mathbf{a}_{ms}\mathbf{b}_{s1} + \cdots & \mathbf{a}_{m1}\mathbf{b}_{1m} + \mathbf{a}_{ms}\mathbf{b}_{sm} + \cdots & \cdot & \cdot \\ \cdot & \cdot & & \cdot \end{pmatrix} \tag{4-90}$$

where each repeated index takes on just *one value* (that assigned in the partitioning). The numbers of rows in each submatrix sum of (4-90) equals the number of rows in the first factor of each term. The number of columns is determined by the number of columns in the second factor of each term.

When

$$m \le j < n, \quad \text{and} \quad 1 \le l < m, \tag{4-91}$$

all terms of the element $[\mathbf{A}(\text{block})\mathbf{B}]_{jl}$ are in

$$\mathbf{a}_{m1}\mathbf{b}_{11} + \mathbf{a}_{ms}\mathbf{b}_{s1} + \cdots. \tag{4-92}$$

Furthermore, the part of the total in $\mathbf{a}_{m1}\mathbf{b}_{11}$ is

$$\sum_{k=1}^{s-1} A_{jk}B_{kl}, \tag{4-93}$$

while the part in $\mathbf{a}_{ms}\mathbf{b}_{s1}$ is

$$\sum_{k=s}^{t-1} A_{jk}B_{kl}, \tag{4-94}$$

and so on.

Adding these parts yields the form in (4-7):

$$[\mathbf{A}(\text{block})\mathbf{B}]_{jl} = \sum_{\text{all } k} A_{jk}B_{kl}. \tag{4-95}$$

The argument proceeds similarly for all other possible ranges of j and l. Consequently, block multiplication produces the same result as ordinary matrix multiplication:

$$\mathbf{A}(\text{block})\mathbf{B} = \mathbf{AB}. \tag{4-96}$$

When we apply this rule to two reduced matrices that represent reorientations of a given system, we obtain

$$\mathbf{CD} = \begin{pmatrix} \mathbf{c}_{11} & 0 & 0 & \cdot \\ 0 & \mathbf{c}_{mm} & 0 & \cdot \\ 0 & 0 & \mathbf{c}_{nn} & \cdot \\ \cdot & \cdot & \cdot & \cdot \end{pmatrix} \begin{pmatrix} \mathbf{d}_{11} & 0 & 0 & \cdot \\ 0 & \mathbf{d}_{mm} & 0 & \cdot \\ 0 & 0 & \mathbf{d}_{nn} & \cdot \\ \cdot & \cdot & \cdot & \cdot \end{pmatrix}$$

$$= \begin{pmatrix} \mathbf{c}_{11}\mathbf{d}_{11} & 0 & 0 & \cdot \\ 0 & \mathbf{c}_{mm}\mathbf{d}_{mm} & 0 & \cdot \\ 0 & 0 & \mathbf{c}_{nn}\mathbf{d}_{nn} & \cdot \\ \cdot & \cdot & \cdot & \cdot \end{pmatrix}. \tag{4-97}$$

The product, which also represents a reorientation, has the same reduced form. Reduction of all matrices in a closed set is allowed by the multiplication rule. Formulas governing the reduction will be developed in a chapter on group theory.

4.10   Diagonalizing matrices

When a given square matrix acts on vectors pointing in certain directions, the line of action of the vector is unaffected. Thus, the moment of inertia $\mathbf{I}$ acting on an ω lying along a principal axis of a body yields an angular momentum $\mathbf{L}$ along this axis. Such a column matrix is called an *eigenvector* of the square matrix.

When the given matrix possesses enough independent eigenvectors, the latter make up a square matrix for which an inverse exists. A similarity transformation of the given matrix by this eigenvector matrix then yields a matrix in which zeros appear everywhere except on the principal diagonal, as we will now see.

An *eigenvalue* of a given square matrix $\mathbf{A}$ is a real or complex number λ that satisfies the equation

$$\mathbf{Au} = \lambda\mathbf{u} = \mathbf{u}\lambda. \tag{4-98}$$

The vector $\mathbf{u}$ whose line of action is unaffected by $\mathbf{A}$ is the corresponding eigenvector of $\mathbf{A}$.

Applying the rule for matrix multiplication to (4-98) yields

$$A_{jk}u_k = \lambda\delta_{jk}u_k \tag{4-99}$$

or

$$(A_{jk} - \lambda\delta_{jk})u_k = 0. \tag{4-100}$$

These equations have the same form as those in example 4.9, with all b_j's zero. The numerator in Cramer's formula is therefore zero. For at least one u_k to be different from zero, the denominator D must also be zero:

$$\begin{vmatrix} A_{11} - \lambda & A_{12} & \cdot & \cdot \\ A_{21} & A_{22} - \lambda & \cdot & \cdot \\ \cdot & \cdot & \cdot & \cdot \end{vmatrix} = 0. \tag{4-101}$$

Equation (4-101) is called the *secular equation*.

When **A** is an $n \times n$ matrix, this equation has n roots, $\lambda_1, \lambda_2, \ldots, \lambda_n$, not necessarily all distinct. But if the elements in **A** are sufficiently diverse, there is an independent **u** for each root. Let us label these $\mathbf{u}_1, \mathbf{u}_2, \ldots, \mathbf{u}_n$ and form the square matrices

$$\mathbf{U} = (\mathbf{u}_1 \quad \mathbf{u}_2 \quad \cdot \quad \cdot \quad \cdot \quad \mathbf{u}_n), \tag{4-102}$$

$$\mathbf{\Lambda} = \begin{pmatrix} \lambda_1 & 0 & \cdot & \cdot & \cdot \\ 0 & \lambda_2 & \cdot & \cdot & \cdot \\ \cdot & & \cdot & & \cdot \\ \cdot & & & & \lambda_n \end{pmatrix}. \tag{4-103}$$

The set of satisfied eigenvalue equations can then be written as

$$\mathbf{AU} = \mathbf{U\Lambda}, \tag{4-104}$$

whence

$$\mathbf{U}^{-1}\mathbf{AU} = \mathbf{\Lambda}. \tag{4-105}$$

The eigenvectors may be arranged in any order in **U**; so the eigenvalues may appear in any order in **Λ**. But the step from (4-104) to (4-105) depends on the existence of an inverse $\mathbf{U}^{-1}$, and thus on the presence of n distinct independent eigenvectors.

To determine whether a similarity transformation alters eigenvalues, consider

$$\mathbf{V}^{-1}\mathbf{AV} = \mathbf{B} \tag{4-106}$$

where **V** is any suitable matrix. Multiply (4-106) from the left by **V** and from the right by $\mathbf{V}^{-1}$ to get a substitution for **A**

$$\mathbf{A} = \mathbf{VBV}^{-1} \tag{4-107}$$

in equation (4-105)

$$\mathbf{U}^{-1}(\mathbf{VBV}^{-1})\mathbf{U} = (\mathbf{V}^{-1}\mathbf{U})^{-1}\mathbf{BV}^{-1}\mathbf{U} = \mathbf{\Lambda}. \tag{4-108}$$

The final equation has the form

$$\mathbf{W}^{-1}\mathbf{BW} = \mathbf{\Lambda} \tag{4-109}$$

where **W** is $\mathbf{V}^{-1}\mathbf{U}$. Consequently, **B** reduces to the same diagonal matrix as **A**. But since this contains $\lambda_1, \lambda_2, \ldots, \lambda_n$, an arbitrary similarity transformation does leave the eigenvalues unchanged.

From formula (4-97), multiplying **Λ** by itself once and then by itself $p - 2$ more times yields

$$\mathbf{\Lambda}^2 = \begin{pmatrix} \lambda_1{}^2 & 0 & \cdot & \cdot & \cdot \\ 0 & \lambda_2{}^2 & \cdot & \cdot & \cdot \\ \cdot & \cdot & \cdot & \cdot & \cdot \\ \cdot & \cdot & \cdot & \cdot & \lambda_n{}^2 \end{pmatrix} \tag{4-110}$$

and

$$\mathbf{\Lambda}^p = \begin{pmatrix} \lambda_1{}^p & 0 & \cdot & \cdot & \cdot \\ 0 & \lambda_2{}^p & \cdot & \cdot & \cdot \\ \cdot & \cdot & \cdot & \cdot & \cdot \\ \cdot & \cdot & \cdot & \cdot & \lambda_n{}^p \end{pmatrix}. \tag{4-111}$$

Equation (4-111) enables us to raise a given square matrix **A** to any power. The pth power of

$$\boldsymbol{\Lambda} = \mathbf{U}^{-1}\mathbf{A}\mathbf{U} \tag{4-112}$$

is

$$\boldsymbol{\Lambda}^p = (\mathbf{U}^{-1}\mathbf{A}\mathbf{U})(\mathbf{U}^{-1}\mathbf{A}\mathbf{U}) \cdots (\mathbf{U}^{-1}\mathbf{A}\mathbf{U}) = \mathbf{U}^{-1}\mathbf{A}^p\mathbf{U}, \tag{4-113}$$

whence

$$\mathbf{U}\boldsymbol{\Lambda}^p\mathbf{U}^{-1} = \mathbf{A}^p. \tag{4-114}$$

EXAMPLE 4.12   Diagonalize the reorientation matrix for the C_n operation.
A radius vector **r** is rotated $1/n$ turn about the z axis by the 3×3 matrix in (4-29) when

$$\phi = \frac{2\pi}{n} \text{ radians.}$$

To construct the corresponding secular equation (4-101), subtract the unknown λ from each diagonal element, take the determinant of the matrix, and set it equal to zero. The result is

$$\begin{vmatrix} \cos\phi - \lambda & -\sin\phi & 0 \\ \sin\phi & \cos\phi - \lambda & 0 \\ 0 & 0 & 1-\lambda \end{vmatrix} = 0$$

or

$$(\cos\phi - \lambda)^2(1-\lambda) + \sin^2\phi(1-\lambda) = 0.$$

This equality is satisfied when

$$\lambda = 1$$

or

$$(\cos\phi - \lambda)^2 = -\sin^2\phi.$$

The square root of the latter

$$\cos\phi - \lambda = \mp i\sin\phi,$$

yields the second and third eigenvalues

$$\lambda = \cos\phi \pm i\sin\phi = e^{\pm i\phi}.$$

Since the three eigenvalues are distinct, the corresponding eigenvectors are independent and diagonal form (4-103) exists:

$$\begin{pmatrix} 1 & 0 & 0 \\ 0 & e^{i\phi} & 0 \\ 0 & 0 & e^{-i\phi} \end{pmatrix}.$$

EXAMPLE 4.13   Diagonalize a permutation matrix for C_n.
Let n similar particles be arranged at equidistant intervals around a ring as the four are in figure 4.6. A rotation by C_n then causes the radius vector r_j locating the jth particle to become r_{j+1} (if $n + 1$ is 1):

$$r_j' \equiv C_n r_j = r_{j+1}.$$

A matrix equation summarizing the changes is

$$\begin{pmatrix} r_1' \\ r_2' \\ \cdot \\ r_{n-1}' \\ r_n' \end{pmatrix} = \begin{pmatrix} 0 & 1 & 0 & \cdot & 0 \\ 0 & 0 & 1 & \cdot & 0 \\ \cdot & \cdot & \cdot & \cdot & \cdot \\ 0 & 0 & 0 & \cdot & 1 \\ 1 & 0 & 0 & \cdot & 0 \end{pmatrix} \begin{pmatrix} r_1 \\ r_2 \\ \cdot \\ r_{n-1} \\ r_n \end{pmatrix}$$

or

$$\mathbf{r}' = \mathbf{C}_n \mathbf{r}.$$

The elements in the $n \times n$ matrix representing C_n yield the secular equation

$$\begin{vmatrix} -\lambda & 1 & 0 & \cdot & 0 \\ 0 & -\lambda & 1 & \cdot & 0 \\ \cdot & \cdot & \cdot & \cdot & \cdot \\ 0 & 0 & 0 & \cdot & 1 \\ 1 & 0 & 0 & \cdot & -\lambda \end{vmatrix} = 0.$$

The first term in sum D for the determinant is the product (4-59)

$$(-\lambda)^n,$$

while the other terms come from permutations of the second indices on the elements. Those permutations that leave the first factor as $-\lambda$ all include 0 as a factor. Indeed, if we start multiplying elements along the principal diagonal as in (4-59), but at a certain column, skip more than one row down, we find a zero.

Increasing each of the second indices in (4-59) by 1 moves the path of multiplication to the diagonal of 1's. This product being obtained on $n - 1$ exchanges of indices, it has the form

$$(-1)^{n-1}(1)^n.$$

All permutations of the second indices leaving the first factor unchanged include 0 as a factor. (At a certain column, we can skip down to a $-\lambda$. But then, the only elements in the skipped row that are in columns not used by earlier factors are zeros. One of these has to be in the product.)

All permutations that increase the second index on the first term beyond 2 lead to 0 because the first factor, at least, is 0. Consequently, the secular equation reduces to

$$(-\lambda)^n + (-1)^{n-1}(1)^n = 0.$$

This equation is easily solved. First, rearrange it and write -1 as $e^{\pi i}$, $+1$ as $e^{2m\pi i}$, where m is an integer:

$$(-\lambda)^n = (-1)^n = e^{2m\pi i} e^{n\pi i}.$$

Then take the nth root:

$$-\lambda = e^{2m\pi i/n} e^{\pi i}$$

or

$$\lambda = e^{2m\pi i/n} = e^{im\phi}$$

with

$$m = 0, 1, 2, \ldots, n - 1, \quad \text{and} \quad \phi = \frac{2\pi}{n}.$$

The resulting λ's give matrix (4-103) the form

$$
\begin{pmatrix}
1 & 0 & 0 & \cdot & 0 \\
0 & e^{i\phi} & 0 & \cdot & 0 \\
0 & 0 & e^{i2\phi} & \cdot & 0 \\
\cdot & \cdot & \cdot & \cdot & \cdot \\
0 & 0 & 0 & \cdot & e^{-i\phi}
\end{pmatrix}.
$$

4.11 Matrices with real eigenvalues

Since measurements in the laboratory yield real numbers, each observable scalar is real. Solutions of the secular equation may be complex, however; so if they are to represent possible values of a physical property, an additional condition must be imposed. One that is sufficient will be described here. Some additional background will be presented first, though.

Reflecting the elements of a matrix across the principal diagonal, that is, interchanging its rows and columns, produces a new matrix called the *transpose* of the original one. Letting $\tilde{\mathbf{A}}$ represent the transpose of $\mathbf{A}$, we have

$$
(\tilde{\mathbf{A}})_{jk} = A_{kj}. \tag{4-115}
$$

When the transposed matrix is the same as the original one, the matrix is said to be *symmetric*. The moment-of-inertia $\mathbf{I}$ is such a matrix.

Equation (4-115) tells us that an element of the transpose of $\mathbf{AB}$ is

$$
(\widetilde{\mathbf{AB}})_{jl} = A_{lk}B_{kj}. \tag{4-116}
$$

But

$$
A_{lk} = (\tilde{\mathbf{A}})_{kl} \quad \text{and} \quad B_{kj} = (\tilde{\mathbf{B}})_{jk} \tag{4-117}
$$

combine to yield

$$
A_{lk}B_{kj} = (\tilde{\mathbf{A}})_{kl}(\tilde{\mathbf{B}})_{jk} = (\tilde{\mathbf{B}})_{jk}(\tilde{\mathbf{A}})_{kl}. \tag{4-118}
$$

So

$$
(\widetilde{\mathbf{AB}})_{jl} = (\tilde{\mathbf{B}})_{jk}(\tilde{\mathbf{A}})_{kl}, \tag{4-119}
$$

and

$$
\widetilde{\mathbf{AB}} = \tilde{\mathbf{B}}\tilde{\mathbf{A}}. \tag{4-120}
$$

The transpose of the product of two matrices equals the product of the transposed matrices in reverse order.

Transposing a matrix and taking the complex conjugate of each element produces a new matrix called the *Hermitian adjoint* of the original one. Letting $\mathbf{A}^{\dagger}$ represent this adjoint of $\mathbf{A}$, we have

$$
(\mathbf{A}^{\dagger})_{jk} = A_{kj}^{*}. \tag{4-121}
$$

When the adjoint matrix is the same as the original one, the matrix is said to be *Hermitian*. Note that $\mathbf{I}$ is Hermitian.

Applying (4-121) to the complex conjugate of each side of (4-119) yields

$$[(\mathbf{AB})^{\dagger}]_{jl} = (\mathbf{B}^{\dagger})_{jk}(\mathbf{A}^{\dagger})_{kl} \qquad (4\text{-}122)$$

whence

$$(\mathbf{AB})^{\dagger} = \mathbf{B}^{\dagger}\mathbf{A}^{\dagger}. \qquad (4\text{-}123)$$

The Hermitian adjoint of a product is the product of the adjoints in the opposite order.

Let us now apply this rule to the eigenvalue equation

$$\mathbf{Au} = \lambda\mathbf{u} \qquad (4\text{-}124)$$

to get

$$\mathbf{u}^{\dagger}\mathbf{A}^{\dagger} = \mathbf{u}^{\dagger}\lambda^{*}. \qquad (4\text{-}125)$$

Then multiply (4-124) by $\mathbf{u}^{\dagger}$ from the left and (4-125) by $\mathbf{u}$ from the right:

$$\mathbf{u}^{\dagger}\mathbf{Au} = \mathbf{u}^{\dagger}\lambda\mathbf{u}, \qquad (4\text{-}126)$$

$$\mathbf{u}^{\dagger}\mathbf{A}^{\dagger}\mathbf{u} = \mathbf{u}^{\dagger}\lambda^{*}\mathbf{u}. \qquad (4\text{-}127)$$

When the eigenvalue is *real*,

$$\lambda = \lambda^{*}, \qquad (4\text{-}128)$$

the right sides of (4-126) and (4-127) are the same. Then the left sides are also equal:

$$\mathbf{u}^{\dagger}\mathbf{Au} = \mathbf{u}^{\dagger}\mathbf{A}^{\dagger}\mathbf{u}. \qquad (4\text{-}129)$$

If result (4-129) is to hold for the complete set of eigenvectors, we must have

$$\mathbf{A} = \mathbf{A}^{\dagger}. \qquad (4\text{-}130)$$

That is, matrix $\mathbf{A}$ must be Hermitian.

4.12 Properties of some eigenvectors

A few manipulations suffice to show that the eigenvectors corresponding to different eigenvalues are orthogonal when the operating matrix is Hermitian. When $\mathbf{A}$ is also real, all eigenvectors can be made real. But they do not have to be, because a complex constant times an eigenvector is still an eigenvector.

Let $\mathbf{u}_1$ and $\mathbf{u}_2$ be eigenvectors corresponding to the two eigenvalues, λ_1 and λ_2, of $\mathbf{A}$:

$$\mathbf{Au}_1 = \lambda_1\mathbf{u}_1, \qquad (4\text{-}131)$$

$$\mathbf{Au}_2 = \lambda_2\mathbf{u}_2. \qquad (4\text{-}132)$$

Multiply (4-131) from the left by $\mathbf{u}_2^{\dagger}$:

$$\mathbf{u}_2^{\dagger}\mathbf{Au}_1 = \lambda_1\mathbf{u}_2^{\dagger}\mathbf{u}_1. \qquad (4\text{-}133)$$

Assume that $\mathbf{A}$ is Hermitian and take the adjoint of (4-132):

$$\mathbf{u}_2^{\dagger}\mathbf{A}^{\dagger} = \mathbf{u}_2^{\dagger}\mathbf{A} = \lambda_2\mathbf{u}_2^{\dagger}. \qquad (4\text{-}134)$$

Then multiply the result from the right by $\mathbf{u}_1$:

$$\mathbf{u}_2{}^t\mathbf{A}\mathbf{u}_1 = \lambda_2\mathbf{u}_2{}^t\mathbf{u}_1. \qquad (4\text{-}135)$$

Since the left sides of (4-133) and (4-135) are the same, the right sides must be equal:

$$\lambda_1\mathbf{u}_2{}^t\mathbf{u}_1 = \lambda_2\mathbf{u}_2{}^t\mathbf{u}_1. \qquad (4\text{-}136)$$

When λ_1 equals λ_2, this equation is an identity that does not restrict $\mathbf{u}_2{}^t$ or $\mathbf{u}_1$; but when λ_1 is different from λ_2, the equation can only be satisfied if

$$\mathbf{u}_2{}^t\mathbf{u}_1 = 0. \qquad (4\text{-}137)$$

Hence any two eigenvectors corresponding to different eigenvalues of a Hermitian $\mathbf{A}$ must be *orthogonal*.

If the given matrix $\mathbf{A}$ is both real and symmetric, it is Hermitian and its eigenvalues are real. The complex conjugate of

$$\mathbf{A}\mathbf{u}_1 = \lambda_1\mathbf{u}_1 \qquad (4\text{-}138)$$

is then

$$\mathbf{A}\mathbf{u}_1{}^* = \lambda_1\mathbf{u}_1{}^*. \qquad (4\text{-}139)$$

Let us add these equations and factor $\mathbf{A}$ from the left side:

$$\mathbf{A}\mathbf{u}_1 + \mathbf{A}\mathbf{u}_1{}^* = \mathbf{A}(\mathbf{u}_1 + \mathbf{u}_1{}^*) = \lambda_1(\mathbf{u}_1 + \mathbf{u}_1{}^*). \qquad (4\text{-}140)$$

We see that the real vector $\mathbf{u}_1 + \mathbf{u}_1{}^*$ satisfies the eigenvalue equation. Since λ_1 may be any of the eigenvalues, there is a *real eigenvector* for each eigenvalue of a real symmetric matrix.

EXAMPLE 4.14   What properties do eigenvectors and eigenvalues of the moment-of-inertia matrix generally exhibit?

Combining the eigenvalue equation for the moment-of-inertia matrix

$$\mathbf{I}\omega = \lambda\omega$$

with the equation defining this matrix, (3-81),

$$\mathbf{I}\omega = \mathbf{L}$$

yields

$$\mathbf{L} = \lambda\omega.$$

When the angular velocity ω is an eigenvector for $\mathbf{I}$, the angular momentum $\mathbf{L}$ has the same direction as ω.

Since $\mathbf{I}$ is real and symmetric, a real eigenvector exists for each λ. Any two of these real ω's belonging to different λ's must satisfy equation (4-137)

$$\tilde{\omega}_j\omega_k = 0$$

since $\mathbf{I}$ is Hermitian. As a consequence, the eigenvectors are mutually perpendicular vectors in ordinary space:

$$\omega_j \cdot \omega_k = 0.$$

These eigenvectors define a set of Cartesian axes along which **L** and ω may point simultaneously. The eigenvalues of **I** are the moments of inertia about these axes. The corresponding products of inertia are zero.

4.13 Transformations that preserve Hermitian scalar products

Whenever a homogeneous linear transformation distorts some aspect of a physical system, the transformed system is different from the original one and various scalar products of physically significant vectors are altered. But when the operation merely reorients the system, these products remain invariant.

Some of the pertinent vectors differ from those we have considered, being abstractions in an n-dimensional mathematical space. When these have complex components, we take the complex conjugate of the first factor before dot multiplying it with the second. Such a combination is called the *Hermitian scalar product*.

As examples, we have

$$x_1{}^2 + x_2{}^2 + x_3{}^2 \tag{4-141}$$

where (x_1, x_2, x_3) locates a given element of the system,

$$\tilde{\mathbf{F}} \, d\mathbf{r}, \tag{4-142}$$

where **F** is the force acting on the element and **r** is the conventional radius vector drawn from a point that is not changed to the given element,

$$r_1{}^2 + r_2{}^2 + \cdots + r_n{}^2, \tag{4-143}$$

where component r_j is the 3-dimensional radius vector drawn from the invariant point to the jth part that is permuted,

$$\phi_1{}^*\phi_1 + \phi_2{}^*\phi_2 + \cdots + \phi_n{}^*\phi_n, \tag{4-144}$$

where component ϕ_j is a function associated with the jth part that is permuted, and

$$\phi_1{}^*\psi_1 + \phi_2{}^*\psi_2 + \cdots + \phi_n{}^*\psi_n, \tag{4-145}$$

where functions ϕ_j and ψ_j are both associated with the jth part. The complex conjugate of the first factor is taken so the result is real when the vector is multiplied by itself, as in expression (4-144).

Let us represent (4-141) through (4-145), and similar forms, as $\mathbf{r}^\dagger\mathbf{s}$ before the transformation and as $\mathbf{r}'^\dagger\mathbf{s}'$ after. Being vectors in the same n-dimensional mathematical space, **r** and **s** obey the same transformation law:

$$\mathbf{r}' = \mathbf{A}\mathbf{r}, \tag{4-146}$$

$$\mathbf{s}' = \mathbf{A}\mathbf{s}. \tag{4-147}$$

But if the transformation does not distort the system, it leaves the Hermitian scalar product invariant:

$$\mathbf{r}'^\dagger\mathbf{s}' = \mathbf{r}^\dagger\mathbf{s}. \tag{4-148}$$

Combining these equations to eliminate the primed vectors then yields

$$\mathbf{r}^\dagger\mathbf{A}^\dagger\mathbf{A}\mathbf{s} = \mathbf{r}^\dagger\mathbf{s}. \tag{4-149}$$

For (4-149) to hold regardless of how $\mathbf{r}$ and $\mathbf{s}$ are constructed in their space, we must have

$$\mathbf{A}^\dagger\mathbf{A} = \mathbf{E}. \tag{4-150}$$

Any transformation that obeys this equation is said to be *unitary*. Transforming operator $\mathbf{A}$ is called a unitary matrix.

Note that as long as $\mathbf{A}$ is unitary, the Hermitian scalar product between any two vectors $\mathbf{r}$ and $\mathbf{s}$ on which it acts is invariant:

$$\mathbf{r}^\dagger\mathbf{s} = \mathbf{r}^\dagger\mathbf{E}\mathbf{s} = \mathbf{r}^\dagger\mathbf{A}^\dagger\mathbf{A}\mathbf{s} = (\mathbf{A}\mathbf{r})^\dagger\mathbf{A}\mathbf{s} = \mathbf{r}'^\dagger\mathbf{s}'. \tag{4-151}$$

Consequently, a unitary transformation leaves the magnitude of each vector in the space unchanged.

Some properties of unitary matrices will now be considered. Rewrite equation (4-150) in index notation

$$(\mathbf{A}^\dagger)_{jk}A_{kl} = \delta_{jl} \tag{4-152}$$

and introduce defining relationship (4-121):

$$A_{kj}{}^*A_{kl} = \delta_{jl}. \tag{4-153}$$

Let A_{kl} be the kth component of vector $\mathbf{a}_l$. Equation (4-153) then yields

$$\mathbf{a}_j{}^* \cdot \mathbf{a}_l = \delta_{jl}. \tag{4-154}$$

The vectors formed from the columns of a unitary matrix are orthogonal and of unit magnitude (orthonormal).

Next, multiply (4-150) from the right by $\mathbf{A}^{-1}$,

$$\mathbf{A}^\dagger\mathbf{A}\mathbf{A}^{-1} = \mathbf{E}\mathbf{A}^{-1}, \tag{4-155}$$

and reduce

$$\mathbf{A}^\dagger = \mathbf{A}^{-1}. \tag{4-156}$$

The Hermitian adjoint of a unitary matrix is its inverse.

Finally, multiply (4-156) from the left by $\mathbf{A}$ to get

$$\mathbf{A}\mathbf{A}^\dagger = \mathbf{E}, \tag{4-157}$$

or

$$A_{jk}(\mathbf{A}^\dagger)_{kl} = \delta_{jl}. \tag{4-158}$$

Defining relationship (4-121) converts this to

$$A_{jk}A_{lk}{}^* = \delta_{jl}. \tag{4-159}$$

Letting A_{jk} be the kth component of vector $\mathbf{a}_j$ then yields

$$\mathbf{a}_j \cdot \mathbf{a}_l{}^* = \delta_{jl} \tag{4-160}$$

or

$$\mathbf{a}_j{}^* \cdot \mathbf{a}_l = \delta_{jl}. \tag{4-161}$$

The vectors formed from the rows of $\mathbf{A}$ are also orthonormal.

EXAMPLE 4.15 Show that the elements in matrix S_3 of example 4.4 obey the orthogonality conditions (4-159) and (4-153).

Take the elements in each row as corresponding components of a separate vector

$$\mathbf{a} = -\tfrac{1}{2}\mathbf{i} - \frac{\sqrt{3}}{2}\mathbf{j} + 0\mathbf{k},$$

$$\mathbf{b} = \frac{\sqrt{3}}{2}\mathbf{i} - \tfrac{1}{2}\mathbf{j} + 0\mathbf{k},$$

$$\mathbf{c} = 0\mathbf{i} + 0\mathbf{j} - 1\mathbf{k},$$

and form the dot products:

$$\mathbf{a} \cdot \mathbf{a} = \tfrac{1}{4} + \tfrac{3}{4} + 0 = 1,$$
$$\mathbf{b} \cdot \mathbf{b} = \tfrac{3}{4} + \tfrac{1}{4} + 0 = 1,$$
$$\mathbf{c} \cdot \mathbf{c} = 0 + 0 + 1 = 1,$$

$$\mathbf{a} \cdot \mathbf{b} = -\frac{\sqrt{3}}{4} + \frac{\sqrt{3}}{4} + 0 = 0,$$

$$\mathbf{a} \cdot \mathbf{c} = 0 + 0 + 0 = 0,$$
$$\mathbf{b} \cdot \mathbf{c} = 0 + 0 + 0 = 0.$$

Vectors $\mathbf{a}$, $\mathbf{b}$, $\mathbf{c}$ form an orthonormal set, as (4-159) and (4-161) state.

Then treat each column as containing components of a separate vector

$$\mathbf{d} = -\tfrac{1}{2}\mathbf{i} + \frac{\sqrt{3}}{2}\mathbf{j} + 0\mathbf{k},$$

$$\mathbf{e} = -\frac{\sqrt{3}}{2}\mathbf{i} - \tfrac{1}{2}\mathbf{j} + 0\mathbf{k},$$

$$\mathbf{f} = 0\mathbf{i} + 0\mathbf{j} - 1\mathbf{k},$$

and form the dot products:

$$\mathbf{d} \cdot \mathbf{d} = \tfrac{1}{4} + \tfrac{3}{4} + 0 = 1,$$
$$\mathbf{e} \cdot \mathbf{e} = \tfrac{3}{4} + \tfrac{1}{4} + 0 = 1,$$
$$\mathbf{f} \cdot \mathbf{f} = 0 + 0 + 1 = 1,$$

$$\mathbf{d} \cdot \mathbf{e} = \frac{\sqrt{3}}{4} - \frac{\sqrt{3}}{4} + 0 = 0,$$

$$\mathbf{d} \cdot \mathbf{f} = 0 + 0 + 0 = 0,$$
$$\mathbf{e} \cdot \mathbf{f} = 0 + 0 + 0 = 0.$$

Vectors $\mathbf{d}$, $\mathbf{e}$, $\mathbf{f}$ also form an orthonormal set.

4.14   Recapitulation

A rectangular array of numbers is a matrix if and only if it combines with conformable arrays by formulas (4-3) and (4-7). A column of numbers that is a matrix represents a vector, while a square matrix represents a dyad. The operator that transforms an n-dimensional vector linearly is an $n \times n$ square matrix. When such an operator transforms two vectors linked by a dyad, the dyad itself undergoes a similarity transformation.

Formulas (4-41) and (4-71) do not expand or distort a physical system as long as the transforming matrix **A** is unitary. Conversely, any reorientation or permutation that does not change the physical system in any essential way is represented by such a matrix. A dyad, on the other hand, does alter the vector on which it acts in a fundamental manner. (Recall how **I** changes ω to **L**.)

Two sets of matrices related by a similarity transformation possess a similar algebraic structure. Applying the same transformation to **A** and **B** in

$$\mathbf{AB} = \mathbf{C} \tag{4-162}$$

yields

$$\mathbf{U}^{-1}\mathbf{AUU}^{-1}\mathbf{BU} = \mathbf{U}^{-1}\mathbf{ABU} = \mathbf{U}^{-1}\mathbf{CU} \tag{4-163}$$

or

$$\mathbf{A'B'} = \mathbf{C'}. \tag{4-164}$$

Any set of matrices that multiply as the symmetry operations for a figure can be divided into classes. In each class, typical members **A** and **B** are related by a similarity transformation

$$\mathbf{U}^{-1}\mathbf{AU} = \mathbf{B} \tag{4-165}$$

with other members **U** of the complete set.

DISCUSSION QUESTIONS

4.1 Discuss how rules of combination give meaning to sets of numbers. When are the sets matrices?

4.2 Express multiplication by a scalar as matrix multiplication.

4.3 How are multiple products formed? Why does matrix multiplication follow the associative law?

4.4 Show why **A** and **B** may not be conformable in the product **BA** even though they are conformable in **AB**.

4.5 Explain why

$$(\mathbf{AB})^m \neq \mathbf{A}^m\mathbf{B}^m.$$

4.6 Use matrices to describe the effect of (a) inversion, (b) reflection, and (c) rotation on a radius vector. What are the two different interpretations of each operation?

4.7 Does rotation proceed about a definite axis in 4-dimensional space?

4.8 Explain symmetry operations and list pertinent examples. Show how they are represented by (a) reorientation and (b) permutation matrices.

4.9 Why does a physical system appear the same after a symmetry operation as before?

4.10 Why is the product of two homogeneous linear transformations a homogeneous linear transformation?

4.11 Discuss why a given matrix may not have an inverse. Must $\mathbf{A}$ be square for $\mathbf{A}^{-1}$ to exist?

4.12 Show that all symmetry operations have inverses.

4.13 Prove that the inverse of a product is the product of the inverse matrices in opposite order.

4.14 Construct a prefactor (a multiplier from the left) that converts

$$\begin{pmatrix} x_1 \\ x_2 \\ x_3 \end{pmatrix}$$

to unity. Is this inverse unique?

4.15 Show how a dyadic is reoriented.

4.16 Prove that a similarity transformation does not destroy a representation of a set of symmetry operations—the converted matrices multiply as the original ones.

4.17 How are classes of symmetry operations identified?

4.18 What symmetry operations cause C_3 and C_3^{-1} to belong to the same class?

4.19 What is the trace? Show that the trace of a product is independent of the order of factors.

4.20 What is the character? Why is it invariant to similarity transformations?

4.21 What property is the same for each matrix in a given representation and class?

4.22 Why does block multiplication produce the same result as ordinary matrix multiplication?

4.23 How do reduced matrices multiply?

4.24 What significance do the eigenvectors of the moment-of-inertia matrix have?

4.25 How is the secular equation derived?

4.26 Explain diagonalization. How can a matrix be raised to an infinite power?

4.27 Why are some of the eigenvalues of the permutation matrix C_n eigenvalues for the reorientation matrix for C_n?

4.28 How many independent elements are there in an $n \times n$ symmetric matrix? Why is the moment-of-inertia matrix symmetric?

4.29 Give reasons why a dyadic may be represented as a vector times the transpose of a vector.

4.30 Prove that the transpose of a product equals the product of the transposed matrices in reverse order. Prove that the Hermitian adjoint of a product equals the product of the adjoints in the opposite order.

4.31 When must eigenvectors of a Hermitian matrix be orthogonal?

4.32 When must real eigenvectors exist?

4.33 Explain how a matrix can transform the components of complex vectors without changing their magnitudes at equivalent points.

4.34 Show that symmetry operations are represented by unitary matrices.

4.35 Indicate how orthogonal unit vectors may be formed in two different ways from the elements of a unitary matrix.

PROBLEMS

4.1 Show that an arbitrary 2×2 matrix can be expressed as a linear combination of the appropriate unit matrix and the Pauli spin matrices

$$\sigma_1 = \begin{pmatrix} 0 & 1 \\ 1 & 0 \end{pmatrix}, \sigma_2 = \begin{pmatrix} 0 & -i \\ i & 0 \end{pmatrix}, \sigma_3 = \begin{pmatrix} 1 & 0 \\ 0 & -1 \end{pmatrix}.$$

4.2 Express $\sigma_j \sigma_k$ as the linear combination in problem 4.1. How is $\sigma_k \sigma_j$ related to $\sigma_j \sigma_k$?

4.3 Prove that if matrix **A** commutes with matrix **B**, the matrices have the same number of rows and columns—that is, both are square and of the same size.

4.4 Set up the matrix that rotates radius vector **r** counterclockwise $2\pi/5$ radians in the yz plane.

4.5 What symmetry operations do (a) the permutation matrix

$$\begin{pmatrix} 0 & 0 & 0 & 0 & 1 \\ 1 & 0 & 0 & 0 & 0 \\ 0 & 1 & 0 & 0 & 0 \\ 0 & 0 & 1 & 0 & 0 \\ 0 & 0 & 0 & 1 & 0 \end{pmatrix}$$

and (b) the square of this permutation matrix represent?

4.6 Set up σ_v for reflection in the xz plane and C_3 for rotation by $2\pi/3$ radians about the z axis. Then employ matrix multiplication to determine how (a) $C_3\sigma_v$ and (b) $\sigma_v C_3$ transform the radius vector

$$\begin{pmatrix} 1 \\ 1 \\ 1 \end{pmatrix}.$$

4.7 Construct the moment-of-inertia matrix for a uniform rectangular parallelepiped, employing axes that pass through the center of mass parallel to the edges. Show how this $\mathbf{I}$ transforms when the coordinate system is rotated clockwise by 45° about the z axis. When are the new axes also principal axes?

4.8 The symmetry operations for a molecule are E, C_n, $C_n{}^2$, ..., C_{-n}, C_2', C_2'', ... Show that including C_2', C_2'', ... whose axes are perpendicular to the principal axis causes $C_n{}^m$ and $C_{-n}{}^m$ to belong to the same class.

4.9 Diagonalize the matrix

$$\begin{pmatrix} \dfrac{1}{2} & -\dfrac{1}{\sqrt{2}} & \dfrac{1}{2} \\[3mm] -\dfrac{1}{2} & -\dfrac{1}{\sqrt{2}} & -\dfrac{1}{2} \\[3mm] -\dfrac{1}{\sqrt{2}} & 0 & \dfrac{1}{\sqrt{2}} \end{pmatrix}.$$

4.10 The simultaneous equations (4-100) relate the components of $\mathbf{u}$ for each eigenvalue. Employ these in constructing a set of eigenvectors for the reorientation matrix $\mathbf{C}_n$ in example 4.12.

4.11 Compute the trace for the matrix in problem 4.9 before and after diagonalization. Prove that the character for reorientation $C_n{}^m$ equals that for $C_{-n}{}^m$.

4.12 Show that the matrix in problem 4.9 represents a rotation-reflection.

4.13 Prove that

$$\begin{pmatrix} 1 & 0 \\ 0 & 1 \end{pmatrix} \quad \text{and} \quad \begin{pmatrix} 0 & 1 \\ -1 & 0 \end{pmatrix}$$

have the algebraic properties of 1 and $\sqrt{-1}$, respectively. Then show how a complex number can be represented as a real linear combination of these matrices.

4.14 If

$$\mathbf{Ax} = \mathbf{Bx}$$

for any conformable column matrix $\mathbf{x}$ in which the first component x_1 is 0, what can be said about $\mathbf{A}$ and $\mathbf{B}$?

4.15 Establish a formula for differentiating the product $\mathbf{AA}$ with respect to the independent variable t.

4.16 Find the matrix that rotates radius vector $\mathbf{r}$ counterclockwise by $2\pi/6$ radians in the zx plane. Determine how it changes $\mathbf{r}$ when the initial components of the radius vector are $(1, -1, -1)$.

4.17 Construct matrices representing (a) $C_6\sigma_v'$ and (b) $\sigma_v'C_6$ where the z axis is the C_6 axis and σ_v' is reflection in the yz plane.

4.18 The digits in 312 are placed in a column and the matrix is coded through the multiplication

$$\begin{pmatrix} 1 & 3 & 2 \\ 5 & 1 & 0 \\ 0 & 2 & 2 \end{pmatrix} \begin{pmatrix} 3 \\ 1 \\ 2 \end{pmatrix}.$$

By solving the pertinent simultaneous equations, obtain the matrix that decodes the result, giving again 312 in a column.

4.19 Rotate the coordinate axes for the moment of inertia

$$\frac{m}{72} \begin{pmatrix} 4s^2 - b^2 & 0 & 0 \\ 0 & 3b^2 & 0 \\ 0 & 0 & 4s^2 + 2b^2 \end{pmatrix}$$

clockwise by $120°$ around the z axis. When are the new axes also principal axes?

4.20 If the symmetry operations for a given molecule are E, C_4, C_2, $C_4{}^3$, and four vertical σ's whose planes meet on the fourfold axis at $45°$ and $90°$ angles, what are the classes?

4.21 Diagonalize the matrix

$$\begin{pmatrix} 0 & \dfrac{1}{\sqrt{2}} & -\dfrac{1}{\sqrt{2}} \\[2mm] \dfrac{1}{\sqrt{3}} & \dfrac{1}{\sqrt{3}} & \dfrac{1}{\sqrt{3}} \\[2mm] \dfrac{\sqrt{2}}{\sqrt{3}} & -\dfrac{1}{\sqrt{6}} & -\dfrac{1}{\sqrt{6}} \end{pmatrix}.$$

4.22 Recall the simultaneous equations that relate the components of eigenvector $\mathbf{u}$ for each eigenvalue λ. Employ these in constructing a set of eigenvectors for

$$\begin{pmatrix} 0 & 1 & 0 \\ 0 & 0 & 1 \\ 1 & 0 & 0 \end{pmatrix}.$$

4.23 If

$$\mathbf{A} = \begin{pmatrix} 0 & A_{12} & A_{13} \\ 0 & 0 & A_{23} \\ 0 & 0 & 0 \end{pmatrix},$$

what is the exponential of $\mathbf{A}$, defined as the series

$$e^{\mathbf{A}} = \mathbf{E} + \mathbf{A} + \tfrac{1}{2}\mathbf{A}^2 + \tfrac{1}{6}\mathbf{A}^3 + \cdots ?$$

4.24 Show that the matrix in problem 4.21 represents a rotation.

REFERENCES

BOOKS

Bishop, R. E. D., Gladwell, G. M. L., and Michaelson, S. *The Matrix Analysis of Vibration*, pp. 1–35. Cambridge University Press, London, 1965.

Boas, M. L. *Mathematical Methods in the Physical Sciences*, pp. 76–120. John Wiley & Sons, Inc., New York, 1966.

Bradbury, T. C. *Theoretical Mechanics*, pp. 7–37, 92–121. John Wiley & Sons, Inc., New York, 1968.

Eisenman, R. L. *Matrix Vector Analysis*, pp. 24–45, 153–169, 185–246. McGraw-Hill Book Company, New York, 1963.

Goldstein, H. *Classical Mechanics*, pp. 93–142. Addison-Wesley Publishing Company, Inc., Reading, Mass., 1950.

Hollingsworth, C. A. *Vectors, Matrices, and Group Theory for Scientists and Engineers*, pp. 55–183. McGraw-Hill Book Company, New York, 1967.

Margenau, H., and Murphy, G. M. *The Mathematics of Physics and Chemistry*, 2nd ed., pp. 301–332. D. Van Nostrand Company, Inc., Princeton, N.J., 1956.

ARTICLES

Best, G. C. Two Theorem Tables of Matrix Algebra. *Math. Comput.*, **15**, 19 (1961).

Carroll, P. J., Jr., and Kyame, J. J. Matrix Representation of Thermodynamics of Multicomponent Systems. *Am. J. Phys.*, **30**, 282 (1962).

Chen, F. Y. Similarity Transformation and the Eigenvibration Problem of Certain Far-Coupled Systems. *Am. J. Phys.*, **38**, 1036 (1970).

Doughty, S. P., Jr., and Infante, E. F. Matrix Proof of the Theorem of Rodrigues and Hamilton. *Am. J. Phys.*, **32**, 712 (1964).

Eakin, D. M., and Davis, S. P. An Application of Matrix Optics. *Am. J. Phys.*, **34**, 758 (1966).

Halbach, K. Matrix Representation of Gaussian Optics. *Am. J. Phys.*, **32**, 90 (1964).

Hilborn, R. C. A Note on Euler Angle Rotations. *Am. J. Phys.*, **40**, 1036 (1972).

Howard, J. E. Singular Cases of the Inertia Tensor. *Am. J. Phys.*, **35**, 281 (1967).

NiCastro, J. R. A. J. Invariant Variables under Similarity Transformations. *Am. J. Phys.*, **39**, 1311 (1971).

Romer, R. H. Matrix Description of Collisions on an Air Track. *Am. J. Phys.*, **35**, 862 (1967).

Schwartz, H. M. Derivation of the Matrix of Rotation about a Given Direction as a Simple Exercise in Matrix Algebra. *Am. J. Phys.*, **31**, 730 (1963).

Strobel, G. L. Matrices and Superballs. *Am. J. Phys.*, **36**, 834 (1968).

Thurnauer, P. G. Kinematics of Finite, Rigid-Body Displacements. *Am. J. Phys.*, **35**, 1145 (1967).

5 / *Generalized Coordinates, Velocities, Forces*

5.1 Restructuring the laws of motion

The bodies or particles in a physical system generally interact with each other. As a consequence, the individual Newtonian equations governing their movements are not independent.

The set of equations can nevertheless be solved when enough constraints and/or symmetries are present. A *constraint* is a physical restriction on the freedom of motion, imposed by properties and arrangements of material. *Symmetry* is said to be present when operations in addition to the identity operation transform the system into itself.

Constraints are described by relationships among the Cartesian coordinates and velocities of particles in the given system. Such relationships serve to simplify the equations of motion. Furthermore, considerations of symmetry indicate how the coordinates may be combined to form more appropriate independent variables.

Constraints that merely restrict the Cartesian coordinates can be introduced by allowing only certain functions of these coordinates to vary. The constraints are then said to be present *implicitly*. Symmetries can also be allowed for by transforming to appropriate functions of the Cartesian coordinates. To facilitate the introduction of constraints and the use of symmetry-adapted independent variables, let us recast the equations of motion so they involve scalars depending on such functions, or on arbitrary functions, of the Cartesian coordinates. These functions are called *generalized coordinates* and are labeled $q_1, q_2, \ldots, q_n$.

Now, the work done on a given system for *any* infinitesimal change in coordinates can be determined in principle. The infinitesimal work δW can be expressed as a linear function of the δq_k's and of δt. The coefficient of δq_k is called the *generalized force Q_k*.

This work δW is also the sum over the particles of the conventional force acting on each, dotted with the infinitesimal displacement of the particle. It is appropriate to introduce Newton's second law here and transform the differential form to the system of generalized coordinates. Equating the result to that involving the generalized forces and identifying a certain expression as the kinetic energy T leads to a set of equations obtained by Joseph L. Lagrange.

When the generalized forces can be derived from a potential U by the same differentiations that cause them to come from T, the scalar $T - U$ governs the system. This scalar is called the *Lagrangian*, in honor of its inventor. The derivative of $T - U$ with respect to a generalized velocity $\dot{q}_k$ is called the *generalized momentum* p_k.

In carrying out the development of these ideas, we will need the well-known formulas for expanding infinitesimal quantities and derivatives of functions. A review of pertinent material follows.

EXAMPLE 5.1   If a property f depends on expressions $q_1, q_2, \ldots$ independently, how does it change when the q_k's change by infinitesimal amounts? The q_k's may include the Cartesian coordinates, or generalized coordinates, and time, or they may be determined by some prescription applied over a range of the independent variables —as in chapter 11.

The relationship between f and the q_k's is represented by a formula

$$f = f(q_1, q_2, \ldots)$$

in which each q_k enters independently. We assume that all expressions vary smoothly, q_1 changing by Δq_1, q_2 by $\Delta q_2, \ldots$, to produce change Δf in f:

$$
\begin{aligned}
\Delta f &= f(q_1 + \Delta q_1, q_2 + \Delta q_2, \ldots) - f(q_1, q_2, \ldots) \\
&= f(q_1 + \Delta q_1, q_2 + \Delta q_2, \ldots) - f(q_1, q_2 + \Delta q_2, \ldots) \\
&\quad + f(q_1, q_2 + \Delta q_2, q_3 + \Delta q_3, \ldots) - f(q_1, q_2, q_3 + \Delta q_3, \ldots) \\
&\quad + \cdots \\
&= \frac{f(q_1 + \Delta q_1, q_2 + \Delta q_2, \ldots) - f(q_1, q_2 + \Delta q_2, \ldots)}{\Delta q_1} \Delta q_1 \\
&\quad + \frac{f(q_1, q_2 + \Delta q_2, q_3 + \Delta q_3, \ldots) - f(q_1, q_2, q_3 + \Delta q_3, \ldots)}{\Delta q_2} \Delta q_2 \\
&\quad + \cdots .
\end{aligned}
$$

When $\Delta f, \Delta q_1, \Delta q_2, \ldots$ become infinitesimal, they become the first *variations* $\delta f, \delta q_1, \delta q_2, \ldots$. Furthermore, the ratios of increments become partial derivatives as follows:

$$
\lim_{\substack{\Delta q_1 \to \delta q_1 \\ \Delta q_2 \to \delta q_2 \\ \cdots}} \frac{f(q_1 + \Delta q_1, q_2 + \Delta q_2, \ldots) - f(q_1, q_2 + \Delta q_2, \ldots)}{\Delta q_1}
$$

$$
= \lim_{\substack{\Delta q_2 \to \delta q_2 \\ \cdots}} f_{q_1}{}'(q_1, q_2 + \Delta q_2, \ldots) = f_{q_1}{}'(q_1, q_2, \ldots) = \frac{\partial f}{\partial q_1}
$$

$$
\lim_{\substack{\Delta q_2 \to \delta q_2 \\ \cdots}} \frac{f(q_1, q_2 + \Delta q_2, q_3 + \Delta q_3, \ldots) - f(q_1, q_2, q_3 + \Delta q_3, \ldots)}{\Delta q_2}
$$

$$
= f_{q_2}{}'(q_1, q_2, q_3, \ldots) = \frac{\partial f}{\partial q_2} .
$$

Thus variation δf is given by

$$\delta f = \frac{\partial f}{\partial q_1} \delta q_1 + \frac{\partial f}{\partial q_2} \delta q_2 + \cdots.$$

When $q_1, q_2, \ldots$ are ordinary variables, this equation is written in the form

$$df = \frac{\partial f}{\partial q_1} dq_1 + \frac{\partial f}{\partial q_2} dq_2 + \cdots$$

and the infinitesimals $df, dq_1, \ldots$ are called first differentials.

EXAMPLE 5.2 Expand $f(q_1, q_2, \ldots)$ about the point where all coordinates in the given set are zero.

When the q_k's are small, products containing more than a certain number of these variables as factors are negligible with respect to those containing smaller numbers. So we are concerned with the expansion

$$f(q_1, q_2, \ldots) = A + B_k q_k + C_{kl} q_k q_l + \cdots$$

in which terms beyond a certain power are to be neglected. As before, the summation convention applies to each index that is repeated.

Differentiate the assumed form and let the coefficients be symmetric in their indices:

$$\frac{\partial f}{\partial q_r} = B_r + C_{kr} q_k + C_{rl} q_l + \cdots$$

$$= B_r + 2C_{rl} q_l + \cdots.$$

Differentiate again:

$$\frac{\partial^2 f}{\partial q_r \, \partial q_s} = 2C_{rs} + \cdots.$$

The process can be repeated any number of times to get as many derivatives as needed.

In each of these equations let all q_k's equal zero and solve for a coefficient. Substitute the results back into the original form to get the series

$$f = (f)_0 + \left(\frac{\partial f}{\partial q_k}\right)_0 q_k + \frac{1}{2}\left(\frac{\partial^2 f}{\partial q_k \, \partial q_l}\right)_0 q_k q_l + \cdots$$

in which the subscript 0 indicates the point, the origin, where the derivatives of f are evaluated. The series is called a *Taylor expansion* about the origin. However, it is readily generalized to an expansion about any given point. By subtracting the value of f at the point from both sides of this new expansion, we obtain a useful expansion for Δf.

5.2   Transforming to generalized coordinates

A separate radius vector $\mathbf{r}_j$ is needed to describe the position of each particle, or body, in a system. The velocity of each is given by the derivative $\dot{\mathbf{r}}_j$.

At an arbitrary time t, all these vectors can be assigned without violating Newton's laws; for the second law governs $\ddot{\mathbf{r}}_j$. However, there may be other conditions relating the $\mathbf{r}_j$'s, $\dot{\mathbf{r}}_j$'s, and t, each of which is a constraint on the system. Specific examples will be considered later.

Transforming the $\mathbf{r}_j$'s automatically transforms the $\dot{\mathbf{r}}_j$'s; for the latter are derivatives of the former. A homogeneous linear transformation would have the form (4-44). A general transformation, on the other hand, is effected by the substitution

$$x_j = x_j(t, q_1, \ldots, q_n), \tag{5-1}$$

$$y_j = y_j(t, q_1, \ldots, q_n), \tag{5-2}$$

$$z_j = z_j(t, q_1, \ldots, q_n), \tag{5-3}$$

in which x_j, y_j, z_j are inertial rectangular coordinates of the jth particle, or body. The new variables $q_1, \ldots, q_n$ are generalized coordinates. They may not all be independent; some of the constraints may impose relationships on them, their first temporal derivatives, and the time.

The number n of generalized coordinates is less than the number of Cartesian coordinates by the number of constraints that have been introduced implicitly. The number of *degrees of freedom*, on the other hand, is integer n minus the number of additional constraints that are present.

For convenience, let us construct an n-dimensional space in which $q_1, \ldots, q_n$ are Cartesian components of a single radius vector $\mathbf{q}$:

$$\mathbf{q} = q_k \mathbf{e}_k. \tag{5-4}$$

The base vectors $\mathbf{e}_1, \ldots, \mathbf{e}_n$ are mutually perpendicular in this space. The derivative

$$\dot{\mathbf{q}} = \frac{d\mathbf{q}}{dt} \tag{5-5}$$

is then called the *generalized velocity*, while

$$\dot{q}_k = \frac{dq_k}{dt} \tag{5-6}$$

is called the kth component of the generalized velocity, or simply the kth generalized velocity.

In physical space, $\dot{q}_k$ may be a linear velocity, or an angular velocity, or something more complicated. Since velocities can be assigned arbitrarily even when the positions of all particles are given, the independent variables include $t, q_1, q_2, \ldots, \dot{q}_1, \dot{q}_2, \ldots$.

Usually, the functional relationships in equations (5-1), (5-2), (5-3) are not linear and equations (4-44) do not apply, but the theory in examples 5.1 and 5.2 indicates that differentials of the coordinates do transform linearly. Indeed, we have

$$dx_j = \frac{\partial x_j}{\partial t} dt + \frac{\partial x_j}{\partial q_k} dq_k, \tag{5-7}$$

whence

$$\dot{x}_j = \frac{dx_j}{dt} = \frac{\partial x_j}{\partial t} + \frac{\partial x_j}{\partial q_k} \dot{q}_k. \tag{5-8}$$

In this equation all partial derivatives of x_j depend only on t and the q_k's because x_j itself does; therefore, differentiating it with respect to the independent variable $\dot{q}_l$ yields just one term:

$$\frac{\partial \dot{x}_j}{\partial \dot{q}_l} = 0 + \frac{\partial x_j}{\partial q_k} \frac{\partial \dot{q}_k}{\partial \dot{q}_l} = \frac{\partial x_j}{\partial q_k} \delta_{kl} = \frac{\partial x_j}{\partial q_l}. \tag{5-9}$$

Thus the partial derivative of $\dot{x}_j$ with respect to $\dot{q}_l$ is the coefficient of dq_l in the equation transforming the set $dt, dq_1, \ldots, dq_n$ to dx_j.

Furthermore, partial differentiation and total differentiation of x_j commute. Treating $\partial x_j / \partial q_k$ as a function of t and the q_k's yields

$$\frac{d}{dt} \frac{\partial x_j}{\partial q_k} = \frac{\partial^2 x_j}{\partial t \, \partial q_k} + \frac{\partial^2 x_j}{\partial q_l \, \partial q_k} \dot{q}_l \tag{5-10}$$

(as before, the formula in example 5.1 is used). On the other hand, differentiating (5-8) with respect to q_k gives us

$$\frac{\partial \dot{x}_j}{\partial q_k} = \frac{\partial^2 x_j}{\partial q_k \, \partial t} + \frac{\partial^2 x_j}{\partial q_k \, \partial q_l} \dot{q}_l. \tag{5-11}$$

Therefore,

$$\frac{d}{dt} \frac{\partial x_j}{\partial q_k} = \frac{\partial \dot{x}_j}{\partial q_k}. \tag{5-12}$$

These formulas will enable us to introduce general derivatives of kinetic energy into the expression for work done on the system.

We can often construct formulas for pertinent small displacements of parts of a setup from the way the corresponding coordinate system is defined. Then we may be able to formulate T (and Q_k) directly, without employing the transformation equations (5-1), (5-2), and (5-3). Consider the following examples:

EXAMPLE 5.3 Compose the displacement ds from the movements that occur when each cylindrical coordinate changes infinitesimally. Associate mass m with the moving point, interval dt with the displacement, and formulate the corresponding kinetic energy T.

Construct reference rectangular axes in some inertial frame. Let coordinate r be the distance of the mass point from the z axis, ϕ be the angle between the zx plane and the perpendicular dropped on the z axis from the point, and z be the distance of the point above the xy plane.

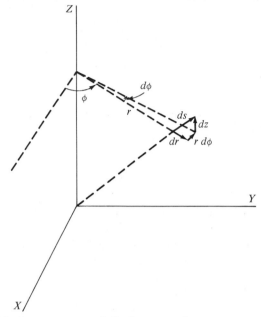

FIGURE 5.1   Cylindrical components of displacement *d*s.

Then if *r* increases infinitesimally and no other changes occur, the point moves out distance *dr* (see figure 5.1). When ϕ alone changes, it moves at distance *r* around the *z* axis. If ϕ is in radians and the increase is small, the distance traced out is *r dϕ*. When *z* alone changes infinitesimally, the point moves up distance *dz*.

The net effect of these three motions executed simultaneously is element *d*s. Since the constituents are mutually perpendicular, the Pythagorean theorem applies:

$$ds^2 = dr^2 + r^2\,d\phi^2 + dz^2.$$

For the square of the speed, we then have

$$\left(\frac{ds}{dt}\right)^2 = \left(\frac{dr}{dt}\right)^2 + r^2\left(\frac{d\phi}{dt}\right)^2 + \left(\frac{dz}{dt}\right)^2$$

or

$$\dot{s}^2 = \dot{r}^2 + r^2\dot{\phi}^2 + \dot{z}^2.$$

When mass *m* moves with the point, (2-101) yields

$$T = \tfrac{1}{2}m\dot{r}^2 + \tfrac{1}{2}mr^2\dot{\phi}^2 + \tfrac{1}{2}m\dot{z}^2.$$

EXAMPLE 5.4   Resolve *d*s into the displacements that occur when each spherical coordinate changes by itself. Associate mass *m* with the point and derive the corresponding *T*.

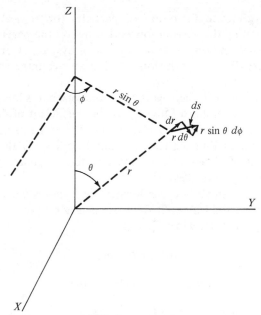

FIGURE 5.2 Spherical components of displacement d**s**.

Construct rectangular axes in some inertial frame. Then let coordinate r be the distance of the mass point from the origin, θ be the angle between the radius vector **r** and the reference z axis, and ϕ be the angle between the zx plane and the plane through **r** and the z axis (see figure 5.2).

An increase in r by dr causes the point to move dr further from the origin. An increase in θ by $d\theta$ causes it to move perpendicular to **r** along the arc $r\,d\theta$. Both of these displacements are in the vertical plane through **r**. But when ϕ increases by $d\phi$, the point moves perpendicular to this plane along the arc of length $r \sin \theta\, d\phi$ (the lever arm is the projection of **r** on the perpendicular dropped on the z axis).

Since these elements are mutually perpendicular, the square of the magnitude of the resultant is

$$ds^2 = dr^2 + r^2\, d\theta^2 + r^2 \sin^2 \theta\, d\phi^2,$$

and the square of the speed is

$$\dot{s}^2 = \dot{r}^2 + r^2\dot{\theta}^2 + r^2 \sin^2 \theta\dot{\phi}^2.$$

The kinetic energy of mass m moving with the point is then

$$T = \tfrac{1}{2}m\dot{r}^2 + \tfrac{1}{2}mr^2\dot{\theta}^2 + \tfrac{1}{2}mr^2 \sin^2 \theta\dot{\phi}^2.$$

5.3   Generalized forces

The evolution of a given configuration of particles (or bodies) with given initial velocities is determined by the varying forces that act on the parts. We presume that these forces are known in some inertial Cartesian system. Consequently, we can construct a form for δW, the work done during an arbitrary small change in the configuration.

A transformation of this δW then leads to a function that is linear in δq_1, δq_2, . . . , δt. We will define the generalized force Q_k as the coefficient of δq_k in this function.

Since work is a scalar, the transformation does not alter the value of δW. The work done on a set of N interacting particles (or small bodies) by the forces during the change is independent of choice of coordinate system. It corresponds to the same increase in kinetic energy, regardless.

Now, each particle, or small body, is located by a radius vector $\mathbf{r}_j$ drawn from a reference point in an inertial frame. If the force acting on the jth particle is $\mathbf{F}_j$ while it moves distance $\delta \mathbf{r}_j$, the total work done is

$$\delta W = \mathbf{F}_j \cdot \delta \mathbf{r}_j. \tag{5-13}$$

Introducing Newton's second law converts (5-13) to

$$\begin{aligned} \delta W &= m_j \ddot{\mathbf{r}}_j \cdot \delta \mathbf{r}_j \\ &= m_j(\ddot{x}_j\, \delta x_j + \ddot{y}_j\, \delta y_j + \ddot{z}_j\, \delta z_j). \end{aligned} \tag{5-14}$$

Since x_j is a function of $q_1, \ldots, q_n$, t, the formula in example 5.1 yields the expansion

$$\delta x_j = \frac{\partial x_j}{\partial q_k}\, \delta q_k + \frac{\partial x_j}{\partial t}\, \delta t. \tag{5-15}$$

When this is substituted into equation (5-14), we get terms containing the factor

$$\begin{aligned} \ddot{x}_j \frac{\partial x_j}{\partial q_k} &= \frac{d}{dt}\left(\dot{x}_j \frac{\partial x_j}{\partial q_k} \right) - \dot{x}_j \frac{d}{dt}\frac{\partial x_j}{\partial q_k} \\ &= \frac{d}{dt}\left(\dot{x}_j \frac{\partial \dot{x}_j}{\partial \dot{q}_k} \right) - \dot{x}_j \frac{\partial \dot{x}_j}{\partial q_k} = \frac{d}{dt}\frac{\partial}{\partial \dot{q}_k}\left(\frac{\dot{x}_j{}^2}{2} \right) - \frac{\partial}{\partial q_k}\left(\frac{\dot{x}_j{}^2}{2} \right) \\ &= \frac{1}{m_j}\left[\frac{d}{dt}\frac{\partial}{\partial \dot{q}_k}\left(\frac{m_j \dot{x}_j{}^2}{2} \right) - \frac{\partial}{\partial q_k}\left(\frac{m_j \dot{x}_j{}^2}{2} \right) \right]. \end{aligned} \tag{5-16}$$

The first equality follows from the formula for differentiating a product, the second equality from (5-9) and (5-12), and the third one from the formula for differentiating a square. Similar factors appear in the other two groups of terms.

Introducing the kinetic energy

$$T = \tfrac{1}{2}m_j \dot{x}_j{}^2 + \tfrac{1}{2}m_j \dot{y}_j{}^2 + \tfrac{1}{2}m_j \dot{z}_j{}^2 \tag{5-17}$$

simplifies the resulting expression for the infinitesimal work done on the system to

$$\delta W = \left(\frac{d}{dt} \frac{\partial T}{\partial \dot{q}_k} - \frac{\partial T}{\partial q_k} \right) \delta q_k$$

$$+ m_j \left(\ddot{x}_j \frac{\partial x_j}{\partial t} + \ddot{y}_j \frac{\partial y_j}{\partial t} + \ddot{z}_j \frac{\partial z_j}{\partial t} \right) \delta t. \tag{5-18}$$

This formula is valid whether all the δq_k's are independent and whether the variation is along a possible path of the system or not. A variation that is hypothetical, that does not actually take place, is said to be *virtual*.

From section 5.2, the vector $\mathbf{q}$ represents the configuration of the system at a given time t. Let us define the *generalized force* $\mathbf{Q}$ as the vector whose dot product with the displacement $\delta \mathbf{q}$ yields the corresponding work done on the system. When δt is not zero, we set the coefficient of δt equal to Q_t and write

$$\delta W = \mathbf{Q} \cdot \delta \mathbf{q} + Q_t \, \delta t = Q_k \, \delta q_k + Q_t \, \delta t. \tag{5-19}$$

This equation enables expressions for each component of the generalized force, $Q_1, Q_2, \ldots, Q_n$, to be deduced from the complete expression for δW.

Let us now consider all the q_k's to be physically independent (explicit constraints will be considered later). Since the variations may be virtual, δt and all except *any* one δq_k can be made zero. Dividing the resulting forms in (5-18), (5-19) by the remaining, nonzero, δq_k, and equating what is left gives us

$$\frac{d}{dt} \frac{\partial T}{\partial \dot{q}_k} - \frac{\partial T}{\partial q_k} = Q_k \tag{5-20}$$

a general formulation of Newton's second law.

Let us call (5-20) the *Lagrange force equation*. A second Lagrange equation, in which $T - U$ is the dependent variable, will be established later. In either formula, index k may be any positive integer up to and including n. Consequently, we have n simultaneous differential equations to solve (one for each degree of freedom when all constraints have been introduced implicitly).

EXAMPLE 5.5 Apply Lagrange's method to Atwood's machine, where mass m_1 is supported by a string that passes over a pulley to mass m_2. Neglect the mass of the string and pulley. Also, neglect the friction between the pulley and its axle.

Let the distance x of the mass m_1 below the center of the pulley be the generalized coordinate for the problem (see figure 5.3). Then (5-20) becomes

$$\frac{d}{dt} \frac{\partial T}{\partial \dot{x}} - \frac{\partial T}{\partial x} = Q.$$

When mass m_1 moves distance δx down, mass m_2 moves δx up, and the work done by gravity is

$$\delta W = m_1 g \, \delta x - m_2 g \, \delta x.$$

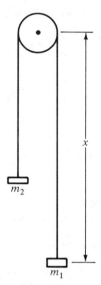

FIGURE 5.3   Coordinate system for Atwood's machine.

Comparing this with formula (5-19) yields

$$Q = (m_1 - m_2)g.$$

Since the constraint makes m_2 move at the same speed as m_1, the kinetic energy is given by

$$T = \tfrac{1}{2}m_1\dot{x}^2 + \tfrac{1}{2}m_2\dot{x}^2.$$

Differentiating T with respect to x and with respect to its time derivative $\dot{x}$ yields

$$\frac{\partial T}{\partial x} = 0, \qquad \frac{\partial T}{\partial \dot{x}} = (m_1 + m_2)\dot{x},$$

so Lagrange's equation becomes

$$\frac{d}{dt}(m_1 + m_2)\dot{x} - 0 = (m_1 - m_2)g,$$

whence

$$\ddot{x} = \frac{m_1 - m_2}{m_1 + m_2}\,g.$$

To calculate tension τ in the string, consider the mass m_1 by itself. The forces doing work on m_1 include τ and $m_1 g$:

$$Q\,\delta x = (m_1 g - \tau)\,\delta x.$$

Now

$$T = \tfrac{1}{2}m_1\dot{x}^2,$$

so Lagrange's equation,

$$\frac{d}{dt}\,m_1\dot{x} - 0 = m_1g - \tau,$$

yields

$$\tau = m_1g - m_1\ddot{x}.$$

Finally, introducing the previous result for $\ddot{x}$ gives

$$\tau = \frac{2m_1m_2}{m_1 + m_2}\,g.$$

EXAMPLE 5.6   Obtain the Lagrangian equations for a particle constrained by a string to move at a varying distance r from a fixed point on a smooth horizontal table.

Assume that the point and the table are at rest in an inertial frame. Draw on the table a reference line, from which the angle of rotation θ is measured as in figure 5.4. Then the square of an element of length along the particle's path is

$$ds^2 = dr^2 + r^2\,d\theta^2$$

in the inertial frame, whence

$$v^2 = \left(\frac{ds}{dt}\right)^2 = \left(\frac{dr}{dt}\right)^2 + r^2\left(\frac{d\theta}{dt}\right)^2 = \dot{r}^2 + r^2\dot{\theta}^2$$

and

$$T = \tfrac{1}{2}m(\dot{r}^2 + r^2\dot{\theta}^2)$$

where m is the mass of the particle.

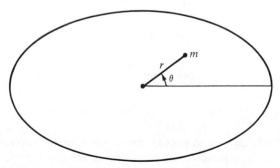

FIGURE 5.4   Coordinate system for a particle on a horizontal table.

Differentiating the kinetic energy with respect to the generalized coordinates θ, r, and their derivatives gives

$$\frac{\partial T}{\partial \theta} = 0, \qquad \frac{\partial T}{\partial r} = mr\dot{\theta}^2,$$

$$\frac{\partial T}{\partial \dot{\theta}} = mr^2\dot{\theta}, \qquad \frac{\partial T}{\partial \dot{r}} = m\dot{r}.$$

Because the string exerts its force along the radius vector r, no work is done on the particle as long as r is constant. That is,

$$\delta W_\theta = Q_\theta \, \delta\theta = 0$$

and

$$Q_\theta = 0.$$

On the other hand, work may be done as r is altered; so

$$Q_r \neq 0.$$

Substituting these expressions into the Lagrange equation for θ,

$$\frac{d}{dt}\frac{\partial T}{\partial \dot{\theta}} - \frac{\partial T}{\partial \theta} = Q_\theta$$

yields

$$\frac{d}{dt}(mr^2\dot{\theta}) = 0,$$

whence

$$r^2\dot{\theta} = A$$

a constant. Thence

$$r\dot{\theta}^2 = \frac{r^4\dot{\theta}^2}{r^3} = \frac{A^2}{r^3}.$$

Substituting into the equation for r,

$$\frac{d}{dt}\frac{\partial T}{\partial \dot{r}} - \frac{\partial T}{\partial r} = Q_r$$

gives

$$m\ddot{r} - mr\dot{\theta}^2 = Q_r$$

and

$$m\ddot{r} - m\frac{A^2}{r^3} = Q_r.$$

If Q_r or $\delta W_r = Q_r \, \delta r$ were given, this equation might be solved.

EXAMPLE 5.7 A body subject to gravity moves on a smooth sphere along a vertical great circle. Find its angular acceleration assuming the great circle to be at rest in an inertial frame.

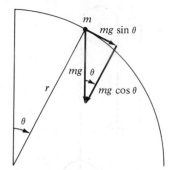

FIGURE 5.5 Components of the gravitational force acting on a particle at angle θ along the arc of a vertical circle with radius r.

For generalized coordinates, choose colatitude θ and radius r as in figure 5.5. Since only θ varies, we need the Lagrange equation

$$\frac{d}{dt}\frac{\partial T}{\partial \dot{\theta}} - \frac{\partial T}{\partial \theta} = Q_\theta$$

for which (as in example 5.6)

$$T = \tfrac{1}{2}m(\dot{r}^2 + r^2\dot{\theta}^2)$$

and

$$\frac{\partial T}{\partial \theta} = 0, \qquad \frac{\partial T}{\partial \dot{\theta}} = mr^2\dot{\theta}.$$

From figure 5.5, the work done by gravity when the particle moves $\delta\theta$ is

$$\delta W = (mg \sin\theta)r\,\delta\theta = Q_\theta\,\delta\theta,$$

so

$$Q_\theta = mgr \sin\theta.$$

Therefore, Lagrange's equation becomes

$$\frac{d}{dt}(mr^2\dot{\theta}) - 0 = mgr \sin\theta$$

or

$$\ddot{\theta} = \frac{g}{r}\sin\theta.$$

EXAMPLE 5.8   Indicate how the Lagrangian kinetic energy and generalized forces are obtained for the pulley complex in figure 5.6.

Introduce generalized coordinate x as the distance of mass m_1 below the center of the fixed pulley and generalized coordinate y as the distance of mass m_3 below the center of the mobile pulley.

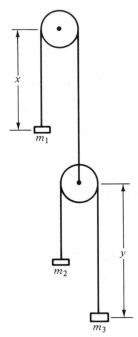

FIGURE 5.6   Coordinates for a pulley complex.

Whenever mass m_1 moves down distance δx, the mobile pulley moves up distance δx. And if y also increases by δy, mass m_2 moves up distance $\delta y + \delta x$ while mass m_3 moves down $\delta y - \delta x$. The downward velocities are $\dot{x}$ for m_1, $-\dot{x}$ for the movable pulley, $-\dot{y} - \dot{x}$ for m_2, and $\dot{y} - \dot{x}$ for m_3. Since these are linear velocities, they are used directly in T.

When y is fixed and x changes by δx, m_1 moves down distance δx while m_2 and m_3 move up distance δx. The work done by gravity on the masses is then

$$Q_1 \, \delta q_1 = [m_1 g - (m_2 + m_3)g] \, \delta x.$$

When x is fixed and y changes by δy, m_3 moves down distance δy while m_2 moves up the same distance. The work done by gravity is

$$Q_2 \, \delta q_2 = (m_3 g - m_2 g) \, \delta y.$$

5.4 Functions and variables in the Lagrange force equations

Before applying equations (5-20), we should select generalized coordinates that respect the known constraints and forces in the region under study. The kinetic energy and the known generalized forces are formulated as functions of these co-ordinates, their temporal derivatives, and time. Derivatives of T are calculated and the results are inserted into the Lagrange equations. Finally, the set is manipulated to produce the desired solutions.

In constructing T, we start with definitions of the generalized coordinates. Pertinent line elements are identified and the square of each possible displacement of a particle or body is determined. Forming the corresponding square of speed, placing it in equation (2-101), and combining the result with those for the other elements yields the kinetic energy expression.

Generalized forces can be obtained from how portions of the system move, if the known information determines the derivatives that enter the pertinent Lagrange equations. Conversely, the paths traversed by different parts of a system can be obtained from the forces. Expressions for the Q_k's are placed on the right sides of the Lagrange equations, the appropriate expressions for T on the left, and formulas for the coordinates are obtained by integration.

Equation (5-19) states that the force acting in the direction of a motion described by δq_k can be obtained from $\delta W_k / \delta q_k$, where δW_k is the work done when only q_k varies. The common force that does not behave in this way is the magnetic force acting on a moving charge. Such a force always points perpendicular to the direction of motion.

An interpretation of the partial derivative in the first term of the Lagrange equation comes from its meaning in a special case. When rectangular coordinates are employed, the kinetic energy T is given by (5-17). Differentiating this form with respect to a typical velocity component yields the momentum on which the corresponding component of force acts:

$$\frac{\partial T}{\partial \dot{x}_1} = m_1 \dot{x}_1 = (p_1)_x. \tag{5-21}$$

By analogy, we call

$$\frac{\partial T}{\partial \dot{q}_k} = p_k \tag{5-22}$$

the momentum on which the force Q_k acts. The generalized momentum to be defined later is obtained from $T - U$ in the same manner.

The ordinary derivative in (5-20) is the time rate of change in the momentum on which Q_k acts. The second term $\partial T / \partial q_k$ is a correction term that allows for the presence of q_k in T. It is a fictitious force that has to be subtracted from dp_k/dt to get Q_k. The force Q_k is real because it equals $\delta W / \delta q_k$ when only q_k varies.

EXAMPLE 5.9   If

$$x = x(u, v),$$
$$y = y(u, v),$$

how is the element of length ds related to du and dv? What condition makes ds at constant u orthogonal to ds at constant v?

From the general form for the differential of a function, the dependence of x on u and v implies that

$$dx = \frac{\partial x}{\partial u} du + \frac{\partial x}{\partial v} dv,$$

while the dependence of y on u and v implies that

$$dy = \frac{\partial y}{\partial u} du + \frac{\partial y}{\partial v} dv.$$

Applying the Pythagorean theorem to the perpendicular elements dx and dy then yields

$$ds^2 = dx^2 + dy^2$$

$$= \left(\frac{\partial x}{\partial u} du + \frac{\partial x}{\partial v} dv\right)^2 + \left(\frac{\partial y}{\partial u} du + \frac{\partial y}{\partial v} dv\right)^2$$

$$= \left[\left(\frac{\partial x}{\partial u}\right)^2 + \left(\frac{\partial y}{\partial u}\right)^2\right] du^2 + 2\left[\frac{\partial x}{\partial u}\frac{\partial x}{\partial v} + \frac{\partial y}{\partial u}\frac{\partial y}{\partial v}\right] du\, dv$$

$$+ \left[\left(\frac{\partial x}{\partial v}\right)^2 + \left(\frac{\partial y}{\partial v}\right)^2\right] dv^2.$$

When v is constant, dv is zero and the last two terms drop out. The line element itself then has the form

$$ds_u = A(u, v)\, du.$$

Similarly when u is constant, du is zero and we have

$$ds_v = B(u, v)\, dv.$$

When ds_v is perpendicular to ds_u, the Pythagorean theorem applies

$$ds^2 = ds_u^2 + ds_v^2 = A^2\, du^2 + B^2\, dv^2.$$

and there is no cross term in ds^2. The term containing $du\, dv$ in the preceding equation for ds^2 drops out if and only if

$$\frac{\partial x}{\partial u}\frac{\partial x}{\partial v} + \frac{\partial y}{\partial u}\frac{\partial y}{\partial v} = 0.$$

5.5   Handling explicit constraints

Each independent constraint on a physical system restricts movements of mass elements in a different way. Indeed, it links possible infinitesimal variations of the coordinates by an independent linear equation. Such an equation may have one form for certain configurations and disappear for others (consider a particle constrained to move on a sphere *or* outside of it). We can then solve the problem with the constraint and determine from the solution when it is no longer applicable, or solve the problem without the constraint and determine when it must be introduced.

Some of the equations of constraint can be integrated individually. The resulting integrals may be used to eliminate coordinates and their derivatives from the kinetic energy expression and from the two formulas for δW. Such constraints are then implicitly present. In simple cases, the q_k's are defined to make the elimination trivial.

Other constraining equations are not integrable by themselves. Each of these must be considered as an additional condition linking the δq_k's. Furthermore, we may wish to consider some of the integrable constraints as conditions on the δq_k's.

Let us suppose that m independent constraints are to be considered. The corresponding equations need involve only some of the variations in coordinates. But for the m equations to be consistent, they must make m of the variations depend on other variations. Let us pick out a possible dependent set and label them as the first m δq_k's.

The explicit relationships among the variations invalidate the derivation of (5-20). Nevertheless, (5-18) and (5-19) can be combined at any given time t to yield the equation

$$\left(\frac{d}{dt} \frac{\partial T}{\partial \dot{q}_k} - \frac{\partial T}{\partial q_k} \right) \delta q_k = Q_k \, \delta q_k, \tag{5-23}$$

which rearranges to

$$\left(\frac{d}{dt} \frac{\partial T}{\partial \dot{q}_k} - \frac{\partial T}{\partial q_k} - Q_k \right) \delta q_k = 0. \tag{5-24}$$

The explicit constraints relate the variations in some region:

$$a_{1k} \, \delta q_k = 0,$$
$$\vdots$$
$$a_{mk} \, \delta q_k = 0. \tag{5-25}$$

To allow for these equations, we multiply each by a factor to be determined when the nature of the solution becomes evident. The results are added to (5-24) and the choice is made.

Indeed, we multiply the first of equations (5-25) by $-\lambda_1$, the second by $-\lambda_2, \ldots,$ the mth by $-\lambda_m$. Each of the resulting equations is then added to (5-24) in turn to produce

$$\left(\frac{d}{dt} \frac{\partial T}{\partial \dot{q}_k} - \frac{\partial T}{\partial q_k} - Q_k - \lambda_j a_{jk} \right) \delta q_k = 0. \tag{5-26}$$

All the variations would be independent if m, the upper limit on j, were zero. Each constraining equation (5-25) reduces the independent δq_k's by 1 and increases the needed λ_j's by 1.

Let us choose the $m\lambda_j$'s so the first m parenthetical expressions in (5-26) are zero; that is, so

$$\frac{d}{dt} \frac{\partial T}{\partial \dot{q}_k} - \frac{\partial T}{\partial q_k} - Q_k - \lambda_j a_{jk} = 0 \qquad \text{for} \quad 1 \le k \le m. \tag{5-27}$$

Since the first m δq_k's have been made dependent on the other variations, these others are independent. We can choose to let all but one equal zero. Then if k is l in the remaining term (the one different from zero), we have

$$\frac{d}{dt} \frac{\partial T}{\partial \dot{q}_l} - \frac{\partial T}{\partial q_l} - Q_l - \lambda_j a_{jl} = 0 \qquad \text{for} \quad m < l \le n \tag{5-28}$$

after the δq_l has been canceled.

Equations (5-27) and (5-28) can be rewritten in the form

$$\frac{d}{dt}\frac{\partial T}{\partial \dot{q}_k} - \frac{\partial T}{\partial q_k} = Q_k + \lambda_j a_{jk}. \tag{5-29}$$

On comparing this form with (5-20), we see that the sum

$$\lambda_j a_{jk} \tag{5-30}$$

acts as the contribution of the constraints to the kth generalized force. Indeed, the constraints could be replaced by this set of forces without altering the motion.

Because the λ_j's are chosen at a given time, they can vary with t. In working an actual problem, do not worry about this dependence. Instead, eliminate each λ_j from the set (5-29). The $n - m$ relationships remaining can then be combined with equations (5-25).

Because the method of undetermined multipliers was introduced by Lagrange, the λ_j's are often called *Lagrange multipliers*.

EXAMPLE 5.10   Obtain a second-order differential equation describing the way a homogeneous sphere of mass m and radius r rolls without slippage on a great circle of a sphere of radius R as in figure 5.7.

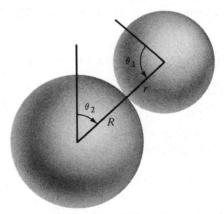

FIGURE 5.7   Coordinates for a mobile sphere rolling along a vertical great circle of a fixed sphere.

The center of mass of the moving sphere travels along an arc of length $(R + r)\theta_2$ while the sphere turns through the angle $\theta_1 + \theta_2$. At any given instant of time, the translational speed equals $(R + r)\dot{\theta}_2$ while the angular velocity equals $\dot{\theta}_1 + \dot{\theta}_2$. The moment of inertia is $(\frac{2}{5})mr^2$; and the kinetic energy is

$$T = \tfrac{1}{2}m(R + r)^2\dot{\theta}_2{}^2 + \tfrac{1}{5}mr^2(\dot{\theta}_1 + \dot{\theta}_2)^2.$$

On the other hand, the work done by gravity when the angle to the point of contact changes by $\delta\theta_2$ is

$$\delta W = mg \sin \theta_2 (R + r)\, \delta\theta_2.$$

Hence the generalized forces are

$$Q_1 = 0$$
$$Q_2 = mg(R + r)\sin \theta_2.$$

Differentiating the kinetic energy, we obtain

$$\frac{\partial T}{\partial \theta_1} = 0, \qquad \frac{\partial T}{\partial \theta_2} = 0,$$

$$\frac{\partial T}{\partial \dot\theta_1} = \tfrac{2}{5}mr^2(\dot\theta_1 + \dot\theta_2), \qquad \frac{\partial T}{\partial \dot\theta_2} = m(R + r)^2\dot\theta_2 + \tfrac{2}{5}mr^2(\dot\theta_1 + \dot\theta_2),$$

$$\frac{d}{dt}\frac{\partial T}{\partial \dot\theta_1} = \tfrac{2}{5}mr^2(\ddot\theta_1 + \ddot\theta_2), \qquad \frac{d}{dt}\frac{\partial T}{\partial \dot\theta_2} = m(R + r)^2\ddot\theta_2 + \tfrac{2}{5}mr^2(\ddot\theta_1 + \ddot\theta_2).$$

Since there is no slippage, the length of path traced out on one sphere equals that on the other; for an infinitesimal shift

$$r\,\delta\theta_1 - R\,\delta\theta_2 = 0$$

whence

$$a_{11} = r, \qquad a_{12} = -R.$$

Substituting these expressions into equations (5-27) and (5-28),

$$\tfrac{2}{5}mr^2(\ddot\theta_1 + \ddot\theta_2) - \lambda r = 0,$$
$$m(R + r)^2\ddot\theta_2 + \tfrac{2}{5}mr^2(\ddot\theta_1 + \ddot\theta_2) - mg(R + r)\sin \theta_2 + \lambda R = 0;$$

and eliminating $\ddot\theta_1$ in the first equation with the formula

$$r\ddot\theta_1 = R\ddot\theta_2$$

derived from the equation of constraint, results in

$$\tfrac{2}{5}m(R\ddot\theta_2 + r\ddot\theta_2) - \lambda = 0.$$

Subtracting the first equation from the second,

$$m(R + r)^2\ddot\theta_2 - mg(R + r)\sin \theta_2 + \lambda(R + r) = 0,$$

canceling $(R + r)$ and substituting in the expression for λ,

$$m(R + r)\ddot\theta_2 - mg \sin \theta_2 + \tfrac{2}{5}m(R + r)\ddot\theta_2 = 0,$$

and reducing, we end up with

$$\ddot\theta_2 - \frac{5g}{7(R + r)} \sin \theta_2 = 0$$

a form of the pendulum equation.

The δW for this system can be integrated. Setting it equal to $-dV$,

$$dV = -mg(R + r)\sin \theta_2 \, d\theta_2,$$

we find that

$$V = mg(R + r)\cos \theta_2.$$

Function V is said to be the potential from which the forces are derived. Indeed here

$$-\frac{\partial V}{\partial \theta_1} = Q_1, \qquad -\frac{\partial V}{\partial \theta_2} = Q_2.$$

5.6   Potentials

Let us return to systems with no explicit constraints, those governed by Lagrange force equation (5-20). Each component of the generalized force is related to δW by (5-19), while δW is related to the conventional forces by (5-13). However, some of these forces are conservative, derivable from a potential V (see section 3.7). We will now determine how this V may enter (5-20). We will also consider how the potential can be generalized to represent magnetic forces. Dissipative forces will be considered in section 5.8.

A part of each term on the right of equation (5-13) is generally integrable without specifying the path. The corresponding infinitesimal work δW_{pot} is an exact differential. Consequently, it defines the change in a function

$$\delta W_{\text{pot}} \equiv -\delta V \tag{5-31}$$

that depends on the position of each mass element in a reference inertial frame:

$$V = V(x_1, y_1, z_1, \ldots, x_N, y_N, z_N). \tag{5-32}$$

Since this function acts as a source of kinetic energy, it is referred to as *potential energy*. The corresponding forces are called *conservative* or *potential forces*, as in section 3.7.

Any given transformation (5-1), (5-2), (5-3) converts this V to a function of t, $q_1, \ldots, q_n$. The formula in example 5.1 then shows

$$\delta V = \frac{\partial V}{\partial q_k} \delta q_k + \frac{\partial V}{\partial t} \delta t. \tag{5-33}$$

When other forces are absent, δW_{pot} is δW and (5-31) converts (5-19) to

$$\delta V = -Q_k \, \delta q_k - Q_t \, \delta t. \tag{5-34}$$

Since the right sides of (5-33), (5-34) are equal,

$$Q_k = -\frac{\partial V}{\partial q_k}. \tag{5-35}$$

Since V does not depend on any $\dot{q}_k$,

$$\frac{\partial V}{\partial \dot{q}_k} = 0. \tag{5-36}$$

Using (5-35) to eliminate Q_k from (5-20) and adding the total derivative of expression (5-36) to the result leads to

$$\frac{d}{dt}\frac{\partial T}{\partial \dot{q}_k} - \frac{\partial T}{\partial q_k} = \frac{d}{dt}\frac{\partial V}{\partial \dot{q}_k} - \frac{\partial V}{\partial q_k}, \tag{5-37}$$

or

$$\frac{d}{dt}\frac{\partial L}{\partial \dot{q}_k} - \frac{\partial L}{\partial q_k} = 0, \tag{5-38}$$

where

$$L = T - V. \tag{5-39}$$

Scalar L is called the Lagrangian for the physical system.

The part of quantity $-\delta W$ that does not contribute to δV is dissipated. It cannot reappear as kinetic energy while a cycle of changes is being completed. Instead, it increases the temperature of parts of the system. The corresponding forces are the well-known frictional forces that oppose motion in macroscopic systems.

The final part of each term on the right of (5-13) involves forces that are always normal to the direction of motion. These deflect the mass elements without doing work on them. The only known forces of this kind are magnetic.

Since magnetic forces are a relativistic consequence of Coulomb's law, we might expect them to be represented as potential forces are. Indeed, such is the case.

It is possible to construct a generalized potential

$$U = U(t, q_1, \ldots, q_n, \dot{q}_1, \ldots, \dot{q}_n) \tag{5-40}$$

in such a way that the formula

$$Q_k = \frac{d}{dt}\frac{\partial U}{\partial \dot{q}_k} - \frac{\partial U}{\partial q_k} \tag{5-41}$$

yields the electric and magnetic forces acting on given charged mass elements, besides the potential or conservative forces present. Equation (5-20) then becomes

$$\frac{d}{dt}\frac{\partial T}{\partial \dot{q}_k} - \frac{\partial T}{\partial q_k} = \frac{d}{dt}\frac{\partial U}{\partial \dot{q}_k} - \frac{\partial U}{\partial q_k}. \tag{5-42}$$

If the Lagrangian is redefined as

$$L = T - U, \tag{5-43}$$

equation (5-38) is still obeyed.

5.7   Time rate of change in the canonical momentum

When force component Q_k comes from a generalized potential

$$U = U(t, q_1, \ldots, q_n, \dot{q}_1, \ldots, \dot{q}_n) \tag{5-44}$$

by the same differentiations that relate it to T, movement of the mass elements is governed by the Lagrange equations

$$\frac{d}{dt} \frac{\partial L}{\partial \dot{q}_k} = \frac{\partial L}{\partial q_k} \tag{5-45}$$

in which the Lagrangian is the scalar

$$L = T - U. \tag{5-46}$$

As a generalization of (5-22), we now let

$$\frac{\partial L}{\partial \dot{q}_k} = p_k. \tag{5-47}$$

And in the n-dimensional space of formula (5-4), we construct the vector

$$\mathbf{p} = p_k \mathbf{e}_k. \tag{5-48}$$

Expression p_k is called the kth component of the *canonical* or *generalized momentum* $\mathbf{p}$. Furthermore, we define the *del operator*

$$\nabla = \mathbf{e}_k \frac{\partial}{\partial q_k} \tag{5-49}$$

so that

$$\nabla L = \mathbf{e}_k \frac{\partial L}{\partial q_k}. \tag{5-50}$$

In the abstract n-dimensional space, equation (5-45) is then the kth component of

$$\frac{d\mathbf{p}}{dt} = \nabla L. \tag{5-51}$$

This vector form of Lagrange's equations is convenient for generally treating systems in which dissipative forces are absent or negligible.

5.8   Rayleigh's dissipation function

Whenever frictional losses occur, additional forces not derivable from a generalized potential U appear. Indeed, the difference between Q_k and the right side of (5-42) becomes a quantity Q_k' called the *residual force*. Introducing L as before then yields the equation

$$\frac{d}{dt} \frac{\partial L}{\partial \dot{q}_k} - \frac{\partial L}{\partial q_k} = Q_k'. \tag{5-52}$$

A model for dissipation consists of a spherical body moving slowly through a viscous fluid. For such a system, the time rate of energy dissipation is a linear function of the square of each component of velocity, according to *Stokes' law*.

Each element of work done against friction, by a set of small bodies dissipating energy in this way, is

$$-\delta W' = (a_j \dot{x}_j{}^2 + b_j \dot{y}_j{}^2 + c_j \dot{z}_j{}^2)\, \delta t = 2\mathscr{F}\, \delta t. \qquad (5\text{-}53)$$

Here a_j, b_j, c_j are proportionality constants, index j designates the pertinent small body, and $\mathscr{F}$ is the *Rayleigh dissipation function*, defined as

$$\mathscr{F} = -\frac{1}{2} \frac{\delta W'}{\delta t} \qquad (5\text{-}54)$$

one half the time rate of energy dissipation.

If X_j', Y_j', Z_j' are Cartesian components of the residual force acting on the jth particle and δx_j, δy_j, δz_j are real or virtual displacements, the corresponding work is

$$\begin{aligned}
\delta W' &= X_j'\, \delta x_j + Y_j'\, \delta y_j + Z_j'\, \delta z_j \\
&= (X_j' \dot{x}_j + Y_j' \dot{y}_j + Z_j' \dot{z}_j)\, \delta t \qquad (5\text{-}55)
\end{aligned}$$

according to (5-13).

On comparing equations (5-53) and (5-55), we find that

$$\begin{aligned}
X_1' &= -a_1 \dot{x}_1 & Y_1' &= -b_1 \dot{y}_1 & Z_1' &= -c_1 \dot{z}_1 \\
&= -\frac{\partial \mathscr{F}}{\partial \dot{x}_1}, & &= -\frac{\partial \mathscr{F}}{\partial \dot{y}_1}, & &= -\frac{\partial \mathscr{F}}{\partial \dot{z}_1}, \qquad (5\text{-}56)
\end{aligned}$$

and so on. These components, together with expansions of the variations from formula (5-7) transform equation (5-55) to

$$\begin{aligned}
\delta W' &= -\left(\frac{\partial \mathscr{F}}{\partial \dot{x}_j}\, \delta x_j + \frac{\partial \mathscr{F}}{\partial \dot{y}_j}\, \delta y_j + \frac{\partial \mathscr{F}}{\partial \dot{z}_j}\, \delta z_j\right) \\
&= -\left(\frac{\partial \mathscr{F}}{\partial \dot{x}_j}\frac{\partial x_j}{\partial q_k} + \frac{\partial \mathscr{F}}{\partial \dot{y}_j}\frac{\partial y_j}{\partial q_k} + \frac{\partial \mathscr{F}}{\partial \dot{z}_j}\frac{\partial z_j}{\partial q_k}\right)\delta q_k \\
&\quad -\left(\frac{\partial \mathscr{F}}{\partial \dot{x}_j}\frac{\partial x_j}{\partial t} + \frac{\partial \mathscr{F}}{\partial \dot{y}_j}\frac{\partial y_j}{\partial t} + \frac{\partial \mathscr{F}}{\partial \dot{z}_j}\frac{\partial z_j}{\partial t}\right)\delta t. \qquad (5\text{-}57)
\end{aligned}$$

Introducing formula (5-9) and the analogous formulas into the coefficient of δq_k and identifying this coefficient as Q_k' in

$$\delta W' = Q_k'\, \delta q_k + Q_t\, \delta t \qquad (5\text{-}58)$$

leads to

$$Q_k' = -\frac{\partial \mathscr{F}}{\partial \dot{x}_j}\frac{\partial \dot{x}_j}{\partial \dot{q}_k} - \frac{\partial \mathscr{F}}{\partial \dot{y}_j}\frac{\partial \dot{y}_j}{\partial \dot{q}_k} - \frac{\partial \mathscr{F}}{\partial \dot{z}_j}\frac{\partial \dot{z}_j}{\partial \dot{q}_k} = -\frac{\partial \mathscr{F}}{\partial \dot{q}_k}. \qquad (5\text{-}59)$$

This result reduces equation (5-52) to

$$\frac{d}{dt}\frac{\partial L}{\partial \dot{q}_k} - \frac{\partial L}{\partial q_k} + \frac{\partial \mathscr{F}}{\partial \dot{q}_k} = 0. \tag{5-60}$$

Being scalars, L and $\mathscr{F}$ are invariant to any change in coordinates. However, either may be subjected to gauge transformations that have no physical consequences. Such transformations will be discussed in section 5-9.

Some systems not ordinarily considered in mechanics are governed by (5-60). Indeed, a Lagrangian L can be set up as long as the total energy can be divided into a part analogous to the kinetic energy and a part analogous to the generalized potential of a mechanical system. A dissipation function $\mathscr{F}$ can be formulated if the rate at which energy is dissipated increases linearly with the square of each component of a generalized velocity. Then $\mathscr{F}$ is one half this rate.

EXAMPLE 5.11   During each time interval dt, a generator expends work $\mathscr{E}\, dq$ in pumping charge dq from one of its terminals to the other. The charge passes through a resistor to a capacitor. An equal amount of charge moves from the other side of the capacitor through an inductor to the original terminal, as in figure 5.8. The capacitor stores energy $\frac{1}{2}q^2/\mathscr{C}$, while the resistor dissipates energy at the rate $\mathscr{R}\dot{q}^2$ and the inductor holds energy $\frac{1}{2}\mathscr{L}\dot{q}^2$ as a flywheel would. Identify the energy quantities and construct the corresponding Lagrange equation.

Since it is analogous to potential energy, the energy in the capacitor minus that supplied by the generator for flow of charge q is equated to V:

$$V = \frac{1}{2}\frac{q^2}{\mathscr{C}} - \int \mathscr{E}\, dq.$$

Since it resembles kinetic energy, we set the energy associated with the inductor equal to T:

$$T = \tfrac{1}{2}\mathscr{L}\dot{q}^2.$$

Furthermore, $\mathscr{F}$ is one-half the rate of energy dissipation:

$$\mathscr{F} = \tfrac{1}{2}\mathscr{R}\dot{q}^2.$$

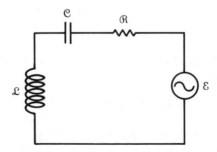

FIGURE 5.8   Circuit with resistance, capacitance, and inductance in series.

Calculating the pertinent derivatives,

$$\frac{\partial L}{\partial \dot{q}} = \mathscr{L}\dot{q}, \qquad \frac{d}{dt}\frac{\partial L}{\partial \dot{q}} = \mathscr{L}\ddot{q},$$

$$\frac{\partial L}{\partial q} = -\frac{q}{\mathscr{C}} + \mathscr{E}, \qquad \frac{\partial \mathscr{F}}{\partial \dot{q}} = \mathscr{R}\dot{q},$$

and substituting the results into (5-60) then yields

$$\mathscr{L}\ddot{q} + \frac{q}{\mathscr{C}} - \mathscr{E} + \mathscr{R}\dot{q} = 0,$$

or

$$\mathscr{L}\frac{d^2 q}{dt^2} + \mathscr{R}\frac{dq}{dt} + \frac{q}{\mathscr{C}} = \mathscr{E}.$$

5.9 Gauge transformations

Equation (5-41) does not uniquely define the generalized potential. Consequently, the Lagrangian in (5-45) is not uniquely defined. Any expression that satisfies this equation identically can be added to the L for a given problem, with no effect on the motion.

Let us take an arbitrary function of $t, q_1, \ldots, q_n$ and differentiate it with respect to time, following the formula in example 5.1:

$$\frac{df}{dt} = \frac{\partial f}{\partial t} + \frac{\partial f}{\partial q_l}\frac{dq_l}{dt} = \frac{\partial f}{\partial t} + \frac{\partial f}{\partial q_l}\dot{q}_l. \qquad (5\text{-}61)$$

Then differentiate with respect to $\dot{q}_k$

$$\frac{\partial}{\partial \dot{q}_k}\frac{df}{dt} = \frac{\partial f}{\partial q_k} \qquad (5\text{-}62)$$

and again with respect to t

$$\frac{d}{dt}\frac{\partial}{\partial \dot{q}_k}\frac{df}{dt} = \frac{d}{dt}\frac{\partial f}{\partial q_k} = \frac{\partial}{\partial q_k}\frac{df}{dt}. \qquad (5\text{-}63)$$

In the last step, we have interchanged the order of differentiation, as in (5-12).

The overall equality in (5-63) has the same form as equation (5-45); the expression

$$\frac{df(t, q_1, \ldots, q_n)}{dt} \qquad (5\text{-}64)$$

satisfies the Lagrange equation identically. Consequently, we can add (5-64) to L with no change in the physical results. Since the resulting transformation

$$L' = L + \frac{df(t, q_1, \ldots, q_n)}{dt} \qquad (5\text{-}65)$$

does regauge the dependence of the Lagrangian on the independent variables, it is called a *gauge transformation*.

The canonical momenta are also altered. Indeed, applying formula (5-47) to the transformed Lagrangian yields

$$p_k{}' = \frac{\partial L'}{\partial \dot{q}_k} = \frac{\partial L}{\partial \dot{q}_k} + \frac{\partial}{\partial \dot{q}_k} \frac{df}{dt} = p_k + \frac{\partial f}{\partial q_k}. \tag{5-66}$$

In the last step, equations (5-47) and (5-62) have been employed.

We can introduce a coordinate transformation

$$q_k = q_k(t, q_1{}', \dots, q_n{}') \tag{5-67}$$

along with a gauge transformation. The expression

$$\frac{df(t, q_1{}', \dots, q_n{}')}{dt} \tag{5-68}$$

then satisfies the final Lagrange equation identically. Corresponding to (5-65), we have

$$L'(t, q_1{}', \dots, q_n{}', \dot{q}_1{}', \dots, \dot{q}_n{}') = L(t, q_1(t, q_1{}', \dots, q_n{}'), \dots, \dot{q}_1(t, q_1{}', \dots, q_n{}'), \dots)$$
$$+ \frac{df(t, q_1{}', \dots, q_n{}')}{dt}. \tag{5-69}$$

We could start with an arbitrary function of time, the initial coordinates, and the final coordinates such that

$$F(t, q_1, \dots, q_n, q_1{}', \dots, q_n{}')$$
$$= F(t, q_1(t, q_1{}', \dots, q_n{}'), \dots, q_n(t, q_1{}', \dots, q_n{}'), q_1{}', \dots, q_n{}')$$
$$= f(t, q_1{}', \dots, q_n{}'). \tag{5-70}$$

Then (5-69) could be rewritten in the form

$$L'(t, q_1{}', \dots, q_n{}', \dot{q}_1{}', \dots, \dot{q}_n{}')$$
$$= L(t, q_1, \dots, q_n, \dot{q}_1, \dots, \dot{q}_n) + \frac{dF(t, q_1, \dots, q_n, q_1{}', \dots, q_n{}')}{dt}. \tag{5-71}$$

Besides relating the new Lagrangian to the old, this equation relates the new coordinates to the old ones. In a sense, function F generates the transformation.

EXAMPLE 5.12   Transform the Lagrangian

$$L = \tfrac{1}{2} m(\dot{x}^2 - \omega_0{}^2 x^2)$$

with the generating function

$$f = \tfrac{1}{2} i m \omega_0 x^2.$$

Since f is a function of x alone, it has the required form. Differentiating it

$$\frac{df}{dt} = im\omega_0 x\dot{x}$$

and adding the result to the given L, as in (5-65), leads to

$$L' = \tfrac{1}{2}m(\dot{x}^2 - \omega_0^2 x^2) + im\omega_0 x\dot{x}$$
$$= \tfrac{1}{2}m(\dot{x} + i\omega_0 x)^2.$$

5.10 Suitable independent variables

All aspects of the motion of a physical system are determined by given conditions and interactions. Any symmetry in space that these influences possess can with advantage be represented by generalized coordinates. Indeed, we have already used the r, θ, ϕ of figure 5.2 in treating motion within a central field. Only one vertical distance x was needed in discussing Atwood's machine of figure 5.3.

Lagrange's equations facilitate the use of such pertinent coordinates, since they do not require us to resolve any accelerations. Furthermore, the kinetic energy T and forces Q_k can often be set up directly.

Additional simplification of the equations can result when both the coordinates and momenta are transformed. In effect, a more subtle symmetry than any we have so far considered is then present. An introduction to the procedure will appear in chapter 10.

DISCUSSION QUESTIONS

5.1 Why does the expansion of df in terms of dq_1, dq_2, ... hold even when q_1, q_2, ... are not physically independent variables? What does use of this expansion imply about the continuity of f, q_1, q_2, ...?

5.2 Rearrange the Taylor series expansion of f about a given point to get a series for Δf. Show what happens to this expression when all increments become infinitesimal.

5.3 How does the transformation to generalized coordinates differ from a linear transformation?

5.4 Find the matrix for transforming

$$\begin{pmatrix} dq_1 \\ dq_2 \\ \vdots \\ dq_n \\ dt \end{pmatrix} \text{ to } \begin{pmatrix} dx_1 \\ \vdots \\ dx_N \\ dy_1 \\ \vdots \\ dz_1 \\ \vdots \end{pmatrix}.$$

5.5 Show that

$$\frac{\partial \dot{x}_j}{\partial \dot{q}_k} = \frac{\partial x_j}{\partial q_k} \quad \text{and} \quad \frac{d}{dt}\frac{\partial x_j}{\partial q_k} = \frac{\partial \dot{x}_j}{\partial q_k}.$$

5.6 Why is δW invariant? How can we use δW in defining the generalized force Q_k?

5.7 Derive Lagrange's force equations from Newton's laws. Why is this derivation more than just a mathematical exercise?

5.8 What coordinate systems fit the symmetry of the paths of (a) a simple-pendulum bob, (b) a mass point constrained to a sphere, (c) a projectile, (d) a weight in Atwood's machine, (e) the electron in $H_2{}^+$, (f) a nonperiodic comet? How does proper choice of a coordinate system simplify calculations?

5.9 When is a generalized force the same as the corresponding component of conventional force? When is it different?

5.10 In working a problem with forces given in some inertial frame, how do we obtain appropriate expressions for (a) the Lagrangian kinetic energy and (b) the generalized forces in the desired system?

5.11 When is the fictitious-force term different from zero?

5.12 Describe various constraints. When and how can constraints be introduced implicitly into the equations of motion?

5.13 How is the derivation of Lagrange's force equation altered when explicit constraints exist among the δq_k's?

5.14 How are Lagrange multipliers used?

5.15 What do conservative forces contribute to δW? How does an evernormal force affect the form, but not the magnitude, of δW? In what way do dissipative forces affect δW?

5.16 How is the generalized potential U defined? What advantage does it have over the conventional potential V?

5.17 Obtain the differential equation governing

$$L = T - U.$$

5.18 How can the set of Lagrange equations for a system be represented by a single vector equation in configuration space (the space in which the q_k's are Cartesian components)?

5.19 Describe the use of Rayleigh's dissipation function.

5.20 How can the laws of mechanics govern the behavior of an electric circuit?

5.21 What arbitrariness exists in the definition of L?

5.22 What can one do to prevent mathematical complexities from obscuring the physical relationships in a given problem?

PROBLEMS

5.1 (a) Constraints cause a small body to move on a spherical surface fixed in an inertial frame. Using fundamental definitions and the Pythagorean theorem, express the square of a small displacement of the particle as a function of constants and angles. Then construct the corresponding kinetic energy T. (b) Express the kinetic energy of a particle constrained to a fixed plane as a function of one distance and one angle.

5.2 Set up Lagrange's force equations in polar coordinates and use them to find the force law for a particle moving along the orbit

$$\frac{1}{r} = ae^{b\phi}$$

in an inertial frame with the radius vector r sweeping out area at a constant rate.

5.3 Use Lagrange's force equation to find an expression for the angular acceleration of a pendulum bob moving in a vertical plane. Also, find an expression for the tension in the string supporting the bob.

5.4 The string from a $5m$ mass passes over a fixed pulley to a mass m, to which is attached a pulley. Over this movable pulley passes a string supporting a $2m$ mass on one side and a $3m$ mass on the other. Choose two generalized coordinates, find expressions for kinetic energy T and generalized forces Q_1, Q_2, and write the equations of motion from them. Finally calculate how fast the $5m$ mass accelerates.

5.5 Using Lagrange's force equation, obtain the acceleration of a rough plank moving on a smooth plane inclined at 30° from the horizontal.

5.6 The plank in problem 5.5 extends 16 feet down the inclined plane. A dog weighing as much as the plank runs down it just fast enough to slow its acceleration to $(\frac{1}{10})g$. Calculate his speed when he reaches the lower end of the plank, if both dog and plank were at rest when the dog was at the upper end of the plank.

5.7 Throughout its movement on a smooth horizontal table, a particle is constrained by a string that extends from it, through a hole in the table, to a like particle hanging vertically. Put suitable generalized coordinates and forces into the corresponding Lagrange force equations. Integrate the simpler equation and substitute the result into the other one to get a differential equation in a single coordinate (dependent variable).

5.8 A particle is connected to point P on a smooth horizontal table by a string of length a. The table rotates at constant angular velocity ω about an axis distance b from P. Assume that at $t = 0$ the particle is distance $a + b$ from the axis and find the acceleration of θ, the angle between the string and a line drawn on the table from P.

5.9 For Atwood's machine formulate the kinetic energy T and forces Q_k as if the weights were independent. Then introduce a differential constraining relationship and solve using the undetermined (Lagrange) multiplier method.

5.10 What are the differential equations of constraint for a disk rolling on a plane? Let the point of contact be (x, y), while a is the radius of the disk, ϕ the angle it rotates through, and θ the angle between the projection of its axis on the xy plane and the x axis.

5.11 A hoop is released from rest at the top of a plane of length l inclined at angle ϕ from the horizontal. If it rolls directly down without slipping, what are its kinetic energy, potential energy, and differential equation of constraint? Using the Lagrange multiplier method, find the acceleration of the hoop down the plane.

5.12 The viscous force F retarding a sphere of radius r in a fluid with viscosity η is given by Stokes' law

$$F = -6\pi\eta r v$$

where v is the velocity of the sphere. Set up the corresponding dissipation function. Then set up Lagrange's equation for the sphere falling vertically under influence of gravity in the fluid. Solve for the limiting velocity.

5.13 Formulate a Lagrangian function for an inductor and a capacitor connected in parallel to a varying emf.

5.14 A circuit containing a capacitor and a secondary coil is coupled to a primary coil across which a source of emf acts. Set up the Lagrangian if the energy of mutual interaction of the two circuits is $\mathcal{M}_{12}\dot{q}_1\dot{q}_2$.

5.15 Derive the equation of motion from each Lagrangian in Example 5.12. Does the gauge transformation alter the result?

5.16 Each infinitesimal displacement of a mass point can be resolved into the components

$$ds_1 = h_1(q_1, q_2, q_3)\, dq_1,$$
$$ds_2 = h_2(q_1, q_2, q_3)\, dq_2,$$
$$ds_3 = h_3(q_1, q_2, q_3)\, dq_3,$$

in a given inertial frame. If these components are mutually perpendicular, what is the particle's kinetic energy?

5.17 Show that if

$$x = a \cosh u \cos v,$$
$$y = a \sinh u \sin v,$$

point (x, y) moves along an ellipse when u is constant and along a hyperbola when v is constant. Show that each of these ellipses meets each of the hyperbolas at right angles. Then express the kinetic energy of a particle in the uv system.

5.18 Set up Langrange's force equations in cylindrical coordinates and use them to find the force law for a particle moving along the orbit

$$r\phi = \text{constant}$$

in an inertial frame with the radius vector r sweeping out equal areas in equal times.

5.19 Using Lagrange's force equations, obtain the central-force law that causes a particle to move along the circle

$$r = A \cos \phi.$$

5.20 Find out how the period of motion in problem 5.19 varies with the diameter A of the orbit.

5.21 The string from a mass m passes over a fixed pulley to a mass $2m$ moving without friction on a plane inclined at 60° from the horizontal. Set up Lagrange's force equation for the system and obtain the acceleration of the $2m$ mass.

5.22 A mass m is suspended by a string of length l from a support at x which moves back and forth horizontally following the equation

$$x = a \cos \omega t.$$

Assuming the string to remain in a single vertical plane, set up Lagrange's equation for the motion. If the angle of oscillation is small, what does the equation reduce to? Integrate this approximate equation for the final steady-state motion.

5.23 Mass m moves without friction on a straight wire inclined at angle α from a vertical z axis. Now if the wire revolves about the z axis with constant angular velocity ω, what is Lagrange's equation for the motion and its solution?

5.24 For problem 5.21 formulate the kinetic energy T and forces Q_k as if the weights were independent. Then introduce a differential constraining relationship and solve using the undetermined (Lagrange) multiplier method.

5.25 With the Lagrange multiplier method, find a single differential equation describing how a hoop of radius r rolls inside a fixed hoop of radius R.

5.26 A uniform plank of length $2l$, thickness $2h$, and mass m is set crosswise on a fixed horizontal roller of radius a. If the center of mass is directly over the line of contact when the plank is horizontal and if the amplitude of oscillation is small, what Lagrange equation governs its seesaw motion near either turning point, where θ is small? Under what conditions is acceleration of the inclination angle negative near the turning point where θ is positive and positive near the one where θ is negative? When is the equilibrium position stable?

5.27 Formulate Lagrangian and dissipation functions for a circuit in which an inductor and a resistor are connected in parallel to a varying emf. Then set up the corresponding Lagrange equations.

5.28 Show that the generating function

$$F = q_k q_k'$$

transforms the Lagrangian, through (5-71), so each new momentum equals an old coordinate and the negative of each new coordinate equals an old momentum.

REFERENCES

BOOKS

Greenwood, D. T. *Principles of Dynamics*, pp. 229–280. Prentice-Hall, Inc., Englewood Cliffs, N.J., 1965.

Hauser, W. *Introduction to the Principles of Mechanics*, pp. 142–163. Addison-Wesley Publishing Company, Inc., Reading, Mass., 1965.

Huang, T. C. *Engineering Mechanics*, pp. 639–647, 702–707, 809–813. Addison-Wesley Publishing Company, Inc., Reading, Mass., 1967.

Symon, K. R. *Mechanics*, 2nd ed., pp. 354–405. Addison-Wesley Publishing Company, Inc., Reading, Mass., 1960.

ARTICLES

Andersen, C. M., and **von Baeyer, H. C.** Theory of a Ball Rolling on a $1/\rho$ Surface of Revolution. *Am. J. Phys.*, **38**, 140 (1970).

Davis, W. R. Constraints in Particle Mechanics and Geometry. *Am. J. Phys.*, **35**, 916 (1967).

Denman, H. H. On Linear Friction in Lagrange's Equation. *Am. J. Phys.*, **34**, 1147 (1966).

Frank, W., and **Trigg, G. L.** Momentum and Conservation Laws in Newtonian and Canonical Formalisms. *Am. J. Phys.*, **28**, 315 (1960).

Iddings, C. K., and **Teague, M.** Potentials as the Result of Unobserved Degrees of Freedom. *Am. J. Phys.*, **39**, 619 (1971).

Kelly, E. M. Lagrange's Equations from Intrinsic Geometry of Coordinate Systems. *Am. J. Phys.*, **36**, 908 (1968).

Levy-Leblond, J. M. Conservation Laws for Gauge-Variant Lagrangians in Classical Mechanics. *Am. J. Phys.*, **39**, 502 (1971).

Plybon, B. F. Conservation Laws for Undergraduates. *Am. J. Phys.*, **39**, 1372 (1971).

Solbrig, A. W., Jr. Momentum in Newtonian and Lagrangian Formalisms. *Am. J. Phys.*, **28**, 680 (1960).

6 / *Vibrational Coordinates, Frequencies, and Amplitudes*

6.1 Resolving the collective motions of the particles in a given system Influences on the resulting modes

Whenever coordinates can be chosen so that equations (5-20) separate into independent (or nearly independent) subsets, the corresponding motions are independent (or nearly independent) of each other. The divisions must appear in both the kinetic energy T and the generalized forces $Q_1, \ldots, Q_n$.

According to example 3.6, the kinetic energy of the mass elements in a system does split into a part $\frac{1}{2} M \dot{\mathbf{R}} \cdot \dot{\mathbf{R}}$ associated with translation and a part $\frac{1}{2} m_j \dot{\mathbf{r}}_j{}^i \cdot \dot{\mathbf{r}}_j{}^i$ associated with the other motions. In free space, or in a uniform gravitational field, the force components split similarly. Then there are three equations governing the translation and $3N - 3$ governing the rotational, vibrational, and random relative motions.

Translations and rotations have already been discussed. Let us now focus our attention on vibrations. We will assume that conservative forces bind the system together. Furthermore, we will consider the corresponding potential energy V and the kinetic energy T to be quadratic functions of the same generalized coordinates. For most systems, such forms are valid only when various amplitudes are small.

In the absence of other forces, the resulting Lagrange equations are homogeneous and linear. The condition that a linear combination of these equations has the form of the harmonic-oscillator equation is a set of simultaneous equations with the multipliers as unknowns. The condition that these simultaneous equations be consistent is the secular equation. Solving the secular equation yields the frequency for each oscillation.

Each frequency converts the simultaneous equations into a set relating the original coordinates. Variation of one of the resulting combinations, keeping all other such combinations fixed, then represents progression of the chosen oscillation, or *mode* of motion.

A periodic driving force that produces steady-state oscillation can be introduced. A type of dissipative force will be studied and its effect determined. Some aspects of nonlinear vibrations are also to be treated.

Rotation modifies vibrational modes because it introduces centripetal and Coriolis accelerations. Indeed, the kinetic energy $\frac{1}{2}m_j\dot{\mathbf{r}}_j{}^i \cdot \dot{\mathbf{r}}_j{}^i$ splits into a rotational term, a vibrational term, and an interaction term. The nature of each will be considered.

6.2   A model oscillator

A model for a particular mode of vibration is a mass m supported by a rubber band or a steel spring as in figure 6.1.

Let x be the distance of a given point on the weight from a reference level while x_e is this distance when the system is at equilibrium. Let the displacement of the weight from its equilibrium position be q at time t:

$$q = x - x_e. \tag{6-1}$$

For the common band or spring, observers find that the displacement is represented approximately by a sinusoidal function of time

$$q = A \sin (2\pi vt - b) \tag{6-2}$$

in which v is the frequency, A the amplitude, and b a phase angle.

Differentiating (6-2) twice yields the result

$$\ddot{q} = -4\pi^2 v^2 A \sin (2\pi vt - b) = -4\pi^2 v^2 q. \tag{6-3}$$

This has the form

$$\ddot{q} + \omega^2 q = 0 \tag{6-4}$$

in which ω equals $2\pi v$ and is called the *angular frequency*.

When the frequency is zero, equation (6-4) reduces to

$$\ddot{q} = 0 \tag{6-5}$$

a relationship requiring q to change at a constant rate. Such behavior is associated with free translation in a particular direction and with free rotation about a given

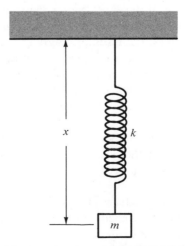

FIGURE 6.1   Mass supported by a spring in a gravitational field.

axis, as well as with a vibration that has degenerated into uniform motion in one direction only during the time the equation is valid.

When ω is positive, equation (6-4) describes vibration at a definite frequency. Because such motion is presumably associated with each mode, let us seek to combine the displacements of particles in a given system so that generalized coordinates obeying this equation are obtained. But we will first have to consider the nature of T and V.

6.3   Kinetic energy

If m_j is the mass of the jth particle (or element) and x_j, y_j, z_j are its Cartesian coordinates in an inertial or center-of-mass frame, then the kinetic energy is

$$T = \tfrac{1}{2}m_j\dot{x}_j^{\,2} + \tfrac{1}{2}m_j\dot{y}_j^{\,2} + \tfrac{1}{2}m_j\dot{z}_j^{\,2} \tag{6-6}$$

as long as the summation indicated by repetition of the indices is carried out over all the particles or elements in the given system (remember the definition in section 3.3 and the result in example 3.6).

When transformation equations (5-1) do not contain t explicitly, equations (5-8) reduce to equalities

$$\dot{x}_j = \frac{\partial x_j}{\partial q_k}\,\dot{q}_k \tag{6-7}$$

in which the partial derivatives vary with the q_k's but *not* with the $\dot{q}_k$'s. The square of $\dot{x}_j$ is then a homogeneous quadratic form in the $\dot{q}_k$'s, containing terms such as $A_{11}\dot{q}_1\dot{q}_1$, $A_{12}\dot{q}_1\dot{q}_2$, $A_{21}\dot{q}_2\dot{q}_1, \ldots$.

This form and the corresponding forms for $\dot{y}_j^{\,2}$, $\dot{z}_j^{\,2}$ change equation (6-6) to

$$T = \dot{q}_j A_{jk}\dot{q}_k = \dot{\mathbf{q}}^{\dagger}\mathbf{A}\dot{\mathbf{q}}. \tag{6-8}$$

Each expression A_{jk} depends on the q_k's but not on the $\dot{q}_k$'s. But as long as the displacements from an equilibrium configuration remain small, the q_k's do not vary appreciably and each A_{jk} is approximately constant.

The A_{jk}'s are also constant, regardless of any displacement, for certain coordinates. Indeed, if we let

$$
\begin{aligned}
x_1 - (x_1)_0 &= \eta_1, \\
y_1 - (y_1)_0 &= \eta_2, \\
z_1 - (z_1)_0 &= \eta_3, \\
x_2 - (x_2)_0 &= \eta_4, \\
&\ \vdots \\
z_N - (z_N)_0 &= \eta_{3N},
\end{aligned} \tag{6-9}
$$

with $(x_1)_0$, $(y_1)_0$, $(z_1)_0$, $(x_2)_0, \ldots, (z_N)_0$ all constant, the derivatives are not affected:

$$
\begin{aligned}
\dot{x}_1 &= \dot{\eta}_1, \\
\dot{y}_1 &= \dot{\eta}_2, \\
\dot{z}_1 &= \dot{\eta}_3, \\
\dot{x}_2 &= \dot{\eta}_4, \\
&\ \vdots \\
\dot{z}_N &= \dot{\eta}_{3N}.
\end{aligned} \tag{6-10}
$$

The nonzero A_{jk}'s in

$$T = \dot{\eta}_j A_{jk} \dot{\eta}_k = \dot{\boldsymbol{\eta}}^\dagger \mathbf{A} \dot{\boldsymbol{\eta}} \tag{6-11}$$

are then the half masses.

In any case, let us divide the terms in T so that $\mathbf{A}$ is symmetric:

$$A_{jk} = A_{kj}. \tag{6-12}$$

Thus, the quantity $f(q_1, \dots)\dot{q}_j\dot{q}_k$ in T is split equally, with the coefficient in one half being labeled A_{jk}, that in the other half A_{kj}.

6.4   A suitable potential

When the forces binding mass elements together are conservative, they are derivable from a potential function V. Let us now consider how such a function is represented by a quadratic form similar to that for T.

We first suppose that the function is analytic in the coordinates:

$$V = b + b_j q_j + b_{jk} q_j q_k + b_{jkl} q_j q_k q_l + \cdots. \tag{6-13}$$

Since each Q_m is $-\partial V/\partial q_m$, we may choose

$$b = 0 \tag{6-14}$$

without affecting any forces.

Furthermore, we may construct the coordinates so that each q_j vanishes when the system is in an equilibrium configuration. But in this configuration, the expression

$$\frac{\partial V}{\partial q_m} = b_m + b_{mk} q_k + b_{jm} q_j + \cdots \tag{6-15}$$

must also vanish. Then we have

$$b_m = 0. \tag{6-16}$$

When there is only one mode of vibration, the spring may be chosen so that terms in V beyond the second degree vanish. Anharmonicities are said to be absent and the spring is neither soft nor hard. When there are several modes, the use of such springs does not ensure the vanishing of all cubic and higher terms. But such terms will be negligible if the pertinent displacements are small.

When such terms can be neglected, equations (6-14) and (6-16) reduce (6-13) to

$$V = b_{jk} q_j q_k = q_j b_{jk} q_k = \mathbf{q}^\dagger \mathbf{B} \mathbf{q}. \tag{6-17}$$

When using the η_j's, we generally let $(x_1)_0, (y_1)_0, (z_1)_0, (x_2)_0, \dots, (z_N)_0$ be the Cartesian coordinates of an equilibrium configuration. Then

$$V = \eta_j B_{jk} \eta_k = \boldsymbol{\eta}^\dagger \mathbf{B} \boldsymbol{\eta}. \tag{6-18}$$

As with T, the quantities in the sum can be distributed so that $\mathbf{B}$ is symmetric $(B_{jk} = B_{kj})$.

6.5   The secular equation

An unplanned solution of the equations of motion for a given system of interacting particles may be very difficult. But if the kinetic and potential energies are *quadratic forms* with effectively constant coefficients, we can combine the equations to get an equation (6-4) for each kind of vibration in the system.

When coordinates describing all displacements in the system are employed, a zero frequency appears for each rotational and translational mode. However, the translational energy is distinct from the remaining kinetic energy, and the rotational energy can also be separated out as a first approximation. As a consequence, we may employ coordinates that do not measure the position of the center of mass in an inertial frame, and, as a first approximation, coordinates that do not measure the orientation of the system in space.

Our method does require using coordinates $\zeta_1, \zeta_2, \ldots, \zeta_n$ for which the pertinent kinetic energy has the form $\dot{\zeta}_j A_{jk} \dot{\zeta}_k$ and the pertinent potential energy has the form $\zeta_j B_{jk} \zeta_k$, with all A_{jk}'s and B_{jk}'s constant. We also symmetrize matrices **A** and **B** as equation (6-12) and the corresponding equation $B_{jk} = B_{kj}$ indicate.

Differentiating the quadratic T and V with respect to ζ_l and $\dot{\zeta}_l$ proceeds as follows:

$$\frac{\partial T}{\partial \zeta_l} = 0, \tag{6-19}$$

$$\frac{\partial T}{\partial \dot{\zeta}_l} = \frac{\partial}{\partial \dot{\zeta}_l} (\dot{\zeta}_j A_{jk} \dot{\zeta}_k) = A_{lk} \dot{\zeta}_k + \dot{\zeta}_j A_{jl} = 2A_{lk} \dot{\zeta}_k, \tag{6-20}$$

$$\frac{\partial V}{\partial \zeta_l} = \frac{\partial}{\partial \zeta_l} (\zeta_j B_{jk} \zeta_k) = B_{lk} \zeta_k + \zeta_j B_{jl} = 2B_{lk} \zeta_k, \tag{6-21}$$

$$\frac{\partial V}{\partial \dot{\zeta}_l} = 0. \tag{6-22}$$

Motion in conservative systems is governed by Lagrange equations (5-38). Substituting the final expressions of (6-19) through (6-22) into these equations yields

$$A_{lk} \ddot{\zeta}_k + B_{lk} \zeta_k = 0, \tag{6-23}$$

whence

$$\mathbf{A}\ddot{\zeta} + \mathbf{B}\zeta = 0. \tag{6-24}$$

If any additional forces acted on the particles, besides those due to their mutual interactions, they would appear on the right, as in equation (5-52). Such forces will be considered later.

Let us change the free index in (6-23) to j, multiply the result by C_{lj}, and sum over j. We thus obtain the general linear combination

$$C_{lj} A_{jk} \ddot{\zeta}_k + C_{lj} B_{jk} \zeta_k = 0, \tag{6-25}$$

or

$$\mathbf{CA}\ddot{\zeta} + \mathbf{CB}\zeta = 0. \tag{6-26}$$

If it is possible to replace the first term of (6-25) by $\ddot{q}_l$ and the second one by $\omega^2 q_l$, the desired reduction of (6-23) to

$$\ddot{q}_l + \omega^2 q_l = 0 \tag{6-27}$$

would be effected. The substitutions are possible if the identity

$$q_l = C_{lj} A_{jk} \zeta_k = \frac{1}{\omega^2} C_{lj} B_{jk} \zeta_k \tag{6-28}$$

exists. That is, if

$$C_{lj}(\omega^2 A_{jk} - B_{jk}) = 0. \tag{6-29}$$

However, a nontrivial solution, with some $C_{lj} \neq 0$, can be found only if the determinant of the coefficients of the C_{lj}'s is zero:

$$\begin{vmatrix} \omega^2 A_{11} - B_{11} & \omega^2 A_{21} - B_{21} & \cdot & \cdot & \cdot \\ \omega^2 A_{12} - B_{12} & \omega^2 A_{22} - B_{22} & \cdot & \cdot & \cdot \\ \cdot & \cdot & \cdot & \cdot & \cdot \end{vmatrix} = 0. \tag{6-30}$$

This equation in ω^2 is called the *secular equation*. It has n roots, not necessarily distinct. The roots are real, though, for each real *normal frequency* $\omega/2\pi$. A corresponding q is called a *normal coordinate*.

EXAMPLE 6.1   Two identical simple pendulums interact because of a spring joining the bobs as in figure 6.2. Find their normal frequencies of vibration, if the spring does not affect the equilibrium positions of the bobs.

We assume that each bob acts as if its mass m were concentrated at its center, distance r from the corresponding point of support. If at a given instant the angular

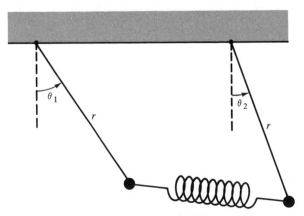

FIGURE 6.2   Coordinate systems for the coupled pendulums.

velocity of the first bob is $\dot\theta_1$ and that of the second one $\dot\theta_2$, the kinetic energy of the bobs is then

$$T = \tfrac{1}{2}mr^2\dot\theta_1{}^2 + \tfrac{1}{2}mr^2\dot\theta_2{}^2.$$

All other kinetic energies will be neglected.

From the angles, we find that the height of the first bob above its equilibrium position is $r(1 - \cos\theta_1)$, while that of the second bob is $r(1 - \cos\theta_2)$. So at the given instant, the gravitational potential of these is

$$mgr(1 - \cos\theta_1) + mgr(1 - \cos\theta_2) = 2mgr - mgr\cos\theta_1 - mgr\cos\theta_2$$
$$\simeq \tfrac{1}{2}mgr\theta_1{}^2 + \tfrac{1}{2}mgr\theta_2{}^2.$$

In the approximation, each trigonometric function has been expanded, and terms higher than the quadratic are neglected.

On the other hand, the horizontal component of distance between the bobs is

$$x_2 - x_1 = a + r\sin\theta_2 - r\sin\theta_1$$

while the vertical component is

$$y_2 - y_1 = -r\cos\theta_2 + r\cos\theta_1.$$

The square of the distance between them is

$$
\begin{aligned}
l^2 &= (x_2 - x_1)^2 + (y_2 - y_1)^2 \\
&= a^2 + r^2\sin^2\theta_2 + r^2\sin^2\theta_1 + 2ar\sin\theta_2 - 2ar\sin\theta_1 \\
&\quad - 2r^2\sin\theta_1\sin\theta_2 + r^2\cos^2\theta_2 + r^2\cos^2\theta_1 - 2r^2\cos\theta_2\cos\theta_1 \\
&\simeq a^2 + r^2\theta_2{}^2 + r^2\theta_1{}^2 + 2ar(\theta_2 - \theta_1) - 2r^2\theta_1\theta_2 \\
&= (a + r\theta_2 - r\theta_1)^2.
\end{aligned}
$$

Here again, the trigonometric functions have been expanded and higher terms neglected.

The displacement from the equilibrium distance is

$$l - a \simeq r\theta_2 - r\theta_1.$$

If the spring is neither soft nor hard, the corresponding potential energy is

$$\tfrac{1}{2}k(l - a)^2 = \tfrac{1}{2}k(r\theta_2 - r\theta_1)^2.$$

Neglecting all other potential energies, we have

$$
\begin{aligned}
V &\simeq \tfrac{1}{2}mgr\theta_1{}^2 + \tfrac{1}{2}mgr\theta_2{}^2 + \tfrac{1}{2}kr^2\theta_1{}^2 - kr^2\theta_1\theta_2 + \tfrac{1}{2}kr^2\theta_2{}^2 \\
&= \tfrac{1}{2}(mgr + kr^2)\theta_1{}^2 - \tfrac{1}{2}kr^2\theta_1\theta_2 - \tfrac{1}{2}kr^2\theta_2\theta_1 + \tfrac{1}{2}(mgr + kr^2)\theta_2{}^2.
\end{aligned}
$$

Substituting the coefficients for T and V into the secular equation now yields

$$
\tfrac{1}{2}r^2
\begin{vmatrix}
m\omega^2 - \dfrac{mg}{r} - k & k \\[2ex]
k & m\omega^2 - \dfrac{mg}{r} - k
\end{vmatrix}
= 0,
$$

whence

$$\omega_1 = \sqrt{\frac{g}{r}}, \qquad \omega_2 = \sqrt{\frac{g}{r} + 2\frac{k}{m}}.$$

The corresponding frequencies are

$$\nu_1 = \frac{1}{2\pi}\sqrt{\frac{g}{r}}, \qquad \nu_2 = \frac{1}{2\pi}\sqrt{\frac{g}{r} + 2\frac{k}{m}}.$$

EXAMPLE 6.2   Obtain normal coordinates for the coupled pendulums of figure 6.2.
   The coefficients of the unknowns in equation (6-29) are the elements in the determinant of (6-30). Thus from the secular equation in example 6.1, we get the set

$$C_{11}\left(m\omega^2 - \frac{mg}{r} - k\right) + C_{12}k = 0.$$

$$C_{11}k + C_{12}\left(m\omega^2 - \frac{mg}{r} - k\right) = 0.$$

For the solution

$$\omega^2 = \frac{g}{r},$$

the first equation of the set becomes

$$C_{11}\left(m\frac{g}{r} - \frac{mg}{r} - k\right) + C_{12}k = 0,$$

whence

$$-C_{11} + C_{12} = 0 \qquad \text{or} \qquad C_{11} = C_{12}.$$

But since the coefficients in the kinetic energy expression are

$$A_{11} = \tfrac{1}{2}mr^2, \qquad A_{12} = 0,$$
$$A_{21} = 0, \qquad A_{22} = \tfrac{1}{2}mr^2,$$

equation (6-28) yields

$$q_1 = C_{11}\tfrac{1}{2}mr^2\theta_1 + C_{12}\tfrac{1}{2}mr^2\theta_2.$$

Choosing

$$C_{11} = \frac{2}{mr^2},$$

and using $C_{12} = C_{11}$, we get

$$q_1 = \theta_1 + \theta_2.$$

   Similarly, when

$$\omega^2 = \frac{g}{r} + 2\frac{k}{m},$$

the first equation (6-29) becomes

$$C_{21}\left(\frac{mg}{r} + 2k - \frac{mg}{r} - k\right) + C_{22}k = 0.$$

Then

$$C_{21} + C_{22} = 0 \quad \text{or} \quad C_{21} = -C_{22}.$$

With

$$C_{21} = \frac{2}{mr^2},$$

as before, equation (6-28) gives us

$$q_2 = \theta_1 - \theta_2.$$

To determine what these generalized coordinates stand for, consider the motion when *only one* varies from zero. Indeed if $q_1 \neq 0$ and $q_2 = 0$, we have

$$\theta_1 = \theta_2.$$

Then the pendulums move in phase. The spring exerts no force and the frequency is the same as for a single isolated pendulum:

$$\nu_1 = \frac{1}{2\pi} \sqrt{\frac{g}{r}}.$$

By letting $q_2 \neq 0$ and $q_1 = 0$, we obtain

$$\theta_1 = -\theta_2.$$

Then the pendulums are 180° out of phase. The spring exerts its maximum effect, causing the pendulums to slow down faster as they move apart from their equilibrium positions and causing them to return faster after they stop. Thus the frequency of vibration is increased. Our analysis showed that

$$\nu_2 = \frac{1}{2\pi} \sqrt{\frac{g}{r} + 2\frac{k}{m}}.$$

The general motion is a superposition of the two modes. Note that combining the equations for q_1 and q_2 yields the angular displacements

$$\theta_1 = \frac{q_1 + q_2}{2},$$

$$\theta_2 = \frac{q_1 - q_2}{2}.$$

Thus, an increase in q_1 increases θ_1 and θ_2 by the same amount, while an increase in q_2 increases θ_1 and decreases θ_2 by the same amount.

6.6 Simultaneous diagonalization of matrices A and B

By transforming the variables so $\mathbf{A}$ in the kinetic energy expression becomes $\mathbf{E}$, the unit matrix, one can reduce the computation of section 6.5 to the diagonalization carried out in section 4.10.

First, multiply each equation (6-9) by the square root of half the mass whose displacement is being described. Let the result equal a new variable

$$\xi_k = \sqrt{\frac{m_j}{2}}\, \eta_k \tag{6-31}$$

in which j labels the pertinent particle. Differentiate this variable:

$$\dot{\xi}_1 = \sqrt{\frac{m_1}{2}}\, \dot{x}_1,$$

$$\dot{\xi}_2 = \sqrt{\frac{m_1}{2}}\, \dot{y}_1.$$

$$\vdots \tag{6-32}$$

Then substitute into equation (6-6) to get

$$T = \dot{\xi}_k \dot{\xi}_k = \dot{\xi}_k \delta_{kl} \dot{\xi}_l \tag{6-33}$$

and into (6-17) to obtain

$$V = \xi_k B_{kl}{}' \xi_l. \tag{6-34}$$

The argument in section 6.5 now yields the secular equation

$$\begin{vmatrix} \omega^2 - B_{11}{}' & -B_{21}{}' & \cdot & \cdot & \cdot \\ -B_{12}{}' & \omega^2 - B_{22}{}' & \cdot & \cdot & \cdot \\ \cdot & \cdot & \cdot & \cdot & \cdot \end{vmatrix} = 0. \tag{6-35}$$

Since matrix $\mathbf{B}'$ is symmetric, equation (6-35) does not differ essentially from equation (4-101), which is the condition that equations (4-100) have nontrivial solutions. A matrix form of the pertinent simultaneous equations is

$$\mathbf{B}'\mathbf{u} = \omega^2 \mathbf{u} = \mathbf{u}\omega^2 \tag{6-36}$$

where ω^2 is the lth eigenvalue while $\mathbf{u}$ is the lth eigenvector composed of the components $C_{l1}, C_{l2}, \ldots, C_{ln}$.

With our substitutions, the first equality in (6-28) reduces to the form

$$q_l = C_{lj} \delta_{jk} \xi_k = C_{lj} \xi_j. \tag{6-37}$$

The components of the eigenvector are the coefficients that multiply the ξ_j's in the expression for the lth generalized coordinate.

6.7 Pure forced vibrations

The movements we have discussed so far occur when the system vibrates freely after being initially disturbed in some way. If external influences continue to act, additional forces must be included beyond those coming from spatial variations in the potential V. These may be added on the right as in formula (5-52).

The additional force acting on the jth particle can be resolved into x, y, and z components F_{3j-2}, F_{3j-1}, F_{3j}. Multiplying each by the corresponding displacement and adding yields the work

$$\delta W = F_l \, \delta \eta_l. \tag{6-38}$$

The Lagrange equations now become

$$A_{jk}\ddot{\eta}_k + B_{jk}\eta_k = \tfrac{1}{2}F_j. \tag{6-39}$$

Combining them as in (6-25) yields

$$C_{lj}A_{jk}\ddot{\eta}_k + C_{lj}B_{jk}\eta_k = \tfrac{1}{2}C_{lj}F_j. \tag{6-40}$$

With the generalized force defined by the sum

$$Q_l = \tfrac{1}{2}C_{lj}F_j \tag{6-41}$$

and each generalized coordinate introduced as in (6-27), equation (6-40) reduces to

$$\ddot{q}_l + \omega^2 q_l = Q_l. \tag{6-42}$$

The effectiveness of a set of forces in exciting a normal mode, that is, in causing q_l to vary, depends on the combination Q_l. If this combination is sinusoidal,

$$Q_l = (Q_l)_a \cos (\omega_a t - \delta_a), \tag{6-43}$$

the equation of motion becomes

$$\ddot{q}_l + \omega^2 q_l = (Q_l)_a \cos (\omega_a t - \delta_a) \tag{6-44}$$

where $(Q_l)_a$ is the amplitude of the applied force of angular frequency ω_a and phase δ_a. A particular solution of (6-44) is

$$q_l = \frac{(Q_l)_a \cos (\omega_a t - \delta_a)}{\omega^2 - \omega_a^2}. \tag{6-45}$$

Since equations (6-28) are linear and do not eliminate any information about the coordinates, they invert to a linear form. With η_k replacing ζ_k, we have

$$\eta_k = h_{kl}q_l \tag{6-46}$$

where expression h_{kl} is the coefficient of q_l obtained in the inversion process. Substituting in solution (6-45) and labeling each angular frequency ω obtained from the secular equation by the subscript l gives us

$$\eta_k = \frac{h_{kl}(Q_l)_a \cos (\omega_a - \delta_a)}{\omega_l^2 - \omega_a^2}. \tag{6-47}$$

In general, the solutions of the homogeneous equations have to be added to this particular solution. However, in the presence of damping, the corresponding free vibrations do not last.

Note that the amplitude of a particular vibration increases as its ω_a approaches any ω_l. Indeed, solution (6-47) predicts an infinite amplitude at

$$\omega_a = \omega_l, \tag{6-48}$$

but in practice, dissipative forces limit the amplitude.

EXAMPLE 6.3   What periodic torques, acting on the coupled pendulums of figure 6.2, cause only the first mode q_1 to be excited?

In the pendulum system, each θ_l plays the role of η_l in the text. Formula (6-38) is applicable if F_l is the conventional torque acting in the direction of increasing θ_l. Example 6.2 tells us that the coefficients which the Lagrange equations should be multiplied by are

$$C_{11} = C_{12} = \frac{2}{mr^2},$$

$$C_{21} = -C_{22} = \frac{2}{mr^2}.$$

Therefore formula (6-41) yields

$$Q_1 = \frac{1}{mr^2}(F_1 + F_2),$$

$$Q_2 = \frac{1}{mr^2}(F_1 - F_2).$$

For

$$Q_2 = 0$$

and the particular solution q_2 of (6-42) to vanish, we must have

$$F_2 = F_1.$$

That is, the torque acting on the second pendulum must be equal in magnitude and phase to the torque acting on the first one. The angular frequency ω_a may have any value, however.

6.8   Damped free vibrations

Any oscillation that involves (a) the movement of a macroscopic body through a fluid, (b) the flexing of a condensed phase, or (c) the rubbing of one body on another is damped. When the frictional force acting on each particle in a given system is a symmetric linear function of the particle velocities, as when Stokes' law applies, the effects can be treated without too much difficulty.

For whenever the residual force acting on the jth particle is

$$F_j' = -\frac{\partial \mathscr{F}}{\partial \dot{\eta}_j} = -2D_{jk}\dot{\eta}_k \quad \text{with} \quad D_{jk} = D_{kj} = \text{constant}, \tag{6-49}$$

integration yields the Rayleigh function

$$\mathscr{F} = \dot{\eta}_j D_{jk} \dot{\eta}_k. \tag{6-50}$$

Lagrange equation (5-60) then becomes

$$A_{lk}\ddot{\eta}_k + D_{lk}\dot{\eta}_k + B_{lk}\eta_k = 0. \tag{6-51}$$

When the force acting on each particle is proportional to *both* the particle's velocity and mass, matrix $\mathbf{D}$ has the same shape as $\mathbf{A}$. Each element of $\mathbf{D}$ is then parameter f times the corresponding element of $\mathbf{A}$ and (6-51) reduces to the equation

$$A_{lk}\ddot{\eta}_k + f A_{lk}\dot{\eta}_k + B_{lk}\eta_k = 0. \tag{6-52}$$

That transformation (6-28) simplifies to

$$\ddot{q}_l + f\dot{q}_l + \omega^2 q_l = 0. \tag{6-53}$$

Introducing the trial solution $e^{-i\omega' t}$,

$$\omega'^2 + i\omega' f - \omega^2 = 0, \tag{6-54}$$

and solving for ω' yields

$$\omega' = -i\frac{f}{2} \pm \left(\omega^2 - \frac{f^2}{4}\right)^{1/2}. \tag{6-55}$$

Hence

$$q_l = Ce^{-ft/2} \cos\left[\left(\omega^2 - \frac{f^2}{4}\right)^{1/2} t + \delta\right]. \tag{6-56}$$

When $\frac{1}{2}f < \omega$, the trigonometric factor oscillates with angular frequency $(\omega^2 - f^2/4)^{1/2}$. The amplitude of motion is given by the other factor $C\exp(-ft/2)$. The time $2/f$ for this to fall to $1/e$ of its initial value is called the *time constant* for the mode. Increasing $\frac{1}{2}f$ to ω reduces the frequency of oscillation to zero. When $\frac{1}{2}f > \omega$, the frequency is imaginary and q_l approaches zero without successive changes in sign.

When matrix $\mathbf{D}$ is not a constant times matrix $\mathbf{A}$, equation (6-51) does not reduce to (6-53) and the damped motion may differ fundamentally from the undamped motion.

EXAMPLE 6.4 How does solution (6-56) behave when $\frac{1}{2}f > \omega$?

Making f larger than 2ω causes the angle on which the cosine acts in (6-56) to be imaginary. Then the result behaves as a hyperbolic function. Indeed, the identity

$$\cos x = \cosh ix$$

converts (6-55) to

$$q_l = Ce^{-ft/2} \cosh\left[\left(\frac{f^2}{4} - \omega^2\right)^{1/2} t + \delta'\right].$$

Constants C and δ' are determined by the values of q_l and $\dot{q}_l$ at $t = 0$.

The damping causes the coordinate q_l to decrease eventually (if not initially). Thus when t is large enough, the formula for q_l reduces to

$$q_l \simeq \tfrac{1}{2}C \exp\left[-\frac{f}{2} + \left(\frac{f^2}{4} - \omega^2\right)^{1/2}\right] t$$

and the coefficient of t in the exponent is negative.

EXAMPLE 6.5   For what damping coefficient f does the amplitude of a given free oscillation decrease to a small value most rapidly?

Increasing f from zero increases the rate at which the exponential factor $\exp(-ft/2)$ in (6-56) becomes small. As long as $f < 2\omega$, the amplitude of the other factor in (6-56) is constant. But when $f > 2\omega$ and t is large enough, this amplitude increases.

When $f > 2\omega$, we also have

$$\frac{f^2}{4} - 2\frac{f}{2}\omega + \omega^2 < \frac{f^2}{4} - \omega^2$$

and

$$\frac{f}{2} - \omega < \left(\frac{f^2}{4} - \omega^2\right)^{1/2}$$

or

$$\frac{f}{2} - \left(\frac{f^2}{4} - \omega^2\right)^{1/2} < \omega.$$

As a consequence, coordinate q_l in the final approximation in example 6.4 does not approach a given small value as fast as it would if

$$q_l = \tfrac{1}{2}C \exp(-\omega t),$$

that is, when $f = 2\omega$.

The damping is most rapid when parameter f is twice the angular frequency of free vibration. Under these circumstances, the oscillation is said to be damped *critically*.

6.9   Damped forced vibrations

Friction reduces the amplitude of each forced vibration. The effect is readily calculable if it is governed by the Rayleigh dissipation function employed in equation (6-52). For then, (6-44) is replaced by

$$\ddot{q}_l + f\dot{q}_l + \omega^2 q_l = (Q_l)_a \cos(\omega_a t - \delta_a)$$
$$= A_a \cos \omega_a t + B_a \sin \omega_a t. \tag{6-57}$$

In solving this equation, we first suppose that each forced vibration has the same frequency but not the same phase as its cause. Thus, we look for solutions of the type

$$q_l = C_a \cos \omega_a t + D_a \sin \omega_a t. \tag{6-58}$$

Substituting (6-58) into equation (6-57) and rearranging yields

$$(-C_a\omega_a{}^2 + D_a\omega_a f + C_a\omega^2 - A_a)\cos \omega_a t$$
$$+ (-D_a\omega_a{}^2 - C_a\omega_a f + D_a\omega^2 - B_a)\sin \omega_a t = 0. \tag{6-59}$$

This equation is not valid for all t's unless the coefficients are zero:

$$C_a(\omega^2 - \omega_a{}^2) + D_a f\omega_a = A_a, \tag{6-60}$$

$$-C_a f\omega_a + D_a(\omega^2 - \omega_a{}^2) = B_a. \tag{6-61}$$

Solving these simultaneous equations gives us

$$C_a = \frac{A_a(\omega^2 - \omega_a{}^2) - B_a f\omega_a}{(\omega^2 - \omega_a{}^2)^2 + (f\omega_a)^2}, \tag{6-62}$$

$$D_a = \frac{B_a(\omega^2 - \omega_a{}^2) + A_a f\omega_a}{(\omega^2 - \omega_a{}^2)^2 + (f\omega_a)^2}. \tag{6-63}$$

The amplitude of the corresponding q_l is

$$r = (C_a{}^2 + D_a{}^2)^{1/2}$$

$$= \frac{(A_a{}^2 + B_a{}^2)^{1/2}}{[(\omega^2 - \omega_a{}^2)^2 + (f\omega_a)^2]^{1/2}}. \tag{6-64}$$

The amplitude varies with the applied angular frequency ω_a. On differentiating equation (6-64) and setting the derivative equal to zero, we find that a maximum occurs where

$$\omega^2 - \omega_a{}^2 = \tfrac{1}{2}f^2 \tag{6-65}$$

or

$$\omega_a = (\omega^2 - \tfrac{1}{2}f^2)^{1/2}. \tag{6-66}$$

The corresponding frequency is referred to as a *resonant frequency*.

EXAMPLE 6.6   What is the complete solution of equation (6-57)?

Equation (6-58) describes the steady state, after transients have disappeared. In general, we must add to it the solution of the homogeneous equation, which is (6-55). Thus

$$q_l = C_a \cos \omega_a t + D_a \sin \omega_a t + Ce^{-ft/2} \cos\left[\left(\omega^2 - \frac{f^2}{4}\right)^{1/2} t + \delta\right].$$

Coefficients C_a and D_a are given by (6-62) and (6-63).

In the design of equipment, possible transients must be allowed for because they can cause failure.

6.10   Anharmonic motion

We have just seen how the steady-state amplitude of a mode rises to a maximum and then falls as the driving frequency is increased. Now, the resonance peak can be made to *lean* toward higher or lower frequencies by hardening or softening the springs.

For simplicity, let us consider a one-mode system for which the potential energy

$$V = ab\eta^2 + \tfrac{1}{2}ac\eta^4, \tag{6-67}$$

the kinetic energy

$$T = \tfrac{1}{2}m\dot\eta^2 \equiv a\dot\eta^2, \tag{6-68}$$

and the driving force

$$Q' = 2aA \cos \omega t. \tag{6-69}$$

These expressions, with η as q, convert Lagrange equation (5-52) to

$$\ddot\eta + b\eta + c\eta^3 = A \cos \omega t. \tag{6-70}$$

The third term on the left makes this equation of motion nonlinear and difficult to solve. Consequently, an approximation procedure will be followed.

First, we try a function having the same frequency as the driving force but differing in phase from it by $0°$ or $180°$, as the particular solution in section 6.7 does:

$$\eta \simeq B \cos \omega t. \tag{6-71}$$

This cannot satisfy the differential equation, however, because

$$c\eta^3 = cB^3 \cos^3 \omega t = c(\tfrac{3}{4}B^3 \cos \omega t + \tfrac{1}{4}B^3 \cos 3\omega t). \tag{6-72}$$

Indeed, the $c\eta^3$ in (6-70) yields a $\cos 3\omega t$ term which cannot be balanced against anything else in the equation.

To allow for this, add a third harmonic to the trial solution. But when it is added, the $c\eta^3$ expression yields terms in $\cos \omega t$, $\cos 3\omega t$, $\cos 5\omega t$, $\cos 7\omega t$, and $\cos 9\omega t$. Additional terms must be added until we end up with a solution that contains all odd harmonics of the forcing frequency.

To outline the behavior near resonance, however, we need only consider the two terms

$$\eta \simeq B \cos \omega t + C \cos 3\omega t. \tag{6-73}$$

Then we have

$$\ddot{\eta} \simeq -\omega^2 B \cos \omega t - 9\omega^2 C \cos 3\omega t \tag{6-74}$$

and

$$c\eta^3 \simeq c\tfrac{3}{4}B^3 \cos \omega t + c\tfrac{1}{4}B^3 \cos 3\omega t \tag{6-75}$$

if we drop terms containing c times a power of C. This approximation is justified if c and C are both small, as we presume.

Substituting these expressions into the differential equation yields

$$\begin{aligned} -\omega^2 B \cos \omega t - 9\omega^2 C \cos 3\omega t + bB \cos \omega t + bC \cos 3\omega t \\ + c\tfrac{3}{4}B^3 \cos \omega t + c\tfrac{1}{4}B^3 \cos 3\omega t = A \cos \omega t. \end{aligned} \tag{6-76}$$

Since this is an identity in t, the coefficients of $\cos \omega t$ and of $\cos 3\omega t$ must be the same on each side. Thus

$$\tfrac{3}{4}cB^3 + (b - \omega^2)B - A = 0, \tag{6-77}$$

$$C = \frac{\tfrac{1}{4}cB^3}{9\omega^2 - b}, \tag{6-78}$$

whence

$$\eta \simeq B \cos \omega t + \frac{cB^3}{4(9\omega^2 - \omega_0{}^2)} \cos 3\omega t. \tag{6-79}$$

By definition,

$$\omega_0{}^2 = b, \tag{6-80}$$

so ω_0 is the resonant angular frequency when $c = 0$.

Since form (6-78) is not small when $\omega^2 \simeq \omega_0{}^2/9$, the approximations break down in that neighborhood. Higher terms must then be considered. In general, C is of the same order as c.

Thus when c is small, the main contribution to the amplitude of vibration is from B. Rearranging (6-77),

$$\left(1 - \frac{\omega^2}{\omega_0^2}\right) B - \frac{A}{\omega_0^2} = - \frac{3cB^3}{4\omega_0^2}, \tag{6-81}$$

shows us that the allowed values of B can be obtained from the intersections of the straight line

$$z = \left(1 - \frac{\omega^2}{\omega_0^2}\right) B - \frac{A}{\omega_0^2} \tag{6-82}$$

with the cubic parabola

$$z = - \frac{3cB^3}{4\omega_0^2}. \tag{6-83}$$

Now, the straight line always passes through the point $(0, -A/\omega_0^2)$. When ω is 0, the slope is 1. As ω increases, the slope decreases—until it is $-\infty$ when ω is ∞. The cubic parabola, on the other hand, is independent of ω (see figure 6.3).

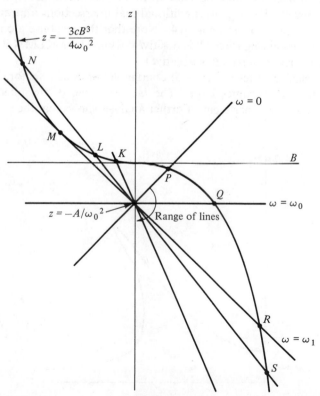

FIGURE 6.3   Straight lines and curve whose intersections yield the possible amplitudes.

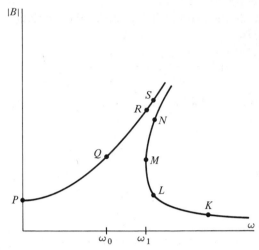

FIGURE 6.4   Variation of the amplitude $|B|$ with angular frequency of the driving force.

Typical magnitudes $|B|$ may be read from the intersections. When the driving angular frequency reaches ω_1, two additional real intersections appear. Thus, a plot of $|B|$ against ω appears as in figure 6.4. Note that the response along curve PQRS is in phase with the driving force (B is positive), while along KLMN it is 180° behind the phase of the driving force (B is negative).

Some of the neglected terms in (6-75) contain B^2 or B as a factor. These become appreciable when $|B|$ becomes large. The fact that two of the intersections go to $\pm \infty$ as ω increases is not pertinent. Further analysis shows that the oscillation is self-limiting.

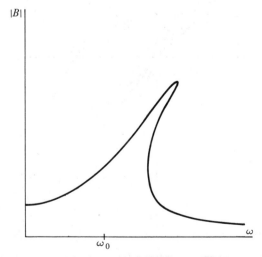

FIGURE 6.5   Response curve, with the anharmonicity coefficient c positive and with some damping present.

In an actual system, some damping is always present. This tends to reduce the amplitude and does alter the phase relationships, as in the linear problem discussed previously (see section 6.9). Then the first branch joins smoothly onto the other, as in figure 6.5, with the phase relationships changing continuously from one extreme to the other. When c is negative, the resonance curve leans toward lower frequencies, as in figure 6.6.

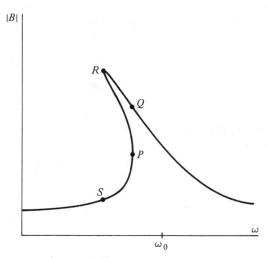

FIGURE 6.6   Response curve for c negative.

Either kind of curve leads to an amplitude jump at some ω when the angular driving frequency is changed. Thus in figure 6.6 an increase in ω from 0 will lead to an increase of $|B|$ along the curve SP. From P, the amplitude has to jump up to Q. Then it follows the resonance curve smoothly down. Similarly, a decrease in ω from a large value will cause the amplitude $|B|$ to move smoothly up the curve through Q to R. At R it has to jump down to S and then continue smoothly on.

6.11   Excitation of a subharmonic

Anharmonicities can also cause a system to oscillate at some unit fraction of the driving frequency, say ω/n, if this angular frequency is close to the resonant frequency ω_0. If the nonlinearity is cubic, as in section 6.10, odd multiples of this frequency also appear.

Thus, the first two terms of the series representing the $\omega/3$ solution of (6-70) are

$$\eta \simeq D \cos \frac{\omega}{3} t + E \cos \omega t. \qquad (6\text{-}84)$$

Then

$$\ddot{\eta} \simeq -\tfrac{1}{9}\omega^2 D \cos \frac{\omega}{3} t - \omega^2 E \cos \omega t, \qquad (6\text{-}85)$$

and

$$\eta^3 \simeq \tfrac{3}{4}D^3 \cos \frac{\omega}{3} t + \tfrac{1}{4}D^3 \cos \omega t + \tfrac{3}{2}D^2 E \cos \omega t$$

$$+ \tfrac{3}{4}D^2 E \cos \frac{\omega}{3} t + \tfrac{3}{2}DE^2 \cos \frac{\omega}{3} t + \tfrac{3}{4}E^3 \cos \omega t \qquad (6\text{-}86)$$

if higher harmonics are neglected.

These expressions convert the differential equation (6-70) to an identity in *t*. Equating coefficients of $\cos (\omega/3)t$ and of $\cos \omega t$ then gives us

$$(b - \tfrac{1}{9}\omega^2)D + \tfrac{3}{4}c(D^3 + D^2 E + 2DE^2) = 0, \qquad (6\text{-}87)$$

$$(b - \omega^2)E + \tfrac{1}{4}c(D^3 + 6D^2 E + 3E^3) = A. \qquad (6\text{-}88)$$

Note that when *D* is 0, (6-88) reduces to the same form as (6-77). A first approximation to $|E|$ is that obtained before for $|B|$, as illustrated in figures 6.5 and 6.6.

Solving (6-87) for ω^2 yields

$$\omega^2 = 9b + \tfrac{27}{4}c(D^2 + DE + 2E^2). \qquad (6\text{-}89)$$

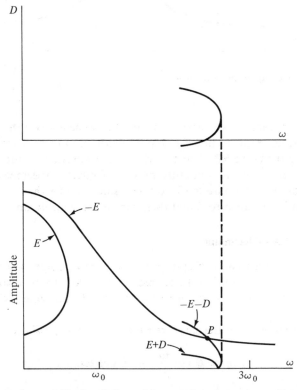

FIGURE 6.7   How the amplitude of the subharmonic response combines with that of the harmonic response when *c* is negative.

Now, E is nearly constant for frequencies in a small range far from the resonance peak. Over this range, a plot of (6-89) is nearly elliptical when c is negative and nearly hyperbolic when c is positive (see the upper graphs in figures 6.7 and 6.8).

Differentiating equation (6-89) with respect to D and setting the result equal to 0 locates the extremum of each curve at

$$D = -\tfrac{1}{2}E, \tag{6-90}$$

or

$$\omega^2 = 9(b + \tfrac{21}{16}cE^2). \tag{6-91}$$

Thus, the subharmonic vibrations exist only when

$$\omega < 3(b + \tfrac{21}{16}cE^2)^{1/2} \tag{6-92}$$

if c is negative, and only when

$$\omega > 3(b + \tfrac{21}{16}cE^2)^{1/2} \tag{6-93}$$

if c is positive.

In the allowed range, a point P is found where D is zero and the subharmonic vanishes. This appears as a bifurcation point in the lower graphs of figures 6.7 and 6.8.

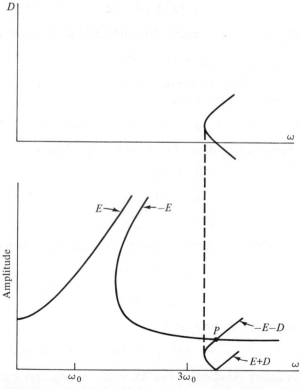

FIGURE 6.8   Amplitudes of the subharmonic and harmonic responses for c positive.

6.12   Effect of asymmetry in the potential

So far, each system has been governed by a potential that is symmetric about the positions of equilibrium. A given displacement from such a point in either direction increases the potential energy by the same amount. But asymmetry often exists in practice. Thus, the V of a diatomic molecule rises faster as the molecule is compressed than as it is stretched from the equilibrium configuration. Energy sufficient to dissociate the molecule will only compress it a finite amount.

The asymmetry introduces into the free oscillation all higher harmonics. Only the second one will be approximated here, however, because of computational difficulties. The method to be employed is known as the *perturbation method*.

We consider the system governed by the potential energy

$$V = ab\eta^2 - \tfrac{2}{3}a\lambda\eta^3,$$ (6-94)

the kinetic energy

$$T = \tfrac{1}{2}m\dot{\eta}^2 = a\dot{\eta}^2,$$ (6-95)

and the driving force

$$Q' = 0.$$ (6-96)

Lagrange's equation then takes on the form

$$\ddot{\eta} + b\eta - \lambda\eta^2 = 0.$$ (6-97)

We assume that the term $-\lambda\eta^2$ perturbs the solution in an analytic manner; so

$$\eta = \eta_0 + \lambda\eta_1 + \lambda^2\eta_2 + \cdots$$ (6-98)

where η_0 is the solution for the simple oscillator.

This series converts Lagrange's equation to

$$\ddot{\eta}_0 + \lambda\ddot{\eta}_1 + \cdots + b\eta_0 + b\lambda\eta_1 + \cdots - \lambda\eta_0{}^2 - \cdots = 0.$$ (6-99)

For this result to be valid regardless of λ, it must be an identity in λ. Therefore

$$\ddot{\eta}_0 + b\eta_0 = 0,$$ (6-100)

$$\ddot{\eta}_1 + b\eta_1 = \eta_0{}^2,$$ (6-101)

$$\vdots \quad .$$ (6-102)

The solution of (6-100) with the phase chosen so η_0 is a maximum at $t = 0$ is

$$\eta_0 = A \cos \omega_0 t$$ (6-103)

where, as before,

$$\omega_0{}^2 = b.$$ (6-104)

Substituting this result into (6-101) yields

$$\ddot{\eta}_1 + \omega_0{}^2\eta_1 = A^2 \cos^2 \omega_0 t = \tfrac{1}{2}A^2 + \tfrac{1}{2}A^2 \cos 2\omega_0 t.$$ (6-105)

Since the complementary function of (6-105), with the arbitrary constants, already occurs in the zeroth-order solution η_0, we need only the particular integral.

Introducing the trial form

$$\eta_1 = B \cos 2\omega_0 t + C \tag{6-106}$$

converts (6-105) to

$$-4\omega_0{}^2 B \cos 2\omega_0 t + \omega_0{}^2 B \cos 2\omega_0 t + \omega_0{}^2 C = \tfrac{1}{2}A^2 + \tfrac{1}{2}A^2 \cos 2\omega_0 t, \tag{6-107}$$

whence

$$B = -\frac{A^2}{6\omega_0{}^2}, \tag{6-108}$$

$$C = \frac{A^2}{2\omega_0{}^2}. \tag{6-109}$$

Substituting these constants into (6-106), multiplying by λ, and then adding to (6-103) as in series (6-98) gives us

$$\eta \simeq A \cos \omega_0 t - \lambda \frac{A^2}{6\omega_0{}^2} (\cos 2\omega_0 t - 3). \tag{6-110}$$

Higher terms in the series, containing higher-harmonic contributions, can be obtained on solving equations (6-102) in order.

6.13 Incomplete separation of rotational motion from the vibrational modes

In deriving the secular equation (section 6.5), we assumed that T and V appear as in (6-11) and (6-17), with $\mathbf{A}$ and $\mathbf{B}$ constant. This constancy implies that certain displacements in the given system are small. The equation, therefore, may not allow for rotation at an appreciable rate, and does not show how rotation interacts with the vibrations.

But if the system is in free space, or in a uniform gravitational field, as we assume, the potential V does allow translatory motion to separate from the other motions (section 6.1). So let us consider only movements with respect to the center of mass.

We also assume that the system is rotating freely. The position about which each particle oscillates, the position that it would occupy if the system were not vibrating, is found, and vector $\mathbf{a}_j$ drawn from the center of mass to this point. The displacement of the pertinent particle from this point is labeled ρ_j (see figure 6.9).

Since the array of equilibrium positions acts as a rigid body turning at angular velocity ω, equation (3-73) shows that

$$\dot{\mathbf{a}}_j = \omega \times \mathbf{a}_j. \tag{6-111}$$

The velocity of mass element m_j with respect to a nonrotating frame based on the center of mass is the time derivative of

$$\mathbf{r}_j{}^i = \mathbf{a}_j + \rho_j. \tag{6-112}$$

The internal kinetic energy is

$$T = \tfrac{1}{2} m_j \dot{\mathbf{r}}_j{}^i \cdot \dot{\mathbf{r}}_j{}^i. \tag{6-113}$$

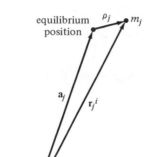

FIGURE 6.9   Vectors used to locate the element of mass m_j with respect to the center of mass and the equilibrium position.

Combining (6-112) and (6-113) yields

$$T = \tfrac{1}{2}m_j(\dot{\mathbf{a}}_j + \dot{\boldsymbol{\rho}}_j) \cdot (\dot{\mathbf{a}}_j + \dot{\boldsymbol{\rho}}_j)$$
$$= \tfrac{1}{2}m_j\dot{\mathbf{a}}_j \cdot \dot{\mathbf{a}}_j + \tfrac{1}{2}m_j\dot{\boldsymbol{\rho}}_j \cdot \dot{\boldsymbol{\rho}}_j + m_j\dot{\mathbf{a}}_j \cdot \dot{\boldsymbol{\rho}}_j. \qquad (6\text{-}114)$$

The first sum in the final expression is the rotational kinetic energy assuming all mass elements are at their equilibrium positions:

$$T_{\text{rot}} = \tfrac{1}{2}m_j\dot{\mathbf{a}}_j \cdot \dot{\mathbf{a}}_j. \qquad (6\text{-}115)$$

Because these positions are determined for the given rotational state (not for the system at rest), this sum includes the effect of centrifugal distortion. The second sum would be the vibrational kinetic energy if the coordinate system rotating with the equilibrium positions were an inertial frame:

$$T_{\text{vib}} = \tfrac{1}{2}m_j\dot{\boldsymbol{\rho}}_j \cdot \dot{\boldsymbol{\rho}}_j. \qquad (6\text{-}116)$$

The last sum results from a coupling between the rotation and vibrations and is called the *Coriolis energy*. Because $\dot{\rho}_j$ is generally larger than $\dot{a}_j$, this Coriolis energy is usually much larger than the contribution of centrifugal distortion to T_{rot}.

Introducing relationship (6-111) and the definition

$$\mathbf{L} = m_j\mathbf{a}_j \times \dot{\boldsymbol{\rho}}_j \qquad (6\text{-}117)$$

into the last sum of (6-114) yields

$$m_j\dot{\mathbf{a}}_j \cdot \dot{\boldsymbol{\rho}}_j = m_j\boldsymbol{\omega} \times \mathbf{a}_j \cdot \dot{\boldsymbol{\rho}}_j = \boldsymbol{\omega} \cdot m_j\mathbf{a}_j \times \dot{\boldsymbol{\rho}}_j$$
$$= \boldsymbol{\omega} \cdot \mathbf{L}. \qquad (6\text{-}118)$$

Then (6-114) can be rewritten as

$$T = T_{\text{rot}} + T_{\text{vib}} + \boldsymbol{\omega} \cdot m_j\mathbf{a}_j \times \dot{\boldsymbol{\rho}}_j$$
$$= T_{\text{rot}} + T_{\text{vib}} + \boldsymbol{\omega} \cdot \mathbf{L}. \qquad (6\text{-}119)$$

Note that **L** is here the angular momentum due to the vibrational motion at the given instant of time.

When a rotation couples with a vibrational mode to produce some motion characteristic of another vibrational mode, the two vibrations interact. In general, the frequency of the lower one is depressed, that of the upper one raised.

6.14 Additional remarks

In general, coordinates cannot be chosen so that V is a homogeneous quadratic function of the coordinates while T is a homogeneous quadratic function of the temporal derivatives of the coordinates. But when the amplitudes of oscillation are small enough, such quadratic forms suffice.

Then the equations of motion are linear and solution of the corresponding secular equation leads to the normal frequencies and normal coordinates. However, extensive calculations are required unless the system is very small or symmetric. External forces that depend on the time alone, and dissipative forces whose matrix is diagonalized when the other terms in the matrix equation of motion are, do not destroy the linearity.

Retaining additional terms in the potential V beyond those quadratic in the coordinates does destroy the linearity. The resulting equations may be linearized by the perturbation method. But even for a single normal coordinate, the procedure is tedious (consider determining η_4 in section 6.12).

In a nonlinear system, some resonance occurs between any two modes whose frequencies are commensurable. Such resonance can lead to an instability. Furthermore, increasing the energy in a mode of motion may cause it to become unstable. Discussion of these effects is beyond the scope of the present text.

DISCUSSION QUESTIONS

6.1 What motions remain after the translational and rotational ones are separated out?

6.2 Explain why the kinetic energy T has the form

$$\dot{\mathbf{q}}^{\dagger}\mathbf{A}\dot{\mathbf{q}}.$$

6.3 When are the A_{jk}'s constant? Does setting

$$A_{jk} = A_{kj}$$

limit T?

6.4 Under what circumstances does the potential energy V have the form

$$\boldsymbol{\eta}^{\dagger}\mathbf{B}\boldsymbol{\eta}?$$

6.5 Formulate Lagrange equations for a set of particles governed by the quadratic potential.

6.6 What is the model system for a mode of vibration?

6.7 Show how the equation of motion of the model system is obtained from the Lagrange equations for the set of particles.

6.8 Derive the secular equation.

6.9 What are normal coordinates and how are they obtained?

6.10 Why are the Lagrange equations for a set of particles generally intractable?

6.11 In what sense is a steady-state forced vibration governed by the secular equation?

6.12 How must the additional forces combine if a mode is to remain unexcited?

6.13 For what kind of damping is the reduction of the equations of motion to normal form by the secular equation possible?

6.14 How is the resonant frequency affected by damping?

6.15 How does the amplitude vary around a resonance?

6.16 What causes a resonance peak to lean toward (a) higher frequencies, (b) lower frequencies?

6.17 Explain why the amplitude may jump as a particular frequency is reached.

6.18 Discuss how higher harmonics arise.

6.19 Explain the appearance of subharmonics.

6.20 How can a problem be linearized by the perturbation technique?

PROBLEMS

6.1 Formulate and solve the secular equation for vibration of the diatomic molecule AB. Neglect any anharmonicity in the system.

6.2 Find normal coordinates for the vibrating AB.

6.3 Formulate and solve the secular equation for vibration of a linear symmetrical triatomic molecule BAB. Neglect any bending of the system.

6.4 Obtain normal coordinates for the longitudinal vibrations of BAB. Show how the atoms move when each of these coordinates changes by itself.

6.5 Set up and solve the secular equation representing small oscillations of a double pendulum consisting of mass m suspended by a string of length l from a second mass m which in turn is suspended from a fixed point by a string of length l.

6.6 Derive normal coordinates for the double pendulum in problem 6.5. What relationship exists between the instantaneous angular displacements in each of the pure modes?

6.7 What periodic forces, acting on the double pendulum of problem 6.6, cause the two bobs to move exactly in phase?

6.8 The magnitude of a free oscillation falls to $(1/e)$th of its initial value in $\frac{1}{2}$ period of oscillation. Relate its actual frequency and its resonant frequency to the frequency of oscillation if no damping were present.

6.9 Show to what approximation the simple-pendulum equation has the form (6-70). Use the results from section 6.10 to obtain roughly how the frequency of oscillation varies with its amplitude.

6.10 A two-dimensional oscillator is governed by the Lagrangian

$$L = \tfrac{1}{2}m\dot{x}^2 + \tfrac{1}{2}M\dot{y}^2 - \tfrac{1}{2}ax^2 - \frac{1}{2}\frac{aM}{m}y^2 - cxy.$$

Find the frequencies of its normal modes from its secular equation.

6.11 Derive normal coordinates for the oscillator in problem 6.10.

6.12 Consider the bending of linear BAB to be independent of the motions studied in problem 6.3. Furthermore, assume the potential to be a quadratic function of the deviation of <BAB from 180°, set up the secular equation, and solve.

6.13 Obtain normal coordinates for the transverse movements of BAB. Show which represent (a) translation, (b) rotation, (c) real vibration.

6.14 Set up and solve the secular equation for *small* displacements in isosceles-triangular AAA, if its potential is a quadratic function of the deviation of <AAA from 90° and a quadratic function of the deviation of each leg from its equilibrium length. Neglect any motion out of the plane of the molecule, for simplicity.

6.15 For the four rational roots in problem 6.14, formulate normal coordinates.

6.16 What periodic forces, acting on the oscillator in problem 6.10, cause only one mode to be excited?

6.17 Show that solution (6-56) is not complete when the damping is critical. Find the additional term that is then necessary.

6.18 At opposing points on two horizontal fixed planes are attached two equivalent springs. The other ends of the springs are joined to a mass m. Neglect gravity and obtain the differential equation describing the mass's transverse movement. When is the equation intrinsically nonlinear?

REFERENCES

BOOKS

Bradbury, T. C. *Theoretical Mechanics*, pp. 222–250. John Wiley & Sons, Inc., New York, 1968.

Goldstein, H. *Classical Mechanics*, pp. 318–346. Addison-Wesley Publishing Company, Inc., Reading, Mass., 1950.

Goodman, L. E., and Warner, W. H. *Dynamics*, pp. 512–559. Wadsworth Publishing Company, Inc., Belmont, Calif., 1963.

Greenwood, D. T. *Principles of Dynamics*, pp. 446–497. Prentice-Hall, Inc., Englewood Cliffs, N.J., 1965.

Stoker, J. J. *Nonlinear Vibrations in Mechanical and Electrical Systems*, pp. 1–18, 81–117. Interscience Publishers, New York, 1950.

Symon, K. R. *Mechanics*, 2nd ed., pp. 473–512. Addison-Wesley Publishing Company, Inc., Reading, Mass., 1960.

Volterra, E., and Zachmanoglou, E. C. *Dynamics of Vibrations*, pp. 1–256, 439–512. Charles E. Merrill Books, Inc., Columbus, Ohio, 1965.

ARTICLES

Edwards, T. W., and Hultsch, R. A. Mass Distribution and Frequencies of a Vertical Spring. *Am. J. Phys.*, **40**, 445 (1972).

Lee, S. M. Exact Normal-Mode Analysis for a Linear Lattice with "Periodic Impurities." *Am. J. Phys.*, **37**, 888 (1969).

Lee, S. M. The Isochronous Problem Inside the Spherically Uniform Earth. *Am. J. Phys.*, **40**, 315 (1972).

Levenson, M. E. A Numerical Determination of Subharmonic Response for the Duffing Equation $\ddot{x} + \alpha x + \beta x^3 = F \cos \omega t$. *Quart. Appl. Math.*, **25**, 11 (1967).

Musa, S. A., and Kronauer, R. E. Sub- and Superharmonic Synchronization in Weakly Nonlinear Systems: Integral Constraints and Duality. *Quart. Appl. Math.*, **25**, 399 (1968).

Parker, L. Adiabatic Invariance in Simple Harmonic Motion. *Am. J. Phys.*, **39**, 24 (1971).

Phelps, F. M., III, and Hunter, J. H., Jr. Reply to Joshi's Comments on a Damping Term in the Equations of Motion of the Inverted Pendulum. *Am. J. Phys.*, **34**, 533 (1966).

Slee, F. W. The Prediction of the Frequency of a Nonlinear Oscillator. *Am. J. Phys.*, **39**, 578 (1971).

Stockard, D. P., Johnson, T. L., and Sears, F. W. Study of Amplitude Jumps. *Am. J. Phys.*, **35**, 961 (1967).

Tanttila, W. H. Rattles, Swings, and Subharmonics. *Am. J. Phys.*, **35**, 543 (1967).

Weinstock, R. Normal-Mode Frequencies of Finite One-Dimensional Lattices with Single Mass Defect: Exact Solutions. *Am. J. Phys.*, **39**, 484 (1971).

7 / Symmetry in Physical Systems

7.1 Consistency and continuity of behavior

Employing the classical laws of mechanics and known properties of materials, engineers can design rockets and a support system that enable two men to go to the neighborhood of a given spot on the moon. Using the statistical laws governing nuclear behavior and the known properties of the pertinent substances, engineers can design a nuclear reactor to supply the power consumed by a large city.

The physical world is indeed consistent. Each identifiable part subject to a given environment behaves in a reproducible manner. Each cause acting within the region produces a definite spectrum of effects.

A symmetry operation, an operation that changes a small region into an equivalent system while leaving the environment unchanged, does not essentially alter causes. So it does not alter the spectrum of effects that the original situation allowed.

Furthermore, consequences are not isolated. Changing the environment (boundary conditions) slightly produces a small change in the behavior of the region, as long as the system is not carried across a discontinuity or a near-discontinuity. As an example of a discontinuity, recall the jump in amplitude of a nonlinear vibrating system when the frequency of a driving force passes a critical value (section 6.10). A near-discontinuity is a precipitous rise or fall in certain properties associated, for instance, with a shock front.

Distorting a symmetric region generally involves changing the causes and the consequent effects continuously. The final states are essentially perturbations of states of the symmetric system. In many regions that we will consider, the perturbations are small and can be neglected. Otherwise, we might first calculate properties of the approximate symmetric system and then estimate the effects of perturbations that are present.

7.2 Symmetry operations

Whenever a model system possesses symmetry, it possesses elements with respect to which symmetry operations can be carried out. Examples of such elements include the space the system occupies, a mirror plane, a proper axis (for rotation), an improper

TABLE 7.1   Common Elements of Symmetry and the Symmetry Operations Associated with Them

Geometric entity left unchanged	A corresponding geometric operation		The geometric operation plus the pertinent change in nature (color, spin, or quark)	
	Description	Symbols*	Description	Symbols
The system	Identity	E or 1	One of the conversions	$\mathcal{O}$ or $\underline{1}$
Mirror plane	Reflection in the plane	σ_v, σ_d σ_h m $/m$	The reflection followed by the conversion	$\mathcal{O}\sigma_v, \mathcal{O}\sigma_d$ $\mathcal{O}\sigma_h$ $\underline{m}$ $\underline{/m}$
Line (proper axis)	Rotation by $2\pi/n$ about the axis	C_n $(2, 3, .)$	The rotation followed by the conversion	$\mathcal{O}C_n$ $\underline{n}$ $(\underline{2}, \underline{3}, .)$
Point (as center of symmetry)	Inversion (reflection) through the center	i $\bar{1}$	The inversion followed by the conversion	$\mathcal{O}i$ $\underline{\bar{1}}$
Point (with an improper axis)	Rotation by $2\pi/n$ followed by inversion	iC_n $\bar{n}$	The rotoinversion followed by the conversion	$\mathcal{O}iC_n$ $\underline{\bar{n}}$
Point (with an improper axis)	Reflection in plane perpendicular to axis followed by rotation by $2\pi/n$	S_n n/m	The rotoreflection followed by the conversion	$\mathcal{O}S_n$ $\underline{n/m}$
None	Translation	t t	The translation followed by the conversion	$\mathcal{O}t$ $\underline{t}$

*The Schönflies notation appears in the first column, the Hermann-Manguin notation in the second.

axis with a unique plane perpendicular to it (for rotation-reflection), a point (for inversion), and arrays of these.

Geometric operations that can serve as symmetry operations have been discussed in sections 4.3 and 4.4. These can be combined with nongeometric operations as indicated in table 7.1. The nongeometric operations are called conversions.

In color groups, a conversion involves changing one color to another in a cycle. Thus, $\mathcal{O}$ may shift black to white and white to black. In magnetic groups, the conversion involves reversing the spins associated with various parts of the system. In particle theory, we deal with operations that permute nearly equivalent fundamental states (the quarks).

7.3 Symmetry groups

The complete set of reorientations and/or conversions that transform a model into equivalent systems is characterized by four properties.

Because each transformation in the set changes the pertinent region into a system like the original one, any of the transformations converts this second system into a third one that is also similar to the initial one. Consequently, the transformation from the initial to the final form is one of the operations in the set. Any two symmetry operations combine to give a symmetry operation for the given system. The set is therefore said to be *closed*.

In a series of three transformations, we may replace the first two by the equivalent single transformation, or the last two by the equivalent, without affecting the results. The *associative* law is therefore obeyed.

Furthermore, no transformation is one-way. For each, there is a reverse transformation that converts the second form back to the original form. This is called the *inverse* of the given transformation. And the set always contains the *identity* operation, which leaves the system as is.

Any set of operators meeting these conditions is said to form a *group*. If part of the operators also meet these conditions, they are said to form a *subgroup* of the set.

7.4 Classifying geometric groups

Because of the way the operations combine, the presence or absence of certain symmetry operations determines whether others are present. We can identify the complete group of operations that leave a symmetric system essentially unchanged by answering only a few questions.

A convenient sequence appears in figure 7.1. First, decide whether the covering operations (the operations that convert the system into an equivalent system) include at least two C_n with $n > 2$. If the answer is yes, then decide whether the group contains any reflections. If the answer is yes, then determine whether a fivefold rotation axis is present. If it is, the group is $\mathbf{I}_h$.

If the answer to the first question is no, decide whether the group contains any C_n with $n > 1$. If the answer is yes, find out whether the group contains any twofold

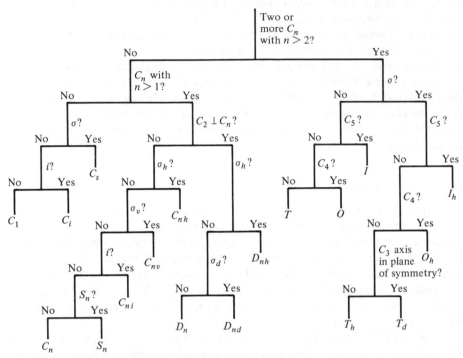

FIGURE 7.1   Successive questions for identifying geometric groups.

rotation axes perpendicular to the principal axis. If the answer is yes, then decide whether σ_h is present. If yes, the group is $\mathbf{D}_{nh}$.

Proceed similarly for the other groups. Note that each is a subgroup of the group of operations that transform a sphere centered on the origin into an equivalent sphere. The groups that do not contain two or more C_n with $n > 2$ are also subgroups of the group of operations that transform a circular cylinder centered on the principal axis into itself.

Some of these groups combine with certain translatory operations to yield 3-dimensional lattices.

7.5   Lattices for describing crystals

A homogeneous region possesses translatory symmetry and is called a crystal if it can be divided into segments that are the same in content, size, shape, and orientation. In general, three sets of dividing surfaces are required. Successive surfaces in the first set differ by the same translation $\mathbf{a}_1$, successive surfaces in the second set by $\mathbf{a}_2$, successive surfaces in the third set by $\mathbf{a}_3$.

A specific division of the system arises if planes are drawn through a given position of these vectors and translated by repetitive $\mathbf{a}_1$'s, $\mathbf{a}_2$'s, and $\mathbf{a}_3$'s. Each part of the

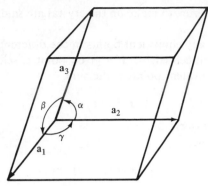

FIGURE 7.2   Vectors defining a unit cell.

crystal then appears as in figure 7.2, with the intervector angles γ, α, β. Such a segment is called a *unit cell* for the system. Possible three-dimensional unit cells are described in table 7.2.

For a given crystal, the unit may be chosen in many different ways. The smallest possible unit, of any shape, is called a *primitive* cell (P). Many of the conventional unit cells are, for convenience, not primitive (see table 7.3).

A division into primitive cells can be partially described by the vertices at which the cells meet. The array of vertices is called the crystal *lattice*. The atoms or ions that

TABLE 7.2   Properties of Possible Unit Cells

Lattice	Unit-cell edges	Angles
Triclinic	$a_1 \neq a_2 \neq a_3$	$\alpha \neq \beta \neq \gamma \neq 90°$
Monoclinic	$a_1 \neq a_2 \neq a_3$	$\alpha = \gamma = 90°, \beta \neq 90°$
Orthorhombic	$a_1 \neq a_2 \neq a_3$	$\alpha = \beta = \gamma = 90°$
Tetragonal	$a_1 = a_2 \neq a_3$	$\alpha = \beta = \gamma = 90°$
Cubic	$a_1 = a_2 = a_3$	$\alpha = \beta = \gamma = 90°$
Hexagonal	$a_1 = a_2 \neq a_3$	$\alpha = \beta = 90°, \gamma = 120°$
Rhombohedral	$a_1 = a_2 = a_3$	$\alpha = \beta = \gamma \neq 90°$

TABLE 7.3   Bravais Lattices

Locations of equivalent points	Symbol	Where observed
Corners only	P	All lattices
Corners and center of one face	C	Monoclinic and orthorhombic
Corners and centers of all faces	F	Orthorhombic and cubic
Corners and center of body	I	Orthorhombic, tetragonal, and cubic

are associated with each vertex to make up the crystal are said to form a *basis* for the crystal.

The lattice points, and any equivalent points in the different cells, are separated by an integral number of unit-cell edges. If a given point is selected as the origin, the radius vector locating equivalent points is the sum

$$\mathbf{r} = n_j \mathbf{a}_j \tag{7-1}$$

in which n_1, n_2, n_3 are integers.

If a sinusoidal disturbance

$$\mathbf{F} = \mathbf{F}_0 \sin(\mathbf{k} \cdot \mathbf{r} - \alpha) \sin \omega t \tag{7-2}$$

is to affect each of these positions in the same manner, the propagation vector $\mathbf{k}$ must be chosen so that $\mathbf{k} \cdot \mathbf{r}$ increases by an integral number of 2π radians for each $\mathbf{r}$ given by (7-1). This condition is satisfied when $\mathbf{k}$ is an integral combination

$$\mathbf{k} = h_j \mathbf{A}_j \tag{7-3}$$

of the *reciprocal vectors*

$$\mathbf{A}_1 = 2\pi \frac{\mathbf{a}_2 \times \mathbf{a}_3}{\mathbf{a}_1 \cdot \mathbf{a}_2 \times \mathbf{a}_3}, \tag{7-4}$$

$$\mathbf{A}_2 = 2\pi \frac{\mathbf{a}_3 \times \mathbf{a}_1}{\mathbf{a}_1 \cdot \mathbf{a}_2 \times \mathbf{a}_3}, \tag{7-5}$$

$$\mathbf{A}_3 = 2\pi \frac{\mathbf{a}_1 \times \mathbf{a}_2}{\mathbf{a}_1 \cdot \mathbf{a}_2 \times \mathbf{a}_3}. \tag{7-6}$$

For then,

$$\mathbf{A}_1 \cdot \mathbf{a}_1 = 2\pi, \quad \mathbf{A}_1 \cdot \mathbf{a}_2 = 0, \quad \mathbf{A}_1 \cdot \mathbf{a}_3 = 0, \tag{7-7}$$

$$\mathbf{A}_2 \cdot \mathbf{a}_1 = 0, \quad \mathbf{A}_2 \cdot \mathbf{a}_2 = 2\pi, \quad \mathbf{A}_2 \cdot \mathbf{a}_3 = 0, \tag{7-8}$$

$$\mathbf{A}_3 \cdot \mathbf{a}_1 = 0, \quad \mathbf{A}_3 \cdot \mathbf{a}_2 = 0, \quad \mathbf{A}_3 \cdot \mathbf{a}_3 = 2\pi, \tag{7-9}$$

and

$$\mathbf{k} \cdot \mathbf{r} = (h_j \mathbf{A}_j) \cdot (n_l \mathbf{a}_l) = 2\pi(h_j n_j) = 2\pi \text{ (integer)}. \tag{7-10}$$

The disturbance may be provided by a beam of neutrons or of X rays.

The possible propagation vectors

$$\mathbf{k} = h_j \mathbf{A}_j \tag{7-11}$$

with each h_j an integer define an array of points called the *reciprocal lattice*. In a sense, this array describes the symmetry within the crystal as well as the lattice generated by (7-1).

EXAMPLE 7.1   What lattice is reciprocal to a body-centered cubic lattice?
From figure 7.3, a primitive cell of the body-centered cubic (bcc) lattice is bounded by the vectors

$$\mathbf{a}_1 = \frac{d}{2}(\mathbf{i} + \mathbf{j} - \mathbf{k}),$$

$$\mathbf{a}_2 = \frac{d}{2}(-\mathbf{i} + \mathbf{j} + \mathbf{k}),$$

$$\mathbf{a}_3 = \frac{d}{2}(\mathbf{i} - \mathbf{j} + \mathbf{k}),$$

where d is the distance between opposite faces of the unit cube. Equation (1-34) then shows that

$$\mathbf{a}_1 \cdot \mathbf{a}_2 \times \mathbf{a}_3 = \frac{d^3}{8} \begin{vmatrix} 1 & 1 & -1 \\ -1 & 1 & 1 \\ 1 & -1 & 1 \end{vmatrix} = \frac{d^3}{2},$$

and equations (7-4), (7-5), (7-6) yield

$$\mathbf{A}_1 = \frac{2\pi}{d^3/2} \frac{d^2}{4} \begin{vmatrix} \mathbf{i} & \mathbf{j} & \mathbf{k} \\ -1 & 1 & 1 \\ 1 & -1 & 1 \end{vmatrix} = \frac{2\pi}{d}(\mathbf{i} + \mathbf{j}),$$

$$\mathbf{A}_2 = \frac{2\pi}{d^3/2} \frac{d^2}{4} \begin{vmatrix} \mathbf{i} & \mathbf{j} & \mathbf{k} \\ 1 & -1 & 1 \\ 1 & 1 & -1 \end{vmatrix} = \frac{2\pi}{d}(\mathbf{j} + \mathbf{k}),$$

$$\mathbf{A}_3 = \frac{2\pi}{d^3/2} \frac{d^2}{4} \begin{vmatrix} \mathbf{i} & \mathbf{j} & \mathbf{k} \\ 1 & 1 & -1 \\ -1 & 1 & 1 \end{vmatrix} = \frac{2\pi}{d}(\mathbf{i} + \mathbf{k}).$$

According to figure 7.4, a primitive cell of the face-centered cubic (fcc) lattice is bounded by

$$\mathbf{a}_1 = \frac{e}{2}(\mathbf{i} + \mathbf{j}),$$

$$\mathbf{a}_2 = \frac{e}{2}(\mathbf{j} + \mathbf{k}),$$

$$\mathbf{a}_3 = \frac{e}{2}(\mathbf{i} + \mathbf{k})$$

where e is the length of an edge of the pertinent unit cube. Thus, the reciprocal lattice is a face-centered cubic lattice with cube edge equal to

$$e = \frac{4\pi}{d}.$$

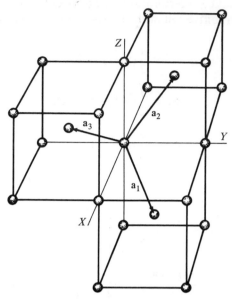

FIGURE 7.3   Primitive base vectors for a body-centered cubic lattice.

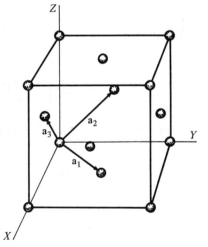

FIGURE 7.4   Primitive base vectors for a face-centered cubic lattice.

7.6   Subatomic constituents of matter

Very interesting conversional symmetries occur among the so-called fundamental particles of nature. Before we can study these, though, we need to survey pertinent properties of the particles.

On bombarding matter with very small projectiles, we find that most of the mass is concentrated in small nuclei. The particles easily removed from matter, the electrons, occupy the nearly void regions that are left, in cloudlike fashion.

The nuclei are positively charged; they are presumably formed from common hydrogen nuclei and neutrons. To explain how such compact structures can be stable against the large electromagnetic forces tending to break them apart, we must postulate the existence of nuclear (or strong-interaction) forces.

Neutron-rich and proton-rich nuclei emit beta particles and neutrinos to gain stability. In explaining such decay, we postulate the existence of weak (or Fermi) forces.

Table 7.4 lists properties of particles with the smallest rest masses. The photon and the graviton are the quanta associated with electromagnetic and gravitational waves, respectively. In a sense, they represent means by which energy may be transferred.

From an unstable particle, energy may also be emitted as an electron-neutrino pair or as a muon-neutrino pair. Because the electron and the muon carry either positive or negative charge, such emission does alter the charge on the source. The two neutrinos, the two electrons, and the two muons are called *leptons* and are not subject to nuclear forces.

Other physical particles are subject to strong nuclear interactions. The lighter of these with zero spin are listed in table 7.5. The pions, the kaons, and the eta particle are treated as stable particles because they have relatively long mean lives.

The lighter strongly interacting particles with spin one are listed in table 7.6. These have such short mean lives that they appear as *resonances* in experiments. The integral spin particles that interact strongly with nuclear particles are called *mesons*.

The lightest of the known strongly interacting particles with one half unit of spin appear in table 7.7. Since these have considerable mean lives, they are treated as stable particles.

Table 7.8 describes the lightest particles with three half units of spin. All of these appear as resonances in bombardment experiments except omega-minus, which is relatively stable. The half-integral spin strongly interacting particles are called *baryons*.

7.7 Hypercharge and isotopic spin

Each of the tables has been constructed by bringing together particles with the same spin, similar masses, and similar mean lives or resonance widths. Within each set are subsets of more closely related particles.

Furthermore, the η and K particles in table 7.5 behave as heavy pions; the Ξ, Σ, and Λ particles in table 7.7 as heavy nucleons. The particles in table 7.6 behave as excited versions of those in table 7.5; the particles in table 7.8 as excited forms of those in table 7.7. The *antiparticle*, listed for each, is the entity that can combine with the particle to yield energy, or its equivalent, on collision with the particle. In tables 7.7 and 7.8, the antiparticles form distinct sets.

Let us consider each set of related particles as a *multiplet*. Each subset, whose members differ from their neighbors by one unit of electric charge and a consequent small variation in mass is called a *submultiplet*.

TABLE 7.4 Physical Particles with the Smallest Rest Masses*

Particle	Symbol	Electric charge	Rest mass, MeV	Spin	Antiparticle	Mean life, sec	Decay products
Photon	γ	0	0	1	γ (Same)	Stable	—
Graviton	—	0	0	2	Same	Stable	—
Electron neutrino	ν_e	0	0	$\frac{1}{2}$	$\tilde{\nu}_e$	Stable	—
Muon neutrino	ν_μ	0	0	$\frac{1}{2}$	$\tilde{\nu}_\mu$	Stable	—
Electron	β^-, e^-	−1	0.511004	$\frac{1}{2}$	β^+, e^+	Stable	—
Muon	μ^-	−1	105.660	$\frac{1}{2}$	μ^+	2.198×10^{-6}	$e^- + \tilde{\nu}_e + \nu_\mu$

*Particle Data Group: A. Rittenberg, A. Barbaro-Galtieri, T. Lasinski, A. H. Rosenfeld, T. G. Trippe, M. Roos, C. Bricman, P. Söding, N. Barash-Schmidt, and C. G. Wohl, Review of Particle Properties, *Rev. Mod. Phys.*, **43**, S1 (1971).

TABLE 7.5 The Lightest Strongly Interacting Particles with no Spin*

Particle	Symbol and charge	Rest mass, MeV	Spin	Quark composition	Antiparticle	Mean life, sec	Decay products
Pion							
Neutral	π^0	134.97	0	$\frac{1}{2}p\tilde{p} + \frac{1}{2}n\tilde{n}$	π^0	0.84×10^{-16}	2γ
Positive	π^+	139.58	0	$p\tilde{n}$	π^-	2.60×10^{-8}	$\mu^+ + \nu_\mu$
Negative	π^-	139.58	0	$n\tilde{p}$	π^+	2.60×10^{-8}	$\mu^- + \tilde{\nu}_\mu$
Kaon							
Positive	K^+	493.8	0	$p\tilde{\lambda}$	K^-	1.24×10^{-8}	$\left\{ \begin{array}{l} \mu^\pm + \nu_\mu \cdots \\ \pi^\pm + \pi^0 \\ \pi^\pm + \pi^+ + \pi^- \end{array} \right.$
Negative	K^-	493.8	0	$\lambda\tilde{p}$	K^+	1.24×10^{-8}	
Neutral	K^0	497.8	0	$n\tilde{\lambda}$	$\tilde{K}^0$	50% K_S^0, 50% K_L^0	
	K_S^0					0.86×10^{-10}	$\pi^{+,0} + \pi^{-,0}$
	K_L^0					5.17×10^{-8}	$\left\{ \begin{array}{l} \pi^\pm + e^\mp + \nu_e \cdots \\ \pi^\pm + \mu^\mp + \nu_\mu \cdots \\ \pi^{+,0} + \pi^{-,0} + \pi^0 \end{array} \right.$
Eta	η^0	548.8	0	$\frac{1}{6}p\tilde{p} + \frac{1}{6}n\tilde{n} + \frac{2}{3}\lambda\tilde{\lambda}$	η^0	2.5×10^{-19}	$\left\{ \begin{array}{l} 2\gamma \\ \pi^{+,0} + \pi^{-,0} + \pi^0 \end{array} \right.$
Ex	X^0	957.5	0	$\frac{1}{3}p\tilde{p} + \frac{1}{3}n\tilde{n} + \frac{1}{3}\lambda\tilde{\lambda}$	X^0	$>1.6 \times 10^{-22}$	$\left\{ \begin{array}{l} \pi^{+,0} + \pi^{-,0} + \eta^0 \\ \pi^+ + \pi^- + \gamma \\ 2\gamma \end{array} \right.$

*Particle Data Group, Review of Particle Properties, *Rev. Mod. Phys.*, **43**, S1 (1971).

TABLE 7.6 The Light Strongly Interacting Particles with Spin 1*

Particle	Symbol and charge	Rest mass, MeV	Spin	Quark composition	Antiparticle	Resonance width, MeV	Decay products
Rho							
Positive	ρ^+	765	1	$\underline{p\bar{n}}$	ρ^-	125	$\pi^+ + \pi^0$
Negative	ρ^-	765	1	$\underline{n\bar{p}}$	ρ^+	125	$\pi^- + \pi^0$
Neutral	ρ^0	767	1	$\frac{1}{2}\underline{p\bar{p}} + \frac{1}{2}\underline{n\bar{n}}$	ρ^0	125	$\left\{\begin{array}{l}\pi^0 + \pi^0 \\ \pi^+ + \pi^-\end{array}\right.$
Omega	ω^0	783.9	1	$\frac{1}{6}\underline{p\bar{p}} + \frac{1}{6}\underline{n\bar{n}} + \frac{2}{3}\underline{\lambda\bar{\lambda}}$	ω^0	11.4	$\pi^+ + \pi^- + \pi^0$
Kay Star							
Positive	K^{*+}	892	1	$\underline{p\bar{\lambda}}$	K^{*-}	50.3	$\left\{\begin{array}{l}K^+ + \pi^0 \\ K^0 + \pi^+\end{array}\right.$
Negative	K^{*-}	892	1	$\underline{\lambda\bar{p}}$	K^{*+}	50.3	$\left\{\begin{array}{l}K^- + \pi^0 \\ \bar{K}^0 + \pi^-\end{array}\right.$
Neutral	K^{*0}	899	1	$\underline{n\bar{\lambda}}$	$\tilde{K}^{*0}$	52.2	$\left\{\begin{array}{l}K^0 + \pi^0 \\ K^+ + \pi^- \\ K^- + \pi^+\end{array}\right.$
Phi	ϕ^0	1019.5	1	$\frac{1}{3}\underline{p\bar{p}} + \frac{1}{3}\underline{n\bar{n}} + \frac{1}{3}\underline{\lambda\bar{\lambda}}$	ϕ^0	4.0	$\left\{\begin{array}{l}K^+ + K^- \\ K^0_L + K^0_S \\ \pi^+ + \pi^- + \pi^0\end{array}\right.$

*Particle Data Group, Review of Particle Properties, *Rev. Mod. Phys.*, **43**, S1 (1971).

TABLE 7.7   The Lightest Strongly Interacting Particles with Spin $\frac{1}{2}$*

Particle	Symbol and charge	Rest mass, MeV	Spin	Quark composition	Antiparticle	Mean life, sec	Decay products
Proton	p^+	938.26	$\frac{1}{2}$	$\overline{ppn}$	$\bar{p}^-$	Stable	$p + \beta^- + \tilde{\nu}_e$
Neutron	n^0	939.55	$\frac{1}{2}$	$\overline{pnn}$	$\tilde{n}^0$	0.93×10^3	
Lambda	Λ^0	1115.6	$\frac{1}{2}$	$\overline{pn\lambda}$	$\tilde{\Lambda}^0$	2.52×10^{-10}	$\begin{cases} p + \pi^- \\ n + \pi^0 \end{cases}$
Sigma							
Positive	Σ^+	1189.4	$\frac{1}{2}$	$\overline{pp\lambda}$	$\tilde{\Sigma}^-$	0.80×10^{-10}	$\begin{cases} p + \pi^0 \\ n + \pi^+ \end{cases}$
Neutral	Σ^0	1192.5	$\frac{1}{2}$	$\overline{pn\lambda}$	$\tilde{\Sigma}^0$	$<1.0 \times 10^{-14}$	$\Lambda^0 + \gamma$
Negative	Σ^-	1197.4	$\frac{1}{2}$	$\overline{nn\lambda}$	$\tilde{\Sigma}^+$	1.49×10^{-10}	$n + \pi^-$
Xi							
Neutral	Ξ^0	1314.7	$\frac{1}{2}$	$\overline{p\lambda\lambda}$	$\tilde{\Xi}^0$	3.0×10^{-10}	$\Lambda^0 + \pi^0$
Negative	Ξ^-	1321.3	$\frac{1}{2}$	$\overline{n\lambda\lambda}$	$\tilde{\Xi}^+$	1.66×10^{-10}	$\Lambda^0 + \pi^-$

*Particle Data Group, Review of Particle Properties, *Rev. Mod. Phys.*, **43**, S1 (1971).

TABLE 7.8   The Light Strongly Interacting Particles with Spin $\frac{3}{2}$*

Particle	Symbol and charge	Rest mass, MeV	Spin	Quark composition	Antiparticle	Resonance width, MeV	Decay products
Delta							
Doubly positive	Δ^{++}	1233	$\frac{3}{2}$	ppp	$\tilde{\Delta}^{--}$	120	$p^+ + \pi^+$
Singly positive	Δ^+	1233	$\frac{3}{2}$	ppn	$\tilde{\Delta}^-$	120	$\begin{cases} p^+ + \pi^0 \\ n^0 + \pi^+ \end{cases}$
Neutral	Δ^0	1234	$\frac{3}{2}$	pnn	$\tilde{\Delta}^0$	120	$\begin{cases} p^+ + \pi^- \\ n^0 + \pi^0 \end{cases}$
Negative	Δ^-	1241	$\frac{3}{2}$	nnn	$\tilde{\Delta}^+$	120	$n^0 + \pi^-$
Sigma Star							
Positive	Σ^{*+}	1383	$\frac{3}{2}$	ppλ	$\tilde{\Sigma}^{*-}$	36	$\Lambda^0 + \pi^+$
Neutral	Σ^{*0}	1385	$\frac{3}{2}$	pnλ	$\tilde{\Sigma}^{*0}$	36	$\Lambda^0 + \pi^0$
Negative	Σ^{*-}	1386	$\frac{3}{2}$	nnλ	$\tilde{\Sigma}^{*+}$	36	$\Lambda^0 + \pi^-$
Xi Star							
Neutral	Ξ^{*0}	1529	$\frac{3}{2}$	p$\lambda\lambda$	$\tilde{\Xi}^{*0}$	7	$\begin{cases} \Xi^0 + \pi^0 \\ \Xi^- + \pi^+ \end{cases}$
Negative	Ξ^{*-}	1534	$\frac{3}{2}$	n$\lambda\lambda$	$\tilde{\Xi}^{*+}$	7	$\begin{cases} \Xi^0 + \pi^- \\ \Xi^- + \pi^0 \end{cases}$
Omega						Mean life, sec	
Negative	Ω^-	1672	$\frac{3}{2}$	$\lambda\lambda\lambda$	$\tilde{\Omega}^+$	1×10^{-10}	$\begin{cases} \Xi^0 + \pi^- \\ \Xi^- + \pi^0 \end{cases}$

*Particle Data Group, Review of Particle Properties, *Rev. Mod. Phys.*, **43**, S1 (1971).

Each submultiplet can similarly be considered to be one unit away from its neighboring submultiplets by one unit of *hypercharge*. The origin for the hypercharge scale is located so that the average hypercharge of the particles in the complete set is zero. The signs are chosen so that the numbers for the common baryons decrease from plus to minus as we go from submultiplets containing lighter to those containing heavier particles. For the corresponding antiparticles, the order is reversed; the numbers increase with increasing mass.

Let us label each particle in a submultiplet by integers or half integers that increase by one unit with each one-unit increase in electric charge. The origin is chosen at the middle; so for each positive observed number, there is also a negative one. The maximum number in the subset is called the *isotopic spin*; the labeling integer or half-integer is the third component of isotopic spin.

Let Y be the hypercharge and I_3 the pertinent isotopic spin component for a given particle. Then the particle is represented by a point in the $I_3 Y$ plane. A multiplet is represented by an array of such points.

A translation from one point of the array to another represents the transmutation of one particle of the multiplet into another. A second translation represents conversion of the second particle into a third. A system can go directly from the first to the third point, and the over-all process is still a transmutation. To each transmutation there is an inverse. In a series of transmutations, the way the conversions are associated is unimportant. The set of operations linking members of a multiplet constitutes a group.

Transmutation of a neutron to a proton increases the electric charge by 1 but does not change the submultiplet. It increases the isotopic spin I_3 by 1 and leaves the hypercharge unchanged. Transmuting a lambda to a proton changes the submultiplet into its next higher form in which there is one more possible isotopic spin I_3. It increases Y by 1 and I_3 by $\frac{1}{2}$. The operation that transmutes a lambda to a neutron increases Y by 1 and I_3 by $-\frac{1}{2}$. An inverse for each of these also exists.

Let us construct the $I_3 Y$ plane so that all vectors representing these elementary operations are of the same length. Then the particles lie on a regular hexagonal lattice. Let us label the operations $A_\pm$, $B_\pm$, $C_\pm$ as in figure 7.5.

7.8 Hypothetical fundamental states for strongly interacting particles

Electromagnetic effects distinguish between particles connected by an integral number of A_+'s or A_-'s. Furthermore, weak interactions distinguish between states with different Y's. In a multiplet for which these effects are negligible, except insofar as they label the particles, translations A_+, B_+, C_+, A_-, B_-, C_- are equivalent. If one of them is present, the others presumably should also be present.

When none of the operations can act on a given subatomic particle, the particle stands by itself as in figure 7.6. When one translation is present, with its inverse, a second particle exists. A third particle arranged symmetrically with respect to the first two then allows all six transmutations to be present singly.

Because the step-up and -down vectors are of equal length, by choice, the three particles must lie at the corners of an equilateral triangle. For A_+ and A_- to be

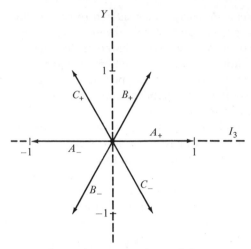

FIGURE 7.5   Basic step-up and step-down operations of the group.

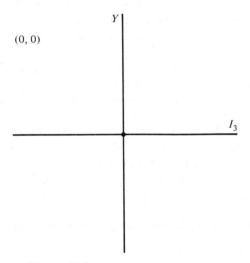

(0, 0)

FIGURE 7.6   A vacuum-like multiplet.

present, two must lie on a horizontal line. These two may be either above or below the I_3 axis. The zero for Y is chosen so the third particle lies on the other side of the I_3 axis at twice the distance (see figures 7.7 and 7.8).

The particles in figure 7.7 are called *quarks*. Those in figure 7.8 are the corresponding antiparticles. None of them exists individually, insofar as we know; but the larger multiplets are constructed from them. For convenience, quark properties are listed in table 7.9.

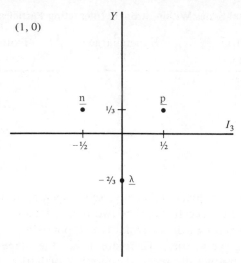

FIGURE 7.7   The quark multiplet.

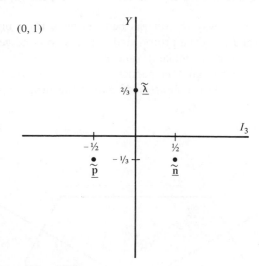

FIGURE 7.8   The antiquark multiplet.

Neglect of the electromagnetic field except for its effect on labeling the particles makes the different members of a row of a given array equivalent. Consequently, the array is symmetric with respect to reflection in the Y axis.

Neglect of both electromagnetism and weak interaction makes the λ quark equivalent to the $\underset{\sim}{n}$ quark and to the p quark. But reflection of figure 7.7 (or 7.8) in axes passing through the origin, inclined at $\pm 60°$ from the Y axis, permutes these particles (scales are chosen so the step-up and -down vectors are all the same length). As long as the larger arrays are built up from these triplets, reflections in their inclined axes also

TABLE 7.9   Fundamental States Within Strongly Interacting Particles

Quark	Electric charge q	Hypercharge Y	Isotopic spin I_3	Spin s_z
p	$\frac{2}{3}$	$\frac{1}{3}$	$\frac{1}{2}$	$\pm\frac{1}{2}$
n	$-\frac{1}{3}$	$\frac{1}{3}$	$-\frac{1}{2}$	$\pm\frac{1}{2}$
λ	$-\frac{1}{3}$	$-\frac{2}{3}$	0	$\pm\frac{1}{2}$
$\tilde{p}$	$-\frac{2}{3}$	$-\frac{1}{3}$	$-\frac{1}{2}$	$\mp\frac{1}{2}$
$\tilde{n}$	$\frac{1}{3}$	$-\frac{1}{3}$	$\frac{1}{2}$	$\mp\frac{1}{2}$
$\tilde{\lambda}$	$\frac{1}{3}$	$\frac{2}{3}$	0	$\mp\frac{1}{2}$

replace λ quarks by n or p quarks and conversely. Such multiplets are consequently symmetric with respect to reflections in the two inclined axes.

A boundary drawn through the extremal lattice points as in figure 7.9, must also exhibit these reflection symmetries. Therefore, it has the shape shown in the figure. This is characterized by two distances, the number of lattice sections at maximum and at minimum Y, λ and μ. We call the set of particles the (λ, μ) multiplet.

EXAMPLE 7.2   Indicate the way the quarks behave when (a) a proton collides with another proton to form a π^+, (b) a proton strikes a neutron and generates a π^-, (c) a proton collides with a proton to make π^+ and π^-.

Whenever one particle hits another particle, it does work on the struck particle. And as long as the projectile is compressed against the target particle, some of the

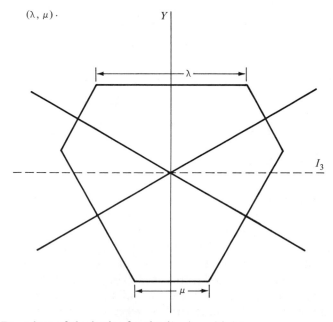

FIGURE 7.9   Boundary of the lattice for the (λ, μ) multiplet.

initial kinetic energy must be present as potential energy. When the mass of this potential energy is larger than the rest mass of a quark-antiquark pair in combination, the pair may appear and become distributed among the other quarks in a particular way.

Tables 7.5 and 7.7 show the quarks that each particle contains. They also show the maximum z component of spin that each structure can possess. For bookkeeping purposes, consider that this spin arises as the sum of the s_z values of the constituent quarks. Then add to the formula for the particle an upward pointing arrow ↑ for every $s_z = +\frac{1}{2}$ quark and a downward pointing arrow ↓ for every $s_z = -\frac{1}{2}$ quark.

Reaction (a) involves the materialization of energy to $\underline{n}^\uparrow$ and $\underline{\tilde{n}}^\downarrow$. The new $\underline{n}$ displaces a $\underline{p}$ from one of the protons while the $\underline{\tilde{n}}$ pairs with this $\underline{p}$:

$$\underline{ppn}^{\uparrow\uparrow\downarrow} + \underline{ppn}^{\uparrow\uparrow\downarrow} \rightarrow \underline{ppn}^{\uparrow\uparrow\downarrow} + \underline{pnn}^{\uparrow\uparrow\downarrow} + \underline{p\tilde{n}}^{\uparrow\downarrow}.$$

In reaction (b) potential energy arising from the collision appears as $\underline{p}^\uparrow$ and $\underline{\tilde{p}}^\downarrow$. The new $\underline{p}$ displaces an $\underline{n}$ from the neutron while the $\underline{\tilde{p}}$ pairs with this $\underline{n}$:

$$\underline{ppn}^{\uparrow\uparrow\downarrow} + \underline{pnn}^{\uparrow\uparrow\downarrow} \rightarrow \underline{ppn}^{\uparrow\uparrow\downarrow} + \underline{ppn}^{\uparrow\uparrow\downarrow} + \underline{n\tilde{p}}^{\uparrow\downarrow}.$$

Potential energy from the collision of two protons materializes as $\underline{n}^\uparrow$ and $\underline{\tilde{n}}^\downarrow$ and as $\underline{p}^\uparrow$, $\underline{\tilde{p}}^\downarrow$ in reaction (c). The new $\underline{p}$ pairs with the $\underline{\tilde{n}}$; the new $\underline{n}$ with the $\underline{\tilde{p}}$:

$$\underline{ppn}^{\uparrow\uparrow\downarrow} + \underline{ppn}^{\uparrow\uparrow\downarrow} \rightarrow \underline{ppn}^{\uparrow\uparrow\downarrow} + \underline{ppn}^{\uparrow\uparrow\downarrow} + \underline{p\tilde{n}}^{\uparrow\downarrow} + \underline{n\tilde{p}}^{\uparrow\downarrow}.$$

EXAMPLE 7.3   What happens to the quarks when a positive pion strikes a proton to produce a Σ^+ and a K^+?

Production of the so-called strange particles is associated with conversion of energy to a $\underline{\lambda}$, $\underline{\tilde{\lambda}}$ pair. In this particular reaction, some of the energy comes from annihilation of $\underline{n}$ with $\underline{\tilde{n}}$. The rest of the energy comes from kinetic energy of the bombarding particle. The $\underline{\lambda}$ produced replaces the $\underline{n}$ in the proton; the $\underline{\tilde{\lambda}}$ replaces the $\underline{\tilde{n}}$ in the pion:

$$\underline{p\tilde{n}}^{\uparrow\downarrow} + \underline{ppn}^{\uparrow\uparrow\downarrow} \rightarrow \underline{pp\lambda}^{\uparrow\uparrow\downarrow} + \underline{p\tilde{\lambda}}^{\uparrow\downarrow}.$$

EXAMPLE 7.4   How do the quarks behave when (a) a neutron decays to a proton, (b) a sigma-plus decays to a nucleon and a pion?

In beta decay, excess energy is lost as an electron-neutrino pair. The change in electric charge alters the third component of isotopic spin, converting one ordinary quark into another. Thus, reaction (a) involves an $\underline{n}$ going to a $\underline{p}$:

$$\underline{pnn}^{\uparrow\uparrow\downarrow} \rightarrow \underline{ppn}^{\uparrow\uparrow\downarrow} + \beta^{-\uparrow} + \tilde{\nu}^\downarrow.$$

Associated with the extreme hypercharge in a $\underline{\lambda}$ is excess energy. This may materialize as a π^0 meson,

$$\underline{pp\lambda}^{\uparrow\uparrow\downarrow} \rightarrow \underline{ppn}^{\uparrow\uparrow\downarrow} + (\tfrac{1}{2}\underline{p\tilde{p}} + \tfrac{1}{2}\underline{n\tilde{n}})^{\uparrow\downarrow},$$

or as an $\underline{n}^\uparrow$, $\underline{\tilde{n}}^\downarrow$ pair. If the $\underline{n}$ replaces a $\underline{p}$ in the Σ^+, we have

$$\underline{pp\lambda}^{\uparrow\uparrow\downarrow} \rightarrow \underline{pnn}^{\uparrow\uparrow\downarrow} + \underline{p\tilde{n}}^{\uparrow\downarrow}.$$

The change in hypercharge converts the $\underline{\lambda}$ to an $\underline{n}$.

7.9   Exchange properties

According to the quark model, every member of a multiplet consists of a combination of p, n, and λ particles. We may consider each of these constituents as involving a fundamental particle in a different state. The indistinguishability of the different fundamental particles then leads to a generalized Pauli principle limiting the combinations that can be formed and the number of multiplet members to be associated with each point of the hexagonal lattice in the $I_3 Y$ plane.

In subatomic structures, the state of a particle is described by a function of its coordinates. For the jth particle in fundamental state p, we write

$$u_{\text{p}} \text{ (coordinates of } j) \equiv u_{\text{I}}(j). \tag{7-12}$$

For the same particle in state n, we write

$$u_{\text{n}} \text{ (coordinates of } j) \equiv u_{\text{II}}(j). \tag{7-13}$$

For the particle in state λ, we write

$$u_{\lambda} \text{ (coordinates of } j) \equiv u_{\text{III}}(j). \tag{7-14}$$

Composite states are described by products of functions. For particles 1 and 2 in states I and II, we might form the following:

$$u_{\text{I}}(1)u_{\text{I}}(2), \tag{7-15}$$

$$u_{\text{II}}(1)u_{\text{II}}(2), \tag{7-16}$$

$$u_{\text{I}}(1)u_{\text{II}}(2), \tag{7-17}$$

$$u_{\text{II}}(1)u_{\text{I}}(2). \tag{7-18}$$

But if the fundamental constituents are really indistinguishable, each composite state must exhibit a definite symmetry under interchange of any two of the particles.

Functions (7-17) and (7-18) as they stand lack this property. On the other hand, the renormalized sum

$$\frac{1}{\sqrt{2}} \left[u_{\text{I}}(1)u_{\text{II}}(2) + u_{\text{II}}(1)u_{\text{I}}(2) \right] \tag{7-19}$$

does not change, while the renormalized difference

$$\frac{1}{\sqrt{2}} \left[u_{\text{I}}(1)u_{\text{II}}(2) - u_{\text{II}}(1)u_{\text{I}}(2) \right] = \frac{1}{\sqrt{2}} \begin{vmatrix} u_{\text{I}}(1) & u_{\text{II}}(1) \\ u_{\text{I}}(2) & u_{\text{II}}(2) \end{vmatrix} \tag{7-20}$$

merely changes sign when particles 1 and 2 are interchanged. Function (7-19) is said to be *symmetric*, (7-20) *antisymmetric*, with respect to the exchange.

States constructed from more than two quarks may be symmetric with respect to interchange of the particles in some sets and antisymmetric with respect to interchange in other sets.

7.10 Young tables and diagrams

Each member of a multiplet is described by a function that does not change when some of its constituent particles are permuted but does change sign when some of these particles are exchanged with others. This property can be represented by a *Young table*, in which integers describing states linked symmetrically appear in a row and those describing a set of states linked antisymmetrically appear in a column. If a square is placed around each integer, an array of squares characterizing the multiplet is obtained. It is called a *Young diagram*.

The hypothetical quark multiplet consists of states p, n̲, and λ̲ for a single fundamental particle. These are described by functions (7-12), (7-13), (7-14). For bookkeeping purposes, each is represented by its Roman index within a square. Thus, we let

$$\boxed{\text{I}} \qquad \text{stand for} \qquad u_{\text{I}}(1), \tag{7-21}$$

$$\boxed{\text{II}} \qquad \text{stand for} \qquad u_{\text{II}}(1), \tag{7-22}$$

$$\boxed{\text{III}} \qquad \text{stand for} \qquad u_{\text{III}}(1). \tag{7-23}$$

The single square

$$\boxed{} \tag{7-24}$$

characterizes all three states.

Because the fundamental particles are indistinguishable, each composite system must contain all possible permutations of the arrangement occurring in the first term of the governing function, with equal weight. In one of these permutations, the Roman indices either increase or remain unchanged in each step through a set of factors that combine symmetrically, but increase step by step through a set of factors that combine antisymmetrically. The system itself is labeled by the corresponding Young table.

A multiplet containing two quarks bound symmetrically is described by (7-21), (7-22), (7-23) combined as in (7-15), (7-16), (7-19). We have

$$\boxed{\text{I} \mid \text{I}} \quad : \ u_{\text{I}}(1)u_{\text{I}}(2), \tag{7-25}$$

$$\boxed{\text{II} \mid \text{II}} \quad : \ u_{\text{II}}(1)u_{\text{II}}(2), \tag{7-26}$$

$$\boxed{\text{III} \mid \text{III}} \quad : \ u_{\text{III}}(1)u_{\text{III}}(2), \tag{7-27}$$

$$\boxed{\text{I} \mid \text{II}} \quad : \ \frac{1}{\sqrt{2}}\,[u_{\text{I}}(1)u_{\text{II}}(2) + u_{\text{II}}(1)u_{\text{I}}(2)], \tag{7-28}$$

$$\boxed{\text{I} \mid \text{III}} \quad : \ \frac{1}{\sqrt{2}}\,[u_{\text{I}}(1)u_{\text{III}}(2) + u_{\text{III}}(1)u_{\text{I}}(2)], \tag{7-29}$$

$$\boxed{\text{II} \mid \text{III}} \quad : \ \frac{1}{\sqrt{2}}\,[u_{\text{II}}(1)u_{\text{III}}(2) + u_{\text{III}}(1)u_{\text{II}}(2)]. \tag{7-30}$$

Each Young table lists the Roman indices from the term in which the numerals either increase or remain unchanged. All possible combinations meeting this condition appear. Indeed, we can obtain every member of the set by placing each Roman numeral I, II, III into the first box in turn and placing the same one or a higher one in the second box. The diagram

$$\boxed{} \qquad (7\text{-}31)$$

characterizes the multiplet.

When any two quarks are combined as in (7-20), we obtain the antisymmetric combinations

$$\begin{array}{c}\boxed{\text{I}}\\\boxed{\text{II}}\end{array} \quad : \quad \frac{1}{\sqrt{2}}\,[u_{\text{I}}(1)u_{\text{II}}(2) - u_{\text{II}}(1)u_{\text{I}}(2)], \qquad (7\text{-}32)$$

$$\begin{array}{c}\boxed{\text{I}}\\\boxed{\text{III}}\end{array} \quad : \quad \frac{1}{\sqrt{2}}\,[u_{\text{I}}(1)u_{\text{III}}(2) - u_{\text{III}}(1)u_{\text{I}}(2)], \qquad (7\text{-}33)$$

$$\begin{array}{c}\boxed{\text{II}}\\\boxed{\text{III}}\end{array} \quad : \quad \frac{1}{\sqrt{2}}\,[u_{\text{II}}(1)u_{\text{III}}(2) - u_{\text{III}}(1)u_{\text{II}}(2)]. \qquad (7\text{-}34)$$

Each Young table lists the Roman indices from the term in which the numerals increase. We can obtain every member of the set by placing I in the upper square and the higher number, II or III, in the lower one, or II in the upper square and the higher number III in the lower one. The skeleton

$$\begin{array}{c}\boxed{}\\\boxed{}\end{array} \qquad (7\text{-}35)$$

characterizes the given multiplet.

For three quarks, there is only one nonzero antisymmetric combination

$$\begin{array}{c}\boxed{\text{I}}\\\boxed{\text{II}}\\\boxed{\text{III}}\end{array} \quad : \quad \frac{1}{\sqrt{6}}\begin{vmatrix} u_{\text{I}}(1) & u_{\text{II}}(1) & u_{\text{III}}(1) \\ u_{\text{I}}(2) & u_{\text{II}}(2) & u_{\text{III}}(2) \\ u_{\text{I}}(3) & u_{\text{II}}(3) & u_{\text{III}}(3) \end{vmatrix}. \qquad (7\text{-}36)$$

The reference term is the one in which the numerals increase from I to III. The corresponding Young diagram is

$$\begin{array}{c}\boxed{}\\\boxed{}\\\boxed{}\end{array}. \qquad (7\text{-}37)$$

Since a multiplet containing only one member must lie at the origin of the $I_3 Y$ plane, the net hypercharge and isotopic spin here are zero.

Larger determinants constructed from the quarks in the same way all contain duplicate columns. Consequently, they vanish; only three numerals can appear in the column of the Young table. The numerals must increase, square by square, and beyond III there is no quark.

Any number of quarks can be combined symmetrically, however. Each combination is labeled by the Roman indices in a reference term chosen as before. In this term, the numerals either remain the same or increase, from factor to factor.

A system may involve as many as three symmetric sets whose members combine antisymmetrically with those of the other sets. The three-member columns are written first, then the two-member columns, and finally the one-member columns:

$$(7\text{-}38)$$

Each three-member column can only be filled in one way, as in (7-36). Since the result is vacuum-like, none of these columns contributes to Y and I_3. The other columns increase Y the most when they do not contain III. The two-member columns would then contain ṇ and p with their isotopic spins paired. To make I_3 as large as possible for the maximum Y, we would finally have to place p quarks in all one-member columns.

Consequently, the system in the upper right corner of figure 7.9 is represented by

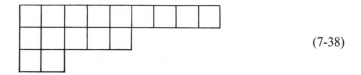

$$(7\text{-}39)$$

Each p contributing $\frac{1}{2}$ to I_3, the number of single-member columns is double the third component of isotopic spin at the corner. This number is the number λ of lattice sections at maximum Y.

The systems with the smallest Y arise when the table contains as many threes as possible. But III can be only at the bottom of each column. To make I_3 as large as possible for this Y, I must be present in as many columns as contain openings:

$$(7\text{-}40)$$

Now each two-member column contributes $\frac{1}{2}$ to I_3, but the other columns contribute nothing. The number of two-member columns yields two times the third component of isotopic spin. Since the system is at the lower right corner of figure 7.9, this number equals the number μ of lattice sections at minimum Y.

7.11   Baryon multiplets

We have seen how all particles in a multiplet are connected by step-up and step-down operations of a group. These operations link the points of a hexagonal lattice in the $I_3 Y$ plane. The boundary of this lattice is characterized by the number of lattice sections at maximum and at minimum Y. The number of particles at each point is given by the number of Young tables corresponding to it.

For half-integral-spin systems, we must combine an odd number of quarks. The one-quark multiplets have already been described. Next in complexity are the three-quark multiplets defined by the diagrams

(7-41)

(7-42)

(7-43)

Diagram (7-41) corresponds to a symmetric relationship among the quarks that allows all spins to be parallel. Applying the rule that numbers in a row may either remain the same or increase in any step from left to right, the following combinations are possible:

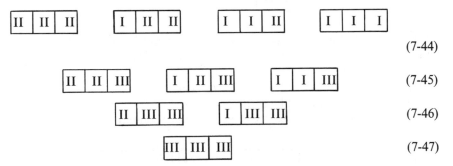

(7-44)

(7-45)

(7-46)

(7-47)

Row (7-44) corresponds to the delta particles, row (7-45) to the sigma star particles, row (7-46) to the xi star particles, and row (7-47) to the omega minus particle. Plotting these in the $I_3 Y$ plane yields figure 7.10. Note that each lattice point occurs just once.

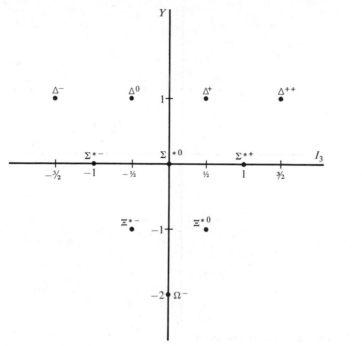

FIGURE 7.10 The common baryon decuplet.

Diagram (7-42) corresponds to two quarks in a symmetric configuration and one in antisymmetric relationship to them, thus allowing a spin $\frac{1}{2}$ arrangement. Applying the rules that numbers must increase step by step in descending a column, while they may either remain the same or increase on going from left to right in a row, yields the following combinations:

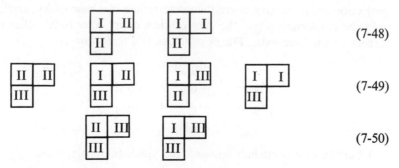

Row (7-48) corresponds to the nucleons, the triplet in row (7-49) to the sigma particles, the singlet to the Λ^0, and row (7-50) to the xi particles. In the $I_3 Y$ plane, these combinations form figure 7.11. Note how two particles are associated with the origin, one particle with all other positions.

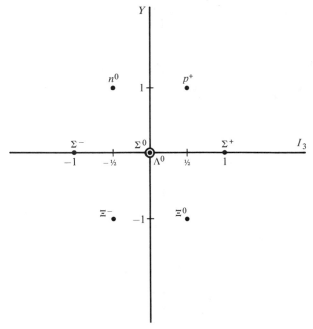

FIGURE 7.11   The common baryon octet.

Three antiquarks combine similarly. Corresponding to figures 7.10 and 7.11 are the antibaryon decuplet and antibaryon octet from tables 7.8 and 7.7.

EXAMPLE 7.5   From the possible independent spin states, determine how many baryons have the composition (a) pnλ, (b) ppn, (c) ppp.

We assume that orbital motions of the three quarks cancel. The possible spin projections of each baryon are then determined by those of its quarks.

In the combination pnλ the spin of each quark may have either orientation with respect to a unique axis. The possibilities add as follows:

$$p^{\uparrow}n^{\uparrow}\lambda^{\uparrow} \qquad\qquad s_z = \tfrac{3}{2}$$

$$p^{\uparrow}n^{\uparrow}\lambda^{\downarrow} \quad p^{\uparrow}n^{\downarrow}\lambda^{\uparrow} \quad p^{\downarrow}n^{\uparrow}\lambda^{\uparrow} \qquad s_z = \tfrac{1}{2}$$

$$p^{\uparrow}n^{\downarrow}\lambda^{\downarrow} \quad p^{\downarrow}n^{\downarrow}\lambda^{\uparrow} \quad p^{\downarrow}n^{\uparrow}\lambda^{\downarrow} \qquad s_z = -\tfrac{1}{2}$$

$$p^{\downarrow}n^{\downarrow}\lambda^{\downarrow} \qquad\qquad s_z = -\tfrac{3}{2}.$$

A baryon with three half units of spin yields the projections

$$s_z = \tfrac{3}{2}, \qquad \tfrac{1}{2}, \qquad -\tfrac{1}{2}, \qquad -\tfrac{3}{2}$$

while one with one half unit yields

$$s_z = \tfrac{1}{2}, \qquad -\tfrac{1}{2}.$$

The projections for $\underline{pn\lambda}$ are obtained if just one of the former baryons and two of the latter contribute. Thus, the (0, 0) position in the spin-$\frac{3}{2}$ multiplet is singly occupied; but in the spin-$\frac{1}{2}$ multiplet it is doubly occupied. The pertinent particles are Σ^{*0} and Σ^0, Λ^0.

We also assume that identical particles are indistinguishable. There is no difference between

$$p^\uparrow p^\downarrow n^\uparrow \qquad \text{and} \qquad p^\downarrow p^\uparrow n^\uparrow$$

or between

$$p^\downarrow p^\uparrow n^\downarrow \qquad \text{and} \qquad p^\uparrow p^\downarrow n^\downarrow.$$

Consequently, two fewer states arise when two of the quarks are the same. For the $\underline{ppn}$, we find only six distinct states:

$$
\begin{array}{ccl}
p^\uparrow p^\uparrow n^\uparrow & & s_z = \frac{3}{2} \\
p^\uparrow p^\uparrow n^\downarrow \qquad p^\uparrow p^\downarrow n^\uparrow & & s_z = \frac{1}{2} \\
p^\downarrow p^\downarrow n^\uparrow \qquad p^\downarrow p^\uparrow n^\downarrow & & s_z = -\frac{1}{2} \\
p^\downarrow p^\downarrow n^\downarrow & & s_z = -\frac{3}{2}.
\end{array}
$$

These correspond to the Δ^+ with spin $\frac{3}{2}$, yielding $s_z = \frac{3}{2}, \frac{1}{2}, -\frac{1}{2}, -\frac{3}{2}$, and the proton with spin $\frac{1}{2}$, yielding $s_z = \frac{1}{2}, -\frac{1}{2}$. Therefore, the $(\frac{1}{2}, 1)$ position in the spin-$\frac{3}{2}$ multiplet, and the $(\frac{1}{2}, 1)$ position in the spin-$\frac{1}{2}$ multiplet, are each singly occupied.

Similarly, we cannot distinguish between

$$p^\uparrow p^\uparrow p^\downarrow \qquad p^\uparrow p^\downarrow p^\uparrow \qquad p^\downarrow p^\uparrow p^\uparrow$$

or between

$$p^\uparrow p^\downarrow p^\downarrow \qquad p^\downarrow p^\uparrow p^\downarrow \qquad p^\downarrow p^\downarrow p^\uparrow.$$

Four fewer states arise when all the quarks are the same.

With the $\underline{ppp}$ combination, there are only four independent states:

$$
\begin{array}{cl}
p^\uparrow p^\uparrow p^\uparrow & s_z = \frac{3}{2} \\
p^\uparrow p^\uparrow p^\downarrow & s_z = \frac{1}{2} \\
p^\uparrow p^\downarrow p^\downarrow & s_z = -\frac{1}{2} \\
p^\downarrow p^\downarrow p^\downarrow & s_z = -\frac{3}{2}.
\end{array}
$$

These come from the Δ^{++} with spin equal to $\frac{3}{2}$. The $(\frac{3}{2}, 1)$ position is singly occupied in the spin-$\frac{3}{2}$ multiplet. It is vacant in the spin-$\frac{1}{2}$ multiplet.

7.12 Meson multiplets

Each meson is produced when the appropriate baryon strikes another baryon with enough force. In the process, sufficient kinetic energy is transformed to potential energy and materialized as quark-antiquark pairs. A new quark may displace an old quark from a baryon. When more than one pair is produced, exchange among the new pairs may occur. Consequently, all possible combinations of quark and anti-quark may appear.

In determining the corresponding multiplets, we employ the terminology of sections 7.9 and 7.10. We also identify an antiquark as a fundamental antiparticle in one of the fundamental states. The jth antiparticle in the first state is denoted by

$$u_{\tilde{\mathbf{P}}} \text{ (coordinates of } j) \equiv u_{\tilde{\mathbf{I}}}(j) \quad \text{or} \quad \boxed{\tilde{\text{I}}}\,. \tag{7-51}$$

For the same antiparticle in the second fundamental state, we write

$$u_{\tilde{\mathbf{n}}} \text{ (coordinates of } j) \equiv u_{\tilde{\mathbf{II}}}(j) \quad \text{or} \quad \boxed{\tilde{\text{II}}}\,. \tag{7-52}$$

For the antiparticle in the third fundamental state, we have

$$u_{\tilde{\mathbf{\lambda}}} \text{ (coordinates of } j) \equiv u_{\widetilde{\mathbf{III}}}(j) \quad \text{or} \quad \boxed{\widetilde{\text{III}}}\,. \tag{7-53}$$

An antiquark can react with a quark of the same Roman numeral to yield energy; a vacuum-like configuration then results. But

$$\boxed{\begin{array}{c} \\ \hline \\ \end{array}} \tag{7-54}$$

reacts in this way with

$$\boxed{} \tag{7-55}$$

in the same column. Consequently

$$\boxed{\begin{array}{c} \text{I} \\ \hline \text{II} \end{array}} \quad \text{behaves as} \quad \boxed{\widetilde{\text{III}}} \tag{7-56}$$

$$\boxed{\begin{array}{c} \text{II} \\ \hline \text{III} \end{array}} \quad \text{behaves as} \quad \boxed{\tilde{\text{I}}} \tag{7-57}$$

$$\boxed{\begin{array}{c} \text{I} \\ \hline \text{III} \end{array}} \quad \text{behaves as} \quad \boxed{\tilde{\text{II}}}\,. \tag{7-58}$$

Replacing each two-member column in (7-48), (7-49), (7-50) by the antiquark box it behaves as, yields the array:

$$\tag{7-59}$$

$$\tag{7-60}$$

$$\tag{7-61}$$

Similarly substituting into (7-36) leads to the one possibility

$$\boxed{\begin{array}{c}\widetilde{\mathrm{III}} \\ \hline \mathrm{III}\end{array}} \, . \tag{7-62}$$

Because a λ can annihilate an $\tilde{\lambda}$ and appear as either $\mathrm{n\tilde{n}}$ or $\mathrm{p\bar{p}}$, an n can annihilate an $\tilde{\mathrm{n}}$ and appear as either $\mathrm{p\bar{p}}$ or $\lambda\tilde{\lambda}$, a p can annihilate an $\bar{\mathrm{p}}$ and appear as either $\mathrm{n\tilde{n}}$ or $\lambda\tilde{\lambda}$, the two inner states in row (7-60) and the state corresponding to row (7-62) involve $\mathrm{p\bar{p}}$, $\mathrm{n\tilde{n}}$, and $\lambda\tilde{\lambda}$. Since (7-62) represents the vacuum-like singlet, each of the pertinent forms contributes equally to the corresponding function:

$$\frac{1}{\sqrt{3}}\left[u_{\mathrm{I}}(1)u_{\tilde{\mathrm{I}}}(2) + u_{\mathrm{II}}(1)u_{\tilde{\mathrm{II}}}(2) + u_{\mathrm{III}}(1)u_{\widetilde{\mathrm{III}}}(2)\right]. \tag{7-63}$$

Going with the two outer states in row (7-60) is the combination in which λ and $\tilde{\lambda}$ do not contribute:

$$\frac{1}{\sqrt{2}}\left[u_{\mathrm{I}}(1)u_{\tilde{\mathrm{I}}}(2) - u_{\mathrm{II}}(1)u_{\tilde{\mathrm{II}}}(2)\right]. \tag{7-64}$$

The other member of the octet consists of what remains:

$$\frac{1}{\sqrt{6}}\left[u_{\mathrm{I}}(1)u_{\tilde{\mathrm{I}}}(2) + u_{\mathrm{II}}(1)u_{\tilde{\mathrm{II}}}(2) - 2u_{\mathrm{III}}(1)u_{\widetilde{\mathrm{III}}}(2)\right]. \tag{7-65}$$

The spin of the antiquark may be either parallel or antiparallel to that of the quark. The parallel arrangement leads to the mesons with spin equal to 1 (see figure 7.12) for the octet. The antiparallel configuration corresponds to the spin being 0, as in figure 7.13. The vacuum-like singlets occur in mesons X^0 and ϕ^0.

7.13 Pertinent points

A given region or system is said to be symmetric whenever there are reorientations, translations, or similar operations that convert it into various equivalent forms. The complete set of operations (and each of its closed subsets) forms a group.

The reorientations by themselves may make up any of the geometric groups listed in figure 7.1. The translations by themselves generate one or more of the lattices described in section 7.5. The so-called internal, or dynamic, operations relate subatomic particles. For each set of these, a reference lattice exists in the $I_3 Y$ plane.

A given setup may deviate from a symmetric system because of (a) variations in supposedly equivalent masses, (b) variations in supposedly equivalent distances, (c) variations in parameters for supposedly equivalent springs, (d) less symmetric surroundings, and so on. Electromagnetic and weak interactions break the symmetry among the particles of a multiplet, causing the observed variations in mass.

In general, a distortion from a symmetric model perturbs physical properties continuously. Consequently, the changes may be estimated.

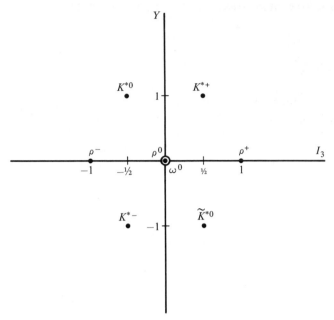

FIGURE 7.12   The octet of common spin-1 mesons.

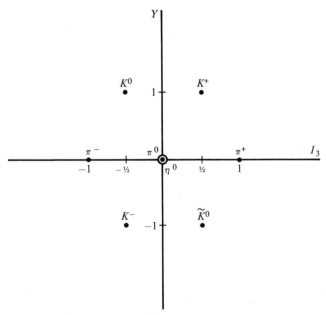

FIGURE 7.13   The octet of common spin-0 mesons.

DISCUSSION QUESTIONS

7.1 In what way do symmetry arguments apply to a system that is not quite symmetric?

7.2 Distinguish between symmetry elements and symmetry operations.

7.3 List the operations that transform a checkerboard into itself.

7.4 What binary operation connecting vectors does not obey the associative law?

7.5 Cite transformations that do not have inverses.

7.6 Why would the presence of any covering operations for a given physical system imply the presence of other covering operations? Give examples.

7.7 What may be associated with each point of a crystal lattice? How are the different parts of the basis oriented with respect to the lattice?

7.8 When are the vectors A_1, A_2, A_3 for a reciprocal lattice mutually orthogonal?

7.9 What lattice is reciprocal to a face-centered cubic lattice?

7.10 Discuss the forces needed to explain physical phenomena.

7.11 How are (a) multiplets and (b) submultiplets of particles identified?

7.12 Describe the transmutations that relate subatomic particles. How do these form groups?

7.13 What elementary operations are always present in these groups? To what approximation are they equivalent?

7.14 When do these operations trace out a regular hexagonal lattice? Show that two parameters characterize the boundary of this lattice.

7.15 In what kind of processes do mesons appear? Does the nature of these support the view that mesons consist of quark-antiquark pairs?

7.16 Must the step-up and step-down operations be equivalent for the exchange properties to hold rigorously?

7.17 Cite an alternate way for listing the Roman indices in a Young table.

7.18 Why can a three-member column be filled in only one way? Why do the other columns increase Y the most when they do not contain III?

7.19 Why must III be at the bottom of a column? Why must I be present in as many columns as contain openings, for I_3 to be as large as possible? Why does a column containing I and III contribute $\frac{1}{2}$ to the total I_3?

7.20 Why does λ in figure 7.9 equal the number of one-member columns in the corresponding Young diagram? Why is μ the number of two-member columns?

7.21 How do we obtain the number of independent states corresponding to each lattice point in the $I_3 Y$ plane?

7.22 Use Young tables to explain the nature of (a) the baryon multiplets, (b) the meson multiplets.

PROBLEMS

7.1 Show that the operators multiplication by each of the different nth roots of unity constitute a group. To what geometric group does the set correspond faithfully?

7.2 By a systematic procedure, determine the group of covering operations that transform each regular polyhedron into itself.

7.3 If (a) rotation C_4, (b) operation S_3 transforms a small region into an equivalent physical system, what other reorientations must be symmetry operations for the system?

7.4 What group of operations leaves (a) $zR(r)$, (b) $(x^2 - y^2)R(r)$, (c) $xyzR(r)$ unchanged?

7.5 Show why a crystal cannot contain C_5, C_7, and C_8 axes of symmetry.

7.6 Identify the lattice for which

$$\mathbf{a}_1 = \frac{d}{2}(\mathbf{i} + \sqrt{3}\,\mathbf{j}),$$

$$\mathbf{a}_2 = \frac{d}{2}(-\mathbf{i} + \sqrt{3}\,\mathbf{j}),$$

$$\mathbf{a}_3 = e\mathbf{k}.$$

Find the vectors bounding a unit cell of the reciprocal lattice and identify it.

7.7 Use quark theory to show what else must result if a proton and a K^+ come from a collision of a proton with a proton.

7.8 From how the quark spins may be aligned with respect to a unique direction, determine the nature of the two basic baryon multiplets.

7.9 Construct all allowed Young tables for two quarks and an antiquark combined symmetrically. Arrange them to indicate the nature of the multiplet.

7.10 Use Young tables to determine the number of states in the (2, 2) multiplet.

7.11 What are alternate symbols for $3/m$, $\bar{2}$, $\bar{4}$, $\bar{6}$? What group contains operations $\bar{4}$, 3, and m?

7.12 To what groups do the covering operations for (a) NH_3 (ammonia), (b) C_6H_6 (benzene), (c) PCl_5 (phosphorus pentachloride), (d) C_2H_6 (ethane) belong?

7.13 Prove that the transformations

$$x' = x, \qquad x' = \frac{1}{1 - x}, \qquad x' = \frac{x - 1}{x},$$

form a group. Find the symmetry group to which the set corresponds faithfully.

7.14 What group of operations leaves

$$A + B(x^2 + y^2 + z^2) + Cz^2 + Di[(x + iy)^3 - (x - iy)^3]$$

unchanged?

7.15 If a crystal lattice possesses sixfold symmetry about an axis and if **a** is one of the smallest nonvanishing allowed translations perpendicular to the axis, what other translations of the same magnitude also convert the crystal into itself? Show that any pair of these add to give allowed translations.

7.16 Construct the lattice that is reciprocal to the primitive lattice for which

$$\begin{aligned} \mathbf{a}_1 &= d\mathbf{i}, \\ \mathbf{a}_2 &= e\mathbf{i} + f\mathbf{j}, \\ \mathbf{a}_3 &= g\mathbf{k}. \end{aligned}$$

Classify both lattices.

7.17 Use quark theory to determine four different collision reactions that lead to a Σ^0 and a K^0. Employ only common particles for reactants.

7.18 Explain how a triplet of particles can arise from combinations of three occupied fundamental states.

7.19 Construct all allowed Young tables for the (1, 2) multiplet.

7.20 Determine the number of states at each lattice point of the multiplet

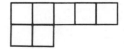

REFERENCES

BOOKS

Cotton, F. A. *Chemical Applications of Group Theory*, pp. 6–49. Interscience Publishers, New York, 1963.

Eisenman, R. L. *Matrix Vector Analysis*, pp. 247–258. McGraw-Hill Book Company, New York, 1963.

Kittel, C. *Introduction to Solid State Physics*, 3rd ed., pp. 3–76. John Wiley & Sons, Inc., New York, 1966.

Lipkin, H. J. *Lie Groups for Pedestrians*, 2nd ed., pp. 33–56, 96–102, 123–136. North-Holland Publishing Co., Amsterdam, 1966.

ARTICLES

Atoji, M. Graphical Representations of Magnetic Space Groups. *Am. J. Phys.*, **33**, 212 (1965).

Duley, W. W. Symmetry and Multipole Expansions in Electrostatics. *Am. J. Phys.*, **39**, 1087 (1971).

Gatto, R. Symmetries in Fundamental Interactions. *Nuovo Cimento, Suppl.*, **4**, 414 (1966).

McMillan, J. A. Stereographic Projections of the Colored Crystallographic Point Groups. *Am. J. Phys.*, **35**, 1049 (1967).

McMillan, J. A. Symmetry and Properties of Crystals: Theorem of Group Intersection. *Am. J. Phys.*, **37**, 793 (1969).

Metzger, E. Musical Scales and Algebraic Groups. *Am. J. Phys.*, **35**, 441 (1967).

Park, D. Resource Letter SP-1 on Symmetry in Physics. *Am. J. Phys.*, **36**, 577 (1968).

Zeldovich, Ya. B. Classification of Elementary Particles and Quarks "For the Layman." *Soviet Phys. Uspekhi.*, **8**, 489 (1965).

8 / *Matrices for Describing a Group*

8.1   Constructing matrix representations

Symmetry in a given region is manifested by the existence of reorientations and/or conversions that change the region into equivalent systems. Any closed set of the symmetry operations constitutes a group.

Such a closed set is characterized by how its elements combine with each other. Now, explicit mathematical operators that combine as the elements do can be found for any given group. A complete set of such operators corresponds to the group under consideration and so *represents* it.

In many systems, a certain property or attribute is measured through expressions that are *linearly* mixed by each covering operation. By trial, coefficients in the linear series can be determined and then used to make up a matrix representing the operation.

Elements of such matrices, and the traces constructed from them, obey relationships to be determined in this chapter. These relationships can be used to break a small movement down into modes with elementary symmetry properties. Since modes belonging to different symmetry species do not interact, this analysis can greatly simplify secular equations. The relationships also help correlate states before and after a change in symmetry.

8.2   Groups and bases

Before developing this theory, we need to recall and restate the fundamental definitions.

A *group* is a set of elements $R, S, T, \ldots$ with a prescription for combining pairs that meets the two conditions:

1. *Associativity*. The way elements are associated in forming a result is unimportant:

$$(RS)T = R(ST) = RST. \tag{8-1}$$

Multiple combinations depend only on the elements and the order in which they are arranged.

2. *Solvability.* Equations containing unknowns X, Y, Z

$$RS = X, \tag{8-2}$$

$$RY = S, \tag{8-3}$$

$$ZR = S, \tag{8-4}$$

are solvable with elements in the set.

Since X is in the group as long as R and S are, equation (8-2) shows that the set is *closed* under the combining operation (often called multiplication). Since S may be the same as R, (8-3) and (8-4) imply the existence of an *identity*. If the set contains the identity, S may be selected as this element, and the equations also imply the existence of an *inverse*. The number of distinct elements in the group is called the *order* of the group.

Operators that combine as the group elements combine form a *representation* of the group. Operands that the operators mix linearly constitute a *basis* for the representation.

Each geometric group is represented by innumerable sets of matrices. The bases for these include

  (a) vectors,
  (b) products of vectors,
  (c) functions,
  (d) products of functions.

In the following discussion, the bases from a given set need only be distinguished from each other. Therefore, each is marked by a single index. To remind ourselves that an operand need not be a scalar, we set it in boldface type.

EXAMPLE 8.1 Identify the finite groups that systems of particles can generate.

The particles may be arranged so that the pertinent symmetry operations form one or more of the following:

1. A group containing only the identity. Since the E operation corresponds to rotation by $2\pi/1$, it is designated C_1 and the group is called $\mathbf{C_1}$.

2. A group containing only rotation-reflections. When n in the S_n is 1, the group contains E, σ_h, and is called $\mathbf{C_s}$; when n is 2, the group contains E, i, and is called $\mathbf{C_i}$. For larger n's, it is labeled $\mathbf{S_n}$.

3. Simple rotation groups. Each of these contains rotation C_n and its integral powers.

 (a) If the group contains no other elements, it is called the $\mathbf{C_n}$ group. Since $C_n{}^n$ and larger rotations merely duplicate the effects of smaller rotations, the group consists of n distinct elements.

 (b) If reflections in n planes containing the principal axis are also present, v is added to the subscript. Note that the angle between successive σ_v planes is π/n radians. $\mathbf{C_{2v}}$, $\mathbf{C_{3v}}$, $\mathbf{C_{4v}}$, $\mathbf{C_{5v}}$, $\mathbf{C_{6v}}$ are common.

 (c) If instead, reflection in a plane perpendicular to the principal axis is present, h is added to the subscript. Examples include $\mathbf{C_{1h}}$, $\mathbf{C_{2h}}$, $\mathbf{C_{3h}}$, $\mathbf{C_{4h}}$, $\mathbf{C_{5h}}$, $\mathbf{C_{6h}}$.

4. Dihedral groups. Besides C_n and its integral powers, each of these contains twofold rotations about n axes perpendicular to the principal axis (C_2', C_2'', ...).

(a) If the group contains no other elements, it is identified as $\mathbf{D}_n$.

(b) If reflections in n planes containing the principal axis are also present, d is added to the subscript. $\mathbf{D}_{2d}$ and $\mathbf{D}_{3d}$ are common.

(c) If the dihedral group includes reflection in a horizontal plane passing through the twofold axes, h is added to the subscript. Some σ_v's are of necessity also present. $\mathbf{D}_{2h}$, $\mathbf{D}_{3h}$, $\mathbf{D}_{4h}$, $\mathbf{D}_{5h}$, $\mathbf{D}_{6h}$ are common.

5. Groups of higher symmetry. Each of these contains more than one C_n operation with $n > 2$.

(a) The $\mathbf{T}$ group contains the C_n operations that rotate a regular tetrahedron into itself.

(b) If we include the other covering operations that also send a regular tetrahedron into itself, we get the $\mathbf{T}_d$ group.

(c) The $\mathbf{O}$ group contains the C_n operations that rotate a regular octahedron, or cube, into itself.

(d) If we include the other covering operations that also send a regular octahedron into itself, we get the $\mathbf{O}_h$ group.

(e) The $\mathbf{I}$ group consists of the C_n operations that rotate a regular icosahedron, or regular dodecahedron, into itself.

(f) If we include the other covering operations that also send a regular icosahedron into itself, we get the $\mathbf{I}_h$ group.

EXAMPLE 8.2 Identify the infinite groups that are generated by arrangements of particles and/or continua.

Matter may be placed so that the symmetry operations make up one or more of the following:

1. Groups containing all two-dimensional rotations. Each of these contains all rotations that change a cylindrically symmetric body into itself.

(a) If there are no other elements in the group, it is designated $\mathbf{C}_\infty$.

(b) If reflection in any vertical plane containing the axis is also a symmetry operation, the group is labeled $\mathbf{C}_{\infty v}$.

(c) Inclusion of the inversion operation changes $\mathbf{C}_{\infty v}$ to the dihedral group $\mathbf{D}_{\infty h}$.

2. Groups containing all three-dimensional rotations. Each of these contains all rotations that change a spherically symmetric body into itself.

(a) If there are no other elements in the group, it is known as $\mathbf{N}_{3p}$.

(b) If the reflections and inversions that also change a sphere into itself are included, the group is called $\mathbf{N}_3$.

8.3 How the operands generate representations

Bases consist of similar, but independent, expressions

$$\mathbf{e}_1, \mathbf{e}_2, \ldots, \mathbf{e}_r \tag{8-5}$$

associated with equivalent parts, orientations, or aspects of a physical system. They may be vectors, functions, or products of these.

As long as each symmetry operation

$$R, S, SR, \ldots \tag{8-6}$$

transforms each base linearly, we can write

$$R\mathbf{e}_k = \mathbf{e}_j D_{jk}(R), \tag{8-7}$$

$$S\mathbf{e}_j = \mathbf{e}_i D_{ij}(S), \tag{8-8}$$

and

$$SR\mathbf{e}_k = \mathbf{e}_i D_{ik}(SR). \tag{8-9}$$

As before, each term in which an index appears more than once is summed over the allowed values of the index.

Introducing substitutions (8-7) and (8-8) into the left side of (8-9) leads to an alternate form:

$$SR\mathbf{e}_k = S\mathbf{e}_j D_{jk}(R) = \mathbf{e}_i D_{ij}(S) D_{jk}(R). \tag{8-10}$$

Setting it equal to the right side of (8-9) yields

$$D_{ij}(S) D_{jk}(R) = D_{ik}(SR) \tag{8-11}$$

and

$$\mathbf{D}(S)\mathbf{D}(R) = \mathbf{D}(SR). \tag{8-12}$$

The $r \times r$ matrices constructed from the coefficients combine as elements of the group do and form a representation. The number r of independent bases is called the *dimensionality* of the representation.

A typical coefficient $D_{jk}(R)$ is a number depending only on operation R and on the bases $\mathbf{e}_j$ and $\mathbf{e}_k$. The subscript on $\mathbf{e}_j$ labels the *row* of $\mathbf{D}(R)$ by which the base is multiplied in transformation equations (8-7). We say that operand $\mathbf{e}_j$ belongs to the *j*th row of the representation.

8.4   Suitable operands for a vibrating system

Small displacements, or arrays of unit vectors along which they may be resolved, and linear combinations of these, can serve as bases.

Consider a system of particles oscillating about N equilibrium points. Place an origin at each equilibrium position and draw a vector to the instantaneous location of the corresponding particle. Let each displacement vector be resolved along three mutually perpendicular unit vectors also drawn from the equilibrium point.

A suitable set of these unit vectors is then labeled

$$\mathbf{u}_1, \ldots, \mathbf{u}_{3j-2}, \mathbf{u}_{3j-1}, \mathbf{u}_{3j}, \ldots, \mathbf{u}_{3N} \tag{8-13}$$

where j is the number of the pertinent origin (or particle). Let these $3N$ $\mathbf{u}_k$'s be arranged perpendicular to each other in a $3N$-dimensional mathematical space.

We can then compose $3N$ new base vectors,

$$\mathbf{e}_m = \mathbf{u}_k a_{km}, \tag{8-14}$$

representing independent cooperative movements of the particles. Note that the $\mathbf{e}_m$'s are linear combinations of the $\mathbf{u}_k$'s.

Because symmetry operations permute at least some of the equilibrium positions, they permute appropriately arranged $\mathbf{u}_k$'s and so transform the $\mathbf{e}_m$'s *linearly*. The $\mathbf{e}_m$'s are consequently suitable bases for representations.

EXAMPLE 8.3 Set up appropriate unit vectors for oscillations of the particles in figure 8.1.

On each mass m_j, erect a unit vector $\mathbf{r}_j$ pointing inward toward the center of mass, a unit vector $\mathbf{t}_j$ perpendicular to $\mathbf{r}_j$, pointing counterclockwise in the plane of the equilibrium positions, and a unit vector $\mathbf{s}_j$ that is equal to $\mathbf{t}_j \times \mathbf{r}_j$.

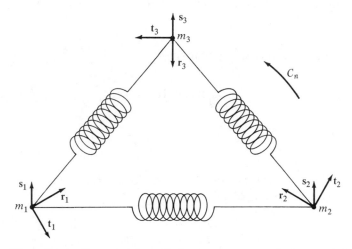

FIGURE 8.1 Three interacting particles. If the masses are equivalent and the springs are equivalent, the system is regulated by the symmetry group $\mathbf{D}_{3h}$, and its subgroups $\mathbf{C}_{3h}$, $\mathbf{D}_3$, $\mathbf{C}_{3v}$, $\mathbf{C}_3$, $\mathbf{C}_{2v}$, $\mathbf{C}_2$, and $\mathbf{C}_s$.

EXAMPLE 8.4 Show that the combination

$$\mathbf{e}_3 = \frac{1}{\sqrt{3}} (\mathbf{r}_1 + \mathbf{r}_2 + \mathbf{r}_3)$$

of vectors from figure 8.1 is a complete basis for the pertinent $\mathbf{C}_3$ group.

Applying each symmetry operation of $\mathbf{C}_3$ to $\mathbf{r}_j$ produces the following effects

$$E\mathbf{r}_j = \mathbf{r}_j,$$
$$C_3\mathbf{r}_j = \mathbf{r}_{j+1},$$
$$C_3{}^2\mathbf{r}_j = \mathbf{r}_{j+2},$$

if the numbers 1, 2, 3 form a cyclic system.

Such permutations of the $\mathbf{r}_j$'s merely convert $\mathbf{e}_3$ into itself:

$$Ee_3 = E \frac{1}{\sqrt{3}} (\mathbf{r}_1 + \mathbf{r}_2 + \mathbf{r}_3) = \frac{1}{\sqrt{3}} (\mathbf{r}_1 + \mathbf{r}_2 + \mathbf{r}_3) = \mathbf{e}_3,$$

$$C_3 \mathbf{e}_3 = C_3 \frac{1}{\sqrt{3}} (\mathbf{r}_1 + \mathbf{r}_2 + \mathbf{r}_3) = \frac{1}{\sqrt{3}} (\mathbf{r}_2 + \mathbf{r}_3 + \mathbf{r}_1) = \mathbf{e}_3,$$

$$C_3{}^2 \mathbf{e}_3 = C_3{}^2 \frac{1}{\sqrt{3}} (\mathbf{r}_1 + \mathbf{r}_2 + \mathbf{r}_3) = \frac{1}{\sqrt{3}} (\mathbf{r}_3 + \mathbf{r}_1 + \mathbf{r}_2) = \mathbf{e}_3.$$

Therefore, no other combination of the $\mathbf{r}_j$'s is necessarily involved with $\mathbf{e}_3$ as a base vector.

EXAMPLE 8.5   Show that the two combinations

$$\mathbf{e}_1 = \frac{\sqrt{2}}{\sqrt{3}} (\mathbf{r}_1 - \tfrac{1}{2}\mathbf{r}_2 - \tfrac{1}{2}\mathbf{r}_3),$$

$$\mathbf{e}_2 = \frac{1}{\sqrt{2}} (\mathbf{r}_3 - \mathbf{r}_2),$$

of vectors from figure 8.1 form a complete basis for the $\mathbf{C}_3$ group.
    Using the formulas in example 8.4,

$$Ee_1 = \mathbf{e}_1,$$
$$Ee_2 = \mathbf{e}_2,$$

$$C_3 \mathbf{e}_1 = C_3 \frac{\sqrt{2}}{\sqrt{3}} (\mathbf{r}_1 - \tfrac{1}{2}\mathbf{r}_2 - \tfrac{1}{2}\mathbf{r}_3) = \frac{\sqrt{2}}{\sqrt{3}} (\mathbf{r}_2 - \tfrac{1}{2}\mathbf{r}_3 - \tfrac{1}{2}\mathbf{r}_1)$$

$$= -\frac{1}{2} \frac{\sqrt{2}}{\sqrt{3}} (\mathbf{r}_1 - \tfrac{1}{2}\mathbf{r}_2 - \tfrac{1}{2}\mathbf{r}_3) - \frac{\sqrt{3}}{2} \frac{1}{\sqrt{2}} (\mathbf{r}_3 - \mathbf{r}_2)$$

$$= -\frac{1}{2} \mathbf{e}_1 - \frac{\sqrt{3}}{3} \mathbf{e}_2,$$

$$C_3 \mathbf{e}_2 = C_3 \frac{1}{\sqrt{2}} (\mathbf{r}_3 - \mathbf{r}_2) = \frac{1}{\sqrt{2}} (\mathbf{r}_1 - \mathbf{r}_3)$$

$$= +\frac{\sqrt{3}}{2} \frac{\sqrt{2}}{\sqrt{3}} (\mathbf{r}_1 - \tfrac{1}{2}\mathbf{r}_2 - \tfrac{1}{2}\mathbf{r}_3) - \frac{1}{2} \frac{1}{\sqrt{2}} (\mathbf{r}_3 - \mathbf{r}_2)$$

$$= +\frac{\sqrt{3}}{2} \mathbf{e}_1 - \tfrac{1}{2}\mathbf{e}_2.$$

Likewise,

$$C_3{}^2\mathbf{e}_1 = -\tfrac{1}{2}\mathbf{e}_1 + \frac{\sqrt{3}}{2}\,\mathbf{e}_2,$$

$$C_3{}^2\mathbf{e}_2 = -\frac{\sqrt{3}}{2}\,\mathbf{e}_1 - \tfrac{1}{2}\mathbf{e}_2.$$

Each operation of group $\mathbf{C}_3$ converts bases $\mathbf{e}_1$ and $\mathbf{e}_2$ into a mixture of the two. Nothing is left over. We may conveniently summarize the results in the following matrix equations:

$$E\begin{pmatrix}\mathbf{e}_1\\\mathbf{e}_2\end{pmatrix} = \begin{pmatrix}1 & 0\\0 & 1\end{pmatrix}\begin{pmatrix}\mathbf{e}_1\\\mathbf{e}_2\end{pmatrix},$$

$$C_3\begin{pmatrix}\mathbf{e}_1\\\mathbf{e}_2\end{pmatrix} = \begin{pmatrix}-\tfrac{1}{2} & -\dfrac{\sqrt{3}}{2}\\[2mm]\dfrac{\sqrt{3}}{2} & -\tfrac{1}{2}\end{pmatrix}\begin{pmatrix}\mathbf{e}_1\\\mathbf{e}_2\end{pmatrix},$$

$$C_3{}^2\begin{pmatrix}\mathbf{e}_1\\\mathbf{e}_2\end{pmatrix} = \begin{pmatrix}-\tfrac{1}{2} & \dfrac{\sqrt{3}}{2}\\[2mm]-\dfrac{\sqrt{3}}{2} & -\tfrac{1}{2}\end{pmatrix}\begin{pmatrix}\mathbf{e}_1\\\mathbf{e}_2\end{pmatrix}.$$

EXAMPLE 8.6 Prove that the combination

$$\mathbf{e}_4 = \frac{1}{\sqrt{3}}\,(\mathbf{r}_1 + \omega\mathbf{r}_2 + \omega^2\mathbf{r}_3)$$

of vectors from figure 8.1 is a complete basis for a representation of the $\mathbf{C}_3$ group. Here ω is exp $(2\pi i/3)$.

Using the formulas from example 8.4 to express the effect of each symmetry operation on each term in $\mathbf{e}_4$ leads to

$$E\mathbf{e}_4 = \mathbf{e}_4,$$

$$C_3\mathbf{e}_4 = \frac{1}{\sqrt{3}}\,(\mathbf{r}_2 + \omega\mathbf{r}_3 + \omega^2\mathbf{r}_1) = \omega^2\,\frac{1}{\sqrt{3}}\,(\omega\mathbf{r}_2 + \omega^2\mathbf{r}_3 + \mathbf{r}_1)$$

$$= \omega^2\mathbf{e}_4,$$

$$C_3{}^2\mathbf{e}_4 = \frac{1}{\sqrt{3}}\,(\mathbf{r}_3 + \omega\mathbf{r}_1 + \omega^2\mathbf{r}_2) = \omega\,\frac{1}{\sqrt{3}}\,(\omega^2\mathbf{r}_3 + \mathbf{r}_1 + \omega\mathbf{r}_2)$$

$$= \omega\mathbf{e}_4.$$

Since each operation transforms $\mathbf{e}_4$ into a constant times $\mathbf{e}_4$, no other vector is needed to complete the basis. Another mixture of $\mathbf{r}_1$, $\mathbf{r}_2$, $\mathbf{r}_3$ that is transformed into a power of ω times itself by each operation of C_3 is

$$\mathbf{e}_5 = \frac{1}{\sqrt{3}} (\mathbf{r}_1 + \omega^2 \mathbf{r}_2 + \omega \mathbf{r}_3).$$

8.5   Reducibility and the lemmas of Schur

To a given expression $\mathbf{u}_1$ associated with a part of a physical system, we must add similar expressions $\mathbf{u}_2$, $\mathbf{u}_3$, ..., $\mathbf{u}_r$ associated with all equivalent parts in order to produce a set that is always transformed within itself by symmetry operations of the group. Such a set is said to be complete.

When the expressions are chosen so that these transformations are linear, the set generates a representation. When n linear combinations of the $\mathbf{u}_k$'s also form a complete set, with $n < r$, the original representation is *reducible* to one based on this set and the residue. Whenever such smaller complete sets cannot be formed from a given representation, the representation is said to be *irreducible*.

Only certain kinds of irreducible representations can be constructed for any given geometric group. Since these types of representations characterize the group, their study is rewarding.

Consider a set of independent operands

$$\mathbf{u}_1, \mathbf{u}_2, \ldots, \mathbf{u}_r \tag{8-15}$$

that each operation R of the group transforms linearly:

$$R\mathbf{u}_k = \mathbf{u}_j D_{jk}(R) \qquad \text{where} \quad j = 1, 2, \ldots, r. \tag{8-16}$$

If the representation is reducible, there exist n operands

$$\mathbf{e}_m = \mathbf{u}_k a_{km} \qquad \text{where} \quad k = 1, 2, \ldots, r \tag{8-17}$$

with $n < r$ such that

$$R\mathbf{e}_m = \mathbf{e}_l D_{lm}'(R) \qquad \text{where} \quad l = 1, 2, \ldots, n \tag{8-18}$$

for each R in the group.

Operate on each side of (8-17) with R and introduce expansion (8-16) into the result:

$$R\mathbf{e}_m = R\mathbf{u}_k a_{km} = \mathbf{u}_j D_{jk}(R) a_{km}. \tag{8-19}$$

Also expand (8-18) with (8-17):

$$R\mathbf{e}_m = \mathbf{e}_l D_{lm}'(R) = \mathbf{u}_j a_{jl} D_{lm}'(R). \tag{8-20}$$

Thus, we obtain two different expressions for the same thing:

$$\mathbf{u}_j D_{jk}(R) a_{km} = \mathbf{u}_j a_{jl} D_{lm}'(R). \tag{8-21}$$

Equating coefficients of $\mathbf{u}_j$ then yields

$$D_{jk}(R)a_{km} = a_{jl}D_{lm}'(R) \tag{8-22}$$

or

$$\mathbf{D}(R)\mathbf{A} = \mathbf{A}\mathbf{D}'(R). \tag{8-23}$$

Note that $\mathbf{A}$ contains r rows and n columns, while $\mathbf{D}$ is an $r \times r$ matrix and $\mathbf{D}'$ an $n \times n$ matrix. $\mathbf{A}$ does not depend on R; it is constant.

This argument can be reversed. Suppose that each matrix $\mathbf{D}(R)$ of an r-dimensional representation times one $r \times n$ matrix $\mathbf{A}$ equals $\mathbf{A}$ times the corresponding matrix $\mathbf{D}'(R)$ of an n-dimensional representation:

$$\mathbf{D}(R)\mathbf{A} = \mathbf{A}\mathbf{D}'(R). \tag{8-24}$$

We then have

$$D_{jk}(R)a_{km} = a_{jl}D_{lm}'(R) \qquad \text{where} \quad \begin{array}{l} k \text{ runs from 1 to } r, \\ l \text{ runs from 1 to } n. \end{array} \tag{8-25}$$

Let us multiply both sides by the jth base for the $\mathbf{D}(R)$'s and sum from 1 to r:

$$\mathbf{u}_j D_{jk}(R)a_{km} = \mathbf{u}_j a_{jl}D_{lm}'(R). \tag{8-26}$$

Then reduce with relationship (8-16):

$$R\mathbf{u}_k a_{km} = \mathbf{u}_k a_{kl}D_{lm}'(R). \tag{8-27}$$

On the right, the dummy index j has been changed to k. This substitution does not affect the result because k denotes the same integers as j.

Equations (8-27) shows that the operands

$$\mathbf{u}_k a_{km} = \mathbf{e}_m \tag{8-28}$$

form a basis for the primed representation $\mathbf{D}'(R)$, $\mathbf{D}'(S)$, ... as long as matrix $\mathbf{A}$ is not zero. Then if $n < r$, the unprimed representation $\mathbf{D}(R)$, $\mathbf{D}(S)$, ... is reducible. Note particularly the following situations:

1. If the $\mathbf{D}$ and $\mathbf{D}'$ representations are irreducible and $n < r$, the only matrix $\mathbf{A}$ satisfying (8-24) is zero. This statement forms part of *Schur's first lemma*.

When $n = r$, the $\mathbf{D}$ and $\mathbf{D}'$ matrices have the same dimensionality. And if the $\mathbf{D}$ representation is irreducible, the $r\mathbf{e}_m$'s must be linearly independent (if any one of them were a linear function of the others, the operations R, S, ... would mix only s independent functions where $s < n$ and $\mathbf{D}$ would be reducible, contrary to assumption). Therefore, (8-28) can be inverted. The $\mathbf{u}_k$'s can be expressed in terms of the $\mathbf{e}_m$'s and $\mathbf{A}^{-1}$ exists, except when $\mathbf{A} = 0$.

2. If the $\mathbf{D}$ representation is irreducible and $n = r$, *either* (8-24) can be rewritten in the form

$$\mathbf{A}^{-1}\mathbf{D}(R)\mathbf{A} = \mathbf{D}'(R) \tag{8-29}$$

that is, $\mathbf{D}$ and $\mathbf{D}'$ are related by a similarity transformation, *or* $\mathbf{A}$ is zero. This is the second part of *Schur's first lemma*.

Finally, suppose that one matrix $\mathbf{A}$ commutes with all the matrices in an irreducible representation of dimensionality r:

$$\mathbf{D}(R)\mathbf{A} = \mathbf{AD}(R). \qquad (8\text{-}30)$$

What can we then deduce concerning the nature of $\mathbf{A}$?

Construct the eigenvalue equation

$$\mathbf{Av} = \lambda\mathbf{v} \qquad (8\text{-}31)$$

with some eigenvector $\mathbf{v}$. Act on it with $\mathbf{D}(R)$,

$$\mathbf{D}(R)\mathbf{Av} = \mathbf{D}(R)\lambda\mathbf{v}, \qquad (8\text{-}32)$$

and commute the operators on each side,

$$\mathbf{AD}(R)\mathbf{v} = \lambda\mathbf{D}(R)\mathbf{v}. \qquad (8\text{-}33)$$

We find that each $\mathbf{D}(R)\mathbf{v}$ is also an eigenvector.

Since $\mathbf{A}$ is an $r \times r$ matrix, it has r independent eigenvectors. If all of them are obtained with the forms $\mathbf{D}(R)\mathbf{v}$, then only one eigenvalue exists. But if all are not obtainable in this way, operations of the group, through $\mathbf{D}(R)$ acting on $\mathbf{v}$, would mix only n vectors where $n < r$. Then $\mathbf{D}$ would be reducible contrary to assumption.

Any $\mathbf{A}$ that commutes with all matrices in an irreducible representation has r equal eigenvalues. Substituting these into equation (4-105),

$$\mathbf{U}^{-1}\mathbf{AU} = \begin{pmatrix} \lambda & 0 & \cdot & 0 \\ 0 & \lambda & \cdot & 0 \\ \cdot & \cdot & \cdot & \cdot \\ 0 & 0 & \cdot & \lambda \end{pmatrix} = \lambda\mathbf{E}, \qquad (8\text{-}34)$$

and multiplying the result from the left by $\mathbf{U}$, from the right by $\mathbf{U}^{-1}$ gives

$$\mathbf{A} = \mathbf{U}\lambda\mathbf{EU}^{-1} = \lambda\mathbf{EUU}^{-1} = \lambda\mathbf{E}. \qquad (8\text{-}35)$$

3. If $\mathbf{D}$ and $\mathbf{D}'$ are identical irreducible representations in (8-24), then

$$\mathbf{A} = \lambda\mathbf{E}. \qquad (8\text{-}36)$$

That is, $\mathbf{A}$ is a number times the unit matrix. This statement constitutes *Schur's second lemma*.

EXAMPLE 8.7   One irreducible representation of the $\mathbf{C}_{3v}$ group consists of the matrices

$$\begin{pmatrix} 1 & 0 \\ 0 & 1 \end{pmatrix}, \quad \begin{pmatrix} -\frac{1}{2} & -\frac{\sqrt{3}}{2} \\ \frac{\sqrt{3}}{2} & -\frac{1}{2} \end{pmatrix}, \quad \begin{pmatrix} -\frac{1}{2} & \frac{\sqrt{3}}{2} \\ -\frac{\sqrt{3}}{2} & -\frac{1}{2} \end{pmatrix},$$

$$\begin{pmatrix} 1 & 0 \\ 0 & -1 \end{pmatrix}, \quad \begin{pmatrix} -\frac{1}{2} & -\frac{\sqrt{3}}{2} \\ -\frac{\sqrt{3}}{2} & \frac{1}{2} \end{pmatrix}, \quad \begin{pmatrix} -\frac{1}{2} & \frac{\sqrt{3}}{2} \\ \frac{\sqrt{3}}{2} & \frac{1}{2} \end{pmatrix}.$$

Show that the 6-dimensional vector constructed from the 12 elements is $\sqrt{3}$ units long and is perpendicular to the vector constructed from the 21 elements.

Multiply each 12 element by a different base vector in the 6-dimensional space and add:

$$\mathbf{A} = 0\mathbf{e}_1 - \frac{\sqrt{3}}{2}\mathbf{e}_2 + \frac{\sqrt{3}}{2}\mathbf{e}_3 + 0\mathbf{e}_4 - \frac{\sqrt{3}}{2}\mathbf{e}_5 + \frac{\sqrt{3}}{2}\mathbf{e}_6.$$

Similarly treat each 21 element:

$$\mathbf{B} = 0\mathbf{e}_1 + \frac{\sqrt{3}}{2}\mathbf{e}_2 - \frac{\sqrt{3}}{2}\mathbf{e}_3 + 0\mathbf{e}_4 - \frac{\sqrt{3}}{2}\mathbf{e}_5 + \frac{\sqrt{3}}{2}\mathbf{e}_6.$$

Then dot multiply $\mathbf{A}$ with itself:

$$\mathbf{A} \cdot \mathbf{A} = 0 + \tfrac{3}{4} + \tfrac{3}{4} + 0 + \tfrac{3}{4} + \tfrac{3}{4} = 3.$$

Dot multiply $\mathbf{A}$ with $\mathbf{B}$:

$$\mathbf{A} \cdot \mathbf{B} = 0 - \tfrac{3}{4} - \tfrac{3}{4} + 0 + \tfrac{3}{4} + \tfrac{3}{4} = 0.$$

Vectors can likewise be set up from the 11 elements and the 22 elements. Each of the vectors has magnitude $\sqrt{3}$ and is perpendicular to the others. The number 3 equals the number of elements in the group, 6, divided by the dimensionality of the representation, 2.

8.6   Restrictions on matrix elements of representations

Example 8.7 showed how corresponding elements of an irreducible matrix representation produce vectors of definite magnitude mutually orthogonal to each other. Similar relationships exist among vectors from different irreducible representations. The pertinent formulas come from applying Schur's lemmas in the following manner:

We consider a group of symmetry operations

$$S = R_1, R_2, \ldots, R_g \tag{8-37}$$

represented irreducibly by the matrices

$$\mathbf{D}(R_1), \mathbf{D}(R_2), \ldots, \mathbf{D}(R_g). \tag{8-38}$$

The sum

$$\mathbf{A} = \mathbf{D}(R_i)\mathbf{X}\mathbf{D}(R_i^{-1}) \tag{8-39}$$

containing the arbitrary conformable matrix $\mathbf{X}$, and the product

$$\mathbf{D}(S)\mathbf{A} = \mathbf{D}(S)\mathbf{D}(R_i)\mathbf{X}\mathbf{D}(R_i^{-1}) \tag{8-40}$$

are constructed. The summation proceeds over all operations of the group:

$$i = 1, 2, \ldots, g. \tag{8-41}$$

We multiply the right side of (8-40) by the unit matrix $\mathbf{D}(S^{-1})\mathbf{D}(S)$:

$$\mathbf{D}(S)\mathbf{A} = \mathbf{D}(S)\mathbf{D}(R_i)\mathbf{X}\mathbf{D}(R_i^{-1})\mathbf{D}(S^{-1})\mathbf{D}(S). \tag{8-42}$$

Equation (8-12) is introduced, together with the relationship

$$R_i^{-1}S^{-1} = (SR_i)^{-1}, \tag{8-43}$$

derived as in section 4.5, to cause the two factors before $\mathbf{X}$ and the two after $\mathbf{X}$ to combine:

$$\mathbf{D}(S)\mathbf{A} = \mathbf{D}(SR_i)\mathbf{X}\mathbf{D}[(SR_i)^{-1}]\mathbf{D}(S). \tag{8-44}$$

Since the summation is over all operations of the group, the same terms are obtained, in a different order, if SR_i is replaced with R_i. Then introducing $\mathbf{A}$ leads to

$$\mathbf{D}(S)\mathbf{A} = \mathbf{D}(R_i)\mathbf{X}\mathbf{D}(R_i^{-1})\mathbf{D}(S) = \mathbf{A}\mathbf{D}(S) \tag{8-45}$$

the equation to which Schur's second lemma applies; so

$$\mathbf{A} = \lambda\mathbf{E}. \tag{8-46}$$

The number λ depends on how $\mathbf{X}$ is chosen.

To obtain properties of the vectors constructed from different elements of $\mathbf{D}(R_i)$, we let

$$X_{lm} = 1 \tag{8-47}$$

and all other elements of $\mathbf{X}$ be zero. The corresponding lambda is labeled λ_{lm}. Equations (8-39) and (8-46) then combine to yield

$$A_{jk} = D_{jl}(R_i)D_{mk}(R_i^{-1}) = \lambda_{lm}\delta_{jk} \tag{8-48}$$

since the jkth element of $\mathbf{E}$ is the Kronecker delta δ_{jk}.

Let $j = k$ and sum over the dimensionality n:

$$D_{kl}(R_i)D_{mk}(R_i^{-1}) = n\lambda_{lm}. \tag{8-49}$$

Transpose the factors on the left side, introduce (8-12), and reduce:

$$D_{mk}(R_i^{-1})D_{kl}(R_i) = D_{ml}(R_i^{-1}R_i) = gD_{ml}(E) = g\delta_{ml}. \tag{8-50}$$

Then equate the result to the right side of (8-49) and rearrange to get

$$\lambda_{lm} = \frac{g}{n}\,\delta_{ml}. \tag{8-51}$$

Substituting this λ_{lm} into the right side of (8-48) yields

$$D_{jl}(R_i)D_{mk}(R_i^{-1}) = \frac{g}{n}\,\delta_{ml}\delta_{jk}. \tag{8-52}$$

The scalar product of the vector constructed from the jl elements of an irreducible representation with the vector from the mk elements of the inverse matrices from the same representation equals zero unless l is m and j is k. Then it equals the order of the group g divided by the dimensionality n.

The second representation may be made different from the first. Thus, we may start with the two irreducible representations

$$\mathbf{D}^1(R_1), \mathbf{D}^1(R_2), \ldots, \mathbf{D}^1(R_g) \tag{8-53}$$

$$\mathbf{D}^2(R_1), \mathbf{D}^2(R_2), \ldots, \mathbf{D}^2(R_g) \tag{8-54}$$

with the superscript on each **D** identifying the representation.

As in the earlier discussion, we construct the sum

$$\mathbf{A} = \mathbf{D}^2(R_i)\mathbf{X}\mathbf{D}^1(R_i{}^{-1}) \tag{8-55}$$

and the product

$$\mathbf{D}^2(S)\mathbf{A} = \mathbf{D}^2(S)\mathbf{D}^2(R_i)\mathbf{X}\mathbf{D}^1(R_i{}^{-1}). \tag{8-56}$$

The right side is multiplied by a product equal to the unit matrix:

$$\mathbf{D}^2(S)\mathbf{A} = \mathbf{D}^2(S)\mathbf{D}^2(R_i)\mathbf{X}\mathbf{D}^1(R_i{}^{-1})\mathbf{D}^1(S^{-1})\mathbf{D}^1(S). \tag{8-57}$$

Equation (8-12) is employed to combine factors:

$$\mathbf{D}^2(S)\mathbf{A} = \mathbf{D}^2(SR_i)\mathbf{X}\mathbf{D}^1[(SR_i)^{-1}]\mathbf{D}^1(S). \tag{8-58}$$

Since the summation is over all matrices in each representation, replacing SR_i with R_i merely rearranges terms. Equation (8-58) is equivalent to

$$\mathbf{D}^2(S)\mathbf{A} = \mathbf{D}^2(R_i)\mathbf{X}\mathbf{D}^1(R_i{}^{-1})\mathbf{D}^1(S) = \mathbf{A}\mathbf{D}^1(S), \tag{8-59}$$

a form of (8-24). Consequently, Schur's first lemma applies. We have

$$\mathbf{A} = 0 \tag{8-60}$$

as long as the two irreducible representations are not linked by a similarity transformation.

As before, we let X_{lm} be 1 and all other elements in **X** be 0. Then equations (8-55) and (8-60) yield

$$A_{jk} = D_{jl}{}^2(R_i)D_{mk}{}^1(R_i{}^{-1}) = 0 \tag{8-61}$$

when the two representations are not equivalent.

8.7 Magnitudes and mutual orthogonality of the element vectors

Equation (8-61) is valid for any two different irreducible matrix representations of the group (8-37). Number 2 may be replaced by μ; number 1 by ν. Then the equation is combined with (8-52) to get

$$D_{jl}{}^\mu(R_i)D_{mk}{}^\nu(R_i{}^{-1}) = \frac{g}{n}\,\delta_{ml}\delta_{jk}\delta_{\mu\nu} \tag{8-62}$$

where g is the order of the group and n the dimensionality of either irreducible representation μ or ν.

In section 4.13, it was shown that each symmetry operation is effected by a matrix whose inverse is its Hermitian adjoint. Since we are concerned with groups of such operations, we assume the same relationship to exist for the **D**'s. Equation (8-12) indicates that $\mathbf{D}^v(R_i^{-1})$ is the inverse of $\mathbf{D}^v(R_i)$. Setting this inverse equal to the Hermitian adjoint converts (8-62) to

$$D_{jl}{}^{\mu}(R_i)D_{km}{}^{v*}(R_i) = \frac{g}{n}\,\delta_{jk}\delta_{lm}\delta_{\mu v}. \tag{8-63}$$

The Hermitian scalar product between the vector constructed from the *jl* elements of one irreducible representation with the vector constructed from the *km* elements of another irreducible representation is zero unless the representations and the elements are the same. Then the product equals the order of the group divided by the dimensionality. This result is referred to as the *orthogonality theorem* for representations.

EXAMPLE 8.8   Formulate one vector from the characters of the representation in example 8.7 and a second one from the characters of the following representation:

$$(1),\ (1),\ (1),\ (-1),\ (-1),\ (-1).$$

Show that these both have magnitude $\sqrt{6}$ and are orthogonal to each other.

The character of a matrix representing an element of a group equals its trace. The 6-dimensional vector constructed from the traces of the matrices in example 8.7 is

$$\mathbf{A} = 2\mathbf{e}_1 - \mathbf{e}_2 - \mathbf{e}_3 + 0\mathbf{e}_4 + 0\mathbf{e}_5 + 0\mathbf{e}_6,$$

while that constructed from those above is

$$\mathbf{B} = \mathbf{e}_1 + \mathbf{e}_2 + \mathbf{e}_3 - \mathbf{e}_4 - \mathbf{e}_5 - \mathbf{e}_6.$$

The magnitude squares are given by

$$\mathbf{A} \cdot \mathbf{A} = 4 + 1 + 1 + 0 + 0 + 0 = 6,$$
$$\mathbf{B} \cdot \mathbf{B} = 1 + 1 + 1 + 1 + 1 + 1 = 6,$$

while the projection of **A** on **B** is

$$\frac{1}{B}\,\mathbf{A} \cdot \mathbf{B} = \frac{1}{\sqrt{6}}\,(2 - 1 - 1 + 0 + 0 + 0) = 0.$$

8.8   Magnitude and mutual orthogonality of the character vectors

The theory in section 4.8 and example 4.11 enables us to partition a group like (8-37) into classes. Since the matrices in a representation correspond to elements in the group, they divide up in the same way. A similarity transformation of all members in a representation does not change the algebraic properties and consequently does not change the partitioning.

The similarity transformation does alter some matrix elements, but it leaves the trace of each matrix (its character χ) unchanged. To restate theorem (8-63) in terms of this invariant property, we set $j = l$, $k = m$, and sum over the repeated indices:

$$D_{ll}{}^{\mu}(R_i)D_{mm}{}^{v*}(R_i) = \frac{g}{n}\,\delta_{ml}\delta_{lm}\delta_{\mu v}. \tag{8-64}$$

Introducing the definition of character on the left and carrying out the summations on the right lead to

$$\chi^{\mu}(R_i)\chi^{\nu*}(R_i) = g\delta_{\mu\nu}. \tag{8-65}$$

The Hermitian scalar product between vectors constructed from the characters of two irreducible representations is zero unless the representations are the same. Then it equals the number of elements in the group g.

From section 4.8, the different matrices of a representation that belong to a class all have the same character. The number of independent characters in the representation equals the number of classes c. We may consider the character for a class times the number of group elements in the class as the component of a character vector in a hypothetical space. But in this space only c mutually perpendicular vectors can be erected. Therefore, the number of distinct irreducible representations equals c or less. Actually, the latter possibility can be ruled out; but that will not be done here.

Characters for the irreducible representations of a group can often be obtained from explicit matrices representing the group operations. In addition, various theorems such as (8-65) and the one for the number of irreducible representations can be employed. Results for common groups appear in the appendix.

EXAMPLE 8.9   Show that group C_{3v} has 3 irreducible representations.

Proceeding as in examples 4.10 and 4.11, we find that E is in a class by itself, while C_3 and $C_3{}^2$ are in a second class, and σ_v, σ_v', and σ_v'' are in a third class; there are 3 classes in the group. Consequently, there are no more than 3 irreducible representations.

Two of the irreducible representations appear in examples 8.7 and 8.8. A third one is the completely symmetric one

$$(1), (1), (1), (1), (1), (1).$$

8.9   The irreducible representations in a given reducible representation

Let us now consider how to find the irreducible representations that contribute to a given reducible one. As in section 8.5, we start with r independent operands

$$\mathbf{u}_1, \mathbf{u}_2, \ldots, \mathbf{u}_r \tag{8-66}$$

and their transformations by the symmetry operations

$$R\mathbf{u}_k = \mathbf{u}_j D_{jk}(R) \qquad \text{where} \quad j = 1, 2, \ldots, r. \tag{8-67}$$

Reduction of the representation is possible if there are n independent operands

$$\mathbf{e}_m = \mathbf{u}_k a_{km} \qquad \text{where} \qquad k = 1, 2, \ldots, r, \tag{8-68}$$

with $n < r$ such that

$$R\mathbf{e}_m = \mathbf{e}_l D_{lm}'(R) \qquad \text{where} \qquad \begin{matrix} l = 1, 2, \ldots, n, \\ m = 1, 2, \ldots, n, \end{matrix} \tag{8-69}$$

for each R in the group. If (8-68) have been properly constructed (as we will assume), this n-dimensional representation is irreducible. The numbers $D_{lm}'(R)$ then form a submatrix on the diagonal of the $\mathbf{D}'$ matrix to be set up later.

Since the original $\mathbf{u}_j$'s are linearly independent, we can form $r - n$ operands independent of each other and of the $\mathbf{e}_i$'s:

$$\mathbf{v}_m = \mathbf{u}_k b_{km}. \tag{8-70}$$

No operation of the group can convert any of a $\mathbf{v}_m$ into some of the $\mathbf{e}_i$'s of (8-68) because each of these operations has an inverse in the group and none of the operations converted an $\mathbf{e}_m$ of (8-68) into something containing a contribution from the $\mathbf{v}_m$'s. The n $\mathbf{e}_i$'s in (8-68) exhaust a subspace. Consequently,

$$R\mathbf{v}_m = \mathbf{v}_i D_{im}''(R) \qquad \text{where} \quad i = 1, 2, \ldots, r - n. \tag{8-71}$$

If further reduction is possible, there are additional $\mathbf{e}_m$'s,

$$\mathbf{e}_m = \mathbf{v}_i b_{im} \equiv \mathbf{u}_k a_{km} \tag{8-72}$$

such that

$$R\mathbf{e}_m = \mathbf{e}_l D_{lm}'(R) \qquad \text{where} \quad \begin{aligned} l &= n + 1, \ldots, o, \\ m &= n + 1, \ldots, o, \end{aligned} \tag{8-73}$$

and there is a smaller set of independent operands. From these we may get additional $\mathbf{e}_m$'s, and so on. Each set of $D_{lm}'(R)$'s forms a submatrix.

In the end, there are r linearly independent $\mathbf{e}_i$'s subdivided into irreducible sets. Any given R mixes the $\mathbf{e}_i$'s only within each irreducible set. If we form the row matrices

$$\mathbf{u} = (\mathbf{u}_1 \quad \mathbf{u}_2 \quad \cdot \quad \cdot \quad \cdot \quad \mathbf{u}_r), \tag{8-74}$$

$$\mathbf{e} = (\mathbf{e}_1 \quad \mathbf{e}_2 \quad \cdot \quad \cdot \quad \cdot \quad \mathbf{e}_r), \tag{8-75}$$

then the transformation equations can be summarized as

$$R\mathbf{u} = \mathbf{u}\mathbf{D}(R) \tag{8-76}$$

and

$$R\mathbf{e} = \mathbf{e}\mathbf{D}'(R) \tag{8-77}$$

where $\mathbf{D}(R)$ and $\mathbf{D}'(R)$ are $r \times r$ square matrices.

Because operation R merely mixes the $\mathbf{e}_i$'s in each subset, matrix $\mathbf{D}'(R)$ is of block form, with the $n \times n$, $(o - n) \times (o - n), \ldots$ submatrices on the diagonal and zeros everywhere else:

$$\mathbf{D}'(R) = \begin{pmatrix} \square & 0 & 0 & \cdot \\ 0 & \square & 0 & \cdot \\ 0 & 0 & \square & \cdot \\ \cdot & \cdot & \cdot & \cdot \end{pmatrix}. \tag{8-78}$$

The trace of $\mathbf{D}'(R)$ is the sum of the traces of the submatrices.

Equations (8-68), (8-72), ... can be summarized as

$$\mathbf{e} = \mathbf{u}\mathbf{A} \tag{8-79}$$

or

$$\mathbf{e}\mathbf{A}^{-1} = \mathbf{u}. \tag{8-80}$$

Substituting (8-80) into (8-76),

$$Ru = eA^{-1}D(R), \tag{8-81}$$

and acting on this result with the constant (unaffected by R) matrix $\mathbf{A}$ from the right yields

$$RuA = eA^{-1}D(R)A \tag{8-82}$$

or

$$Re = eA^{-1}D(R)A. \tag{8-83}$$

Comparing this equation with (8-77) leads to the result

$$\mathbf{D}'(R) = \mathbf{A}^{-1}\mathbf{D}(R)\mathbf{A} \tag{8-84}$$

that each reduced matrix is obtained by the same similarity transformation of the original matrix. Since this process does not affect the trace, the character of each $\mathbf{D}(R)$ equals that of the corresponding $\mathbf{D}'(R)$. If a_μ is the number of times the submatrix of the μth representation occurs in $\mathbf{D}'(R)$ and if $\chi^\mu(R)$ is its character, the character of the reducible representation is

$$\chi(R) = a_\mu \chi^\mu(R). \tag{8-85}$$

The character $\chi(R)$ equals the sum of terms on the diagonal of $\mathbf{D}(R)$. To find it, we need only determine coefficient b in

$$Ru_k = bu_k + cu_{k+1} + \cdots \tag{8-86}$$

for each $\mathbf{u}_k$ and add the results. The other elements in $\mathbf{D}(R)$ are not needed. To find a_μ, determine $\chi(R)$ for an R in each class. Then, use the table listing the characters of the irreducible representations to see how they can add to produce those of the reducible representation.

When the breakdown is not evident, use the following formula: Construct the sum

$$\chi(R_i)\chi^{\nu*}(R_i) = a_\mu\chi^\mu(R_i)\chi^{\nu*}(R_i). \tag{8-87}$$

Introduce orthogonality relation (8-65) in the form

$$a_\mu\chi^\mu(R_i)\chi^{\nu*}(R_i) = a_\mu g \delta_{\mu\nu} = a_\nu g \tag{8-88}$$

and solve for a_ν:

$$a_\nu = \frac{1}{g} \chi(R_i)\chi^{\nu*}(R_i). \tag{8-89}$$

EXAMPLE 8.10   Find all characters of the reducible representation of $\mathbf{D}_{3h}$ for which the $\mathbf{t}_j$ vectors in figure 8.1 are bases.

Apply a symmetry operation from each class to each $\mathbf{t}_j$ vector:

$$
\begin{array}{lll}
E\mathbf{t}_j = \mathbf{t}_j, & S_3\mathbf{t}_j = \mathbf{t}_{j+1}, & \sigma_v\mathbf{t}_1 = -\mathbf{t}_1, \\
\sigma_h\mathbf{t}_j = \mathbf{t}_j, & C_2'\mathbf{t}_1 = -\mathbf{t}_1, & \sigma_v\mathbf{t}_2 = -\mathbf{t}_3, \\
C_3\mathbf{t}_j = \mathbf{t}_{j+1}, & C_2'\mathbf{t}_2 = -\mathbf{t}_3, & \sigma_v\mathbf{t}_3 = -\mathbf{t}_2, \\
& C_2'\mathbf{t}_3 = -\mathbf{t}_2. &
\end{array}
$$

Then substitute the results into matrix equation (8-76):

$$E\mathbf{t} = (\mathbf{t}_1 \quad \mathbf{t}_2 \quad \mathbf{t}_3) \begin{pmatrix} 1 & 0 & 0 \\ 0 & 1 & 0 \\ 0 & 0 & 1 \end{pmatrix},$$

$$\sigma_h\mathbf{t} = (\mathbf{t}_1 \quad \mathbf{t}_2 \quad \mathbf{t}_3) \begin{pmatrix} 1 & 0 & 0 \\ 0 & 1 & 0 \\ 0 & 0 & 1 \end{pmatrix},$$

$$C_3\mathbf{t} = (\mathbf{t}_1 \quad \mathbf{t}_2 \quad \mathbf{t}_3) \begin{pmatrix} 0 & 0 & 1 \\ 1 & 0 & 0 \\ 0 & 1 & 0 \end{pmatrix},$$

$$S_3\mathbf{t} = (\mathbf{t}_1 \quad \mathbf{t}_2 \quad \mathbf{t}_3) \begin{pmatrix} 0 & 0 & 1 \\ 1 & 0 & 0 \\ 0 & 1 & 0 \end{pmatrix},$$

$$C_2'\mathbf{t} = (\mathbf{t}_1 \quad \mathbf{t}_2 \quad \mathbf{t}_3) \begin{pmatrix} -1 & 0 & 0 \\ 0 & 0 & -1 \\ 0 & -1 & 0 \end{pmatrix},$$

$$\sigma_v\mathbf{t} = (\mathbf{t}_1 \quad \mathbf{t}_2 \quad \mathbf{t}_3) \begin{pmatrix} -1 & 0 & 0 \\ 0 & 0 & -1 \\ 0 & -1 & 0 \end{pmatrix}.$$

Read off the sum of terms along the principal diagonal of each $\mathbf{D}(R)$ and record the results in a table:

Base vectors $(\mathbf{t}_1, \mathbf{t}_2, \mathbf{t}_3)$	Representation Γ_t	E	σ_h	$2C_3$	$2S_3$	$3C_2'$	$3\sigma_v$
		3	3	0	0	-1	-1

Column header line: "Character for each class"

EXAMPLE 8.11   What irreducible representations are based on linear combinations of the $\mathbf{t}_j$ vectors?

Equation (8-85) states that

$$\chi(R) = a_1\chi^1(R) + a_2\chi^2(R) + \cdots$$

the character for a representation of an operation equals the number of times the first irreducible representation appears times its character for the same operation plus similar terms for each of the other irreducible representations.

But from the appendix, the $\mathbf{D}_{3h}$ group contains the following characters:

Representation	E	σ_h	$2C_3$	$2S_3$	$3C_2'$	$3\sigma_v$
A_2'	1	1	1	1	-1	-1
E'	2	2	-1	-1	0	0
$A_2' + E'$	3	3	0	0	-1	-1

Column header line: "Characters for each class"

Note how the characters for one A_2' and one E' add to give those for the representation based on $(\mathbf{t}_1, \mathbf{t}_2, \mathbf{t}_3)$. Therefore, linear combinations of the $\mathbf{t}_j$'s form bases for A_2' and for E':

$$\Gamma_t = A_2' + E'.$$

8.10 Constructing bases for irreducible representations

The similar, independent expressions

$$\mathbf{u}_1, \mathbf{u}_2, \ldots, \mathbf{u}_r \qquad (8\text{-}90)$$

associated with equivalent parts of a system form a basis for a representation whenever every symmetry operation in a group transforms each into a linear combination of the others. In practice, the expressions may measure standard causes or effects. Thus, they may describe generalized forces or the generalized accelerations that they produce.

Combinations of the expressions that belong to a certain row of an irreducible representation are particularly simple. Since a cause of one type tends to produce an effect of the same type, we need to determine how such combinations can be found.

We start with the independent operands (8-90) and the way they transform:

$$R\mathbf{u}_k = \mathbf{u}_j D_{jk}(R) \qquad \text{where} \quad j = 1, 2, \ldots, r. \qquad (8\text{-}91)$$

Bases for each irreducible representation μ that occurs are unknown linear functions of the $\mathbf{u}_k$'s:

$$\mathbf{e}_m = \mathbf{u}_k a_{km} \qquad \text{where} \quad \begin{array}{l} k = 1, 2, \ldots, r, \\ m = 1, \ldots, h+1, \ldots, h+n_\mu, \ldots, r. \end{array} \qquad (8\text{-}92)$$

There are r linearly independent $\mathbf{e}_m$'s, n_μ in each occurrence of μ.

Any given symmetry operation R_i mixes the bases within each irreducible representation:

$$R_i \mathbf{e}_m = \mathbf{e}_l D_{lm}(R_i) \qquad \text{where} \quad \begin{array}{l} m = h+1, \ldots, h+n_\mu, \\ l = h+1, \ldots, h+n_\mu. \end{array} \qquad (8\text{-}93)$$

For the μth representation the rows and columns of $\mathbf{D}(R_i)$ are now numbered from $h+1$ instead of from 1.

Let us represent the inversion of set (8-92) as

$$\mathbf{u}_j = b_{jm} \mathbf{e}_m \qquad \text{where} \quad m = 1, 2, \ldots, r. \qquad (8\text{-}94)$$

Then operate on the result with R_i and use (8-93) to expand:

$$R_i \mathbf{u}_j = b_{jm} R_i \mathbf{e}_m = b_{jm} \mathbf{e}_l D_{lm}(R_i) = b_{jp}{}^\mu \mathbf{e}_o{}^\mu D_{op}{}^\mu(R_i) \qquad (8\text{-}95)$$

where

$$o = 1, 2, \ldots, n_\mu,$$
$$p = 1, 2, \ldots, n_\mu,$$

while μ labels the irreducible representations that are summed over.

Let us multiply (8-95) by the complex conjugate of the character of the νth representation,

$$\chi_\nu^*(R_i) = D_{qq}^{\ \nu*}(R_i) \tag{8-96}$$

and sum over all operations of the group:

$$\chi_\nu^*(R_i)R_i\mathbf{u}_j = D_{qq}^{\ \nu*}(R_i)D_{op}^{\ \mu}(R_i)b_{jp}^{\ \mu}\mathbf{e}_o^{\ \mu}$$

$$= \frac{g}{n}\delta_{oq}\delta_{pq}\delta_{\mu\nu}b_{jp}^{\ \mu}\mathbf{e}_o^{\ \mu} = \frac{g}{n}b_{jm}\mathbf{e}_m \quad \text{(in ν only).} \tag{8-97}$$

In the second equality, orthogonality relationship (8-63) has been introduced. We have

$$\frac{n}{g}\chi_\nu^*(R_i)R_i\mathbf{u}_j = b_{jm}\mathbf{e}_m \quad \text{(in ν only).} \tag{8-98}$$

The result contains *only* those terms of (8-94) that are fractions of bases for the νth representation. Consequently, the operator sum

$$\frac{n}{g}\chi_\nu^*(R_i)R_i \tag{8-99}$$

projects $\mathbf{u}_j$ onto the space of these bases; it is a *projection operator*.

A constant times the projection is a base when the representation occurs just once. When it occurs more than once, the projection may contain parts from more than one occurrence. To eliminate all but one part, combine the sum with combinations generated from other $\mathbf{u}_h$'s:

$$\mathbf{f}_j = c_h\chi_\nu^*(R_i)R_i\mathbf{u}_h. \tag{8-100}$$

Suitable c_h's may be determined from the way each operation of the group transforms each combination.

EXAMPLE 8.12   Construct bases for irreducible representations of $\mathbf{D}_{3h}$ from the $\mathbf{t}_j$ vectors in figure 8.1.

According to the results in example 8.11, combinations of the $\mathbf{t}_j$'s form bases for A_2' and E'. Since neither representation occurs more than once, the left side of (8-97) yields base vectors for each:

$$\mathbf{f}_k = \chi_\nu^*(R_i)R_i\mathbf{u}_j.$$

When the various symmetry operations act on the array of vectors, the following results are obtained:

Effect on operand of operator

Operand	E	σ_h	C_3	$C_3^{\ 2}$	S_3	$S_3^{\ 2}$	C_2'	C_2''	C_2'''	σ_v	σ_v'	σ_v''
$\mathbf{t}_1$	$\mathbf{t}_1$	$\mathbf{t}_1$	$\mathbf{t}_2$	$\mathbf{t}_3$	$\mathbf{t}_2$	$\mathbf{t}_3$	$-\mathbf{t}_1$	$-\mathbf{t}_3$	$-\mathbf{t}_2$	$-\mathbf{t}_1$	$-\mathbf{t}_3$	$-\mathbf{t}_2$
$\mathbf{t}_2$	$\mathbf{t}_2$	$\mathbf{t}_2$	$\mathbf{t}_3$	$\mathbf{t}_1$	$\mathbf{t}_3$	$\mathbf{t}_1$	$-\mathbf{t}_3$	$-\mathbf{t}_2$	$-\mathbf{t}_1$	$-\mathbf{t}_3$	$-\mathbf{t}_2$	$-\mathbf{t}_1$
$\mathbf{t}_3$	$\mathbf{t}_3$	$\mathbf{t}_3$	$\mathbf{t}_1$	$\mathbf{t}_2$	$\mathbf{t}_1$	$\mathbf{t}_2$	$-\mathbf{t}_2$	$-\mathbf{t}_1$	$-\mathbf{t}_3$	$-\mathbf{t}_2$	$-\mathbf{t}_1$	$-\mathbf{t}_3$

Now take the complex conjugate of the character for each operation from the A_2' row of the table in example 8.11 and multiply it by the corresponding $R_i \mathbf{u}_1$ from the first row in the table above and add (following the formula for $\mathbf{f}_k$):

$$\mathbf{f}_1 = \mathbf{t}_1 + \mathbf{t}_1 + \mathbf{t}_2 + \mathbf{t}_3 + \mathbf{t}_2 + \mathbf{t}_3 + \mathbf{t}_1 + \mathbf{t}_3 + \mathbf{t}_2 + \mathbf{t}_1 + \mathbf{t}_3 + \mathbf{t}_2$$
$$= 4(\mathbf{t}_1 + \mathbf{t}_2 + \mathbf{t}_3).$$

The $R_i\mathbf{u}_2$'s from the second row of the table above and the $R_i\mathbf{u}_3$'s from the third row yield the same $\mathbf{f}_1$ for representation A_2'.

Take the complex conjugate of each character from the E' row, multiply it by each corresponding $R_i\mathbf{u}_1$, and add, following the formula for $\mathbf{f}_k$:

$$\mathbf{f}_2 = 2\mathbf{t}_1 + 2\mathbf{t}_1 - \mathbf{t}_2 - \mathbf{t}_3 - \mathbf{t}_2 - \mathbf{t}_3 + 0$$
$$= 4\mathbf{t}_1 - 2\mathbf{t}_2 - 2\mathbf{t}_3.$$

Alternatively, multiply by each corresponding $R_i\mathbf{u}_2$ and add,

$$\mathbf{f}_3 = 4\mathbf{t}_2 - 2\mathbf{t}_3 - 2\mathbf{t}_1,$$

or by each corresponding $R_i\mathbf{u}_3$ and add,

$$\mathbf{f}_4 = 4\mathbf{t}_3 - 2\mathbf{t}_1 - 2\mathbf{t}_2.$$

Since

$$\mathbf{f}_2 + \mathbf{f}_3 = -\mathbf{f}_4$$

only two of these are independent. Because E' is a 2-dimensional representation, it is spanned by two independent vectors.

A vector that is orthogonal (perpendicular in a space in which $\mathbf{t}_1$, $\mathbf{t}_2$, $\mathbf{t}_3$ are mutually perpendicular) to $\mathbf{f}_2$ and thus completely independent of it is constructed by inspection:

$$\mathbf{f}_3 - \mathbf{f}_4 = 6\mathbf{t}_2 - 6\mathbf{t}_3.$$

Each of the orthogonal vectors can be reduced to unit length (normalized) with the following results:

$$\mathbf{e}_1 = \frac{1}{\sqrt{3}} (\mathbf{t}_1 + \mathbf{t}_2 + \mathbf{t}_3) \qquad \text{for} \quad A_2',$$

$$\mathbf{e}_2 = \frac{1}{\sqrt{6}} (2\mathbf{t}_1 - \mathbf{t}_2 - \mathbf{t}_3)$$
$$\text{for} \quad E'.$$
$$\mathbf{e}_3 = \frac{1}{\sqrt{2}} (\mathbf{t}_2 - \mathbf{t}_3)$$

Since there are the same number of base vectors as $\mathbf{t}$'s, they span the space—no more are needed.

8.11   Mutual orthogonality of base operands from irreducible sets

We have just considered how similar, independent expressions associated with equivalent parts of a physical system yield bases

$$\mathbf{e}_1{}^1, \ldots, \mathbf{e}_{n_1}{}^1, \mathbf{e}_1{}^2, \ldots, \mathbf{e}_{n_2}{}^2, \ldots \qquad (8\text{-}101)$$

for irreducible representations $1, 2, \ldots$. Let us now determine how such bases are related when any two of them combine to form a scalar

$$\mathbf{e}_j{}^{\mu*} \cdot \mathbf{e}_k{}^{\nu}. \qquad (8\text{-}102)$$

Since a physical scalar is not associated with directions, it cannot be altered by a covering operation, an operation that merely reorients the given physical system. If R is a typical reorientation operation, we must have

$$R(\mathbf{e}_j{}^{\mu*} \cdot \mathbf{e}_k{}^{\nu}) = (R\mathbf{e}_j{}^{\mu})^* \cdot (R\mathbf{e}_k{}^{\nu}) = \mathbf{e}_j{}^{\mu*} \cdot \mathbf{e}_k{}^{\nu}. \qquad (8\text{-}103)$$

If each operation in the pertinent group is applied and the results averaged, the same scalar is obtained:

$$\frac{1}{g} (R_i\mathbf{e}_j{}^{\mu})^* \cdot (R_i\mathbf{e}_k{}^{\nu}) = \mathbf{e}_j{}^{\mu*} \cdot \mathbf{e}_k{}^{\nu}. \qquad (8\text{-}104)$$

Each factor in each term can be expanded with (8-18):

$$\mathbf{e}_j{}^{\mu*} \cdot \mathbf{e}_k{}^{\nu} = \frac{1}{g} \left[\mathbf{e}_l D_{li}{}^{\mu}(R_i) \right]^* \cdot \left[\mathbf{e}_m D_{mk}{}^{\nu}(R_i) \right]$$

$$= \frac{1}{g} \mathbf{e}_l{}^* \cdot \mathbf{e}_m D_{lj}{}^{\mu*}(R_i) D_{mk}{}^{\nu}(R_i), \qquad (8\text{-}105)$$

and the result reduced with orthogonality theorem (8-63):

$$\mathbf{e}_j{}^{\mu*} \cdot \mathbf{e}_k{}^{\nu} = \left(\frac{1}{n} \mathbf{e}_l{}^* \cdot \mathbf{e}_m \delta_{lm} \right) \delta_{\mu\nu} \delta_{jk} = N \delta_{\mu\nu} \delta_{jk}. \qquad (8\text{-}106)$$

Parameter N represents the Hermitian scalar product of a base with itself.

Note that the scalar combination is zero and the bases $\mathbf{e}_j{}^{\mu}$ and $\mathbf{e}_k{}^{\nu}$ are orthogonal when $\mu \neq \nu$ (when they are in different irreducible representations) or when $j \neq k$ (when they are in entirely different rows of the same representation).

EXAMPLE 8.13   How can the different bases for a given representation be constructed from one $\mathbf{e}_l$?

First determine the effect of each operation of the group on the given base using (8-18):

$$R_i\mathbf{e}_l = \mathbf{e}_j D_{jl}{}^{\mu}(R_i).$$

Then multiply each result by $D_{km}^{v*}(R_i)$, sum over all operations of the group, and reduce:

$$D_{km}^{v*}(R_i)R_i e_l = e_j D_{jl}^{\mu}(R_i)D_{km}^{v*}(R_i) = e_j \frac{g}{n} \delta_{kj}\delta_{ml}\delta_{v\mu}$$

$$= \frac{g}{n} \delta_{v\mu}\delta_{ml} e_k.$$

The orthogonality theorem is used in the next to the last step.

We see that the operator

$$\frac{n}{g} D_{km}^{v*}(R_i)R_i$$

acting on e_l yields zero when $v \neq \mu$ and $m \neq l$. But when the two representations are the same and we use the lth column from $\mathbf{D}^v(R_i)$, the process yields e_k.

If bases independent of the e_l's obey the same transformation law

$$R_i \mathbf{f}_l = \mathbf{f}_j D_{jl}^{\mu}(R_i),$$

the representation is said to occur again. Operation

$$\frac{n}{g} D_{km}^{v*}(R_i)R_i$$

acting on $\mathbf{f}_l$ will then produce

$$\mathbf{f}_k$$

when $v = \mu$ and $m = l$. This operand belongs to the same row of the representation as e_k.

8.12 Summary

Elements in a set form a group with respect to a combining operation if and only if (a) the result of the operation on any two ordered members of the set is a member of the set, (b) one element, the identity, does not alter any other member on combining with it, (c) for each element, there is an inverse that combines with the given element to yield the identity, (d) the result from combining three or more elements does not depend on which two consecutive elements are combined first, whether the preceding member or the following member is combined with this product next, and so on.

Any two members R and S that are linked through a similarity transformation by some element T in the group

$$S = T^{-1}RT \tag{8-107}$$

are said to belong to the same class. Specific operators combining as the group elements do constitute a representation of the group. Operands for such operators form a basis for the representation.

Matrix representations arise when the group elements transform the given bases linearly. Such a representation is reducible if the operands combine linearly to yield two or more smaller sets each of which is a complete basis for a representation. It is irreducible when such reduction is not possible.

The *jl* elements of the *g* irreducible matrices representing a group are components of a vector orthogonal, in the Hermitian sense, to any other similarly constituted vector. The magnitude of the vector equals $\sqrt{g/n}$. As a consequence, the characters of each irreducible representation form a vector orthogonal to each other similarly constituted vector for the group.

The part of a given base in the space of a given irreducible representation can be obtained by subjecting the base to each operation of the group multiplied by n/g times the complex conjugate of the corresponding character and adding the results.

DISCUSSION QUESTIONS

8.1 How do symmetry operations affect the symmetry and energy of a system? How do they affect characteristics that are not uniquely determined by the symmetry and energy? Give examples.

8.2 What is a group? What is a representation? What is a basis?

8.3 Show when independent expressions associated with equivalent parts of a system form a basis for a representation.

8.4 How can arrays of vectors serve as bases? How can linear combinations of such vectors serve as bases?

8.5 When is a basis complete?

8.6 When is a representation (a) reducible, (b) irreducible?

8.7 Why must the **A** in

$$\mathbf{A}\mathbf{D}(R) = \mathbf{D}'(R)\mathbf{A}$$

be zero if **D** and a dissimilar **D'** representation are both irreducible?

8.8 Why is any matrix that commutes with all matrices in an irreducible representation a number times the unit matrix?

8.9 Construct the reorientation matrices that represent the $\mathbf{C}_{3v}$ group. From these obtain the matrices in example 8.7.

8.10 Show that the vector formed from the 11 elements in the set of matrices from example 8.7 is orthogonal to the vector constructed from the 22 elements.

8.11 Explain how Schur's lemmas lead to restrictions on the matrix elements of irreducible representations.

8.12 What characterizes each class in a representation?

8.13 Why can't the number of distinct irreducible representations exceed the number of classes?

8.14 Derive the orthogonality theorem for the characters.

8.15 How are the characters for a given set of operands obtained? Illustrate.

8.16 What can we deduce when the symmetry operations R_i acting on a given $\mathbf{u}_j$ do not yield in turn all the $\mathbf{u}_k$'s, or combinations of all of them?

8.17 How do we combine the operands of a reducible representation to get those for a constituent irreducible representation? Give examples.

8.18 Explain the formulas that tell one how to get the bases corresponding to given rows of an irreducible representation.

8.19 Can we picture bases of representations as mutually perpendicular equivalent vectors in an r-dimensional mathematical space?

PROBLEMS

8.1 Four identical particles are bound together by similar springs so they oscillate about the four corners of a square. Number these corners consecutively in the counterclockwise direction. Then from each, draw a unit vector $\mathbf{r}_j$ pointing directly towards the center, a unit vector $\mathbf{t}_j$ in the same plane but perpendicular to $\mathbf{r}_j$ and pointing in the counterclockwise direction, and a unit vector $\mathbf{s}_j$ equal to $\mathbf{t}_j \times \mathbf{r}_j$ (that is, pointing up perpendicular to the plane).

8.2 Prove that the vectors

$$\mathbf{e}_1 = \frac{1}{\sqrt{2}} (\mathbf{r}_1 - \mathbf{r}_3)$$

$$\mathbf{e}_2 = \frac{1}{\sqrt{2}} (\mathbf{r}_2 - \mathbf{r}_4)$$

form a complete basis for a representation of the C_4 subgroup of the system in problem 8.1.

8.3 Show that the vector

$$\mathbf{e}_3 = \tfrac{1}{2}(\mathbf{r}_1 - i\mathbf{r}_2 - \mathbf{r}_3 + i\mathbf{r}_4)$$

is a complete basis for a representation of the C_4 subgroup of the system in problem 8.1. Express $\mathbf{e}_3$ as a linear function of $\mathbf{e}_1$ and $\mathbf{e}_2$. What kind of representation are $\mathbf{e}_1$ and $\mathbf{e}_2$ the basis for?

8.4 From the appropriate reorientation matrices deduce the following representation

$$\begin{pmatrix} 1 & 0 \\ 0 & 1 \end{pmatrix}, \begin{pmatrix} 0 & - \\ 1 & 0 \end{pmatrix}, \begin{pmatrix} -1 & 0 \\ 0 & -1 \end{pmatrix}, \begin{pmatrix} 0 & 1 \\ -1 & 0 \end{pmatrix},$$

of the C_4 group. Explain why the 4-dimensional vector constructed from the 12 elements is *not* perpendicular to the vector constructed from the 21 elements.

8.5 If C_4 is represented by (-1), what are E, C_2, and C_4^{-1} represented by? Also, what is the completely symmetric 1-dimensional representation of the C_4 group?

8.6 Determine all characters of the reducible representation of D_{4h} for which the r_j vectors in problem 8.1 are bases. Let the planes for σ_v, σ_v' and the axes for $(C_2')_a$, $(C_2')_b$ pass through the midpoints of opposite edges of the square while the planes for σ_d, σ_d' and the axes for $(C_2'')_a$, $(C_2'')_b$ pass through opposite corners. The classes in the group include E, $2C_4$, C_2, $2\sigma_v$, $2\sigma_d$, σ_h, $2S_4$, i, $2C_2'$, $2C_2''$.

8.7 Find the irreducible representations that are based on linear combinations of the r_j vectors of problem 8.1.

8.8 Generate bases for irreducible representations of D_{4h} from the r_j vectors of problem 8.1. Show that these bases are mutually perpendicular in the space in which the r_j's are base vectors.

8.9 Show that, for each irreducible representation but one in any given group, the sum of the characters over all operations is zero.

8.10 Prove that the function

$$f = r_1 \cdot r_1 - r_2 \cdot r_2 + r_3 \cdot r_3 - r_4 \cdot r_4$$

is a complete basis for the C_4 subgroup of the system in problem 8.1. To what representation does it belong?

8.11 Four equivalent particles oscillate about alternate corners of a cube, the equilibrium positions forming a tetrahedron. Let the three cube edges from the first occupied corner on the base be x, y, z axes, with corresponding unit vectors u_1, u_2, u_3. Erect similar unit vectors u_4, u_5, u_6 along cube edges from the opposite corner of the base, with u_4 pointing in the $-x$ direction, u_5 in the $-y$ direction, and u_6 in the z direction. Go to the occupied top corner from which the edges proceed in the $-x$, y, $-z$ directions and label the corresponding unit vectors u_7, u_8, u_9. From the last occupied corner construct unit vectors u_{10}, u_{11}, u_{12} pointing in the x, $-y$, $-z$ directions.

8.12 Prove that the vectors

$$e_1 = \tfrac{1}{2}(u_1 + u_4 + u_7 + u_{10})$$
$$e_2 = \tfrac{1}{2}(u_2 + u_5 + u_8 + u_{11})$$
$$e_3 = \tfrac{1}{2}(u_3 + u_6 + u_9 + u_{12})$$

form a complete basis for a representation of the T subgroup of the system in problem 8.11. Note that the axis for a C_2 passes through the middle of opposite edges of the tetrahedron, while the axis for a C_3 passes through a vertex to the middle of the opposite face.

8.13 Six identical particles are bound together with the symmetry of a regular hexagon. On each equilibrium position, construct a unit vector $\mathbf{r}_j$ pointing toward the center. Then prove that the vector

$$\mathbf{e} = \frac{1}{\sqrt{6}}(\mathbf{r}_1 + \omega\mathbf{r}_2 + \omega^2\mathbf{r}_3 + \omega^3\mathbf{r}_4 + \omega^4\mathbf{r}_5 + \omega^5\mathbf{r}_6),$$

where $\omega = e^{2\pi i/6}$ is a complete basis for a representation of the C_6 subgroup.

8.14 If C_6 is represented by (-1), what are E, C_3, C_2, C_3^{-1}, C_6^{-1} represented by? What other 1-dimensional representation does the C_6 group possess? Why can't C_5 be represented by (-1)?

8.15 From the appropriate reorientation matrices deduce the following representation

$$\begin{pmatrix} 1 & 0 & 0 \\ 0 & 1 & 0 \\ 0 & 0 & 1 \end{pmatrix}, \begin{pmatrix} -1 & 0 & 0 \\ 0 & -1 & 0 \\ 0 & 0 & 1 \end{pmatrix}, \begin{pmatrix} 1 & 0 & 0 \\ 0 & -1 & 0 \\ 0 & 0 & -1 \end{pmatrix}, \begin{pmatrix} -1 & 0 & 0 \\ 0 & 1 & 0 \\ 0 & 0 & -1 \end{pmatrix},$$

of the D_2 group. From this deduce three 1-dimensional representations. Also write the completely symmetric representation.

8.16 Determine all characters of the reducible representation of T for which the $\mathbf{u}_j$ vectors in problem 8.11 are bases. The classes of the group include E, $3C_2$, $4C_3$, $4C_3^2$.

8.17 Find the irreducible representations that are based on linear combinations of the $\mathbf{u}_j$ vectors of problem 8.11.

8.18 Generate bases for irreducible representations of T from the $\mathbf{u}_j$ vectors of problem 8.11.

8.19 Obtain the characters and the composition of the representation of which vectors $\mathbf{e}_1$, $\mathbf{e}_2$, $\mathbf{e}_3$ of problem 8.12 are the basis. Then show how these vectors are related to the bases found in problem 8.18.

8.20 If the expressions

$$\mathbf{e}_1, \mathbf{e}_2, \ldots, \mathbf{e}_r$$

constitute a basis for an irreducible representation, to what representation does the sum

$$\mathbf{e}_k{}^*\mathbf{e}_k$$

belong?

REFERENCES

BOOKS

Cotton, F. A.   *Chemical Applications of Group Theory*, pp. 50–76.   Interscience Publishers, New York, 1963.

Cracknell, A. P.   *Applied Group Theory*, pp. 3–41.   Pergamon Press, Oxford, 1968.

Hall, G. G.   *Applied Group Theory*, pp. 1–65.   American Elsevier Publishing Company, Inc., New York, 1967.

Hollingsworth, C. A.   *Vectors, Matrices, and Group Theory for Scientists and Engineers*, pp. 187–281.   McGraw-Hill Book Company, New York, 1967.

Tinkham, M.   *Group Theory and Quantum Mechanics*, pp. 6–49.   McGraw-Hill Book Company, New York, 1964.

ARTICLES

Bradley, C. J., and Davies, B. L.   Magnetic Groups and Their Corepresentations. *Rev. Mod. Phys.*, **40**, 359 (1968).

Brown, E.   A Simple Alternative to Double Groups.   *Am. J. Phys.*, **38**, 704 (1970).

Gamba, A.   Representations and Classes in Groups of Finite Order.   *J. Math. Phys.*, **9**, 186 (1968).

9 / *Modes of Motion*

9.1 Separating the equations of motion

A set of Lagrange equations governs the classical movements in any given region.

Any symmetry, or near-symmetry, in the region permits the equations to be combined into noninteracting, or weakly interacting, subsets. Group-theoretical considerations enable us to separate the equations until each subset corresponds to just one row of an irreducible representation. Equations describing pure translations and rotations can then be removed from the appropriate simultaneous equations.

The resulting subsets are independent when the amplitudes of motion are small. When the amplitudes are appreciable, various interactions arise. Even when a vibrational mode is harmonic, the effective moments of inertia of the system change with the amplitude.

Coriolis interaction exists between representations whose bases multiply to form bases for rotational representations. Furthermore, large amplitudes lead to appreciable distortion of the system during certain phases of the motion.

Group-theoretical considerations indicate when distortions cause (a) a degenerate representation to split into two or more distinct representations and (b) two or more distinct representations to coalesce. Where the splitting occurs, the pertinent frequencies diverge continuously with increasing asymmetry. Where the combination occurs, interaction appears and increases continuously, driving the pertinent frequencies farther apart.

For simplicity, our main concern will be with relationships that are valid when the amplitudes of motion are small.

9.2 Suitable coordinates and forces

Different parts of a symmetric region, and the forces acting thereon, may very well be referred to different coordinate systems.

Suppose that a set of positions is known where all particles or small masses in the region can simultaneously rest without accelerating. Erect Cartesian axes at each of these equilibrium positions and orient the axes so they point in equivalent directions.

Then each symmetry operation merely permutes some coordinate axes and reorients others.

Let the unit vectors defining the axes at the jth position be

$$\mathbf{u}_{3j-2}, \mathbf{u}_{3j-1}, \mathbf{u}_{3j} \tag{9-1}$$

as in section 8.4. Also let the instantaneous position of the jth mass be measured by its coordinates

$$\eta_{3j-2}, \eta_{3j-1}, \eta_{3j}, \tag{9-2}$$

while the force acting on it is measured by its components

$$\phi_{3j-2}, \phi_{3j-1}, \phi_{3j} \tag{9-3}$$

in the local Cartesian frame.

Multiplying each force by the displacement over which it acts during any arbitrary small movement, and adding, then yields the work

$$\delta W = \phi_k \, \delta \eta_k. \tag{9-4}$$

According to sections 8.10 and 8.11, the r $\mathbf{u}_k$'s combine linearly to yield r mutually orthogonal $\mathbf{e}_m$'s. Inverting the set yields each $\mathbf{u}_k$ as a linear sum of the $\mathbf{e}_m$'s. Each sum can be plotted in the r-dimensional Euclidean space defined by the $\mathbf{e}_m$'s. Linear combinations of the $\mathbf{u}_k$'s then appear as vectors in this hypothetical space. We can construct

$$\mathbf{q} = \eta_k \mathbf{u}_k, \tag{9-5}$$

$$\mathbf{Q} = \phi_k \mathbf{u}_k. \tag{9-6}$$

The $\mathbf{u}_k$'s must be mutually perpendicular to cause an arbitrary $\mathbf{Q}$ dot an arbitrary $\delta \mathbf{q}$ to give δW:

$$\mathbf{Q} \cdot \delta \mathbf{q} = \phi_k \mathbf{u}_k \cdot \delta \eta_l \mathbf{u}_l = \phi_k \, \delta \eta_k = \delta W. \tag{9-7}$$

We can project the coordinate and force vectors on the bases

$$q_l = \mathbf{q} \cdot \mathbf{e}_l, \tag{9-8}$$

$$Q_l = \mathbf{Q} \cdot \mathbf{e}_l, \tag{9-9}$$

and construct the expansions

$$\mathbf{q} = q_l \mathbf{e}_l, \tag{9-10}$$

$$\mathbf{Q} = Q_l \mathbf{e}_l. \tag{9-11}$$

The scalar product for work done can then be rewritten in the form

$$\delta W = \mathbf{Q} \cdot \delta \mathbf{q} = Q_l \, \delta q_l, \tag{9-12}$$

in which each term represents the contribution from a different row of an irreducible representation.

Combining equations (9-8), (9-5), and (8-92),

$$q_l = \mathbf{q} \cdot \mathbf{e}_l = \eta_k \mathbf{u}_k \cdot \mathbf{u}_m a_{ml} = \eta_k a_{kl} \tag{9-13}$$

shows that each q_l is a linear combination of the η_k's. Since the number of independent q_l's equals the number of η_k's, the relationships can be inverted. Substituting the results into equation (6-11) and collecting like terms then produces the double sum

$$T = \dot{q}_j A_{jk} \dot{q}_k \qquad (9\text{-}14)$$

in which the A_{jk}'s are linear functions of the masses.

In practice, $\mathbf{e}_l$ may be replaced by a convenient multiple

$$\mathbf{f}_l = k\mathbf{e}_l. \qquad (9\text{-}15)$$

The generalized coordinate q_l may then be taken as

$$q_l = \mathbf{q} \cdot \mathbf{f}_l. \qquad (9\text{-}16)$$

As long as Q_l is obtained from a potential that is expressed properly in terms of these q_k's, and as long as T is correctly expressed, no error is introduced.

9.3 Symmetry-adapted equations of motion

The kinetic energy expression (9-14) and the force expressions from work (9-12) simplify Lagrange equations (5-20),

$$\frac{d}{dt}\frac{\partial T}{\partial \dot{q}_l} - \frac{\partial T}{\partial q_l} = Q_l. \qquad (9\text{-}17)$$

to

$$2A_{lk}\ddot{q}_k = Q_l. \qquad (9\text{-}18)$$

The generalized coordinates from formula (9-16) lead to similar equations, but with different A_{lk} and Q_l.

Because any pertinent symmetry operation converts the set of particles into an equivalent set, it cannot destroy these equalities. So long as the right side of (9-18) belongs to one row of a single representation, called a species, the left side must also belong to the species.

Since any of the $\ddot{q}_k$'s may differ from zero, all A_{lk}'s must be zero except for those multiplying the $\ddot{q}_k$'s belonging to the same species as Q_l. The equations corresponding to individual species are consequently independent. If a species occurs n times, its study requires the solution of *only n* simultaneous equations.

Introducing the symmetry-adapted coordinates $\mathbf{q} \cdot \mathbf{e}_l$ (or $\mathbf{q} \cdot \mathbf{f}_l$) can thus simplify the equations of motion. In practice, we set up Lagrange's equations in the conventional manner and then combine them to separate the representations, the rows of representations, the equations describing translations, and those describing rotations.

9.4 Base vectors for a square array

Let us amplify these ideas by treating a particular system, four identical particles bound together so they oscillate about the corners of a square.

The equilibrium positions may translate or rotate; so we first pick out a particular set and label the corresponding particles in counterclockwise order, 1, 2, 3, 4. Unit

TABLE 9.1   Effect of Symmetry Operations on Representative Unit Vectors

Operand	Operator															
	E	C_4	C_4^{-1}	C_2	$(C_2')_a$	$(C_2')_b$	$(C_2'')_a$	$(C_2'')_b$	i	S_4	S_4^{-1}	σ_h	σ_v	σ_v'	σ_d	σ_d'
$\mathbf{u}_1$	$\mathbf{u}_1$	$\mathbf{u}_4$	$\mathbf{u}_{10}$	$\mathbf{u}_7$	$\mathbf{u}_5$	$\mathbf{u}_{11}$	$\mathbf{u}_2$	$\mathbf{u}_8$	$\mathbf{u}_7$	$\mathbf{u}_4$	$\mathbf{u}_{10}$	$\mathbf{u}_1$	$\mathbf{u}_5$	$\mathbf{u}_{11}$	$\mathbf{u}_2$	$\mathbf{u}_8$
$\mathbf{u}_2$	$\mathbf{u}_2$			$\mathbf{u}_8$					$\mathbf{u}_8$			$\mathbf{u}_2$				
$\mathbf{u}_3$	$\mathbf{u}_3$	$\mathbf{u}_6$	$\mathbf{u}_{12}$	$\mathbf{u}_9$	$-\mathbf{u}_6$	$-\mathbf{u}_{12}$	$-\mathbf{u}_3$	$-\mathbf{u}_9$	$-\mathbf{u}_9$	$-\mathbf{u}_6$	$-\mathbf{u}_{12}$	$-\mathbf{u}_3$	$\mathbf{u}_6$	$\mathbf{u}_{12}$	$\mathbf{u}_3$	$\mathbf{u}_9$
$\mathbf{u}_4$	$\mathbf{u}_4$			$\mathbf{u}_{10}$					$\mathbf{u}_{10}$			$\mathbf{u}_4$				
$\mathbf{u}_5$	$\mathbf{u}_5$			$\mathbf{u}_{11}$					$\mathbf{u}_{11}$			$\mathbf{u}_5$				
$\mathbf{u}_6$	$\mathbf{u}_6$			$\mathbf{u}_{12}$					$-\mathbf{u}_{12}$			$-\mathbf{u}_6$				

displacement of the *j*th particle, from its corner of the square, along the edge to the right (looking in toward the center), is labeled $\mathbf{u}_{3j-2}$, that along the edge to the left $\mathbf{u}_{3j-1}$, and that directly up $\mathbf{u}_{3j}$. Furthermore, axes and planes for symmetry operations are defined as figure 9.1 indicates.

By trial, we then find that each symmetry operation converts pertinent $\mathbf{u}_k$'s as table 9.1 shows.

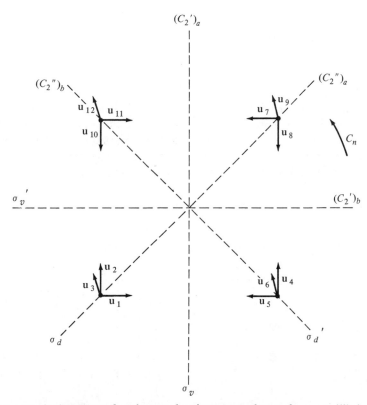

FIGURE 9.1 Equivalent sets of orthogonal unit vectors drawn from equilibrium positions at the corners of a square. The symmetry of the system is described by group $\mathbf{D}_{4h}$ and its subgroups, which include $\mathbf{C}_{4h}$, $\mathbf{D}_4$, $\mathbf{C}_{4v}$, and $\mathbf{C}_4$.

Combining the $R_i\mathbf{u}_k$'s with characters from table 9.2, as we did in example 8.12, and reducing yields the vectors:

$$A_{1g}: \quad \mathbf{f}_1 = \mathbf{u}_1 + \mathbf{u}_2 + \mathbf{u}_4 + \mathbf{u}_5 + \mathbf{u}_7 + \mathbf{u}_8 + \mathbf{u}_{10} + \mathbf{u}_{11}, \quad (9\text{-}19)$$

$$A_{2g}: \quad \mathbf{f}_2 = \mathbf{u}_1 - \mathbf{u}_2 + \mathbf{u}_4 - \mathbf{u}_5 + \mathbf{u}_7 - \mathbf{u}_8 + \mathbf{u}_{10} - \mathbf{u}_{11}, \quad (9\text{-}20)$$

$$A_{2u}: \quad \mathbf{f}_3 = \mathbf{u}_3 + \mathbf{u}_6 + \mathbf{u}_9 + \mathbf{u}_{12}, \quad (9\text{-}21)$$

$$B_{1g}: \quad \mathbf{f}_4 = \mathbf{u}_1 - \mathbf{u}_2 - \mathbf{u}_4 + \mathbf{u}_5 + \mathbf{u}_7 - \mathbf{u}_8 - \mathbf{u}_{10} + \mathbf{u}_{11}, \qquad (9\text{-}22)$$

$$B_{1u}: \quad \mathbf{f}_5 = \mathbf{u}_3 - \mathbf{u}_6 + \mathbf{u}_9 - \mathbf{u}_{12}, \qquad (9\text{-}23)$$

$$B_{2g}: \quad \mathbf{f}_6 = \mathbf{u}_1 + \mathbf{u}_2 - \mathbf{u}_4 - \mathbf{u}_5 + \mathbf{u}_7 + \mathbf{u}_8 - \mathbf{u}_{10} - \mathbf{u}_{11}, \qquad (9\text{-}24)$$

$$E_g: \quad \mathbf{f}_7 = \mathbf{u}_3 - \mathbf{u}_9, \qquad (9\text{-}25)$$

$$\mathbf{f}_8 = \mathbf{u}_6 - \mathbf{u}_{12}, \qquad (9\text{-}26)$$

$$E_u: \quad \mathbf{f}_9 = \mathbf{u}_1 - \mathbf{u}_7, \qquad (9\text{-}27)$$

$$\mathbf{f}_{10} = \mathbf{u}_{11} - \mathbf{u}_5, \qquad (9\text{-}28)$$

$$\mathbf{f}_{11} = \mathbf{u}_2 - \mathbf{u}_8, \qquad (9\text{-}29)$$

$$\mathbf{f}_{12} = \mathbf{u}_4 - \mathbf{u}_{10}. \qquad (9\text{-}30)$$

From equation (8-98), each of these vectors is a linear combination of bases for the pertinent irreducible representation. Since the number of independent vectors for A_{1g}, A_{2g}, A_{2u}, B_{1g}, B_{1u}, B_{2g}, E_g is equal to the dimensionality of each representation, these vectors are bases as they stand. But four independent combinations arise when v is E_u. These require two occurrences of the doubly degenerate representation.

TABLE 9.2   Characters for the Irreducible Representations of the $\mathbf{D}_{4h}$ Group

Representation	E	$2C_4$	C_2	$2C_2'$	$2C_2''$	i	$2S_4$	σ_h	$2\sigma_v$	$2\sigma_d$
A_{1g}	1	1	1	1	1	1	1	1	1	1
A_{2g}	1	1	1	-1	-1	1	1	1	-1	-1
B_{1g}	1	-1	1	1	-1	1	-1	1	1	-1
B_{2g}	1	-1	1	-1	1	1	-1	1	-1	1
E_g	2	0	-2	0	0	2	0	-2	0	0
A_{1u}	1	1	1	1	1	-1	-1	-1	-1	-1
A_{2u}	1	1	1	-1	-1	-1	-1	-1	1	1
B_{1u}	1	-1	1	1	-1	-1	1	-1	-1	1
B_{2u}	1	-1	1	-1	1	-1	1	-1	1	-1
E_u	2	0	-2	0	0	-2	0	2	0	0

The vectors $\mathbf{f}_9$, $\mathbf{f}_{10}$, $\mathbf{f}_{11}$, $\mathbf{f}_{12}$ correspond to the displacements in figure 9.2. By trial, we find that operations of the group transform them as table 9.3 indicates. The arrays form bases for the *reducible* representation $2E_u$.

To separate the two occurrences, we note that the first two entries in each column are

$$\mathbf{f}_9, \mathbf{f}_{10}, \quad \text{or} \quad -\mathbf{f}_9, -\mathbf{f}_{10}, \quad \text{or} \quad \mathbf{f}_{11}, \mathbf{f}_{12}, \quad \text{or} \quad -\mathbf{f}_{11}, -\mathbf{f}_{12}.$$

The other two entries also consist of one of these pairs.

TABLE 9.3 Effect of Each Symmetry Operation on the Combinations $\pm\frac{1}{4}\chi_{E_u}*(R_i)R_i\mathbf{u}_k$ listed

Operand	Operator															
	E	C_4	$C_4{}^{-1}$	C_2	$(C_2')_a$	$(C_2')_b$	$(C_2'')_a$	$(C_2'')_b$	i	S_4	$S_4{}^{-1}$	σ_h	σ_v	σ_v'	σ_d	σ_d'
$\mathbf{f}_9$	$\mathbf{f}_9$	$\mathbf{f}_{12}$	$-\mathbf{f}_{12}$	$-\mathbf{f}_9$	$-\mathbf{f}_{10}$	$\mathbf{f}_{10}$	$\mathbf{f}_{11}$	$-\mathbf{f}_{11}$	$-\mathbf{f}_9$	$\mathbf{f}_{12}$	$-\mathbf{f}_{12}$	$\mathbf{f}_9$	$-\mathbf{f}_{10}$	$\mathbf{f}_{10}$	$\mathbf{f}_{11}$	$-\mathbf{f}_{11}$
$\mathbf{f}_{10}$	$\mathbf{f}_{10}$	$\mathbf{f}_{11}$	$-\mathbf{f}_{11}$	$-\mathbf{f}_{10}$	$-\mathbf{f}_9$	$\mathbf{f}_9$	$\mathbf{f}_{12}$	$-\mathbf{f}_{12}$	$-\mathbf{f}_{10}$	$\mathbf{f}_{11}$	$-\mathbf{f}_{11}$	$\mathbf{f}_{10}$	$-\mathbf{f}_9$	$\mathbf{f}_9$	$\mathbf{f}_{12}$	$-\mathbf{f}_{12}$
$\mathbf{f}_{11}$	$\mathbf{f}_{11}$	$-\mathbf{f}_{10}$	$\mathbf{f}_{10}$	$-\mathbf{f}_{11}$	$\mathbf{f}_{12}$	$-\mathbf{f}_{12}$	$\mathbf{f}_9$	$-\mathbf{f}_9$	$-\mathbf{f}_{11}$	$-\mathbf{f}_{10}$	$\mathbf{f}_{10}$	$\mathbf{f}_{11}$	$\mathbf{f}_{12}$	$-\mathbf{f}_{12}$	$\mathbf{f}_9$	$-\mathbf{f}_9$
$\mathbf{f}_{12}$	$\mathbf{f}_{12}$	$-\mathbf{f}_9$	$\mathbf{f}_9$	$-\mathbf{f}_{12}$	$\mathbf{f}_{11}$	$-\mathbf{f}_{11}$	$\mathbf{f}_{10}$	$-\mathbf{f}_{10}$	$-\mathbf{f}_{12}$	$-\mathbf{f}_9$	$\mathbf{f}_9$	$\mathbf{f}_{12}$	$\mathbf{f}_{11}$	$-\mathbf{f}_{11}$	$\mathbf{f}_{10}$	$-\mathbf{f}_{10}$

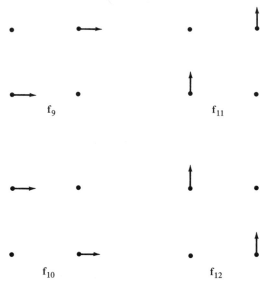

FIGURE 9.2   Displacements described by $\mathbf{f}_9, \mathbf{f}_{10}, \mathbf{f}_{11}$, and $\mathbf{f}_{12}$.

Consequently, operation R_i would cause

$$\mathbf{f}_9 + \mathbf{f}_{10} \quad \text{to go into} \quad \pm(\mathbf{f}_9 + \mathbf{f}_{10}) \quad \text{or} \quad \pm(\mathbf{f}_{11} + \mathbf{f}_{12}),$$

and

$$\mathbf{f}_{11} + \mathbf{f}_{12} \quad \text{to go into} \quad \pm(\mathbf{f}_{11} + \mathbf{f}_{12}) \quad \text{or} \quad \pm(\mathbf{f}_9 + \mathbf{f}_{10}).$$

It would make

$$\mathbf{f}_9 - \mathbf{f}_{10} \quad \text{go into} \quad \pm(\mathbf{f}_9 - \mathbf{f}_{10}) \quad \text{or} \quad \pm(\mathbf{f}_{11} - \mathbf{f}_{12}),$$

and

$$\mathbf{f}_{11} - \mathbf{f}_{12} \quad \text{go into} \quad \pm(\mathbf{f}_{11} - \mathbf{f}_{12}) \quad \text{or} \quad \pm(\mathbf{f}_9 - \mathbf{f}_{10}).$$

One occurrence of the representation is associated with the combinations

$$E_u: \quad \mathbf{f}_{13} = \mathbf{f}_9 + \mathbf{f}_{10} = \mathbf{u}_1 - \mathbf{u}_5 - \mathbf{u}_7 + \mathbf{u}_{11}, \tag{9-31}$$

$$\mathbf{f}_{14} = \mathbf{f}_{11} + \mathbf{f}_{12} = \mathbf{u}_2 + \mathbf{u}_4 - \mathbf{u}_8 - \mathbf{u}_{10}, \tag{9-32}$$

the other with the combinations

$$E_u: \quad \mathbf{f}_{15} = \mathbf{f}_9 - \mathbf{f}_{10} = \mathbf{u}_1 + \mathbf{u}_5 - \mathbf{u}_7 - \mathbf{u}_{11}, \tag{9-33}$$

$$\mathbf{f}_{16} = \mathbf{f}_{11} - \mathbf{f}_{12} = \mathbf{u}_2 - \mathbf{u}_4 - \mathbf{u}_8 + \mathbf{u}_{10}. \tag{9-34}$$

9.5   The corresponding motions

The particles in a cluster may move so that just one q_l varies from zero. The composite $\mathbf{q}$ then lies along the corresponding $\mathbf{f}_l$, and $\mathbf{q}$ itself equals a scalar function of time multiplied by $\mathbf{f}_l$. At any given instant of time, $\mathbf{q}$ is plus or minus some quantity times $\mathbf{f}_l$.

Accordingly, a typical phase in an elementary motion of an array is represented by a plot of the corresponding $\mathbf{f}_l$. Formulas (9-19)–(9-26) and (9-31)–(9-34) yield the arrangements in figure 9.3. The corresponding motions may be described as follows.

In the A_{1g} motion, the four particles move symmetrically in and out along radii, executing a breathing movement. The A_{2g} type involves the particles moving perpendicular to radii in one direction around a circle; it is a rotation. In the A_{2u} motion, the particles move together out of the paper, in a translatory motion. The B_{1g} kind involves the particles moving perpendicular to the radii in alternating directions.

The B_{1u} type involves concerted motion of opposite corners up and down; it is an undulatory movement. In the B_{2g} motion, opposite corners move in and out. The distorted form here is diamond-shaped. In the E_g movement, the particles rotate about axes in the plane of the reference positions.

The first kind of E_u motion is a translation in the plane of the equilibrium positions. The second kind involves a scissors movement of opposite sides.

EXAMPLE 9.1 For the square array of particles, set up a linear combination of displacements that represent each symmetry-adapted generalized coordinate q_l.

In the r-dimensional hypothetical space in which the $\mathbf{u}_k$'s are mutually perpendicular unit vectors, the Cartesian displacements are components of a single vector,

$$\mathbf{q} = \eta_k \mathbf{u}_k,$$

while the base for a row of each irreducible representation is another vector,

$$\mathbf{f}_l = a_{lm}\mathbf{u}_m.$$

Substituting these forms into equation (9-16) yields

$$q_l = \mathbf{q} \cdot \mathbf{f}_l = a_{lk}\eta_k$$

a variable containing the η_k's combined as the $\mathbf{u}_m$'s are in the formula for $\mathbf{f}_l$. Since this variable is proportional to the projection of $\mathbf{q}$ on $\mathbf{f}_l$, it has the same symmetry properties; it belongs to the same row of the irreducible representation as $\mathbf{f}_l$ does.

Equations (9-19) through (9-26) and (9-31) through (9-34) thus yield the following symmetry-adapted coordinates:

$$
\begin{aligned}
A_{1g}: \quad & q_1 = \mathbf{q} \cdot \mathbf{f}_1 = \eta_1 + \eta_2 + \eta_4 + \eta_5 + \eta_7 + \eta_8 + \eta_{10} + \eta_{11}, \\
A_{2g}: \quad & q_2 = \mathbf{q} \cdot \mathbf{f}_2 = \eta_1 - \eta_2 + \eta_4 - \eta_5 + \eta_7 - \eta_8 + \eta_{10} - \eta_{11}, \\
A_{2u}: \quad & q_3 = \mathbf{q} \cdot \mathbf{f}_3 = \eta_3 + \eta_6 + \eta_9 + \eta_{12}, \\
B_{1g}: \quad & q_4 = \mathbf{q} \cdot \mathbf{f}_4 = \eta_1 - \eta_2 - \eta_4 + \eta_5 + \eta_7 - \eta_8 - \eta_{10} + \eta_{11}, \\
B_{1u}: \quad & q_5 = \mathbf{q} \cdot \mathbf{f}_5 = \eta_3 - \eta_6 + \eta_9 - \eta_{12}, \\
B_{2g}: \quad & q_6 = \mathbf{q} \cdot \mathbf{f}_6 = \eta_1 + \eta_2 - \eta_4 - \eta_5 + \eta_7 + \eta_8 - \eta_{10} - \eta_{11}, \\
E_g: \quad & q_7 = \mathbf{q} \cdot \mathbf{f}_7 = \eta_3 - \eta_9, \\
& q_8 = \mathbf{q} \cdot \mathbf{f}_8 = \eta_6 - \eta_{12}, \\
E_u: \quad & q_{13} = \mathbf{q} \cdot \mathbf{f}_{13} = \eta_1 - \eta_5 - \eta_7 + \eta_{11}, \\
& q_{14} = \mathbf{q} \cdot \mathbf{f}_{14} = \eta_2 + \eta_4 - \eta_8 - \eta_{10}, \\
& q_{15} = \mathbf{q} \cdot \mathbf{f}_{15} = \eta_1 + \eta_5 - \eta_7 - \eta_{11}, \\
& q_{16} = \mathbf{q} \cdot \mathbf{f}_{16} = \eta_2 - \eta_4 - \eta_8 + \eta_{10}.
\end{aligned}
$$

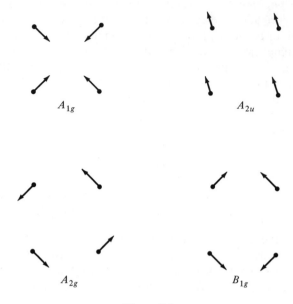

Figure 9.3a

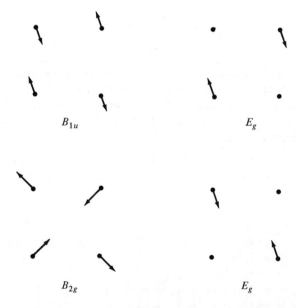

Figure 9.3b

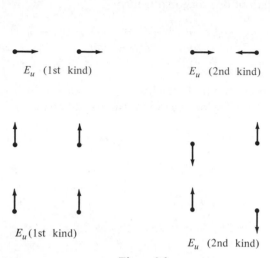

E_u (1st kind)
E_u (2nd kind)
E_u(1st kind)
E_u (2nd kind)

Figure 9.3c

FIGURE 9.3 Arrangements of vectors that act as bases for the (a) A_{1g}, (b) A_{2g}, (c) A_{2u}, (d) B_{1g}, (e) B_{1u}, (f) B_{2g}, (g) E_g, (h) E_u representations of the $\mathbf{D}_{4h}$ group.

EXAMPLE 9.2 Linearize the formula for the extension of a diagonal spring in figure 9.4.

The displacements are labeled so they correspond to the unit vectors in figure 9.1.

$$\boldsymbol{\eta}_1 = \eta_1 \mathbf{u}_1, \boldsymbol{\eta}_2 = \eta_2 \mathbf{u}_2, \ldots .$$

k_1
m
m
k_2
k_1
k_2
k_1
m
m
k_1

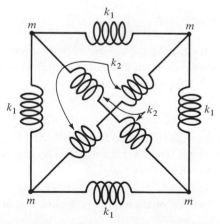

FIGURE 9.4 $\mathbf{D}_{4h}$ arrangement of masses and springs to be considered.

The square of the length of one diagonal is then

$$l^2 = (a - \eta_7 - \eta_1)^2 + (a - \eta_8 - \eta_2)^2 + (\eta_9 - \eta_3)^2$$
$$= 2a^2 - 2a\eta_1 - 2a\eta_2 - 2a\eta_7 - 2a\eta_8 + 2\eta_1\eta_7 + 2\eta_2\eta_8$$
$$- 2\eta_3\eta_9 + \eta_1{}^2 + \eta_2{}^2 + \eta_7{}^2 + \eta_8{}^2 + \eta_9{}^2 + \eta_3{}^2$$

where a is the equilibrium length of an edge.

To find coefficients for a Taylor expansion, we first differentiate

$$2l\, \frac{\partial l}{\partial \eta_1} = -2a + 2\eta_1 + 2\eta_7,$$

$$2l\, \frac{\partial l}{\partial \eta_2} = -2a + 2\eta_2 + 2\eta_8,$$

$$2l\, \frac{\partial l}{\partial \eta_7} = -2a + 2\eta_7 + 2\eta_1,$$

$$2l\, \frac{\partial l}{\partial \eta_8} = -2a + 2\eta_8 + 2\eta_2,$$

$$2l\, \frac{\partial l}{\partial \eta_3} = 2\eta_3 - 2\eta_9,$$

$$2l\, \frac{\partial l}{\partial \eta_9} = 2\eta_9 - 2\eta_3.$$

Substituting the values

$$l = \sqrt{2}a, \qquad \eta_1 = \eta_2 = \eta_7 = \eta_8 = \eta_9 = \eta_3 = 0$$

that hold at the equilibrium position yields

$$\left(\frac{\partial l}{\partial \eta_k} \right)_0 = -\frac{1}{\sqrt{2}}$$

when k is 1, 2, 7, 8. Otherwise, the partial derivative is zero.

Consequently, the Taylor expansion about the equilibrium length is

$$l = \sqrt{2}a - \frac{1}{\sqrt{2}} (\eta_1 + \eta_2 + \eta_7 + \eta_8) + \cdots.$$

The extension is obtained on shifting $\sqrt{2}a$ to the left side.

In a linear theory, the higher terms are dropped. The extension of an edge spring is linearized in a similar manner.

9.6 Simultaneous equations for small-amplitude oscillations of the masses

Let us next represent the interaction between each pair of masses by a spring consistent with the $\mathbf{D}_{4h}$ symmetry. For simplicity, only small displacements are considered.

By the procedure in example 9.2, we then find that the instantaneous extension of each edge is

$$l_1 - a \simeq -\eta_1 - \eta_5, \tag{9-35}$$

$$l_2 - a \simeq -\eta_4 - \eta_8, \tag{9-36}$$

$$l_3 - a \simeq -\eta_7 - \eta_{11}, \tag{9-37}$$

$$l_4 - a \simeq -\eta_{10} - \eta_2, \tag{9-38}$$

while the extension of each diagonal is

$$l_5 - \sqrt{2}a \simeq -\frac{1}{\sqrt{2}}(\eta_1 + \eta_2 + \eta_7 + \eta_8), \tag{9-39}$$

$$l_6 - \sqrt{2}a \simeq -\frac{1}{\sqrt{2}}(\eta_4 + \eta_5 + \eta_{10} + \eta_{11}). \tag{9-40}$$

Remember how the displacements are labeled to correspond with the unit vectors of figure 9.1.

Letting the potential energy stored in a spring be one-half the *spring constant* times the square of the extension then yields

$$V = \tfrac{1}{2}k_1[(\eta_1 + \eta_5)^2 + (\eta_4 + \eta_8)^2 + (\eta_7 + \eta_{11})^2 + (\eta_{10} + \eta_2)^2]$$
$$+ \tfrac{1}{2}k_2[\tfrac{1}{2}(\eta_1 + \eta_2 + \eta_7 + \eta_8)^2 + \tfrac{1}{2}(\eta_4 + \eta_5 + \eta_{10} + \eta_{11})^2], \tag{9-41}$$

a quadratic form like (6-18).

As in (6-11), the kinetic energy expression is

$$T = \tfrac{1}{2}m[(\dot{\eta}_1^2 + \dot{\eta}_2^2 + \dot{\eta}_3^2) + (\dot{\eta}_4^2 + \dot{\eta}_5^2 + \dot{\eta}_6^2)$$
$$+ (\dot{\eta}_7^2 + \dot{\eta}_8^2 + \dot{\eta}_9^2) + (\dot{\eta}_{10}^2 + \dot{\eta}_{11}^2 + \dot{\eta}_{12}^2)]. \tag{9-42}$$

These produce a 12×12 determinant in the secular equation.

Since the η_3, η_6, η_9, and η_{12} displacements are not present in (9-41), the one vibration that involves these, B_{1u}, must behave as a translation or rotation when its amplitude is small; it must be intrinsically nonlinear (a similar situation was considered in problem 6.18).

The Lagrange equation for each of these transverse displacements reduces to the kinetic energy part, in the approximation we are considering. It does not involve the other coordinates and so does not influence their behavior. In considering them, we may drop the terms containing $\dot{\eta}_3^2$, $\dot{\eta}_6^2$, $\dot{\eta}_9^2$, and $\dot{\eta}_{12}^2$ from T.

The matrix of the coefficients in T then reduces to

$$\mathbf{A} = \tfrac{1}{2}m \begin{pmatrix} 1 & 0 & 0 & 0 & 0 & 0 & 0 & 0 \\ 0 & 1 & 0 & 0 & 0 & 0 & 0 & 0 \\ 0 & 0 & 1 & 0 & 0 & 0 & 0 & 0 \\ 0 & 0 & 0 & 1 & 0 & 0 & 0 & 0 \\ 0 & 0 & 0 & 0 & 1 & 0 & 0 & 0 \\ 0 & 0 & 0 & 0 & 0 & 1 & 0 & 0 \\ 0 & 0 & 0 & 0 & 0 & 0 & 1 & 0 \\ 0 & 0 & 0 & 0 & 0 & 0 & 0 & 1 \end{pmatrix}. \tag{9-43}$$

Multiplying (9-41) out yields the corresponding matrix for the coefficients in V:

$$\mathbf{B} = \tfrac{1}{2} \times \tag{9-44}$$

$$
\begin{pmatrix}
k_1 + \tfrac{1}{2}k_2 & \tfrac{1}{2}k_2 & 0 & k_1 & \tfrac{1}{2}k_2 & \tfrac{1}{2}k_2 & 0 & 0 \\
\tfrac{1}{2}k_2 & k_1 + \tfrac{1}{2}k_2 & 0 & 0 & \tfrac{1}{2}k_2 & \tfrac{1}{2}k_2 & k_1 & 0 \\
0 & 0 & k_1 + \tfrac{1}{2}k_2 & \tfrac{1}{2}k_2 & 0 & k_1 & \tfrac{1}{2}k_2 & \tfrac{1}{2}k_2 \\
k_1 & 0 & \tfrac{1}{2}k_2 & k_1 + \tfrac{1}{2}k_2 & 0 & 0 & \tfrac{1}{2}k_2 & \tfrac{1}{2}k_2 \\
\tfrac{1}{2}k_2 & \tfrac{1}{2}k_2 & 0 & 0 & k_1 + \tfrac{1}{2}k_2 & \tfrac{1}{2}k_2 & 0 & k_1 \\
\tfrac{1}{2}k_2 & \tfrac{1}{2}k_2 & k_1 & 0 & \tfrac{1}{2}k_2 & k_1 + \tfrac{1}{2}k_2 & 0 & 0 \\
0 & k_1 & \tfrac{1}{2}k_2 & \tfrac{1}{2}k_2 & 0 & 0 & k_1 + \tfrac{1}{2}k_2 & \tfrac{1}{2}k_2 \\
0 & 0 & \tfrac{1}{2}k_2 & \tfrac{1}{2}k_2 & k_1 & 0 & \tfrac{1}{2}k_2 & k_1 + \tfrac{1}{2}k_2
\end{pmatrix}
$$

Substituting the elements of $\mathbf{A}$ and $\mathbf{B}$ into (6-23) and multiplying each equation by 2 yields

$$m\ddot{\eta}_1 + (k_1 + \tfrac{1}{2}k_2)\eta_1 + \tfrac{1}{2}k_2\eta_2 + 0 + k_1\eta_5$$
$$+ \tfrac{1}{2}k_2\eta_7 + \tfrac{1}{2}k_2\eta_8 + 0 + 0 = 0, \tag{9-45}$$

$$m\ddot{\eta}_2 + (k_1 + \tfrac{1}{2}k_2)\eta_2 + 0 + 0 + \tfrac{1}{2}k_2\eta_7$$
$$+ \tfrac{1}{2}k_2\eta_8 + k_1\eta_{10} + 0 + \tfrac{1}{2}k_2\eta_1 = 0, \tag{9-46}$$

$$m\ddot{\eta}_4 + (k_1 + \tfrac{1}{2}k_2)\eta_4 + \tfrac{1}{2}k_2\eta_5 + 0 + k_1\eta_8$$
$$+ \tfrac{1}{2}k_2\eta_{10} + \tfrac{1}{2}k_2\eta_{11} + 0 + 0 = 0, \tag{9-47}$$

$$m\ddot{\eta}_5 + (k_1 + \tfrac{1}{2}k_2)\eta_5 + 0 + 0 + \tfrac{1}{2}k_2\eta_{10}$$
$$+ \tfrac{1}{2}k_2\eta_{11} + k_1\eta_1 + 0 + \tfrac{1}{2}k_2\eta_4 = 0, \tag{9-48}$$

$$m\ddot{\eta}_7 + (k_1 + \tfrac{1}{2}k_2)\eta_7 + \tfrac{1}{2}k_2\eta_8 + 0 + k_1\eta_{11}$$
$$+ \tfrac{1}{2}k_2\eta_1 + \tfrac{1}{2}k_2\eta_2 + 0 + 0 = 0, \tag{9-49}$$

$$m\ddot{\eta}_8 + (k_1 + \tfrac{1}{2}k_2)\eta_8 + 0 + 0 + \tfrac{1}{2}k_2\eta_1$$
$$+ \tfrac{1}{2}k_2\eta_2 + k_1\eta_4 + 0 + \tfrac{1}{2}k_2\eta_7 = 0, \tag{9-50}$$

$$m\ddot{\eta}_{10} + (k_1 + \tfrac{1}{2}k_2)\eta_{10} + \tfrac{1}{2}k_2\eta_{11} + 0 + k_1\eta_2$$
$$+ \tfrac{1}{2}k_2\eta_4 + \tfrac{1}{2}k_2\eta_5 + 0 + 0 = 0, \tag{9-51}$$

$$m\ddot{\eta}_{11} + (k_1 + \tfrac{1}{2}k_2)\eta_{11} + 0 + 0 + \tfrac{1}{2}k_2\eta_4$$
$$+ \tfrac{1}{2}k_2\eta_5 + k_1\eta_7 + 0 + \tfrac{1}{2}k_2\eta_{10} = 0, \tag{9-52}$$

if the term containing $\ddot{\eta}_k$ is written first, with the other terms following it in order.

Next, equations (9-45) through (9-52) will be combined as the formulas in example 9.1, or equations (9-19)–(9-26) and (9-31)–(9-34), indicate. Each resulting equation will govern a single normal motion because in the given system no more than one vibration belongs to a row of an irreducible representation.

EXAMPLE 9.3   What differential equation governs the large-amplitude breathing motion of a square array held together by springs that are neither soft nor hard?

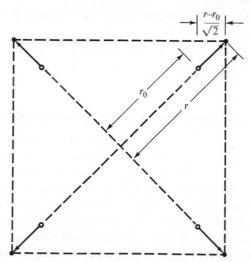

FIGURE 9.5   A typical stage in the breathing (A_{1g}) motion of a square array.

The first diagram of figure 9.3 shows how each mass moves in and out together, preserving the square arrangement at all times, in the A_{1g} motion. Now, let r be the distance of a corner from the center of mass at time t and r_0 be the equilibrium position, as in figure 9.5. Set

$$q = r - r_0.$$

Then the kinetic energy of the system is

$$T = 4(\tfrac{1}{2}m\dot{r}^2) = 2m\dot{q}^2,$$

while the potential energy stored in the diagonal springs at any instant of time is

$$V_{\text{diag}} = 2\{\tfrac{1}{2}k_2[2(r - r_0)]^2\} = 4k_2(r - r_0)^2 = 4k_2q^2,$$

and that in the springs along the edges is

$$V_{\text{edge}} = 4\left\{\tfrac{1}{2}k_1\left[2\left(\frac{r - r_0}{\sqrt{2}}\right)\right]^2\right\} = 4k_1(r - r_0)^2 = 4k_1q^2.$$

The Lagrangian,

$$L = T - V = 2m\dot{q}^2 - 4(k_1 + k_2)q^2,$$

converts the equation of motion,

$$\frac{d}{dt}\frac{\partial L}{\partial \dot{q}} - \frac{\partial L}{\partial q} = 0,$$

to

$$4m\ddot{q} + 8(k_1 + k_2)q = 0$$

or

$$m\ddot{q} + 2(k_1 + k_2)q = 0,$$

the same form that a harmonic oscillator obeys. The solutions are sinusoidal functions of frequency

$$v = \frac{1}{2\pi} \left[\frac{2(k_1 + k_2)}{m} \right]^{1/2}$$

(see section 6.2).

9.7 Separating representations from the linearized equations of motion

A linear combination of equations (9-45) through (9-52) will be constructed to make the resulting derivative term a basis for each irreducible representation that occurs, in turn. The rest of each equation must have a symmetry compatible with the derivative part; it must collectively belong to the same representation (recall the discussion in section 9.3). If the representation occurs but once, these terms make up a parameter times the pertinent generalized coordinate, and a solution can be written immediately. If it occurs more than once, but only one occurrence involves a vibration, the same procedure can be followed. Otherwise, the equation must be considered simultaneously with the other independent vibrational equations for the same row of the irreducible representation.

The procedure is simple because each of the starting equations contains just one derivative term, which is m times an $\ddot{\eta}_k$. For these to add up to an $m\ddot{q}_1$, the equations need only be combined as the η_k's are in example 9.1.

In the basis for the A_{1g} representation, all the η_k's are added. Consequently, we add equations (9-45) through (9-52) as they stand. Then we let

$$\eta_1 + \eta_2 + \eta_4 + \eta_5 + \eta_7 + \eta_8 + \eta_{10} + \eta_{11} = q_1 \tag{9-53}$$

and

$$\ddot{\eta}_1 + \ddot{\eta}_2 + \ddot{\eta}_4 + \ddot{\eta}_5 + \ddot{\eta}_7 + \ddot{\eta}_8 + \ddot{\eta}_{10} + \ddot{\eta}_{11} = \ddot{q}_1 \tag{9-54}$$

to get the result,

$$m\ddot{q}_1 + 2(k_1 + k_2)q_1 = 0, \tag{9-55}$$

which agrees with that in example 9.3. On comparing it with equations (6-4) and (6-3), we see that the breathing frequency is

$$v_1 = \frac{1}{2\pi} \left[\frac{2(k_1 + k_2)}{m} \right]^{1/2}. \tag{9-56}$$

For the A_{2g} representation, we have

$$\eta_1 - \eta_2 + \eta_4 - \eta_5 + \eta_7 - \eta_8 + \eta_{10} - \eta_{11} = q_2. \tag{9-57}$$

Combining equations (9-45) through (9-52) with alternating signs yields a form that reduces to

$$m\ddot{q}_2 = 0. \tag{9-58}$$

Integrating this equation yields a constant $\dot{q}_2$ over any small interval of time. The A_{2g} rotation proceeds at a fixed angular velocity.

Taking equation (9-45) minus (9-46) minus (9-47) plus (9-48) plus (9-49) minus (9-50) minus (9-51) plus (9-52) and letting

$$\eta_1 - \eta_2 - \eta_4 + \eta_5 + \eta_7 - \eta_8 - \eta_{10} + \eta_{11} = q_4 \tag{9-59}$$

leads to

$$m\ddot{q}_4 + 2k_1 q_4 = 0. \tag{9-60}$$

Consequently, the frequency of the B_{1g} small oscillation is

$$\nu_4 = \frac{1}{2\pi}\left(\frac{2k_1}{m}\right)^{1/2}. \tag{9-61}$$

Since this motion does not cause the length of a diagonal to vary, in the approximation that the amplitude is small, the frequency is independent of the spring constant k_2.

Equation (9-24) shows that the normal coordinate for the B_{2g} oscillation is

$$\eta_1 + \eta_2 - \eta_4 - \eta_5 + \eta_7 + \eta_8 - \eta_{10} - \eta_{11} = q_6. \tag{9-62}$$

From equation (9-45) plus (9-46) minus (9-47) minus (9-48) plus (9-49) plus (9-50) minus (9-51) minus (9-52), we get

$$m\ddot{q}_6 + 2k_2 q_6 = 0, \tag{9-63}$$

whence

$$\nu_6 = \frac{1}{2\pi}\left(\frac{2k_2}{m}\right)^{1/2}. \tag{9-64}$$

This result does not involve k_1 because the small distortion takes place without any change in length of the edges.

A generalized coordinate for the E_u translation is obtained from equation (9-31) as

$$\eta_1 - \eta_5 - \eta_7 + \eta_{11} = q_{13}. \tag{9-65}$$

Combining (9-45) minus (9-48) minus (9-49) plus (9-52) leaves

$$m\ddot{q}_{13} = 0, \tag{9-66}$$

whence

$$\dot{q}_{13} = \text{constant}. \tag{9-67}$$

Similarly, with

$$\eta_2 + \eta_4 - \eta_8 - \eta_{10} = q_{14}, \tag{9-68}$$

we find that

$$\dot{q}_{14} = \text{constant}. \tag{9-69}$$

To separate out an E_u scissors motion, use the combination in (9-33), taking equation (9-45) plus (9-48) minus (9-49) minus (9-52). Then letting

$$\eta_1 + \eta_5 - \eta_7 - \eta_{11} = q_{15} \tag{9-70}$$

leaves

$$m\ddot{q}_{15} + 2k_1 q_{15} = 0. \tag{9-71}$$

For the independent motion of the same type, use the combination in (9-34). Then

$$\eta_2 - \eta_4 - \eta_8 + \eta_{10} = q_{16}, \tag{9-72}$$

and

$$m\ddot{q}_{16} + 2k_1 q_{16} = 0. \tag{9-73}$$

Equations (9-71) and (9-73) contain the same coefficients as (9-60) does; so the frequency of the E_u oscillation is the same:

$$\nu_{15} = \nu_{16} = \frac{1}{2\pi}\left(\frac{2k_1}{m}\right)^{1/2}. \tag{9-74}$$

Different modes of motion are said to be *degenerate* if they appear with the same frequency. The degeneracy between the two E_u modes is due to the symmetry of the system in space, as reflected in the 2-dimensionality of the representation itself. The degeneracy with the B_{1g} mode, on the other hand, is not required by the spatial symmetry. Insofar as the $\mathbf{D}_{4h}$ group is concerned, it is *accidental*.

This degeneracy is caused by equivalences between spring constants and between masses in the Lagrange equations. The $\mathbf{D}_{4h}$ operations do not include possible permutations of the equivalent springs with respect to the masses. Nevertheless, the pertinent reduced equations of motion, and the governing Lagrangian, are invariant to these additional permutations. We say there is a *form invariance* present which the group has not taken into account.

Oftentimes, an accidental degeneracy that is parameter-independent indicates the existence of more symmetry operations than those being considered.

9.8   Reducing the symmetry

Let us next consider what happens if symmetry in a system is reduced by altering masses, and/or spring constants, and/or equilibrium positions. General principles will be developed first.

Any reduction in symmetry eliminates some of the symmetry operations. The $\mathbf{D}_{2h}$ subgroup of $\mathbf{D}_{4h}$, for instance, does not contain the two C_4 operations, the two S_4 operations, the two σ_d operations, or the two C_2'' operations. In the reduction, the distinctions between some representations disappear. Furthermore, degenerate representations may be split.

The matrices needed to effect the remaining symmetry operations on a given basis are the same as before, so the characters for the surviving classes of a representation are not altered by the reduction. We need only compare these characters with some of those for the reduced group to identify the representation. Each initially degenerate representation that is split, ends up as a reducible representation. The theory for breaking down a reducible representation, developed in section 8.9, applies.

Tables 9.2 and 9.4 contain the information needed to correlate the modes of $\mathbf{D}_{4h}$ and $\mathbf{D}_{2h}$ oscillators. First, delete the columns in table 9.2 for $2C_4$, $2S_4$, $2\sigma_d$, and $2C_2''$. Notice that $(C_2')_b$, $(C_2')_a$ become $C_2(x)$, $C_2(y)$ while the σ_v's become $\sigma(zx)$ and $\sigma(yz)$.

TABLE 9.4 Characters for Representations of the $\mathbf{D}_{2h}$ Group

Representation	Class							
	E	$C_2(z)$	$C_2(y)$	$C_2(x)$	i	$\sigma(xy)$	$\sigma(zx)$	$\sigma(yz)$
A_g	1	1	1	1	1	1	1	1
B_{1g}	1	1	-1	-1	1	1	-1	-1
B_{2g}	1	-1	1	-1	1	-1	1	-1
B_{3g}	1	-1	-1	1	1	-1	-1	1
A_u	1	1	1	1	-1	-1	-1	-1
B_{1u}	1	1	-1	-1	-1	-1	1	1
B_{2u}	1	-1	1	-1	-1	1	-1	1
B_{3u}	1	-1	-1	1	-1	1	1	-1
Γ_{E_g}	2	-2	0	0	2	-2	0	0
Γ_{E_u}	2	-2	0	0	-2	2	0	0

Then the remaining characters for A_{1g} and B_{1g} are those of A_g, those remaining for B_{1u} of $\mathbf{D}_{4h}$ are those of A_u in $\mathbf{D}_{2h}$, and so on. For the degenerate E_g and E_u of $\mathbf{D}_{4h}$, we are left with the characters of the reducible representations Γ_{E_g} and Γ_{E_u}. These break down into B_{2g}, B_{3g} and B_{2u}, B_{3u}, respectively. The results in table 9.5 are thus obtained.

TABLE 9.5 Correlating Modes of $\mathbf{D}_{4h}$ and $\mathbf{D}_{2h}$

$\mathbf{D}_{4h}$	$\mathbf{D}_{2h}$
A_{1g}	A_g
B_{1g}	A_g
B_{1u}	A_u
A_{2g}	B_{1g}
B_{2g}	B_{1g}
A_{2u}	B_{1u}
E_g	B_{2g}, B_{3g}
E_u	B_{2u}, B_{3u}

9.9 Small-amplitude oscillations of a $\mathbf{D}_{2h}$ array

Let us now reduce the symmetry of the system in figure 9.4 by altering the two springs on opposite sides by the same amount. The configuration becomes that described by figure 9.6, and equation (9-41) is replaced with

$$V = \tfrac{1}{2}k_1[(\eta_1 + \eta_5)^2 + (\eta_7 + \eta_{11})^2]$$
$$+ \tfrac{1}{2}k_2[(\eta_4 + \eta_8)^2 + (\eta_{10} + \eta_2)^2]$$
$$+ \tfrac{1}{2}k_3[\tfrac{1}{2}(\eta_1 + \eta_2 + \eta_7 + \eta_8)^2 + \tfrac{1}{2}(\eta_4 + \eta_5 + \eta_{10} + \eta_{11})^2]. \qquad (9\text{-}75)$$

The change causes parameter k_2 in equations (9-45) through (9-52) to become k_3, and k_1 in (9-46), (9-47), (9-50), and (9-51) to become k_2. Since the same basis vectors

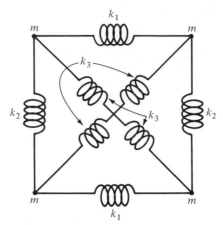

FIGURE 9.6   A $\mathbf{D}_{2h}$ arrangement of masses and springs.

are still valid, we may combine equations as the formulas in example 9.1 indicate to separate the representations.

The combinations for vibrational modes in the same representation must be considered together, however, since they are generally not independent. The interaction causes the original basis vectors to mix, so in practice, it is well to choose convenient linear mixtures of them. The corresponding combinations of equations are then formed.

As an example, we have the two A_g combinations that came from the A_{1g} and B_{1g} modes of $\mathbf{D}_{4h}$. Instead of using bases $\mathbf{f}_1$ and $\mathbf{f}_4$, we can employ the orthogonal set

$$\mathbf{f}_{17} = \tfrac{1}{2}(\mathbf{f}_1 + \mathbf{f}_4) = \mathbf{u}_1 + \mathbf{u}_5 + \mathbf{u}_7 + \mathbf{u}_{11}, \tag{9-76}$$

$$\mathbf{f}_{18} = \tfrac{1}{2}(\mathbf{f}_1 - \mathbf{f}_4) = \mathbf{u}_2 + \mathbf{u}_4 + \mathbf{u}_8 + \mathbf{u}_{10}. \tag{9-77}$$

First, note that matrix $2\mathbf{B}$ is $\hspace{3cm}$ (9-78)

$$\begin{pmatrix}
k_1 + \tfrac{1}{2}k_3 & \tfrac{1}{2}k_3 & 0 & k_1 & \tfrac{1}{2}k_3 & \tfrac{1}{2}k_3 & 0 & 0 \\
\tfrac{1}{2}k_3 & k_2 + \tfrac{1}{2}k_3 & 0 & 0 & \tfrac{1}{2}k_3 & \tfrac{1}{2}k_3 & k_2 & 0 \\
0 & 0 & k_2 + \tfrac{1}{2}k_3 & \tfrac{1}{2}k_3 & 0 & k_2 & \tfrac{1}{2}k_3 & \tfrac{1}{2}k_3 \\
k_1 & 0 & \tfrac{1}{2}k_3 & k_1 + \tfrac{1}{2}k_3 & 0 & 0 & \tfrac{1}{2}k_3 & \tfrac{1}{2}k_3 \\
\tfrac{1}{2}k_3 & \tfrac{1}{2}k_3 & 0 & 0 & k_1 + \tfrac{1}{2}k_3 & \tfrac{1}{2}k_3 & 0 & k_1 \\
\tfrac{1}{2}k_3 & \tfrac{1}{2}k_3 & k_2 & 0 & \tfrac{1}{2}k_3 & k_2 + \tfrac{1}{2}k_3 & 0 & 0 \\
0 & k_2 & \tfrac{1}{2}k_3 & \tfrac{1}{2}k_3 & 0 & 0 & k_2 + \tfrac{1}{2}k_3 & \tfrac{1}{2}k_3 \\
0 & 0 & \tfrac{1}{2}k_3 & \tfrac{1}{2}k_3 & k_1 & 0 & \tfrac{1}{2}k_3 & k_1 + \tfrac{1}{2}k_3
\end{pmatrix},$$

while matrix $\mathbf{A}$ is still given by (9-43).

For simplicity, number the rows and columns as the η_k's are numbered: 1, 2, 4, 5, 7, 8, 10, 11. Substitute the elements from the jth row of $\mathbf{A}$, $\mathbf{B}$ into equation (6-23) and multiply by 2. The coefficient of η_k in the jth simultaneous equation is then $2B_{jk}$, which can be obtained directly from matrix (9-78). The coefficient of $\ddot{\eta}_j$, in the jth equation, is m as before.

Equations (9-76) and (9-77) lead to the generalized coordinates

$$\eta_1 + \eta_5 + \eta_7 + \eta_{11} = q_{17}, \tag{9-79}$$

$$\eta_2 + \eta_4 + \eta_8 + \eta_{10} = q_{18}, \tag{9-80}$$

for the A_g motions. Corresponding to q_{17}, we add the 1st, 5th, 7th, 11th equations to derive

$$m\ddot{q}_{17} + (2k_1 + k_3)q_{17} + k_3 q_{18} = 0. \tag{9-81}$$

Following the combinations in q_{18}, we add the 2nd, 4th, 8th, and 10th equations to get

$$k_3 q_{17} + m\ddot{q}_{18} + (2k_2 + k_3)q_{18} = 0. \tag{9-82}$$

(Remember, the 3rd, 6th, 9th, and 12th equations are missing from the set.)
A sinusoidal solution of angular frequency ω satisfies equation (6-4):

$$\ddot{q}_l = -\omega^2 q_l. \tag{9-83}$$

This substitution changes the differential equations to algebraic ones,

$$[m\omega^2 - (2k_1 + k_3)]q_{17} - k_3 q_{18} = 0, \tag{9-84}$$

$$-k_3 q_{17} + [m\omega^2 - (2k_2 + k_3)]q_{18} = 0, \tag{9-85}$$

that are homogeneous.
For either q_{17} or q_{18} to differ from zero, we must consequently have

$$\begin{vmatrix} m\omega^2 - (2k_1 + k_3) & -k_3 \\ -k_3 & m\omega^2 - (2k_2 + k_3) \end{vmatrix} = 0. \tag{9-86}$$

This 2×2 secular determinant will be solved in example 9.4.
The A_u mode comes from the B_{1u} mode in $\mathbf{D}_{4h}$, for which

$$\eta_3 - \eta_6 + \eta_9 - \eta_{12} = q_5. \tag{9-87}$$

This is a vibrational mode. Since $\eta_3, \eta_6, \eta_9, \eta_{12}$ are absent from potential (9-75), no restoring force appears in the linear approximation. The vibration is intrinsically nonlinear.
Although there are two B_{1g} modes, they do not lead to a secular equation. The coordinate

$$\eta_1 - \eta_2 + \eta_4 - \eta_5 + \eta_7 - \eta_8 + \eta_{10} - \eta_{11} = q_2 \tag{9-88}$$

obtained from (9-20) causes us to combine the simultaneous equations with alternating signs. We then find

$$m\ddot{q}_2 = 0 \tag{9-89}$$

as in equation (9-58). The motion is the rotation described before.
The other B_{1g} motion correlates with the B_{2g} motion in the $\mathbf{D}_{4h}$ structure, for which

$$\eta_1 + \eta_2 - \eta_4 - \eta_5 + \eta_7 + \eta_8 - \eta_{10} - \eta_{11} = q_6. \tag{9-90}$$

Combining the simultaneous equations in the same manner, plus the 1st and 2nd, minus the 4th and 5th, plus the 7th and 8th, minus the 10th and 11th, yields

$$m\ddot{q}_6 + 2k_3 q_6 = 0, \tag{9-91}$$

whence

$$v_6 = \frac{1}{2\pi} \left(\frac{2k_3}{m} \right)^{1/2}. \tag{9-92}$$

The B_{2g} and B_{3g} modes of $\mathbf{D}_{2h}$ are rotations, as the E_g modes of the $\mathbf{D}_{4h}$ system were. The translations of the $\mathbf{D}_{2h}$ system include the B_{1u}, B_{2u}, and B_{3u} modes, corresponding to the A_{2u} and the first kind of E_u in the $\mathbf{D}_{4h}$ system.

Correlating with the E_u vibrations of $\mathbf{D}_{4h}$ are the B_{2u} and B_{3u} nontranslatory motions of $\mathbf{D}_{2h}$. The combination

$$\eta_1 + \eta_5 - \eta_7 - \eta_{11} = q_{15} \tag{9-93}$$

formed from (9-33) behaves as a base for the B_{2u} representation, while the one

$$\eta_2 - \eta_4 - \eta_8 + \eta_{10} = q_{16} \tag{9-94}$$

constructed from (9-34) transforms as a base for B_{3u}.

Combining plus the 1st and 5th, minus the 7th and 11th simultaneous equation, as (9-93) indicates, yields

$$m\ddot{q}_{15} + 2k_1 q_{15} = 0. \tag{9-95}$$

The frequency for the B_{2u} vibration is therefore

$$v_{15} = \frac{1}{2\pi} \left(\frac{2k_1}{m} \right)^{1/2}. \tag{9-96}$$

Equation (9-96) involves only the spring constant of the two springs that vary appreciably during the motion.

Similarly, from (9-94) we find

$$m\ddot{q}_{16} + 2k_2 q_{16} = 0 \tag{9-97}$$

and

$$v_{16} = \frac{1}{2\pi} \left(\frac{2k_2}{m} \right)^{1/2} \tag{9-98}$$

for the B_{3u} vibration. Observe how the E_u vibrational level has split.

EXAMPLE 9.4   What are the normal frequencies for the A_g oscillations of the $\mathbf{D}_{2h}$ structure?

First multiply out the determinant in equation (9-86):

$$m^2 \omega^4 - 2(k_1 + k_2 + k_3) m \omega^2 + 4k_1 k_2 + 2k_2 k_3 + 2k_3 k_1 + k_3{}^2 - k_3{}^2 = 0.$$

Then solve the quadratic for ω^2, take the square root, and divide by 2π:

$$v = \frac{1}{2\pi}\left[\frac{1}{m}(k_1 + k_2 + k_3 \pm \sqrt{k_1{}^2 + k_2{}^2 + k_3{}^2 - 2k_1k_2})\right]^{1/2}.$$

Notice that when $k_1 = k_2$, the solutions reduce to (9-56) and (9-61), for now k_3 represents the constant for a diagonal spring.

EXAMPLE 9.5   Show that the interaction introduced between the A_{1g} and B_{1g} modes by reducing the symmetry from $\mathbf{D}_{4h}$ causes the two normal frequencies to move farther apart.

For convenience, work with $m\omega^2 = m(2\pi v)^2$ rather than with v itself. The upper root of the secular equation in example 9.4 is then

$$m\omega^2 = k_1 + k_2 + k_3 + (k_3{}^2 + k_1{}^2 - 2k_1k_2 + k_2{}^2)^{1/2}$$
$$= k_1 + k_2 + k_3 + [k_3{}^2 + (k_1 - k_2)^2]^{1/2}$$
$$= k_1 + k_2 + k_3 + k_3 + \frac{(k_1 - k_2)^2}{2k_3} + \cdots,$$

while the lower root is

$$m\omega^2 = k_1 + k_2 + k_3 - [k_3{}^2 + (k_1 - k_2)^2]^{1/2}$$
$$= k_1 + k_2 + k_3 - k_3 - \frac{(k_1 - k_2)^2}{2k_3} - \cdots.$$

The radical has been expanded using the binomial theorem.

Note how increasing $|k_1 - k_2|$ while keeping the average of k_1 and k_2 constant causes the upper root to move higher, the lower one to move still lower.

9.10   Principles employed

The spatial symmetry of an interacting set of particles is specified by the complete list of symmetry operations that transform the set into other sets that appear the same. The list of operations forms a group. Any displacement of the particles, from an equilibrium configuration, forms a basis for a matrix representation of the group. The resultant restoring forces also make up a basis.

Each of these can be broken down into elementary contributions belonging to rows of irreducible representations. We found that the force belonging to a given row of a given representation tends to produce a displacement that belongs to the same row. Consequently, the equations of motion can be combined to yield separate equations for different representations and rows.

Reducing the symmetry causes some representations to merge. Interaction then appears between the occurrences, and their normal frequencies are driven further apart.

Lowering the symmetry also introduces reducibility into some representations. Each new distinct irreducible representation yields a distinct mode with a distinct normal frequency.

Symmetry arguments are more general than those of any particular physical theory. They endure even where Newtonian mechanics does not. Even though a molecule is governed by quantum mechanics, its motions can be broken down into the same species as the corresponding classical system. These species are employed in interpreting the vibrational and vibronic spectra of the molecule.

DISCUSSION QUESTIONS

9.1 Why are the base vectors $\mathbf{u}_1$, $\mathbf{u}_2$, ... mutually perpendicular in the r-dimensional mathematical space?

9.2 How do given generalized coordinates and generalized forces appear in this space?

9.3 How does the basis for a row of a given representation appear? How do projections of generalized coordinates and generalized forces on this row appear?

9.4 Why does a given linear combination of displacements (η_k's) yield the same q_l even when other kinds of motion are present?

9.5 How are contributions from different representations separated in the virtual work δW?

9.6 Why does $\ddot{q}_k$ belong to the same representation as q_k?

9.7 Explain the equation

$$2A_{jk}\ddot{q}_k = Q_j$$

9.8 Why does A_{jk} differ from zero only when q_j and q_k belong to the same species—that is, only when the corresponding $\mathbf{f}_j$ and $\mathbf{f}_k$ are bases for the same row of the same irreducible representation?

9.9 How does group theory help us identify the normal vibrations of a symmetric mechanical system?

9.10 How does knowledge of the normal vibrations aid in solving the equations of motion?

9.11 Explain why linearity of each spring in a system does not necessarily make higher terms in its potential V vanish. How does small-vibration theory consider these terms?

9.12 What differential equation does small-vibration theory yield for an intrinsically nonlinear mode of vibration?

9.13 When can the Lagrange equation be set up directly for one of the vibrational modes by itself?

9.14 Consider four equivalent masses joined along the edges of a square by equivalent springs and along the diagonals by two springs that are alike and explain how the normal frequencies arise.

9.15 Why does the distinction between some representations disappear when the symmetry of the given system is lowered?

9.16 How are degenerate representations split? Why is the theory for breaking down reducible representations applicable?

9.17 Explain the effects of reducing the symmetry of an array from $\mathbf{D}_{4h}$ to $\mathbf{D}_{2h}$.

9.18 How general are symmetry arguments?

PROBLEMS

9.1 Find the irreducible representations for which displacements in figure 9.7 are bases. Then set up base vectors for each one. Assume the motion to be planar and oscillatory. A mass m_1 is located at each joint and mass m_2 at the center. The constraining bars, connected by pin joints at the corners, are effectively massless.

9.2 Assume that in problem 9.1, the springs exert no force when the system is at rest. Also, let the spring constants be k_1 and k_2 as labeled. Then formulate the Lagrange equation for each representation and solve for the normal frequencies when the amplitudes are small.

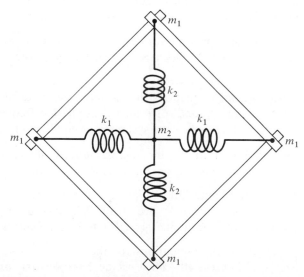

FIGURE 9.7 A $\mathbf{C}_{2v}$ arrangement of masses, springs, and constraining rods. Assume that each rod is the same and that each spring has the same equilibrium length.

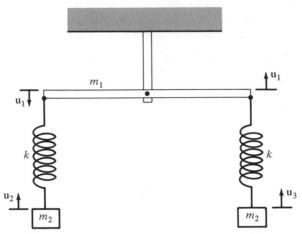

FIGURE 9.8   A C_{2v} arrangement oscillating in the gravitational field of the earth. The mass m_1 of the beam is distributed uniformly along its length.

9.3 Let k_3 be the spring constant for one-half of a diagonal spring in figure 9.4. Then relate k_3 to k_2, the constant for the whole spring.

9.4 Find the irreducible representations of which the displacements shown in figure 9.8 are bases. Then construct base vectors for the representations.

9.5 Formulate Lagrange equations for actual displacements in the figure 9.8 system, separate the representations, and solve for the normal frequencies in the approximation that the center of mass of the beam is at the support.

9.6 Find the irreducible representations for the allowed motions in the C_{6v} symmetric system of figure 9.9. Describe a typical stage in the motion of each type.

9.7 Construct the Lagrange equations that govern back-and-forth motions of the central mass in figure 9.9, and solve for the oscillation frequency. Assume the springs to exert no force when the system is at equilibrium.

9.8 For the other allowed irreducible representations of the C_{6v} structure, set up the Lagrange equations and obtain the normal frequencies.

9.9 Construct an approximate differential equation governing the large amplitude B_{2g} vibration of figure 9.4. See figure 9.3. Assume the springs are neither soft nor hard and keep terms through the third power in q.

9.10 Find the irreducible representations for the allowed motions in the C_{6v} symmetric systems of figure 9.10. Then set up base vectors for each species and describe a typical stage in the corresponding motion. Assume that the pin joint in the middle keeps the rods all in one plane.

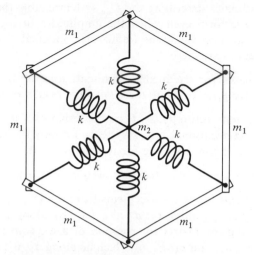

FIGURE 9.9 A $\mathbf{C}_{6v}$ arrangement of rods, springs, and a mass m_2. The mass m_1 of each rod is distributed uniformly along its edge of the hexagon to a good approximation.

9.11 Suppose that in figure 9.10, the springs exert no forces when the system is at rest. Then formulate Lagrange equations for it, assuming the amplitudes are small.

9.12 Combine the equations in problem 9.11 so the representations are separated and solve for the normal frequencies.

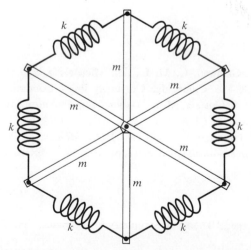

FIGURE 9.10 Six equivalent rods arranged as spokes to the ends of six equivalent springs. The mass m of each rod is distributed uniformly over its length.

9.13 What coordinates describing the C_{6v} system enable the kinetic energy to have the quadratic form even when the amplitudes of oscillation are finite? What potential function yields the frequencies obtained in problem 9.12 even when the amplitudes are large?

9.14 Generate bases for irreducible representations of D_{3h} from the r_j and s_j vectors of figure 8.1. Tabulate these with the results of example 8.12.

9.15 Project the unit vectors r_j, s_j, t_j on Cartesian axes u_{3j-2}, u_{3j-1}, u_{3j} that are parallel to a conventional x, y, z system. Employ the results to separate a base vector for vibration from one for translation in the E' representation.

9.16 Sketch the independent elementary motions in the D_{3h} structure.

9.17 Let components of the displacements from the D_{3h} equilibrium configuration be ρ_1 along r_1, ρ_2 along r_2, σ_1 along s_1, σ_2 along s_2, τ_1 along t_1, and τ_2 along t_2. Using the matrix for rotation of axes, convert these to the displacements η_1, η_2, η_3, and η_4, η_5, η_6 which lie along u_1, u_2, u_3, and u_4, u_5, u_6, respectively. Recall the approximate formula for the extension of an edge and construct the potential energy of the spring along the first edge as a quadratic form. Set up similar forms for the other two edges. Finally set up the Lagrange equations linear in ρ_j, σ_j, τ_j.

9.18 Use the results from problems 9.14, 9.15 to separate the representations in these Lagrange equations and solve for the normal frequencies of the D_{3h} structure.

9.19 Set up a differential equation governing the large amplitude breathing motion of the triangular array assuming the springs to be neither soft nor hard.

REFERENCES

BOOKS

Bishop, R. E. D., Gladwell, G. M. L., and Michaelson, S. *The Matrix Analysis of Vibration*, pp. 36–175. Cambridge University Press, London, 1965.

Hurty, W. C., and Rubinstein, M. F. *Dynamics of Structures*, pp. 110–415. Prentice-Hall, Inc., Englewood Cliffs, N.J., 1964.

Wilson, E. B., Jr., Decius, J. C., and Cross, P. C. *Molecular Vibrations: The Theory of Infrared and Raman Vibrational Spectra*, pp. 102–145. McGraw-Hill Book Company, New York, 1955.

ARTICLES

Wilde, R. E. A General Theory of Symmetry Coordinates. *Am. J. Phys.*, **32**, 45 (1964).

10 / *Generalized Momentum*

10.1 Continuous symmetry

Whenever the forces acting on the parts of a given system are derivable from a generalized potential U, possible motions in the system are determined by the Lagrangian function

$$L = T - U. \tag{10-1}$$

As a consequence, any change in a generalized coordinate that does not alter possible appearances of the system does not by itself affect L. Such a change is in a sense a symmetry operation.

Furthermore, when a system moves from one state to a neighboring state, a given Lagrangian shifts continuously and uniquely. So if each *infinitesimal* change in a q_l, over a certain range, effects a symmetry operation in the above sense, it cannot alter L and

$$\frac{\partial L}{\partial q_l} = 0 \tag{10-2}$$

throughout the range.

An indicator for the presence of such a region is a variable that becomes constant wherever equation (10-2) applies. According to the pertinent Lagrange equation, the derivative $\partial L/\partial \dot{q}_l \equiv p_l$ is such a variable.

Since this derivative is a function of $q_1, \ldots, q_n, \dot{q}_1, \ldots, \dot{q}_n, t$, it can be introduced as an independent variable in place of a generalized velocity. Introducing the p_l corresponding to each q_l allows us to eliminate all $\dot{q}_l$'s and enables us to work with $p_1, \ldots, p_n, q_1, \ldots, q_n, t$ as the independent variables. The appropriate energy function and the appropriate equations of motion are then the Hamiltonian function and the Hamiltonian equations.

The behavior of the system is represented by the motion of a point in a mathematical $p_1, \ldots, p_n, q_1, \ldots, q_n$ space. The behavior of an ensemble of systems is represented by the movement of a swarm of points in this space. We will consider what condition this swarm must satisfy.

Finally, both coordinates and momenta will be transformed. While this step does not aid in solving specific problems, it can nevertheless bring out previously hidden relationships in the system.

10.2   Generalized momentum

As in section 5.7, we are considering a system free of dissipative effects. Motion in the system must then satisfy equation (5-45)

$$\frac{d}{dt}\frac{\partial L}{\partial \dot{q}_l} = \frac{\partial L}{\partial q_l} \tag{10-3}$$

with L given by (10-1).

But in a region where continuous symmetry exists, equation (10-2) holds for some q_l, and

$$\frac{d}{dt}\frac{\partial L}{\partial \dot{q}_l} = 0. \tag{10-4}$$

Integrating equation (10-4) yields

$$\frac{\partial L}{\partial \dot{q}_l} = \text{constant.} \tag{10-5}$$

Thus, a variable that is constant wherever (10-2) is valid is the *generalized momentum* introduced in equation (5-47):

$$\frac{\partial L}{\partial \dot{q}_l} = p_l. \tag{10-6}$$

To recognize this as a generalization of conventional momentum, let U be a function of position only, and T be a function of Cartesian velocities. Then the Lagrangian has the form

$$L = \tfrac{1}{2}m_1(\dot{x}_1{}^2 + \dot{y}_1{}^2 + \dot{z}_1{}^2) + \tfrac{1}{2}m_2(\dot{x}_2{}^2 + \dot{y}_2{}^2 + \dot{z}_2{}^2) + \cdots - V(x_1, y_1, z_1, \ldots) \tag{10-7}$$

and the partial derivatives,

$$\frac{\partial L}{\partial \dot{x}_1} = m_1\dot{x}_1 = (p_x)_1 \cdots, \tag{10-8}$$

are the conventional components of the linear momenta.

The n simultaneous equations (10-6) contain the q_l's, $\dot{q}_l$'s, and t as independent variables on the left. With no relationship among the p_l's, the equations would generally allow us to solve for the $\dot{q}_l$'s individually. These expressions could then be substituted into the equations of motion.

EXAMPLE 10.1   If the potential energy of a particle varies only with its distance from a point fixed in an inertial frame, what are its conserved momenta?

Let us place the origin of a spherical coordinate system at the center of the field, so the potential has the simple form

$$V = V(r).$$

In such a system, the kinetic energy is

$$T = \tfrac{1}{2}m(\dot{r}^2 + r^2\dot{\theta}^2 + r^2\sin^2\theta\dot{\phi}^2),$$

where θ is the colatitude and ϕ the longitude.

From the Lagrangian

$$L = T - V$$

we then derive the generalized momentum

$$p_\phi = \frac{\partial L}{\partial \dot{\phi}} = mr^2\sin^2\theta\dot{\phi} = (r\sin\theta)(mr\sin\theta\dot{\phi})$$

$$= (r\sin\theta)(mv_\phi).$$

Since L does not contain ϕ explicitly, equation (10-3) shows that p_ϕ is constant. Since L does contain r and θ, p_r and p_θ vary.

The same form is obtained for p_ϕ regardless of how the z axis is oriented. Consequently, the angular momentum is conserved about any axis passing through the center.

10.3 The Hamiltonian

We have considered the Lagrangian L for a given system to be a smoothly varying function of $\dot{q}_1, \ldots, \dot{q}_n, q_1, \ldots, q_n, t$. Consequently, along any possible element of path, differential dL is a linear function of $d\dot{q}_1, \ldots, d\dot{q}_n, dq_1, \ldots, dq_n, dt$. A Legendre transformation, in which dL is subtracted from the differential of an appropriate sum, leads to a linear function of $dp_1, \ldots, dp_n, dq_1, \ldots, dq_n, dt$. This linear function is the differential of an energy function, the Hamiltonian H. Identifying coefficients in the expression for dH as partial derivatives of H leads to new equations of motion. Under special circumstances, the Hamiltonian is $T + V$.

Since L is a function of the $\dot{q}_k$'s, the q_k's, and t, the result in example 5.1 implies that

$$dL = \frac{\partial L}{\partial \dot{q}_k}\,d\dot{q}_k + \frac{\partial L}{\partial q_k}\,dq_k + \frac{\partial L}{\partial t}\,dt. \tag{10-9}$$

The generalized momentum,

$$p_k = \frac{\partial L}{\partial \dot{q}_k}, \tag{10-10}$$

converts Lagrange equations (10-3) to

$$\dot{p}_k \equiv \frac{d}{dt}\,p_k = \frac{\partial L}{\partial q_k}. \tag{10-11}$$

Equation (10-9) can therefore be rewritten in the form

$$dL = p_k \, d\dot{q}_k + \dot{p}_k \, dq_k + \frac{\partial L}{\partial t} \, dt. \tag{10-12}$$

We can free the differential of terms containing $d\dot{q}_k$ by subtracting dL from the differential of $p_k\dot{q}_k$. Alternatively, we can start with the *Hamiltonian H*

$$H \equiv p_k\dot{q}_k - L \tag{10-13}$$

and differentiate

$$dH = p_k \, d\dot{q}_k + \dot{q}_k \, dp_k - dL. \tag{10-14}$$

Then (10-12) can be introduced and the expression reduced:

$$dH = p_k \, d\dot{q}_k + \dot{q}_k \, dp_k - p_k \, d\dot{q}_k - \dot{p}_k \, dq_k - \frac{\partial L}{\partial t} \, dt$$

$$= \dot{q}_k \, dp_k - \dot{p}_k \, dq_k - \frac{\partial L}{\partial t} \, dt. \tag{10-15}$$

Since we consider H to be a function of $p_1, \ldots, p_n, q_1, \ldots, q_n, t$, its differential is also written as

$$dH = \frac{\partial H}{\partial p_k} \, dp_k + \frac{\partial H}{\partial q_k} \, dq_k + \frac{\partial H}{\partial t} \, dt. \tag{10-16}$$

For the two linear expressions representing dH to be the same, we must have

$$\dot{q}_k = \frac{\partial H}{\partial p_k}, \tag{10-17}$$

$$\dot{p}_k = -\frac{\partial H}{\partial q_k}, \tag{10-18}$$

$$\frac{\partial H}{\partial t} = -\frac{\partial L}{\partial t}. \tag{10-19}$$

Equations (10-17), (10-18), (10-19) are called *Hamilton's equations* of motion. Note how p_k and q_k enter in a nearly symmetrical manner into them.

The expression for the Hamiltonian can be simplified when the potential is independent of the particle velocities and no time-dependent constraints exist. For then, the q_k's can be related to inertial Cartesian coordinates by transformation equations free of the time t, and the derivation of equation (6-8) is valid:

$$T = \dot{q}_j A_{jk} \dot{q}_k. \tag{10-20}$$

Differentiating this equation,

$$\frac{\partial T}{\partial \dot{q}_l} = \dot{q}_j A_{jl} + A_{lk}\dot{q}_k, \tag{10-21}$$

multiplying by $\dot{q}_l$ and summing, we obtain

$$\dot{q}_l \frac{\partial T}{\partial \dot{q}_l} = \dot{q}_j A_{jl}\dot{q}_l + \dot{q}_l A_{lk}\dot{q}_k = 2T. \tag{10-22}$$

If the potential does not depend on the $\dot{q}_k$'s, we have

$$\frac{\partial T}{\partial \dot{q}_l} = \frac{\partial}{\partial \dot{q}_l}(T - V) = \frac{\partial L}{\partial \dot{q}_l} = p_l. \tag{10-23}$$

The last step follows from the definition of p_l (section 10.2).
  Combining equations (10-22) and (10-23),

$$\dot{q}_l p_l = 2T, \tag{10-24}$$

and substituting the result into the definition of H yields

$$H = p_k \dot{q}_k - (T - V) = 2T - T + V$$
$$= T + V. \tag{10-25}$$

  When the potential does not vary with any $\dot{q}_l$ and the constraints (if present) do not vary with time, the Hamiltonian H equals the kinetic energy plus the potential energy.

EXAMPLE 10.2 Construct Hamilton's equations for a particle moving along the x axis in the parabolic potential $\frac{1}{2}kx^2$.
  The kinetic energy,

$$T = \tfrac{1}{2}m\dot{x}^2,$$

and the potential energy,

$$V = \tfrac{1}{2}kx^2,$$

yield the Lagrangian,

$$L = T - V = \tfrac{1}{2}m\dot{x}^2 - \tfrac{1}{2}kx^2.$$

  Differentiate to find the momentum,

$$p = \frac{\partial L}{\partial \dot{x}} = m\dot{x},$$

and rewrite T in terms of this p,

$$T = \frac{m^2\dot{x}^2}{2m} = \frac{p^2}{2m}.$$

Since the potential depends on the coordinate, and no time-dependent constraints exist, equation (10-25) is valid:

$$H = T + V = \frac{p^2}{2m} + \tfrac{1}{2}kx^2.$$

  Substitute the coordinate x, momentum p, and this Hamiltonian H into equations (10-17) and (10-18):

$$\dot{x} = \frac{\partial H}{\partial p} = \frac{p}{m},$$

$$\dot{p} = -\frac{\partial H}{\partial x} = -kx.$$

The first of these Hamiltonian equations would follow from the customary definition of the momentum while the second would follow directly from Newton's second law.

10.4   Conservation laws

Associated with the presence of continuous symmetry (see section 10.1) are absence from the Lagrangian of the corresponding variable, and a consequent constancy of a conjugate variable.

Indeed wherever successive small changes in a generalized coordinate q_l cause no change in the governing function L, we have the condition

$$\frac{\partial L}{\partial q_l} = 0. \tag{10-26}$$

This converts Lagrange equation (10-3) to

$$\frac{d}{dt} \frac{\partial L}{\partial \dot{q}_l} = 0. \tag{10-27}$$

Equation (10-27) makes the variable

$$\frac{\partial L}{\partial \dot{q}_l} = p_l \tag{10-28}$$

constant. Presence of the symmetry described by (10-26) causes the quantity p_l to be conserved during the pertinent motion.

Absence of time t from the Lagrangian

$$\frac{\partial L}{\partial t} = 0 \tag{10-29}$$

also leads to a conservation law. Equation (10-16) yields the derivative relationship

$$\frac{dH}{dt} = \frac{\partial H}{\partial p_k} \frac{dp_k}{dt} + \frac{\partial H}{\partial q_k} \frac{dq_k}{dt} + \frac{\partial H}{\partial t}. \tag{10-30}$$

Combining this with Hamilton's equations (10-17), (10-18), (10-19), and with condition (10-29), we find,

$$\frac{dH}{dt} = - \frac{\partial H}{\partial p_k} \frac{\partial H}{\partial q_k} + \frac{\partial H}{\partial q_k} \frac{\partial H}{\partial p_k} - \frac{\partial L}{\partial t}$$

$$= 0 - 0 = 0, \tag{10-31}$$

that the Hamiltonian H is constant.

The continuous symmetry in time expressed by (10-29) leads to this conservation-of-energy law.

10.5   Simple phase diagrams

An instantaneous state of a system may be referred to as a phase in its life. Such a state is described by the pertinent generalized coordinates and momenta, $q_1, \ldots,$ $q_n, p_1, \ldots, p_n$. These may be plotted in the Cartesian manner in a $2n$-dimensional Euclidean space called the *phase space* for the system.

Each point in the space then represents a distinct configuration changing in a particular manner. A succession of points, a trajectory in phase space, represents an evolution of the given system. A family of trajectories describes various possible evolutions succinctly.

For simplicity, let us here limit ourselves to a consideration of 1-dimensional motion. We also assume that the force acting on the given particle is the sum of a component depending on the position and a component depending on the speed or momentum

$$p = m\dot{x}. \tag{10-32}$$

Letting $-f(x)/m$ be the first component and $-\phi(p)/m$ be the second component reduces Newton's force equation to the form

$$\dot{p} = \frac{-f(x) - \phi(p)}{m}. \tag{10-33}$$

Each element of a given trajectory then has the slope

$$\frac{dp}{dx} = \frac{dp/dt}{dx/dt} = \frac{\dot{p}}{\dot{x}} = \frac{-f(x) - \phi(p)}{m\dot{x}} = \frac{-f(x) - \phi(p)}{p}. \tag{10-34}$$

This slope is defined at all points except where both the numerator and denominator are zero, and the particle is at rest. Such a point is called a *singular point*.

When the spring force $-f(x)/m$ is linear in x, units may be chosen so that equation (10-34) becomes

$$\frac{dp}{dx} = \frac{-\phi(p) - x}{p}. \tag{10-35}$$

Then a simple construction, due to Alfred Lienard, yields the trajectories.

If the slope at P(x, p) is desired, first plot

$$x = -\phi(p) \tag{10-36}$$

and locate on this curve the point Q that is at the same height p as P (see figure 10.1). Point R is located on the x axis at the same distance from the p axis as Q. The trajectory is then drawn through P *perpendicular* to PR. This result follows because the length of QP is $x + \phi(p)$ and the slope of RP is

$$\frac{p}{x + \phi(p)}. \tag{10-37}$$

To plot the corresponding trajectory in phase space, continue the line through P a short distance. Then the construction is repeated to get a new slope at the end point, and so on. The construction can be made sufficiently accurate by taking the segment lengths small enough.

EXAMPLE 10.3   Construct a phase diagram for a harmonic oscillator retarded by a frictional force of constant magnitude.

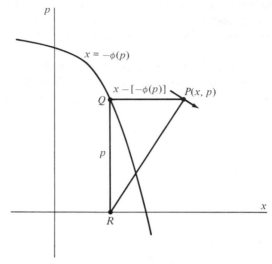

FIGURE 10.1   Lienard construction for obtaining the direction of the trajectory at a given point in phase space.

Lienard's procedure is applicable because the oscillation is harmonic, and, consequently, the spring force is linear in x. First choose units to make this force equal $-x/m$ as in equation (10-35). Then the retarding force is formulated to be constant for each sign of $\dot{p}$:

$$-\frac{\phi(p)}{m} = \begin{cases} -a/m & \text{when} \quad \dot{p} > 0 \\ a/m & \text{when} \quad \dot{p} < 0. \end{cases}$$

The first equation yields the half-line $x = -a$, drawn up from S in figure 10.2. The second equation yields the half-line $x = a$, drawn down from T in figure 10.2. Because the first half-line is vertical, point R of the Lienard construction stays at S as long as p is positive. Therefore, the trajectory element perpendicular to each position of PR lies along a circle centered at S as shown. But when p becomes negative, the point R shifts to T and the solution lies along the smaller semicircle centered on T.

At point U, the momentum p is 0, so the mass has no kinetic energy. Furthermore, since the spring force $-x/m$ is smaller than the frictional force a/m, the mass can no longer move.

10.6   Liouville's theorem

Conditions in a specific system at a distinct time are represented by a single point in its phase space. A set of possible instantaneous states is represented by a swarm of points. The swarm may distort as it moves along, but its density is conserved, as we will now see.

The swarm describes a set, or *ensemble*, of systems composed of identical particles obeying the same force laws. However, each member of the set is evolving from an initial state slightly different from those of its immediate neighbors.

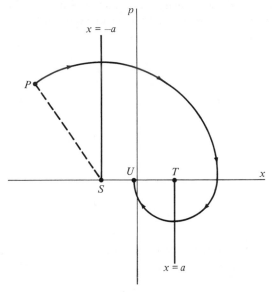

FIGURE 10.2 Phase diagram for a mass subject to a linear restoring force and a constant frictional retarding force.

We presume that the points are numerous enough to let ρ be their density and to allow the number of systems having points in a volume $d\tau$ at time t be

$$dN = \rho \, d\tau = \rho \, dp_1 \cdots dp_n \, dq_1 \cdots dq_n. \qquad (10\text{-}38)$$

The components of velocity of a given point are $\dot{p}_1, \ldots, \dot{p}_n, \dot{q}_1, \ldots, \dot{q}_n$. If these were constant over a local region, all points within a slanting hyperprism of height $\dot{p}_1$ would pass through the hypersurface,

$$dp_2 \cdots dp_n \, dq_1 \cdots dq_n, \qquad (10\text{-}39)$$

in unit time (see figure 10.3).

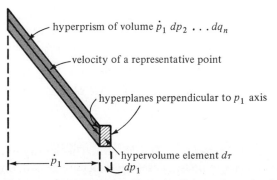

FIGURE 10.3 Volume containing the points that would pass through a bounding hyperplane in unit time if the velocity of every approaching point were constant.

Generally, the velocity and density vary from point to point. But the transfer depends on ρ and $\dot{p}_1$ *at the surface*; the number of points passing through the surface per unit time equals the density there times the volume of the hyperprism:

$$\rho\dot{p}_1 \, dp_2 \cdots dp_n \, dq_1 \cdots dq_n. \tag{10-40}$$

This number of points passes into the volume $d\tau$. The rate of flow out through the parallel surface on the opposite side of $d\tau$ is quantity (10-40) plus the pertinent differential:

$$\left[\rho\dot{p}_1 + \frac{\partial(\rho\dot{p}_1)}{\partial p_1} \, dp_1\right] dp_2 \cdots dp_n \, dq_1 \cdots dq_n. \tag{10-41}$$

Subtracting (10-41) from (10-40) gives the net rate of accumulation between the two faces:

$$-\frac{\partial(\rho\dot{p}_1)}{\partial p_1} \, d\tau. \tag{10-42}$$

A similar term exists for the rate of accumulation between each set of parallel faces. Setting the total equal to the rate at which the number of points within $d\tau$ increases,

$$\left[-\frac{\partial}{\partial p_k}(\rho\dot{p}_k) - \frac{\partial}{\partial q_k}(\rho\dot{q}_k)\right] d\tau = \frac{\partial dN}{\partial t} = \frac{\partial\rho}{\partial t} \, d\tau, \tag{10-43}$$

and rearranging the result, we get

$$\frac{\partial\rho}{\partial t} + \frac{\partial\rho}{\partial p_k}\dot{p}_k + \frac{\partial\rho}{\partial q_k}\dot{q}_k + \rho\frac{\partial\dot{p}_k}{\partial p_k} + \rho\frac{\partial\dot{q}_k}{\partial q_k} = 0, \tag{10-44}$$

or

$$\frac{d\rho}{dt} + \rho\frac{\partial\dot{p}_k}{\partial p_k} + \rho\frac{\partial\dot{q}_k}{\partial q_k} = 0. \tag{10-45}$$

Since neighboring points describe neighboring states of the same physical system, they are governed by the same Hamiltonian equations. Differentiating (10-18), (10-17), summing, rearranging one second derivative,

$$\frac{\partial\dot{p}_k}{\partial p_k} = -\frac{\partial^2 H}{\partial p_k \, \partial q_k}, \tag{10-46}$$

$$\frac{\partial\dot{q}_k}{\partial q_k} = \frac{\partial^2 H}{\partial q_k \, \partial p_k} = \frac{\partial^2 H}{\partial p_k \, \partial q_k} = -\frac{\partial\dot{p}_k}{\partial p_k}, \tag{10-47}$$

and combining the result with equation (10-45), we obtain

$$\frac{d\rho}{dt} = 0. \tag{10-48}$$

Equation (10-48) is a form of *Liouville's theorem*: The density of points representing an ensemble of like systems, in phase space, does not change with time. This invariance depends on employing p_k's with q_k's; it depends on using variables properly conjugate to the coordinates in place of the generalized velocities.

EXAMPLE 10.4   Observing a definite system at a specific time yields the coordinate q with a standard deviation of Δq_0 and the conjugate p with a standard deviation of Δp_0. If the coordinate q is effectively independent of the other coordinates, how is the uncertainty in p related to that of q at some later time?

The data are the same as those we would obtain if we were averaging corresponding coordinates and momenta of a set of independent systems, each similar to the original system, but distributed over a range of q's and p's to produce the observed means and standard deviations. To this ensemble, Liouville's theorem applies.

Indeed, because the measured q and p are essentially independent of the other variables, the theorem is valid in the pq plane. Thus, the phase-space points initially in the area $\Delta p_0 \, \Delta q_0$ remain in an area that large, and the product of the standard deviations does not change with time.

If Δq is the standard deviation in the coordinate and Δp the standard deviation in the corresponding momentum at some later time, we have

$$\Delta p \, \Delta q = \Delta p_0 \, \Delta q_0 = \text{constant}$$

(see figure 10.4).

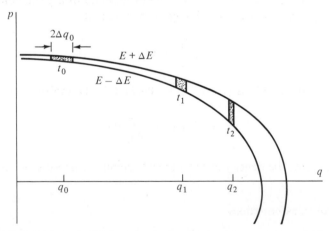

FIGURE 10.4   Evolution of the phase area containing the systems within ΔE and Δq_0 of E and q_0 initially. The magnitude of the area remains constant to satisfy Liouville's theorem.

10.7   Thermodynamic equilibrium

In thermodynamic studies, we are concerned with the collective behavior of a large number of particles. We are not interested in individual coordinates and momenta, since there are too many of them to cope with in a typical system.

Instead, a thermodynamic system is represented by the average behavior of *all mechanical systems* consistent with the applied constraints. These limitations may include the amount of each susbtance, the total energy, and the volume. Strictly, each of these is specified within limits.

Some aspects of this average behavior can be deduced from Liouville's theorem. Imagine a large number of systems consistent with the constraints but possessing differing detailed coordinates and momenta at a specific time.

Each system is represented by a moving point in phase space. If the boundary conditions of the various systems are chosen so that ρ is constant throughout the allowed region of phase space at, say, $t = 0$, then in

$$\frac{d\rho}{dt} = \frac{\partial \rho}{\partial t} + \frac{\partial \rho}{\partial p_k} \dot{p}_k + \frac{\partial \rho}{\partial q_k} \dot{q}_k, \tag{10-49}$$

each term in the two series is zero, and

$$\frac{d\rho}{dt} = \frac{\partial \rho}{\partial t}. \tag{10-50}$$

But equation (10-48) demands that $d\rho/dt$ be zero. Therefore, the partial derivative $\partial \rho/\partial t$ is zero and the density at any distinct point in phase space does not change with time. The chosen distribution describes *equilibrium* conditions.

On the other hand, if equilibrium exists, the probability of a given configuration of coordinates and momenta does not change with time and

$$\frac{\partial \rho}{\partial t} = 0 \tag{10-51}$$

at each point. With equations (10-48) and (10-49), (10-51) implies a zero

$$d\rho = \frac{\partial \rho}{\partial p_k} dp_k + \frac{\partial \rho}{\partial q_k} dq_k. \tag{10-52}$$

But if this $d\rho$ is zero, the density in the allowed regions of phase space is constant.

10.8   Canonical transformations

The coordinate and gauge transformations of section 5.9 do not alter the form of the Hamiltonian equations. As a consequence, they are said to be canonical.

For convenience, let us represent the initial sets of generalized coordinates and momenta by **q** and **p**, the final, or primed, sets by **Q** and **P**. Also let us replace L' by $\mathscr{L}$. Then equation (5-71) becomes

$$\mathscr{L}(t, \mathbf{Q}, \dot{\mathbf{Q}}) = L(t, \mathbf{q}, \dot{\mathbf{q}}) + \frac{d}{dt} F(t, \mathbf{q}, \mathbf{Q}), \tag{10-53}$$

whence

$$\mathscr{L}(t, \mathbf{Q}, \dot{\mathbf{Q}}) = L(t, \mathbf{q}, \dot{\mathbf{q}}) + \frac{\partial F}{\partial t} + \frac{\partial F}{\partial q_k} \dot{q}_k + \frac{\partial F}{\partial Q_k} \dot{Q}_k. \tag{10-54}$$

Differentiating (10-54) with respect to $\dot{q}_l$ and $\dot{Q}_l$ yields

$$0 = \frac{\partial L}{\partial \dot{q}_l} + \frac{\partial F}{\partial q_l} \quad \text{or} \quad \frac{\partial L}{\partial \dot{q}_l} = -\frac{\partial F}{\partial q_l} \tag{10-55}$$

and

$$\frac{\partial \mathscr{L}}{\partial \dot{Q}_l} = \frac{\partial F}{\partial Q_l}. \tag{10-56}$$

But the appropriate generalization of equation (10-6) is

$$\frac{\partial \mathscr{L}}{\partial \dot{Q}_l} = P_l, \tag{10-57}$$

so equation (10-56) is rewritten as

$$P_l = \frac{\partial F}{\partial Q_l}. \tag{10-58}$$

Variable P_l is the transformed momentum.

By analogy with equation (10-13), the transformed Hamiltonian is defined as

$$\mathscr{H} = P_k \dot{Q}_k - \mathscr{L}. \tag{10-59}$$

Differentiating, we get

$$\frac{\partial \mathscr{H}}{\partial P_l} = \dot{Q}_l, \tag{10-60}$$

$$\frac{\partial \mathscr{H}}{\partial Q_l} = -\frac{\partial \mathscr{L}}{\partial Q_l} = -\frac{d}{dt}\frac{\partial \mathscr{L}}{\partial \dot{Q}_l}$$

$$= -\dot{P}_l. \tag{10-61}$$

The second equality in (10-61) follows because $\mathscr{L}$ satisfies the transformed Lagrange equation, the third equality from the generalization of (10-6).

Since equations (10-60) and (10-61) involve $\mathscr{H}$, Q_l, P_l as (10-17) and (10-18) involve H, q_l, p_l, the transformation has not altered the form of the Hamiltonian equations. In effect, the new Hamiltonian of $Q_1, \ldots, Q_n, P_1, \ldots, P_n, t$ behaves as the old Hamiltonian of the coordinates and conjugate momenta could. We refer to such a transformation as *canonical*.

Substituting (10-58), (10-54), (10-55), (10-6) into (10-59) and reducing, we find

$$\mathscr{H} = \frac{\partial F}{\partial Q_k} \dot{Q}_k - L - \frac{\partial F}{\partial t} + \frac{\partial L}{\partial \dot{q}_k} \dot{q}_k - \frac{\partial F}{\partial Q_k} \dot{Q}_k$$

$$= p_k \dot{q}_k - L - \frac{\partial F}{\partial t} = H - \frac{\partial F}{\partial t}, \tag{10-62}$$

a relationship between the transformed and the original Hamiltonian.

10.9   Hamilton's principal function

The variables appearing in the generating function may be changed by a Legendre transformation.

Let

$$P_k Q_k - F(t, \mathbf{q}, \mathbf{Q}) = S. \tag{10-63}$$

On differentiating both sides with respect to Q_l, we obtain

$$P_l - \frac{\partial F}{\partial Q_l} = \frac{\partial S}{\partial Q_l}. \tag{10-64}$$

Equation (10-58) shows that the left side of (10-64) is zero. Therefore, function S, which is called *Hamilton's principal function*, does not depend on Q_l; but rather

$$S = S(t, \mathbf{q}, \mathbf{P}) \tag{10-65}$$

only.

Differentiating definition (10-63) with respect to q_l and P_l yields

$$\frac{\partial S}{\partial q_l} = -\frac{\partial F}{\partial q_l} = \frac{\partial L}{\partial \dot{q}_l} = p_l, \tag{10-66}$$

$$\frac{\partial S}{\partial P_l} = Q_l, \tag{10-67}$$

the corresponding initial generalized momentum and final coordinate, if (10-55) and (10-6) are introduced.

Differentiating with respect to t yields

$$\frac{\partial S}{\partial t} = -\frac{\partial F}{\partial t}, \tag{10-68}$$

so equation (10-62) can be rewritten as

$$\mathcal{H} = H + \frac{\partial S}{\partial t}. \tag{10-69}$$

We may construct a canonical transformation by (a) choosing either

$$F(t, \mathbf{q}, \mathbf{Q}) \qquad \text{or} \qquad S(t, \mathbf{q}, \mathbf{P}),$$

and (b) applying equations (10-58), (10-55), (10-6), (10-62) or equations (10-66), (10-67), (10-69).

EXAMPLE 10.5   Find a principal function S that generates the transformation from rectangular coordinates x, y to cylindrical coordinates r, ϕ. From this function derive the momenta conjugate to r and ϕ.

It is given that

$$\begin{pmatrix} q_1 \\ q_2 \end{pmatrix} = \begin{pmatrix} x \\ y \end{pmatrix}, \qquad \begin{pmatrix} Q_1 \\ Q_2 \end{pmatrix} = \begin{pmatrix} r \\ \phi \end{pmatrix},$$

$$\begin{pmatrix} p_1 \\ p_2 \end{pmatrix} = \begin{pmatrix} p_x \\ p_y \end{pmatrix}, \qquad \begin{pmatrix} P_1 \\ P_2 \end{pmatrix} = \begin{pmatrix} p_r \\ p_\phi \end{pmatrix}.$$

First substitute the appropriate elements into (10-67) and introduce the familiar relationships between the coordinate systems:

$$\frac{\partial S}{\partial p_r} = r = (x^2 + y^2)^{1/2},$$

$$\frac{\partial S}{\partial p_\phi} = \phi = \tan^{-1} \frac{y}{x}.$$

Integrate these equations and combine the results to obtain

$$S(x, y, p_r, p_\phi) = p_r(x^2 + y^2)^{1/2} + p_\phi \tan^{-1} \frac{y}{x}.$$

Substitute into equation (10-66),

$$p_x = \frac{\partial S}{\partial x} = \frac{x}{r} p_r - \frac{y}{r^2} p_\phi,$$

$$p_y = \frac{\partial S}{\partial y} = \frac{y}{r} p_r + \frac{x}{r^2} p_\phi,$$

and solve for the transformed momenta,

$$p_r = \frac{1}{r}(x p_x + y p_y),$$

$$p_\phi = x p_y - y p_x.$$

10.10 Transforming the harmonic oscillator to a free translator

Because a canonical transformation does not alter the form of Hamilton's equations, it can be used to relate dynamical systems that are similar. A nontrivial example will be considered here.

Let us start with the Hamiltonian

$$H = \frac{p^2}{2\mu} + \frac{kq^2}{2} \tag{10-70}$$

which converts equations (10-17), (10-18) to

$$\dot{q} = \frac{p}{\mu}, \tag{10-71}$$

$$\dot{p} = -kq. \tag{10-72}$$

On differentiating (10-71), combining the result with (10-72),

$$\ddot{q} = \frac{\dot{p}}{\mu} = -\frac{k}{\mu} q, \tag{10-73}$$

and integrating, we find that

$$q = A \sin(\omega t + \alpha) \tag{10-74}$$

where

$$\omega = \left(\frac{k}{\mu}\right)^{1/2}. \tag{10-75}$$

Equation (10-74) describes the motion of a *harmonic oscillator*.

Substituting (10-74) into (10-71), and the resulting p together with (10-74) and (10-75) into (10-70) yields

$$H = \frac{1}{2\mu} [\mu A \omega \cos(\omega t + \alpha)]^2 + \frac{k}{2} [A \sin(\omega t + \alpha)]^2$$

$$= \frac{k}{2} A^2, \tag{10-76}$$

a constant. But since the potential is independent of $\dot{q}$, the Hamiltonian is the energy E and

$$A = \left(\frac{2E}{k}\right)^{1/2}. \tag{10-77}$$

Now, *uniform rectilinear motion* in the Q, P system is described by the equations

$$Q = \omega t + \alpha, \tag{10-78}$$

$$P = \text{constant.} \tag{10-79}$$

Substituting these into formulas (10-60), (10-61) leads to

$$\frac{\partial \mathcal{H}}{\partial P} = \dot{Q} = \omega, \tag{10-80}$$

$$\frac{\partial \mathcal{H}}{\partial Q} = -\dot{P} = 0. \tag{10-81}$$

We also assume

$$\frac{\partial \mathcal{H}}{\partial t} = 0. \tag{10-82}$$

The differential of the transformed Hamiltonian $\mathcal{H}$ then reduces to one term

$$d\mathcal{H} = \frac{\partial \mathcal{H}}{\partial P} dP + \frac{\partial \mathcal{H}}{\partial Q} dQ + \frac{\partial \mathcal{H}}{\partial t} dt = \omega \, dP \tag{10-83}$$

that integrates to

$$\mathcal{H} = \omega P \tag{10-84}$$

if the constant of integration is set equal to zero. If this Hamiltonian is also set equal to the energy E, equation (10-84) rearranges to

$$P = \frac{E}{\omega}. \tag{10-85}$$

Equations (10-78), (10-79) now show how the harmonic oscillator is transformed into the free translator. Substituting (10-85) into (10-77),

$$A = \left(\frac{2P\omega}{k}\right)^{1/2}, \tag{10-86}$$

and (10-86), (10-78) into (10-74) yields

$$q = \left(\frac{2P\omega}{k}\right)^{1/2} \sin Q, \tag{10-87}$$

while combining this result with (10-71) gives

$$p = (2P\mu\omega)^{1/2} \cos Q. \tag{10-88}$$

Transforming the oscillatory motion to uniform rectilinear motion does not aid in solving the harmonic oscillator problem. It merely indicates a similarity in behavior. The phase angle in the sinusoidal function for the oscillator increases uniformly with time as the distance of the translator does from its starting position.

10.11 Infinitesimal transformations

The question of invariance of an expression during a finite transformation can be broken down into that of invariance during each infinitesimal step of the process. If the overall transformation is a canonical transformation, the process breaks down into small steps generated by S or F.

Consider a canonical transformation generated by the Hamiltonian principal function

$$S(t, \mathbf{q}, \mathbf{P}) = q_k P_k + \varepsilon G(t, \mathbf{q}, \mathbf{P}) \tag{10-89}$$

in which ε is an arbitrary small parameter and G a well-behaved function. As long as the operation differs infinitesimally from the identity operation, $\mathbf{P}$ differs only infinitesimally from $\mathbf{p}$ and $G(t, \mathbf{q}, \mathbf{p})$ differs only infinitesimally from $G(t, \mathbf{q}, \mathbf{P})$. Then equation (10-89) can be rewritten as

$$S(t, \mathbf{q}, \mathbf{P}) = q_k P_k + \varepsilon G(t, \mathbf{q}, \mathbf{p}). \tag{10-90}$$

Applying equation (10-66) to (10-90) yields

$$p_l = \frac{\partial S}{\partial q_l} = P_l + \varepsilon \frac{\partial G}{\partial q_l}, \tag{10-91}$$

while applying equation (10-67) leads to

$$Q_l = \frac{\partial S}{\partial P_l} = q_l + \varepsilon \frac{\partial G}{\partial p_l}. \tag{10-92}$$

Each of these differs by an amount of order ε from the identity relationship; so our assumption about **P** is valid.

The expressions whose invariance we are interested in are certain fundamental Poisson brackets.

10.12   Poisson brackets

A Poisson bracket is a combination of derivatives of two functions in phase space, which turns out to be invariant to a canonical transformation from one phase space to another.

In particular, the Poisson bracket of functions g and h is defined as

$$[g, h] = \frac{\partial g}{\partial q_k} \frac{\partial h}{\partial p_k} - \frac{\partial g}{\partial p_k} \frac{\partial h}{\partial q_k} \tag{10-93}$$

in the space of the q_l's and p_l's.

When one of these functions is a coordinate or a momentum, we get a single derivative:

$$[q_j, h] = \frac{\partial q_j}{\partial q_k} \frac{\partial h}{\partial p_k} - \frac{\partial q_j}{\partial p_k} \frac{\partial h}{\partial q_k} = \frac{\partial h}{\partial p_j}, \tag{10-94}$$

$$[p_j, h] = \frac{\partial p_j}{\partial q_k} \frac{\partial h}{\partial p_k} - \frac{\partial p_j}{\partial p_k} \frac{\partial h}{\partial q_k} = -\frac{\partial h}{\partial q_j}. \tag{10-95}$$

When both are coordinates or momenta, we get either zero or one:

$$[q_j, q_k] = \frac{\partial q_k}{\partial p_j} = 0, \tag{10-96}$$

$$[p_j, p_k] = -\frac{\partial p_k}{\partial q_j} = 0, \tag{10-97}$$

$$[q_j, p_k] = \frac{\partial p_k}{\partial p_j} = \delta_{jk}. \tag{10-98}$$

Now, differentiating equations (10-91) and (10-92) lead to

$$\frac{\partial Q_j}{\partial q_k} = \delta_{jk} + \varepsilon \frac{\partial^2 G}{\partial q_k \, \partial p_j}, \tag{10-99}$$

$$\frac{\partial Q_j}{\partial p_k} = \varepsilon \frac{\partial^2 G}{\partial p_k \, \partial p_j}, \tag{10-100}$$

$$\frac{\partial P_j}{\partial q_k} = -\varepsilon \frac{\partial^2 G}{\partial q_k \, \partial q_j}, \tag{10-101}$$

$$\frac{\partial P_j}{\partial p_k} = \delta_{jk} - \varepsilon \frac{\partial^2 G}{\partial p_k \, \partial q_j}. \tag{10-102}$$

Substituting these expressions into the defining relationship (10-93) and dropping terms in ε^2 yields

$$[Q_j, P_k] = \left(\delta_{jl} + \varepsilon \frac{\partial^2 G}{\partial q_l \, \partial p_j}\right)\left(\delta_{kl} - \varepsilon \frac{\partial^2 G}{\partial p_l \, \partial q_k}\right) - \left(\varepsilon \frac{\partial^2 G}{\partial p_l \, \partial p_j}\right)\left(-\varepsilon \frac{\partial^2 G}{\partial q_l \, \partial q_k}\right)$$

$$= \delta_{jl} \delta_{kl} + \varepsilon \delta_{kl} \frac{\partial^2 G}{\partial q_l \, \partial p_j} - \varepsilon \delta_{jl} \frac{\partial^2 G}{\partial p_l \, \partial q_k}$$

$$= \delta_{jk} + \varepsilon \left(\frac{\partial^2 G}{\partial q_k \, \partial p_j} - \frac{\partial^2 G}{\partial p_j \, \partial q_k}\right), \tag{10-103}$$

whence

$$[Q_j, P_k] = \delta_{jk}. \tag{10-104}$$

Similarly, we can show that

$$[Q_j, Q_k] = 0, \tag{10-105}$$

$$[P_j, P_k] = 0. \tag{10-106}$$

The infinitesimal transformation has not changed the fundamental Poisson brackets evaluated in the initial phase space.

Thus an integral of such transformations leaves them invariant. We have

$$[Q_j, P_k] = \delta_{jk}, \tag{10-107}$$

$$[Q_j, Q_k] = 0, \tag{10-108}$$

$$[P_j, P_k] = 0, \tag{10-109}$$

when Q_j and P_k differ by any expression from q_j and p_k, as long as the two sets of variables are related by a canonical transformation.

These equations can be used to prove that any Poisson bracket is invariant to such a transformation. We will not pursue this development here, however.

EXAMPLE 10.6 Show that the Cartesian components of angular momentum of a particle obey the cyclic relationship

$$[L_1, L_2] = L_3.$$

From equation (2-107),

$$L_1 = x_2 p_3 - x_3 p_2,$$
$$L_2 = x_3 p_1 - x_1 p_3,$$
$$L_3 = x_1 p_2 - x_2 p_1.$$

Substitute the first two components into the Poisson bracket equation (10-93). Then carry out the indicated differentiations and identify the result:

$$[L_1, L_2] = \frac{\partial L_1}{\partial x_k} \frac{\partial L_2}{\partial p_k} - \frac{\partial L_1}{\partial p_k} \frac{\partial L_2}{\partial x_k}$$

$$= 0 + 0 + (-p_2)(-x_1) - 0 - 0 - x_2 p_1$$

$$= x_1 p_2 - x_2 p_1 = L_3.$$

Similarly, we can show that

$$[L_2, L_3] = L_1,$$
$$[L_3, L_1] = L_2.$$

EXAMPLE 10.7   Why can no more than one Cartesian component of the angular momentum serve as a canonical variable at one time?

Equation (10-109) shows that the Poisson bracket between two canonically transformed momenta must be 0. In example 10.6, we saw that each combination gave plus (or minus) the remaining component.

10.13   General remarks

In chapter 5, we recast the equations of motion in a form suitable for any generalized coordinate system. Gauge and coordinate transformations that leave the motion unchanged were also introduced.

In this chapter, we have replaced the generalized velocities with variables that become constant whenever the system is as symmetric as possible with respect to the corresponding coordinates. These are the generalized momenta.

A Legendre transformation that introduces the p_i's in place of the $\dot{q}_i$'s converts the Lagrangian L to the Hamiltonian H. The equations of motion then become

$$\dot{q}_k = \frac{\partial H}{\partial p_k},$$ (10-110)

$$\dot{p}_k = -\frac{\partial H}{\partial q_k}.$$ (10-111)

The motion itself can be represented by a single point moving in a phase space for the system. An ensemble of systems that differ only in their initial q_k's and p_k's is represented by a swarm of points moving through the phase space. The density of these points around a given moving point is invariant, according to Liouville's theorem.

Any transformation

$$P_k = P_k(p_1, \ldots, p_n, q_1, \ldots, q_n, t),$$ (10-112)

$$Q_k = Q_k(p_1, \ldots, p_n, q_1, \ldots, q_n, t),$$ (10-113)

$$t = t,$$ (10-114)

that allows a new function,

$$\mathscr{H} = \mathscr{H}(P_1, \ldots, P_n, Q_1, \ldots, Q_n, t),$$ (10-115)

to be set up that satisfies equations of the Hamiltonian form

$$\dot{Q}_k = \frac{\partial \mathscr{H}}{\partial P_k},$$ (10-116)

$$\dot{P}_k = -\frac{\partial \mathscr{H}}{\partial Q_k},$$ (10-117)

is called a *canonical transformation*. Such a transformation can be generated by the functions F and S of sections 10.8 and 10.9.

DISCUSSION QUESTIONS

10.1 What does absence of a q_l from L indicate?

10.2 How may the definition of p_l be related to this absence? When is equation (5-47), or (10-6), not a generalization of equation (5-22)?

10.3 Cite examples of conserved generalized momenta.

10.4 By differentiating

$$H = p_k \dot{q}_k - L,$$

show that the Hamiltonian is not a function of any $\dot{q}_l$.

10.5 Explain Hamilton's equations.

10.6 When is H not conserved? When is H different from $T + V$?

10.7 How does the evolution of a given system appear in phase space?

10.8 When is the force on a particle divisible into a part depending on position and a part depending on momentum?

10.9 How does a system evolve when it is near a singular point in its phase diagram?

10.10 Explain what density is conserved in phase space.

10.11 Discuss the relationship that exists between the uncertainty in p and the uncertainty in q when Liouville's theorem can be applied to the pq plane by itself.

10.12 Explain what happens in the pertinent phase space when thermodynamic equilibrium exists in the given system.

10.13 Cite reasons for considering the p_l's to be more fundamental than the $\dot{q}_l$'s.

10.14 Show how the gauge and coordinate transformations do not alter the form of Hamilton's equations. How may the variables in the generating function be changed?

10.15 How can a canonical transformation reveal symmetry in a physical system? How can a canonical transformation reduce the number of independent variables?

10.16 How can a canonical transformation relate two kinds of motion?

10.17 What principal function S generates the identity transformation?

10.18 Why can't more than one Cartesian component of angular momentum be a generalized momentum in a problem?

10.19 Are Q_l and P_l necessarily coordinate and momentum?

10.20 Since the Hamiltonian H of a free particle is its kinetic energy T, equation (10-18) then reduces to

$$\dot{p}_k = -\frac{\partial T}{\partial q_k}.$$

The Lagrangian L of such a particle is also T and equation (10-3) leads to

$$\frac{\partial T}{\partial q_k} = \frac{d}{dt}\frac{\partial L}{\partial \dot{q}_k} = \frac{d}{dt}p_k = \dot{p}_k.$$

Reconcile these two different formulas for $\dot{p}_k$.

PROBLEMS

10.1 Set up Hamilton's equations for a mass m falling along the spiral

$$z = -k\phi, \qquad r = \text{constant},$$

under the influence of a constant gravitational field directed along the minus z axis. Combine the equations and integrate to obtain ϕ as a function of t.

10.2 Set up Hamilton's equations for a mass m that moves without friction on a circular wire that rotates at constant angular velocity ω about a vertical diameter. Find the equilibrium position and the frequency of small oscillation about this position.

10.3 Formulate the Hamiltonian function for a uniform hemisphere rocking back and forth along a straight line on a horizontal plane. Find the frequency of small oscillation for the system.

10.4 Construct Hamilton's equations for a body falling freely through a straight hole bored between two points on the surface of a homogeneous gravitating sphere. Solve for the time of transit from surface to surface if the radius of the sphere is R and the acceleration due to gravity at its surface is g.

10.5 Sketch a phase diagram showing typical motions of a mass m that is constrained to move without friction on a circular wire fixed upright in a gravitational field.

10.6 Successive turning points for a mass subject to a linear restoring force and a frictional force of constant magnitude are $+14$ cm and -10 cm from the point at which no forces act. Construct the corresponding trajectory in phase space and continue it to locate the position where the mass comes to rest.

10.7 A set of identical particles is initially distributed in the earth's gravitational field uniformly through heights lying between z_0 and $z_0 + a$ and through vertical momenta lying between p_0 and $p_0 + b$. Multiply the initial range in coordinate by the initial range in momentum. Then calculate the heights and momenta of extreme particles in the set at time t, assuming the particles to move freely, and obtain the area occupied in phase space at this time. Does Liouville's theorem apply?

10.8 If the equations

$$Q_k = a_{kl}q_l,$$
$$P_k = a_{kl}p_l,$$
$$K = H,$$

are to define a canonical transformation of the generalized coordinates q_l and generalized momenta p_l, what relationship must the a_{kl}'s obey?

10.9 Identify the canonical transformation generated by the function

$$F = q_k Q_k.$$

10.10 Prove that

$$[q_j, H] = \dot{q}_j,$$
$$[p_j, H] = \dot{p}_j,$$
$$\frac{dg}{dt} = [g, H] + \frac{\partial g}{\partial t},$$

if H is the Hamiltonian for the system.

10.11 Set up Hamilton's equations for a mass m sliding smoothly over the inner surface of the upper half of a cone, fixed so the axis is vertical and the only external force acting is that of gravity. Solve for the angular velocity of rotation about the axis and for the distance of the particle from the apex when the particle traverses a horizontal path. Assume the particle to be disturbed from this path and solve for the frequency of small oscillation about it. When is the frequency of oscillation equal to that of rotation about the axis?

10.12 The governor for a steam engine consists of two balls, each of mass m, attached by light arms to sleeves on a rotating rod, as in figure 10.5. The upper

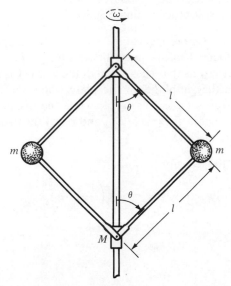

FIGURE 10.5 Essentials of a flyball governor.

sleeve is fixed to the rod, while the lower one, of mass M, is free to move up or down. Assume the arms to be massless and the angular velocity ω to be constant. Then set up Hamilton's equations describing the system, solve for the angle θ at which the arms would stop waving, and obtain the frequency of small oscillation about this steady value.

10.13 A mass m is attached by a string of length l to a support at rest in the earth's field. While the mass is swinging in a plane, the string is shortened at the rate

$$\frac{dl}{dt} = -a$$

where a is constant. Formulate the Hamiltonian function and equations. Then solve for the frequency of small oscillation.

10.14 Construct the Hamiltonian function for a uniform rectangular slab of thickness a seesawing back and forth on a fixed cylinder of radius R. The axis of the cylinder is horizontal while the axis of the slab rocks in a vertical plane perpendicular to the cylinder axis.

10.15 Sketch phase diagrams showing typical motions of a simple pendulum swinging from a point of support on a wall. Assume that the wall is perpendicular to the plane of motion but is inclined at angle α from the vertical. Suppose that in each collision with the wall the bob dissipates half of its incident kinetic energy.

10.16 Calculate successive turning points for a 0.100 kg mass drawn over a surface with a friction coefficient of 0.10 by a spring with $k = 10.0$ N m^{-1}, if the initial displacement from the point at which the spring is relaxed is 10 cm. Then construct the trajectory in phase space.

10.17 An electron microscope causes electrons scattered from an object of length x_1 to be focused in an image of length x_2. Formulate the phase area occupied by the electrons that pass through the lens (a) as the electrons leave the object and (b) as they impinge on the plate where the image is formed. Let the effective aperture of the lens be A, while the distance of the object from the lens is D_1 and that of the image D_2, and relate the two phase areas. Assume that $x_1 \ll D_1$ and $x_2 \ll D_2$.

10.18 Show that the transformation

$$Q = \log\left(\frac{1}{q}\sin p\right),$$

$$P = q \cot p,$$

$$K = H,$$

is canonical.

10.19 Show that the equations

$$Q = q^{1/2} \cos 2p,$$
$$P = q^{1/2} \sin 2p,$$
$$K = H,$$

describe a canonical transformation.

10.20 Prove that

$$[A, [B, C]] + [B, [C, A]] + [C, [A, B]] = 0.$$

REFERENCES

BOOKS

Goldstein, H. *Classical Mechanics*, pp. 215–247. Addison-Wesley Publishing Company, Inc., Reading, Mass., 1950.

Konopinski, E. J. *Classical Descriptions of Motion*, pp. 188–233. W. H. Freeman and Company, San Francisco, 1969.

Marion, J. B. *Classical Dynamics of Particles and Systems*, 2nd ed., pp. 220–233. Academic Press, Inc., New York, 1970.

Stoker, J. J. *Nonlinear Vibrations in Mechanical and Electrical Systems*, pp. 19–80. Interscience Publishers, New York, 1950.

Symon, K. R. *Mechanics*, 2nd ed., pp. 396–400. Addison-Wesley Publishing Company, Inc., Reading, Mass., 1960.

ARTICLES

Anderson, J. L. Symmetries and Invariances of Canonical Theories. *Am. J. Phys.*, **40**, 541 (1972).

Beran, M. J. Use of the Characteristic Function in the Derivation of Liouville's Equation. *Am. J. Phys.*, **35**, 242 (1967).

Campbell, P. M. Classical Integrals Not Associated with Symmetry Transformations. *Am. J. Phys.*, **37**, 1161 (1969).

Campbell, P. M. Classical Transformation Operators and the *S* Matrix. *Am. J. Phys.*, **36**, 931 (1968).

Chamorro, A. On the Conservation Laws in Classical Mechanics. *Am. J. Phys.*, **37**, 610 (1969).

Cheng, C. General Born Diagram and Legendre Transformation. *Am. J. Phys.*, **38**, 956 (1970).

Chern, B., and **Tubis, A.** Invariance Principles in Classical and Quantum Mechanics. *Am. J. Phys.*, **35**, 254 (1967).

Deprit, A. Free Rotation of a Rigid Body Studied in the Phase Plane. *Am. J. Phys.*, **35**, 424 (1967).

Hurley, J. Necessary and Sufficient Conditions for a Canonical Transformation. *Am. J. Phys.*, **40,** 533 (1972).

Jeffers, W. A., Jr. Phase Diagrams for the Overdamped Oscillator. *Am. J. Phys.*, **39,** 1210 (1971).

Levitas, A. D. Algebraic Equivalence of the Lagrangian and of the Hamiltonian Forms of the Equations of Motion. *Am. J. Phys.*, **36,** 1144 (1968).

Ludford, G. S. S., and **Yannitell, D. W.** Canonical Transformations without Hamilton's Principle. *Am. J. Phys.*, **36,** 231 (1968).

Rodrigues, L. M. C. S., and **Rodrigues, P. R.** Further Developments in Generalized Classical Mechanics. *Am. J. Phys.*, **38,** 557 (1970).

Sciamanda, R. J. Expansion of Available Phase Space and Approach to Equilibrium. *Am. J. Phys.*, **37,** 808 (1969).

11 / Hamilton's Principle and the Calculus of Variations

11.1 Differential and integral forms for laws

A law of nature may appear as a relationship among neighboring and/or consecutive states of a given system. Particular physical properties are then dependent variables in a set of *differential* equations. Specifying the properties at appropriate limits allows the equations to be integrated and the properties to be obtained throughout a given region or interval.

On the other hand, general features of the over-all behavior can be picked out and formulated as a principle. When the statement imposes a condition on an integral, the law is said to be in its *integral* form. When this condition implies no more nor less than the differential laws imply, the differential equations can be deduced from the integral formulation and the integral formulation can be deduced from the differential laws.

So far, we have employed differential statements of Newtonian mechanics. In this chapter, we will develop the corresponding integral statement, Hamilton's principle, according to which the time integral of the difference between kinetic energy T and generalized potential U is stationary. Particles seem to behave as purposeful agents, choosing paths that make this time integral extremal.

But according to Newton's laws, the behavior of a system is determined at any given time by the existing state and by the forces acting. There is no leeway for choice. Accelerations occur at the same time that the pertinent forces act—not at some earlier or later time. Thus, they satisfy the *causal principle* in its strongest form. Not only are the effects determined by the causes (here the forces) but they are *not* separated from them in time.

In this chapter, Hamilton's principle will be derived from Newton's laws, *and* Lagrange's equations, a form of Newton's laws, will be derived from Hamilton's principle. As a consequence, the principle implies no more nor less than the differential laws; it does not violate the strong causality condition.

In deriving the integral statement from the differential law and in carrying out the inverse transformation, we will employ properties of variations. Because these properties and their uses may be unfamiliar, they will be studied next. Furthermore,

319

the scope and power of the calculus of variations will be illustrated with examples and problems in which geodesics, brachistochrones, minimal surfaces, and other stationary solutions will be determined. Later, in chapter 20, we will find that the geodesic problem has not only mathematical, but also physical significance.

11.2 Pertinent operations

A typical problem in the calculus of variations requires a definite integral of some function of unknown functions to be stationary to small variations in the unknown functions.

Let a particular manifestation of one of these unknown functions be q while a suitable neighboring example is Q. A plot of the corresponding property against the independent variables appears as the paths in figure 11.1. Now, some criterion can be employed to associate points on one of these paths with those on the other and the difference between the two functions can be obtained at corresponding points. Whenever this difference is infinitesimal, it is called the *variation*

$$\delta q = Q - q. \tag{11-1}$$

In most of our work, such variations are taken at the same stage in the integration (exceptions arise in some of the problems). When the integration proceeds over t alone, corresponding points are then associated with the same t and the variation of t itself is zero:

$$\delta t = 0. \tag{11-2}$$

When the integration proceeds over several independent variables, all of these variables are similarly held constant in calculating a δq.

Since Q and q represent possible paths, they are well-behaved functions, and equation (11-1) can be differentiated. When Q and q depend only on t, we have

$$\frac{d}{dt}\delta q = \frac{dQ}{dt} - \frac{dq}{dt}. \tag{11-3}$$

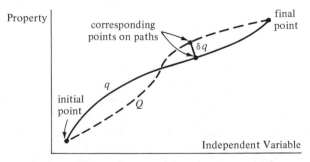

FIGURE 11.1   Examples from the continuum of functions representing q.

But from the definition of variation, entity $\delta\dot{q}$ is the difference between the derivatives at corresponding points on the two curves:

$$\delta\dot{q} = \frac{dQ}{dt} - \frac{dq}{dt}.$$ (11-4)

Because the right sides of equations (11-3) and (11-4) are the same, the left sides are equal:

$$\frac{d}{dt}\,\delta q = \delta\dot{q}.$$ (11-5)

A similar derivation can be carried out for the variation of a function of several variables, and differentiation with respect to any one of them. The general result is that variation and differentiation commute.

The variation of a function of the q's is defined as the change in the function when the q's are varied. Thus

$$\delta f = f(Q_1,\ldots) - f(q_1,\ldots)$$ (11-6)

and

$$\delta I = \delta \int_{t_1}^{t_2} f\,dt$$

$$= \int_{t_1}^{t_2} f(Q_1,\ldots)\,dt - \int_{t_1}^{t_2} f(q_1,\ldots)\,dt$$

$$= \int_{t_1}^{t_2} [f(Q_1,\ldots) - f(q_1,\ldots)]\,dt$$

$$= \int_{t_1}^{t_2} \delta f\,dt.$$ (11-7)

Note that the integration may be either single or multiple; in any case, the above steps can be carried out. The variation operator can be brought inside an integral. The final integrand can then be expanded by means of the procedure in example 5.1.

Existence of the final form in (11-7) does not imply existence of the initial form. If the infinitesimal δf is not exact, a function f does not exist and δf is not expressible as $f(Q_1,\ldots) - f(q_1,\ldots)$. The work δW may be such an infinitesimal.

In the problems to be discussed, the *continuity principle* is valid. By this, we mean that a small cause produces a small effect. Any allowable infinitesimal variation in q away from a given path then produces an infinitesimal change in I. Furthermore, successive similar variations in q produce similar effects.

Thus the variation in I plotted against some parameter measuring possible variations in a q is a smooth curve. Near an *extremum*, therefore, we must have

$$\delta I = 0.$$ (11-8)

Substituting equation (11-8) into (11-7) then yields

$$\int_{t_1}^{t_2} \delta f \, dt = 0 \tag{11-9}$$

as the condition that the integral

$$I = \int_{t_1}^{t_2} f \, dt \tag{11-10}$$

must meet.

EXAMPLE 11.1   Show that the variation $A\varepsilon \sin(t/\varepsilon)$, where ε is infinitesimal, changes a function of t that is well-behaved into one that is not.

Let the original function be q and the variation be

$$\delta q = A\varepsilon \sin \frac{t}{\varepsilon}.$$

The derivative of this is

$$\frac{d}{dt} \delta q = A \cos \frac{t}{\varepsilon} = \delta \dot{q}.$$

The frequency of oscillation of both δq and $\delta \dot{q}$ is $1/(2\pi)$ times the coefficient of t in the argument:

$$\nu = \frac{1}{2\pi\varepsilon}.$$

As ε approaches zero, this frequency increases without limit. Furthermore, the amplitude $A\varepsilon$ of δq becomes infinitesimal, while $\delta \dot{q}$ oscillates within its extremes

$$-A \leq \delta \dot{q} \leq A.$$

Although

$$Q = q + \delta q$$

differs only infinitesimally from q, its slope oscillates from $\dot{q} + A$ to $\dot{q} - A$ at an unbounded rate with change of t. Thus, the varied function Q does not have a definite derivative and is not well-behaved. Following the spirit of the continuity principle, we reject such functions.

11.3   Single unknown function

Let us first consider how the well-behaved q that passes through given points $q(t_1)$ and $q(t_2)$ while making the integral

$$I = \int_{t_1}^{t_2} f(q, \dot{q}, t) \, dt \tag{11-11}$$

extreme can be found. We assume that the continuity principle, as reflected in equation (11-9), is valid.

Because f is here a function of q, $\dot{q}$, and t, the procedure in example 5.1 yields

$$\delta f = \frac{\partial f}{\partial q}\, \delta q + \frac{\partial f}{\partial \dot{q}}\, \delta \dot{q} + \frac{\partial f}{\partial t}\, \delta t. \tag{11-12}$$

Choosing corresponding points at a given t eliminates the last term. Substituting the result into (11-7) and introducing condition (11-8) leads to

$$\delta I = \int_{t_1}^{t_2} \delta f\, dt$$

$$= \int_{t_1}^{t_2} \left(\frac{\partial f}{\partial q}\, \delta q + \frac{\partial f}{\partial \dot{q}}\, \delta \dot{q} \right) dt = 0. \tag{11-13}$$

Since variation and differentiation commute,

$$\delta \dot{q}\, dt = \left(\frac{d}{dt}\, \delta q \right) dt = d(\delta q), \tag{11-14}$$

the second term of the integral can be expanded by use of the formula for differentiating a product:

$$\int_{t_1}^{t_2} \frac{\partial f}{\partial \dot{q}}\, \delta \dot{q}\, dt = \int_{t_1}^{t_2} \frac{\partial f}{\partial \dot{q}}\, d(\delta q) = \frac{\partial f}{\partial \dot{q}}\, \delta q \,\Big|_{t_1}^{t_2} - \int_{t_1}^{t_2} d\left(\frac{\partial f}{\partial \dot{q}} \right) \delta q$$

$$= 0 - \int_{t_1}^{t_2} \frac{d}{dt} \left(\frac{\partial f}{\partial \dot{q}} \right) \delta q\, dt. \tag{11-15}$$

Note that the integrated part is zero because δq is zero at the end points.

This result reduces equation (11-13) to

$$\int_{t_1}^{t_2} \left(\frac{\partial f}{\partial q} - \frac{d}{dt} \frac{\partial f}{\partial \dot{q}} \right) \delta q\, dt = 0. \tag{11-16}$$

Now, variation δq is arbitrary inside the limits of integration. We may make it zero everywhere except in a given small interval where it is assumed to be positive. Then to satisfy equation (11-16), we must set

$$\frac{\partial f}{\partial q} - \frac{d}{dt} \frac{\partial f}{\partial \dot{q}} = 0 \tag{11-17}$$

in the interval. Since the interval may be anywhere within the end points, equation (11-17) is valid throughout the range of integration. This equation is named after its deviser *Euler*.

An alternate form of Euler's equation is useful when $\partial f / \partial t$ is zero. We multiply (11-17) by $\dot{q}$,

$$\dot{q}\, \frac{\partial f}{\partial q} - \dot{q}\, \frac{d}{dt} \frac{\partial f}{\partial \dot{q}} = 0, \tag{11-18}$$

and rewrite (11-12) in derivative form

$$\frac{df}{dt} = \frac{\partial f}{\partial q}\dot{q} + \frac{\partial f}{\partial \dot{q}}\frac{d\dot{q}}{dt} + \frac{\partial f}{\partial t}. \qquad (11\text{-}19)$$

Adding (11-18), (11-19), and rearranging then yields

$$\frac{df}{dt} - \frac{\partial f}{\partial \dot{q}}\frac{d}{dt}\dot{q} - \dot{q}\frac{d}{dt}\frac{\partial f}{\partial \dot{q}} - \frac{\partial f}{\partial t} = 0 \qquad (11\text{-}20)$$

or

$$\frac{d}{dt}\left(f - \dot{q}\frac{\partial f}{\partial \dot{q}}\right) - \frac{\partial f}{\partial t} = 0. \qquad (11\text{-}21)$$

Our derivation shows that equation (11-17), and its alternate form (11-21), must be satisfied by a physically useful (well-behaved) $q(t)$ making δI zero; Euler's equation is a *necessary condition* for extrema. In practice, one of the forms is integrated and the result examined. If this is not flexible enough to pass through the given points $q(t_1)$, $q(t_2)$, then no solution exists. Otherwise, the result may either maximize or minimize I.

EXAMPLE 11.2   Elaborate on the step from equation (11-16) to (11-17).

Let the limits of the small interval in which δq is different from zero be t_A and t_B, as in figure 11.2. Then break equation (11-16) down as follows:

$$\int_{t_1}^{t_A} (\ldots) \, \delta q \, dt + \int_{t_A}^{t_B} (\ldots) \, \delta q \, dt + \int_{t_B}^{t_2} (\ldots) \, \delta q \, dt = 0.$$

The first and last integrals are zero because of the way δq is chosen. From the middle integral, the quantity in parenthesis may be factored, if t_B is close enough to t_A. We are left with the term

$$\left(\frac{\partial f}{\partial q} - \frac{d}{dt}\frac{\partial f}{\partial \dot{q}}\right)_{t=t_A} \int_{t_A}^{t_B} \delta q \, dt = 0$$

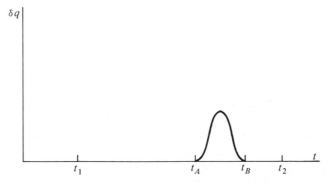

FIGURE 11.2   A variation that is different from zero only within the interval $t_A < t < t_B$.

from which the integral can be canceled, leaving

$$\left(\frac{\partial f}{\partial q} - \frac{d}{dt} \frac{\partial f}{\partial \dot{q}} \right)_{t=t_A} = 0.$$

EXAMPLE 11.3 Show that we may let the variation be $\varepsilon(t - t_A)^2(t - t_B)^2$ inside the interval t_A to t_B and zero outside.

Differentiating the form

$$\delta q = \varepsilon(t - t_A)^2(t - t_B)^2$$

for inside the interval yields

$$\delta\dot{q} = \frac{d}{dt} \delta q = 4\varepsilon \left(t - \frac{t_A + t_B}{2} \right)(t - t_A)(t - t_B)$$

while differentiating the expression

$$\delta q = 0$$

for outside the interval yields

$$\delta\dot{q} = \frac{d}{dt} \delta q = 0.$$

Except about the points t_A and t_B, the variations are analytical and clearly suitable. Now at these points the fourth degree parabola and its derivative are both zero. The variations δq and $\delta\dot{q}$ are thus also continuous here and do meet the required standards of behavior.

Furthermore, a continuous curve q having continuous $\dot{q}$ and $\ddot{q}$ may be chosen. Then, the expression

$$\frac{d}{dt} \frac{\partial f}{\partial \dot{q}} = \frac{\partial^2 f}{\partial t \, \partial \dot{q}} + \frac{\partial^2 f}{\partial q \, \partial \dot{q}} \dot{q} + \frac{\partial^2 f}{\partial \dot{q}^2} \ddot{q}$$

is continuous as required in the integration by parts.

EXAMPLE 11.4 Find the shortest path between two points on a sphere.

A convenient coordinate system is the spherical system based on the center of the sphere as origin and a reference axis and plane passing through the center. With respect to the latter, θ is the colatitude and ϕ the longitude. Then if a is the radius of the sphere, the element of distance along the path is

$$ds = a(d\theta^2 + \sin^2\theta \, d\phi^2)^{1/2} = a(1 + \sin^2\theta \, \phi'^2)^{1/2} \, d\theta$$

where $\phi' = d\phi/d\theta$. The length of the path is the integral of this element from the initial point to the final point:

$$s = \int_{\theta_1}^{\theta_2} a(1 + \sin^2\theta \, \phi'^2)^{1/2} \, d\theta = \int f(\phi, \phi', \theta) \, d\theta.$$

Since this integral has the same form as integral (11-11), with ϕ replacing q and θ replacing t, the theory in section 11.3 applies.

As the integrand does not contain ϕ, the derivative with respect to ϕ is zero

$$\frac{\partial f}{\partial \phi} = 0.$$

This value reduces Euler's equation (11-17)

$$\frac{\partial f}{\partial \phi} - \frac{d}{d\theta}\frac{\partial f}{\partial \phi'} = 0$$

to

$$\frac{d}{d\theta}\frac{\partial f}{\partial \phi'} = 0.$$

Integration then yields

$$\frac{\partial f}{\partial \phi'} = ca.$$

Now, differentiate the integrand with respect to ϕ' and substitute the result into the integrated equation

$$\frac{\partial f}{\partial \phi'} = \frac{a \sin^2 \theta \, \phi'}{(1 + \sin^2 \theta \, \phi'^2)^{1/2}} = ca.$$

Square and rearrange,

$$\left(1 - \frac{c^2}{\sin^2 \theta}\right)\phi'^2 = \frac{c^2}{\sin^4 \theta},$$

introduce the pertinent trigonometric identities,

$$(1 - c^2 - c^2 \cot^2 \theta)\phi'^2 = c^2 \csc^4 \theta,$$

solve for $d\phi$, replacing $c/(1 - c^2)^{1/2}$ with k,

$$d\phi = \frac{k \csc^2 \theta \, d\theta}{(1 - k^2 \cot^2 \theta)^{1/2}},$$

and integrate again to get

$$\phi = \alpha - \sin^{-1}(k \cot \theta)$$

or

$$\sin(\alpha - \phi) = k \cot \theta,$$

whence

$$a \sin \theta \cos \phi \sin \alpha - a \sin \theta \sin \phi \cos \alpha = ka \cos \theta.$$

Since a is the distance of a point from the origin, the conventional formulas for transforming to rectangular coordinates convert this result to

$$\begin{cases} x \sin \alpha - y \cos \alpha = kz \\ x^2 + y^2 + z^2 = a^2, \end{cases}$$

a plane passing through the center intersecting the sphere along a great circle. In applications, constants α and k are chosen to make the great circle pass through the initial and final points.

EXAMPLE 11.5   Find the vertical curve along which a particle will most quickly fall through a constant gravitational field from the origin to a lower point.   Neglect friction.

Since velocity v of the particle is ds/dt, the integral for the time of descent is

$$t = \int_A^B dt = \int \frac{ds}{ds/dt} = \int \frac{ds}{v}.$$

If axes are chosen as in figure 11.3, the element of distance is

$$ds = (dx^2 + dy^2)^{1/2} = (1 + y'^2)^{1/2} \, dx$$

where y' represents dy/dx.

Choosing the energy E as zero initially,

$$E = \tfrac{1}{2}mv^2 - mgy = 0,$$

solving for v,

$$v = (2gy)^{1/2},$$

and substituting these expressions for v, ds into the integral for time, we get

$$t = \int_A^B \frac{(1 + y'^2)^{1/2}}{(2gy)^{1/2}} \, dx = \int_A^B f \, dx.$$

This integral has the same form as (11-11), with y replacing q and x replacing t.

The integrand f does not contain x explicitly, so $\partial f/\partial x$ is zero and the second form of Euler's equation (11-21) integrates to yield

$$f - y' \frac{\partial f}{\partial y'} = \text{constant.}$$

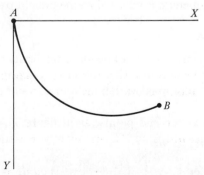

FIGURE 11.3   The curve of quickest descent, which is called the brachistochrone.

Combining this result with the derivative of the integrand with respect to y',

$$\frac{\partial f}{\partial y'} = \frac{y'}{(2gy)^{1/2}(1 + y'^2)^{1/2}},$$

we obtain

$$\frac{(1 + y'^2)^{1/2}}{(2gy)^{1/2}} - \frac{y'^2}{(2gy)^{1/2}(1 + y'^2)^{1/2}} = \frac{1}{(2gy)^{1/2}(1 + y'^2)^{1/2}} = \text{constant}.$$

whence

$$(2gy)^{1/2}(1 + y'^2)^{1/2} = \frac{1}{\text{constant}}$$

or

$$y(1 + y'^2) = 2a$$

where a is a constant. Thus

$$y \, dy^2 = (2a - y) \, dx^2.$$

Let us introduce the substitution

$$y = a(1 - \cos \theta).$$

Then

$$dy = a \sin \theta \, d\theta$$

and the differential equation becomes

$$(1 - \cos \theta)a^2 \sin^2 \theta \, d\theta^2 = (1 + \cos \theta) \, dx^2.$$

Multiply by $(1 - \cos \theta)$, cancel $\sin^2 \theta$, and take the square root:

$$a(1 - \cos \theta) \, d\theta = dx.$$

Finally, integrate and let the constant of integration be zero (so x is 0 when y is 0):

$$x = a(\theta - \sin \theta).$$

This equation together with the definition of θ,

$$y = a(1 - \cos \theta),$$

describes a cycloid. Parameter a is adjusted so the cycloid passes through point B.

11.4 More general paths

In some problems the quantity q is not known at the limits of integration. In others, we may wish to consider a solution with a corner—a discontinuity in the derivative $\dot{q}$. These possibilities allow additional variations and so yield additional conditions on the solutions.

Indeed when q varies at the end points, as in figure 11.4, the integrated part of equation (11-15) no longer drops out. Consequently, equation (11-16) now becomes

$$\frac{\partial f}{\partial \dot{q}} \, \delta q \, \bigg|_{t_1}^{t_2} + \int_{t_1}^{t_2} \left(\frac{\partial f}{\partial q} - \frac{d}{dt} \frac{\partial f}{\partial \dot{q}} \right) \delta q \, dt = 0. \tag{11-22}$$

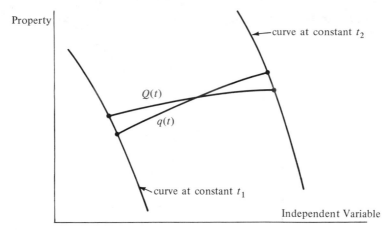

FIGURE 11.4   Unvaried path $q(t)$ and varied path $Q(t)$ with different end points.

Variation δq is arbitrary. As long as it is zero at the limits, the argument proceeds as before, yielding Euler's equation as a necessary condition. So in a variation from an extremum, the quantity in parenthesis is zero and equation (11-22) reduces to

$$\frac{\partial f}{\partial \dot{q}} \, \delta q \, \bigg|_{t_1}^{t_2} = 0. \tag{11-23}$$

The variation δq can be set equal to zero at one limit and different from zero at the other. The nonzero variation is then canceled to get

$$\frac{\partial f}{\partial \dot{q}} = 0 \quad \text{at } t_1 \text{ and } t_2. \tag{11-24}$$

If q is not known at the end points, we employ condition (11-24) as well as Euler's equation in solving the problem.

If the solution is to have a discontinuous slope at some t_2 between the end points t_1 and t_3, break the integration down into two terms

$$I = \int_{t_1}^{t_2} f(q, \dot{q}, t) \, dt + \int_{t_2}^{t_3} f(r, \dot{r}, t) \, dt. \tag{11-25}$$

Here $q(t)$ and $r(t)$ are the two smooth parts of the path, as in figure 11.5. Then we assume that an extremum exists and we consider special variations from it.

Thus, the first integral may be held fixed and the second integral varied. We then obtain as a necessary condition

$$\frac{\partial f}{\partial r} - \frac{d}{dt} \frac{\partial f}{\partial \dot{r}} = 0 \tag{11-26}$$

over the second interval. In like manner, the first integral alone may be varied. The necessary condition that $q(t)$ be a solution of Euler's equation is obtained.

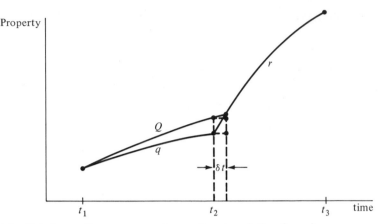

FIGURE 11.5   Solution with a corner and a pertinent variation from it.

Next, the junction at t_2 is allowed to vary up or down. The integrated part of the first integral becomes

$$\left(\frac{\partial f}{\partial \dot{q}}\right)_{t_2} \delta q(t_2) \tag{11-27}$$

while the integrated part of the second one is

$$-\left(\frac{\partial f}{\partial \dot{r}}\right)_{t_2} \delta r(t_2) = -\left(\frac{\partial f}{\partial \dot{r}}\right)_{t_2} \delta q(t_2). \tag{11-28}$$

The variation of the sum in (11-25) now yields two integrals of form (11-16) plus these two integrated parts. At an extremum, the Euler equations (11-17) and (11-26) are valid, and the integrals vanish. We are left with the equation

$$\left[\left(\frac{\partial f}{\partial \dot{q}}\right)_{t_2} - \left(\frac{\partial f}{\partial \dot{r}}\right)_{t_2}\right] \delta q(t_2) = 0. \tag{11-29}$$

Since $\delta q(t_2)$ need not be zero, we must also have

$$\left(\frac{\partial f}{\partial \dot{q}}\right)_{t_2} = \left(\frac{\partial f}{\partial \dot{r}}\right)_{t_2}. \tag{11-30}$$

Along the over-all path, the first derivative is evaluated just left of the junction and the second derivative on the immediate right.

Finally, we may vary the junction to left or right. For convenience, we allow it to run up $r(t)$ as in figure 11.5. Then the variation in integral I is

$$\delta I = \int_{t_1}^{t_2+\delta t} f(Q, \dot{Q}, t)\, dt - \int_{t_1}^{t_2} f(q, \dot{q}, t)\, dt - \int_{t_2}^{t_2+\delta t} f(r, \dot{r}, t)\, dt. \tag{11-31}$$

The difference between the integrals for the two paths up to t_2 is given by the left side of (11-22) with the integrated part at t_1 zero. For the part at t_2, the variation of the end point is

$$(\delta q)_{t_2} \simeq \frac{dr}{dt} \delta t - \frac{dQ}{dt} \delta t \simeq \frac{dr}{dt} \delta t - \frac{dq}{dt} \delta t. \tag{11-32}$$

The difference between the integrals from t_2 to $t_2 + \delta t$ is approximately

$$f(Q, \dot{Q}, t) \, \delta t - f(r, \dot{r}, t) \, \delta t. \tag{11-33}$$

The approximations become exact as the shift δt is made small.

Thus, equation (11-31) reduces to

$$\delta I = \int_{t_1}^{t_2} \left(\frac{\partial f}{\partial q} - \frac{d}{dt} \frac{\partial f}{\partial \dot{q}} \right) \delta q \, dt + \frac{\partial f}{\partial \dot{q}} \left(-\frac{dq}{dt} \delta t + \frac{dr}{dt} \delta t \right)$$
$$+ f(Q, \dot{Q}, t) \, \delta t - f(r, \dot{r}, t) \, \delta t$$
$$= 0. \tag{11-34}$$

Since the variation is away from an extremum, the net integral from t_1 to t_2 is zero. Cancelling expression δt from the remaining terms, replacing $f(Q, \dot{Q}, t)$ by its approximate equivalent $f(q, \dot{q}, t)$, replacing the coefficient of $\dot{r}$ by its equivalent from (11-30) $\partial f / \partial \dot{r}$, and rearranging yields

$$-\frac{\partial f}{\partial \dot{q}} \dot{q} + f(q, \dot{q}, t) = -\frac{\partial f}{\partial \dot{r}} \dot{r} + f(r, \dot{r}, t) \qquad \text{at the corner.} \tag{11-35}$$

The solution with one corner is composed of curves $q(t)$ and $r(t)$ satisfying Euler's equations (11-17) and (11-26), the continuity condition

$$q(t_2) = r(t_2), \tag{11-36}$$

and the change-in-slope conditions (11-30) and (11-35).

11.5   Several unknown functions

Let us now seek the functions $q_1(t)$, $q_2(t)$, ... that pass through given end points $q_1(t_1)$, $q_1(t_2)$, $q_2(t_1)$, $q_2(t_2)$, ... while making the integral

$$I = \int_{t_1}^{t_2} f(q_1, q_2, \ldots, \dot{q}_1, \dot{q}_2, \ldots, t) \, dt \tag{11-37}$$

extreme. Because we are interested in physically realizable situations, we employ the same continuity requirements as before.

Around an extremum, the variation of the integral vanishes:

$$\delta I = \delta \int_{t_1}^{t_2} f \, dt = \int_{t_1}^{t_2} \delta f \, dt = 0. \tag{11-38}$$

Expanding δf by the procedure in example 5.1 and eliminating the term in δt because variations are taken at a given t, we obtain

$$\int_{t_1}^{t_2} \left(\frac{\partial f}{\partial q_k} \delta q_k + \frac{\partial f}{\partial \dot{q}_k} \delta \dot{q}_k \right) dt = 0. \tag{11-39}$$

Let us commute the variation and differentiation operations in the second sum, integrate the result by parts, and introduce zero for each δq_k at the end points:

$$\int_{t_1}^{t_2} \frac{\partial f}{\partial \dot{q}_k} \delta \dot{q}_k \, dt = \int_{t_1}^{t_2} \frac{\partial f}{\partial \dot{q}_k} d(\delta q_k) = \frac{\partial f}{\partial \dot{q}_k} \delta q_k \Big|_{t_1}^{t_2} - \int_{t_1}^{t_2} d\left(\frac{\partial f}{\partial \dot{q}_k} \right) \delta q_k$$

$$= 0 - \int_{t_1}^{t_2} \frac{d}{dt} \left(\frac{\partial f}{\partial \dot{q}_k} \right) \delta q_k \, dt. \tag{11-40}$$

Thus, equation (11-39) reduces to

$$\int_{t_1}^{t_2} \left(\frac{\partial f}{\partial q_k} - \frac{d}{dt} \frac{\partial f}{\partial \dot{q}_k} \right) \delta q_k \, dt = 0. \tag{11-41}$$

Note that $\delta q_1, \delta q_2, \ldots, \delta q_k, \ldots$ are arbitrary and independent functions of t. Let us make only one of them, δq_k say, different from zero anywhere within the limits of integration. Furthermore, let us make it appreciable only within an arbitrary small interval. Then for the left side of (11-41) to vanish, we must have

$$\frac{\partial f}{\partial q_k} - \frac{d}{dt} \frac{\partial f}{\partial \dot{q}_k} = 0 \tag{11-42}$$

in the interval. Since both k and the interval are arbitrary, this result must hold throughout the range of integration for each q_k.

11.6   Hamilton's principle

Note that equations (11-42) and (5-38) are the same when f is L. If the particles of a system move along paths satisfying the law

$$\delta \int_{t_1}^{t_2} L \, dt = 0, \tag{11-43}$$

the paths are governed by the Lagrange equations. But if residual forces are present, equations (5-38) are not valid. A more general integral law is then needed. This law will now be constructed from Newton's equations. We will see how it reduces to (11-43) when each Q_k' is zero.

We start with an inertial Cartesian system in which coordinates of the jth particle are x_j, y_j, z_j at time t while the mass of this particle is m_j. By definition, the kinetic energy T of the set is the sum

$$T = \tfrac{1}{2} m_j (\dot{x}_j{}^2 + \dot{y}_j{}^2 + \dot{z}_j{}^2). \tag{11-44}$$

A variation of the paths alters this expression by the amount

$$\delta T = m_j(\dot{x}_j \, \delta\dot{x}_j + \dot{y}_j \, \delta\dot{y}_j + \dot{z}_j \, \delta\dot{z}_j)$$

$$= m_j\left(\dot{x}_j \frac{d}{dt} \delta x_j + \dot{y}_j \frac{d}{dt} \delta y_j + \dot{z}_j \frac{d}{dt} \delta z_j\right). \tag{11-45}$$

Let us expand each term in the sums with the formula for differentiating a product

$$m_j\dot{x}_j \frac{d}{dt} \delta x_j = m_j \frac{d}{dt}(\dot{x}_j \, \delta x_j) - m_j\ddot{x}_j \, \delta x_j$$

$$\vdots \tag{11-46}$$

and introduce Newton's second law

$$m_j\dot{x}_j \frac{d}{dt} \delta x_j = \frac{d}{dt}(m_j\dot{x}_j \, \delta x_j) - X_j \, \delta x_j$$

$$\vdots \tag{11-47}$$

Then let us transfer the terms containing a force times a distance to the left side:

$$\delta T + X_j \, \delta x_j + Y_j \, \delta y_j + Z_j \, \delta z_j = \frac{d}{dt} m_j(\dot{x}_j \, \delta x_j + \dot{y}_j \, \delta y_j + \dot{z}_j \, \delta z_j). \tag{11-48}$$

The transferred terms add up to the virtual work done on the system during the variation. Equation (5-19) allows this work to be expressed in terms of the generalized forces

$$X_j \, \delta x_j + Y_j \, \delta y_j + Z_j \, \delta z_j = \delta W = Q_k \, \delta q_k \tag{11-49}$$

if corresponding points are chosen so δt is zero.

Integrating the resulting form of (11-48) over time gives us

$$\int_{t_1}^{t_2} (\delta T + \delta W) \, dt = \int_{t_1}^{t_2} (\delta T + Q_k \, \delta q_k) \, dt$$

$$= m_j(\dot{x}_j \, \delta x_j + \dot{y}_j \, \delta y_j + \dot{z}_j \, \delta z_j)\Big|_{t_1}^{t_2} = 0. \tag{11-50}$$

The integrated right side is zero because the variations are zero at the end points. The over-all relationship

$$\int_{t_1}^{t_2} (\delta T + \delta W) \, dt = 0 \tag{11-51}$$

is known as *Hamilton's principle*.

Since T is a function of the q's, $\dot{q}$'s, and t, the integral of δT can be expanded as the integral of δf was in equation (11-41). But the work W done on the particles need not be such a function. The integral of its variation expands similarly only when the forces are derivable from the generalized potential U by differentiation:

$$Q_k = \frac{d}{dt} \frac{\partial U}{\partial \dot{q}_k} - \frac{\partial U}{\partial q_k}. \tag{11-52}$$

We then obtain

$$\int_{t_1}^{t_2} \left(\frac{\partial T}{\partial q_k} - \frac{d}{dt} \frac{\partial T}{\partial \dot{q}_k} - \frac{\partial U}{\partial q_k} + \frac{d}{dt} \frac{\partial U}{\partial \dot{q}_k} \right) \delta q_k \; dt = 0 \qquad (11\text{-}53)$$

or

$$\int_{t_1}^{t_2} \left(\frac{\partial L}{\partial q_k} - \frac{d}{dt} \frac{\partial L}{\partial \dot{q}_k} \right) \delta q_k \; dt = 0 \qquad (11\text{-}54)$$

where the Lagrangian

$$L = T - U. \qquad (11\text{-}55)$$

Equation (11-54) implies that

$$\delta \int_{t_1}^{t_2} L \; dt = 0. \qquad (11\text{-}56)$$

Any small well-behaved variation in a q_k from its correct path must not alter the integral of L. Indeed, it must not violate equation (11-54). Consequently, each expression in parenthesis in the integrand equals zero. The calculus of variations thus yields Lagrange's equation (5-38) for each q_k.

11.7   Maximizing or minimizing a multiple integral

Many problems involve extremizing a multiple integral. To see how these may be treated, let us seek the function

$$u(x, y, z) \qquad (11\text{-}57)$$

that is known on a surface enclosing a region in which the integral

$$I = \int_{z_1}^{z_2} \int_{y_1}^{y_2} \int_{x_1}^{x_2} f(u, u_x, u_y, u_z, x, y, z) \; dx \; dy \; dz \qquad (11\text{-}58)$$

is to be made as large or small as possible. The symbols u_x, u_y, u_z represent $\partial u/\partial x$, $\partial u/\partial y$, and $\partial u/\partial z$.

The standard continuity requirements are employed, and the function is chosen to make any allowed variation in the integral zero:

$$\delta I = \delta \int \int \int f \; dx \; dy \; dz = \int \int \int \delta f \; dx \; dy \; dz = 0. \qquad (11\text{-}59)$$

At any stage of the integration, x, y, z are fixed and the procedure in example 5.1 gives us

$$\delta f = \frac{\partial f}{\partial u} \delta u + \frac{\partial f}{\partial u_x} \delta u_x + \frac{\partial f}{\partial u_y} \delta u_y + \frac{\partial f}{\partial u_z} \delta u_z. \qquad (11\text{-}60)$$

Thus, the variation δI contains three integrals like

$$\int \int \left(\int \frac{\partial f}{\partial u_x} \delta u_x \; dx \right) dy \; dz. \qquad (11\text{-}61)$$

But variation at a fixed x, y, z and differentiation with respect to x, y, or z commute (see section 11.2). Therefore we can integrate each of the three integrals by parts and introduce the zero variations at the limits:

$$\int_{x_1}^{x_2} \frac{\bar{\partial} f}{\partial u_x} \, \delta u_x \, dx = \int \frac{\partial f}{\partial u_x} \frac{\partial}{\partial x} (\delta u) \, dx = \int \frac{\partial f}{\partial u_x} \, d(\delta u)$$

$$= \frac{\partial f}{\partial u_x} \, \delta u \Big|_{x_1}^{x_2} - \int_{x_1}^{x_2} d\left(\frac{\partial f}{\partial u_x}\right) \delta u = 0 - \int_{x_1}^{x_2} \frac{\partial}{\partial x}\left(\frac{\partial f}{\partial u_x}\right) \delta u \, dx$$

$$\vdots \qquad\qquad (11\text{-}62)$$

Equation (11-59) thus becomes

$$\int\int\int \left(\frac{\partial f}{\partial u} - \frac{\bar{\partial}}{\partial x} \frac{\partial f}{\partial u_x} - \frac{\partial}{\partial y} \frac{\partial f}{\partial u_y} - \frac{\partial}{\partial z} \frac{\partial f}{\partial u_z} \right) \delta u \, dx \, dy \, dz = 0. \qquad (11\text{-}63)$$

Since the variation δu is arbitrary within the limits of integration, we may choose it different from zero in an arbitrary small volume and let it vanish everywhere else. Then to keep relationship (11-63) intact, we must set

$$\frac{\partial f}{\partial u} - \frac{\partial}{\partial x} \frac{\partial f}{\partial u_x} - \frac{\partial}{\partial y} \frac{\partial f}{\partial u_y} - \frac{\partial}{\partial z} \frac{\partial f}{\partial u_z} = 0 \qquad (11\text{-}64)$$

in the small volume. But this may be chosen at will inside the limits; hence equation (11-64) must hold throughout the enclosed region.

EXAMPLE 11.6   What differential equation does an analytical u satisfy when the average of $u_x{}^2 + u_y{}^2 + u_z{}^2$ within a closed surface is as small as a given $u(x, y, z)$ pattern on the surface allows? The average is given by the ratio

$$\frac{I}{V} = \frac{\int\int\int (u_x{}^2 + u_y{}^2 + u_z{}^2) \, dx \, dy \, dz}{\int\int\int dx \, dy \, dz}.$$

Since the denominator is the volume within the given surface, it is constant during the pertinent variations and we need only extremize the numerator I.

Because the integrand in I is

$$f = x_x{}^2 + u_y{}^2 + u_z{}^2,$$

the partial derivatives are

$$\frac{\partial f}{\partial u} = 0 \qquad\qquad \frac{\partial}{\partial x} \frac{\partial f}{\partial u_x} = 2u_{xx}$$

$$\frac{\partial}{\partial y} \frac{\partial f}{\partial u_y} = 2u_{yy} \qquad \frac{\partial}{\partial z} \frac{\partial f}{\partial u_z} = 2u_{zz}$$

and condition (11-64) becomes

$$0 - 2u_{xx} - 2u_{yy} - 2u_{zz} = 0$$

or

$$\frac{\partial^2 u}{\partial x^2} + \frac{\partial^2 u}{\partial y^2} + \frac{\partial^2 u}{\partial z^2} = 0.$$

This result is called *Laplace's equation*.

Decreasing the average $u_x{}^2 + u_y{}^2 + u_z{}^2$ corresponds to smoothing the function. In practice, heat flow tends to do this to the temperature within a body. Thus, the steady-state distribution allowed by any given surface pattern is described by Laplace's equation.

EXAMPLE 11.7   The ends of an elastic string having the relaxed length *a* are fixed to posts distance *a* apart. Show how Hamilton's principle governs its motion.

Let dx be the length of an element of the string when the string is in its equilibrium position at rest. Also, let **u** be the vector describing the displacement of the left edge of the element and $\mathbf{u} + d\mathbf{u}$ that of the right edge at time t, as figure 11.6 shows. In moving from the equilibrium position, the element stretches by the amount

$$\left(\mathbf{i}\, dx + \frac{\partial \mathbf{u}}{\partial x}\, dx \right) - \mathbf{i}\, dx = \frac{\partial \mathbf{u}}{\partial x}\, dx.$$

We expect any tension in the element to be an analytic function of the ratio of this extension to the equilibrium length dx. The linear term in the series expansion is

$$\tau \frac{\partial \mathbf{u}}{\partial x}$$

with τ constant. In approximately linear materials, higher terms are small, even where the derivative $\partial \mathbf{u}/\partial x$ is finite.

The integral of force acting on an element times the corresponding component of displacement is then

$$dV = \int \tau \frac{\cdot\, \mathbf{u}}{\partial x} \cdot d\left(\frac{\partial \mathbf{u}}{\partial x} \right) dx = \tfrac{1}{2}\tau \frac{\partial \mathbf{u}}{\partial x} \cdot \frac{\partial \mathbf{u}}{\partial x}\, dx \equiv \tfrac{1}{2}\tau \mathbf{u}_x \cdot \mathbf{u}_x\, dx.$$

The potential energy stored in the whole string by such movement is

$$V = \tfrac{1}{2}\tau \int_0^a \mathbf{u}_x \cdot \mathbf{u}_x\, dx.$$

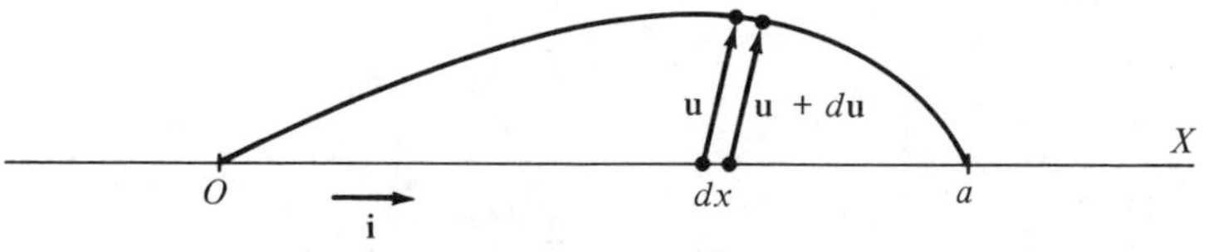

FIGURE 11.6   Configuration of a vibrating string at time t.

The kinetic energy of an element is one-half its mass times the velocity of the element squared:

$$dT = \tfrac{1}{2}\rho\, dx\, \frac{\partial \mathbf{u}}{\partial t} \cdot \frac{\partial \mathbf{u}}{\partial t} \equiv \tfrac{1}{2}\rho\, dx\, \mathbf{u}_t \cdot \mathbf{u}_t.$$

Here ρ is the linear density of the relaxed string. Integrating over the string yields

$$T = \tfrac{1}{2}\rho \int_0^a \mathbf{u}_t \cdot \mathbf{u}_t\, dx.$$

Constructing the Lagrangian

$$L = T - V = \int_0^a (\tfrac{1}{2}\rho\mathbf{u}_t \cdot \mathbf{u}_t - \tfrac{1}{2}\tau\mathbf{u}_x \cdot \mathbf{u}_x)\, dx$$

and substituting it into equation (11-56) gives us

$$\delta \int_{t_1}^{t_2} \int_0^a (\tfrac{1}{2}\rho\mathbf{u}_t \cdot \mathbf{u}_t - \tfrac{1}{2}\tau\mathbf{u}_x \cdot \mathbf{u}_x)\, dx\, dt = 0$$

or

$$\delta \int_{t_1}^{t_2} \int_0^a (\tfrac{1}{2}\rho u_{j,t} u_{j,t} - \tfrac{1}{2}\tau u_{j,x} u_{j,x})\, dx\, dt = 0.$$

Note that the summation convention applies to j but *not* to t and x. The latter are variables, not indices.

We now have to determine three unknown functions—the components of $\mathbf{u}$. Following the procedures in sections 11.5 and 11.7, we find that an equation like (11-64) governs each of these:

$$\frac{\partial f}{\partial u_j} - \frac{\partial}{\partial t}\frac{\partial f}{\partial u_{j,t}} - \frac{\partial}{\partial x}\frac{\partial f}{\partial u_{j,x}} = 0.$$

Applying this equation to the integrand above yields

$$0 - \rho u_{j,tt} + \tau u_{j,xx} = 0$$

or

$$-\rho\mathbf{u}_{tt} + \tau\mathbf{u}_{xx} = 0$$

and

$$\frac{\partial^2 \mathbf{u}}{\partial x^2} = \frac{\rho}{\tau}\frac{\partial^2 \mathbf{u}}{\partial t^2}.$$

11.8 Local constraints

So far, we have assumed that the properties $q_1, q_2, \ldots, q_n$ can vary independently. But constraints that restrict the q's may be present. Lagrange's method of undetermined multipliers is then used to combine the constraining equations with the integral condition.

An important example is afforded by a mechanical system subject to the independent relationships

$$a_{1k}(\mathbf{q}, t)\dot{q}_k + b_1(\mathbf{q}, t) = 0$$
$$\vdots$$
$$a_{mk}(\mathbf{q}, t)\dot{q}_k + b_m(\mathbf{q}, t) = 0. \tag{11-65}$$

Since these may be rewritten in the form

$$a_{jk}(\mathbf{q}, t)\,\delta q_k + b_j(\mathbf{q}, t)\,\delta t = 0, \tag{11-66}$$

and variations are taken at a given t, we have

$$a_{jk}\,\delta q_k = 0. \tag{11-67}$$

The set of particles is governed by Hamilton's principle (11-50)

$$\int_{t_1}^{t_2} (\delta T + Q_k\,\delta q_k)\,dt = 0. \tag{11-68}$$

As in section 5.5, each sum $a_{jk}\,\delta q_k$ is multiplied by a factor $\lambda_j(t)$, and the products are added to the integrand in (11-68). Because each sum is zero, the integrand is not decreased thereby. Then expanding δT as δf was in section 11.5 yields

$$\int_{t_1}^{t_2} \left(\frac{\partial T}{\partial q_k} - \frac{d}{dt}\frac{\partial T}{\partial \dot{q}_k} + Q_k + \lambda_j a_{jk} \right) \delta q_k\,dt = 0. \tag{11-69}$$

Now, only $n - m$ of the δq_k's are independent at any given t. But m independent λ_j's exist. Choose these so that the coefficients of the first m δq_k's are zero:

$$\frac{d}{dt}\frac{\partial T}{\partial \dot{q}_k} - \frac{\partial T}{\partial q_k} - Q_k - \lambda_j a_{jk} = 0 \qquad \text{for } 1 \le k \le m. \tag{11-70}$$

These δq_k's are then selected to be the dependent ones. Choose all but one of the others to be zero everywhere and let δq_l be different from zero only in an arbitrary small interval.

Then for the left side of (11-69) to vanish, we must have

$$\frac{d}{dt}\frac{\partial T}{\partial \dot{q}_l} - \frac{\partial T}{\partial q_l} - Q_l + \lambda_j a_{jl} = 0 \tag{11-71}$$

in the arbitrary interval within the limits of integration. Because l may be any number from $m + 1$ to n, these results agree with (5-27) and (5-28).

11.9 Integral constraints

The conditions that restrict variations may also appear as integrals of functions of the q's, $\dot{q}$'s, and t. As long as each such constraint involves the same integration as the integral to be extremized, constant Lagrange multipliers are sufficient.

The problem is to make

$$\int f\, dt = I \tag{11-72}$$

stationary when

$$\int f_1\, dt = c_1$$

$$\vdots$$

$$\int f_m\, dt = c_m. \tag{11-73}$$

Here all integrals have the same limits t_1 and t_2, all integrands are functions of $q_1, \ldots, q_n, \dot{q}_1, \ldots, \dot{q}_n, t$, and all c's are constant.

First, we multiply equations (11-73) by the undetermined constants $\lambda_1, \lambda_2, \ldots, \lambda_m$ and add the resulting equations to equation (11-72). On the left side, we obtain

$$\int g\, dt \tag{11-74}$$

where

$$g = f + \lambda_j f_j. \tag{11-75}$$

On the right, we have

$$I + \lambda_j c_j. \tag{11-76}$$

Because the sum over j in (11-76) is constant, the variation of the complete expression reduces to the variation of I. But since δI is zero about a stationary value, we seek the q_k's that make

$$\delta \int g\, dt = 0. \tag{11-77}$$

Proceeding as in section 11.5, let us put this into the form

$$\int_{t_1}^{t_2} \left(\frac{\partial g}{\partial q_k} - \frac{d}{dt}\frac{\partial g}{\partial \dot{q}_k} \right) \delta q_k\, dt = 0 \tag{11-78}$$

and investigate its properties.

For the purpose of discussion, we break the range of integration down into p small intervals and consider the average values of the δq_k's in these intervals as "variables." Since the constraining equations impose m relationships among these, any m of the "variables" may be considered as dependent.

We choose λ's that make the corresponding factors in (11-78) zero:

$$\frac{\partial g}{\partial q_k} - \frac{d}{dt}\frac{\partial g}{\partial \dot{q}_k} = 0. \tag{11-79}$$

Since the other "variables" are now arbitrary, they may all be set equal to zero except one labeled $\delta q_l\ (t = a)$. Then for (11-78) to hold, the corresponding factor in the integrand must be zero:

$$\frac{\partial g}{\partial q_l} - \frac{d}{dt}\frac{\partial g}{\partial \dot{q}_l} = 0 \qquad \text{when} \quad t = a. \tag{11-80}$$

Thus, formula (11-79) is valid regardless of the value of the index and t.

When n is one, the alternate form

$$\frac{d}{dt}\left(g - \dot{q}\,\frac{\partial g}{\partial \dot{q}}\right) - \frac{\partial g}{\partial t} = 0 \tag{11-81}$$

can be derived, as in section 11.3.

EXAMPLE 11.8 Maximize the volume within a solid of revolution, keeping the area constant.

The outside area of the ring in figure 11.7 is obtained when its width

$$ds = (dx^2 + dy^2)^{1/2} = (1 + y'^2)^{1/2}\,dx$$

is multiplied by its length

$$2\pi y.$$

The latter is generated by rotation of a point of the element about the x axis. Integrating this differential area over the range of x, we find that

$$A = 2\pi \int y\,ds = 2\pi \int_0^a y(1 + y'^2)^{1/2}\,dx.$$

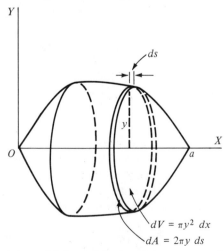

FIGURE 11.7 Element of the solid formed by revolving a curve about the x axis.

When the cross sectional area

$$\pi y^2$$

is multiplied by the thickness

$$dx$$

of the volume element bounded by the ring and parallel plane faces, the volume of the solid element is obtained. Integrating this, we find for the total volume

$$V = \pi \int_0^a y^2 \, dx.$$

Let us extremize $V/\pi = I$ subject to the constraint that $A/2\pi = c_1$ be constant. Then f is y^2, f_1 is $y(1 + y'^2)^{1/2}$, and

$$g = y^2 + \lambda y(1 + y'^2)^{1/2}.$$

Also, x is the variable of integration, corresponding to t in section 11.9.
    Since

$$\frac{\partial g}{\partial x} = 0,$$

the alternate Euler equation (11-81) reduces to

$$\frac{d}{dx}\left(g - y' \frac{\partial g}{\partial y'}\right) = 0$$

whence

$$g - y' \frac{\partial g}{\partial y'} = c.$$

Substituting our g into this equation, we get

$$y^2 + \lambda y(1 + y'^2)^{1/2} - y'\lambda y \frac{y'}{(1 + y'^2)^{1/2}} = c.$$

For the surface to be closed (as we assume), y must be zero at two points (if not more). Substituting zero for y makes c zero, even when y' is infinite. In the latter case, the equation reduces to

$$y^2 + \lambda yy' - \lambda yy' = c$$

and the second and third terms cancel.
    Setting c equal to zero, canceling y, and clearing of fractions, we obtain

$$y(1 + y'^2)^{1/2} + \lambda = 0$$

whence

$$y^2 y'^2 = \lambda^2 - y^2$$

or

$$\frac{dy}{dx} = \pm \frac{(\lambda^2 - y^2)^{1/2}}{y}.$$

Integrating gives the equation for a circle

$$\lambda^2 = (x - x_0)^2 + y^2.$$

Revolving it about the x axis produces a sphere.

Parameter λ is calculated from the given area. Parameter x_0 may be chosen so that the circle passes through the origin. If the limit a is then taken to be greater than the diameter 2λ, the complete curve consists of the semicircle from $x = 0$ to $x = 2\lambda$ and the line $y = 0$ from $x = 2\lambda$ to $x = a$.

Because the curve contains a corner, the additional conditions of section 11.4 need to be introduced (with g in place of f). Because c is 0, the section $y = 0$ does satisfy the pertinent Euler equation. Because the corner is at $y = 0$, the change-in-slope conditions are also met.

11.10   Change of coordinates in a multiple integral

The argument in section 11.7 does not depend on the nature of the variables of integration—as long as they are well-behaved. In general, they would be chosen to fit the symmetry of the problem. But the basic law may be known only in the Cartesian system. How is it to be transformed?

Suppose the law involves extremizing the double integral

$$I = \int \int f(x, y) \, dx \, dy \tag{11-82}$$

and we wish to employ the coordinates u, v defined by

$$x = P(u, v), \qquad y = Q(u, v). \tag{11-83}$$

In the integral, $f(x, y)$ is to be replaced with

$$f[P(u, v), Q(u, v)] \tag{11-84}$$

and $dx \, dy$ by

$$J \, du \, dv. \tag{11-85}$$

Note that the new element of area is generally different from the old. This difference does not matter as long as the total area is covered in the integration.

Let the new element be bounded by a vector $d\mathbf{a}$ along which only u changes and a vector $d\mathbf{b}$ along which only v changes (see figure 11.8). From the differentials of equations (11-83)

$$dx = \frac{\partial x}{\partial u} \, du + \frac{\partial x}{\partial v} \, dv, \tag{11-86}$$

$$dy = \frac{\partial y}{\partial u} \, du + \frac{\partial y}{\partial v} \, dv, \tag{11-87}$$

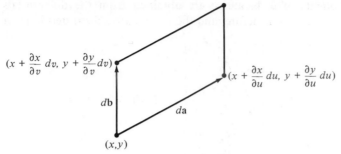

FIGURE 11.8 Vectors bounding an element of area when u and v are the coordinates. Along $d\mathbf{a}$ from point (x, y) u alone varies; along $d\mathbf{b}$, v alone varies.

we then obtain the Cartesian components

$$da_x = \frac{\partial x}{\partial u}\, du, \qquad da_y = \frac{\partial y}{\partial u}\, du, \qquad (11\text{-}88)$$

and

$$db_x = \frac{\partial x}{\partial v}\, dv, \qquad db_y = \frac{\partial y}{\partial v}\, dv. \qquad (11\text{-}89)$$

The area of the element equals the magnitude of the cross product of the two bounding vectors:

$$dS = |d\mathbf{a} \times d\mathbf{b}| = \begin{vmatrix} da_x & da_y \\ db_x & db_y \end{vmatrix}. \qquad (11\text{-}90)$$

With (11-88), (11-89) this reduces to

$$dS = \begin{vmatrix} \dfrac{\partial x}{\partial u}\, du & \dfrac{\partial y}{\partial u}\, du \\[2ex] \dfrac{\partial x}{\partial v}\, dv & \dfrac{\partial y}{\partial v}\, dv \end{vmatrix} = \begin{vmatrix} \dfrac{\partial x}{\partial u} & \dfrac{\partial y}{\partial u} \\[2ex] \dfrac{\partial x}{\partial v} & \dfrac{\partial y}{\partial v} \end{vmatrix}\, du\, dv$$

$$= J(x, y/u, v)\, du\, dv. \qquad (11\text{-}91)$$

The determinant labeled $J(x, y/u, v)$ is called the *Jacobian* of x and y with respect to u and v.

Substituting the transformed f and differential of area into the original integral (11-82) yields

$$I = \iint f[P(u, v), Q(u, v)]J(x, y/u, v)\, du\, dv. \qquad (11\text{-}92)$$

We handle triple integrals similarly. The new element of volume is bounded by line segments along which only u, only v, and only w vary. If the corresponding vectors are $d\mathbf{a}$, $d\mathbf{b}$, $d\mathbf{c}$, the magnitude dV is given by $d\mathbf{a} \cdot d\mathbf{b} \times d\mathbf{c}$.

The components of **a**, **b**, and **c** are obtained from the differentials $dx(u, v, w)$, $dy(u, v, w)$, $dz(u, v, w)$, as before, and the results are substituted into formula (1-34):

$$dV = \begin{vmatrix} \dfrac{\partial x}{\partial u} & \dfrac{\partial y}{\partial u} & \dfrac{\partial z}{\partial u} \\[2mm] \dfrac{\partial x}{\partial v} & \dfrac{\partial y}{\partial v} & \dfrac{\partial z}{\partial v} \\[2mm] \dfrac{\partial x}{\partial w} & \dfrac{\partial y}{\partial w} & \dfrac{\partial z}{\partial w} \end{vmatrix} du\, dv\, dw = J\, du\, dv\, dw. \qquad (11\text{-}93)$$

When

$$x = P(u, v, w), \qquad y = Q(u, v, w), \qquad z = R(u, v, w), \qquad (11\text{-}94)$$

we have

$$\int\!\!\int\!\!\int f(x, y, z)\, dx\, dy\, dz = \int\!\!\int\!\!\int f(P, Q, R)J\, du\, dv\, dw. \qquad (11\text{-}95)$$

The 3×3 determinant is called the *Jacobian* $J(x, y, z/u, v, w)$.

Transformation of the integral is generally easier than transforming the Euler equation. The latter can be done, of course, by direct application of the formula in example 5.1.

EXAMPLE 11.9   Obtain Laplace's equation in polar coordinates by transforming the variables of

$$I = \int\!\!\int (u_x^2 + u_y^2)\, dx\, dy$$

and extremizing the result.

The distance of a point from the z axis is labeled r and the angle between the corresponding radius vector and the x axis is labeled ϕ. Projecting r on the x and y axes then gives us

$$x = r \cos \phi,$$
$$y = r \sin \phi,$$

while treating u as a function of r and ϕ yields

$$du = \frac{\partial u}{\partial r}\, dr + \frac{\partial u}{\partial \phi}\, d\phi = u_r\, dr + u_\phi\, d\phi,$$

whence

$$u_x = \frac{\partial u}{\partial x} = u_r \frac{\partial r}{\partial x} + u_\phi \frac{\partial \phi}{\partial x},$$

$$u_y = \frac{\partial u}{\partial y} = u_r \frac{\partial r}{\partial y} + u_\phi \frac{\partial \phi}{\partial y}.$$

Differentiate the transformation equations

$$dx = -r \sin \phi \, d\phi + \cos \phi \, dr,$$
$$dy = r \cos \phi \, d\phi + \sin \phi \, dr,$$

and solve the simultaneous set for dr and $d\phi$

$$dr = \cos \phi \, dx + \sin \phi \, dy,$$

$$d\phi = - \frac{\sin \phi}{r} dx + \frac{\cos \phi}{r} dy.$$

Comparing these with the general forms

$$dr = \frac{\partial r}{\partial x} dx + \frac{\partial r}{\partial y} dy,$$

$$d\phi = \frac{\partial \phi}{\partial x} dx + \frac{\partial \phi}{\partial y} dy,$$

yields the pertinent partial derivatives. Substitute these derivatives into the equations for u_x, u_y:

$$u_x = u_r \cos \phi - u_\phi \frac{\sin \phi}{r},$$

$$u_y = u_r \sin \phi + u_\phi \frac{\cos \phi}{r}.$$

Square and add:

$$u_x{}^2 + u_y{}^2 = u_r{}^2 \cos^2 \phi - 2u_r u_\phi \frac{\sin \phi \cos \phi}{r} + u_\phi{}^2 \frac{\sin^2 \phi}{r^2}$$

$$+ u_r{}^2 \sin^2 \phi + 2u_r u_\phi \frac{\sin \phi \cos \phi}{r} + u_\phi{}^2 \frac{\cos^2 \phi}{r^2}$$

$$= u_r{}^2 + \frac{1}{r^2} u_\phi{}^2.$$

Now, the element of area in polar coordinates is

$$dS = r \, dr \, d\phi.$$

Substitute these into the given integral:

$$I = \int\int (u_x{}^2 + u_y{}^2) \, dx \, dy = \int\int \left(u_r{}^2 + \frac{1}{r^2} u_\phi{}^2 \right) r \, dr \, d\phi.$$

Differentiate the new integrand

$$\frac{\partial f}{\partial u_r} = 2ru_r, \qquad \frac{\partial}{\partial r}\frac{\partial f}{\partial u_r} = 2u_r + 2ru_{rr},$$

$$\frac{\partial f}{\partial u_\phi} = \frac{2}{r}u_\phi, \qquad \frac{\partial}{\partial \phi}\frac{\partial f}{\partial u_\phi} = \frac{2}{r}u_{\phi\phi}.$$

Equation (11-64) thus becomes

$$0 - 2\left(ru_{rr} + u_r + \frac{1}{r}u_{\phi\phi}\right) = 0$$

or

$$u_{rr} + \frac{1}{r}u_r + \frac{1}{r^2}u_{\phi\phi} = 0,$$

whence

$$\frac{1}{r}\frac{\partial}{\partial r}\left(r\frac{\partial u}{\partial r}\right) + \frac{1}{r^2}\frac{\partial^2 u}{\partial \phi^2} = 0.$$

The element of area $r\, d\phi\, dr$ comes directly from a geometric description of the coordinates and the definition of radian measure. Indeed, we need not calculate the Jacobian whenever the area or volume element in (11-92) or (11-95) can be found by elementary means.

While we will not do it here, we can also obtain the expression for the Laplacian $u_{xx} + u_{yy} + u_{zz}$ in any orthogonal coordinate system by considering the diffusion of a hypothetical substance into and out of a typical volume element.

11.11   Extensions of the theory

When the integrand f depends on the nth derivative of q, expanding δf yields a term containing the variation of this derivative. To handle it, we integrate the term by parts n times, obtaining an integral with δq as a factor. The integrated parts vanish as long as the variations of the lower derivatives are zero at the limits. Other terms are reduced in like manner. The complete integral δI is then treated as the integral in (11-16).

We may allow the independent variables, such as t, to shift during variations. When the limits are fixed, results equivalent to those obtained before are found. When the limits are mobile, further conditions in addition to Euler's equation are found.

We may remove the restriction that the derivative $(d/dt)\,\delta q$ be the same order as δq; that is, that δq be *weak*. If the derivative is not necessarily small when the variation is small, the variation is said to be *strong*. A minimum for weak small variations that fails to be a minimum for some strong small variations is not a true minimum, even though it be the physically significant one. Indeed, the result obtained in example 11.6 exhibits this defect. When strong variations are admitted, the triple integral is found to allow lower values than those given by Laplace's equation. The corresponding solutions are, of course, not well-behaved.

DISCUSSION QUESTIONS

11.1 How may a physical law be expressed in both differential and integral forms?

11.2 State the causality condition. Show that Hamilton's principle is consistent with it.

11.3 Show that variation and differentiation commute. Show when variation and integration commute.

11.4 State the continuity principle. Why should it be valid in physical problems? How does it imply that any δI from an extremum be zero?

11.5 Construct a variation whose derivative is not small at points where the variation itself is infinitesimal.

11.6 A student stated that at an end point where $\delta q = 0$, one has

$$\delta \dot{q} = \frac{d}{dt}\, \delta q = \frac{d}{dt}\, (0) = 0.$$

What is wrong with his argument? What variations can be different from zero at such an end point?

11.7 Explain how δq varies when

$$\int_{t_1}^{t_3} M(t)\, \delta q\, dt = M(t_2) \int_{t_1}^{t_3} \delta q\, dt.$$

How may this formula help one in going from equation (11-16) to (11-17)?

11.8 Review the derivation of Euler's equation.

11.9 (a) Why is formula (11-17) particularly useful when $\partial f / \partial q$ vanishes everywhere while $\partial f / \partial t$ does not? (b) Why is formula (11-21) better when $\partial f / \partial t$ is zero and $\partial f / \partial q$ is not?

11.10 How are problems in the calculus of variations like those in partial differential equations?

11.11 Show that the equation defining θ in example 11.5 does not introduce a restriction on x or y.

11.12 Describe extremization problems where q is mobile at the end points. What condition in addition to Euler's equation must then be satisfied?

11.13 Derive the conditions that govern the change in slope at a corner in a solution.

11.14 In what respect are $q_1, q_2, \ldots, \dot{q}_1, \dot{q}_2, \ldots, t$ all independent variables in equation (11-37)? In what respect is only t independent?

11.15 Derive Hamilton's principle from Newton's laws.

11.16 Does Hamilton's principle govern the behavior of an electric circuit?

11.17 In the variation of the integral in section 11.7, are derivatives of u at the limits known or fixed?

11.18 What kind of function does Laplace's equation describe?

11.19 Use Hamilton's principle in deriving the differential equation for a vibrating string.

11.20 Distinguish between local constraints and integral constraints.

11.21 How are the Lagrange multipliers defined? How are they evaluated in an actual problem?

11.22 Compare the results from (a) making an integral I extreme while keeping an integral J constant and (b) making J extreme while keeping I constant.

11.23 Why doesn't $\int f\, dA$ change with the shape of dA?

11.24 Obtain Laplace's equation in cylindrical coordinates by constructing and varying the appropriate integral.

PROBLEMS

11.1 Use calculus of variations to find the equation for the shortest path between two points in a plane.

11.2 Connect the two given points by vector **c** directly, and then by **a** drawn from the initial point to an intermediate point followed by **b** from this intermediate point to the final point. With vector analysis, show that the broken line is longer than the single straight line between the end points.

11.3 Both segments of the broken line **a** + **b** satisfy Euler's equation in problem 11.1. When do they meet the change-in-slope conditions?

11.4 Use the theory for movable end points in determining the shortest paths between two parallel straight lines.

11.5 Find the brachistochrone for a field in which the potential energy of the particle decreases as the square of the vertical distance fallen from rest.

11.6 Determine the curve joining two given points which produces the least area when it is revolved about an axis lying in the plane of the curve.

11.7 Minimize the path which links two points on a cone of revolution.

11.8 Derive the Euler equation for making

$$\int_{t_1}^{t_2} f(q, \dot{q}, \ddot{q}, t)\, dt$$

stationary.

11.9 Set up both Euler equations that result from applying Hamilton's principle to a particle moving in the potential

$$V = \tfrac{1}{2}kx^2.$$

11.10 Obtain the partial differential equation describing the surface assumed by a soap film held by a wire bent in the form of a simple closed curve. The surface tension acts to minimize the area of the film.

11.11 Find the planar curve that generates the solid of revolution of given mass and uniform density having the greatest gravitational attraction on a point mass located at the origin—a point on its axis.

11.12 A string lies in the xy plane above the x axis. While keeping its end points fixed, vary its shape until the area under it is a minimum or maximum.

11.13 Using the identities

$$\delta \, dq = d \, \delta q \qquad \text{and} \qquad \delta \, dt = d \, \delta t,$$

show that the integral

$$\int_A^B \Phi(q, t, dq, dt)$$

is stationary when

$$\Phi_q' - d\Phi_{dq}' = 0$$

and

$$\Phi_t' - d\Phi_{dt}' = 0.$$

Using the identity

$$\Phi = f \, dt,$$

show that these equations are equivalent to the Euler equation (11-17).

11.14 Derive the Euler equations for making

$$\int_{t_1}^{t_2} f(q_1, \ldots, q_n, \dot{q}_1, \ldots, \dot{q}_n, t) \, dt = I$$

stationary when the $m < n$ conditions

$$f_1(q_1, \ldots, \dot{q}_1, \ldots, t) = 0,$$
$$\vdots$$
$$f_m(q_1, \ldots, \dot{q}_1, \ldots, t) = 0$$

are valid.

11.15 Find the q that minimizes the integral

$$\int_{t_1}^{t_2} \dot{q}^2 \, dt.$$

11.16 Show when the condition

$$\delta \int_{t_1}^{t_2} f(\dot{q}) \, dt = 0$$

gives the same function q as problem 11.15.

11.17 Find the shortest paths between two concentric coplanar circles.

11.18 Find the shortest curve between two points on a right circular cylinder. In the solution express the rectangular coordinates of the moving point in terms of the angle of rotation about the axis of the cylinder.

11.19 Minimize the potential energy of a chain of uniform density hanging between fixed supports.

11.20 Of all the planar curves of constant density joining two fixed points that subtend a small angle at the origin, choose the one having the smallest moment of inertia with respect to the origin.

11.21 Derive the Euler equation for making

$$\int_{t_1}^{t_2} f\left(q, \frac{d^3 q}{dt^3}, t\right) dt$$

stationary.

11.22 Find the shortest curve between two points in space.

11.23 Minimize the integral

$$\int_{t_1}^{t_2} f(l\dot{q}_1 + m\dot{q}_2 + n\dot{q}_3) \, dt.$$

11.24 Form $\int L \, dt$ for a particle on which no forces act in an inertial system. Then use the calculus of variations to show that the motion is rectilinear.

11.25 Obtain the differential equation that makes the integral

$$\int \int \int (u_x^2 + u_y^2 + u_z^2 + k^2 u^2) \, dx \, dy \, dz$$

stationary.

11.26 Find the planar curve of given length which encloses the greatest area.

11.27 Using the identities

$$\delta \, dq = d \, \delta q \qquad \text{and} \qquad \delta \, dt = d \, \delta t$$

show that in general

$$\frac{d}{dt} \delta q = \delta \dot{q} + \dot{q} \frac{d \, \delta t}{dt}.$$

REFERENCES

BOOKS

Bolza, O. *Lectures on the Calculus of Variations*, pp. 1–114. Dover Publications, Inc., New York, 1961.

Dettman, J. W. *Mathematical Methods in Physics and Engineering*, pp. 68–136. McGraw-Hill Book Company, New York, 1962.

Gelfand, I. M., and **Fomin, S. V.** *Calculus of Variations* (Translated and edited by R. A. Silverman), pp. 1–226. Prentice-Hall, Inc., Englewood Cliffs, N.J., 1963.

Weinstock, R. *Calculus of Variations*, pp. 1–198. McGraw-Hill Book Company, New York, 1952.

ARTICLES

Bochner, S. The Role of Mathematics in the Rise of Mechanics. *Am. Scientist*, **50,** 294 (1962).

Candotti, E., Palmieri, C., and **Vitale, B.** On the Inversion of Noether's Theorem in Classical Dynamical Systems. *Am. J. Phys.*, **40,** 424 (1972).

Keskinen, R. Complex Potentials in Classical Mechanics and Geometrical Optics. *Am. J. Phys.*, **40,** 418 (1972).

Leitmann, G. Some Remarks on Hamilton's Principle. *J. Appl. Mech.*, **30** (*Trans. ASME*, **85**), 623 (1963).

Prather, J. L. Undetermined Multipliers. *Am. J. Phys.*, **33,** 657 (1965).

Ray, J. R. Nonholonomic Constraints and Gauss's Principle of Least Constraint. *Am. J. Phys.*, **40,** 179 (1972).

Riahi, F. On Lagrangians with Higher Order Derivatives. *Am. J. Phys.*, **40,** 386 (1972).

Saletan, E. J., and **Cromer, A. H.** A Variational Principle for Nonholonomic Systems. *Am. J. Phys.*, **38,** 892 (1970).

Smith, M. J. Variational Derivation of Young's Law and Laplace's Capillary Equation. *Am. J. Phys.*, **38,** 1153 (1970).

van der Vaart, H. R. Variational Principle for Certain Nonconservative Systems. *Am. J. Phys.*, **35,** 419 (1967).

Venezian, G. Terrestrial Brachistochrone. *Am. J. Phys.*, **34,** 701 (1966).

12 / Common Vector Operators

12.1 Fields

We have seen how study of the solar system led to a very good mathematical law governing gravitation, while study of the interaction between charges with a torsion balance led to the fundamental law for electrostatics. The study of the interactions among moving charges in neighboring wires also leads to a law governing magnetism.

In these situations, one body or particle affects distant bodies or particles. But how can one directly act on another removed from it? Must not each mass and charge affect the surrounding space in some manner so that each other mass or charge is no longer in an isotropic region?

Indeed, we suppose that each mass and charge sets up a field that influences the other bodies. Mathematical techniques for dealing with such fields will be developed in this chapter. For the sake of generality, some of these will be worked out for n-dimensional, rather than 3-dimensional space.

12.2 The gradient of a scalar function

Let us first consider a function ϕ that associates a real scalar with each point in an n-dimensional space. The scalar presumably varies continuously with position, so the derivative $d\phi/ds$ exists for any line element $d\mathbf{s}$ drawn from any point, except where the function goes bad.

At each point, let orthogonal unit vectors

$$\mathbf{e}_1, \mathbf{e}_2, \ldots, \mathbf{e}_n \tag{12-1}$$

be defined, with the corresponding Cartesian coordinates

$$x_1, x_2, \ldots, x_n. \tag{12-2}$$

Line element $d\mathbf{s}$ drawn from the point then expands as

$$d\mathbf{s} = dx_j \mathbf{e}_j. \tag{12-3}$$

352

The unit vector having the direction of $d\mathbf{s}$ is

$$\mathbf{f} = \frac{d\mathbf{s}}{ds} = \frac{dx_j}{ds}\,\mathbf{e}_j = a_{rj}\mathbf{e}_j \tag{12-4}$$

where a_{rj} is introduced as the direction cosine dx_j/ds. From example 5.1, the rate of change in ϕ along this path is

$$\frac{d\phi}{ds} = \frac{\partial\phi}{\partial x_j}\frac{dx_j}{ds} = a_{rj}\frac{\partial\phi}{\partial x_j}. \tag{12-5}$$

The final form in (12-5) can be interpreted as the scalar product of

$$\frac{\partial\phi}{\partial x_j}\,\mathbf{e}_j \tag{12-6}$$

with the unit vector $\mathbf{f}$. So $d\phi/ds$ is the projection of (12-6) in the direction of $d\mathbf{s}$.

Since this projection is largest when $d\mathbf{s}$ lies along (12-6), vector (12-6) gives the direction in which $d\phi/ds$ is largest, and the rate of change $d\phi/ds$ in that direction. Vector (12-6) is called the *gradient* of ϕ.

We write

$$\text{grad } \phi \equiv \nabla\phi \equiv \mathbf{e}_j\frac{\partial}{\partial x_j}\,\phi \tag{12-7}$$

where

$$\nabla = \mathbf{e}_j\frac{\partial}{\partial x_j} \tag{12-8}$$

We call ∇ the *del* operator.

By definition, the gradient of a real function is the vector function giving, at each point, the direction in which the scalar function changes most rapidly, and the rate of change in that direction. If the function ϕ were complex, the arguments above would apply to the real part and the imaginary part separately.

EXAMPLE 12.1   Show graphically how $d\phi/ds$ varies with the direction of $d\mathbf{s}$ from a given point.

Sketch the constant ϕ surface through the given point and a neighboring surface on which ϕ is larger by amount $d\phi$, as in figure 12.1. If ϕ is well-behaved, these are nearly parallel planes in the pertinent neighborhood. (Deviations from parallelism and planarity introduce higher order differentials into the formulas.)

Then the shortest distance from the given point to the other surface is perpendicular to the surfaces, as $d\mathbf{r}$ in figure 12.1 is. Since $d\phi/dr$ is larger than any other $d\phi/ds$, $\nabla\phi$ has the magnitude $d\phi/dr$ and the direction of $d\mathbf{r}$.

The definition of the cosine tells us that

$$ds = \frac{dr}{\cos\theta},$$

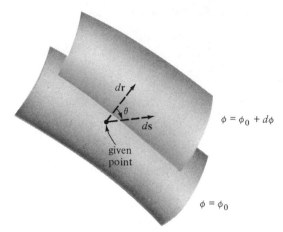

FIGURE 12.1   Normal and general displacement vectors drawn from a given point to a neighboring constant ϕ surface.

so we have the result

$$\frac{d\phi}{ds} = \cos \theta \, \frac{d\phi}{dr}$$

in which the magnitude of the gradient is projected in the direction of $d\mathbf{s}$. This can be written in the form

$$\frac{d\phi}{ds} = \mathbf{f} \cdot \nabla\phi$$

if $\mathbf{f}$ lies along $d\mathbf{s}$ and is of unit magnitude.

12.3   The divergence of a vector function

Let us next consider a function $\mathbf{A}$ that associates a real vector with each point in an n-dimensional space. Such a function possesses real components

$$A_1, A_2, \ldots, A_n \tag{12-9}$$

with respect to given unit vectors (12-1). The components presumably vary with position so that all derivatives

$$\frac{dA_1}{ds}, \frac{dA_2}{ds}, \ldots, \frac{dA_n}{ds} \tag{12-10}$$

exist everywhere except where the function goes bad.

   To study this function, we construct a hypothetical fluid whose velocity at any given point is $\mathbf{A}$ divided by its mass density ρ:

$$\mathbf{v} = \frac{\mathbf{A}}{\rho} \quad \text{or} \quad \mathbf{A} = \rho\mathbf{v}. \tag{12-11}$$

Since **v** equals the volume of the hypothetical fluid passing by the point, per unit cross sectional area, per unit time, multiplying it by ρ yields the corresponding rate of mass flow **A**.

Let us suppose that the fluid is conserved. Then any change in density within an infinitesimal volume element is due to movement of the fluid in and out. Let us also choose the element to be rectangular, with edges

$$dx_1', dx_2', \ldots, dx_n' \tag{12-12}$$

and with its center at $(x_1, x_2, \ldots, x_n)$. The primed axes are obtained from the unprimed axes by the appropriate rotation (see figure 12.2).

The rate at which hypothetical-fluid mass flows by the center point in the direction of increasing x_1' is

$$\rho v_1' \, dx_2' \, dx_3' \, \ldots. \tag{12-13}$$

The rate at which it flows in through the back face is

$$\rho v_1' \, dx_2' \, dx_3' \, \cdots + d(\rho v_1' \, dx_2' \, dx_3' \, \cdots)$$

$$= \left[\rho v_1' + \frac{\partial(\rho v_1')}{\partial x_1'} \left(-\frac{dx_1'}{2} \right) \right] dx_2' \, dx_3' \, \ldots, \tag{12-14}$$

while the rate at which it flows out through the parallel front face is

$$\rho v_1' \, dx_2' \, dx_3' \, \cdots + d(\rho v_1' \, dx_2' \, dx_3' \, \cdots)$$

$$= \left[\rho v_1' + \frac{\partial(\rho v_1')}{\partial x_1'} \left(+\frac{dx_1'}{2} \right) \right] dx_2' \, dx_3' \, \cdots. \tag{12-15}$$

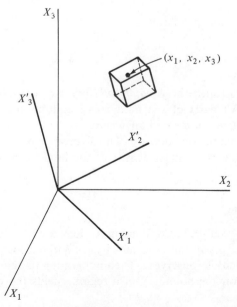

FIGURE 12.2   Reference axes for a general rectangular element of 3-dimensional space.

In unit time, the mass within the element decreases by expression (12-15) minus expression (12-14)

$$\frac{\partial(\rho v_1')}{\partial x_1'} \, dx_1' \, dx_2' \, dx_3' \cdots \tag{12-16}$$

from movement through these two faces. We have similar terms for each of the other pairs of faces. Adding these and dividing by the volume to obtain the rate of decrease in density yields

$$\frac{\partial(\rho v_1')}{\partial x_1'} + \frac{\partial(\rho v_2')}{\partial x_2'} + \cdots + \frac{\partial(\rho v_n')}{\partial x_n'} = -\frac{\partial \rho}{\partial t} \tag{12-17}$$

or

$$\frac{\partial(\rho v_j')}{\partial x_j'} = -\frac{\partial \rho}{\partial t}. \tag{12-18}$$

Whenever the mass moving into an element does not replace all that moves out, the fluid is said to diverge from the element. Since $-\partial \rho/\partial t$ is a measure of this divergence that is independent of the size of the element, we call the sum on the left of (12-18) the *divergence* in the rate of mass flow.

In this sum, each independent differentiating operator of del acts on the corresponding component of $\rho \mathbf{v}$. Indeed

$$\frac{\partial(\rho v_j')}{\partial x_j'} = \mathbf{e}_j' \cdot \mathbf{e}_k' \frac{\partial}{\partial x_j'} (\rho v_k') = \mathbf{e}_j' \frac{\partial}{\partial x_j'} \cdot (\rho v_k' \mathbf{e}_k')$$

$$= \nabla \cdot \rho \mathbf{v}, \tag{12-19}$$

and equation (12-18) can be rewritten as

$$\frac{\partial \rho}{\partial t} + \nabla \cdot \rho \mathbf{v} = 0. \tag{12-20}$$

Equation (12-20) is called the *equation of continuity* because it applies to a fluid that is conserved. All parts of the fluid move continuously from point to point without being annihilated or created anywhere.

Notice that $\rho \mathbf{v}$ is a momentum density. The divergence of a vector function is minus the time rate of change in the mass density of the fluid for which the function is the momentum density.

12.4 Gauss's theorem

Corresponding to any well-behaved function $\mathbf{A}$, there are hypothetical fluids for which the function is the density ρ times the velocity $\mathbf{v}$. Without loss of generality, we may assume that such a fluid is conserved. Then integrating its rate of mass flow through unit cross section, over the boundary of a region, yields the volume integral of the rate of density decrease.

For simplicity, let us assume the space to be 3-dimensional. The total amount of the hypothetical fluid in a given volume is then

$$m = \int_V \rho \, dV. \tag{12-21}$$

Differentiating this expression with respect to time and introducing the equation of continuity yields

$$\frac{\partial m}{\partial t} = \int_V \frac{\partial \rho}{\partial t} \, dV = -\int_V \nabla \cdot (\rho \mathbf{v}) \, dV. \tag{12-22}$$

The rate of flow of the hypothetical fluid through element $d\mathbf{S}$ of the bounding surface is

$$(\rho \mathbf{v}) \cdot d\mathbf{S} \tag{12-23}$$

if the sign of $d\mathbf{S}$ is chosen as in figure 12.3. Integrating over the surface enclosing volume V gives us

$$-\frac{\partial m}{\partial t} = \int_S (\rho \mathbf{v}) \cdot d\mathbf{S}. \tag{12-24}$$

Let us eliminate $\partial m / \partial t$ from (12-22) and (12-24),

$$\int_V \nabla \cdot (\rho \mathbf{v}) \, dV = \int_S (\rho \mathbf{v}) \cdot d\mathbf{S}, \tag{12-25}$$

and replace $\rho \mathbf{v}$ by $\mathbf{A}$,

$$\int_V \nabla \cdot \mathbf{A} \, dV = \int_S \mathbf{A} \cdot d\mathbf{S}. \tag{12-26}$$

The resulting relationship is called *Gauss's theorem*.

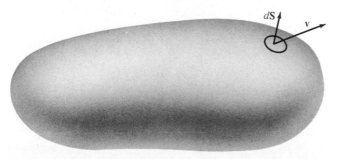

FIGURE 12.3   The surface surrounding volume V and the velocity $\mathbf{v}$ of the hypothetical fluid through an element $d\mathbf{S}$ of it.

EXAMPLE 12.2   Find a differential equation that governs an inverse-square field.
In section 2.9, we saw how an inverse-square field leads to Gauss's law

$$\int_S \mathbf{I} \cdot d\mathbf{S} = -4\pi G \int_V \rho \, dV.$$

Let us employ identity (12-26) to change the left side to a volume integral:

$$\int_V \nabla \cdot \mathbf{I} \, dV = -4\pi G \int_V \rho \, dV.$$

This relationship must hold regardless of what volume is taken. Therefore, we must have

$$\nabla \cdot \mathbf{I} = -4\pi G \rho.$$

12.5   Circulation around a line

A scalar field ρ may be considered as defining the density of a hypothetical (or real) fluid at each point in space. An accompanying vector field $\mathbf{v}$ then defines the velocity of the fluid at each point. Sources may exist where elements of the fluid appear, and sinks where they disappear. In absence of sources or sinks, ρ and $\mathbf{v}$ are related through the equation of continuity (12-20).

The velocity field may contain regions in which the fluid moves as it does near a vortex. In 3-dimensional space, the effect is measured by another vector field. Before studying it in detail, let us consider some general features of the situation.

Suppose a uniform 3-dimensional fluid is circulating about a line, whose positive direction is indicated by unit vector $\mathbf{n}$, at the constant angular velocity $\dot{\phi}$, as in figure 12.4. Let us integrate the projection of the velocity along a circular path distance r from the center:

$$\oint r\dot{\phi} \, ds = r\dot{\phi} \oint ds = r\dot{\phi}2\pi r = 2\dot{\phi}A. \tag{12-27}$$

If we divide the integral by the enclosed area A, we get the magnitude $2\dot{\phi}$, a quantity independent of r. Thus, this quotient, together with $\mathbf{n}$, characterize the rotation.

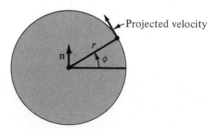

FIGURE 12.4   Velocity of an element of the hypothetical (or real) fluid projected on a circle centered on vector $\mathbf{n}$ on the vortex line.

12.6 The curl of a vector function

This result leads us to consider the line integral of the projection of any given vector field **v** along any convenient infinitesimal planar path about a given point as center divided by the area enclosed.

Let us choose a point (x, y, z), construct an arbitrary rectangle as in figure 12.5, and orient the primed coordinate system so the x' axis is parallel to the first edge and the y' axis parallel to the second edge. Then dy' is zero along the first and third edges, while dx' is zero along the second and fourth edges. Also, both 1 and 3 are dx' long; 2 and 4 dy' long.

At the middle of the first edge, vector **v** is then

$$\mathbf{v}(x,\ y,\ z) + \frac{\partial \mathbf{v}}{\partial y'}\left(-\frac{dy'}{2}\right), \tag{12-28}$$

while at the middle of the second edge, it is

$$\mathbf{v}(x,\ y,\ z) + \frac{\partial \mathbf{v}}{\partial x'}\left(\frac{dx'}{2}\right), \tag{12-29}$$

at the middle of the third edge, it is

$$\mathbf{v}(x,\ y,\ z) + \frac{\partial \mathbf{v}}{\partial y'}\left(\frac{dy'}{2}\right), \tag{12-30}$$

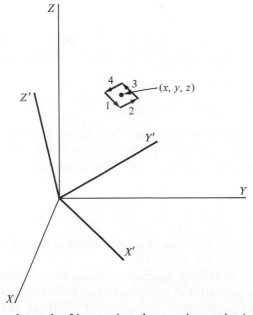

FIGURE 12.5 Rectangular path of integration about a given point (x, y, z).

and at the middle of the fourth edge, it is

$$\mathbf{v}(x, y, z) + \frac{\partial \mathbf{v}}{\partial x'}\left(-\frac{dx'}{2}\right). \tag{12-31}$$

Except for differentials of higher order, $\int \mathbf{v} \cdot d\mathbf{s}$ along each small edge equals $\mathbf{v}$ at the middle dot the pertinent $d\mathbf{s}$. For the first edge, we have the dot product of (12-28) with $\mathbf{i}' \, dx'$, for the second edge, (12-29) dot $\mathbf{j}' \, dy'$, for the third edge, (12-30) dot $-\mathbf{i}' \, dx'$, for the fourth edge, (12-31) dot $-\mathbf{j}' \, dy'$. Thus

$$\oint \mathbf{v} \cdot d\mathbf{s} = \left(v_{x'} - \frac{1}{2}\frac{\partial v_{x'}}{\partial y'} \, dy'\right) dx' + \left(v_{y'} + \frac{1}{2}\frac{\partial v_{y'}}{\partial x'} \, dx'\right) dy'$$

$$- \left(v_{x'} + \frac{1}{2}\frac{\partial v_{x'}}{\partial y'} \, dy'\right) dx' - \left(v_{y'} - \frac{1}{2}\frac{\partial v_{y'}}{\partial x'} \, dx'\right) dy'$$

$$= \left(\frac{\partial v_{y'}}{\partial x'} - \frac{\partial v_{x'}}{\partial y'}\right) dx' \, dy'. \tag{12-32}$$

The first factor in the final expression is the z' component of

$$\begin{vmatrix} \mathbf{i}' & \mathbf{j}' & \mathbf{k}' \\ \dfrac{\partial}{\partial x'} & \dfrac{\partial}{\partial y'} & \dfrac{\partial}{\partial z'} \\ v_{x'} & v_{y'} & v_{z'} \end{vmatrix} = \mathbf{V}, \tag{12-33}$$

while the second factor is the magnitude of the rectangular area, which we label $\Delta\mathbf{S}$. If the positive direction of $\Delta\mathbf{S}$ is taken along $\mathbf{k}'$, overall equation (12-32) can be replaced with

$$\oint \mathbf{v} \cdot d\mathbf{s} = \mathbf{V} \cdot \Delta\mathbf{S}. \tag{12-34}$$

This integral is largest for a given area when the vector $\Delta\mathbf{S}$ points along $\mathbf{V}$. Then magnitude V equals $\oint \mathbf{v} \cdot d\mathbf{s}$ divided by ΔS. Vector $\mathbf{V}$ is called the curl of "velocity" $\mathbf{v}$.

Thus, the *curl* of a vector $\mathbf{v}$ is the vector whose magnitude is the maximum limit of the ratio

$$\frac{\oint \mathbf{v} \cdot d\mathbf{s}}{\Delta S} \tag{12-35}$$

as planar area ΔS becomes small, and whose direction is that of the right hand normal to the planar area yielding the maximum.

In the determinant of (12-33), each independent differentiating operator acts as the component of the first factor of a vector product whose second factor is $\mathbf{v}$. Furthermore, each is a pertinent part of ∇. So equation (12-33) can be rewritten as

$$\nabla \times \mathbf{v} = \mathbf{V} = \text{curl } \mathbf{v}. \tag{12-36}$$

EXAMPLE 12.3 Determine the curl of the radius vector drawn from the origin to point (x_1, x_2, x_3).

Employ formula (1-70) with the differentiating operator $\partial/\partial x_m$ in place of the *m*th component of **A**:

$$(\nabla \times \mathbf{r})_p = \varepsilon_{pmn} \frac{\partial}{\partial x_m} x_n = \varepsilon_{pmn} \delta_{mn} = 0.$$

Since each component of the curl is 0, we have

$$\nabla \times \mathbf{r} = 0.$$

12.7 Stokes' theorem

We have just seen how the integral of a vector field times the element of distance over which it acts, around a small planar section of a surface, equals the projection of the curl of the field on the vector representing the section times the surface area:

$$\oint_{\text{around } \Delta S} \mathbf{v} \cdot d\mathbf{s} = \nabla \times \mathbf{v} \cdot \Delta \mathbf{S}. \tag{12-37}$$

Any well-behaved surface can be built up from infinitesimal nearly planar sections to which (12-37) applies. Summing the relationship over the constituent elements then yields a very useful formula.

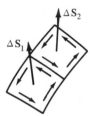

FIGURE 12.6 Neighboring sections of a given surface and the directions in which the bounding line elements are traversed when the two vectors representing the areas are directed upwards out of the paper.

Let us start with one small element and add another, as in figure 12.6. The integral

$$\oint \mathbf{v} \cdot d\mathbf{s} \tag{12-38}$$

for the second section then traverses the common edge in the direction opposite to that followed for the first one. The net contribution from this edge is therefore zero.

Let us keep adding elements until the desired surface is built up, as in figure 12.7. Because the contributions from all common edges drop out, the sum of the integrals on the left of (12-37) is merely the integral around the bounding curve:

$$\sum_{\substack{\text{surface} \\ \text{elements}}} \oint \mathbf{v} \cdot d\mathbf{s} = \oint_c \mathbf{v} \cdot d\mathbf{s}. \tag{12-39}$$

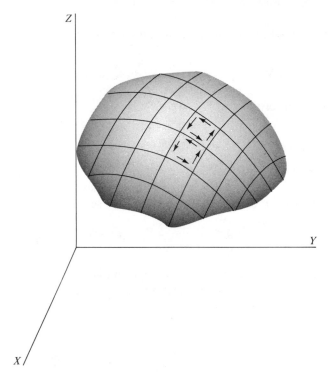

FIGURE 12.7   The complete surface.

But the right side of (12-37) is

$$\nabla \times \mathbf{v} \cdot \Delta \mathbf{S}. \tag{12-40}$$

The sum of these expressions over the infinitesimal surface elements within C is the corresponding integral. This must equal sum (12-39):

$$\int_S \nabla \times \mathbf{v} \cdot d\mathbf{S} = \oint_C \mathbf{v} \cdot d\mathbf{s}. \tag{12-41}$$

Equation (12-41) is called *Stokes' theorem*. It relates a surface integral to a line integral, somewhat as Gauss's theorem relates a volume integral to a surface integral.

12.8   Action of del on composite functions

In various applications, the function whose gradient, divergence, or curl is to be taken is composite. But by manipulation, we can reduce such an expression to a sum of terms in which del acts on a single constituent function at a time.

First, notice that the del operator has both vectorial and differentiating aspects. The vector property may be considered separate from the differentiating action.

The given expression is first expanded into a series of terms in which the differentiation acts on each constituent function by itself. The null contributions are discarded and the remaining forms rearranged by the rules of vector algebra so only the function to be acted on appears to the right of each del. In the work, we may identify the function to be differentiated by an underline, as in example 12.4.

Alternatively, we may introduce a Cartesian coordinate system. The conventions of section 1.8 together with formula (12-8) are then employed (see example 12.5).

EXAMPLE 12.4   By operator methods, show that the identity

$$\nabla \cdot (\phi \mathbf{V}) = \mathbf{V} \cdot \nabla \phi + \phi \nabla \cdot \mathbf{V}$$

is true.

Break the operand down into terms in which the differentiation acts on one factor alone. Distinguish this factor from the others by an underline:

$$\nabla \cdot (\phi \mathbf{V}) = \nabla \cdot (\underline{\phi} \mathbf{V}) + \nabla \cdot (\phi \underline{\mathbf{V}}).$$

Then rearrange each term so only the underlined factor is to the right of ∇:

$$\nabla \cdot (\underline{\phi} \mathbf{V}) + \nabla \cdot (\phi \underline{\mathbf{V}}) = \mathbf{V} \cdot \nabla \phi + \phi \nabla \cdot \mathbf{V}.$$

In this step, the usual rules of vector algebra are employed.

EXAMPLE 12.5   With index procedures, derive the identity in example 12.4.

In a Cartesian system for which the base vectors are $\mathbf{e}_1$, $\mathbf{e}_2$, $\mathbf{e}_3$, del is given by (12-8), while $\mathbf{V}$ is $V_k \mathbf{e}_k$. Consequently,

$$\nabla \cdot (\phi \mathbf{V}) = \mathbf{e}_j \frac{\partial}{\partial x_j} \cdot (\phi V_k \mathbf{e}_k) = \frac{\partial}{\partial x_j} (\phi V_j)$$

$$= V_j \frac{\partial \phi}{\partial x_j} + \phi \frac{\partial V_j}{\partial x_j} = \mathbf{e}_j V_j \cdot \mathbf{e}_k \frac{\partial}{\partial x_k} \phi + \phi \mathbf{e}_j \frac{\partial}{\partial x_j} \cdot V_k \mathbf{e}_k$$

$$= \mathbf{V} \cdot \nabla \phi + \phi \nabla \cdot \mathbf{V}.$$

12.9   General base vectors and the metric tensor

Each point in a physically significant space can be labeled by a set of generalized coordinates. Furthermore, the distance between neighboring points can be determined. Then various operations can be carried out.

In particular, we assume that distance and direction have meaning, so that vectors can be erected at any given point and scalar products taken between any two. If the n independent numbers labeling the point are

$$q^1, q^2, \ldots, q^n, \tag{12-42}$$

we let the direction in which the point moves when only q^j changes be represented by the base vector

$$\mathbf{g}_j. \tag{12-43}$$

The vector whose projection on $\mathbf{g}_j$ is the reciprocal of the magnitude of $\mathbf{g}_j$ and whose direction is perpendicular to that of all base vectors except $\mathbf{g}_j$ is labeled

$$\mathbf{g}^j. \tag{12-44}$$

Thus we have

$$\mathbf{g}^k \cdot \mathbf{g}_j = \delta_j{}^k \tag{12-45}$$

where $\delta_j{}^k$ is the Kronecker delta.

The base vector (12-43) is said to be *covariant*, while the reciprocal vector (12-44) is said to be *contravariant*. Elements (12-42) are also said to be contravariant. Upper indices indicate contravariance; lower indices covariance. The covariant elements will be defined later.

The units and magnitude of each base vector $\mathbf{g}_j$ are chosen so that the vector converts the corresponding dq^j to the distance moved when only q^j varies. When all the coordinates vary, the resulting displacement is the sum of the contributing vectors:

$$d\mathbf{s} = \mathbf{g}_j \, dq^j. \tag{12-46}$$

For the square of the magnitude of this distance, we have

$$ds^2 = d\mathbf{s} \cdot d\mathbf{s} = \mathbf{g}_j \cdot \mathbf{g}_k \, dq^j \, dq^k = g_{jk} \, dq^j \, dq^k. \tag{12-47}$$

In the last step, the jkth covariant component of the *metric tensor* has been introduced by the definition

$$\mathbf{g}_j \cdot \mathbf{g}_k = g_{jk}. \tag{12-48}$$

Since the factors in a scalar product commute, the metric tensor is symmetric:

$$g_{jk} = g_{kj}. \tag{12-49}$$

Differential dq^j is called the jth contravariant component of the displacement. We employ the relationship

$$d\mathbf{s} = \mathbf{g}^j \, dq_j \tag{12-50}$$

in which the reciprocal base vectors are given by (12-45), to define the corresponding covariant components.

Dotting this vector with itself yields

$$ds^2 = d\mathbf{s} \cdot d\mathbf{s} = \mathbf{g}^j \cdot \mathbf{g}^k \, dq_j \, dq_k = g^{jk} \, dq_j \, dq_k. \tag{12-51}$$

Here the jkth contravariant component of the metric tensor is introduced as

$$\mathbf{g}^j \cdot \mathbf{g}^k = g^{jk}. \tag{12-52}$$

Note that

$$g^{jk} = g^{kj}. \tag{12-53}$$

The scalar product of (12-50) with (12-46) is

$$ds^2 = d\mathbf{s} \cdot d\mathbf{s} = \mathbf{g}^j \cdot \mathbf{g}_k \, dq_j \, dq^k = dq_k \, dq^k. \tag{12-54}$$

In the last step, equation (12-45) has been introduced.

For this result to equal that in (12-47), we must have

$$dq_k = g_{jk}\, dq^j = g_{kj}\, dq^j.$$
(12-55)

For it to equal that in (12-51), we must have

$$dq^k = g^{jk}\, dq_j = g^{kj}\, dq_j.$$
(12-56)

Multiplication by g_{jk} lowers the repeated index while multiplication by g^{jk} raises the repeated index on the pertinent vector element.

The procedure can be used to transform a base vector into its reciprocal:

$$\mathbf{g}^k = g^{jk}\mathbf{g}_j.$$
(12-57)

Substituting this result into equation (12-45) leads to the orthogonality relationship:

$$\mathbf{g}^k \cdot \mathbf{g}_l = g^{jk}\mathbf{g}_j \cdot \mathbf{g}_l = g^{jk}g_{jl} = \delta_l^{\ k}.$$
(12-58)

EXAMPLE 12.6 Show how contravariant and covariant components each add to produce a given vector in two dimensions.

Consider a point where the base vectors are $\mathbf{g}_1$, $\mathbf{g}_2$ and the contravariant elements of vector $\mathbf{v}$ are v^1, v^2.

An element times the corresponding base equals what $\mathbf{v}$ would be if all its other components were zero. In figure 12.8, we obtain $v^1\mathbf{g}_1$ and $v^2\mathbf{g}_2$. Note how these add to yield $\mathbf{v}$.

The reciprocal base $\mathbf{g}^1$ must be perpendicular to $\mathbf{g}_2$ and $\mathbf{g}^2$ must be perpendicular to $\mathbf{g}_1$, as shown. The corresponding elements v_1 and v_2 convert these to the vectors $v_1\mathbf{g}^1$ and $v_2\mathbf{g}^2$. Note how these also add to yield $\mathbf{v}$.

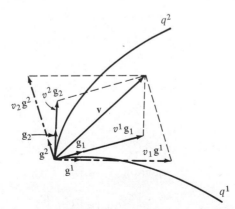

FIGURE 12.8 Contravariant and covariant components of a 2-dimensional vector.

EXAMPLE 12.7   If movement of a particle is being described with the spherical coordinates r, θ, ϕ, what are the components of the corresponding metric tensor?

The square of an element of length along the particle's path is

$$ds^2 = dr^2 + r^2\, d\theta^2 + r^2 \sin^2\theta\, d\phi^2.$$

Comparing this with (12-47) when

$$r = q^1, \qquad \theta = q^2, \qquad \phi = q^3,$$

yields

$$\mathbf{g} = \begin{pmatrix} 1 & 0 & 0 \\ 0 & r^2 & 0 \\ 0 & 0 & r^2 \sin^2\theta \end{pmatrix}.$$

EXAMPLE 12.8   In the spherical coordinate system, how are the base vectors and their reciprocals related to the unit vectors **l**, **m**, **n**?

From figure 1.15 and the definitions of angle, sine, and vector addition, we find that

$$d\mathbf{s} = \mathbf{l}\, dr + \mathbf{m}r\, d\theta + \mathbf{n}r \sin\theta\, d\phi.$$

Comparing this form with equation (12-46) yields

$$\mathbf{g}_1 = \mathbf{l}, \qquad \mathbf{g}_2 = r\,\mathbf{m}, \qquad \mathbf{g}_3 = r \sin\theta\,\mathbf{n}.$$

Since these bases are mutually orthogonal, the reciprocal vectors have the same directions but reciprocal magnitudes:

$$\mathbf{g}^1 = \mathbf{l} \qquad \mathbf{g}^2 = \frac{1}{r}\,\mathbf{m} \qquad \mathbf{g}^3 = \frac{1}{r \sin\theta}\,\mathbf{n}.$$

12.10 Partial derivative of a physical vector

Differentiating the jth component of a vector does not yield the jth component of the derivative when the base vectors vary, for, such variation causes additional terms to appear in the derivative. The nature of these will now be determined.

From equation (12-46), the jth base vector is the coefficient of dq^j in the expression for $d\mathbf{s}$, so

$$\mathbf{g}_j = \frac{\partial \mathbf{s}}{\partial q^j}. \tag{12-59}$$

Let us differentiate this with respect to q^k, interchange the order of differentiation, and reintroduce the appropriate base vector:

$$\frac{\partial \mathbf{g}_j}{\partial q^k} = \frac{\partial}{\partial q^k}\frac{\partial \mathbf{s}}{\partial q^j} = \frac{\partial}{\partial q^j}\frac{\partial \mathbf{s}}{\partial q^k} = \frac{\partial \mathbf{g}_k}{\partial q^j}. \tag{12-60}$$

Next, differentiate one-half of equation (12-48),

$$\frac{1}{2}\frac{\partial g_{jk}}{\partial q^l} = \frac{1}{2}\frac{\partial \mathbf{g}_j}{\partial q^l}\cdot\mathbf{g}_k + \tfrac{1}{2}\mathbf{g}_j\cdot\frac{\partial \mathbf{g}_k}{\partial q^l}, \tag{12-61}$$

and rotate the indices:

$$\frac{1}{2}\frac{\partial g_{jl}}{\partial q^k} = \frac{1}{2}\frac{\partial \mathbf{g}_j}{\partial q^k}\cdot\mathbf{g}_l + \tfrac{1}{2}\mathbf{g}_j\cdot\frac{\partial \mathbf{g}_l}{\partial q^k}, \tag{12-62}$$

$$\frac{1}{2}\frac{\partial g_{kl}}{\partial q^j} = \frac{1}{2}\frac{\partial \mathbf{g}_k}{\partial q^j}\cdot\mathbf{g}_l + \tfrac{1}{2}\mathbf{g}_k\cdot\frac{\partial \mathbf{g}_l}{\partial q^j}. \tag{12-63}$$

Constructing (12-62) plus (12-63) minus (12-61) and introducing the symmetry (12-60) then leads to

$$\frac{1}{2}\left(\frac{\partial g_{jl}}{\partial q^k} + \frac{\partial g_{kl}}{\partial q^j} - \frac{\partial g_{jk}}{\partial q^l}\right) = \frac{\partial \mathbf{g}_j}{\partial q^k}\cdot\mathbf{g}_l. \tag{12-64}$$

Let us next try writing the partial derivative of a base vector as a linear combination of base vectors:

$$\frac{\partial \mathbf{g}_j}{\partial q^k} = \Gamma_{jk}{}^o\mathbf{g}_o. \tag{12-65}$$

Equation (12-64) becomes

$$\frac{1}{2}\left(\frac{\partial g_{jl}}{\partial q^k} + \frac{\partial g_{kl}}{\partial q^j} - \frac{\partial g_{jk}}{\partial q^l}\right) = \Gamma_{jk}{}^o\mathbf{g}_o\cdot\mathbf{g}_l = \Gamma_{jk}{}^o g_{ol}. \tag{12-66}$$

Multiplying through by g^{nl} then gives us an expression for the three-index symbol,

$$\tfrac{1}{2}g^{nl}\left(\frac{\partial g_{jl}}{\partial q^k} + \frac{\partial g_{kl}}{\partial q^j} - \frac{\partial g_{jk}}{\partial q^l}\right) = \Gamma_{jk}{}^n; \tag{12-67}$$

so the procedure is valid. Form (12-67) is referred to as the *Christoffel* three-index symbol of the *second kind*.

An arbitrary vector can be expanded in terms of the base vectors

$$\mathbf{v} = v^j\mathbf{g}_j \tag{12-68}$$

at any given point. Differentiating this equation,

$$\frac{\partial \mathbf{v}}{\partial q^k} = \frac{\partial v^j}{\partial q^k}\mathbf{g}_j + v^j\frac{\partial \mathbf{g}_j}{\partial q^k}, \tag{12-69}$$

and introducing formula (12-65), we finally obtain

$$\frac{\partial \mathbf{v}}{\partial q^k} = \frac{\partial v^j}{\partial q^k}\mathbf{g}_j + v^l\Gamma_{lk}{}^j\mathbf{g}_j$$

$$= \left(\frac{\partial v^j}{\partial q^k} + \Gamma_{lk}{}^j v^l\right)\mathbf{g}_j. \tag{12-70}$$

The coefficient of $\mathbf{g}_j$ is the jth component of the partial derivative of $\mathbf{v}$ with respect to q^k.

12.11   A covariant form of Newton's second law

Properties of the base vectors and the related metric tensor let us derive Lagrange's force equations in a straightforward manner.

We start with generalized coordinates and velocities constructed as in chapter 5, but focus our attention on a single particle (or small body) of the system. From the work done on this unit during a real or virtual change δq^n, the corresponding generalized force Q_{ni} is obtained. If m_i is the mass of the particle and a_n its acceleration, Newton's second law then yields

$$m_i a_n = Q_{ni}. \tag{12-71}$$

Because the acceleration is the time rate of change of the conventional velocity, we have to interpret $d\mathbf{s}$ in equation (12-46) as an actual displacement of the particle. If the time during which this takes place is dt, then we have

$$\frac{d\mathbf{s}}{dt} = \frac{dq^j}{dt} \mathbf{g}_j = \dot{q}^j \mathbf{g}_j. \tag{12-72}$$

Formally, we merely divide equation (12-46) by dt and introduce the definition of $\dot{q}^j$.

Differentiating (12-72) yields the acceleration

$$\mathbf{a} = \frac{d}{dt}(\dot{q}^j \mathbf{g}_j) = \ddot{q}^l \mathbf{g}_l + \dot{q}^j \dot{\mathbf{g}}_j. \tag{12-73}$$

Since the base vector depends only on the generalized coordinates, the time derivative is

$$\dot{\mathbf{g}}_j = \frac{\partial \mathbf{g}_j}{\partial q^k} \dot{q}^k = \Gamma_{jk}{}^l \dot{q}^k \mathbf{g}_l. \tag{12-74}$$

In the second equality, we have introduced relationship (12-65).

Combining (12-73) and (12-74),

$$\mathbf{a} = (\ddot{q}^l + \Gamma_{jk}{}^l \dot{q}^j \dot{q}^k)\mathbf{g}_l, \tag{12-75}$$

we see that

$$a^l = \ddot{q}^l + \Gamma_{jk}{}^l \dot{q}^j \dot{q}^k. \tag{12-76}$$

To convert this element to the covariant component, we multiply by g_{nl}; then we introduce formula (12-66):

$$\begin{aligned}
a_n &= g_{nl}\ddot{q}^l + g_{nl}\Gamma_{jk}{}^l \dot{q}^j \dot{q}^k \\
&= g_{nl}\ddot{q}^l + \frac{1}{2}\left(\frac{\partial g_{jn}}{\partial q^k} + \frac{\partial g_{kn}}{\partial q^j} - \frac{\partial g_{jk}}{\partial q^n}\right)\dot{q}^j \dot{q}^k.
\end{aligned} \tag{12-77}$$

Next, differentiate the product $g_{nk}\dot{q}^k$,

$$\begin{aligned}
\frac{d}{dt}(g_{nk}\dot{q}^k) &= \frac{\partial g_{nk}}{\partial q^j} \dot{q}^j \dot{q}^k + g_{nk}\ddot{q}^k \\
&= \left(\frac{1}{2}\frac{\partial g_{kn}}{\partial q^j} + \frac{1}{2}\frac{\partial g_{jn}}{\partial q^k}\right)\dot{q}^j \dot{q}^k + g_{nl}\ddot{q}^l,
\end{aligned} \tag{12-78}$$

and use the overall equality to simplify (12-77),

$$a_n = \frac{d}{dt}(g_{nk}\dot{q}^k) - \frac{1}{2}\frac{\partial g_{jk}}{\partial q^n}\dot{q}^j\dot{q}^k. \tag{12-79}$$

Multiplying this expression by the mass of the particle yields the corresponding generalized force

$$m_i a_n = \frac{d}{dt}(m_i g_{nk}\dot{q}^k) - \frac{1}{2}m_i\frac{\partial g_{jk}}{\partial q^n}\dot{q}^j\dot{q}^k = Q_{ni}. \tag{12-80}$$

The kinetic energy of the particle is

$$T_i = \frac{1}{2}m_i\left(\frac{ds}{dt}\right)^2 = \frac{1}{2}m_i\dot{q}^j\mathbf{g}_j \cdot \dot{q}^k\mathbf{g}_k = \frac{1}{2}m_i g_{jk}\dot{q}^j\dot{q}^k. \tag{12-81}$$

Here we have employed (12-72) and (12-48).

Differentiating this expression with respect to q^n,

$$\frac{\partial T_i}{\partial q^n} = \frac{1}{2}m_i\frac{\partial g_{jk}}{\partial q^n}\dot{q}^j\dot{q}^k, \tag{12-82}$$

with respect to $\dot{q}^n$,

$$\frac{\partial T_i}{\partial \dot{q}^n} = \frac{1}{2}m_i g_{nk}\dot{q}^k + \frac{1}{2}m_i g_{jn}\dot{q}^j = m_i g_{nk}\dot{q}^k, \tag{12-83}$$

and substituting into equation (12-80), we find

$$\frac{d}{dt}\frac{\partial T_i}{\partial \dot{q}^n} - \frac{\partial T_i}{\partial q^n} = Q_{ni}. \tag{12-84}$$

Since the kinetic energy of the set of particles is the sum of the T_i's, we have

$$T = \sum_i T_i. \tag{12-85}$$

Since the work done on all the particles is the sum of that done on each particle, each term in (5-19) represents such a sum, and we can write

$$Q_n = \sum_i Q_{ni}. \tag{12-86}$$

Summing equation (12-84) over all particles in the system yields the Lagrange force equation

$$\frac{d}{dt}\frac{\partial T}{\partial \dot{q}^n} - \frac{\partial T}{\partial q^n} = Q_n. \tag{12-87}$$

12.12 The del operator in generalized coordinates

When n independent unit vectors obeying the condition

$$\mathbf{e}_j \cdot \mathbf{e}_k = \delta_{jk} \tag{12-88}$$

can be set up at each point of an *n*-dimensional space, the formula

$$\nabla = \mathbf{e}_j \frac{\partial}{\partial x_j} \tag{12-89}$$

defines the del operator.

But if the $\mathbf{e}_k$'s are chosen to be the base vectors and $\mathbf{e}^j$'s their reciprocals, equation (12-45) shows that

$$\mathbf{e}^j \cdot \mathbf{e}_k = \delta_k{}^j. \tag{12-90}$$

This relationship is consistent with (12-88) if,

$$\mathbf{e}^j = \mathbf{e}_j, \tag{12-91}$$

each reciprocal vector equals the corresponding Cartesian base vector.

Then there is no distinction between covariant and contravariant components, and equation (12-89) can be rewritten as

$$\nabla = \mathbf{e}^j \frac{\partial}{\partial x^j}, \tag{12-92}$$

whence

$$\nabla\phi = \mathbf{e}^j \frac{\partial \phi}{\partial x^j}. \tag{12-93}$$

If ϕ is considered a function of the generalized coordinates (12-42), the procedure in example 5.1 yields

$$\nabla\phi = \mathbf{e}^j \frac{\partial q^k}{\partial x^j} \frac{\partial \phi}{\partial q^k}. \tag{12-94}$$

We now have to identify the coefficient of $\partial\phi/\partial q^k$.

Since the Cartesian components of an arbitrary displacement $d\mathbf{s}$ are $dx^1, \ldots, dx^n$, we have

$$d\mathbf{s} = \mathbf{e}_k \, dx^k = \mathbf{e}_k \frac{\partial x^k}{\partial q^l} dq^l = \mathbf{g}_l \, dq^l. \tag{12-95}$$

The second equality involves the formula in example 5.1, the overall equality the formula in (12-46). Consequently,

$$\mathbf{g}_l = \mathbf{e}_k \frac{\partial x^k}{\partial q^l}. \tag{12-96}$$

If this is dot multiplied by

$$\mathbf{e}^i \frac{\partial q^j}{\partial x^i}, \tag{12-97}$$

we obtain

$$\left(\mathbf{e}^i \frac{\partial q^j}{\partial x^i}\right) \cdot \left(\mathbf{e}_k \frac{\partial x^k}{\partial q^l}\right) = \delta_k{}^i \frac{\partial q^j}{\partial x^i} \frac{\partial x^k}{\partial q^l} = \frac{\partial q^j}{\partial q^l} = \delta_l{}^j. \tag{12-98}$$

Since the reciprocal base vector $\mathbf{g}^j$ is defined by the equations

$$\mathbf{g}^j \cdot \mathbf{g}_l = \delta_l{}^j, \tag{12-99}$$

(12-97) is $\mathbf{g}^j$.

Substituting this result into (12-94) gives us

$$\nabla\phi = \mathbf{g}^k \frac{\partial\phi}{\partial q^k}. \tag{12-100}$$

Hence

$$\nabla = \mathbf{g}^k \frac{\partial}{\partial q^k}. \tag{12-101}$$

EXAMPLE 12.9   What form does del assume in spherical coordinates?

Take the reciprocal base vectors from example 12.8 and put them into formula (12-101) to get

$$\nabla = \mathbf{l}\frac{\partial}{\partial r} + \mathbf{m}\frac{1}{r}\frac{\partial}{\partial\theta} + \mathbf{n}\frac{1}{r\sin\theta}\frac{\partial}{\partial\phi}.$$

12.13   A general formula for the divergence

When the base vectors vary from point to point, the formula for the divergence involves the Christoffel three-index symbol.

Let the contravariant components of a vector $\mathbf{v}$ be

$$v^1, v^2, \ldots, v^n \tag{12-102}$$

while the jth base vector is

$$\mathbf{g}_j. \tag{12-103}$$

Then we have

$$\mathbf{v} = v^j\mathbf{g}_j \tag{12-104}$$

and

$$\nabla \cdot \mathbf{v} = \left(\mathbf{g}^j\frac{\partial}{\partial q^j}\right) \cdot (v^k\mathbf{g}_k) = \mathbf{g}^j \cdot \mathbf{g}_k\frac{\partial v^k}{\partial q^j} + v^k\mathbf{g}^j \cdot \frac{\partial\mathbf{g}_k}{\partial q^j}$$

$$= \mathbf{g}^j \cdot \mathbf{g}_k\frac{\partial v^k}{\partial q^j} + v^k\Gamma_{kj}{}^l\mathbf{g}^j \cdot \mathbf{g}_l = \frac{\partial v^j}{\partial q^j} + v^k\Gamma_{kj}{}^j. \tag{12-105}$$

In the third equality, formula (12-65) has been introduced; in the fourth equality, formula (12-45).

Part of the three-index symbol consists of one-half the following expression

$$g^{jl}\left(\frac{\partial g_{kl}}{\partial q^j} - \frac{\partial g_{kj}}{\partial q^l}\right) = g^{jl}\left(\frac{\partial\mathbf{g}_k}{\partial q^j} \cdot \mathbf{g}_l + \mathbf{g}_k \cdot \frac{\partial\mathbf{g}_l}{\partial q^j} - \frac{\partial\mathbf{g}_k}{\partial q^l} \cdot \mathbf{g}_j - \mathbf{g}_k \cdot \frac{\partial\mathbf{g}_j}{\partial q^l}\right)$$

$$= \frac{\partial\mathbf{g}_k}{\partial q^j} \cdot \mathbf{g}^j - \frac{\partial\mathbf{g}_k}{\partial q^l} \cdot \mathbf{g}^l = 0. \tag{12-106}$$

In the first equality, we have employed (12-48); in the second equality, (12-60) and (12-57). The symbol reduces to the remaining part:

$$\Gamma_{kj}{}^j = \tfrac{1}{2}g^{jl}\frac{\partial g_{jl}}{\partial q^k}. \tag{12-107}$$

In 3-dimensional space, the determinant of the metric tensor is

$$g = \varepsilon^{jkl} g_{1j} g_{2k} g_{3l}. \tag{12-108}$$

From (12-58), we have

$$1 = g_{1j} g^{1j} \tag{12-109}$$

or

$$g = g_{1j} g g^{1j}. \tag{12-110}$$

Comparing (12-108) and (12-110), we see that

$$g g^{1j} = \varepsilon^{jkl} g_{2k} g_{3l}. \tag{12-111}$$

Similarly,

$$g g^{2k} = \varepsilon^{jkl} g_{1j} g_{3l}, \tag{12-112}$$

$$g g^{3l} = \varepsilon^{jkl} g_{1j} g_{2k}. \tag{12-113}$$

Differentiating (12-108) then yields

$$\frac{\partial g}{\partial q^i} = \varepsilon^{jkl} g_{2k} g_{3l} \frac{\partial g_{1j}}{\partial q^i} + \varepsilon^{jkl} g_{1j} g_{3l} \frac{\partial g_{2k}}{\partial q^i} + \varepsilon^{jkl} g_{1j} g_{2k} \frac{\partial g_{3l}}{\partial q^i}$$

$$= g g^{1j} \frac{\partial g_{1j}}{\partial q^i} + g g^{2k} \frac{\partial g_{2k}}{\partial q^i} + g g^{3l} \frac{\partial g_{3l}}{\partial q^i} \tag{12-114}$$

or

$$\frac{1}{2g} \frac{\partial g}{\partial q^k} = \tfrac{1}{2} g^{jl} \frac{\partial g_{jl}}{\partial q^k}. \tag{12-115}$$

Hence the divergence of a vector function is given by

$$\nabla \cdot \mathbf{v} = \frac{\partial v^j}{\partial q^j} + \frac{v^k}{2g} \frac{\partial g}{\partial q^k} = \frac{1}{\sqrt{g}} \frac{\partial(\sqrt{g}\, v^j)}{\partial q^j}. \tag{12-116}$$

If the function is expanded in terms of *unit* base vectors $\mathbf{e}_1, \mathbf{e}_2, \mathbf{e}_3$ proportional to $\mathbf{g}_1, \mathbf{g}_2, \mathbf{g}_3$, respectively, we have

$$\mathbf{v} = v^{qj} \mathbf{e}_j = v^{qj} h^j \mathbf{g}_j \tag{12-117}$$

and

$$\nabla \cdot \mathbf{v} = \frac{1}{\sqrt{g}} \frac{\partial(\sqrt{g}\, v^{qj} h^j)}{\partial q^j}. \tag{12-118}$$

Element v^{qj} is called the *physical* component of $\mathbf{v}$ in the direction $\mathbf{e}_j$.

In an *orthogonal* coordinate system

$$g_{jk} = 0 \qquad \text{when} \quad j \neq k, \tag{12-119}$$

so

$$g = g_{11} g_{22} g_{33} \tag{12-120}$$

and

$$\nabla \cdot \mathbf{v} = \frac{1}{(g_{11} g_{22} g_{33})^{1/2}} \frac{\partial(g_{11} g_{22} g_{33})^{1/2} v^j}{\partial q^j}. \tag{12-121}$$

12.14   Formulas for del dot del

In example 11.6, we saw how the requirement that ϕ be as smooth as possible within a given closed surface led to a differential equation that can be written as

$$\nabla \cdot \nabla \phi = 0. \tag{12-122}$$

Let us now obtain alternate forms for the operator

$$\nabla \cdot \nabla = \nabla^2 \tag{12-123}$$

called the *Laplacian*.

Equation (12-100) states that the covariant components of the gradient of ϕ are $\partial\phi/\partial q^k$. Multiplication by g^{jk} changes these to the contravariant form

$$g^{jk} \frac{\partial\phi}{\partial q^k}. \tag{12-124}$$

Substituting (12-124) into formula (12-116) then leads to

$$\nabla \cdot \nabla \phi = \frac{1}{\sqrt{g}} \frac{\partial}{\partial q^j} \left(\sqrt{g}\, g^{jk} \frac{\partial\phi}{\partial q^k} \right). \tag{12-125}$$

In an *orthogonal* coordinate system

$$g_{jl} = 0 \qquad \text{when} \quad j \neq l. \tag{12-126}$$

These zeros reduce equations (12-58) to

$$g^{11}g_{11} = 1, \qquad g^{21}g_{22} = 0, \qquad g^{31}g_{33} = 0, \tag{12-127}$$

$$g^{12}g_{11} = 0, \qquad g^{22}g_{22} = 1, \qquad g^{32}g_{33} = 0, \tag{12-128}$$

$$g^{13}g_{11} = 0, \qquad g^{23}g_{22} = 0, \qquad g^{33}g_{33} = 1, \tag{12-129}$$

whence

$$g^{11} = \frac{1}{g_{11}}, \qquad g^{22} = \frac{1}{g_{22}}, \qquad g^{33} = \frac{1}{g_{33}}, \tag{12-130}$$

and

$$g^{jl} = 0 \qquad \text{when} \quad j \neq l. \tag{12-131}$$

Furthermore, we have

$$g = g_{11}g_{22}g_{33}. \tag{12-132}$$

The Laplacian of ϕ then simplifies to

$$\nabla^2 \phi = \frac{1}{\sqrt{g}} \left[\frac{\partial}{\partial q^1} \left(\sqrt{\frac{g_{22}g_{33}}{g_{11}}} \frac{\partial\phi}{\partial q^1} \right) + \frac{\partial}{\partial q^2} \left(\sqrt{\frac{g_{33}g_{11}}{g_{22}}} \frac{\partial\phi}{\partial q^2} \right) + \frac{\partial}{\partial q^3} \left(\sqrt{\frac{g_{11}g_{22}}{g_{33}}} \frac{\partial\phi}{\partial q^3} \right) \right].$$
$$\tag{12-133}$$

EXAMPLE 12.10   What form does the Laplacian assume in spherical coordinates? The elements of the metric tensor from example 12.7 convert (12-133) to

$$\nabla^2 \phi = \frac{1}{r^2 \sin \theta} \left[\frac{\partial}{\partial r} \left(r^2 \sin \theta \frac{\partial \phi}{\partial r} \right) + \frac{\partial}{\partial \theta} \left(\sin \theta \frac{\partial \phi}{\partial \theta} \right) + \frac{\partial}{\partial \phi} \left(\frac{1}{\sin \theta} \frac{\partial \phi}{\partial \phi} \right) \right]$$

$$= \frac{1}{r^2} \frac{\partial}{\partial r} \left(r^2 \frac{\partial \phi}{\partial r} \right) + \frac{1}{r^2 \sin \theta} \frac{\partial}{\partial \theta} \left(\sin \theta \frac{\partial \phi}{\partial \theta} \right) + \frac{1}{r^2 \sin^2 \theta} \frac{\partial^2 \phi}{\partial \phi^2}.$$

12.15   Additional relationships involving the base vectors

Certain cross products appear in the expression for the curl. These will be reduced in this section.

First, we recall (12-45) relating the reciprocal vectors to the base vectors in a given space:

$$\mathbf{g}^k \cdot \mathbf{g}_l = \delta_l^k. \tag{12-134}$$

Differentiating this equation,

$$\mathbf{g}^k \cdot \frac{\partial \mathbf{g}_l}{\partial q^j} + \frac{\partial \mathbf{g}^k}{\partial q^j} \cdot \mathbf{g}_l = 0, \tag{12-135}$$

and employing (12-65) to introduce the three-index symbol,

$$\mathbf{g}^k \cdot \mathbf{g}_o \Gamma_{lj}{}^o + \frac{\partial \mathbf{g}^k}{\partial q^j} \cdot \mathbf{g}_l = 0, \tag{12-136}$$

we find the first factor in the second term must be

$$\frac{\partial \mathbf{g}^k}{\partial q^j} = -\Gamma_{nj}{}^k \mathbf{g}^n. \tag{12-137}$$

Taking $\mathbf{g}^j$ cross each side of (12-137) and summing, we get

$$\mathbf{g}^j \times \frac{\partial \mathbf{g}^k}{\partial q^j} = -\Gamma_{nj}{}^k \mathbf{g}^j \times \mathbf{g}^n = 0. \tag{12-138}$$

The result is zero because interchanging j and n leaves $\Gamma_{nj}{}^k$ unaltered but changes the sign of $\mathbf{g}^j \times \mathbf{g}^n$. To each positive term in the sum, there is a canceling negative term.

Let us now restrict ourselves to a 3-dimensional space in which Cartesian coordinates x^1, x^2, x^3 can be erected at each point. The square of an element of length is then given by

$$ds^2 = dx^j \, dx^j. \tag{12-139}$$

The procedure in example 5.1 allows each dx^j to be expanded in terms of the dq^k's:

$$ds^2 = \frac{\partial x^j}{\partial q^k} \frac{\partial x^j}{\partial q^l} \, dq^k \, dq^l = g_{kl} \, dq^k \, dq^l. \tag{12-140}$$

The last equality follows from (12-47).

We note that

$$g_{kl} = \frac{\partial x^j}{\partial q^k} \frac{\partial x^j}{\partial q^l}; \tag{12-141}$$

so the determinant of the metric tensor is

$$g = \begin{vmatrix} g_{11} & g_{12} & g_{13} \\ g_{21} & g_{22} & g_{23} \\ g_{31} & g_{32} & g_{33} \end{vmatrix}$$

$$= \begin{vmatrix} \dfrac{\partial x^1}{\partial q^1} & \dfrac{\partial x^2}{\partial q^1} & \dfrac{\partial x^3}{\partial q^1} \\ \dfrac{\partial x^1}{\partial q^2} & \dfrac{\partial x^2}{\partial q^2} & \dfrac{\partial x^3}{\partial q^2} \\ \dfrac{\partial x^1}{\partial q^3} & \dfrac{\partial x^2}{\partial q^3} & \dfrac{\partial x^3}{\partial q^3} \end{vmatrix} \begin{vmatrix} \dfrac{\partial x^1}{\partial q^1} & \dfrac{\partial x^1}{\partial q^2} & \dfrac{\partial x^1}{\partial q^3} \\ \dfrac{\partial x^2}{\partial q^1} & \dfrac{\partial x^2}{\partial q^2} & \dfrac{\partial x^2}{\partial q^3} \\ \dfrac{\partial x^3}{\partial q^1} & \dfrac{\partial x^3}{\partial q^2} & \dfrac{\partial x^3}{\partial q^3} \end{vmatrix}$$

$$= \begin{vmatrix} \dfrac{\partial x^1}{\partial q^1} & \dfrac{\partial x^2}{\partial q^1} & \dfrac{\partial x^3}{\partial q^1} \\ \dfrac{\partial x^1}{\partial q^2} & \dfrac{\partial x^2}{\partial q^2} & \dfrac{\partial x^3}{\partial q^2} \\ \dfrac{\partial x^1}{\partial q^3} & \dfrac{\partial x^2}{\partial q^3} & \dfrac{\partial x^3}{\partial q^3} \end{vmatrix}^2 . \tag{12-142}$$

In the last step, the rows and columns of the second determinant have been interchanged.

The vector drawn from the beginning to the end of the element of length is given by (12-46). But it also has the Cartesian components dx^1, dx^2, dx^3:

$$d\mathbf{s} = \mathbf{g}_k \, dq^k = dx^j \mathbf{e}_j. \tag{12-143}$$

Expanding the Cartesian components as before yields

$$d\mathbf{s} = \frac{\partial x^j}{\partial q^k} \mathbf{e}_j \, dq^k, \tag{12-144}$$

whence

$$\mathbf{g}_k = \frac{\partial x^j}{\partial q^k} \mathbf{e}_j. \tag{12-145}$$

Substituting the components of $\mathbf{g}_1$, $\mathbf{g}_2$, and $\mathbf{g}_3$ into formula (1-34) then gives us

$$\mathbf{g}_1 \cdot \mathbf{g}_2 \times \mathbf{g}_3 = \begin{vmatrix} \dfrac{\partial x^1}{\partial q^1} & \dfrac{\partial x^2}{\partial q^1} & \dfrac{\partial x^3}{\partial q^1} \\[2mm] \dfrac{\partial x^1}{\partial q^2} & \dfrac{\partial x^2}{\partial q^2} & \dfrac{\partial x^3}{\partial q^2} \\[2mm] \dfrac{\partial x^1}{\partial q^3} & \dfrac{\partial x^2}{\partial q^3} & \dfrac{\partial x^3}{\partial q^3} \end{vmatrix}$$

$$= \sqrt{g}. \tag{12-146}$$

The last equality comes from (12-142).

On comparing (12-146) with (12-134), we see that $\mathbf{g}_2 \times \mathbf{g}_3$ is $\sqrt{g}$ times $\mathbf{g}^1$:

$$\mathbf{g}_2 \times \mathbf{g}_3 = \sqrt{g}\, \mathbf{g}^1. \tag{12-147}$$

Similarly,

$$\mathbf{g}_3 \times \mathbf{g}_1 = \sqrt{g}\, \mathbf{g}^2, \tag{12-148}$$

$$\mathbf{g}_1 \times \mathbf{g}_2 = \sqrt{g}\, \mathbf{g}^3. \tag{12-149}$$

Cross multiplying the various pairs of these cross products and reducing then leads to

$$\mathbf{g}_1 = \sqrt{g}\, \mathbf{g}^2 \times \mathbf{g}^3, \tag{12-150}$$

$$\mathbf{g}_2 = \sqrt{g}\, \mathbf{g}^3 \times \mathbf{g}^1, \tag{12-151}$$

$$\mathbf{g}_3 = \sqrt{g}\, \mathbf{g}^1 \times \mathbf{g}^2. \tag{12-152}$$

12.16   The curl of a vector field

Because the cross products we have obtained are between base vectors of the same kind, we expand the given variable vector in terms of the ones that appear in the expression for del, (12-101):

$$\mathbf{v} = v_k \mathbf{g}^k. \tag{12-153}$$

Taking the curl of this vector then leads to

$$\nabla \times \mathbf{v} = \left(\mathbf{g}^j \frac{\partial}{\partial q^j} \right) \times (v_k \mathbf{g}^k)$$

$$= \mathbf{g}^j \times \mathbf{g}^k \frac{\partial v_k}{\partial q^j} + v_k \mathbf{g}^j \times \frac{\partial \mathbf{g}^k}{\partial q^j}$$

$$= \frac{\varepsilon^{jkl}}{\sqrt{g}} \frac{\partial v_k}{\partial q^j} \mathbf{g}_l. \tag{12-154}$$

In the last step, equations (12-150), (12-151), (12-152), and (12-138) have been introduced. Expression ε^{jkl} is the permutation symbol.

12.17 A brief review

When a scalar function is well-behaved, it varies smoothly throughout space at any given time. The maximum rate of change and the direction in which this occurs then define a vector called the *gradient* at each point. From equation (12-101), the general formula for the vector is

$$\text{grad } \phi = \nabla\phi = \mathbf{g}^k \frac{\partial\phi}{\partial q^k}. \tag{12-155}$$

A well-behaved vector function can be represented by the density of momentum of a hypothetical fluid. The flow may cause the density of the fluid to vary with time at any given point. The rate at which this density decreases about the point is called the *divergence* of the vector function there. From equation (12-116), the general formula for this scalar is

$$\text{div } \mathbf{v} = \nabla \cdot \mathbf{v} = \frac{1}{\sqrt{g}} \frac{\partial(\sqrt{g}\, v^j)}{\partial q^j}. \tag{12-156}$$

A well-behaved vector function can also be represented by a hypothetical force field. The work done by the force along a small planar closed path can then be found and the result divided by the area of the plane. For each orientation of the plane, the area may be made small to obtain a limiting ratio. The maximum in this ratio at a given point defines the magnitude of the *curl*; the direction of the right-handed normal to the plane the direction of the curl. From equation (12-154), the general formula for this vector is

$$\text{curl } \mathbf{v} = \nabla \times \mathbf{v} = \frac{\varepsilon^{jkl}}{\sqrt{g}} \frac{\partial v_k}{\partial q^j} \mathbf{g}_l. \tag{12-157}$$

When

$$q^1 = x, \qquad q^2 = y, \qquad q^3 = z, \tag{12-158}$$

the base vectors are the conventional ones of elementary vector algebra:

$$g^1 = \mathbf{i}, \qquad g^2 = \mathbf{j}, \qquad g^3 = \mathbf{k}. \tag{12-159}$$

Then we have

$$g_1 = \mathbf{i}, \qquad g_2 = \mathbf{j}, \qquad g_3 = \mathbf{k}, \tag{12-160}$$

$$v^1 = v_1 = v_x, \qquad v^2 = v_2 = v_y, \qquad v^3 = v_3 = v_z, \tag{12-161}$$

and

$$g = \sqrt{1 \cdot 1 \cdot 1} = 1. \tag{12-162}$$

DISCUSSION QUESTIONS

12.1 How does the observed action of one particle on another at a distance support the view that fields exist?

12.2 With sketches, show how the direction and magnitude of the gradient of a function are related to the function itself.

12.3 What is the significance of the divergence of a vector function?

12.4 How can a distributed physical quantity be described as a fluid?

12.5 Interpret Gauss's theorem.

12.6 What integral per unit area characterizes the behavior of a "fluid" near a vortex?

12.7 Show that if **v** describes the velocity of each material element in a fluid, a small paddle wheel tends to rotate wherever

$$\text{curl } \mathbf{v} \neq 0.$$

12.8 Interpret (a) the curl of a vector function and (b) Stokes' theorem.

12.9 Describe how the direction in which the perimeter is traversed is related to the direction of each element $\Delta \mathbf{S}_j$ of surface within.

12.10 Describe the action of del.

12.11 What is a base vector? Why is it considered covariant?

12.12 How is the reciprocal of a base vector taken? Why is a different reciprocal employed in solid-state physics?

12.13 What is the metric tensor? Does its existence imply that the space is Euclidean?

12.14 Construct the contravariant components for the metric tensor in the spherical coordinate system.

12.15 What is the Christoffel three-index symbol of the second kind? Why does it appear in the formula for the derivative?

12.16 Why is the conventional generalized force covariant? Why is the acceleration in Newton's second law (12-71) covariant?

12.17 Explain how the del operator appears in a generalized coordinate system.

12.18 Distinguish between the physical, the contravariant, and the covariant components of a vector.

12.19 Why does the divergence of **v** differ from

$$\frac{\partial v^j}{\partial q^j} \, ?$$

12.20 Explain the general formula for the Laplacian.

12.21 Why isn't the cross product of two base vectors a constant times the third base vector? What is it instead?

12.22 Show how
$$\Gamma_{nj}{}^{k}\mathbf{g}^{j} \times \mathbf{g}^{n}$$
is zero.

12.23 Derive equations (12-150), (12-151), (12-152) from (12-147), (12-148), (12-149).

12.24 Discuss the general formula for the curl.

PROBLEMS

12.1 Work in Cartesian coordinates and obtain the gradient of (a) $x + y$, (b) r^2, (c) $\ln r$.

12.2 Find the unit vector normal to the surface
$$x^2 - xy + yz = 3$$
at the point $(1, 2, 2)$.

12.3 Employing Cartesian coordinates, find (a) the divergence and (b) the curl of (a) $x^2\mathbf{i}$ and (b) $\boldsymbol{\omega} \times \mathbf{r}$.

12.4 Verify the formulas
$$\nabla \times (\phi\mathbf{V}) = \phi\nabla \times \mathbf{V} - \mathbf{V} \times \nabla\phi,$$
$$\nabla \times (\mathbf{U} \times \mathbf{V}) = (\mathbf{V} \cdot \nabla)\mathbf{U} - \mathbf{V}(\nabla \cdot \mathbf{U}) - (\mathbf{U} \cdot \nabla)\mathbf{V} + \mathbf{U}(\nabla \cdot \mathbf{V}),$$
$$\nabla \times (\nabla \times \mathbf{V}) = \nabla(\nabla \cdot \mathbf{V}) - (\nabla \cdot \nabla)\mathbf{V}.$$

12.5 Calculate the line integral of
$$(x^2 + y^2)\mathbf{i} + xy\mathbf{j}$$
around the square bounded by the coordinate axes and the lines $x = c$, $y = c$ in the xy plane. Check Stokes' theorem by also calculating the surface integral of the curl of the given vector over the square.

12.6 Construct the metric tensor, divergence, Laplacian, curl, velocity, and acceleration for cylindrical coordinates defined by the equations
$$x = r \cos \phi,$$
$$y = r \sin \phi,$$
$$z = z.$$

12.7 Coordinates q^1, q^2, q^3 are related to Cartesian coordinates by the equations
$$x^1 = q^1 + q^2 \cos \theta,$$
$$x^2 = q^2 \sin \theta,$$
$$x^3 = q^3,$$
where θ is a constant angle. Express the base vectors for the q^j's in terms of $\mathbf{i}, \mathbf{j}, \mathbf{k}$ and plot them. Construct the metric tensor. Also obtain the covariant, contravariant, and physical components of the acceleration.

12.8 (a) Determine how the angle between two base vectors g_1 and g_2 depends on elements of the metric tensor. (b) Determine

$$\frac{\partial g_{jk}}{\partial q^l} \quad \text{and} \quad \Gamma_{jk}{}^n$$

when the base vectors do not vary with the coordinates.

12.9 From the properties we have discussed, determine how the base vectors and the metric tensor transform in going from one generalized coordinate system to another.

12.10 Show that

$$\Gamma_{kj}{}^l = \frac{\partial^2 x^i}{\partial q^j \, \partial q^k} \frac{\partial q^l}{\partial x^i}.$$

— — —

12.11 With del in Cartesian coordinates, calculate the gradient of (a) z, (b) $1/r$, (c) r^n.

12.12 If possible, find the function that has the gradient (a) $y\mathbf{i} - x\mathbf{j}$ and (b) $(x\mathbf{i} + y\mathbf{j})/((x^2 + y^2)^{3/2})$.

12.13 Employing Cartesian coordinates, calculate (a) the divergence and (b) the curl of (a) $x^n\mathbf{k}$ and (b) $\mathbf{r}/r^n$.

12.14 Verify the formulas

$$\nabla \cdot (\mathbf{U} \times \mathbf{V}) = \mathbf{V} \cdot \nabla \times \mathbf{U} - \mathbf{U} \cdot \nabla \times \mathbf{V},$$
$$\nabla(\mathbf{U} \cdot \mathbf{V}) = (\mathbf{U} \cdot \nabla)\mathbf{V} + \mathbf{U} \times (\nabla \times \mathbf{V}) + (\mathbf{V} \cdot \nabla)\mathbf{U} + \mathbf{V} \times (\nabla \times \mathbf{U}),$$
$$\nabla \cdot \nabla \times \mathbf{V} = 0.$$

12.15 Show that

$$\int_V (\phi \nabla^2 \psi - \psi \nabla^2 \phi) \, dV = \int_V \nabla \cdot (\phi \nabla \psi - \psi \nabla \phi) \, dV$$

$$= \int_S (\phi \nabla \psi - \psi \nabla \phi) \cdot d\mathbf{S}.$$

12.16 From the results in problem 5.17, construct the velocity, base vectors, metric tensor, divergence, and Laplacian for elliptic coordinates defined by the equations

$$x = a \cosh u \cos v,$$
$$y = a \sinh u \sin v,$$
$$z = z.$$

12.17 Parabolic coordinates are related to conventional spherical coordinates by the equations

$$\xi = [r(1 - \cos \theta)]^{1/2},$$
$$\eta = [r(1 + \cos \theta)]^{1/2},$$
$$\phi = \phi.$$

Calculate the metric tensor and the Laplacian for the parabolic system.

12.18 Construct the metric tensor and the Laplacian for the coordinates defined by the equations

$$x = a(\mu^2 - 1)^{1/2}(1 - v^2)^{1/2} \cos \phi,$$
$$y = a(\mu^2 - 1)^{1/2}(1 - v^2)^{1/2} \sin \phi,$$
$$z = a\mu v.$$

12.19 Show that base vectors for the generalized coordinates u^1, u^2, u^3 are related to those for q^1, q^2, q^3 by the equations

$$\mathbf{h}_k = \frac{\partial q^j}{\partial u^k} \mathbf{g}_j,$$

and that

$$\mathbf{h}_1 \times \mathbf{h}_2 = \frac{\partial q^j}{\partial u^1} \frac{\partial q^k}{\partial u^2} \sqrt{g} \varepsilon_{jkl} \mathbf{g}^l.$$

12.20 Determine how the Christoffel three-index symbol of the second kind transforms from one system of coordinates to another.

REFERENCES

BOOKS

Bradbury, T. C. *Theoretical Mechanics*, pp. 43–91. John Wiley & Sons, Inc., New York, 1968.

Brillouin, L. *Tensors in Mechanics and Elasticity* (Translated from French by R. O. Brennan), pp. 1–156. Academic Press, Inc., New York, 1964.

Fung, Y. C. *Foundations of Solid Mechanics*, pp. 31–57. Prentice-Hall, Inc., Englewood Cliffs, N.J., 1965.

Margenau, H., and **Murphy, G. M.** *The Mathematics of Physics and Chemistry*, 2nd ed., pp. 149–197. D. Van Nostrand Company, Inc., Princeton, N.J., 1956.

ARTICLES

Bevc, V. Vector Differential Operations Derived from Physical Definitions. *Am. J. Phys.*, **34,** 507 (1966).

Brown, B. E. Criterion for a Conservative Field. *Am. J. Phys.*, **35,** 352 (1967).

Burns, J. C. Grad ϕ and Curl v. *Am. J. Phys.*, **34,** 268 (1966).

Chen, F. M. Laplacian Operator in Spherical Coordinates, an Alternative Derivation. *Am. J. Phys.*, **40,** 553 (1972).

Goodman, J. M. Paraboloids and Vortices in Hydrodynamics. *Am. J. Phys.*, **37,** 864 (1969).

Groves, G. W. Acceleration Referred to Moving Curvilinear Coordinates. *Am. J. Phys.*, **35,** 927 (1967).

Khare, P. L. Divergence and Curl in Nonorthogonal Curvilinear Coordinates. *Am. J. Phys.*, **38,** 915 (1970).

McDonald, J. E. Maxwellian Interpretation of the Laplacian. *Am. J. Phys.*, **33,** 706 (1965).

Schilling, R. B. The "Del" Operator in Curvilinear Coordinates. *Am. J. Phys.*, **33,** 241 (1965).

Yang, H. T. A Direct Method of Vector Differentiation in Orthogonal Curvilinear Coordinates. *Am. J. Phys.*, **40,** 109 (1972).

13 / Material Continua—Fluids

13.1 Smoothing out molecular effects

To our eyes and visible probes a physical material appears continuous. From other evidence, though, we know that the substance consists of molecules moving back and forth with varying velocities.

So any visible small part contains a large number of more or less independent particles. During any significant interval of time, many of these molecules bombard a typical bordering region, or enter the region, to be replaced by others with similar properties.

The net result is some transfer of momentum and of material. Furthermore, some energy may be dissipated. A detailed analysis of these processes is impossible; there are simply too many particles.

Instead, let us represent each substance as a medium that varies continuously almost everywhere in time and space. The mass density at each point is assumed to be the mass of a small representative region surrounding the point divided by the volume of the region, the momentum density at the point is defined as the net momentum of matter in the small region divided by its volume, and the drift velocity of the fluid is the momentum density divided by the mass density.

Interactions with neighboring regions not assignable to a net transfer of material are attributed to stresses, or forces per unit area, acting within the material. In the hypothetical medium, the stresses vary smoothly with position, time, and orientation of the pertinent boundary. The interactions that they represent correlate with the strains and the spatial rates of change of strains in the material. Hysteresis effects may also be present; so the behavior of the element can depend on its past history.

The definitions of mass density and momentum density have been chosen so each element of the hypothetical medium obeys the conservation-of-mass law and Newton's force law. But application of these principles leads to simultaneous partial differential equations that are nonlinear.

Solutions are consequently possible only for very simple configurations and models. In general, we must be guided by experimental results. High flow rates by a solid object generally cause turbulence. Whenever this is present, the solution is not unique and the effects have a lower symmetry than their causes.

13.2   Conservation of mass

To a probe that averages over molecular effects, a given physical material appears continuous. The mass and momentum seem to be distributed uniformly over each small region that does not include a surface on which a near-discontinuity is maintained.

Consistent with such observations, each substance is represented as a smoothly varying medium. The mass and momentum of every small visible section are assigned to an element of the medium having the same size and shape. The forces that would cause the observed changes in momentum of the real section, if there were no interchange of molecules across any surface moving with the average mass of the pertinent part, are assumed to be acting on the hypothetical element.

Furthermore, the momentum density across any small surface in the physical region equals that across the corresponding surface in the continuous medium. The mass transported by this vector out of a small physical volume equals that transported out of the corresponding element in the continuous medium during the same interval of time. Conservation of mass of the physical material implies conservation of mass of the hypothetical continuous fluid.

If ρ is the mass density and $\rho\mathbf{v}$ the momentum density within the fluid, the motion must satisfy equation (12-20):

$$\frac{\partial \rho}{\partial t} + \nabla \cdot \rho\mathbf{v} = 0. \tag{13-1}$$

Note that the momentum density divided by the mass density gives the velocity $\mathbf{v}$ at which the material appears to drift.

In any given system, the mass density is a function of t, x_1, x_2, x_3. Therefore, its differential has the form

$$d\rho = \frac{\partial \rho}{\partial t}\, dt + \frac{\partial \rho}{\partial x_j}\, dx_j \tag{13-2}$$

while the corresponding total derivative is

$$\frac{d\rho}{dt} = \frac{\partial \rho}{\partial t} + \frac{dx_j}{dt}\frac{\partial \rho}{\partial x_j}. \tag{13-3}$$

This derivative applies to an infinitesimal element of the fluid moving at its velocity $\mathbf{v}$ when x_1, x_2, x_3 are coordinates locating the element. Then

$$\frac{dx_j}{dt} = v_j \tag{13-4}$$

and

$$\frac{D\rho}{Dt} = \frac{\partial \rho}{\partial t} + v_j \frac{\partial \rho}{\partial x_j} = \frac{\partial \rho}{\partial t} + \mathbf{v} \cdot \nabla\rho. \tag{13-5}$$

The capital D's indicate that differentiation is taken following a given element of matter. The corresponding derivative is called the *material derivative*.

Expanding the second term in (13-1) as in example 12.4 leads to

$$\frac{\partial \rho}{\partial t} + \mathbf{v} \cdot \nabla \rho + \rho \nabla \cdot \mathbf{v} = 0, \tag{13-6}$$

whence

$$\frac{D\rho}{Dt} + \rho \nabla \cdot \mathbf{v} = 0. \tag{13-7}$$

EXAMPLE 13.1   Obtain the continuity equation in orthogonal generalized coordinates by introducing such coordinates into the derivation.

Let independent variables q_1, q_2, and q_3 be chosen so that (a) a given triplet identifies a point in space and (b) a change in the first quantity by dq_1, in the second by dq_2, and in the third by dq_3 produces the displacement

$$d\mathbf{s} = \mathbf{g}_j \, dq_j.$$

(Because we have no need to distinguish between contravariant and covariant expressions, each index will be placed in the same position, the lower one.)

When only q_1 varies, differentials dq_2 and dq_3 are zero and we have

$$d\mathbf{s}_1 = \mathbf{g}_1 \, dq_1 = \mathbf{e}_1 g_1 \, dq_1$$

where $\mathbf{e}_1$ is the unit vector pointing along $\mathbf{g}_1$. Similarly, a change in only q_2 yields

$$d\mathbf{s}_2 = \mathbf{g}_2 \, dq_2 = \mathbf{e}_2 g_2 \, dq_2,$$

while a change in only q_3 gives

$$d\mathbf{s}_3 = \mathbf{g}_3 \, dq_3 = \mathbf{e}_3 g_3 \, dq_3.$$

Since moving each of these line elements a differential distance in any direction changes the element by a differential of higher order, displacing four of each kind symmetrically from point (q_1, q_2, q_3), as figure 13.1 shows, produces an approximate parallelepiped surrounding the point.

Wherever $\mathbf{e}_1$, $\mathbf{e}_2$, and $\mathbf{e}_3$ are mutually perpendicular, the coordinates are said to be *orthogonal*. Then the parallelepiped is rectangular and magnitudes $g_1 \, dq_1$, $g_2 \, dq_2$, $g_3 \, dq_3$ behave as elements dx_1', dx_2', dx_3' do in section 12.3. Furthermore, the physical components of velocity $\mathbf{v}$ from the expansion

$$\mathbf{v} = v_j \mathbf{e}_j$$

behave as v_1', v_2', v_3' do.

Then the rate at which fluid flows by the central point of the volume element in the $\mathbf{e}_1$ direction is

$$\rho v_1 g_2 g_3 \, dq_2 \, dq_3,$$

and the rate at which it flows in through the left face is

$$\left[\rho v_1 g_2 g_3 + \frac{\partial(\rho v_1 g_2 g_3)}{g_1 \, \partial q_1} \left(-\frac{g_1 \, dq_1}{2} \right) \right] dq_2 \, dq_3,$$

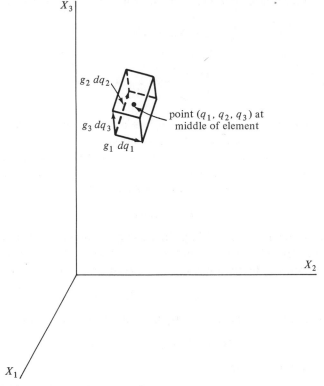

FIGURE 13.1   Volume element bordered by mutually perpendicular line elements.

while the rate at which it flows out through the right face is

$$\left[\rho v_1 g_2 g_3 + \frac{\partial(\rho v_1 g_2 g_3)}{g_1 \, \partial q_1}\left(+\frac{g_1 \, dq_1}{2}\right)\right] dq_2 \, dq_3 .$$

In unit time, the travel through these parallel faces causes the mass decrease

$$\frac{\partial(\rho v_1 g_2 g_3)}{g_1 g_2 g_3 \, \partial q_1} \, g_1 g_2 g_3 \, dq_1 \, dq_2 \, dq_3$$

if the rates do not change. Similar terms exist for the mass decrease from travel through the other two sets of parallel faces. Adding these and dividing by the volume of the element yields the time rate of decrease in density:

$$\frac{1}{g_1 g_2 g_3}\left[\frac{\partial(\rho g_2 g_3 v_1)}{\partial q_1} + \frac{\partial(\rho g_3 g_1 v_2)}{\partial q_2} + \frac{\partial(\rho g_1 g_2 v_3)}{\partial q_3}\right] = -\frac{\partial \rho}{\partial t} .$$

On comparing this equation with (13-1), we see that the left side is a representation of $\nabla \cdot \rho\mathbf{v}$. Note that q_1, q_2, and q_3 are orthogonal generalized coordinates; g_1, g_2, and g_3 are the coefficients of dq_1, dq_2, and dq_3 in the expressions for the perpendicular infinitesimal displacements; v_1, v_2, and v_3 are the corresponding physical components of velocity $\mathbf{v}$; and ρ is the density.

EXAMPLE 13.2   How do (a) steadiness of the flow and (b) incompressibility of the fluid simplify the equation of continuity?

A steady flow does not alter conditions at any fixed point by which the fluid moves. At such a point

$$\frac{\partial \rho}{\partial t} = 0$$

and equation (13-1) becomes

$$\nabla \cdot \rho\mathbf{v} = 0.$$

A fluid is said to be incompressible whenever it moves without appreciable alteration of its density. Then

$$\frac{D\rho}{Dt} = 0$$

and equation (13-7) becomes

$$\nabla \cdot \mathbf{v} = 0.$$

13.3   The vorticity and rate-of-strain dyadics

No matter how small it is, an element of fluid possesses extension. As a consequence, it can undergo rotation, shear deformation, and dilatation, as well as translation (see figures 13.2 through 13.5). These processes involve the spatial rates of change in the components of velocity $\mathbf{v}$.

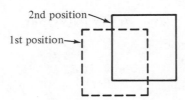

FIGURE 13.2   Translation of the cross section of a rectangular element of fluid in a given time interval.

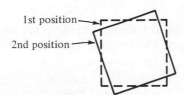

FIGURE 13.3   Rotation of the cross section.

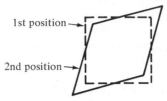

FIGURE 13.4   Shear deformation of the cross section.

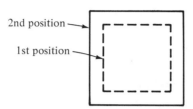

FIGURE 13.5   Dilatation of the cross section.

Let us start with the assumption that the drift velocity $\mathbf{v}$ is a well-behaved function of the coordinates at any given time. Then varying x_1 by dx_1, x_2 by dx_2, and x_3 by dx_3 changes the kth component of $\mathbf{v}$ by the amount

$$dv_k = dx_j \frac{\partial v_k}{\partial x_j} = d\mathbf{r} \cdot \nabla v_k. \tag{13-8}$$

Multiplying each component by the corresponding unit vector and adding yields the increase in vector $\mathbf{v}$ itself

$$d\mathbf{v} = d\mathbf{r} \cdot \nabla \mathbf{v} = (\widetilde{\nabla \mathbf{v}}) \cdot d\mathbf{r}. \tag{13-9}$$

Expression $\nabla \mathbf{v}$ is the dyadic

$$\mathbf{e}_j \frac{\partial}{\partial x_j} v_k \mathbf{e}_k, \tag{13-10}$$

while expression $\widetilde{\nabla \mathbf{v}}$ is the transpose

$$\frac{\partial}{\partial x_j} v_k \mathbf{e}_k \mathbf{e}_j. \tag{13-11}$$

Note how $\widetilde{\nabla \mathbf{v}}$ acts to change the displacement vector $d\mathbf{r}$ to the velocity difference $d\mathbf{v}$. As equation (1-52) was in (3-81), the over-all equation (13-9) can be interpreted as a matrix equation. Then we would write

$$d\mathbf{v} = (\widetilde{\nabla \mathbf{v}}) \, d\mathbf{r}, \tag{13-12}$$

and have

$$d\mathbf{v} = \begin{pmatrix} dv_1 \\ dv_2 \\ dv_3 \end{pmatrix}, \qquad d\mathbf{r} = \begin{pmatrix} dx_1 \\ dx_2 \\ dx_3 \end{pmatrix}, \tag{13-13}$$

$$\widetilde{\nabla \mathbf{v}} = \begin{pmatrix} \dfrac{\partial v_1}{\partial x_1} & \dfrac{\partial v_1}{\partial x_2} & \dfrac{\partial v_1}{\partial x_3} \\[2mm] \dfrac{\partial v_2}{\partial x_1} & \dfrac{\partial v_2}{\partial x_2} & \dfrac{\partial v_2}{\partial x_3} \\[2mm] \dfrac{\partial v_3}{\partial x_1} & \dfrac{\partial v_3}{\partial x_2} & \dfrac{\partial v_3}{\partial x_3} \end{pmatrix}. \tag{13-14}$$

As long as a given matrix is square, we can add it to minus or plus its transpose. We can set

$$\mathbf{A} = \tfrac{1}{2}(\mathbf{A} - \tilde{\mathbf{A}}) + \tfrac{1}{2}(\mathbf{A} + \tilde{\mathbf{A}}). \tag{13-15}$$

The first term on the right forms an antisymmetric matrix; the last term a symmetric matrix. Expression $\widetilde{\nabla \mathbf{v}}$ has the structure

$$\widetilde{\nabla \mathbf{v}} = \mathbf{\Omega} + \boldsymbol{\varepsilon} \tag{13-16}$$

with the antisymmetric part

$$\mathbf{\Omega} = \begin{pmatrix} 0 & \dfrac{1}{2}\left(\dfrac{\partial v_1}{\partial x_2} - \dfrac{\partial v_2}{\partial x_1}\right) & \dfrac{1}{2}\left(\dfrac{\partial v_1}{\partial x_3} - \dfrac{\partial v_3}{\partial x_1}\right) \\[3mm] \dfrac{1}{2}\left(\dfrac{\partial v_2}{\partial x_1} - \dfrac{\partial v_1}{\partial x_2}\right) & 0 & \dfrac{1}{2}\left(\dfrac{\partial v_2}{\partial x_3} - \dfrac{\partial v_3}{\partial x_2}\right) \\[3mm] \dfrac{1}{2}\left(\dfrac{\partial v_3}{\partial x_1} - \dfrac{\partial v_1}{\partial x_3}\right) & \dfrac{1}{2}\left(\dfrac{\partial v_3}{\partial x_2} - \dfrac{\partial v_2}{\partial x_3}\right) & 0 \end{pmatrix} \tag{13-17}$$

and the symmetric part

$$\boldsymbol{\varepsilon} = \begin{pmatrix} \dfrac{\partial v_1}{\partial x_1} & \dfrac{1}{2}\left(\dfrac{\partial v_1}{\partial x_2} + \dfrac{\partial v_2}{\partial x_1}\right) & \dfrac{1}{2}\left(\dfrac{\partial v_1}{\partial x_3} + \dfrac{\partial v_3}{\partial x_1}\right) \\[3mm] \dfrac{1}{2}\left(\dfrac{\partial v_2}{\partial x_1} + \dfrac{\partial v_1}{\partial x_2}\right) & \dfrac{\partial v_2}{\partial x_2} & \dfrac{1}{2}\left(\dfrac{\partial v_2}{\partial x_3} + \dfrac{\partial v_3}{\partial x_2}\right) \\[3mm] \dfrac{1}{2}\left(\dfrac{\partial v_3}{\partial x_1} + \dfrac{\partial v_1}{\partial x_3}\right) & \dfrac{1}{2}\left(\dfrac{\partial v_3}{\partial x_2} + \dfrac{\partial v_2}{\partial x_3}\right) & \dfrac{\partial v_3}{\partial x_3} \end{pmatrix} \tag{13-18}$$

From sections 12.5 and 12.6, one-half the curl of $\mathbf{v}$ equals the angular velocity about the point where the partial derivatives are calculated. Cross multiplying this angular velocity with $d\mathbf{r}$, as equation (3-73) indicates, yields the velocity component involved in the rotation. Since such multiplication is effected by the matrix multiplication

$$\mathbf{\Omega}\,d\mathbf{r}, \tag{13-19}$$

the antisymmetric part of $\widetilde{\nabla \mathbf{v}}$ represents the rate of rotation of the element of fluid; it is called the *vorticity* tensor.

The symmetric part of $\widetilde{\nabla \mathbf{v}}$, on the other hand, is called the *rate-of-strain* tensor. Because its elements are also real, the discussion in example 4.14 applies; matrix $\boldsymbol{\varepsilon}$ has three mutually perpendicular eigenvectors. With respect to the corresponding axes, the matrix has the form

$$\left(\boldsymbol{\varepsilon} = \begin{matrix} \dfrac{\partial v_1'}{\partial x_1'} & 0 & 0 \\ 0 & \dfrac{\partial v_2'}{\partial x_2'} & 0 \\ 0 & 0 & \dfrac{\partial v_3'}{\partial x_3'} \end{matrix}\right) \equiv \begin{pmatrix} \varepsilon_1 & 0 & 0 \\ 0 & \varepsilon_2 & 0 \\ 0 & 0 & \varepsilon_3 \end{pmatrix}. \tag{13-20}$$

The trace is invariant to any change of axes. Therefore

$$\varepsilon_{11} + \varepsilon_{22} + \varepsilon_{33} = \varepsilon_1 + \varepsilon_2 + \varepsilon_3$$

$$= \frac{\partial}{\partial x_j}\, v_j = \frac{\partial}{\partial x_j'}\, v_j' = \nabla \cdot \mathbf{v} = \nabla' \cdot \mathbf{v}'. \tag{13-21}$$

The components ε_1, ε_2, and ε_3 of $\widetilde{\nabla \mathbf{v}}$ are called the *principal rates of strain*. The primed axes, with respect to which $\boldsymbol{\varepsilon}$ takes on form (13-20), are called the *principal axes* of the rate of strain.

EXAMPLE 13.3   How is the vorticity matrix related to the components of angular velocity about a given point?

From formulas (12-27) and (12-35), the angular velocity is one-half the curl of the particle velocity at a given position in a fluid. Therefore,

$$\omega = \dot{\varphi}\mathbf{n} = \tfrac{1}{2}\nabla \times \mathbf{v} = \frac{1}{2} \begin{vmatrix} \mathbf{e}_1 & \mathbf{e}_2 & \mathbf{e}_3 \\ \dfrac{\partial}{\partial x_1} & \dfrac{\partial}{\partial x_2} & \dfrac{\partial}{\partial x_3} \\ v_1 & v_2 & v_3 \end{vmatrix}$$

$$= \frac{1}{2}\left(\frac{\partial v_3}{\partial x_2} - \frac{\partial v_2}{\partial x_3}\right)\mathbf{e}_1 + \frac{1}{2}\left(\frac{\partial v_1}{\partial x_3} - \frac{\partial v_3}{\partial x_1}\right)\mathbf{e}_2 + \frac{1}{2}\left(\frac{\partial v_2}{\partial x_1} - \frac{\partial v_1}{\partial x_2}\right)\mathbf{e}_3$$

$$= \omega_1\mathbf{e}_1 + \omega_2\mathbf{e}_2 + \omega_3\mathbf{e}_3,$$

and formula (13-17) can be rewritten as

$$\Omega = \begin{pmatrix} 0 & -\omega_3 & \omega_2 \\ \omega_3 & 0 & -\omega_1 \\ -\omega_2 & \omega_1 & 0 \end{pmatrix}.$$

EXAMPLE 13.4 How does the matrix product $\Omega\, d\mathbf{r}$ accomplish the cross multiplication of ω and $d\mathbf{r}$?

With the result in example 13.3 and the second matrix in line (13-13), we find that

$$\Omega\, d\mathbf{r} = \begin{pmatrix} 0 & -\omega_3 & \omega_2 \\ \omega_3 & 0 & -\omega_1 \\ -\omega_2 & \omega_1 & 0 \end{pmatrix} \begin{pmatrix} dx_1 \\ dx_2 \\ dx_3 \end{pmatrix} = \begin{pmatrix} \omega_2\, dx_3 - \omega_3\, dx_2 \\ \omega_3\, dx_1 - \omega_1\, dx_3 \\ \omega_1\, dx_2 - \omega_2\, dx_1 \end{pmatrix}.$$

The resulting components agree with those calculated from

$$\omega \times d\mathbf{r} = \begin{vmatrix} \mathbf{e}_1 & \mathbf{e}_2 & \mathbf{e}_3 \\ \omega_1 & \omega_2 & \omega_3 \\ dx_1 & dx_2 & dx_3 \end{vmatrix}.$$

13.4 Force integrals

According to Newtonian theory, the time rate of change in momentum of any given region of a fluid must equal the net force causing the change. The momentum itself appears as a volume integral. However, only part of the force appears as a volume integral; the rest appears as a surface integral.

Forces that act throughout a volume include gravitational and electromagnetic forces. These are presumably mediated by a penetrating nonmaterial field. Forces that work across a surface include hydrostatic and viscous forces.

The mass of an infinitesimal element times its velocity, $(\rho\, dV)\mathbf{v}$, is the momentum of the element. Summing this momentum over the given section of material yields the *net momentum*

$$\int_{\bar{V}} \rho\mathbf{v}\, dV. \tag{13-22}$$

The bar over V indicates the volume is that of a definite amount of material, varying in position, shape, and extent with time.

If $\mathbf{f}$ is the force per unit mass acting on element dV, the *body force*, and $\mathbf{t}$ is the force per unit area, or *stress*, acting on the boundary dS of the pertinent element, the net forces acting are

$$\int_{\bar{V}} \rho\mathbf{f}\, dV + \int_{\bar{S}} \mathbf{t}\, dS. \tag{13-23}$$

Here $\bar{S}$ indicates that the corresponding integration proceeds over the surface of the material being considered. This surface changes, in general, with time.

Introducing Newton's second law then gives us

$$\frac{D}{Dt}\int_V \rho\mathbf{v}\,dV = \int_V \rho\mathbf{f}\,dV + \int_S \mathbf{t}\,dS. \tag{13-24}$$

But since the mass of the material is conserved,

$$\frac{D}{Dt}\int_V \rho\,dV = 0, \tag{13-25}$$

equation (13-24) can be rewritten as

$$\int_V \frac{D\mathbf{v}}{Dt}\rho\,dV = \int_V \rho\mathbf{f}\,dV + \int_S \mathbf{t}\,dS. \tag{13-26}$$

13.5   The stress dyadic

The traction or stress vector $\mathbf{t}$ acting on a given bounding surface $d\mathbf{S}$ is the component of a dyadic in the direction of $d\mathbf{S}$. This dyadic varies from point to point and is symmetric.

Consider an element of the fluid bounded by three mutually perpendicular planes and a fourth plane placed as in figure 13.6. Let $\mathbf{t}_1$ be the stress on bounding plane

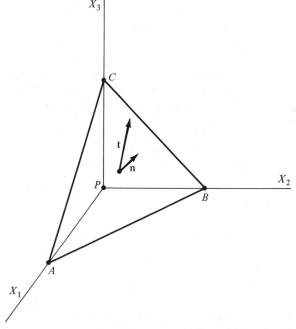

FIGURE 13.6   A tetrahedron of fluid with three mutually perpendicular faces.

PCB, t_2 the stress on PAC, t_3 the stress on PAB, and t the stress on ABC. Also let $-\sigma_{jk}$ be the projection of vector t_j on base vector e_k:

$$t_j = -\sigma_{jk}e_k. \tag{13-27}$$

Let the area of ABC be designated S. Then the area of PCB is the projection of this area on e_1, namely $n \cdot e_1 S$; the area of PAC is $n \cdot e_2 S$; and the area of PAB is $n \cdot e_3 S$. As long as the volume of the tetrahedron is small, the stresses are approximately constant over each face and

$$\int_S t \, dS = tS + n \cdot e_j S t_j \simeq 0. \tag{13-28}$$

In the limit, when the volume is made infinitesimal, the approximations become exact and

$$t = -n \cdot e_j t_j = n \cdot e_j \sigma_{jk} e_k \equiv n \cdot \sigma. \tag{13-29}$$

The final factor is a dyadic with the matrix

$$\begin{pmatrix} \sigma_{11} & \sigma_{12} & \sigma_{13} \\ \sigma_{21} & \sigma_{22} & \sigma_{23} \\ \sigma_{31} & \sigma_{32} & \sigma_{33} \end{pmatrix}. \tag{13-30}$$

From equation (13-29), the kth component of stress that material outside a given element exerts on a small planar face with outward pointing unit normal n is

$$t_k = n \cdot e_j \sigma_{jk}. \tag{13-31}$$

If n is e_j, then the kth component of stress is σ_{jk}, while if n is $-e_j$, the kth component of stress is $-\sigma_{jk}$.

In many discussions, we look at a small rectangular parallelepiped of fluid oriented as in figure 13.7 at a given instant of time. On such an element, components σ_{11}, σ_{12}, and σ_{13} appear as in figure 13.8.

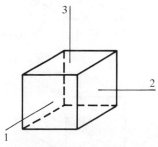

FIGURE 13.7 Cartesian axes for a small rectangular parallelepiped of fluid at a given instant.

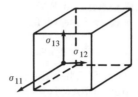

FIGURE 13.8   Components of stress acting on the front face of the parallelepiped.

As long as the σ_{jk}'s vary continuously, matrix σ is symmetric; for, paired with a tangential component of stress on a given face is one on the opposite face. Paired with this couple is a couple on mutually perpendicular faces (see figures 13.9 and 13.10). When the parallelepiped is made small enough, the two stresses in figure 13.9 become equal in magnitude. Similarly, those in figure 13.10 become equal in magnitude. Furthermore, the moment exerted by those in figure 13.9 must be counterbalanced by the moment from figure 13.10.

Such balancing is possible only if

$$\sigma_{21} = \sigma_{12}. \tag{13-32}$$

In general, then, we must have

$$\sigma_{jk} = \sigma_{kj}. \tag{13-33}$$

Since all elements σ_{jk} are real, the discussion in example 4.14 applies. The stress matrix σ has three mutually perpendicular eigenvectors. Transforming to a coordinate system in which such eigenvectors are base vectors, we obtain the diagonal form

$$\sigma = \begin{pmatrix} \sigma_1 & 0 & 0 \\ 0 & \sigma_2 & 0 \\ 0 & 0 & \sigma_3 \end{pmatrix}. \tag{13-34}$$

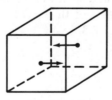

FIGURE 13.9   Component σ_{12} acting on the front face and its twin acting on the back face.

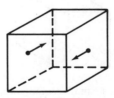

FIGURE 13.10   Component σ_{21} and its twin.

The stresses σ_1, σ_2, and σ_3 that act in the same directions as the normal vectors are called the *principal stresses*. The corresponding Cartesian axes are the principal axes for σ. Since the trace is invariant in any reorientation of axes, we have

$$\sigma_1 + \sigma_2 + \sigma_3 = \sigma_{11} + \sigma_{22} + \sigma_{33}. \tag{13-35}$$

EXAMPLE 13.5   Show that whenever the three principal stresses are equal, matrix σ is diagonal in all rectangular coordinate systems.

Let σ be the stress matrix in an arbitrary Cartesian coordinate system. The set of eigenvalue equations then exhibits form (4-104)

$$\sigma U = U \Lambda$$

in which

$$\Lambda = \begin{pmatrix} \sigma_1 & 0 & 0 \\ 0 & \sigma_2 & 0 \\ 0 & 0 & \sigma_3 \end{pmatrix}.$$

When the eigenvalues are all equal to σ, we then have

$$\Lambda = \begin{pmatrix} \sigma & 0 & 0 \\ 0 & \sigma & 0 \\ 0 & 0 & \sigma \end{pmatrix} = \sigma E.$$

In the physical problem, the eigenvectors are three mutually perpendicular vectors. Consequently, U^{-1} exists and form (4-104) yields

$$\sigma = U \Lambda U^{-1} = U \sigma E U^{-1} = U U^{-1} \sigma E = \sigma E.$$

We see that σ also has the diagonal structure.

13.6   A general equation of motion

The integral of surface forces can now be converted to a volume integral, and a differential equation describing the motion deduced.

Equation (13-29) describes the stress acting on a small element of surface as dot product of the unit normal $\mathbf{n}$ with the stress dyadic σ. Since σ is symmetric, the two factors can be commuted.

$$\mathbf{t} = \mathbf{n} \cdot \sigma = \sigma \cdot \mathbf{n}. \tag{13-36}$$

But $\mathbf{n}$ is an outward pointing normal to the element dS on which $\mathbf{t}$ acts, so

$$\mathbf{t} \, dS = \sigma \cdot \mathbf{n} \, dS = \sigma \cdot d\mathbf{S}. \tag{13-37}$$

To convert the integral of this expression to a volume integral, we first multiply it by an arbitrary constant vector $\mathbf{b}$:

$$\mathbf{b} \cdot \mathbf{t} \, dS = \mathbf{b} \cdot \sigma \cdot d\mathbf{S}. \tag{13-38}$$

The final expression involves vector $\mathbf{b} \cdot \boldsymbol{\sigma}$ dotted with $d\mathbf{S}$, the same kind of expression that occurs in Gauss's theorem. Applying (12-26), we then obtain

$$\int_S \mathbf{b} \cdot \mathbf{t} \, dS = \int_S \mathbf{b} \cdot \boldsymbol{\sigma} \cdot d\mathbf{S} = \int_V \nabla \cdot \mathbf{b} \cdot \boldsymbol{\sigma} \, dV$$

$$= \int_V \mathbf{b} \cdot \nabla \cdot \boldsymbol{\sigma} \, dV. \tag{13-39}$$

Since $\mathbf{b}$ is arbitrary, equation (13-39) demands that

$$\int_S \mathbf{t} \, dS = \int_S \boldsymbol{\sigma} \cdot d\mathbf{S} = \int_V \nabla \cdot \boldsymbol{\sigma} \, dV; \tag{13-40}$$

so equation (13-26) reduces to

$$\int_V \frac{D\mathbf{v}}{Dt} \rho \, dV = \int_V (\rho \mathbf{f} + \nabla \cdot \boldsymbol{\sigma}) \, dV. \tag{13-41}$$

Because the choice of material volume $\overline{V}$ is arbitrary, the integrands must be equal:

$$\rho \frac{D\mathbf{v}}{Dt} = \rho \mathbf{f} + \nabla \cdot \boldsymbol{\sigma}. \tag{13-42}$$

This *equation of motion* was obtained by Augustin L. Cauchy.

Combining the equation of continuity with Cauchy's equation does not generally lead to a solution; for there are too many dependent variables.

In neutral substances, the body force per unit mass is merely $\mathbf{g}$, the acceleration due to gravity. Where charges are separated, $\mathbf{f}$ also involves electric and magnetic forces, but we will not be concerned with these here. The stresses needed to maintain a certain drift-velocity field around a point are determined by properties of the material. A realistic description of these was first given by George G. Stokes.

13.7   Dependence of stresses on rates of strain

Following Stokes, let us consider a moving material with the following properties. Experiments indicate that these are satisfactory as long as conditions are not too extreme.

1. The substance can be treated as a continuous medium governed by an equation of state relating hydrostatic pressure at each point to the density and temperature there.

2. No element of the substance resists deformation when the rate of strain is negligible.

3. Otherwise, the stresses acting on a given element produce unique rates of strain in the element.

4. Therefore, stresses of a given simple symmetry produce rates of strain of the same type.

5. The resting material is isotropic.

6. As a consequence, the principal axes of the stress matrix σ and the rate-of-strain matrix ε coincide.

Invariant to all reorientations of axes are the traces of matrices σ and ε

$$\sigma_{11} + \sigma_{22} + \sigma_{33}, \tag{13-43}$$

$$\varepsilon_{11} + \varepsilon_{22} + \varepsilon_{33}. \tag{13-44}$$

Therefore, forms (13-43) and (13-44) are completely symmetric, as the pressure p is. They belong to the completely symmetric irreducible representation.

According to assumptions 3 and 4, one of these determines the other. But in a resting fluid

$$\varepsilon_{11} = \varepsilon_{22} = \varepsilon_{33} = 0 \tag{13-45}$$

and

$$\sigma_{11} = \sigma_{22} = \sigma_{33} = -p. \tag{13-46}$$

The general relationship is

$$\sigma_{11} + \sigma_{22} + \sigma_{33} = -3p + 3\kappa(\varepsilon_{11} + \varepsilon_{22} + \varepsilon_{33}). \tag{13-47}$$

Coefficient κ is different from zero when dissipation of energy is caused by dilation or compression. It is called the *bulk coefficient of viscosity*.

The array of stresses contains no contributions from other bases only when it is completely symmetric. Such symmetry exists only when

$$\sigma_{11} = \sigma_{22} = \sigma_{33}, \tag{13-48}$$

$$\sigma_{12} = \sigma_{23} = \sigma_{31} = 0, \tag{13-49}$$

and the array appears as in figure 13.11.

The expressions

$$\sigma_{11} - \sigma_{22}, \tag{13-50}$$

$$\sigma_{11} - \sigma_{33}, \tag{13-51}$$

$$\sigma_{22} - \sigma_{33} \tag{13-52}$$

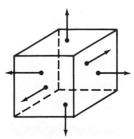

FIGURE 13.11 Array of stresses that is unchanged by each operation of **T** that changes the small cube into itself.

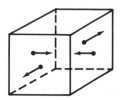

FIGURE 13.12   Array of stresses that is either unchanged or altered in sign by each operation of $\mathbf{D}_4$ that changes the small element into itself.

therefore measure contributions from other bases. These expressions are completely different when they do not contribute to the sum (13-43); that is, when

$$\sigma_{22} = -\sigma_{11}, \tag{13-53}$$

or

$$\sigma_{33} = -\sigma_{11}, \tag{13-54}$$

or

$$\sigma_{33} = -\sigma_{22}. \tag{13-55}$$

A typical array then appears as in figure 13.12. With the small element of fluid a cube, the given array goes into plus or minus itself under each operation of $\mathbf{D}_4$. Therefore, it is a basis for an irreducible representation of the group.

According to assumption 4, such an array determines a similar combination of ε_{jk}'s. We write

$$\sigma_{11} - \sigma_{22} = 2\mu_1(\varepsilon_{11} - \varepsilon_{22}). \tag{13-56}$$

If the resting material is invariant to interchange of axes fixed in it, we also have

$$\sigma_{11} - \sigma_{33} = 2\mu_1(\varepsilon_{11} - \varepsilon_{33}), \tag{13-57}$$

$$\sigma_{22} - \sigma_{33} = 2\mu_1(\varepsilon_{22} - \varepsilon_{33}). \tag{13-58}$$

Coefficient μ_1 is different from zero when dissipation of energy accompanies the motion imposed by the stresses of figure 13.12.

Differences (13-57) and (13-58) are not independent of difference (13-56) and sum (13-47); but the sum of (13-57) and (13-58)

$$\sigma_{11} + \sigma_{22} - 2\sigma_{33} = 2\mu_1(\varepsilon_{11} + \varepsilon_{22} - 2\varepsilon_{33}) \tag{13-59}$$

is independent of (13-56) and (13-47).

Adding twice equation (13-47), thrice equation (13-56), and equation (13-59), dividing by 6,

$$\sigma_{11} = -p + (\kappa - \tfrac{2}{3}\mu_1)(\varepsilon_{11} + \varepsilon_{22} + \varepsilon_{33}) + 2\mu_1\varepsilon_{11}, \tag{13-60}$$

then introducing equation (13-21) leads to

$$\sigma_{11} = -p + (\kappa - \tfrac{2}{3}\mu_1)\nabla \cdot \mathbf{v} + 2\mu_1\varepsilon_{11}, \tag{13-61}$$

while solving for the other diagonal elements yields

$$\sigma_{22} = -p + (\kappa - \tfrac{2}{3}\mu_1)\nabla \cdot \mathbf{v} + 2\mu_1\varepsilon_{22}, \tag{13-62}$$

$$\sigma_{33} = -p + (\kappa - \tfrac{2}{3}\mu_1)\nabla \cdot \mathbf{v} + 2\mu_1\varepsilon_{33}. \tag{13-63}$$

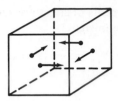

FIGURE 13.13   Another array of stresses that is either unchanged or altered in sign by each operation of the pertinent $\mathbf{D_4}$ group.

Formulas (13-61), (13-62), and (13-63) do not involve the off-diagonal elements. But as long as the stress matrix is symmetric, we have

$$\sigma_{12} = \sigma_{21}, \tag{13-64}$$

$$\sigma_{23} = \sigma_{32}, \tag{13-65}$$

$$\sigma_{31} = \sigma_{13}. \tag{13-66}$$

The array corresponding to (13-64), which is typical of this set, appears in figure 13.13. Since this array goes into plus or minus itself under each operation of $\mathbf{D_4}$, it is a basis for an irreducible representation of the group.

According to assumption 4, such an array determines a similar combination of ε_{jk}'s: We have

$$\sigma_{12} = 2\mu_2\varepsilon_{12}. \tag{13-67}$$

Similarly for the other two configurations, we have

$$\sigma_{23} = 2\mu_2\varepsilon_{23}, \tag{13-68}$$

$$\sigma_{31} = 2\mu_2\varepsilon_{31}. \tag{13-69}$$

The same coefficient μ_2 is involved in all three relationships as long as the resting material is invariant to interchange of axes fixed in it. This coefficient is different from zero when dissipation of energy accompanies the motion imposed by the stresses.

Rotating the stresses in figure 13.12 by 45° around axis 3 converts the array to the type in figure 13.13. If the resting material is symmetric to this transformation, equation (13-56) is not essentially different from (13-67), and

$$\mu_1 = \mu_2 \equiv \mu. \tag{13-70}$$

Coefficient μ is called the *shear coefficient of viscosity*.

In ordinary liquids and gases, each array of stresses that is a basis for a given irreducible representation produces a rate of strain belonging to the same representation. The function of σ_{jk}'s that generates the representation depends only on the corresponding function of ε_{jk}'s when the temperature and pressure are fixed.

EXAMPLE 13.6   What happens to the stress dyadic depicted in figure 13.12 when it is rotated by 45° around the third axis?

In the initial array, all components of stress are zero except σ_{11} and σ_{22}, which are equal in magnitude but opposite in sign

$$\sigma_{11} = -\sigma_{22} = a.$$

The dyadic then has the matrix

$$\sigma = \begin{pmatrix} a & 0 & 0 \\ 0 & -a & 0 \\ 0 & 0 & 0 \end{pmatrix}.$$

To rotate this dyadic by 45°, we apply law (4-71) in the form

$$\sigma' = \mathbf{C}_8 \sigma \mathbf{C}_8{}^{-1}$$

with the matrix for rotating a vector by 45° obtained from equation (4-29)

$$\mathbf{C}_8 = \begin{pmatrix} \dfrac{1}{\sqrt{2}} & -\dfrac{1}{\sqrt{2}} & 0 \\ \dfrac{1}{\sqrt{2}} & \dfrac{1}{\sqrt{2}} & 0 \\ 0 & 0 & 1 \end{pmatrix}.$$

The result is

$$\sigma' = \begin{pmatrix} \dfrac{1}{\sqrt{2}} & -\dfrac{1}{\sqrt{2}} & 0 \\ \dfrac{1}{\sqrt{2}} & \dfrac{1}{\sqrt{2}} & 0 \\ 0 & 0 & 1 \end{pmatrix} \begin{pmatrix} a & 0 & 0 \\ 0 & -a & 0 \\ 0 & 0 & 0 \end{pmatrix} \begin{pmatrix} \dfrac{1}{\sqrt{2}} & \dfrac{1}{\sqrt{2}} & 0 \\ -\dfrac{1}{\sqrt{2}} & \dfrac{1}{\sqrt{2}} & 0 \\ 0 & 0 & 1 \end{pmatrix}$$

$$= \begin{pmatrix} \dfrac{1}{\sqrt{2}} & -\dfrac{1}{\sqrt{2}} & 0 \\ \dfrac{1}{\sqrt{2}} & \dfrac{1}{\sqrt{2}} & 0 \\ 0 & 0 & 1 \end{pmatrix} \begin{pmatrix} \dfrac{a}{\sqrt{2}} & \dfrac{a}{\sqrt{2}} & 0 \\ \dfrac{a}{\sqrt{2}} & -\dfrac{a}{\sqrt{2}} & 0 \\ 0 & 0 & 0 \end{pmatrix}$$

$$= \begin{pmatrix} 0 & a & 0 \\ a & 0 & 0 \\ 0 & 0 & 0 \end{pmatrix}.$$

Note that all elements are zero except σ_{12}' and σ_{21}', which equal a. The corresponding array appears in figure 13.13.

13.8 The Navier-Stokes equation

The formulas just obtained for a Stokesian fluid serve to eliminate $\boldsymbol{\sigma}$ from Cauchy's equation (13-42). Approximations then reduce the equation enough so that some solutions can be obtained.

On examining lines (13-61) through (13-70), we see that the expression

$$-p + (\kappa - \tfrac{2}{3}\mu)\nabla \cdot \mathbf{v} \tag{13-71}$$

contributes only to diagonal elements of $\boldsymbol{\sigma}$. On the other hand, the expression

$$2\mu\varepsilon_{jk} \tag{13-72}$$

contributes to each jkth element. Furthermore, we have

$$2\varepsilon = \nabla\mathbf{v} + \widetilde{\nabla\mathbf{v}}. \tag{13-73}$$

Consequently, the stress dyadic for a Stokesian fluid possesses the form

$$\boldsymbol{\sigma} = [-p + (\kappa - \tfrac{2}{3}\mu)\nabla \cdot \mathbf{v}]\mathbf{E} + \mu(\nabla\mathbf{v} + \widetilde{\nabla\mathbf{v}}). \tag{13-74}$$

In Cauchy's equation (13-42), dyadic $\boldsymbol{\sigma}$ appears as the operand of del dot. But operator $\nabla\cdot$ transforms the first term on the right of (13-74) to a gradient.

$$\nabla \cdot (-p\mathbf{E}) = \mathbf{e}_j \frac{\partial}{\partial x_j} \cdot (-p)\mathbf{e}_k\mathbf{e}_k = -\mathbf{e}_j \cdot \mathbf{e}_k\mathbf{e}_k \frac{\partial p}{\partial x_j}$$

$$= -\mathbf{e}_k \frac{\partial p}{\partial x_k} = -\nabla p. \tag{13-75}$$

Similarly,

$$\nabla \cdot [(\kappa - \tfrac{2}{3}\mu)\nabla \cdot \mathbf{v}\mathbf{E}] = \nabla[(\kappa - \tfrac{2}{3}\mu)\nabla \cdot \mathbf{v}]. \tag{13-76}$$

We also have

$$\nabla \cdot \nabla\mathbf{v} = \nabla^2\mathbf{v} \tag{13-77}$$

and

$$\nabla \cdot \widetilde{\nabla\mathbf{v}} = \mathbf{e}_j \frac{\partial}{\partial x_j} \cdot \mathbf{e}_k \frac{\partial}{\partial x_l} v_k\mathbf{e}_l = \mathbf{e}_j \cdot \mathbf{e}_k\mathbf{e}_l \frac{\partial}{\partial x_j} \frac{\partial}{\partial x_l} v_k$$

$$= \mathbf{e}_l \frac{\partial}{\partial x_l} \frac{\partial}{\partial x_j} v_j = \nabla\nabla \cdot \mathbf{v}. \tag{13-78}$$

Therefore,

$$\nabla \cdot [\mu(\nabla\mathbf{v} + \widetilde{\nabla\mathbf{v}})] = (\nabla\mu) \cdot (\nabla\mathbf{v} + \widetilde{\nabla\mathbf{v}}) + \mu(\nabla^2\mathbf{v} + \nabla\nabla \cdot \mathbf{v}). \tag{13-79}$$

These relationships reduce equation (13-42) to the *Navier-Stokes equation of motion*.

$$\rho \frac{D\mathbf{v}}{Dt} = \rho\mathbf{f} - \nabla p + \nabla[(\kappa - \tfrac{2}{3}\mu)\nabla \cdot \mathbf{v}] + (\nabla\mu) \cdot (\nabla\mathbf{v} + \widetilde{\nabla\mathbf{v}}) + \mu(\nabla^2\mathbf{v} + \nabla\nabla \cdot \mathbf{v}).$$

$$\tag{13-80}$$

In regions where density changes are not fast enough to cause appreciable losses, the coefficient of bulk viscosity is negligible and we may assume

$$\kappa \simeq 0. \tag{13-81}$$

Where temperature and density are not varying enough to affect viscosity μ appreciably, we also have

$$\nabla\mu \simeq 0. \tag{13-82}$$

Furthermore, when the body forces are caused by gravitational effects alone, vector $\mathbf{f}$ is the acceleration due to gravity

$$\mathbf{f} = \mathbf{g}. \tag{13-83}$$

Then equation (13-80) reduces to

$$\rho \frac{D\mathbf{v}}{Dt} = \rho\mathbf{g} - \nabla p + \mu\nabla^2\mathbf{v} + \tfrac{1}{3}\mu\nabla\nabla \cdot \mathbf{v}, \tag{13-84}$$

a result attributed to Louis M. H. Navier.

Because of the highly complicated behavior that equation (13-84) allows, most workers have introduced an additional approximation, that the fluid density is constant. Then the equation of continuity reduces to

$$\nabla \cdot \mathbf{v} = 0, \tag{13-85}$$

and equation (13-84) reduces to

$$\rho \frac{D\mathbf{v}}{Dt} = \rho\mathbf{g} - \nabla p + \mu\nabla^2\mathbf{v}. \tag{13-86}$$

From the third identity in problem 12.4, we know that

$$\nabla^2\mathbf{v} = \nabla(\nabla \cdot \mathbf{v}) - \nabla \times (\nabla \times \mathbf{v}). \tag{13-87}$$

So when equation (13-85) is valid, we also have

$$\rho \frac{D\mathbf{v}}{Dt} = \rho\mathbf{g} - \nabla p - \mu\nabla \times (\nabla \times \mathbf{v}). \tag{13-88}$$

The first term on the right can be expressed as an operation on the height ζ of the pertinent element in the gravitational field. Let $\Delta\zeta$ be the increase in height when the jth Cartesian coordinate increases by Δx_j, as in figure 13.14. The cosine of the angle between the vertical and the direction in which only x_j increases is then

$$\cos\theta = \frac{\Delta\zeta}{\Delta x_j} = \frac{\partial\zeta}{\partial x_j}. \tag{13-89}$$

Since $\mathbf{g}$ points straight down, the angle between $\mathbf{g}$ and $-\mathbf{e}_j$ is θ and

$$\rho\mathbf{g} \cdot \mathbf{e}_j = \rho g\,(-\cos\theta) = -\rho g\,\frac{\partial\zeta}{\partial x_j}\,; \tag{13-90}$$

so the term itself is

$$\rho\mathbf{g} = -\rho g\nabla\zeta. \tag{13-91}$$

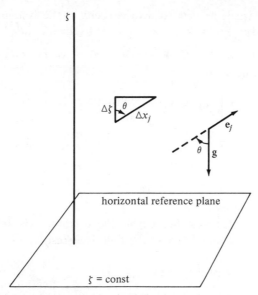

FIGURE 13.14 Increment of height ζ associated with an increment of coordinate x_j.

13.9 Rate of energy dissipation

Newton's second law and the conservation-of-mass law led to equation (13-26), according to which

$$\rho \frac{D\mathbf{v}}{Dt} \tag{13-92}$$

equals the force density acting on an element. Equation (13-88) breaks this force density down into the effects of gravity $\rho\mathbf{g}$, of hydrostatic pressure $-\nabla p$, and of shear viscosity

$$-\mu\nabla \times (\nabla \times \mathbf{v}). \tag{13-93}$$

Dot multiplying a force by the displacement over which it acts yields the work done. Dividing the result by the corresponding interval of time yields the power exerted.

Now, the density of force exerted *by* the element in overcoming the viscous loss is the negative of expression (13-93). Dot multiplying this with

$$\frac{D\mathbf{s}}{Dt} = \mathbf{v} \tag{13-94}$$

yields the density of power dissipated. The power dissipated in volume $\overline{V}$ is thus

$$-\frac{\delta W'}{\delta t} = \mu \int_{\overline{V}} \mathbf{v} \cdot \nabla \times (\nabla \times \mathbf{v}) \, dV. \tag{13-95}$$

This loss may be largely associated with vorticity in the system. The procedure of example 12.4 applied to $\nabla \cdot [\mathbf{v} \times (\nabla \times \mathbf{v})]$ leads to the identity

$$\mathbf{v} \cdot \nabla \times (\nabla \times \mathbf{v}) = (\nabla \times \mathbf{v}) \cdot (\nabla \times \mathbf{v}) - \nabla \cdot [\mathbf{v} \times (\nabla \times \mathbf{v})] \qquad (13\text{-}96)$$

that converts (13-95) to

$$
\begin{aligned}
-\frac{\delta W'}{\delta t} &= \mu \int_V (\nabla \times \mathbf{v})^2 \, dV - \mu \int_V \nabla \cdot [\mathbf{v} \times (\nabla \times \mathbf{v})] \, dV \\
&= \mu \int_V (\nabla \times \mathbf{v})^2 \, dV - \mu \int_S [\mathbf{v} \times (\nabla \times \mathbf{v})] \cdot d\mathbf{S} \\
&= \mu \int_V (\nabla \times \mathbf{v})^2 \, dV. \qquad (13\text{-}97)
\end{aligned}
$$

In the second equality, Gauss's theorem has been employed. In the third equality, we have assumed that the boundary of the fluid is at rest.

13.10   Slow steady flow past a sphere

A fluid that is resting or moving uniformly in one direction does not dissipate energy; all variations in its pressure are counterbalanced by body forces. But slowing down part of the system tends to introduce vorticity and frictional losses.

If the reduction in velocity is made without altering the fluid density to an appreciable extent or at a significant rate anywhere, the approximations in section 13.8 apply and

$$\nabla \cdot \mathbf{v} = 0. \qquad (13\text{-}98)$$

Furthermore, the nonuniformities may be limited enough so that acceleration of each fluid element is relatively small and

$$\rho \frac{D\mathbf{v}}{Dt} \simeq 0. \qquad (13\text{-}99)$$

The Navier-Stokes equation then reduces to

$$0 = \rho \mathbf{g} - \nabla p - \mu \nabla \times (\nabla \times \mathbf{v}). \qquad (13\text{-}100)$$

Here we will suppose that the change in fluid motion is caused by a fixed sphere centered on the origin. The original uniform velocity will be assumed small, so equation (13-100) applies.

Taking the curl of both sides of this equation yields

$$0 = \nabla \times [\nabla \times (\nabla \times \mathbf{v})]. \qquad (13\text{-}101)$$

Now, the reduced equation of continuity (13-98) is satisfied when $\mathbf{v}$ is the curl of an arbitrary vector function:

$$\mathbf{v} = \nabla \times \mathbf{A}. \qquad (13\text{-}102)$$

Substituting (13-102) into (13-101) gives us

$$\nabla \times \{\nabla \times [\nabla \times (\nabla \times \mathbf{A})]\} = 0. \qquad (13\text{-}103)$$

Let us assume that all fluid far away from the sphere moves at a constant velocity **u**. Equation (13-102) can then be integrated for this distant field. Although a unique result is not obtained, the simplest form of all possibilities is

$$\mathbf{A_0} = \tfrac{1}{2}(\mathbf{u} \times \mathbf{r}). \tag{13-104}$$

Since the governing differential equation (13-103) is linear, the effect of the sphere may be introduced by additive terms that vanish at large r. The simplest form related to (13-104) is

$$\mathbf{A}_n = r^{-n}(\mathbf{u} \times \mathbf{r}). \tag{13-105}$$

The possible n's are determined by substitution into (13-103).

The lth Cartesian component of the curl of $\mathbf{A}_n$ is

$$\varepsilon_{lmo}\frac{\partial}{\partial x_m}(r^{-n}\varepsilon_{opq}u_p x_q) = \varepsilon_{olm}\varepsilon_{opq}u_p\left(-nr^{-n-1}\frac{x_m}{r}x_q + r^{-n}\delta_{mq}\right)$$

$$= (2-n)r^{-n}u_l + nr^{-n-2}\mathbf{u}\cdot\mathbf{r}x_l. \tag{13-106}$$

The jth component of the curl of this curl is

$$\{\nabla \times [\nabla \times r^{-n}(\mathbf{u} \times \mathbf{r})]\}_j = \varepsilon_{jkl}\frac{\partial}{\partial x_k}[(2-n)r^{-n}u_l + nr^{-n-2}u_m x_m x_l]$$

$$= n(3-n)r^{-n-2}(\mathbf{u} \times \mathbf{r})_j. \tag{13-107}$$

Applying the curl operation twice has reduced the exponent of r by 2 and multiplied the expression by minus the original exponent times the quantity 3 plus this exponent. Applying it two more times yields

$$\nabla \times \{\nabla \times [\nabla \times (\nabla \times \mathbf{A}_n)]\} = (n+2)(1-n)n(3-n)r^{-n-4}(\mathbf{u} \times \mathbf{r}). \tag{13-108}$$

For this result to equal zero and satisfy equation (13-103), we must have

$$n = -2, 0, 1, \text{ or } 3. \tag{13-109}$$

Since a term with n equal to -2 would dominate at large r and would not reduce to form (13-104), such a term cannot be included. However, arbitrary amounts of the terms with $n = 1$ and $n = 3$ may be present. So

$$\mathbf{A} = \left(\frac{1}{2} + \frac{k_1}{r} + \frac{k_3}{r^3}\right)(\mathbf{u} \times \mathbf{r}). \tag{13-110}$$

Now, the velocity of the fluid is given by the curl of **A**. The curl of expression (13-110) is found by applying formula (13-106):

$$\mathbf{v} = \nabla \times \mathbf{A}$$

$$= (1 + k_1 r^{-1} - k_3 r^{-3})\mathbf{u} + (k_1 r^{-3} + 3k_3 r^{-5})\mathbf{u}\cdot\mathbf{rr}. \tag{13-111}$$

If we let a be the radius of the fixed sphere and assume that the fluid does not slip over the surface of the sphere, we must have $v = 0$ at $r = a$. That is, the coefficients of $\mathbf{u}$ and of $\mathbf{u} \cdot \mathbf{rr}$ must equal zero when r is a:

$$1 + k_1 a^{-1} - k_3 a^{-3} = 0, \tag{13-112}$$

$$k_1 a^{-3} + 3k_3 a^{-5} = 0. \tag{13-113}$$

These equations are satisfied with

$$k_1 = -\tfrac{3}{4}a, \qquad k_3 = \tfrac{1}{4}a^3. \tag{13-114}$$

Form (13-110) does indeed yield a solution with the desired properties.

By means of formula (13-107), the curl of the curl of (13-110) in the field about the sphere is found to be

$$\nabla \times \mathbf{v} = 2k_1 r^{-3} \mathbf{u} \times \mathbf{r} = -\tfrac{3}{2}ar^{-3}\mathbf{u} \times \mathbf{r}. \tag{13-115}$$

If we point the z axis in the direction of $\mathbf{u}$, the magnitude of the curl of $\mathbf{v}$ is

$$|\nabla \times \mathbf{v}| = \frac{3a}{2r^2} u \sin \theta. \tag{13-116}$$

Substituting this quantity into equation (13-97) and integrating yields

$$
\begin{aligned}
-\frac{\delta W'}{\delta t} &= \mu \int_a^\infty \int_0^\pi \int_0^{2\pi} \frac{9a^2}{4r^4} u^2 \sin^2 \theta \, r^2 \sin \theta \, dr \, d\theta \, d\phi \\
&= \tfrac{9}{4}\mu a^2 u^2 \int_a^\infty \frac{dr}{r^2} \int_0^\pi \sin^3 \theta \, d\theta \int_0^{2\pi} d\phi \\
&= 6\pi\mu au^2. \tag{13-117}
\end{aligned}
$$

If we let F be the total drag on the sphere, then Fu equals $-\delta W'/\delta t$ and

$$F = 6\pi\mu au. \tag{13-118}$$

Adding a constant velocity to the system as a whole does not affect any accelerations or any forces. The force F required to drive a sphere of radius a through a medium of viscosity μ at the steady-state (or terminal) velocity u with respect to the fluid is given by equation (13-118). This result is attributed to Stokes.

13.11   Slow steady flow down a symmetric capillary

Let us consider a fluid moving slowly and steadily through a straight capillary or pipe with a constant circular cross section. We assume that the material does not slip over any part of the fixed cylindrical wall.

The approximations made in section 13.8 still apply and

$$\nabla \cdot \mathbf{v} = 0, \tag{13-119}$$

$$\rho \frac{D\mathbf{v}}{Dt} \simeq 0. \tag{13-120}$$

With formula (13-91), the Navier-Stokes equation reduces to

$$0 = -\nabla(p + \rho g \zeta) + \mu \nabla^2 \mathbf{v}. \qquad (13\text{-}121)$$

Let us define cylindrical coordinates r, ϕ, z in the conventional manner with respect to the Cartesian coordinates of figure 13.15. We also assume that the flow is as symmetric as the bounding wall. Thus, the movement is taken to be axial,

$$v_r = v_\phi = 0, \qquad (13\text{-}122)$$

and independent of the azimuthal angle,

$$\frac{\partial v_z}{\partial \phi} = 0. \qquad (13\text{-}123)$$

In the cylindrical coordinate system, the elements of displacement are dr, $r\,d\phi$, and dz. The formula in example 13.1 converts equation (13-119) to

$$\frac{1}{r} \left[\frac{\partial(rv_r)}{\partial r} + \frac{\partial v_\phi}{\partial \phi} + \frac{\partial(rv_z)}{\partial z} \right] = 0. \qquad (13\text{-}124)$$

Conditions (13-122) reduce this equation to

$$\frac{\partial v_z}{\partial z} = 0. \qquad (13\text{-}125)$$

As a consequence, velocity v_z depends only on r.

In cylindrical coordinates, the gradient of a function is

$$\nabla \Phi = \mathbf{l}\, \frac{\partial \Phi}{\partial r} + \mathbf{n}\, \frac{1}{r}\frac{\partial \Phi}{\partial \phi} + \mathbf{k}\, \frac{\partial \Phi}{\partial z} \qquad (13\text{-}126)$$

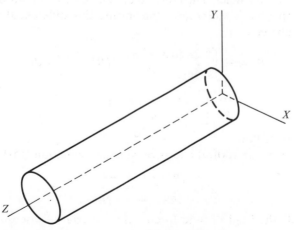

FIGURE 13.15   Reference axes for cylindrically symmetric flow.

according to equation (12-100). Substituting the appropriate elements into the formula for the divergence then yields

$$\nabla^2 \Phi = \frac{1}{r}\left[\frac{\partial}{\partial r}\left(r\frac{\partial \Phi}{\partial r}\right) + \frac{\partial}{\partial \phi}\left(\frac{1}{r}\frac{\partial \Phi}{\partial \phi}\right) + r\frac{\partial}{\partial z}\frac{\partial \Phi}{\partial z}\right]. \tag{13-127}$$

These results let us express the components of equation (13-121) as follows:

$$\frac{\partial(p + \rho g\zeta)}{\partial r} = 0, \tag{13-128}$$

$$\frac{\partial(p + \rho g\zeta)}{\partial \phi} = 0, \tag{13-129}$$

$$\frac{\partial(p + \rho g\zeta)}{\partial z} = \mu\frac{1}{r}\frac{\partial}{\partial r}\left(r\frac{\partial v_z}{\partial r}\right). \tag{13-130}$$

Equations (13-128) and (13-129) show that $p + \rho g\zeta$ depends only on z. Equations (13-123) and (13-125) demonstrate that v_z depends only on r. Therefore, equation (13-130) can be rewritten in the form

$$\frac{d(p + \rho g\zeta)}{dz} = \mu\frac{1}{r}\frac{d}{dr}\left(r\frac{dv_z}{dr}\right). \tag{13-131}$$

Since the right side is not a function of z, the derivative on the left side is a constant. Integrating and setting $v_z = 0$ at $r = R$, the radius of the tube, and $dv_z/dr = 0$ at $r = 0$ leads to

$$v_z = -\frac{1}{4\mu}\frac{d(p + \rho g\zeta)}{dz}(R^2 - r^2). \tag{13-132}$$

The fluid between distance r and distance $r + dr$ from the axis forms a shell that intersects area $2\pi r\, dr$ of a cross section and that travels length v_z perpendicular to this cross section in unit time. Therefore, the volume of this fluid passing a section at given z per unit time is $v_z(2\pi r\, dr)$. Integrating this differential volume over all radii within the tube gives

$$Q = -\int_0^R \frac{2\pi}{4\mu}\frac{(dp + \rho g\zeta)}{dz}(R^2 - r^2)r\, dr$$

$$= \frac{\pi R^4}{8\mu}\left[-\frac{d(p + \rho g\zeta)}{dz}\right], \tag{13-133}$$

the volumetric rate of flow.

When the tube is made vertical and $p \simeq$ constant, equation (15-133) reduces to

$$Q = \frac{\pi R^4 \rho g}{8\mu}\frac{L}{L} = \frac{\pi R^4 H}{8\mu L}. \tag{13-134}$$

Here H represents the head of fluid forcing the material of viscosity μ through the tube of length L, while Q is the volumetric rate of flow. This formula was discovered by Jean L. M. Poiseuille.

13.12 General comments

We often assume that the continuum representing a fluid sticks to any bounding wall. Molecules striking such a surface presumably come into equilibrium with it and then leave as from a layer of fluid having the velocity of the wall.

But at low densities, whenever the molecular mean free paths become comparable to significant dimensions in the system, slip seems to occur. In flow through narrow pores, in the free fall of minute drops, and in the movement of a satellite near the earth, the shear forces acting on the pertinent solid surfaces are less than those predicted by the formulas we have obtained. Apparently, molecules striking walls do retain some of their tangential velocities.

We may also observe a near discontinuity between neighboring flow surfaces (surfaces generated by streamlines). A jet may shoot out through essentially static fluid. A solid object may set up a wake separated from surrounding, rapidly moving fluid by a very thin transition region. Furthermore, a rapidly moving fluid may separate from a curving wall.

The transition layer separating the widely differing flows is said to constitute a free boundary. Unfortunately, such a layer is unstable. The higher the rate of flow on the fast-moving side, the faster the free boundary expands into a mixing zone containing a succession of eddies.

The transition region separating a wall from a fairly homogeneous fluid also becomes unstable at high rates of flow. The point at which erratic behavior appears is measured by the dimensionless Reynolds number

$$R = \frac{\rho v l}{\mu} \tag{13-135}$$

in which ρ is the density of the fluid, v the speed of an element in the body of the fluid, l the pertinent distance, and μ the viscosity.

As a consequence, the flow pattern within or around symmetric boundaries loses symmetry when the appropriate Reynolds number exceeds the critical value. Motions of fluid elements then become irregular and eddies appear. Turbulence is said to exist.

In further developing the mechanics of fluids, we must be guided by these experimental observations.

DISCUSSION QUESTIONS

13.1 How can certain gross mechanical properties of a fluid vary as those of a continuous medium? Cite and discuss other properties that are represented erroneously if at all.

13.2 Explain why the equation of continuity insures conservation of mass in the real material. What is a material derivative?

13.3 How can we employ fluid motion in finding the general expression for ∇ in orthogonal coordinate systems? Why is there no need to distinguish between covariant and contravariant components in this derivation?

13.4 Can $\partial\rho/\partial t$ be zero when the flow is not steady?

13.5 How does an infinitesimal element of fluid differ from a particle without structure (a mass located at a point)?

13.6 Show that $\mathbf{A} - \tilde{\mathbf{A}}$ is antisymmetric while $\mathbf{A} + \tilde{\mathbf{A}}$ is symmetric.

13.7 Explain why the symmetric part of $\widetilde{\nabla\mathbf{v}}$ describes the rate of shear deformation and dilatation.

13.8 Why doesn't application of Newton's second law to an infinitesimal element lead directly to a differential equation for fluid motion?

13.9 Describe the stress dyadic. How does it determine the stress acting on a given $d\mathbf{S}$?

13.10 How can $\boldsymbol{\sigma}$ be diagonalized? Why do the principal stresses act in mutually perpendicular directions? Under what circumstances is $\boldsymbol{\sigma}$ diagonal in all Cartesian coordinate systems?

13.11 How do we reduce the surface integral of surface tractions to a volume integral? Why is this reduction needed in deriving a differential equation for the motion?

13.12 What assumptions characterize a Stokesian fluid?

13.13 Determine the combinations of stresses and of rates of strain that are completely symmetric. What combinations form a basis for the completely symmetric representation? What combinations of $\sigma_{11}, \sigma_{22}, \sigma_{33}$ and of $\varepsilon_{11}, \varepsilon_{22}, \varepsilon_{33}$ measure contributions from other bases?

13.14 Why can we consider the effects of σ_{12}, the effects of σ_{23}, and the effects of σ_{31} separately? Explain when the array represented by σ_{12} is equivalent to the array represented by $\sigma_{11} - \sigma_{22}$.

13.15 Graphically rotate the vectors in figure 13.12 by 45° and resolve the results on an unrotated square prism.

13.16 How is the dyadic $\mathbf{E}$ related to the corresponding unit matrix? Explain the difference between $\nabla \cdot \nabla\mathbf{v}$ and $\nabla \cdot \widetilde{\nabla\mathbf{v}}$. Why does $\frac{1}{2}(\nabla\mathbf{v} + \widetilde{\nabla\mathbf{v}})$ give the symmetric part of both $\nabla\mathbf{v}$ and $\widetilde{\nabla\mathbf{v}}$?

13.17 Explain how the Navier-Stokes equation is set up and simplified.

13.18 Why is the power dissipated largely by the vorticity in a system? How is drag associated with vorticity?

13.19 When is the velocity of each element of a fluid the curl of a vector function? How may this function be constructed to describe uniform flow of the fluid in one direction?

13.20 Why can the velocity **v** be taken zero where the fluid contacts a surface at rest? When does this approximation break down?

13.21 Why is equation (13-132) for v_z invalid near the entrance of a pipe?

13.22 Why must the derivative dv_z/dr be zero at $r = 0$ when v_z does not depend on the azimuthal angle?

13.23 Explain why the symmetry of a flow pattern can be less than that of the boundary conditions.

PROBLEMS

13.1 Formulate the equation of continuity in spherical coordinates.

13.2 Apply the equation of continuity to radially directed motion of an incompressible fluid away from the origin, and integrate. If v_r does not vary with angles θ and ϕ about the center, what source is needed there?

13.3 Determine whether an incompressible fluid can move, so

$$\mathbf{v} = -k \sin \phi \mathbf{i} + k \cos \phi \mathbf{j}.$$

Then calculate the angular velocity of an arbitrary element.

13.4 Show that if a flow is incompressible, irrotational (without rotating elements), and 2-dimensional (dependent only on coordinates x and y), and if the speed of an element is everywhere constant, the flow is uniform in direction.

13.5 In what Cartesian coordinate systems does σ remain diagonal when two of the principal stresses are equal?

13.6 How is the normal component of stress that a fluid exerts on a fixed wall related to the varying density in the fluid?

13.7 Rotate the stress dyadic depicted in figure 13.11 by $2\pi/n$ radians around the third axis.

13.8 Determine how a viscous fluid travels between a fixed horizontal plane and a parallel moving plane.

— — —

13.9 Express the equation of continuity in circular cylindrical coordinates.

13.10 Integrate the equation of continuity to find how the velocity of an incompressible fluid varies when it steadily spreads out as symmetrically as possible from a line source. At what rate does unit length of the source supply fluid?

13.11 Can an incompressible fluid move with the velocity

$$\mathbf{v} = -kr^n \sin \phi \mathbf{i} + kr^n \cos \phi \mathbf{j}?$$

If it can, what is the angular velocity of an element momentarily at a given point?

13.12 When may the expression

$$\mathbf{v} = (ax + by)\mathbf{i} + (cx + dy)\mathbf{j}$$

be the velocity of an incompressible fluid? By integration, determine the path of a given element of such a fluid.

13.13 Show that the antisymmetric nature of $\mathbf{\Omega}$ and the symmetric nature of ε are not affected by any reorientation of axes.

13.14 On a typical small cubic element of fluid at a given instant of time, stress components $\sigma_{11}, \sigma_{22},$ and σ_{33} act. Assume these are all equivalent in magnitude and determine the effect of each operation of group $\mathbf{C_4}$ on the arrays. Then employ the projection-operator technique to construct bases of irreducible representations from these.

13.15 Obtain characters for the reducible representation of $\mathbf{O}_h$ based on equivalent $\sigma_{11}, \sigma_{22},$ and σ_{33} arrays acting on a small cube of the fluid. Find the irreducible representations in this reducible representation.

13.16 A viscous liquid flows slowly but steadily within a pipe whose cross section is the ellipse

$$\frac{x^2}{a^2} + \frac{y^2}{b^2} = 1.$$

Assume that no body forces act and obtain a solution for v_z in which the expression

$$\frac{x^2}{a^2} + \frac{y^2}{b^2} - 1$$

is a factor.

REFERENCES

BOOKS

Birkhoff, G. *Hydrodynamics. A Study in Logic, Fact, and Similitude*, Rev. ed., pp. 1–50. Princeton University Press, Princeton, N.J., 1960.

Chorlton, F. *Textbook of Fluid Dynamics*, pp. 70–135, 310–353. D. Van Nostrand Company, Inc., Princeton, N.J., 1967.

Eskinazi, S. *Vector Mechanics of Fluids and Magnetofluids*, pp. 122–211. Academic Press, Inc., New York, 1967.

Owczarek, J. A. *Introduction to Fluid Mechanics*, pp. 1–63, 101–142, 273–288. International Textbook Company, Scranton, Pa., 1968.

ARTICLES

Dishington, R. H. Rate of Surface-Strain Tensor. *Am. J. Phys.*, **33**, 827 (1965).

Langlois, W. E. An Elementary Proof that the Undetermined Stress in an Incompressible Fluid is of the Form $-p\mathbf{I}$. *Am. J. Phys.*, **39**, 641 (1971).

14 / Electric and Magnetic Field Vectors

14.1 Electric charge

Bringing one material into contact with another by rubbing generally causes each to acquire the ability to attract small bodies. Furthermore, a part of one substance that has acquired this ability repels another part of the same substance that has been treated in the same way, while it attracts the part of the other substance with which it had been rubbed. Rubbing another sample of the first substance with a material different from the second substance yields a body that either repels or attracts the first sample, depending on the choice of material.

A body that has acquired the ability we have just described is said to carry an electric charge. Since a given charged body repels some charged bodies and attracts others, there are two kinds of charge. One of these is always labeled with a positive sign, the other with a negative sign. Consequently, charges repel or attract each other depending on whether they have the same or opposite signs.

In common processes, charge appears to be transported by electrons, holes, and ions. These carriers of charge are tightly bound in the materials called insulators. However, both positive and negative ions travel readily in electrolytes. Ions and electrons are mobile in plasmas. In semiconductors, excited electrons, and the positively charged holes from which these electrons have been excited, move. In solid conductors, the lattice and interstitial ions are relatively fixed, but the valence electrons are highly mobile.

Throughout the body of an ordinary conductor, the density of charge on the valence electrons is nearly counterbalanced by the density of charge on the lattice and interstitial ions. As a consequence, we can readily observe a dynamic interaction between conductors carrying currents. The force that one current element exerts on another is found to vary inversely with the square of the distance between them, as the static forces do. It also varies with the magnitude of each current. Its direction is best described vectorially.

A varying current in one conductor tends to induce a varying current in another that is close by. The induced current is of such direction that it tends to reduce the change of field. A varying electric field is analogous to a current, similarly producing a magnetic field.

413

14.2   Coulomb's law

Experiments of Charles A. de Coulomb and Henry Cavendish showed that the electrostatic forces acting on two relatively small, isolated, charged bodies (a) are directed along the line joining the charges (b) toward each other if the signs on the charges are different and away from each other if their signs are the same, (c) vary directly as the magnitude of each charge, and (d) vary inversely as the square of the distance between them.

Thus, the electric force $\mathbf{F}$ that charge q_1 exerts on charge q distance r away is given by

$$\mathbf{F} = \frac{qq_1}{4\pi\varepsilon_0 r^2}\,\mathbf{l} \tag{14-1}$$

if $\mathbf{l}$ is a unit vector drawn on the straight line from the source charge q_1 to the test charge q and ε_0 is the permittivity of the space between. The parameter ε_0 varies with the system of units employed. When c is the speed of light in free space and

$$\varepsilon_0 = \frac{1}{4\pi c^2 10^{-7}} \simeq 8.854 \times 10^{-12} \text{ coul}^2 \text{ N}^{-1} \text{ m}^{-2}, \tag{14-2}$$

the charges are in coulombs.

In sections 2.1 and 2.2, we noted that forces add as vectors. As a result, the effect of a set of charges $q_1, q_2, \ldots, q_N$ on a test charge q is given by the sum

$$\mathbf{F} = q\,\frac{q_j \mathbf{l}_j}{4\pi\varepsilon_0 r_j^2} \equiv q\mathbf{E}, \tag{14-3}$$

when all charges are at rest.

The force per unit test charge

$$\mathbf{E} = \frac{q_j \mathbf{l}_j}{4\pi\varepsilon_0 r_j^2} \tag{14-4}$$

is called the *electric* field *intensity* at the point where q is located. Multiplying it by the permittivity of free space yields the vector

$$\varepsilon_0\mathbf{E} = \frac{q_j \mathbf{l}_j}{4\pi r_j^2} = \mathbf{D}_{\text{complete}} \tag{14-5}$$

called the complete displacement.

In an actual system, we generally measure only the excess charges resulting from macroscopic movements of electrons, cations, and anions away from a neutral configuration. So only these free charges are included in calculating the conventional *displacement* $\mathbf{D}$:

$$\mathbf{D} = \frac{q_j \mathbf{l}_j}{4\pi r_j^2} \quad \text{where} \quad q_j = j\text{th free source charge.} \tag{14-6}$$

The intensity **E** is always defined as the electric force per unit charge acting on a small free test charge at the given point in the field. As a consequence, the summation in (14-4) must include the bound charges as well as the free charges and **D** is not proportional to **E**. However, we may write

$$\mathbf{D} = \boldsymbol{\varepsilon} \cdot \mathbf{E} \tag{14-7}$$

where ε is a dyadic depending on the material, called the *permittivity tensor*.

Throughout a region containing only free charges, equation (14-7) reduces to over-all equation (14-5) and **D** is proportional to **E**. Introducing an uncharged material that is *isotropic* when neutral cannot, by itself, cause the direction of **E** to differ from that of **D**. It can, of course, alter the magnitude of **E**. So we then have

$$\mathbf{D} = \varepsilon \mathbf{E}. \tag{14-8}$$

Parameter ε is called the permittivity, while $\varepsilon/\varepsilon_0$ is referred to as the *dielectric constant* of the material.

A vector field can be described by hypothetical *lines* that are everywhere tangential to the field vector at a density such that the number of lines per unit cross-sectional area equals the magnitude (or a given constant times the magnitude) of the vector at the corresponding point. Such a description is particularly useful when the lines are either closed (endless) or begin and end on charges causing the field. In the next section, we will see how the lines describing **D** behave.

EXAMPLE 14.1   How does Coulomb's law reduce when SI units are employed?

Since the units in the international system (the SI system) include the meter, the newton, and the coulomb, they require the permittivity of empty space to be given by expression (14-2). This expression converts the scalar form of (14-1) to

$$F = c^2 \times 10^{-7} \frac{qq_1}{r^2} = (2.9979 \times 10^8)^2 \times 10^{-7} \frac{qq_1}{r^2}$$

$$= 8.9876 \times 10^9 \frac{qq_1}{r^2} \text{ N.}$$

Note that the coefficient is nearly 9×10^9 N m^2 coul^{-2}.

14.3   The displacement field

Since the electric field is an inverse-square field, it obeys a differential equation like that derived in example 12.2.

From sum (14-6), the displacement due to one free charge q is

$$\mathbf{D} = \frac{q\mathbf{l}}{4\pi r^2} \tag{14-9}$$

where r is the distance from the charge to the field point and l is a unit vector pointing in this direction. If we construct integral (2-86) with expression (14-9) in place of I and reduce, we get

$$\int_{\substack{\text{surface} \\ \text{enclosing } q}} \mathbf{D} \cdot d\mathbf{S} = \int \frac{q}{4\pi r^2} r^2 \, d\Omega = q \tag{14-10}$$

or

$$\int_{\substack{\text{surface} \\ \text{excluding } q}} \mathbf{D} \cdot d\mathbf{S} = 0. \tag{14-11}$$

Thus, the charge q behaves as a source of q rays or lines and displacement $\mathbf{D}$ behaves as the resulting flux intensity or line density, wherever it is evaluated. Note that the lines do not begin or end anywhere except on the charge. Since $\mathbf{D} \cdot d\mathbf{S}$ is the component of flux intensity normal to the surface dS times the magnitude dS, we call it the flux passing through element dS.

Equation (14-6) implies that displacements due to different charges add. Consequently, the integral of the total flux through a closed surface equals the enclosed free charge. Formula (14-10) applies with $\mathbf{D}$ the net displacement and q the net free charge enclosed; the formula is then called *Gauss's law*. When the charge appears to be distributed continuously at density ρ within a given surface S containing volume V, we write

$$\int_S \mathbf{D} \cdot d\mathbf{S} = \int_V \rho \, dV. \tag{14-12}$$

Gauss's theorem (12-26) changes the left side of (14-12) to a volume integral:

$$\int_V \nabla \cdot \mathbf{D} \, dV = \int_V \rho \, dV. \tag{14-13}$$

For equation (14-13) to be valid regardless of where the closed surface is constructed, we must have

$$\nabla \cdot \mathbf{D} = \rho. \tag{14-14}$$

Here $\mathbf{D}$ is the displacement caused by the free-charge density ρ.

Note how the displacement field emanates from positive charges and terminates on negative charges. Now, such sources or sinks may congregate on surfaces, causing abrupt changes in the normal component of $\mathbf{D}$.

Consider a rectangular area ΔS of the surface on which the charge

$$q = \sigma \, \Delta S, \tag{14-15}$$

where σ is the surface density, is located. Construct a surface parallel to ΔS in the first material and another distance h away in the second material, as in figure 14.1.

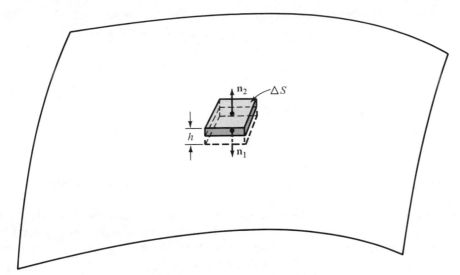

FIGURE 14.1 Rectangular parallelepiped enclosing area ΔS of the surface between two materials.

Complete the enclosure with the side surfaces shown.

The integral of the normal component of **D** over the surface is then equal to charge (14-15), according to Gauss's law. But we may make h so small proportionately that this reduces to the integral over the top and bottom faces. Then if ΔS is chosen small, formula (14-10) reduces to

$$\mathbf{D}_2 \cdot \mathbf{n}_2 \, \Delta S + \mathbf{D}_1 \cdot \mathbf{n}_1 \, \Delta S = \sigma \, \Delta S, \qquad (14\text{-}16)$$

whence

$$\mathbf{n}_2 \cdot (\mathbf{D}_2 - \mathbf{D}_1) = \sigma. \qquad (14\text{-}17)$$

The change in normal component of **D** equals the surface density of charge.

EXAMPLE 14.2   What is the electric field near a plane charged uniformly at surface density σ?

In a uniform medium, an electric field is *uniquely* determined by the source charges. Consequently, the field must be as symmetric as the arrangement of these charges. In other words, the field must be a basis for the completely symmetric representation of the group for the arrangement.

Therefore, in our problem, **D** must be perpendicular to the charged plane everywhere. If we construct a rectangular parallelepiped enclosing area ΔS of the surface, as in figure 14.1, $\mathbf{D} \cdot d\mathbf{S}$ differs from zero only on the top and bottom faces. If each face is the same distance from the charged plane, D must have the same value on each, by symmetry.

Formula (14-10) then yields

$$D \, \Delta S + D \, \Delta S = \sigma \, \Delta S,$$

whence

$$D = \frac{\sigma}{2}.$$

Substituting this into equation (14-8) leads to

$$E = \frac{\sigma}{2\varepsilon}.$$

EXAMPLE 14.3   What is the electric field outside a spherically symmetric charge distribution?

Since the field must have the same symmetry as its source, being uniquely determined, D must be a function of distance r from the center only. And it must lie along $\mathbf{r}$. On a surface at given r, it must be parallel to $d\mathbf{S}$.

Applying Gauss's law (14-10) to such a surface yields

$$\int_S \mathbf{D} \cdot d\mathbf{S} = \int D \, dS = D \int dS = D4\pi r^2 = q$$

where q is the enclosed charge. Then

$$D = \frac{q}{4\pi r^2}$$

and

$$E = \frac{D}{\varepsilon} = \frac{q}{4\pi \varepsilon r^2}.$$

The field at distance r from the center is the same as it would be if all the distributed charge q were at the center.

14.4 Polarization

In the small representative region surrounding a given point in a substance, much of the positive charge is bound to an equal amount of negative charge. However, the average position of such positive charge may be shifted from the average position of the corresponding negative charge because of an inherent anisotropy in the small region. Introducing an electric field introduces opposing electric forces on the positive and negative charges. These forces either alter the existing polarization or introduce some polarization where none existed before. A difference in polarization between neighboring regions causes a net bound charge to appear. This charge contributes to the field.

Displacing a charge q a small distance $\mathbf{s}$ from an opposite charge $-q$ produces a configuration called a dipole. Its strength is measured by the dipole moment

$$q\mathbf{s}. \tag{14-18}$$

Around each point in a polarized material, positive bound charge of density ρ_+ is shifted distance **s** with respect to the corresponding negative charge, producing a dipole-moment density

$$\rho_+\mathbf{s} = \mathbf{P} \qquad (14\text{-}19)$$

that is called *polarization* **P**.

Equidistant to the left and right of such a point let us place dx edges, equidistant to the front and back, dy edges, of a rectangular volume element. The other dx and dy edges are placed up distance dz as in figure 14.2. Multiplying ρ_+ by the z component of **s** by the area of the bottom face yields the positive charge moving into the element through that face as a result of the polarization:

$$\rho_+ s_z \, dx \, dy. \qquad (14\text{-}20)$$

Similarly, the positive charge moving out through the top face is

$$\left[\rho_+ s_z + \frac{\partial}{\partial z} \left(\rho_+ s_z \right) dz \right] dx \, dy. \qquad (14\text{-}21)$$

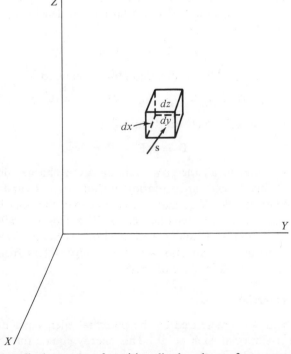

FIGURE 14.2 Mean displacement of positive dipolar charge from negative dipolar charge across the lower face of a rectangular volume element.

The increase of charge in the element due to these movements is the difference

$$-\frac{\partial}{\partial z}(\rho_+ s_z)\, dx\, dy\, dz. \tag{14-22}$$

Like expressions describe the accumulations between the other two pairs of faces. Adding these gives the net bound charge in the element

$$\left[-\frac{\partial}{\partial x}(\rho_+ s_x) - \frac{\partial}{\partial y}(\rho_+ s_y) - \frac{\partial}{\partial z}(\rho_+ s_z)\right] dx\, dy\, dz = \rho_b\, dx\, dy\, dz. \tag{14-23}$$

Introducing the polarization of (14-19) then leads to

$$\left[-\frac{\partial P_x}{\partial x} - \frac{\partial P_y}{\partial y} - \frac{\partial P_z}{\partial z}\right] dx\, dy\, dz = \rho_b\, dx\, dy\, dz, \tag{14-24}$$

whence

$$-\nabla \cdot \mathbf{P} = \rho_b. \tag{14-25}$$

The argument in section 14.3 was carried through for the mobile, or free, charge. Representing the density of this charge as ρ_f alters (14-14) to

$$\nabla \cdot \mathbf{D} = \rho_f. \tag{14-26}$$

However, the argument also describes the displacement due to all charge if the total displacement $\mathbf{D}_{\text{complete}}$ replaces $\mathbf{D}$ and the total charge density $\rho_f + \rho_b$ replaces ρ. So with relationship (14-5), we obtain

$$\nabla \cdot \varepsilon_0 \mathbf{E} = \rho_f + \rho_b. \tag{14-27}$$

Combining equations (14-26), (14-27), and (14-25) leads to

$$\nabla \cdot \mathbf{D} = \rho_f + \rho_b - \rho_b = \nabla \cdot \varepsilon_0 \mathbf{E} + \nabla \cdot \mathbf{P}. \tag{14-28}$$

Consequently, the relationship between $\mathbf{D}$ and $\mathbf{E}$ is

$$\mathbf{D} = \varepsilon_0 \mathbf{E} + \mathbf{P}. \tag{14-29}$$

Introducing a material in all the space among given charges does not alter the conventional $\mathbf{D}$. Instead, such an operation introduces a $\mathbf{P}$ and a compensating change in $\varepsilon_0 \mathbf{E}$. When the field-free material is isotropic, the polarization $\mathbf{P}$ has the same direction as the final force per unit charge $\mathbf{E}$ causing the polarization. Then, $\mathbf{D}$ has the same direction as $\mathbf{E}$ and equation (14-8) is valid. When the field-free dielectric is not isotropic, the direction of $\mathbf{P}$ generally differs from that of $\mathbf{E}$ and equation (14-7) must be used in place of (14-8).

14.5   The scalar potential

A static electric field is characterized by the potential energy per unit charge that a small test charge assumes at each point. This energy equals the work done by the field on the charge, divided by the magnitude of the charge, as the test charge moves from some reference point to the given point.

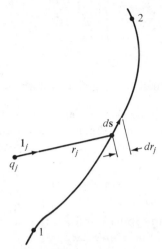

FIGURE 14.3   Scalars and vectors employed in determining the effect of a typical source charge on the work done per unit test charge, moving from point 1 to point 2.

Consider a path from one point (labeled 1) to another (labeled 2) as in figure 14.3. Since the dot product

$$\mathbf{1}_j \cdot d\mathbf{s} = dr_j, \tag{14-30}$$

the pertinent work is given by

$$\int_1^2 \mathbf{E} \cdot d\mathbf{s} = \int_1^2 \frac{q_j}{4\pi\varepsilon r_j{}^2} \mathbf{1}_j \cdot d\mathbf{s} = \frac{q_j}{4\pi\varepsilon} \int_1^2 \frac{dr_j}{r_j{}^2}$$

$$= - \frac{q_j}{4\pi\varepsilon} \left(\frac{1}{r_{j2}} - \frac{1}{r_{j1}} \right). \tag{14-31}$$

If we define the *potential* as the sum

$$\phi = \frac{q_j}{4\pi\varepsilon r_j}, \tag{14-32}$$

then the final expression in (14-31) equals

$$-(\phi_2 - \phi_1) = - \int_1^2 d\phi. \tag{14-33}$$

The result in example 12.1 converts this integral as follows:

$$- \int_1^2 \frac{d\phi}{ds} \, ds = - \int_1^2 \nabla\phi \cdot \mathbf{f} \, ds = - \int_1^2 \nabla\phi \cdot d\mathbf{s}. \tag{14-34}$$

According to (14-31), the initial form was

$$\int_1^2 \mathbf{E} \cdot d\mathbf{s},$$

so we have

$$\mathbf{E} = -\nabla\phi. \tag{14-35}$$

The electric field intensity equals the negative gradient of the potential.

EXAMPLE 14.4   Show that a static electric field is irrotational.

Assume that magnetic effects are negligible. Then the work done by the given field on a unit charge, when the charge moves over a closed path, is 0, according to equation (14-31):

$$\oint \mathbf{E} \cdot d\mathbf{s} = 0.$$

By Stokes' theorem, the line integral around a surface can be converted to the surface integral of the curl. The equation above becomes

$$\int_S \nabla \times \mathbf{E} \cdot d\mathbf{S} = 0.$$

For this formula to hold, regardless of where the bounding line is drawn in the field, we must have

$$\nabla \times \mathbf{E} = 0.$$

14.6   Ampere's law

Andre M. Ampere and later workers have determined how various arrangements of circuits carrying various currents interact with each other. Because of postulates 8 and 9 of section 2.1, we can consider the observed effects as arising from superposition of the forces between small elements of the circuits carrying known currents.

Let us pick out element $d\mathbf{s}_1$ of the curve along which current I_1 travels and element $d\mathbf{s}_2$ of the curve along which current I_2 travels, as in figure 14.4. Differentials $I_1\, d\mathbf{s}_1$ and $I_2\, d\mathbf{s}_2$ are called the corresponding *current elements*. The magnetic experiments indicate that the force $d^2\mathbf{F}_1$ which current element $I_2\, d\mathbf{s}_2$ exerts on $I_1\, d\mathbf{s}_1$ is

$$d^2\mathbf{F}_1 = \frac{\mu_0}{4\pi} \frac{I_1\, d\mathbf{s}_1 \times (I_2\, d\mathbf{s}_2 \times \mathbf{r}/r)}{r^2} \tag{14-36}$$

if $\mathbf{r}$ is the vector extending from $d\mathbf{s}_2$ to $d\mathbf{s}_1$ and μ_0 is the permeability of the space in between. When the force is in newtons and we let

$$\mu_0 = 4\pi \times 10^{-7} \text{ N amp}^{-2}, \tag{14-37}$$

the currents are in amperes.

In words, the magnetic force that a source current element exerts on a test current element (a) is perpendicular to the test element and to the cross product of the source element with the vector joining the elements, (b) varies directly as the sine of the

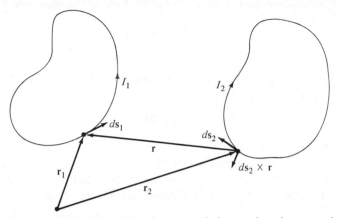

FIGURE 14.4   Vectors used in describing the magnetic interactions between circuits.

angle between the test element and the aforementioned cross product, (c) varies directly as the magnitude of each current and as the length of each element, and (d) varies inversely as the square of the distance between the two elements.

The maximum force per unit current times distance that a magnetic field can exert on a small test element is designated the magnitude of *magnetic induction* **B** at the point where the test element is located. The direction of **B** is taken to be perpendicular to this force and to the most effectively oriented current element in the right-handed sense.

As a consequence, the expression on which $I_1 \, d\mathbf{s}_1$ is cross multiplied in (14-36) is the contribution of element $I_2 \, d\mathbf{s}_2$ to the magnetic induction at the first element. And in general

$$d\mathbf{B} = \frac{\mu_0}{4\pi} \frac{I \, d\mathbf{s} \times \mathbf{r}/r}{r^2} \tag{14-38}$$

where **r** extends from the source element $I \, d\mathbf{s}$ to the point where **B** is located. Being independent, the $d\mathbf{B}$'s from different elements add as separate vectors.

When the field of the source element is written as $d\mathbf{B}$, the subscript 1 is no longer needed on the I and $d\mathbf{s}$ of the test element and equation (14-36) can be written in the form

$$d^2\mathbf{F} = I \, d\mathbf{s} \times d\mathbf{B}$$

which integrates to

$$d\mathbf{F} = I \, d\mathbf{s} \times \mathbf{B}. \tag{14-39}$$

Equivalent to this relationship is the formula

$$\mathbf{F} = q\mathbf{v} \times \mathbf{B} \tag{14-40}$$

for the force **F** acting on a point charge q moving at velocity **v** through a magnetic field.

Now, the contribution of the source element $I\,d\mathbf{s}$ to the *magnetic* field *intensity* $\mathbf{H}$ is taken as

$$d\mathbf{H} = \frac{1}{4\pi} \frac{I\,d\mathbf{s} \times \mathbf{r}/r}{r^2}. \tag{14-41}$$

On comparing (14-41) with (14-38), we see that the relationship between $\mathbf{B}$ and $\mathbf{H}$ is analogous to that between $\mathbf{D}$ and $\mathbf{E}$

$$\mathbf{B} = \mu_0 \mathbf{H} \tag{14-42}$$

when all currents are used in calculating the field vectors.

But in practice, only the currents due to movements of free charge are considered. The effects from the Amperian currents, the uncompensated orbital and spin motions of electrons, are introduced by altering the coefficient of $\mathbf{H}$. Indeed, equation (14-42) is replaced with

$$\mathbf{B} = \boldsymbol{\mu} \cdot \mathbf{H} \tag{14-43}$$

where $\boldsymbol{\mu}$ is a dyadic depending on the material, called the *permeability tensor*.

A medium that is isotropic when free currents are absent cannot, by itself, cause the direction of $\mathbf{B}$ to differ from the direction of $\mathbf{H}$. Then the dyadic acts as a multiplying number μ, the right side of (14-43) reduces to $\mu\mathbf{H}$, and (14-38) is replaced with

$$d\mathbf{B} = \frac{\mu}{4\pi} \frac{I\,d\mathbf{s} \times \mathbf{r}/r}{r^2}, \tag{14-44}$$

where I is the current of free charge. The ratio of μ to μ_0 is called the *relative permeability* or the magnetic constant of the material.

When the Amperian currents reduce the net field, the relative permeability is less than 1 and the substance is said to be *diamagnetic*. When the Amperian currents enhance the field, the relative permeability is greater than 1 and the substance is said to be *paramagnetic*. Substances in which the Amperian currents can cooperate to produce very large enhancements are said to be *ferromagnetic*.

EXAMPLE 14.5   Prove that

$$-\nabla \frac{1}{r} = \frac{\mathbf{r}}{r^3}$$

if ∇ involves differentiation with respect to the coordinates of the end point of $\mathbf{r}$ in figure 14.4.

Vector $\mathbf{r}$ is drawn from the position of source element $I_2\,d\mathbf{s}_2$ to that of test element $I_1\,d\mathbf{s}_1$. If the radius vectors locating these positions are $\mathbf{r}_2$ and $\mathbf{r}_1$, as in figure 14.4, then

$$\mathbf{r} = \mathbf{r}_1 - \mathbf{r}_2 = (x_1 - x_2)\mathbf{i} + (y_1 - y_2)\mathbf{j} + (z_1 - z_2)\mathbf{k}$$

and

$$r = [(x_1 - x_2)^2 + (y_1 - y_2)^2 + (z_1 - z_2)^2]^{1/2}.$$

Subjecting the reciprocal of this expression to the del operator

$$\nabla = \mathbf{i}\,\frac{\partial}{\partial x_1} + \mathbf{j}\,\frac{\partial}{\partial y_1} + \mathbf{k}\,\frac{\partial}{\partial z_1}$$

leads to the desired result:

$$\nabla\frac{1}{r} = \left(\mathbf{i}\,\frac{\partial}{\partial x_1} + \mathbf{j}\,\frac{\partial}{\partial y_1} + \mathbf{k}\,\frac{\partial}{\partial z_1}\right)[(x_1 - x_2)^2 + (y_1 - y_2)^2 + (z_1 - z_2)^2]^{-1/2}$$

$$= -\frac{2(x_1 - x_2)\mathbf{i} + 2(y_1 - y_2)\mathbf{j} + 2(z_1 - z_2)\mathbf{k}}{2[(x_1 - x_2)^2 + (y_1 - y_2)^2 + (z_1 - z_2)^2]^{3/2}}$$

$$= -\frac{\mathbf{r}}{r^3}.$$

14.7 Magnetic induction

Integration of equation (14-44) yields an expression for magnetic induction that can be written as the curl of a vector function. Since the divergence of this curl is zero, the divergence of **B** is zero.

First, let us anticommute the factors in the cross product in expression (14-44) and replace $-\mathbf{r}/r^3$ with its equivalent from example 14.5, $\nabla(1/r)$. Then we note that $d\mathbf{s}$ is constant in the differentiation imposed by ∇; so r can be moved to a position under $d\mathbf{s}$ as long as ∇ still acts on it. The constants can also be moved to the right of $\nabla \times$, with the result:

$$d\mathbf{B} = \frac{\mu I}{4\pi}\left(-\frac{\mathbf{r}}{r^3}\right) \times d\mathbf{s} = \frac{\mu I}{4\pi}\left(\nabla\frac{1}{r}\right) \times d\mathbf{s}$$

$$= \frac{\mu I}{4\pi}\,\nabla \times \frac{d\mathbf{s}}{r} = \nabla \times \frac{\mu I}{4\pi}\frac{d\mathbf{s}}{r}. \tag{14-45}$$

Introducing a *vector potential* **A** by the equation

$$\frac{\mu I}{4\pi}\frac{d\mathbf{s}}{r} = d\mathbf{A} \tag{14-46}$$

reduces (14-45) to

$$d\mathbf{B} = \nabla \times d\mathbf{A}. \tag{14-47}$$

Finally, integrating around all the source circuits yields the net magnetic induction at a given field point:

$$\mathbf{B} = \int \nabla \times d\mathbf{A} = \nabla \times \int d\mathbf{A} = \nabla \times \mathbf{A}. \tag{14-48}$$

The curl operation commutes with the integration since it involves differentiation only with respect to parameters in the integral. Since the divergence of a curl is zero, operating on (14-48) with $\nabla \cdot$ leads to

$$\nabla \cdot \mathbf{B} = \nabla \cdot \nabla \times \mathbf{A} = 0. \tag{14-49}$$

Because **B** forms a vector field, the induction can be described by hypothetical lines that are everywhere tangential to **B** and are, in number per unit cross sectional area, equal to the magnitude of **B** at each point. These lines are analogous to the lines representing **D**. But since the right side of (14-49) is zero, the density in the equations analogous to (14-13) and (14-12) is zero. Hence, there are no magnetic charges (called poles) for the lines of magnetic induction to begin or end on. Instead, these lines appear to be closed curves in space.

14.8   Cyclic integral of H · *dr* in a static field

If a pole were to move continuously in the direction of **H**, from a given point, it would also trace out a closed curve. Consequently, a magnetic field is rotational. To determine exactly how much, when the system is static, let us integrate **H** · *dr* on a loop threading a source circuit.

At a test element, the pertinent field intensity is given by the integral of (14-41) around the source circuit

$$\mathbf{H} = \frac{I}{4\pi} \oint \frac{d\mathbf{s} \times \mathbf{r}/r}{r^2} \, . \tag{14-50}$$

This **H** generally varies with position. Moving the test point by $\delta \mathbf{r}$ is equivalent to holding the point fixed and displacing each circuit element by $-\delta \mathbf{r}$, as in figure 14.5.

The scalar product of the integrand of (14-50) with $\delta \mathbf{r}$ is

$$\frac{\delta \mathbf{r} \cdot d\mathbf{s} \times \mathbf{r}/r}{r^2} = \frac{(\mathbf{r}/r) \cdot \delta \mathbf{r} \times d\mathbf{s}}{r^2} \, . \tag{14-51}$$

But $\delta \mathbf{r} \times d\mathbf{s}$ is the area swept out by $d\mathbf{s}$ during the displacement of the circuit. The dot product with the unit vector $\mathbf{r}/r$ is the projection of this area on a surface

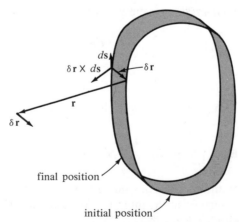

FIGURE 14.5   Area swept out by the circuit displacement that produces the same effect as moving the point of measurement by $\delta \mathbf{r}$, and some pertinent vectors.

perpendicular to $\mathbf{r}$. The ratio of this projection to r^2 is the element of solid angle subtended by the element of the ribbon between the two circuit locations. The integral around the circuit is then the solid angle $\delta\Omega$ subtended by the whole ribbon at the point where $\mathbf{H}$ is being considered.

Dot multiplying both sides of (14-50) with $\delta\mathbf{r}$ consequently yields

$$\mathbf{H} \cdot \delta\mathbf{r} = \frac{I}{4\pi} \oint \frac{\delta\mathbf{r} \cdot d\mathbf{s} \times \mathbf{r}/r}{r^2} = \frac{I}{4\pi} \delta\Omega. \tag{14-52}$$

When the point is taken around a closed path that does not thread the electric circuit, there is a negative change in Ω for each positive change, and

$$\int \delta\Omega = 0. \tag{14-53}$$

When it does thread the circuit once, the total change in solid angle is equal to that surrounding a point:

$$\int \delta\Omega = \pm 4\pi. \tag{14-54}$$

Integrating in the positive direction, as in figure 14.6 yields

$$\oint \mathbf{H} \cdot d\mathbf{r} = \frac{I}{4\pi} \int \delta\Omega = \frac{I}{4\pi} 4\pi = I. \tag{14-55}$$

Addition of another circuit carrying current I' would add another integral on the right of (14-50), and the term $(I'/4\pi)\, \delta\Omega'$ on the right of (14-52). When threaded with the first circuit, we would get $I + I'$ on the right of (14-55). In general, the total

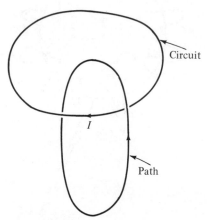

FIGURE 14.6   Path encircling current I.

current threaded would appear, so when the space is filled with current filaments at density **J**, we obtain

$$\oint \mathbf{H} \cdot d\mathbf{r} = \int \mathbf{J} \cdot d\mathbf{S}. \tag{14-56}$$

In section 14.10, we will find that the displacement current $\partial \mathbf{D}/\partial t$ also contributes to the magnetic field. This derivative is, of course, zero in a static field.

EXAMPLE 14.6   Calculate the magnetic field within a toroidal solenoid possessing $\mathbf{C}_\infty$ symmetry.

Since the magnetic field is uniquely determined by the source currents and the ambient medium, it must form a basis for the completely symmetric representation of the group for the system. If a pole were to move continuously in the direction of **H**, from a given point, the pole must trace out a curve with this symmetry.

In our problem, the curve is a circle, perpendicular to the axis and centered on it. Along the circle, the magnitude of **H** must be constant, for the field to be invariant to operation C_n, regardless of n.

If the circle lies within the solenoid, it threads N current loops, where N is the number of turns. Traversing the path in the direction shown in figure 14.7 then leads to

$$\oint \mathbf{H} \cdot d\mathbf{s} = H \oint ds = H 2\pi r = NI,$$

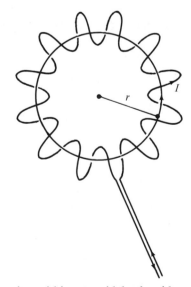

FIGURE 14.7   Path of integration within a toroidal solenoid.

whence

$$H = \frac{NI}{2\pi r}$$

and

$$B = \mu H = \frac{\mu NI}{2\pi r}.$$

If the circle lies without the solenoid, it does not thread any current loops, and

$$H2\pi r = 0,$$

whence

$$H = 0.$$

14.9 Cyclic integral of the work per unit charge acting on a test charge

Michael Faraday's experiments on the currents induced by varying magnetic fields showed that such a current (a) is always directed so its field reduces the rate of change in the magnetic field, as figure 14.8 indicates, (b) is proportional to the area bounded by the circuit, and (c) is proportional to the normal component of the rate of change in the magnetic induction.

Since the current is presumably proportional to the induced emf, the integral of $\mathbf{E} \cdot d\mathbf{s}$, we have the law

$$\oint \mathbf{E} \cdot d\mathbf{s} = - \int \frac{\partial \mathbf{B}}{\partial t} \cdot d\mathbf{S} \tag{14-57}$$

which is similar to formula (14-56).

Stokes' theorem (12-41) changes the left side of (14-57) to a surface integral:

$$\int \nabla \times \mathbf{E} \cdot d\mathbf{S} = - \int \frac{\partial \mathbf{B}}{\partial t} \cdot d\mathbf{S}. \tag{14-58}$$

If this equation is to be valid regardless of how the open surface is chosen, we must have

$$\nabla \times \mathbf{E} = - \frac{\partial \mathbf{B}}{\partial t}. \tag{14-59}$$

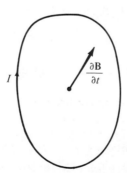

FIGURE 14.8   Electric current induced by a changing magnetic induction.

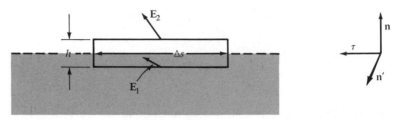

FIGURE 14.9   Path of integration enclosing line Δs on the boundary between two materials and reference unit vectors.

Because of equation (14-57), neighboring parallel components of the electric field intensity differ continuously from each other as long as they lie in a single material, or are tangential components on each side of a boundary between two materials.

Consider such a boundary inscribed with a line of length Δs. Construct a line parallel to it in the first medium, and another distance h away in the second medium, as in figure 14.9. When the ratio of h to Δs is small and Δs itself is small, the pertinent intensities in each medium are approximately equal (constant). Then equation (14-57) reduces to

$$\mathbf{E}_2 \cdot \boldsymbol{\tau} \, \Delta s + \mathbf{E}_1 \cdot (-\boldsymbol{\tau} \, \Delta s) = -\frac{\partial \mathbf{B}}{\partial t} \cdot \mathbf{n}'h \, \Delta s = 0, \qquad (14\text{-}60)$$

whence

$$\boldsymbol{\tau} \cdot (\mathbf{E}_2 - \mathbf{E}_1) = 0. \qquad (14\text{-}61)$$

Note that the direction of $\boldsymbol{\tau}$ on the surface is arbitrary. Furthermore, we may replace $\boldsymbol{\tau}$ with $\mathbf{n}' \times \mathbf{n}$; so equation (14-61) becomes

$$\mathbf{n}' \times \mathbf{n} \cdot (\mathbf{E}_2 - \mathbf{E}_1) = 0 \qquad (14\text{-}62)$$

or

$$\mathbf{n}' \cdot \mathbf{n} \times (\mathbf{E}_2 - \mathbf{E}_1) = 0. \qquad (14\text{-}63)$$

Since the direction of unit vector $\mathbf{n}'$ in the surface is arbitrary, we end up with the condition

$$\mathbf{n} \times (\mathbf{E}_2 - \mathbf{E}_1) = 0, \qquad (14\text{-}64)$$

in which $\mathbf{n}$ is the unit vector normal to the surface.

EXAMPLE 14.7   What principles can be employed to measure $\mathbf{D}$ and $\mathbf{E}$ within a dielectric material?

Assume that small cavities can be constructed in the material without introducing free charges on any new surface. Also assume that a test charge can be placed in the cavity and the force acting on it can be determined.

Around the point to be investigated, cut out a cylinder whose height is very small compared to the diameter of its base, as shown in figure 14.10. By trial and error, orient the cylinder so the force acting on a small charge near the center lies along the

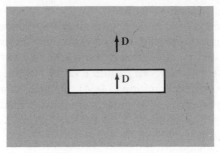

FIGURE 14.10   Very short cylindrical cavity with its axis parallel to **D** in the dielectric.

axis and is maximal. The force per unit charge acting on the test charge is then determined and labeled $\mathbf{E}_c$. Multiplying this vector by the permittivity yields the displacement in the middle of the cavity:

$$\mathbf{D} = \varepsilon_0 \mathbf{E}_c.$$

Equation (14-17) then implies that the axial component of **D** in the dielectric is also $\varepsilon_0 \mathbf{E}_c$. But since the cylinder has been oriented to obtain the maximum E_c, D in the dielectric cannot be larger than $\varepsilon_0 E_c$ and there is no other component at the given point. The displacement **D** in the material is related to the force per unit charge $\mathbf{E}_c$ acting on the charge in the middle of the small oriented cylinder by the equation

$$\mathbf{D} = \varepsilon_0 \mathbf{E}_c.$$

Next form a long needle-shaped cavity as figure 14.11 shows. By trial and error, orient the cavity so the force acting on a small charge near the center lies along the

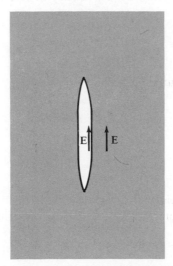

FIGURE 14.11   Long needle-shaped cavity with its axis parallel to **E** in the dielectric.

axis and is maximal. Since in this part of the cavity, effects of the ends are negligible, the field in the center is determined by condition (14-64) alone.

But this condition must apply all the way around the cavity. Since the cross section of the cavity is very small and since $\mathbf{E}$ presumably varies continuously in the original dielectric, we assume that both $\mathbf{E}_1$ and $\mathbf{E}_2$ are constant as $\mathbf{n}$ varies in the cross-sectional plane. Then to satisfy equation (14-64), the axial components of the electric intensity $\mathbf{E}_1 = \mathbf{E}_c$ and that of $\mathbf{E}_2 = \mathbf{E}$ in the neighboring medium must be the same.

Since we have oriented the cavity so $\mathbf{E}_c$ is directed along the axis and is maximal, there is no other component of $\mathbf{E}$ and

$$\mathbf{E} = \mathbf{E}_c.$$

The electric intensity $\mathbf{E}$ in the material equals the electric intensity in the neighboring needle-shaped cavity that points in the direction of $\mathbf{E}$.

14.10   The displacement current

Stokes' theorem also changes the left side of equation (14-56) to a surface integral:

$$\int \nabla \times \mathbf{H} \cdot d\mathbf{S} = \int \mathbf{J} \cdot d\mathbf{S}. \tag{14-65}$$

If this result is to hold true regardless of how the open surface is chosen, we must have

$$\nabla \times \mathbf{H} = \mathbf{J}. \tag{14-66}$$

Taking the divergence of both sides then yields

$$0 = \nabla \cdot \mathbf{J}. \tag{14-67}$$

The current density $\mathbf{J}$ is given by the density of free charges ρ times their mean velocity $\mathbf{v}$:

$$\mathbf{J} = \rho \mathbf{v}. \tag{14-68}$$

Equation (14-67) becomes

$$0 = \nabla \cdot \rho \mathbf{v}, \tag{14-69}$$

and the equation of continuity,

$$\frac{\partial \rho}{\partial t} + \nabla \cdot \rho \mathbf{v} = 0, \tag{14-70}$$

yields

$$\frac{\partial \rho}{\partial t} = 0. \tag{14-71}$$

But charges can accumulate in a region. The free-charge density ρ in equation (14-14) can be different from zero and can vary with time. Equations (14-65) and (14-66) are incomplete.

Differentiating both sides of the displacement equation (14-14),

$$\rho = \nabla \cdot \mathbf{D}, \tag{14-72}$$

with respect to time

$$\frac{\partial \rho}{\partial t} = \nabla \cdot \frac{\partial \mathbf{D}}{\partial t}, \tag{14-73}$$

substituting this result into the conservation equation (14-70), and replacing $\rho \mathbf{v}$ with the current leads to

$$\nabla \cdot \frac{\partial \mathbf{D}}{\partial t} + \nabla \cdot \mathbf{J} = 0. \tag{14-74}$$

If the term to be added to (14-66) is designated $\mathbf{X}$, we have

$$\nabla \times \mathbf{H} = \mathbf{J} + \mathbf{X} \tag{14-75}$$

and

$$0 = \nabla \cdot \nabla \times \mathbf{H} = \nabla \cdot \mathbf{J} + \nabla \cdot \mathbf{X}. \tag{14-76}$$

Equation (14-74) shows that this condition is satisfied if

$$\mathbf{X} = \frac{\partial \mathbf{D}}{\partial t}. \tag{14-77}$$

James C. Maxwell introduced this term, calling it the *displacement current*. The curl of the magnetic field is consequently given by

$$\nabla \times \mathbf{H} = \mathbf{J} + \frac{\partial \mathbf{D}}{\partial t}. \tag{14-78}$$

14.11 Rates at which energy is stored and radiated

Vector $\mathbf{J}$ represents the density of current flow at any given point in a field. Multiplying it by a small cross section perpendicular to its path yields the current, or charge per unit time, passing through the section.

As the charge moves along an element in its path, the field intensity $\mathbf{E}$ acts on it. The component of $\mathbf{E}$ along the path times the elementary distance equals the work done by the field per unit charge. Multiplying this work by the amount of charge per unit time yields the work done by the field on this charge.

If the element of path is dz' and the cross section is $dx'\,dy'$, the scalar

$$E \cos \theta \, dz' J \, dx' \, dy' = \mathbf{E} \cdot \mathbf{J} \, dV \tag{14-79}$$

is this power (see figure 14.12). Since energy is conserved, the negative of this quantity is the rate at which charge passing through does work on the field; it is the power exerted by the current on the element.

Some of the resulting energy may be conducted away, while the rest is stored as electric and magnetic energy in the element. In determining how the energy is partitioned, we start with the divergence of $\mathbf{E} \times \mathbf{H}$. Expand this by the method in section 12.8,

$$\nabla \cdot \mathbf{E} \times \mathbf{H} = \nabla \cdot \underline{\mathbf{E}} \times \mathbf{H} + \nabla \cdot \mathbf{E} \times \underline{\mathbf{H}} = \mathbf{H} \cdot \nabla \times \mathbf{E} - \mathbf{E} \cdot \nabla \times \mathbf{H}, \tag{14-80}$$

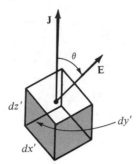

FIGURE 14.12   Element on which the power $-\mathbf{E} \cdot \mathbf{J}\, dx'\, dy'\, dz'$ is being exerted.

introduce field equations (14-59), (14-78),

$$\nabla \cdot \mathbf{E} \times \mathbf{H} = -\mathbf{H} \cdot \frac{\partial \mathbf{B}}{\partial t} - \mathbf{E} \cdot \frac{\partial \mathbf{D}}{\partial t} - \mathbf{E} \cdot \mathbf{J}, \qquad (14\text{-}81)$$

and integrate over an arbitrary volume V,

$$\int_V \nabla \cdot \mathbf{E} \times \mathbf{H}\, dV = -\int_V \mathbf{H} \cdot \frac{\partial \mathbf{B}}{\partial t}\, dV - \int_V \mathbf{E} \cdot \frac{\partial \mathbf{D}}{\partial t}\, dV - \int_V \mathbf{E} \cdot \mathbf{J}\, dV. \qquad (14\text{-}82)$$

Then we employ Gauss's theorem (12-26) to convert the first integral to a surface integral and move the second and third integrals to the left side,

$$\int_S \mathbf{E} \times \mathbf{H} \cdot d\mathbf{S} + \int_V \mathbf{E} \cdot \frac{\partial \mathbf{D}}{\partial t}\, dV + \int_V \mathbf{H} \cdot \frac{\partial \mathbf{B}}{\partial t}\, dV = -\int_V \mathbf{E} \cdot \mathbf{J}\, dV. \qquad (14\text{-}83)$$

Since the right side is the integral of negative (14-79), it is the power exerted by the current flow on the field within V.

The scalar

$$\mathbf{E} \cdot \frac{\partial \mathbf{D}}{\partial t}\, dV \qquad (14\text{-}84)$$

differs from (14-79) only in that $\partial \mathbf{D}/\partial t$ has replaced $\mathbf{J}$. We interpret it as the rate at which work is done on the displacement current in the volume dV. Since the displacement current measures a change in the electric field, (14-84) equals the rate at which energy is stored in this field in dV, and the second integral in (14-83) represents the storage rate in volume V. The product

$$\mathbf{E} \cdot \frac{\partial \mathbf{D}}{\partial t} \qquad (14\text{-}85)$$

is then identified with the rate at which energy is stored in unit volume of the electric field.

Since the third integral involves the analogous magnetic vectors, it is interpreted as the rate at which energy is stored in the magnetic field. The product

$$\mathbf{H} \cdot \frac{\partial \mathbf{B}}{\partial t} \tag{14-86}$$

is the rate at which energy is stored in unit volume of the magnetic field about the given point.

When the only interchange of energy with the outside is by radiation, the first integral measures it. The *Poynting vector*

$$\mathbf{P} = \mathbf{E} \times \mathbf{H} \tag{14-87}$$

is then interpreted as the power flow per unit area in the direction of $\mathbf{P}$. The normal component of $\mathbf{P}$ yields the flow per unit area through the given surface element.

EXAMPLE 14.8   Integrate expression (14-85) to get a formula for the electric energy density.

Consider a process in which the field is built up to the desired extent from zero, integrating the rate at which energy is stored per unit volume of the electric field:

$$\int_{E=0}^{E=E} \mathbf{E} \cdot \frac{\partial \mathbf{D}}{\partial t}\, dt = \int \mathbf{E} \cdot d\mathbf{D}.$$

Then introduce relationship (14-8) with the assumption that ε is constant:

$$\int \mathbf{E} \cdot d\mathbf{D} = \int \mathbf{E} \cdot d(\varepsilon \mathbf{E}) = \varepsilon \int_0^E \mathbf{E} \cdot d\mathbf{E}$$

$$= \frac{\varepsilon \mathbf{E} \cdot \mathbf{E}}{2} = \tfrac{1}{2}\mathbf{E} \cdot \varepsilon \mathbf{E} = \tfrac{1}{2}\mathbf{E} \cdot \mathbf{D}.$$

14.12   Key relationships

Certain combinations of the formulas we have developed are useful in dealing with (a) plasmas and (b) electromagnetic waves.

Between collisions, a charged particle is subject to gravitational and electromagnetic forces. If m is the mass of the particle and $\mathbf{g}$ the acceleration due to gravity alone, the gravitational force is $m\mathbf{g}$. Near a single, spherically symmetric, large mass M, this force is given by (2-93). The electric force is expressed by (14-3) and the magnetic force by (14-40).

According to postulates 8 and 9 of section 2.1, these forces add as independent vectors. When the gravitational force is negligible, only the electric and magnetic forces act and the *Lorentz force law*

$$\mathbf{F} = q(\mathbf{E} + \mathbf{v} \times \mathbf{B}) \tag{14-88}$$

is obtained. Combining this with Newton's second law yields

$$m \frac{d\mathbf{v}}{dt} = q(\mathbf{E} + \mathbf{v} \times \mathbf{B}) \tag{14-89}$$

where m is the mass of the particle carrying charge q at velocity $\mathbf{v}$.

In a given medium, the electric vectors satisfy field equations (14-14), (14-59), constitutive equation (14-7), and a generalized Ohm's law:

$$\nabla \cdot \mathbf{D} = \rho, \tag{14-90}$$

$$\nabla \times \mathbf{E} = -\frac{\partial \mathbf{B}}{\partial t}, \tag{14-91}$$

$$\mathbf{D} = \varepsilon \cdot \mathbf{E}, \tag{14-92}$$

$$\mathbf{J} = \sigma \cdot \mathbf{E}. \tag{14-93}$$

The magnetic vectors obey field equations (14-49), (14-78), and constitutive equation (14-43):

$$\nabla \cdot \mathbf{B} = 0, \tag{14-94}$$

$$\nabla \times \mathbf{H} = \mathbf{J} + \frac{\partial \mathbf{D}}{\partial t}, \tag{14-95}$$

$$\mathbf{B} = \boldsymbol{\mu} \cdot \mathbf{H}. \tag{14-96}$$

The field equations are called *Maxwell's equations.*

In an isotropic medium, the dyadic operators $\varepsilon \cdot$ and $\boldsymbol{\mu} \cdot$ behave as multiplying numbers ε and μ. When the electric and magnetic properties of the medium do not vary with the fields, these multiplying numbers are constant. Then the field equations yield relatively simple wave equations.

Let us take the curl of equation (14-91),

$$\nabla \times (\nabla \times \mathbf{E}) = -\frac{\partial}{\partial t} \nabla \times \mathbf{B}, \tag{14-97}$$

introduce (14-96) with $\boldsymbol{\mu} \cdot$ a constant multiplying number μ, and eliminate $\mathbf{H}$ with (14-95):

$$\nabla \times (\nabla \times \mathbf{E}) = -\frac{\partial}{\partial t} \nabla \times \mu\mathbf{H} = -\mu \frac{\partial}{\partial t} \nabla \times \mathbf{H}$$

$$= -\mu \frac{\partial}{\partial t} \left(\mathbf{J} + \frac{\partial \mathbf{D}}{\partial t} \right). \tag{14-98}$$

Let us also consider $\varepsilon \cdot$ in (14-92) to be a constant multiplying number ε, replace $\mathbf{D}$ in (14-98) with $\varepsilon\mathbf{E}$, and rearrange to get

$$\nabla \times (\nabla \times \mathbf{E}) + \mu\varepsilon \frac{\partial^2 \mathbf{E}}{\partial t^2} = -\mu \frac{\partial \mathbf{J}}{\partial t}, \tag{14-99}$$

or

$$\nabla \times (\nabla \times \mathbf{E}) + \frac{1}{w^2} \frac{\partial^2 \mathbf{E}}{\partial t^2} = -\mu \frac{\partial \mathbf{J}}{\partial t}, \tag{14-100}$$

if we let speed

$$\frac{1}{\sqrt{\mu\varepsilon}} = w. \tag{14-101}$$

Equation (14-100) involves only the electric intensity $\mathbf{E}$ and the current density $\mathbf{J}$.

When the medium is free space, $\mu\varepsilon$ reduces to $\mu_0\varepsilon_0$ and equation (14-101) is replaced by

$$\frac{1}{\sqrt{\mu_0\varepsilon_0}} = c. \tag{14-102}$$

DISCUSSION QUESTIONS

14.1 Compare electromagnetic with gravitational fields.

14.2 Show how Coulomb's law can be stated in (a) integral form and in (b) differential form.

14.3 Which is more fundamental, electric displacement $\mathbf{D}$ or electric field intensity $\mathbf{E}$?

14.4 Why do (a) displacements, (b) intensities, from different sources add?

14.5 What is the field of (a) a planarly symmetric, (b) a spherically symmetric, distribution of charge?

14.6 Explain how a dielectric behaves.

14.7 When does a distribution of equivalent dipoles produce a finite charge density?

14.8 How is a static electric field obtainable from a scalar field?

14.9 Does Ampere's law violate Newton's third law? May work be done on the field by a moving charge?

14.10 Explain how $\mathbf{H}$ may be calculated from (a) all currents, and from (b) only those currents caused by movement of free charges.

14.11 If magnetic poles existed, would the divergence of $\mathbf{B}$ be zero?

14.12 Explain how the solid angle subtended by a current loop at a moving point changes by 4π as the point threads the loop and returns to its starting position.

14.13 What would $\oint \mathbf{H} \cdot d\mathbf{s}$ represent if magnetic poles existed? What does this integral measure?

14.14 When is the electric field rotational? What does the curl of **E** measure?

14.15 Why does the derivation of (14-60) require τ to be directed tangential to the boundary?

14.16 In principle, how may we measure **D** and **E**?

14.17 When is the divergence of the current density **J** zero?

14.18 Explain why the displacement current exists.

14.19 How does an impressed current do work on a field?

14.20 Why is $\frac{1}{2}\mathbf{E} \cdot \mathbf{D}$ interpreted as the electric energy density?

14.21 Why is $\frac{1}{2}\mathbf{H} \cdot \mathbf{B}$ interpreted as the magnetic energy density?

14.22 What is wrong with the following proof that magnetic fields do not exist?

$$\nabla \cdot \mathbf{B} = 0 \qquad\qquad \text{by equation (14-49)}$$
$$\int \mathbf{B} \cdot d\mathbf{S} = \int \nabla \cdot \mathbf{B}\, dV = 0 \qquad \text{Gauss's theorem}$$
$$\mathbf{B} = \nabla \times \mathbf{A} \qquad\qquad \text{by equation (14-48)}$$
$$\int \nabla \times \mathbf{A} \cdot d\mathbf{S} = 0 \qquad\qquad \text{combining}$$
$$\int \mathbf{A} \cdot d\mathbf{r} = \int \nabla \times \mathbf{A} \cdot d\mathbf{S} = 0 \qquad \text{Stokes' theorem}$$
$$\mathbf{A} \cdot d\mathbf{r} = d\phi \qquad \text{or} \qquad \mathbf{A} = \nabla\phi$$
$$\mathbf{B} = \nabla \times \mathbf{A} = \nabla \times \nabla\phi = 0 \qquad \text{curl del identity}$$

14.23 Show that except at $r = 0$, we may have

$$\mathbf{B} = K\,\frac{\mathbf{r}}{r^3}.$$

What kind of source may produce such a field in a limited region?

PROBLEMS

14.1 Equal charges are placed on two similar small balls suspended by strings of equal length from a common point. What relationship governs the angle of inclination from the vertical?

14.2 What charge distribution yields the field

$$\mathbf{D} = A(2z\mathbf{k} - x\mathbf{i} - y\mathbf{j})?$$

14.3 How much charge can be put on a conducting sphere 10.00 cm in radius if the surface field for which corona discharge begins is 3.00×10^6 volt meter^{-1}?

14.4 Sufficiently close to an isolated charge, the electric flux lines spread out symmetrically. If 4 units of charge are placed at A and -1 unit at B, within what angle of line AB do flux lines leave A if they are destined for B?

14.5 Find a potential that describes the field in problem 14.2.

14.6 To what order in a is the potential $\phi(x, y, z)$ in a charge-free region equal to the average of the potentials at points

$$(x \pm a, y, z), \qquad (x, y \pm a, z), \qquad (x, y, z \pm a)?$$

14.7 An infinitely long straight wire of radius a carries a current I distributed uniformly over its cross section. Calculate the magnetic field at distance r from the axis.

14.8 Current I is flowing in a circle that lies in the yz plane and is centered at the origin. From the fact that each current element of given length contributes equally to B_x at any given point on the x axis, obtain a formula for B on this axis.

14.9 Apply the divergence property of $\mathbf{B}$ to get $\partial B_y/\partial y$ from the result in problem 14.8. Then find B_y at a small distance y from the x axis.

14.10 Find the magnetic energy in a toroidal solenoid.

— — —

14.11 A charge q is placed near the midpoint of the straight line joining two similar charges, q and q, and allowed to oscillate back and forth. What is the linear approximation to the restoring force acting on this charge?

14.12 What is the charge on one square kilometer of the earth's surface if an electric field of 300 volts meter^{-1} is directed vertically downward near the surface?

14.13 Find the electric field outside a cylindrically symmetric charge distribution.

14.14 Where is the charge that produces the field

$$\mathbf{E} = A\mathbf{i} + B\mathbf{j} + C\mathbf{k}?$$

What potential describes this field?

14.15 Find a charge distribution that would produce the Yukawa potential

$$\phi = \frac{q}{4\pi\varepsilon_0} \frac{e^{-r/a}}{r}.$$

14.16 Consider a cylindrically symmetric static field for which r is the distance from the axis of symmetry and a is a small quantity such that $r = na$. Determine the order in a such that

$$\phi(r, z) = \frac{1}{8n} \{2n[\phi(r, z + a) + \phi(r, z - a)]$$

$$+ (2n + 1)\phi(r + a, z) + (2n - 1)\phi(r - a, z)\}$$

in a charge-free region of space.

14.17 Find the vector potential due to a pair of parallel, straight, infinite wires carrying the same current in opposite directions. Express the result in terms of the distances from the observation point to each of the wires.

14.18 Let one of the wires in problem 14.17 be removed to a great distance in a convenient direction, and obtain the magnetic induction close to the other one from the vector potential.

14.19 A coil is wound on a nonmagnetic sphere so the turns are in parallel planes with n turns per unit longitudinal length. What is B at the center?

14.20 Steady current I is flowing around a circle in the yz plane at distance a from the origin. From problem 14.9,

$$B_y = \frac{3\mu I a^2 xy}{4(a^2 + x^2)^{5/2}}, \qquad B_z = \frac{3\mu I a^2 xz}{4(a^2 + x^2)^{5/2}}$$

near the x axis. Use these formulas in getting $\partial B_x/\partial y$ and $\partial B_x/\partial z$. With the result from problem 14.8, then find an expression for the longitudinal field B_x at points near the axis.

REFERENCES

BOOKS

Panofsky, W. K. H., and **Phillips, M.** *Classical Electricity and Magnetism*, 2nd ed., pp. 1–60, 95–211. Addison-Wesley Publishing Company, Inc., Reading, Mass., 1962.

Peck, E. R. *Electricity and Magnetism*, pp. 1–136, 197–270, 422–470. McGraw-Hill Book Company, New York, 1953.

Smythe, W. R. *Static and Dynamic Electricity*, 3rd ed., pp. 1–62, 247–367, 415–447. McGraw-Hill Book Company, New York, 1968.

Wangsness, R. K. *Introduction to Theoretical Physics, Classical Mechanics, and Electrodynamics*, pp. 213–341. John Wiley & Sons, Inc., New York, 1963.

ARTICLES

Bork, A. M. Maxwell and the Electromagnetic Wave Equation. *Am. J. Phys.*, **35,** 844 (1967).

Brevik, I. Polar and Axial Vectors in Electrodynamics. *Am. J. Phys.*, **40,** 550 (1972).

Bromberg, J. Maxwell's Electrostatics. *Am. J. Phys.*, **36,** 142 (1968).

Chen, H. S. C. Note on the Magnetic Pole. *Am. J. Phys.*, **33,** 563 (1965).

Franchetti, S. Some Elementary Points of Electrodynamics in the Presence of Magnetized Bodies. *Am. J. Phys.*, **38,** 662 (1970).

Gelman, H. Generalized Conversion of Electromagnetic Units, Measures, and Equations. *Am. J. Phys.*, **34,** 291 (1966).

Greene, J. B., and **Karioris, F. G.** Force on a Magnetic Dipole. *Am. J. Phys.*, **39,** 172 (1971).

Katz, E. Concerning the Number of Independent Variables of the Classical Electromagnetic Field. *Am. J. Phys.*, **33,** 306 (1965).

Klauder, L. T., Jr. A Magnetic Field Diffusion Problem. *Am. J. Phys.*, **37,** 323 (1969).

Meissner, H. Demonstration of Displacement Current. *Am. J. Phys.*, **32,** 916 (1964).

Mello, P. A. A Remark on Maxwell's Displacement Current. *Am. J. Phys.*, **40,** 1010 (1972).

Moorcroft, D. R. Faraday's Law, Potential and Voltage—Discussion of a Teaser. *Am. J. Phys.*, **38,** 376 (1970).

Nicola, M. On the Definition of Electric Charge. *Am. J. Phys.*, **40,** 189 (1972).

Page, C. H. Relations among Systems of Electromagnetic Equations. *Am. J. Phys.*, **38,** 421 (1970).

Phillips, M., French, A. P., and **Rosenfeld, J.** A Magnetic Curl Meter. *Am. J. Phys.*, **40,** 330 (1972).

Pugh, E. M. Poynting Vectors with Steady Currents. *Am. J. Phys.*, **39,** 837 (1971).

Russakoff, G. A Derivation of the Macroscopic Maxwell Equations. *Am. J. Phys.*, **38,** 1188 (1970).

Saslow, W. M., and **Wilkinson, G.** Expulsion of Free Electronic Charge from the Interior of a Metal. *Am. J. Phys.*, **39,** 1244 (1971).

Scanlon, P. J., Henriksen, R. N., and **Allen, J. R.** Approaches to Electromagnetic Induction. *Am. J. Phys.*, **37,** 698 (1969).

Shaw, R. Symmetry, Uniqueness, and the Coulomb Law of Force. *Am. J. Phys.*, **33,** 300 (1965).

Smith, R. L. The Velocities of Light. *Am. J. Phys.*, **38,** 978 (1970).

Whitmer, R. M. Calculation of Magnetic and Electric Fields from Displacement Currents. *Am. J. Phys.*, **33,** 481 (1965).

15 / Plasmas and Electromagnetic Waves

15.1 The nature of a plasma

Any configuration of material containing a significant number of submicroscopic charged particles with movements dominated by extensive electromagnetic interactions is called a *plasma*. The pertinent charged particles are not bound to particular atoms, molecules, or cells, but are relatively free to move over considerable distances. As a consequence, a plasma is generally a very good conductor of electricity.

Examples of plasmas in which the charged particles are electrons and ions include interstellar and interplanetary gases, the ionosphere, vaporized metals, heated gases, products of shocks, electric discharges, and flames. The media in thermonuclear fusion devices and stars are also plasmas. Certain movements in condensed ionized materials are referred to as plasma effects.

The behavior of a plasma depends on how the pertinent particles move. The mean free path of a charged particle, between collisions with other particles, is large when the density of the given system is low. The movements are then relatively simple and will be considered in some detail.

The electric and magnetic fields in a plasma are imposed by external charges and currents, and are altered by currents and accumulations of charge within the system. Particles of a given kind do not move uniformly in any small region, but exhibit a distribution of velocities.

Maxwell's equations, together with the appropriate constitutive equations, govern the propagation of electric and magnetic influences. When the changes in a region are not permanent and not too large, the relationships are approximately linear and we can treat the varying expressions at each frequency, or wave number, separately, as we will do.

15.2 Constituent motions

As a first step toward understanding the behavior of plasmas, let us consider the motion of a single charged particle in simple fields. Equation (14-89) then applies.

442

But in a region where the magnetic field is small enough, the last term drops out and we have

$$m \frac{d\mathbf{v}}{dt} = q\mathbf{E}. \tag{15-1}$$

When the source charges are at rest, intensity $\mathbf{E}$ is given by (14-35) and

$$\nabla \times \mathbf{E} = 0. \tag{15-2}$$

When they move, though, Faraday's law of induction (14-59) applies

$$\nabla \times \mathbf{E} = -\frac{\partial \mathbf{B}}{\partial t}. \tag{15-3}$$

The rate of change of $\mathbf{B}$ may be appreciable even when $\mathbf{B}$ is small. It then contributes to $\mathbf{E}$.

When the electric field is small enough, its effect drops out and we have

$$m \frac{d\mathbf{v}}{dt} = q\mathbf{v} \times \mathbf{B}. \tag{15-4}$$

The acceleration is now perpendicular to $\mathbf{v}$. Speed v is not altered and the kinetic energy of the particle is constant.

If the magnetic induction $\mathbf{B}$ is also constant, the magnitude of the acceleration does not change. Since the acceleration is directed perpendicular to velocity $\mathbf{v}$, it causes the particle to travel around a moving center (the *guiding center*) at a constant rate.

A temporal change in $\mathbf{B}$ can induce appreciable $\mathbf{E}$ along the path of the particle. An estimate of this effect will be made in section 15.4.

15.3 Gyrations caused by a magnetic field

A free charged particle moves in an approximately helical path through any region in which $\mathbf{E}$ is small and $\mathbf{B}$ nearly constant.

Suppose the particle is moving at velocity $\mathbf{v}$ angle θ from the direction of $\mathbf{B}$, as in figure 15.1. If we resolve the velocity along $\mathbf{B}$ and perpendicular to $\mathbf{B}$,

$$v_{\parallel} = v \cos \theta, \tag{15-5}$$

$$v_{\perp} = v \sin \theta, \tag{15-6}$$

we then have

$$v_{\parallel} \simeq \text{constant} \tag{15-7}$$

since $\mathbf{E}$ is small. The magnitude of the cross product involved in equation (15-4) is

$$|\mathbf{v} \times \mathbf{B}| = v_{\perp}B. \tag{15-8}$$

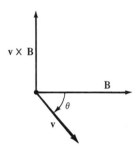

FIGURE 15.1   Velocity of a particle, the magnetic induction to which it is exposed, and their cross product.

Let us take $-\mathbf{l}$ as the instantaneous direction of this vector product:

$$\mathbf{v} \times \mathbf{B} = -v_\perp B\mathbf{l}. \tag{15-9}$$

Then equation (15-4) becomes

$$m\frac{d\mathbf{v}}{dt} = -qv_\perp B\mathbf{l}. \tag{15-10}$$

Since

$$-\frac{q}{|q|}\,\mathbf{l} \tag{15-11}$$

stays perpendicular to $\mathbf{v}_\perp$ (recall the definition of a vector product), the acceleration merely changes the direction of $\mathbf{v}_\perp$. Since $\mathbf{B}$ is nearly fixed and scalar $v_\perp$ does not change appreciably ($\mathbf{E}$ is small), the magnitude of the acceleration is approximately constant. The direction of the particle's velocity is changed at a constant rate perpendicular to $\mathbf{v}_\perp$. Consequently, the particle circles an axis pointing along $\mathbf{B}$ at a distance we label r.

Substituting the radial acceleration given by formula (1-118) into equation (15-10),

$$-m\frac{v_\perp^2}{r}\frac{q}{|q|}\,\mathbf{l} = -qv_\perp B\mathbf{l}, \tag{15-12}$$

and solving for the angular frequency then yields

$$\frac{v_\perp}{r} = \frac{|q|B}{m}. \tag{15-13}$$

This is often called the *cyclotron frequency*.

When q is positive, the force $m\,d\mathbf{v}/dt$ lies along $\mathbf{v} \times \mathbf{B}$ and the guiding center appears as in figure 15.1. The particle then swings around this moving center as in figure 15.2, making the corresponding angular velocity ω point oppositely to $\mathbf{B}$.

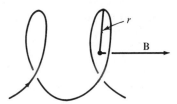

FIGURE 15.2   Helical path of a positively charged particle in a magnetic field.

When q is negative, the center lies on the other side of the particle, and ω points along **B**. Since the magnitude of ω equals $v_\perp/r$, we have

$$\omega = -\frac{q\mathbf{B}}{m} \tag{15-14}$$

and

$$\omega = -\frac{qB}{m}. \tag{15-15}$$

The gyration is oriented so it reduces the field, producing a diamagnetic effect. From equation (15-13), the radius of gyration is

$$r = \frac{mv_\perp}{|q|B}. \tag{15-16}$$

If the particle has some velocity parallel to the field **B**, its guiding center moves at this velocity. The resultant path is a helix, as figure 15.2 shows.

EXAMPLE 15.1   Relate the electron volt (the kinetic energy gained by a free electron in moving from one point to another point 1 volt higher in potential) to the joule.

If the charge on a particle is q while the voltage increment accelerating it is ΔV, the change in potential energy is $q\,\Delta V$. With the total energy constant, the change in kinetic energy is the negative of this value:

$$W = -q\,\Delta V.$$

Inserting the charge on an electron and the 1 volt increment, we see that

$$1\text{ eV} = -(-1.6022 \times 10^{-19}\text{ coulomb})\,(1.0000\text{ volt})$$
$$= 1.6022 \times 10^{-19}\text{ joule}.$$

EXAMPLE 15.2   How is the radius of gyration for (a) an electron, (b) a proton related to the number of electron volts of kinetic energy associated with the perpendicular component of velocity?

The kinetic energy associated with $v_\perp$,

$$W_\perp = \tfrac{1}{2}mv_\perp{}^2,$$

factors out of the numerator of expression (15-16) as follows:

$$r = \frac{mv_\perp}{|q|B} = \frac{(2m)^{1/2}(\tfrac{1}{2}mv_\perp{}^2)^{1/2}}{|q|B} = \frac{(2m)^{1/2}}{|q|} \frac{W_\perp{}^{1/2}}{B}.$$

Let us convert the kinetic energy to the electron-volt scale. Then insert the parameters for an electron,

$$r = \frac{(18.2191 \times 10^{-31} \text{ kg})^{1/2}(1.6022 \times 10^{-19} \text{ joule/ev})^{1/2}}{1.6022 \times 10^{-19} \text{ coulomb}} \frac{W_{ev}{}^{1/2}}{B}$$

$$= 3.372 \times 10^{-6} \frac{W_{ev}{}^{1/2}}{B} \text{ m,}$$

and for a proton,

$$r = 1.445 \times 10^{-4} \frac{W_{ev}{}^{1/2}}{B} \text{ m.}$$

EXAMPLE 15.3   How is the resonant frequency for free electrons related to the prevailing magnetic field?

The cyclic process that can absorb energy is the gyration at angular frequency (15-13). Dividing this by 2π and inserting the appropriate parameters leads to

$$v = \frac{|q|}{2\pi m} B$$

$$= \frac{1.6022 \times 10^{-19} \text{ coulomb}}{(6.2832)9.1096 \times 10^{-31} \text{ kg}} B$$

$$= 2.799 \times 10^{10} B.$$

15.4   Effect of slow temporal variations in B

The acceleration of the charged particle is no longer perpendicular to each element in its path when the magnetic induction varies with time, since equation (15-3) and its integral (14-57) then apply.

The rate at which work is done on the particle equals the conventional frequency of gyration times the charge

$$\frac{\omega}{2\pi} q \tag{15-17}$$

times the work per unit charge done over one turn. This work per turn can be approximated by the cyclic integral

$$\oint \mathbf{E} \cdot d\mathbf{s} \tag{15-18}$$

evaluated along a circular path about the average position of the guiding center.

According to equation (14-57), this cyclic integral equals

$$- \int \frac{\partial \mathbf{B}}{\partial t} \cdot d\mathbf{S}. \tag{15-19}$$

We suppose that $\partial B/\partial t$ is everywhere normal to the plane of the circular path. Replacing it with its average value dB/dt then leads to

$$- \frac{dB}{dt} \pi r^2 \tag{15-20}$$

for (15-19).

But the work done on the particle increases its kinetic energy

$$\tfrac{1}{2} m v_\perp{}^2. \tag{15-21}$$

Setting the time rate of change of (15-21) equal to (15-17) times (15-20) and introducing the *magnetic moment*,

$$M = \pi r^2 \frac{q\omega}{2\pi} = \tfrac{1}{2} r^2 q\omega, \tag{15-22}$$

we find that

$$\frac{d}{dt} (\tfrac{1}{2} m v_\perp{}^2) = \frac{\omega}{2\pi} q \left(- \frac{dB}{dt} \pi r^2 \right) = - M \frac{dB}{dt}. \tag{15-23}$$

Note that M is the area encircled times the current, by definition.

Equations (15-16) and (15-15), which came from the Lorentz force law, can be used to reduce the form for the magnetic moment:

$$M = \tfrac{1}{2} r^2 q\omega = \frac{1}{2} \frac{m^2 v_\perp{}^2}{q^2 B^2} q \left(- \frac{qB}{m} \right) = - \tfrac{1}{2} m v_\perp{}^2 \frac{1}{B}; \tag{15-24}$$

so

$$- MB = \tfrac{1}{2} m v_\perp{}^2. \tag{15-25}$$

Differentiating this result yields

$$\frac{d}{dt} (- MB) = \frac{d}{dt} (\tfrac{1}{2} m v_\perp{}^2). \tag{15-26}$$

For both (15-23) and (15-26) to hold, we must have

$$M = \text{constant}, \tag{15-27}$$

the magnetic moment invariant.

EXAMPLE 15.4   What is the cross product of velocity with magnetic induction for a particle gyrating around a given axis?

Let q be the charge on the particle; and let unit vector **k** represent the direction of the axis, unit vector $(q/|q|)\mathbf{l}$ the radial direction from the axis to the instantaneous position of the particle, and unit vector **n** the product $\mathbf{k} \times \mathbf{l}$. Vectors **k**, **l**, **n** form a right-handed orthogonal system at each point not on the axis.

Assume that within the region of interest magnitude B of the magnetic induction does not change appreciably along a circle at given distance from the axis. Component B_φ in the direction of $(q/|q|)\mathbf{n}$ is then small. Furthermore, B changes slowly along the path of the particle and the radius of gyration r changes very slowly.

If B_z is the axial component and B_r the radial component of $\mathbf{B}$, we have

$$\mathbf{B} \simeq B_z\mathbf{k} + B_r \frac{q}{|q|}\mathbf{l}.$$

If $v_{\parallel}$ is the axial component and $-v_{\perp}$ the angular component of the velocity of the particle, we also have

$$\mathbf{v} \simeq v_{\parallel}\mathbf{k} - v_{\perp}\mathbf{n}.$$

The cross product is

$$\mathbf{v} \times \mathbf{B} = v_{\perp}B_r \frac{q}{|q|}\mathbf{k} - v_{\perp}B_z\mathbf{l} + v_{\parallel}B_r \frac{q}{|q|}\mathbf{n}.$$

The next-to-last term corresponds to the right side of equation (15-9). The preceding term gives the parallel component to be used in equation (15-34).

15.5   Effect of a symmetrically converging magnetic field

In a purely magnetic field, a moving charged particle is acted on by a force $\mathbf{F}$ that is perpendicular to the local field $\mathbf{B}$. As a consequence, convergence of the lines of induction through which the particle passes tilts vector $\mathbf{F}$ back, as figure 15.3 shows. The magnetic force then acquires an axial component that acts to slow down and, if persistent, to reverse the motion of the guiding center. The region is then said to act as a *magnetic mirror*.

Let us assume that within the region of interest magnitude B of the magnetic induction does not change appreciably along a circle any pertinent distance r from a given straight line. If this line is chosen as the axis of a conventional cylindrical coordinate system, then the component of $\mathbf{B}$ in the direction of increasing angle ϕ is negligible.

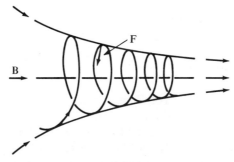

FIGURE 15.3   Path of a positively charged particle around symmetrically converging lines of induction.

The differential equation (14-49) governing changes in magnetic induction,

$$\nabla \cdot \mathbf{B} = 0, \tag{15-28}$$

can then be written in the form

$$\frac{1}{r} \frac{\partial}{\partial r} (rB_r) + \frac{\partial B_z}{\partial z} = 0 \tag{15-29}$$

in which B_r is the radial component of $\mathbf{B}$ and B_z the component along the axis. (Recall the answer to problem 12.6.)

Let us assume that

$$\frac{\partial B_z}{\partial z} = f(z) \tag{15-30}$$

within the region of interest. Then equation (15-29) can be rewritten as

$$d(rB_r) = -\frac{\partial B_z}{\partial z} r \, dr \tag{15-31}$$

and integrated to

$$rB_r = -\frac{\partial B_z}{\partial z} \frac{r^2}{2} \tag{15-32}$$

or

$$B_r = -\frac{r}{2} \frac{\partial B_z}{\partial z}. \tag{15-33}$$

Let us also assume that the given particle of mass m and charge q gyrates around the z axis at velocity $\mathbf{v}$. The calculation in example 15.4 is then applicable. Substituting the expression obtained for the axial component of $\mathbf{v} \times \mathbf{B}$ into that component of equation (15-4) yields

$$m \frac{dv_\parallel}{dt} = \frac{q^2}{|q|} v_\perp B_r. \tag{15-34}$$

From equations (15-13) and (15-15), we have

$$qv_\perp = r|q| \frac{qB_z}{m} = -r|q|\omega; \tag{15-35}$$

so

$$m \frac{dv_\parallel}{dt} = -\frac{q}{|q|} r|q|\omega B_r = -qr\omega \left(-\frac{r}{2} \frac{\partial B_z}{\partial z} \right) = M \frac{\partial B_z}{\partial z} \tag{15-36}$$

where the magnetic moment M has been introduced by equation (15-22).

Let us drop the subscript on B, since the other component of $\mathbf{B}$ is relatively small, and multiply this equation by $v_\parallel = dz/dt$ to get

$$mv_\parallel \frac{dv_\parallel}{dt} = M \frac{\partial B}{\partial z} \frac{dz}{dt} = M \frac{dB}{dt} \tag{15-37}$$

or

$$\frac{d}{dt} (\tfrac{1}{2}mv_\parallel{}^2) = M \frac{dB}{dt}. \tag{15-38}$$

With **E** and $\partial \mathbf{B}/\partial t$ zero, the kinetic energy of the particle is constant:

$$\tfrac{1}{2}mv_{\parallel}^{2} + \tfrac{1}{2}mv_{\perp}^{2} = \text{constant}, \qquad (15\text{-}39)$$

and

$$\frac{d}{dt}\,(\tfrac{1}{2}mv_{\parallel}^{2}) = -\,\frac{d}{dt}\,(\tfrac{1}{2}mv_{\perp}^{2}) = \frac{d}{dt}\,(MB). \qquad (15\text{-}40)$$

The last equality follows from equation (15-26).

On comparing (15-38) and (15-40), we see that

$$M = \text{constant}, \qquad (15\text{-}41)$$

the magnetic moment is again invariant.

EXAMPLE 15.5   If a charged particle is gyrating about the axis of symmetry of a converging magnetic field with a certain $v_{\perp}/v$ at the point where the magnetic induction is B_0, at what point will it be reflected?

Let θ be the angle that the velocity vector makes with an axis parallel to the axis of symmetry; so equation (15-6) applies,

$$\frac{v_{\perp}}{v} = \sin\theta.$$

In a purely magnetic field, the acceleration is perpendicular to **v** and speed v is constant. Then $v_{\perp}$ is proportional to $\sin\theta$.

From equation (15-41), the magnetic moment remains constant. Representing it by the last form in (15-24), we find that

$$\frac{mv_{\perp}^{2}}{2B} = \text{constant},$$

$v_{\perp}^{2}$ is proportional to B.

Combining these two proportionalities to eliminate $v_{\perp}$ yields

$$\sin^{2}\theta = \frac{B}{B_0}\,\sin^{2}\theta_0.$$

Since $\sin\theta$ can get no larger than 1, we have

$$\frac{B}{B_0} \le \frac{1}{\sin^{2}\theta_0}\,.$$

When B/B_0 reaches $1/\sin^{2}\theta_0$, the component $v_{\parallel}$ has reached zero and the particle is reflected.

EXAMPLE 15.6   What fraction of the particles entering a symmetric magnetic mirror is reflected?

If the magnetic field increases from B_0 to B_m as in example 15.5, those particles that initially travel along paths inclined at less than θ_0 from the axis, where

$$\sin^2 \theta_0 = \frac{B_0}{B_m}$$

penetrate the mirror, while those with a θ between θ_0 and $\pi/2$ are reflected.

The number of particles with speed between v and $v + dv$, parallel component of velocity between $v_\parallel$ and $v_\parallel + dv_\parallel$, entering the converging region per unit cross section per unit time equals the number of such particles per unit volume times their $v_\parallel$, which is

$$v_\parallel = v \cos \theta.$$

But if each direction of $\mathbf{v}$ is equally likely, the probability that a speed between v and $v + dv$ gives a parallel component between $v_\parallel$ and $v_\parallel + dv_\parallel$ is proportional to the solid angle

$$d\Omega = 2\pi \sin \theta \, d\theta.$$

The fraction reflected is given by

$$R = \frac{\int_{\theta_0}^{\pi/2} \cos \theta \sin \theta \, d\theta}{\int_0^{\pi/2} \cos \theta \sin \theta \, d\theta}$$

if the constants are canceled from numerator and denominator. Integrating yields

$$R = \frac{\sin^2 \theta|_{\theta_0}^{\pi/2}}{\sin^2 \theta|_0^{\pi/2}} = \frac{1 - \sin^2 \theta_0}{1} = 1 - \frac{B_0}{B_m}.$$

This result does not depend on v. It does presume that the entering particles have an isotropic distribution of velocities, however.

15.6 Variables characterizing a plasma

In an ionized gas or plasma, we must usually consider each kind of particle separately. If the system consists of electrons of charge q_e with mass m_e, ions of charge

$$q_i = -Z_i q_e \tag{15-42}$$

with mass m_i, and possibly neutral molecules, the condition that the system be neutral, on the average, is

$$\bar{n}_e q_e + \bar{n}_i q_i = 0 \tag{15-43}$$

where $\bar{n}_e$ is the mean density of electrons and $\bar{n}_i$ the mean density of ions. When there is more than one species of ion, summation over these is indicated by the repeated i.

If $\bar{\mathbf{v}}_e$ is the average velocity of an electron and $\bar{\mathbf{v}}_i$ the average velocity of an ion, the condition that the system be in equilibrium is

$$\bar{\mathbf{v}}_e = \bar{\mathbf{v}}_i = 0. \tag{15-44}$$

The random motion of each species sets up a kinetic pressure dyad. At equilibrium, these reduce to scalar pressures $\bar{p}_e$ for the electrons and $\bar{p}_i$ for the ions. To the approximation that the ideal gas law applies to each, we have

$$\bar{p}_e = \bar{n}_e k T_e, \tag{15-45}$$

$$\bar{p}_i = \bar{n}_i k T_i. \tag{15-46}$$

The presence of the index on the left indicates no summation on the right. Note that the absolute temperature for the electrons T_e may differ from that for the ions T_i.

A momentary disturbance causes fluctuations in these properties. We let n_e be the change in density of the electrons, n_i the change in density of the ions, v_e the change in velocity of the electrons, v_i the change in velocity of the ions, p_e the change in pressure of the electrons, p_i the change in pressure of the ions.

Because of equation (15-43), the charge density is

$$\rho = n_e q_e + n_i q_i, \tag{15-47}$$

while the current density is

$$\mathbf{J} = (\bar{n}_e + n_e)q_e \mathbf{v}_e + (\bar{n}_i + n_i)q_i \mathbf{v}_i. \tag{15-48}$$

When the disturbances are small, terms involving products of perturbations can be neglected. Then equation (15-48) reduces to

$$\mathbf{J} = \bar{n}_e q_e \mathbf{v}_e + \bar{n}_i q_i \mathbf{v}_i. \tag{15-49}$$

The electrons and the ions are separately conserved. When products of perturbations are neglected, equation (12-20) yields

$$\frac{\partial n_e}{\partial t} + \nabla \cdot \bar{n}_e \mathbf{v}_e = 0, \tag{15-50}$$

$$\frac{\partial n_i}{\partial t} + \nabla \cdot \bar{n}_i \mathbf{v}_i = 0. \tag{15-51}$$

The presence of a single index in the first term indicates no summation in the second term.

EXAMPLE 15.7   A region of a plasma contains 1.00×10^{13} electrons cm^{-3} and a similar concentration of singly charged cations. If the average energy in a translational degree of freedom $\frac{1}{2}kT$ is 10.0 eV, what is the electron pressure in the region?

From example 15.1, we have

$$1 \text{ ev} = 1.6022 \times 10^{-12} \text{ erg.}$$

Equation (15-45) yields

$$\begin{aligned} p &= (1.00 \times 10^{13} \text{ cm}^{-3})(20.0 \text{ eV})(1.602 \times 10^{-12} \text{ erg eV}^{-1}) \\ &\quad \times (9.869 \times 10^{-7} \text{ atm cm}^2 \text{ dyne}^{-1}) \\ &= 3.16 \times 10^{-4} \text{ atm.} \end{aligned}$$

15.7 An approximate equation of motion

In the usual ionized gas or plasma, a given charged particle interacts appreciably with several other particles, each of which interacts appreciably with quite a few other particles, and so on throughout the system. As a consequence, the equation of motion of the particle is not independent of an enormous number of other equations of motion. To make the mathematics tractable, we have to replace these equations with an equation dealing with the average behavior of each species present. The variables defined in the preceding section are applicable and will be employed. We also will assume, for simplicity, that only one species of molecule and one species of ion are present.

The forces acting on the particles of a given kind, per unit volume, include the net Coulomb forces, the net Lorentz force, the negative gradient of the particle pressure (recall a similar term in the Navier-Stokes equation), and the interaction forces with the other species. Indeed, applying Newton's second law to the electrons in the unit volume yields

$$\bar{n}_e m_e \frac{\partial \mathbf{v}_e}{\partial t} = \bar{n}_e q_e (\mathbf{E} + \mathbf{v}_e \times \mathbf{B}_0) - \nabla p_e + \mathbf{P}_{en} + \mathbf{P}_{ei} \qquad (15\text{-}52)$$

if we neglect any change from the imposed magnetic field $\mathbf{B}_0$ because the velocity perturbation $\mathbf{v}_e$ multiplying it is also a small quantity. The similar equation for the ions per unit volume is

$$\bar{n}_i m_i \frac{\partial \mathbf{v}_i}{\partial t} = \bar{n}_i q_i (\mathbf{E} + \mathbf{v}_i \times \mathbf{B}_0) - \nabla p_i + \mathbf{P}_{in} + \mathbf{P}_{ie}. \qquad (15\text{-}53)$$

The last two terms in each equation equal the rate at which momentum is transferred, by interaction, from the pertinent charged species to (a) the neutral species, and to (b) the other charged species. We write

$$\mathbf{P}_{en} = -\bar{n}_e m_e v_{en} \mathbf{v}_e, \qquad (15\text{-}54)$$

$$\mathbf{P}_{in} = -\bar{n}_i m_i v_{in} \mathbf{v}_i, \qquad (15\text{-}55)$$

$$\mathbf{P}_{ei} = -\mathbf{P}_{ie} = -\bar{n}_e m_e v (\mathbf{v}_e - \mathbf{v}_i), \qquad (15\text{-}56)$$

where v_{en} is the frequency of collision between a given electron and any neutral particle times the average fraction of the excess momentum lost in the collision, v_{in} is the same product for a given ion and any neutral particle, while v is the same product for a given electron and any ion.

We will be concerned with changes rapid enough to be approximately adiabatic. Applying the first law of thermodynamics to the ideal electron gas that was hypothesized in (15-45) and letting the increase in internal energy of this gas equal the pressure-volume work done on it leads to the equation of state

$$\frac{\bar{p}_e + p_e}{(\bar{n}_e + n_e)^{\gamma_e}} = \frac{\bar{p}_e}{\bar{n}_e^{\gamma_e}} = \text{constant} \qquad (15\text{-}57)$$

in which γ_e is the heat capacity of the electron gas at constant pressure divided by that at constant volume.

When the perturbation n_e in electron density is small, equation (15-57) yields the linear form

$$p_e = \gamma_e n_e k T_e. \tag{15-58}$$

On comparing results at neighboring points, we then get

$$\nabla p_e = \gamma_e k T_e \, \nabla n_e = m_e V_e^{\,2} \, \nabla n_e. \tag{15-59}$$

Similarly,

$$\nabla p_i = \gamma_i k T_i \, \nabla n_i = m_i V_i^{\,2} \, \nabla n_i. \tag{15-60}$$

The velocities V_e and V_i are defined by the equations

$$V_e^{\,2} = \frac{\gamma_e k T_e}{m_e}, \tag{15-61}$$

$$_i V^2 = \frac{\gamma_i k T_i}{m_i}. \tag{15-62}$$

EXAMPLE 15.8   How much can the average energy associated with one translational degree of freedom separate charges in a plasma?

For simplicity, consider a fluctuation in which all electrons are removed from the slab between

$$x = 0 \quad \text{and} \quad x = h$$

in such a way that the electric field vanishes at $x = 0$. Then an element of the slab of thickness dx, at a given x, contains the charge $-\bar{n}_e q_e \, dx$ per unit area. Applying formula (14-10) to such an element yields

$$dD = -\bar{n}_e q_e \, dx.$$

But

$$\mathbf{E} = \frac{\mathbf{D}}{\varepsilon} \quad \text{and} \quad \nabla \phi = -\mathbf{E}$$

by equations (14-8) and (14-35); so

$$dE = -\frac{\bar{n}_e q_e}{\varepsilon} \, dx$$

and

$$\frac{\partial \phi}{\partial x} = -E = \frac{\bar{n}_e q_e}{\varepsilon} x.$$

Integrating again produces

$$\phi = \frac{\bar{n}_e q_e}{2\varepsilon} x^2.$$

The potential energy of one of the removed electrons is

$$q_e \phi = \frac{\bar{n}_e q_e^{\,2} x^2}{2\varepsilon}.$$

The maximum occurs when $x = h$. If this maximum comes at the expense of the average energy in a translational degree of freedom, then

$$\frac{\bar{n}_e q_e^2 h^2}{2\varepsilon} = \tfrac{1}{2}kT,$$

whence

$$h = \left(\frac{\varepsilon k T}{\bar{n}_e q_e^2} \right)^{1/2}.$$

The quantity h is called the *Debye shielding distance*. It is a measure of the distance over which the electron density can deviate appreciably from $\bar{n}_i Z$.

Because an electron is much lighter than an ion, electrons are much more mobile. A floating solid surface tends to collect them much faster than cations, and a negative potential develops. This can extend into the gas only a few Debye lengths. The region of lowered potential is called a *sheath*. After it is fully developed, electrons and ions are lost to the surface at equal rates.

15.8 Consequences of linearity

We have seen how terms involving products of perturbations can be neglected when disturbances in a plasma are small. The governing equations then become linear, allowing different monochromatic plane waves to propagate independently. Oscillations travelling in one direction at one frequency do not then interact with those that differ in direction or frequency, and can be treated separately.

Let us consider a planar sinusoidal disturbance propagating along the z axis. To the real wave, let us add an imaginary wave with the same amplitude, 90° out of phase. This addition will not cause error because the imaginary wave is independent of the real one.

Each variable then has the form

$$\mathbf{A} = \mathscr{A} \exp \left[j(kz - \omega t) \right] \qquad (15\text{-}63)$$

where $\mathscr{A}$ is a complex constant and j is $\sqrt{-1}$. By carrying out the indicated operations, we find that

$$\nabla \cdot \mathbf{A} = jk A_z, \qquad (15\text{-}64)$$

$$\nabla \times \mathbf{A} = jk(A_x \mathbf{j} - A_y \mathbf{i}), \qquad (15\text{-}65)$$

$$\nabla^2 \mathbf{A} = -k^2 \mathbf{A}, \qquad (15\text{-}66)$$

$$\nabla \times (\nabla \times \mathbf{A}) = k^2 (A_x \mathbf{i} + A_y \mathbf{j}) = k^2 \mathbf{A}_\perp. \qquad (15\text{-}67)$$

Thus, del ∇ produces the same effect as multiplication by $j\mathbf{k}$

$$\nabla = j\mathbf{k} \qquad (15\text{-}68)$$

where $\mathbf{k}$ is the propagation vector in the z direction. Similarly, the effect of differentiation with respect to time is that of multiplication by $-j\omega$

$$\frac{\partial}{\partial t} = -j\omega. \qquad (15\text{-}69)$$

In general, propagation constant k is complex

$$k = \alpha + j\beta. \tag{15-70}$$

Then the velocity at which a given phase propagates is

$$v_\phi = \frac{\omega}{\alpha} \tag{15-71}$$

while the amplitude is

$$\mathscr{A}e^{-\beta z}. \tag{15-72}$$

EXAMPLE 15.9   How do Maxwell's equations reduce when the physical system is linear, and disturbances at one frequency do not excite disturbances at another frequency?

All variables in the equations are and remain real, from the way that they have been defined. A certain real excitation cannot induce imaginary excitations. But mathematically, we may multiply each of these variables by $j = \sqrt{-1}$. Since the original real system does not accumulate imaginary disturbances, the hypothetical imaginary system cannot accumulate real disturbances.

Mathematically, we may add to a real wave, an imaginary wave with the same amplitude, 90° out of phase, and know that this addition will not alter the real part at any future time.

Furthermore, when the system is linear, disturbances at one frequency do not lead to disturbances at any other frequency. Sinusoidal or exponential disturbances, substituted into Maxwell's equations and the auxiliary relationships, never produce a term involving a different frequency. A nonlinear term, however, always expands into linear terms involving other frequencies. Indeed

$$\mathscr{A} \cos^2 \omega t = \mathscr{A}(\tfrac{1}{2} + \tfrac{1}{2} \cos 2\omega t),$$
$$\mathscr{B} \sin^2 \omega t = \mathscr{B}(\tfrac{1}{2} - \tfrac{1}{2} \cos 2\omega t).$$

Consequently, in the linear system we may replace the electric field intensity and displacement with

$$\mathbf{E} = \mathscr{E}e^{-j\omega t},$$
$$\mathbf{D} = \mathscr{D}e^{-j\omega t},$$

where coefficients $\mathscr{E}$ and $\mathscr{D}$ are appropriate complex functions of position. Similarly, we may replace the magnetic field intensity and induction with

$$\mathbf{H} = \mathscr{H}e^{-j\omega t},$$
$$\mathbf{B} = \mathscr{B}e^{-j\omega t}.$$

Differentiating $\mathbf{D}$ and $\mathbf{B}$ with respect to time t then yields

$$\frac{\partial \mathbf{D}}{\partial t} = \mathscr{D}(-j\omega e^{-j\omega t}) = -j\omega \mathbf{D},$$

$$\frac{\partial \mathbf{B}}{\partial t} = \mathscr{B}(-j\omega e^{-j\omega t}) = -j\omega \mathbf{B},$$

and equations (14-91) and (14-95) become

$$\nabla \times \mathbf{E} = j\omega\mathbf{B},$$

$$\nabla \times \mathbf{H} = \mathbf{J} - j\omega\mathbf{D}.$$

The other equations remain unchanged because they do not involve derivatives with respect to time.

15.9   Algebraic equations for small-amplitude waves in the plasma

Maxwell's equations, the conservation laws, and the mechanical force equations govern the propagation of disturbances in a plasma. But when the perturbations are small, a constant times a product of these changes may be neglected with respect to a term linear in one small change. All nonlinear terms then drop out and section 15.8 applies. The differential equations reduce to relatively simple algebraic equations.

Actually, we have already carried out the linearizations. We need only replace the differentiations by their algebraic equivalents and reduce the equations.

We have also eliminated the magnetic field in equation (14-100). Let us now introduce prescriptions (15-67) and (15-69)

$$k^2\mathbf{E}_\perp - \frac{\omega^2}{c^2}\,\mathbf{E} = j\omega\mu\mathbf{J}. \tag{15-73}$$

Then multiply by jc^2/ω^2 and introduce formula (15-49)

$$j\left(\frac{k^2c^2}{\omega^2}\,\mathbf{E}_\perp - \mathbf{E}\right) + \frac{1}{\varepsilon\omega}\,(\bar{n}_e q_e \mathbf{v}_e + \bar{n}_i q_i \mathbf{v}_i) = 0. \tag{15-74}$$

Let us also introduce prescriptions (15-69) and (15-68) into conservation equation (15-50)

$$-j\omega n_e + \bar{n}_e j\mathbf{k} \cdot \mathbf{v}_e = 0 \tag{15-75}$$

and solve for the perturbation of electron density

$$n_e = \frac{\bar{n}_e k v_{ez}}{\omega}. \tag{15-76}$$

Operating on this with ∇ by formula (15-68) yields

$$\nabla n_e = \frac{\bar{n}_e k j\mathbf{k} v_{ez}}{\omega} \tag{15-77}$$

where $\mathbf{k}$ is the propagation vector (assumed to lie along the z axis).

The pressure gradient equation (15-59) then becomes

$$\nabla p_e = \frac{m_e V_e^2 \bar{n}_e j k \mathbf{k} v_{ez}}{\omega}. \tag{15-78}$$

Similarly,

$$\nabla p_i = \frac{m_i V_i^2 \bar{n}_i j k \mathbf{k} v_{iz}}{\omega} .$$ (15-79)

Finally, let us apply the prescriptions to force equation (15-52) and introduce the appropriate substitutions

$$-\bar{n}_e m_e j \omega \mathbf{v}_e = \bar{n}_e q_e (\mathbf{E} + \mathbf{v}_e \times \mathbf{B}_0) - \bar{n}_e m_e j \frac{k\mathbf{k}}{\omega} V_e^2 v_{ez}$$

$$- \bar{n}_e m_e v_{en} \mathbf{v}_e - \bar{n}_e m_e v (\mathbf{v}_e - \mathbf{v}_i).$$ (15-80)

On rearranging this result, we get

$$\bar{n}_e q_e (\mathbf{E} + \mathbf{v}_e \times \mathbf{B}_0) - \bar{n}_e m_e (v + v_{en} - j\omega) \mathbf{v}_e$$

$$- \bar{n}_e m_e j \frac{k\mathbf{k}}{\omega} V_e^2 v_{ez} + \bar{n}_e m_e v \mathbf{v}_i = 0.$$ (15-81)

Similarly from (15-53),

$$\bar{n}_i q_i (\mathbf{E} + \mathbf{v}_i \times \mathbf{B}_0) - [\bar{n}_i m_i (v_{in} - j\omega) + \bar{n}_e m_e v] \mathbf{v}_i$$

$$- \bar{n}_i m_i j \frac{k\mathbf{k}}{\omega} V_i^2 v_{iz} + \bar{n}_e m_e v \mathbf{v}_e = 0.$$ (15-82)

15.10 Condition for existence of small-amplitude waves in the plasma

The final field equation and the two force equations are vector equations that must be satisfied simultaneously. But each has three nontrivial components. In the Cartesian system in which the wave moves parallel to the z axis, we have nine scalar equations, linear and homogeneous in the nine variables

$$E_x, \quad v_{ex}, \quad v_{ix}, \quad E_y, \quad v_{ey}, \quad v_{iy}, \quad E_z, \quad v_{ez}, \quad v_{iz}.$$ (15-83)

The Cramer rule tells us that all of these variables are zero unless the determinant of the coefficients is zero. For a wave to exist in the plasma, this condition must be met.

We have already chosen the z axis to be parallel to the direction of propagation. The imposed magnetic field $\mathbf{B}_0$ then has a longitudinal component B_L in the direction of propagation and a transverse component B_T perpendicular to it. Let us set the y axis in the direction of the transverse component; so we have

$$B_{0x} = 0, \quad B_{0y} = B_T, \quad B_{0z} = B_L.$$ (15-84)

From the x-component of equations (15-74), (15-81), and (15-82), the y-component of these equations, and the z-component of the equations, we then find the coefficients listed in table 15.1.

Multiplying each of the coefficients in a row by a parameter corresponds to multiplying the matching equation by the constant; it does not change the solution. Multiplying each of the coefficients in a column by a parameter corresponds to multiplying the matching variable by the reciprocal of the constant. The determinant must still be zero.

Indeed to simplify the array, we multiply each element by the expression listed at the beginning of its row, and by the expression at the foot of its column, in table 15.1. If we also consider the transfer of momentum between different species to be small, we have

$$v = v_{en} = v_{in} = 0. \tag{15-85}$$

Table 15.1 then yields table 15.2.

Note how we have labeled the square of the plasma frequency for electron and ion

$$\frac{\bar{n}_e q_e^2}{\varepsilon m_e} = \omega_p^2, \tag{15-86}$$

$$\frac{\bar{n}_i q_i^2}{\varepsilon m_i} = \Omega_p^2, \tag{15-87}$$

and the cyclotron frequencies, both longitudinal and transverse, for the electron and the ion

$$\omega_L = -\frac{q_e B_L}{m_e}, \qquad \omega_T = -\frac{q_e B_T}{m_e}, \tag{15-88}$$

$$\Omega_L = \frac{q_i B_L}{m_i}, \qquad \Omega_T = \frac{q_i B_T}{m_i}. \tag{15-89}$$

15.11 Zeroth order dispersion equations

Each nontrivial solution of the field equation and the two force equations causes the determinant of the coefficients listed in table 15.1 to vanish. But the polynomial obtained on multiplying out the determinant is a quartic in k^2. The propagation constant itself equals plus or minus the square root of one of the solutions of this quartic. The four solutions correspond to the four kinds of plasma waves that are possible. The two signs yield the two possible directions of propagation along the z axis, for each type.

Variation of the parameters causes each type to change continuously. Each becomes particularly simple when all off diagonal elements become zero. Then the determinant vanishes only when the propagation constant is a number that causes one of the diagonal elements (one containing the k) to vanish. Let us now determine the nature of these simple waves.

TABLE 15.1   Coefficients of the Nine Simultaneous Homogeneous Equations with Simplifying Multipliers for Each Row and Column

Variables	E_x	v_{ex}	v_{ix}	E_y	v_{ey}
j	$j\left(\dfrac{k^2 c^2}{\omega^2} - 1\right)$	$\dfrac{\bar{n}_e q_e}{\varepsilon\omega}$	$\dfrac{\bar{n}_i q_i}{\varepsilon\omega}$	0	0
$\dfrac{1}{\bar{n}_e q_e}$	$\bar{n}_e q_e$	$-\bar{n}_e m_e(v + v_{en} - j\omega)$	$\bar{n}_e m_e v$	0	$\bar{n}_e q_e B_L$
$\dfrac{1}{\bar{n}_i q_i}$	$\bar{n}_i q_i$	$\bar{n}_e m_e v$	$-\bar{n}_i m_i(v_{in} - j\omega) - \bar{n}_e m_e v$	0	0
-1	0	0	0	$j\left(\dfrac{k^2 c^2}{\omega^2} - 1\right)$	$\dfrac{\bar{n}_e q_e}{\varepsilon\omega}$
$\dfrac{j}{\bar{n}_e q_e}$	0	$-\bar{n}_e q_e B_L$	0	$\bar{n}_e q_e$	$-\bar{n}_e m_e(v + v_{en} - j\omega)$
$\dfrac{j}{\bar{n}_i q_i}$	0	0	$-\bar{n}_i q_i B_L$	$\bar{n}_i q_i$	$\bar{n}_e m_e v$
-1	0	0	0	0	0
$\dfrac{j}{\bar{n}_e q_e}$	0	$\bar{n}_e q_e B_T$	0	0	0
$\dfrac{j}{\bar{n}_i q_i}$	0	0	$\bar{n}_i q_i B_T$	0	0
Multipliers	1	$-j\dfrac{q_e}{m_e\omega}$	$-j\dfrac{q_i}{m_i\omega}$	$-j$	$-\dfrac{q_e}{m_e\omega}$

TABLE 15.1 (continued)

Variables	v_{iy}	E_z	v_{ez}	v_{iz}
j	0	0	0	0
$\dfrac{1}{\bar{n}_e q_e}$	0	0	$-\bar{n}_e q_e B_T$	0
$\dfrac{1}{\bar{n}_i q_i}$	$\bar{n}_i q_i B_L$	0	0	$-\bar{n}_i q_i B_T$
-1	$\dfrac{\bar{n}_i q_i}{\omega}$	0	0	0
$\dfrac{j}{\bar{n}_e q_e}$	$\bar{n}_e m_e \nu$	0	0	0
$\dfrac{j}{\bar{n}_i q_i}$	$-\bar{n}_i m_i(\nu_{in} - j\omega) - \bar{n}_e m_e \nu$	0	0	0
-1	0	$-j$	$\dfrac{\bar{n}_e q_e}{\varepsilon \omega}$	$\dfrac{\bar{n}_i q_i}{\varepsilon \omega}$
$\dfrac{j}{\bar{n}_e q_e}$	0	$\bar{n}_e q_e$	$-\bar{n}_e m_e\left(\nu + \nu_{en} - j\omega + j\dfrac{k^2}{\omega}V_e^2\right)$	$\bar{n}_e m_e \nu$
$\dfrac{j}{\bar{n}_i q_i}$	0	$\bar{n}_i q_i$	$\bar{n}_e m_e \nu$	$-\bar{n}_i m_i\left(\nu_{in} - j\omega + j\dfrac{k^2}{\omega}V_i^2\right) - \bar{n}_e m_e \nu$
Multipliers $-\dfrac{q_i}{m_i \omega}$	$-\dfrac{q_i}{m_i \omega}$	$-j$	$-\dfrac{q_e}{m_e \omega}$	$-\dfrac{q_i}{m_i \omega}$

TABLE 15.2   The Transformed Variables and Elements for the Simplified Consistency Determinant

E_x	$j\dfrac{v_{ex}m_e\omega}{q_e}$	$j\dfrac{v_{ix}m_i\omega}{q_i}$	jE_y	$-\dfrac{v_{ey}m_e\omega}{q_e}$	$-\dfrac{v_{iy}m_i\omega}{q_i}$	jE_z	$-\dfrac{v_{ez}m_e\omega}{q_e}$	$-\dfrac{v_{iz}m_i\omega}{q_i}$
$1-\dfrac{k^2c^2}{\omega^2}$	$\dfrac{\omega_p^2}{\omega^2}$	$\dfrac{\Omega_p^2}{\omega^2}$	0	0	0	0	0	0
1	1	0	0	$\dfrac{\omega_L}{\omega}$	0	0	$-\dfrac{\omega_T}{\omega}$	0
1	0	1	0	0	$-\dfrac{\Omega_L}{\omega}$	0	0	$\dfrac{\Omega_T}{\omega}$
0	0	0	$1-\dfrac{k^2c^2}{\omega^2}$	$\dfrac{\omega_p^2}{\omega^2}$	$\dfrac{\Omega_p^2}{\omega^2}$	0	0	0
0	$\dfrac{\omega_L}{\omega}$	0	1	1	0	0	0	0
0	0	$-\dfrac{\Omega_L}{\omega}$	1	0	1	0	0	0
0	0	0	0	0	0	1	$\dfrac{\omega_p^2}{\omega^2}$	$\dfrac{\Omega_p^2}{\omega^2}$
0	$-\dfrac{\omega_T}{\omega}$	0	0	0	0	1	$1-\dfrac{k^2V_e^2}{\omega^2}$	0
0	0	$\dfrac{\Omega_T}{\omega}$	0	0	0	1	0	$1-\dfrac{k^2V_i^2}{\omega^2}$

When the off diagonal elements are zero and the k makes the 11 and 44 elements zero, we have

$$1 - \frac{k^2 c^2}{\omega^2} = 0, \tag{15-90}$$

whence

$$k = \pm \frac{\omega}{c}. \tag{15-91}$$

Since

$$k = \pm \frac{2\pi}{\lambda} \quad \text{and} \quad \omega = 2\pi v \tag{15-92}$$

where λ is the wavelength and v the conventional frequency, equation (15-91) corresponds to the familiar

$$c = \lambda v. \tag{15-93}$$

Parameter c is the speed at which a given phase of the wave propagates.

Substituting (15-90) into the appropriate simultaneous equations shows that we may have

$$E_x \neq 0 \quad \text{and} \quad E_y = E_z = 0 \tag{15-94}$$

or

$$E_y \neq 0 \quad \text{and} \quad E_x = E_z = 0. \tag{15-95}$$

Consequently, there are two independent transverse electromagnetic waves corresponding to (15-91).

When the off diagonal elements are zero and the k makes the 88 element of the simplified system (table 15-2) vanish, we similarly obtain

$$1 - \frac{k^2 V_e^2}{\omega^2} = 0; \tag{15-96}$$

so

$$k = \pm \frac{\omega}{V_e}. \tag{15-97}$$

The 88 element is the coefficient of a constant times v_{ez} in the pertinent simultaneous equation, whose other terms are now zero. Consequently, we may have

$$v_{ez} \neq 0. \tag{15-98}$$

In the next simultaneous equation, the coefficient of v_{iz} is not zero while the other terms are; so

$$v_{iz} = 0. \tag{15-99}$$

We see that the disturbance is a longitudinal wave in which ion oscillations are negligible. It is called the *electron wave*. From (15–97), the speed at which a given phase travels is V_e.

Likewise from the 99 element of table 15.2, we get

$$1 - \frac{k^2 V_i^2}{\omega^2} = 0, \tag{15-100}$$

$$k = \pm \frac{\omega}{V_i}. \tag{15-101}$$

Combining (15-100) with the appropriate simultaneous equations shows that we now have a wave involving

$$v_{iz} \neq 0 \quad \text{and} \quad v_{ez} = 0, \tag{15-102}$$

traveling at speed V_i. From equations (15-102), it is a longitudinal wave in which ion oscillations dominate. Consequently, it is called the *ion wave*.

An equation linking the propagation constant and the frequency, like (15-91), (15-97), and (15-101) is called a *dispersion equation*.

15.12 Effect of the plasma on a transverse wave

When there is no externally imposed magnetic field, the longitudinal and transverse cyclotron frequencies vanish and the consistency determinant can be factored into three 3 × 3 determinants. Two of these govern the transverse waves, one the longitudinal waves.

Indeed from table 15.2, the simplified consistency equation has the form

$$
\begin{vmatrix}
\begin{array}{ccc} & & \\ & \Delta_T & \\ & & \end{array} &
\begin{array}{ccc} 0 & 0 & 0 \\ 0 & \dfrac{\omega_L}{\omega} & 0 \\ 0 & 0 & -\dfrac{\Omega_L}{\omega} \end{array} &
\begin{array}{ccc} 0 & 0 & 0 \\ 0 & -\dfrac{\omega_T}{\omega} & 0 \\ 0 & 0 & \dfrac{\Omega_T}{\omega} \end{array} \\[4ex]
\begin{array}{ccc} 0 & 0 & 0 \\ 0 & \dfrac{\omega_L}{\omega} & 0 \\ 0 & 0 & -\dfrac{\Omega_L}{\omega} \end{array} &
\begin{array}{ccc} & & \\ & \Delta_T & \\ & & \end{array} &
\begin{array}{ccc} 0 & 0 & 0 \\ 0 & 0 & 0 \\ 0 & 0 & 0 \end{array} \\[4ex]
\begin{array}{ccc} 0 & 0 & 0 \\ 0 & -\dfrac{\omega_T}{\omega} & 0 \\ 0 & 0 & \dfrac{\Omega_T}{\omega} \end{array} &
\begin{array}{ccc} 0 & 0 & 0 \\ 0 & 0 & 0 \\ 0 & 0 & 0 \end{array} &
\begin{array}{ccc} & & \\ & \Delta_L & \\ & & \end{array}
\end{vmatrix} = 0, \tag{15-103}
$$

where

$$\Delta_T = \begin{vmatrix} 1 - \dfrac{k^2 c^2}{\omega^2} & \dfrac{\omega_p^{\,2}}{\omega^2} & \dfrac{\Omega_p^{\,2}}{\omega^2} \\[2mm] 1 & 1 & 0 \\[2mm] 1 & 0 & 1 \end{vmatrix} \qquad (15\text{-}104)$$

and

$$\Delta_L = \begin{vmatrix} 1 & \dfrac{\omega_p^{\,2}}{\omega^2} & \dfrac{\Omega_p^{\,2}}{\omega^2} \\[2mm] 1 & 1 - \dfrac{k^2 V_e^{\,2}}{\omega^2} & 0 \\[2mm] 1 & 0 & 1 - \dfrac{k^2 V_i^{\,2}}{\omega^2} \end{vmatrix}. \qquad (15\text{-}105)$$

When the mean magnetic field $\mathbf{B}_0$ is zero,

$$\omega_L = \omega_T = \Omega_L = \Omega_T = 0, \qquad (15\text{-}106)$$

the subdeterminants off the main diagonal vanish and we must have either

$$\Delta_T = 0 \qquad (15\text{-}107)$$

or

$$\Delta_L = 0. \qquad (15\text{-}108)$$

The former condition yields a purely transverse wave, the latter a purely longitudinal one.

Multiplying out Δ_T and substituting into equation (15-107), we obtain

$$1 - \frac{k^2 c^2}{\omega^2} - \frac{\omega_p^{\,2}}{\omega^2} - \frac{\Omega_p^{\,2}}{\omega^2} = 0, \qquad (15\text{-}109)$$

whence

$$k = \pm \frac{(\omega^2 - \omega_p^{\,2} - \Omega_p^{\,2})^{1/2}}{c}. \qquad (15\text{-}110)$$

Note how the plasma makes k smaller for a given frequency.
Indeed when

$$\omega = (\omega_p^{\,2} + \Omega_p^{\,2})^{1/2} = \omega_{cr}, \qquad (15\text{-}111)$$

the propagation constant vanishes. And when

$$\omega < (\omega_p^{\,2} + \Omega_p^{\,2})^{1/2}, \qquad (15\text{-}112)$$

k is imaginary and the wave is attenuated exponentially.

We call ω_{cr} the *cutoff frequency*. The plasma acts as a high-pass filter, being transparent only to frequencies above the cutoff frequency.

15.13   Effects of the electromagnetic field on the longitudinal waves

The nondiagonal elements in Δ_L similarly affect the electron wave. But the ion wave is influenced differently.

To see why, first multiply out the Δ_L in equation (15-108):

$$\left(1 - \frac{k^2 V_e^2}{\omega^2}\right)\left(1 - \frac{k^2 V_i^2}{\omega^2}\right) - \left(1 - \frac{k^2 V_e^2}{\omega^2}\right)\frac{\Omega_p^2}{\omega^2} - \left(1 - \frac{k^2 V_i^2}{\omega^2}\right)\frac{\omega_p^2}{\omega^2} = 0.$$

(15-113)

Then introduce the new parameters,

$$\omega_0^2 \equiv \omega_p^2 + \Omega_p^2,$$

(15-114)

$$V_s^2 \equiv \frac{\omega_p^2 V_i^2 + \Omega_p^2 V_e^2}{\omega_p^2 + \Omega_p^2},$$

(15-115)

and the dimensionless variables,

$$x \equiv \frac{\omega_p^2 + \Omega_p^2}{\omega^2} \equiv \frac{\omega_0^2}{\omega^2},$$

(15-116)

$$y \equiv \frac{k^2 c^2}{\omega^2},$$

(15-117)

to get the simpler form

$$\left(1 - y\frac{V_e^2}{c^2}\right)\left(1 - y\frac{V_i^2}{c^2}\right) - x\left(1 - y\frac{V_s^2}{c^2}\right) = 0.$$

(15-118)

We recognize this as the equation for a hyperbola.

At very high frequencies, x becomes small and we must have either

$$1 - y\frac{V_e^2}{c^2} = 0$$

(15-119)

or

$$1 - y\frac{V_i^2}{c^2} = 0.$$

(15-120)

These are identical with equations (15-96) and (15-100). The results obtained in the zeroth order calculation are actually satisfied when ω is large enough.

Expressions V_i^2/c^2 and V_s^2/c^2 in equation (15-118) are much smaller than V_e^2/c^2; so when y is very small, yV_i^2/c^2 and yV^2/c^2 are much smaller with respect to 1 than yV_e^2/c^2, and equation (15-118) reduces to

$$1 - y\frac{V_e^2}{c^2} - x = 0.$$

(15-121)

Transforming back to the original variables and parameters changes (15-121) to

$$1 - \frac{k^2 V_e^2}{\omega^2} - \frac{\omega_p^2 + \Omega_p^2}{\omega^2} = 0, \tag{15-122}$$

whence

$$k = \pm \frac{(\omega^2 - \omega_p^2 - \Omega_p^2)^{1/2}}{V_e}. \tag{15-123}$$

This relationship is approximately correct along the electron-wave curve close to the x axis (where y is small). Indeed, it is exact at $x = 0$ where equation (15-121) reduces to (15-119), and is certainly valid at $x = 1$ where y has to vanish and k be zero. At larger x (lower frequencies), k is imaginary and any wave is attenuated rapidly. The cutoff frequency is

$$\omega = (\omega_p^2 + \Omega_p^2)^{1/2} = \omega_{cr} \tag{15-124}$$

as before.

Note that the cutoff occurs on only one of the branches of the hyperbola. On the other, x may increase without limit. Then to satisfy equation (15-118), we must have the coefficient of x vanish:

$$1 - y \frac{V_s^2}{c^2} = 0. \tag{15-125}$$

This condition also insures that

$$\frac{dy}{dx} = 0 \tag{15-126}$$

in equation (15-118). As x increases, the branch of the hyperbola approaches a horizontal asymptote.

Equation (15-125) implies that

$$y = \frac{c^2}{V_s^2}, \tag{15-127}$$

but

$$y = \frac{k^2 c^2}{\omega^2}, \tag{15-128}$$

so

$$k^2 = \frac{\omega^2}{V_s^2} \quad \text{and} \quad k = \pm \frac{\omega}{V_s}. \tag{15-129}$$

When x is large enough, a given phase propagates at speed V_s. This is the *speed of sound* in the plasma.

15.14   Effects from an externally imposed magnetic field

A nonzero longitudinal component in $\mathbf{B}_0$ makes

$$\frac{\omega_L}{\omega} \neq 0 \quad \text{and} \quad \frac{\Omega_L}{\omega} \neq 0. \tag{15-130}$$

If in addition, the transverse component of B_0 is zero, the determinant Δ_L still factors from equation (15-103), and the equations in section 15.13 still hold. But a single Δ_T does not factor out. Consequently, the two transverse waves interact producing two circularly polarized waves.

When the transverse component in B_0 does differ from zero, we have

$$\frac{\omega_T}{\omega} \neq 0 \quad \text{and} \quad \frac{\Omega_T}{\omega} \neq 0. \tag{15-131}$$

The determinant Δ_L no longer factors out of (15-103). As a result, the longitudinal waves mix with the transverse one perpendicular to B_T. If B_L is also different from zero, the other transverse wave is mixed in too.

We will not consider these effects in detail. A few general comments are in order, however.

15.15   The plasma state

Because of its distinctive properties, the state of a substance when it is behaving as a plasma has been described as the fourth state of matter. But the plasma state is not separated from the solid, liquid, and ordinary gaseous states by phase transitions, and it cannot exist in equilibrium with any of the other states of the same material.

A plasma need not be in thermodynamic equilibrium with its surroundings. A substance may be ionized by passing it through an electric discharge, and behave as a plasma for some time after its emergence.

Over any considerable region, a plasma is nearly neutral. In the sheath surrounding a plasma, however, a negative charge accumulates that prevents additional electrons from escaping more rapidly than cations. The thickness of such a charged region is of the order of the Debye length.

In the absence of an imposed magnetic field, two transverse and two longitudinal waves can propagate in a plasma. There is a cutoff frequency below which the transverse and the electron longitudinal waves are attenuated very rapidly, even when momentum interchange between electrons and ions is neglected.

DISCUSSION QUESTIONS

15.1 When is a gas that has moved out of an electric discharge a plasma?

15.2 Show how the rate of change of **B** in a region may be appreciable even though **B** itself remains small or practically constant.

15.3 What determines the motion of the guiding center of a gyrating particle?

15.4 Why is a plasma diamagnetic?

15.5 How can a magnetic field alter the kinetic energy of a charged particle?

15.6 How can a magnetic field reflect some charged particles? Why can't it reflect all of them from a given region?

15.7 Explain what variables and parameters characterize a plasma.

15.8 Why can we average the forces acting on a given species at a given time to obtain the average effect at that time?

15.9 How does the frequency of collision vary with the density of the target species? Explain.

15.10 Derive equation (15-57).

15.11 Is a representative region of a sheath electrically neutral?

15.12 Explain when we can replace each physical variable $\mathbf{A}$ with the complex form

$$\mathbf{A} = \mathscr{A} \exp\left[j(kz - \omega t)\right].$$

15.13 Could we associate the physical variable $\mathbf{A}$ with the imaginary part of the complex form instead of with the real part?

15.14 Why don't we need to add an arbitrary phase to the argument of the exponential function in question 15.12?

15.15 Prove equations (15-64), (15-65), (15-66), and (15-67).

15.16 Combine the results from example 15.9 with the constitutive equations

$$\mathbf{B} = \mu\mathbf{H}, \qquad \mathbf{D} = \varepsilon\mathbf{E}, \qquad c^2 = \frac{1}{\mu\varepsilon},$$

to obtain an algebraic form for $\nabla \times (\nabla \times \mathbf{E})$ that is linear in $\mathbf{J}$ and $\mathbf{E}$.

15.17 Explain the linearized equations that govern the propagation of weak waves in a plasma.

15.18 Why must the determinant of the coefficients in the linearized equations vanish?

15.19 How can a microwave have a longitudinal component when traveling in a wave guide?

15.20 Explain the different kinds of waves that can be excited in a plasma.

PROBLEMS

15.1 Show that a free proton cannot capture a slowly moving electron.

15.2 If protons are to be accelerated in a cyclotron by a 12.0×10^6 sec^{-1} radio-frequency electric field, what magnetic field strength is required?

15.3 What is the approximate radius of gyration of the protons in problem 15.2 when they possess 1.00 Mev?

15.4 Substitute the values of the pertinent fundamental constants into the expression for the Debye shielding distance and reduce the formula.

15.5 If a charged particle enters a converging magnetic field so it spirals symmetrically and if

$$v_\perp = av_\|$$

initially, how strong must the field become to reflect the particle?

15.6 Introduce the magnetic induction at z and at a turning point into the integrand of the invariant action integral

$$J = \oint v_\| \, dz.$$

Is the integral of $v_\| \, dz$ from one turning point to the other invariant?

15.7 For what substance is the cutoff frequency increased most by Ω_p? By what fraction is it increased?

15.8 Estimate the electron concentration in an ionized layer that is transparent to radio waves above 28.0 megahertz.

15.9 Find an expression for the group velocity $d\omega/dk$ of the electron wave when there is no applied magnetic field. What is the product of this with the phase velocity λv?

— — —

15.10 Show what happens to the kinetic energy in the center-of-mass system when an electron is picked up by a polyatomic ion (in the ionosphere, for instance).

15.11 If the effective mass of an electron in a semiconductor subject to a 0.100 volt sec m^{-2} magnetic induction is 0.150 the free electron mass, what is the resonant frequency?

15.12 If the cyclotron resonance in problem 15.11 occurs at 4 °K where the mean velocity for a Maxwellian distribution is 4×10^6 cm sec^{-1}, what is the radius of gyration? About how long must the mean free path of the electron be before this resonance can be observed?

15.13 Calculate the magnetic flux

$$\int \mathbf{B} \cdot d\mathbf{S}$$

that is enclosed by the gyration of a particle. Is it invariant?

15.14 Calculate the Debye length for a rarefied plasma in which T is 1.85×10^4 degrees and n_e is 5.5×10^7 cm^{-3}.

15.15 What is the cutoff frequency in a plasma containing 1.00×10^{12} electrons cm^{-3}?

15.16 At what wavelength of light will a film of sodium become transparent if its density is 0.97 g cm^{-3}?

15.17 Find the equation for the tangent to the ion-wave dispersion curve at $x = 0$.

15.18 What is the group velocity $d\omega/dk$ of the ion wave in the region where the equation in problem 15.17 holds? What is the product of this with the phase velocity λv? What are these quantities in the region of large x, where the curve is very close to its asymptote?

REFERENCES

BOOKS

Delcroix, J. L. *Introduction to the Theory of Ionized Gases* (Translated by M. Clark, Jr., D. J. BenDaniel, and J. M. BenDaniel), pp. 1–134. Wiley-Interscience, New York, 1960.

Denisse, J. F., and **Delcroix, J. L.** *Plasma Waves* (Translated by M. Weinrich and D. J. BenDaniel), pp. 1–77. Wiley-Interscience, New York, 1963.

Rossi, B., and **Olbert, S.** *Introduction to the Physics of Space*, pp. 1–156. McGraw-Hill Book Company, New York, 1970.

Shohet, J. L. *The Plasma State*, pp. 1–12, 39–68. Academic Press, Inc., New York, 1971.

Spitzer, L., Jr. *Physics of Fully Ionized Gases*, pp. 1–48. Wiley-Interscience, New York, 1962.

Tanenbaum, B. S. *Plasma Physics*, pp. 1–138. McGraw-Hill Book Company, New York, 1967.

ARTICLES

Bass, F. G., and **Gurevich, Yu. G.** Nonlinear Theory of the Propagation of Electromagnetic Waves in a Solid-State Plasma and in a Gaseous Discharge. *Soviet Phys. Uspekhi*, **14**, 113 (1971).

Bekefi, G., and **Brown, S. C.** Resource Letter PP-2 on Plasma Physics: Waves and Radiation Processes in Plasmas. *Am. J. Phys.*, **34**, 1001 (1966).

Brown, S. C., *et al.* Outline of a Course in Plasma Physics. *Am. J. Phys.*, **31**, 637 (1963).

Burman, R. A sequel to: "Wave Propagation in Inhomogeneous Gyrotropic Warm Plasmas" by H. Unz. *Am. J. Phys.*, **36**, 635 (1968).

Chen, H. C. Compressivity Tensor and Dispersion Relation for a Plasma. *Am. J. Phys.*, **37**, 1022 (1969).

Collett, E. The Description of Polarization in Classical Physics. *Am. J. Phys.*, **36,** 713 (1968).

French, J. D., Lamb, W. H., Jr., and **Young, P. J.** Students' Calculation of Cosmic Ray Trajectories. *Am. J. Phys.*, **39,** 103 (1971).

Habert, R., and **Samaddar, S. N.** Behavior of the Complex Refractive Index of an Isotropic Plasma. *Am. J. Phys.*, **36,** 1117 (1968).

Hoyaux, M. F. Some Remarks Concerning the Phenomenon of Ambipolar Diffusion in Gaseous Discharges. *Am. J. Phys.*, **35,** 232 (1967).

Hoyaux, M. F. The Low-Pressure Arc Positive Column in a Longitudinal Magnetic Field. *Am. J. Phys.*, **36,** 726 (1968).

Hurt, W. B. The Faraday Dark Space of a Glow Discharge. *Am. J. Phys.*, **37,** 47 (1969).

Kadomtsev, B. B., and **Karpman, V. I.** Nonlinear Waves. *Soviet Phys. Uspekhi*, **14,** 40 (1971).

Maxfield, B. W. Helicon Waves in Solids. *Am. J. Phys.*, **37,** 241 (1969).

Menkes, J. Measurement Objectives in Plasma Physics. *Am. J. Phys.*, **39,** 664 (1971).

Radoski, H. R. Introduction to Geophysical Magnetohydrodynamic Resonances. *Am. J. Phys.*, **35,** 128 (1967).

Scholz, P. D., and **Anderson, T. P.** Four Electromagnetic Shock Tube Experiments to Demonstrate Some High-Temperature High-Speed Flow Phenomena. *Am. J. Phys.*, **38,** 279 (1970).

Schulz, M. Debye Shielding and Virtual Plasma Oscillations. *Am. J. Phys.*, **35,** 117 (1967).

Spavieri, G. Limits of the Adiabatic Approximation. *Am. J. Phys.*, **39,** 599 (1971).

Stern, D. P. Euler Potentials. *Am. J. Phys.*, **38,** 494 (1970).

16 / The Linking of Time to Space

16.1 Concepts that form a basis for Minkowski geometry

Physicists, chemists, and astronomers have long studied the rates of processes on earth and at various places in the universe. They have found that variations in the behavior of a cyclic mechanism are due to changes in the mechanism and in its surroundings. Any number of similar cyclic mechanisms, or clocks, can be constructed; so identical time scales can be established at different positions on extended bodies, or on different particles. The time on one isolated particle can be related to that on another at rest with respect to it by slowly transporting a clock between the places.

1. Indeed, time appears to increase smoothly and uniformly on each isolated small body or particle.

2. The time on one such body can be synchronized with that on another at rest with respect to it.

The possible positions of a free particle at rest with respect to an isolated particle mark out a space erected on the latter. Measurements of distances and angles show that this part of the continuum meets the following conditions.

3. Space is apparently homogeneous and uniform. Variations in the reaction of a test particle with position are presumably caused by variations in one or more field intensities.

4. Space is approximately Euclidean. Within any small, accessible region of the universe, deviations from the Pythagorean theorem are small.

A coordinate system at rest in the nonrotating space based on a given force-free particle is called an *inertial* system or *frame*. Another force-free particle may move at any velocity within limits with respect to the first particle, and an inertial system can be erected on it.

Insofar as a person can tell, all inertial frames are equivalent, for the collection of possible appearances of a physical system is the same in each. No one frame can be picked out and said to be absolutely at rest.

This equivalence of inertial frames was introduced by Albert Einstein in 1905 as the *relativity principle*. The resulting geometry was presented in a simple manner in 1908 by Hermann Minkowski.

16.2   Linearity of the formulas relating different inertial systems

Inertial Cartesian systems that are at rest with respect to each other are linked by one or more of the transformations: a translation, a rotation, a reflection, a scale change. But when the systems are in relative motion, time may also mix with one or more of the coordinates. Since the former transformations are linear, we need only investigate the possible linearity of the latter.

The former transformations are eliminated by constructing the systems as follows: The same units for distance and interval are employed in each. Furthermore, points are measured from the same event, so that when both time coordinates are zero, the two origins coincide.

The path that the origin of the second system Σ' traces out in the first system Σ is made the x axis. The y and z axes are constructed perpendicular to this line and to each other. In Σ', the x' axis is laid out along the x axis, while the y' axis is constructed parallel to the y axis and the z' axis parallel to the z axis, as in figure 16.1.

Let us now consider two different events that occur at a given y and z (implying a given y' and z'). Let the distance between the events be Δx in Σ and $\Delta x'$ in Σ', while the time between them is Δt in Σ and $\Delta t'$ in Σ'.

Because of the homogeneity of space and time in each frame, the relationships among these increments are independent of the positions of either event:

$$\Delta x' = F(\Delta x, \Delta t) \tag{16-1}$$

and

$$\Delta t' = G(\Delta x, \Delta t). \tag{16-2}$$

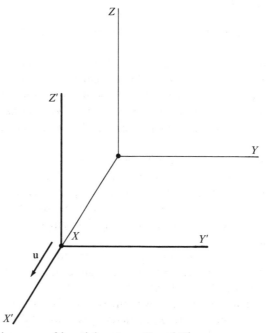

FIGURE 16.1   Cartesian axes of inertial systems Σ and Σ'.

Since the dependences of $\Delta x'$ and $\Delta t'$ on Δx and Δt should be analytic, these equations expand as follows:

$$\Delta x' = a_{00} + a_{10}\,\Delta x + a_{01}\,\Delta t + a_{20}\,\Delta x^2 + a_{11}\,\Delta x\,\Delta t + a_{02}\,\Delta t^2 + \cdots,$$
$$(16\text{-}3)$$
$$\Delta t' = b_{00} + b_{10}\,\Delta x + b_{01}\,\Delta t + b_{20}\,\Delta x^2 + b_{11}\,\Delta x\,\Delta t + b_{02}\,\Delta t^2 + \cdots.$$
$$(16\text{-}4)$$

But since $\Delta x'$ and $\Delta t'$ are zero when Δx and Δt are zero, we have

$$a_{00} = 0, \tag{16-5}$$

$$b_{00} = 0. \tag{16-6}$$

If the events occur in the life of a force-free particle, the same conditions prevail at the end of the interval as at the beginning. Then a second Δt and its accompanying Δx produce the same effect as the first. Doubling Δx and Δt doubles $\Delta x'$ and $\Delta t'$. Such a result is not possible in general unless the higher terms in (16-3) and (16-4) are zero. Therefore,

$$a_{20} = a_{11} = a_{02} = \cdots = 0, \tag{16-7}$$

$$b_{20} = b_{11} = b_{02} = \cdots = 0, \tag{16-8}$$

the coefficients of all nonlinear terms must be zero.

If we take the increments from the time and place the origins were coincident, we obtain x' and t' as linear functions of x and t. Let us write these relationships in the form

$$x' = ax - bct, \tag{16-9}$$

$$ct' = -ex + fct, \tag{16-10}$$

with c a speed to be identified later.

Since Σ' is not moving in the y or z directions, we assume that distances in these directions are invariant:

$$y' = y, \tag{16-11}$$

$$z' = z. \tag{16-12}$$

EXAMPLE 16.1   What forms do equations (16-9) and (16-10) assume if the time in Σ is exactly the same as that in Σ'?

If

$$t' = t,$$

then time does *not* mix in varying ways with space and distance x' is merely distance x minus the distance the origin of Σ' has moved:

$$x' = x - ut.$$

These old formulas are said to describe a *Galilean transformation*.

16.3   The universal speed

Whenever transformation from one inertial frame to another mixes the temporal and spatial coordinates in the linear manner just described, one particular speed is the same in each frame. Constants a, b, e, and f, can be fixed so that parameter c is this invariant speed.

Let us consider a particle that starts from the origin of Σ' at $t' = 0$ and that propagates in a certain direction with the speed c. Let us also assume that the Pythagorean theorem holds. Then the coordinates of the particle obey the equation

$$x'^2 + y'^2 + z'^2 = c^2 t'^2. \tag{16-13}$$

Inserting expressions (16-9), (16-10), (16-11), (16-12) changes this equation to

$$a^2 x^2 - 2abxct + b^2 c^2 t^2 + y^2 + z^2 = e^2 x^2 - 2efxct + f^2 c^2 t^2 \tag{16-14}$$

or

$$(a^2 - e^2)x^2 + 2(ef - ab)xct + y^2 + z^2 = c^2(f^2 - b^2)t^2. \tag{16-15}$$

Equation (16-15) reduces to the form

$$x^2 + y^2 + z^2 = c^2 t^2 \tag{16-16}$$

that describes the particle traveling at speed c in Σ if

$$a^2 - e^2 = 1, \tag{16-17}$$

$$f^2 - b^2 = 1, \tag{16-18}$$

$$ab = ef. \tag{16-19}$$

But our choice of coordinate systems does not really limit how one inertial frame moves with respect to another. The linear transformation equations consequently allow a particle or disturbance to travel at the *same* speed c in all inertial frames. We call this constant the *fundamental speed c*.

Experiments beginning with those of Albert A. Michelson and Edward W. Morley in 1887 indicate the speed of light is this invariant parameter. A spherically symmetric wave front starting at the origin of Σ at $t = 0$ and expanding at speed c in Σ appears as a spherical wave spreading at speed c in Σ' (see figure 16.2).

EXAMPLE 16.2   What is fundamental speed c in a world adhering to the Galilean transformation law inferred in example 16.1?

The transformation equation

$$ct' = -ex + fct$$

reduces to

$$t' = t$$

in a continuum for which either

$$e = 0 \quad \text{and} \quad f = 1$$

or

$$c = \infty.$$

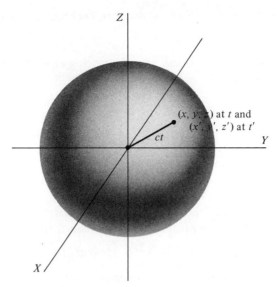

FIGURE 16.2   Coordinates of an element of a spherical wave spreading at the fundamental speed c.

But the law

$$x' = x - ut,$$

discussed in example 16.1, implies that

$$a = 1, \qquad b = \frac{u}{c}.$$

In order for equation (16-19),

$$ef = ab,$$

to hold, constants e and f cannot be 0 and 1, respectively. Hence, speed c is infinite.

16.4   Key covariant functions and differentials

When the transformation describing how one inertial frame moves with respect to another inertial frame leaves a *finite* speed c invariant, the transformation alters the 3-dimensional distance between two given events. However, subtracting the square of the distance that a particle traveling at speed c would have moved during the time between the events from the square of this 3-dimensional distance yields an invariant quantity. The corresponding difference for events infinitesimally far apart will be employed in dealing with continuous changes.

Let us locate the origins of both inertial systems Σ and Σ' on one of the given events. Let us orient the axes as in figure 16.1. Let us also suppose that the second event occurs at position (x, y, z) and time t in Σ and at position (x', y', z') and time t' in Σ'.

In inertial system Σ, the square of the distance between the two events is

$$x^2 + y^2 + z^2, \qquad (16\text{-}20)$$

while the square of the distance that a particle moving at speed c would cover in the time between the two events is

$$c^2t^2. \qquad (16\text{-}21)$$

Subtracting expression (16-21) from (16-20) yields

$$x^2 + y^2 + z^2 - c^2t^2 = r^2. \qquad (16\text{-}22)$$

In inertial system Σ', the similar difference is

$$x'^2 + y'^2 + z'^2 - c^2t'^2 = r'^2. \qquad (16\text{-}23)$$

The coordinates in the two frames are related by equations (16-9) through (16-12). Using these relationships to eliminate the primed coordinates from (16-23) leads to

$$(a^2 - e^2)x^2 + 2(ef - ab)xct + y^2 + z^2 - c^2(f^2 - b^2)t^2 = r'^2. \qquad (16\text{-}24)$$

When c is the fundamental speed, as we assume, equations (16-17) through (16-19) are valid, the left side of (16-24) is identical to the left side of (16-22), and we have

$$r'^2 = x'^2 + y'^2 + z'^2 - c^2t'^2 = x^2 + y^2 + z^2 - c^2t^2 = r^2. \qquad (16\text{-}25)$$

Thus, distance r is invariant:

$$r' = r. \qquad (16\text{-}26)$$

Dividing (16-25) by $-c^2$ and letting $-r'^2/c^2$ be τ'^2, $-r^2/c^2$ be τ^2, yields

$$\tau'^2 = t'^2 - \frac{x'^2 + y'^2 + z'^2}{c^2} = t^2 - \frac{x^2 + y^2 + z^2}{c^2} = \tau^2. \qquad (16\text{-}27)$$

This form is appropriate when r is imaginary. Then we say that time τ is invariant.

But physical laws generally involve infinitesimal changes that do not depend on the choice of inertial frame. The square of one useful differential is analogous to r'^2 in (16-23).

Indeed, let us differentiate the transformation equations (16-9) through (16-12)

$$dx' = a\,dx - bc\,dt, \qquad (16\text{-}28)$$

$$dy' = dy, \qquad (16\text{-}29)$$

$$dz' = dz, \qquad (16\text{-}30)$$

$$c\,dt' = -e\,dx + fc\,dt, \qquad (16\text{-}31)$$

and substitute the results into the expression

$$ds'^2 = dx'^2 + dy'^2 + dz'^2 - c^2\,dt'^2. \qquad (16\text{-}32)$$

We then get

$$
\begin{aligned}
ds'^2 &= (a^2 - e^2)\,dx^2 + 2(ef - ab)\,dx\,c\,dt + dy^2 + dz^2 - c^2(f^2 - b^2)\,dt^2 \\
&= dx^2 + dy^2 + dz^2 - c^2\,dt^2 = ds^2
\end{aligned}
\qquad (16\text{-}33)
$$

when equations (16-17) through (16-19) are satisfied. The *displacement* described by ds is invariant.

Dividing the formulas for ds'^2 and ds^2 by $-c^2$ yields

$$d\tau'^2 = dt'^2 - \frac{dx'^2 + dy'^2 + dz'^2}{c^2}$$

$$= dt^2 - \frac{dx^2 + dy^2 + dz^2}{c^2} = d\tau^2. \tag{16-34}$$

But what time is $d\tau$?

Consider a particle whose coordinates in Σ change by dx, dy, dz during the time dt, as in figure 16.3. Construct Σ' so it moves with the particle during this interval. Then

$$dx' = dy' = dz' = 0 \tag{16-35}$$

and

$$dt'^2 = d\tau^2 \tag{16-36}$$

or

$$dt' = d\tau. \tag{16-37}$$

Now, a clock attached to the particle is momentarily at rest with respect to Σ'. This clock records the interval dt', which by (16-37) equals $d\tau$. Consequently, we

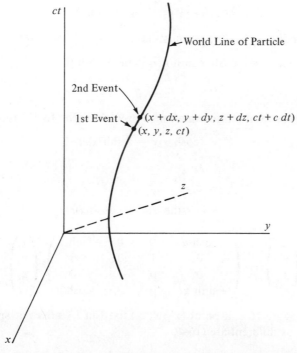

FIGURE 16.3 Successive events in the life of a given particle as observed in Σ.

interpret $d\tau$ as the increase in age of the particle during the small interval under consideration.

Parameter τ is interpreted as the *proper time* or age to be associated with the particle. The path of the particle in the space-time continuum is called its *world line*.

16.5 Coefficients in the transformation law

When Cartesian axes are chosen as in figure 16.1, the primed inertial system is related to the unprimed inertial system by equations (16-9), (16-10), (16-11), and (16-12) with

$$a^2 - e^2 = 1, \tag{16-38}$$

$$f^2 - b^2 = 1, \tag{16-39}$$

$$ab = ef. \tag{16-40}$$

Relationship (16-38) exists between the hyperbolic cosine and the hyperbolic sine of an arbitrary angle

$$\cosh^2 \alpha - \sinh^2 \alpha = 1, \tag{16-41}$$

so we have

$$a = \cosh \alpha, \quad e = \sinh \alpha, \tag{16-42}$$

where α is a parameter to be determined.

Relationship (16-39) is similarly parameterized by setting

$$f = \cosh \beta, \quad b = \sinh \beta. \tag{16-43}$$

Then to meet condition (16-40), we must have

$$\cosh \alpha \sinh \beta = \sinh \alpha \cosh \beta \tag{16-44}$$

or

$$\alpha = \beta. \tag{16-45}$$

Substituting these results into equations (16-9) through (16-12) reduces them to

$$x' = (\cosh \alpha)x - (\sinh \alpha)ct, \tag{16-46}$$

$$y' = y, \tag{16-47}$$

$$z' = z, \tag{16-48}$$

$$ct' = -(\sinh \alpha)x + (\cosh \alpha)ct, \tag{16-49}$$

whence

$$\begin{pmatrix} x' \\ y' \\ z' \\ ct' \end{pmatrix} = \begin{pmatrix} \cosh \alpha & 0 & 0 & -\sinh \alpha \\ 0 & 1 & 0 & 0 \\ 0 & 0 & 1 & 0 \\ -\sinh \alpha & 0 & 0 & \cosh \alpha \end{pmatrix} \begin{pmatrix} x \\ y \\ z \\ ct \end{pmatrix}. \tag{16-50}$$

According to figure 16.1, a point (x', y', z') fixed in Σ' moves at speed u along the x axis of Σ. Let us differentiate (16-46),

$$\frac{dx'}{dt} = (\cosh \alpha)\frac{dx}{dt} - (\sinh \alpha)c, \tag{16-51}$$

substitute 0 for dx'/dt, u for dx/dt,

$$0 = (\cosh \alpha)u - (\sinh \alpha)c, \tag{16-52}$$

and rearrange to get

$$\tanh \alpha = \frac{\sinh \alpha}{\cosh \alpha} = \frac{u}{c}. \tag{16-53}$$

Then solve for the coefficients

$$\sinh \alpha = \frac{u/c}{(1 - u^2/c^2)^{1/2}}, \tag{16-54}$$

$$\cosh \alpha = \frac{1}{(1 - u^2/c^2)^{1/2}}, \tag{16-55}$$

and substitute the results into equations (16-46), (16-47), (16-48), and (16-49), letting

$$\frac{1}{(1 - u^2/c^2)^{1/2}} = \gamma: \tag{16-56}$$

$$x' = \frac{1}{(1 - u^2/c^2)^{1/2}} (x - ut) = \gamma(x - ut), \tag{16-57}$$

$$y' = y, \tag{16-58}$$

$$z' = z, \tag{16-59}$$

$$t' = \frac{1}{(1 - u^2/c^2)^{1/2}} \left(t - \frac{u}{c^2} x \right) = \gamma \left(t - \frac{u}{c^2} x \right). \tag{16-60}$$

Such a change of coordinates was first found by Hendrick A. Lorentz in his electromagnetic investigations based on Maxwell's equations. Equations (16-57) through (16-60) are said to describe a *Lorentz transformation*.

EXAMPLE 16.3   The old idea that simultaneity is absolute implies that distant events behave as if a person can communicate with them instantaneously. What happens to equations (16-56) through (16-60) when fundamental speed c is taken to be infinite?

As c increases without limit, the ratio u/c becomes zero and equations (16-56) through (16-60) reduce to

$$\gamma = \frac{1}{(1 - 0)^{1/2}} = 1,$$

$$x' = x - ut,$$
$$y' = y,$$
$$z' = z,$$
$$t' = t.$$

These formulas that describe the Galilean transformation are used in Newtonian mechanics.

16.6   Projecting (a) a length and (b) a period onto moving inertial frames

When the fundamental speed c is finite, a free measuring rod projects a distance less than its length onto any moving inertial frame. A time interval measured by a clock that is at rest in Σ projects onto Σ' as a greater period plus a distance.

Let us employ the reference systems in figure 16.1. Then equations (16-57) through (16-60) and their increments

$$\Delta x' = \gamma (\Delta x - u \, \Delta t), \tag{16-61}$$

$$\Delta y' = \Delta y, \tag{16-62}$$

$$\Delta z' = \Delta z, \tag{16-63}$$

$$\Delta t' = \gamma \left(\Delta t - \frac{u}{c^2} \, \Delta x \right), \tag{16-64}$$

apply.

Let us also consider a rod fixed in Σ as figure 16.4 shows. Its rest length or *proper length* is the distance between the ends

$$L(0) = \Delta x = x_2 - x_1 \tag{16-65}$$

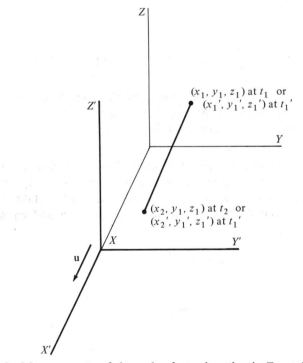

FIGURE 16.4   Measurements of the ends of a rod resting in Σ parallel to the x axis.

determined in the inertial frame in which it is at rest. From the moving system Σ', the two ends are located at a given t':

$$\Delta t' = 0. \tag{16-66}$$

Combining (16-66) and (16-64) yields

$$\Delta t = \frac{u}{c^2} \Delta x, \tag{16-67}$$

the time in Σ between the events observed simultaneously in Σ'. This interval converts equation (16-61) to

$$\Delta x' = \gamma \left(\Delta x - \frac{u^2}{c^2} \Delta x \right) = \frac{1}{(1 - u^2/c^2)^{1/2}} \left(1 - \frac{u^2}{c^2} \right) \Delta x$$

$$= (1 - u^2/c^2)^{1/2} \Delta x, \tag{16-68}$$

whence we obtain

$$\frac{1}{(1 - u^2/c^2)^{1/2}} \Delta x' = \gamma \, \Delta x' = \Delta x \tag{16-69}$$

or

$$\gamma L(u) = L(0). \tag{16-70}$$

Here $L(u)$ is the length measured by the observer moving at speed u with respect to the rod.

Let us next consider the rod fixed in Σ'. The end points are then located in Σ at a given time t:

$$\Delta t = 0. \tag{16-71}$$

This condition causes (16-61) to reduce to

$$\Delta x' = \gamma \, \Delta x, \tag{16-72}$$

whence we obtain

$$L(0) = \gamma L(u), \tag{16-73}$$

the same result as before. We multiply the length of the moving rod by γ, a quantity greater than 1, to get the length of the rod at rest.

In determining how time intervals project, we consider a clock fixed in Σ as figure 16.5 shows. The rest time, or *proper time*, between chosen strokes is then

$$T(0) = \Delta t = t_2 - t_1. \tag{16-74}$$

With respect to Σ, the events occur at the same place:

$$\Delta x = 0. \tag{16-75}$$

In Σ', the events are separated by both $\Delta t'$ and $\Delta x'$. Indeed, substituting (16-75) into equations (16-64) and (16-61) yields

$$\Delta t' = \gamma \, \Delta t, \tag{16-76}$$

$$\Delta x' = -\gamma u \, \Delta t. \tag{16-77}$$

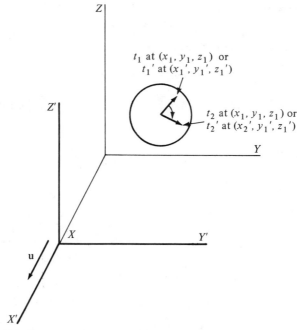

FIGURE 16.5   Measurements on a clock fixed in Σ.

If $T(u)$ is the time between the given strokes that an observer fixed in Σ' would measure, then

$$\Delta t' = T(u). \tag{16-78}$$

Equations (16-78) and (16-74) convert equation (16-76) to

$$T(u) = \gamma T(0). \tag{16-79}$$

When the clock is at rest in Σ', the events are observed with

$$\Delta x' = 0. \tag{16-80}$$

Then

$$\Delta x = u \, \Delta t \tag{16-81}$$

and

$$\Delta t' = \gamma \left(\Delta t - \frac{u^2}{c^2} \Delta t \right) = (1 - u^2/c^2)^{1/2} \, \Delta t, \tag{16-82}$$

whence

$$\gamma \, \Delta t' = \Delta t \tag{16-83}$$

or

$$\gamma T(0) = T(u). \tag{16-84}$$

Period $T(0)$ of a moving clock projects as a longer period $\gamma T(0)$; the clock seems to be running slow. This effect can be observed in a beam of high-energy radioactive particles.

As a demonstration experiment, David H. Frisch and James H. Smith studied the rate of decay of μ mesons associated with cosmic rays. They observed a time dilation of 8.8 ± 0.8, while the γ calculated from the average speed was 8.4 ± 2.

EXAMPLE 16.4   A rocket ship travels to a space station 6 light years away at $\frac{6}{10}c$ and returns immediately at the same speed. Calculate the time for the trip with respect to (a) the ship and (b) the earth.

Since the speed of the ship over the speed of light is

$$\frac{u}{c} = \frac{6}{10},$$

we have

$$1 - \frac{u^2}{c^2} = 1 - \tfrac{36}{100} = 0.64$$

and

$$\gamma = \frac{1}{\sqrt{0.64}} = \frac{1}{0.8}.$$

For the trip out, or the trip back, an observer on earth records the time

$$T(u) = \frac{6 \text{ light years}}{\tfrac{6}{10} \text{ light-year year}^{-1}} = 10 \text{ years.}$$

From equation (16-84), the corresponding time recorded on the ship is

$$T(0) = \frac{T(u)}{\gamma} = \frac{10 \text{ years}}{1/0.8} = 8 \text{ years.}$$

For the passage both ways, we have

$$T(0) = 8 + 8 \text{ years} = 16 \text{ years}$$

and

$$T(u) = 10 + 10 \text{ years} = 20 \text{ years.}$$

Note how different the time is, recorded along two different paths between two given events. The differential of proper time is not exact.

EXAMPLE 16.5   How does the image of a cube moving perpendicular to one of its faces appear a great distance off to one side?

Because light quanta travel at a finite speed c (in free space), the ones that form an image at a given instant of time have left farther points on the given object earlier than they have left nearer points. If the object moves appreciably while earlier reflected quanta reach the distance of the closest points on the object, the image is distorted

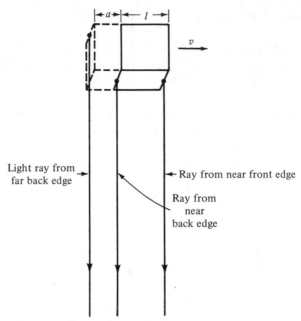

FIGURE 16.6   Rays from a cube moving as the Galilean transformation equations prescribe.

by the motion. This effect is superimposed on the contraction governed by equation (16-73).

Let us consider the side image of a distant cube moving at velocity v perpendicular to one of its faces. If the motion had no effect on the shape of the cube, pertinent rays would travel as figure 16.6 shows.

In time t, the cube would move the distance

$$vt = a.$$

Letting t be the time for light to travel from the far back edge to where the near back edge was earlier,

$$ct = l,$$

we find that

$$\frac{a}{l} = \frac{v}{c}$$

or

$$a = \frac{v}{c} l.$$

A sketch of the image appears in figure 16.7.

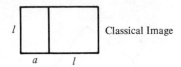

Classical Image

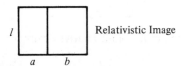

Relativistic Image

FIGURE 16.7   Classical (Galilean) and relativistic (Lorentz) images of the distant moving cube.

If the edges parallel to the direction of motion obey equation (16-70), the contracted length b satisfies

$$\frac{1}{(1 - v^2/c^2)^{1/2}} \, b = l$$

or

$$b = l \left(1 - \frac{v^2}{c^2}\right)^{1/2}.$$

Letting

$$\frac{v}{c} = \sin\theta,$$

we also have

$$\left(1 - \frac{v^2}{c^2}\right)^{1/2} = \cos\theta.$$

The equations for a and b then become

$$a = l \sin\theta,$$
$$b = l \cos\theta.$$

These lengths are the projections of edges of the rotated cube in figure 16.8. Note that the relativistic image is not appreciably distorted but is rotated. This prediction has not been verified experimentally because a person cannot record the image of such a fast moving body.

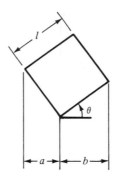

FIGURE 16.8   Apparent rotation of a distant relativistically moving cube.

16.7   Projecting a velocity onto an inertial frame moving in the same direction and the inverse operation

The Lorentz transformation equations yield a law for combining velocities that has experimental support, however.

Consider an effect traveling at velocity **v** in inertial frame Σ during time dt. If the x axis is pointed along the direction of motion, the x coordinate of the effect obeys the equation

$$v = \frac{dx}{dt} \tag{16-85}$$

over the given interval, while the y and z coordinates do not change

$$dy = 0, \qquad dz = 0. \tag{16-86}$$

Inertial frame Σ' moves along the x axis as figure 16.1 shows. With

$$y' = y, \qquad z' = z, \tag{16-87}$$

the effect also moves parallel to the x' axis, and

$$v' = \frac{dx'}{dt'}. \tag{16-88}$$

Equations (16-57) and (16-60) relate x' and t' to x and t. Differentiating these transformation equations yields

$$dx' = d[\gamma(x - ut)] = \gamma (dx - u\, dt), \tag{16-89}$$

$$dt' = d\left[\gamma\left(t - \frac{u}{c^2} x\right)\right] = \gamma\left(dt - \frac{u}{c^2} dx\right). \tag{16-90}$$

Substituting these results into equation (16-88) then leads to the formula

$$v' = \frac{\gamma \, (dx - u \, dt)}{\gamma \, [dt - (u/c^2) \, dx]} = \frac{(dx/dt) - u}{1 - (u/c^2) \, (dx/dt)}$$

$$= \frac{v - u}{1 - uv/c^2} \, . \tag{16-91}$$

In our development of equations (16-57) and (16-60), we chose the parameters so an effect moving at speed c in Σ also moves at speed c in Σ'. We can check equation (16-91) by setting $v = c$ and seeing whether $v' = c$. Indeed, we find that

$$v' = \frac{c - u}{1 - uc/c^2} = \frac{c(c - u)}{c - u} = c. \tag{16-92}$$

Near the middle of the nineteenth century, Armand H. L. Fizeau measured the velocity of light in moving water, obtaining a result that this Einsteinian theory explains.

The speed at which a given phase of light travels within a fluid of refractive index n is

$$v' = \frac{c}{n} \, . \tag{16-93}$$

In the Fizeau experiment, the fluid travels in the same direction at speed u with respect to the observer. The electromagnetic radiation is apparently dragged along by the moving medium.

The resultant velocity is given by equation (16-91) rearranged in the form

$$v = \frac{v' + u}{1 + uv'/c^2} \, . \tag{16-94}$$

Applying the binomial theorem to the denominator

$$v = (v' + u) \left(1 - \frac{uv'}{c^2} + \cdots \right) = v' + u - \frac{uv'^2}{c^2} - \frac{u^2 v'}{c^2} + \cdots , \tag{16-95}$$

introducing equation (16-93),

$$v = \frac{c}{n} + u - u \frac{1}{n^2} - \cdots , \tag{16-96}$$

and dropping higher terms since the fluid speed u is small with respect to the fundamental speed c, we get

$$v = \frac{c}{n} + u \left(1 - \frac{1}{n^2} \right) \tag{16-97}$$

for the phase velocity of the light with respect to the observer. Experimental results do agree with this equation.

EXAMPLE 16.6 What form does equation (16-91) assume when the time in Σ is exactly the same as that in Σ'?

Differentiating the Galilean transformation equations of example 16.1

$$dx' = dx - u\,dt,$$
$$dt' = dt,$$

and forming the derivative, we obtain

$$\frac{dx'}{dt'} = \frac{dx - u\,dt}{dt} = \frac{dx}{dt} - u = v - u.$$

The same result is obtained from equation (16-91) when the final denominator equals one

$$1 - \frac{uv}{c^2} = 1,$$

that is, when

$$c = \infty.$$

16.8   Minkowski coordinate systems

A particle that exists in time is always separated from a reference point in its life by a timelike interval. Each distinct particle is separated from a common reference point by a spacelike interval. Let us here consider how these intervals appear geometrically. Let us also consider the effect of a Lorentz transformation on the reference axes.

When the fundamental speed c is finite, we can associate with the inertial frame Σ a 4-dimensional Euclidean space in which the x, y, z, and ct axes are mutually perpendicular, as figure 16.3 shows. The coordinates of a given particle then trace out a world line.

The velocity of the particle has the elements

$$\frac{dx}{dt} = c\,\frac{dx}{d(ct)}, \qquad \frac{dy}{dt} = c\,\frac{dy}{d(ct)}, \qquad \frac{dz}{dt} = c\,\frac{dz}{d(ct)}, \qquad (16\text{-}98)$$

which equal c times the components of the slope of the world line with respect to the ct axis. When the particle moves at constant velocity, the slope of the world line is constant. When the particle accelerates, the slope changes and the world line is curved.

If the particle moves at speed c all the way to or from the origin, as a photon of light could, its coordinates satisfy

$$x^2 + y^2 + z^2 - c^2t^2 = 0. \qquad (16\text{-}99)$$

But in the 4-space, equation (16-99) describes a hypercone whose generatrix is inclined 45° from the ct axis. This locus is called the *light cone* for the space-time point at the origin. A similar cone is defined for each space-time point.

Any event within a cone can be reached by a particle that passes through the vertex. If this particle travels at *constant velocity* with respect to Σ, it defines the origin of the spatial part of an inertial system Σ'. Then equation (16-27) reduces to

$$-\frac{r^2}{c^2} = t^2 - \frac{x^2 + y^2 + z^2}{c^2} = t'^2 = T^2 \qquad (16\text{-}100)$$

where T is the time recorded by a clock moving with the particle. It is the *proper time* of the event from the vertex. Since a moving clock runs slow, the time recorded by a clock traveling along a different path between the origin and the event (or between the event and the origin) would be less.

The gap between the vertex and an event outside the cone can be spanned by measuring rods moving at the same constant velocity with respect to Σ. Choosing inertial frame Σ' to be the one in which such rods are at rest, we then find that equation (16-25) yields

$$r^2 = x^2 + y^2 + z^2 - c^2t^2 = x'^2 + y'^2 + z'^2 = L^2 > 0 \qquad (16\text{-}101)$$

for the square of their total length. The length itself is called the *proper distance L*. Since moving measuring rods are contracted, the length measured along any other path between the points would be greater.

The locus of points a given proper time T from the vertex form a sheet of a hyperboloid described by equation (16-100). The locus of points a given proper distance from the vertex form a one-sheeted hyperboloid defined by equation (16-101) with L constant. A section across such loci appears in figure 16.9.

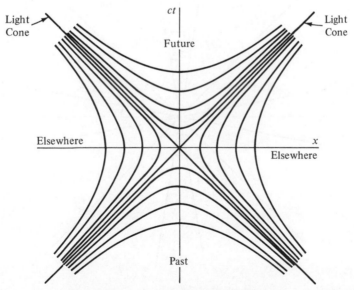

FIGURE 16.9 Curves whose points are a given proper distance or proper time from the origin.

Equations (16-46) through (16-49) relate coordinates and time in two inertial systems oriented as in figure 16.1. Along the ct' axis, coordinate x' is zero, and (16-46) reduces to

$$0 = (\cosh \alpha)x - (\sinh \alpha)ct. \qquad (16\text{-}102)$$

The equation for the ct' axis is

$$x = (\tanh \alpha)ct \qquad (16\text{-}103)$$

in the 4-space. Along the corresponding x' axis, the time t' is zero and (16-49) reduces to

$$0 = -(\sinh \alpha)x + (\cosh \alpha)ct, \qquad (16\text{-}104)$$

whence

$$ct = (\tanh \alpha)x. \qquad (16\text{-}105)$$

From figure 16.10, the tangent of the angle between the ct' axis and the ct axis in the 4-space equals the x coordinate of a point on the ct' axis divided by its ct coordinate. Similarly, the tangent of the angle between the x' axis and the x axis equals the ct coordinate of a point on the x' axis divided by its x coordinate. From equations

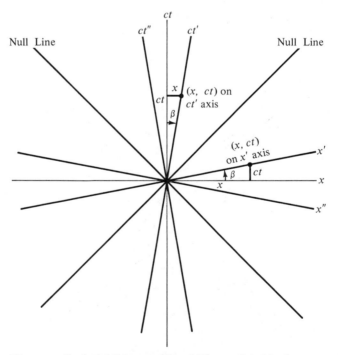

FIGURE 16.10   How axes for inertial frames Σ' and Σ'' are related to the perpendicular set for Σ when the spatial origins of Σ' and Σ'' travel in opposite directions along the x axis.

(16-103) and (16-105), these are the same tanh α. Introducing equation (16-53) then yields

$$\tan \beta = \tanh \alpha = \frac{u}{c}. \tag{16-106}$$

Note that a negative u leads to a negative β, as axes $x''ct''$ illustrate. The angles between each set of axes are bisected by the *null lines*, traces of the light cone based on the origin.

16.9   Significance of a line segment in Minkowski space

The proper length or time represented by a straight line segment of given length depends on its inclination in the 4-dimensional Euclidean space.

Let us suppose that two events are observed in an inertial frame. Through the spatial points where they occur, we pass the x axis for system Σ. The y and z axes are drawn perpendicular to this x axis through a given point on it. The events then map onto the corresponding 4-dimensional Euclidean space at $(x_1, 0, 0, ct_1)$ and $(x_2, 0, 0, ct_2)$. In the xct plane they appear as figure 16.11 shows.

If the measured length of the line drawn between these points is l, while the angle between the line and a horizontal axis parallel to the x axis is β, the trigonometric definitions imply that

$$\Delta x = l \cos \beta, \tag{16-107}$$

$$c\,\Delta t = l \sin \beta. \tag{16-108}$$

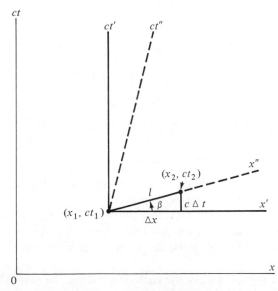

FIGURE 16.11   Straight line segment l and its projections Δx and $c\,\Delta t$ in the hypothetical Euclidean 4-space. The origin of the primed system is placed at (x_1, ct_1) so that l is the radius vector drawn in it to (x_2, ct_2).

A translation of the origins of the x and ct scales to the first event produces inertial frame Σ'. A Lorentz transformation then allows the x'' axis to pass through the second event, if $\beta < 45°$ (see figure 16.11).

In the primed coordinate system increments (16-107), (16-108) are the components of a radius vector drawn from the origin, the first event, to the second event:

$$x' = l \cos \beta, \tag{16-109}$$

$$ct' = l \sin \beta. \tag{16-110}$$

Equation (16-25) applies to a transformation from the primed to the double primed coordinate system

$$r'^2 = x'^2 - c^2t'^2 = x''^2 - c^2t''^2. \tag{16-111}$$

Being on the x'' axis, time t'' at the second event is zero and the proper distance represented by l is x''. Designating this Δs, we have

$$x'^2 - c^2t'^2 = \Delta s^2, \tag{16-112}$$

whence

$$\Delta s^2 = \Delta x^2 - c^2 \, \Delta t^2. \tag{16-113}$$

Let us eliminate Δx and $c \, \Delta t$ with relationships (16-107), (16-108) and reduce the result:

$$\Delta s^2 = l^2(\cos^2 \beta - \sin^2 \beta) = l^2 \frac{\cos^2 \beta - \sin^2 \beta}{\cos^2 \beta + \sin^2 \beta}$$

$$= l^2 \frac{1 - \tan^2 \beta}{1 + \tan^2 \beta}. \tag{16-114}$$

For the proper length of the displayed segment l, we find

$$\Delta s = l \left(\frac{1 - \tan^2 \beta}{1 + \tan^2 \beta} \right)^{1/2}, \tag{16-115}$$

where β is the angle between the line and the x axis.

16.10 Significance of a hyperbolic world line

Let us next consider what variation of acceleration causes a body to move along a hyperbolic world line asymptotic to a light cone.

We suppose that a particle is traveling along an inertial axis with a speed that varies enough to keep the particle a constant proper distance away from a position on the axis. The axis is chosen to be the x axis of Σ, while the reference position is selected to be the origin.

Since the y and z coordinates of the particle are zero, the world line lies in the xct plane of the 4-space. If t is also assumed to be zero when the particle is momentarily at rest in Σ, equation (16-101) is satisfied with y, z zero and with L equal to a constant a:

$$x^2 - c^2t^2 = a^2. \tag{16-116}$$

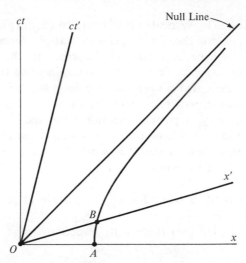

FIGURE 16.12   Hyperbolic path asymptotic to a light cone in the xct plane.

Let the point where this hyperbola crosses the x axis be labeled A, while a point reached later is labeled B, as figure 16.12 illustrates. A Lorentz transformation then takes us to an inertial frame Σ' for which ct' is zero when the particle is at B.

According to equation (16-25), such a transformation changes the form in (16-116) to

$$x'^2 - c^2 t'^2 = a^2 \qquad (16\text{-}117)$$

if y and z are not altered. The resulting curve, in the space in which the x' and ct' axes are perpendicular, is like the initial curve with corresponding points displaced. Compare figure 16.13 with figure 16.12.

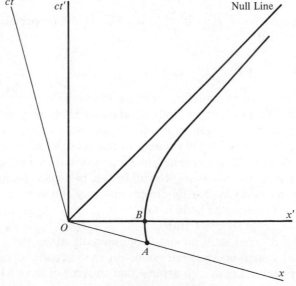

FIGURE 16.13   The path in figure 16.12 plotted with the x' and ct' axes perpendicular. Note that the curve is still a hyperbola with the same semiaxes.

Since the plot of (16-117) is vertical at B in figure 16.13, the particle is then momentarily at rest and d^2x'/dt'^2 is then its proper acceleration. Since this equals d^2x/dt^2 in figure 16.12 where the hyperbola crosses the x axis, with the particle momentarily at rest in Σ, the proper acceleration has not changed in going from A to B.

Such *accelerations* are called *proper* because they are being observed from an inertial frame in which the particle is momentarily at rest. Since B may be chosen anywhere on the trajectory, a particle moving along such a hyperbola is always subject to the same proper acceleration. Conversely, any particle subject to constant proper acceleration traces out a hyperbolic world line asymptotic to the light cone erected on the center of the hyperbola.

EXAMPLE 16.7   What is the acceleration of the particle at point A in figure 16.12 and at point B in figure 16.13?

Differentiate equation (16-116) twice with respect to t and solve for the second derivative of x:

$$\frac{d^2x}{dt^2} = \frac{1}{x}\left[c^2 - \left(\frac{dx}{dt}\right)^2\right].$$

When t is 0, x is a, dx/dt is 0, and

$$\frac{d^2x}{dt^2} = \frac{c^2}{a}.$$

This quantity is the proper acceleration of the particle at A.

Similarly, equation (16-117) yields

$$\frac{d^2x'}{dt'^2} = \frac{c^2}{a}$$

for the second derivative at $t' = 0$, $x' = a$. This is the proper acceleration of the particle when it is at B.

16.11   Rigidity

Because distances depend on the reference frame in which they are measured, no body is absolutely rigid. Two interesting kinds of relative rigidity may be found, however, in nonrotating systems. A given body may translate so that its dimensions are momentarily constant in the inertial frame with respect to which any one of its points is momentarily at rest. The constancy in dimensions then implies that all other points on it have zero instantaneous velocity in this particular frame. Alternatively, the given body may move so that its dimensions stay constant in a certain inertial frame. In the first instance, the body is said to be *properly rigid*; in the second case, it is rigid in the specified reference frame.

Let us consider a meter stick moving longitudinally along the x axis so there is always an inertial system in which all points on the stick are at rest. Such a stick exhibits proper rigidity. Let us also assume that the rear of the stick accelerates at a

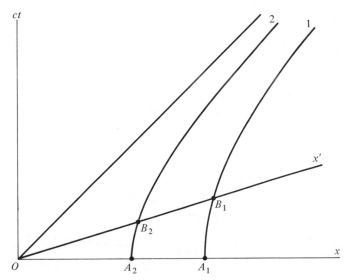

FIGURE 16.14 World lines of the ends of a properly rigid meter stick accelerating longitudinally along the *x* axis from rest at A_1A_2.

constant proper rate from rest in Σ, so its world line plots as hyperbola 2 in figure 16.14. Position B_2 is then the same proper distance from the origin as the initial position A_2.

This distance is displayed as part of the x' axis. Extending this axis one meter farther, the proper length of the stick, we reach the front end. Point B_1 is one meter plus OA_2 in proper distance from the origin.

Because the locus of possible positions of B_1 is a constant proper distance from O, it is a hyperbola having the same center and asymptotes as curve 2. Because its semiaxis is larger, the proper acceleration associated with hyperbola 1 is less, however.

Thus, the front of the stick accelerates at a slower rate than the rear. The meter stick gets shorter and shorter in the original inertial system Σ as t increases. (For the length of a body to remain constant in Σ, its ends would have to accelerate at the same rate.)

Let us next consider two identical rockets initially at rest in Σ, pointing in the same direction along the same line, with the second one linked to the first by a taut *string* of length L. Their identical construction implies that their accelerations will vary in the same manner after firing, so they will trace congruent curves on a Minkowski diagram.

We assume that the rockets are fired at the same time $t = 0$, so an observer in Σ sees them keep a constant distance apart. We also assume that the proper acceleration on each is the same until cutoff. The resulting world lines are hyperbolic, as figure 16.15 shows.

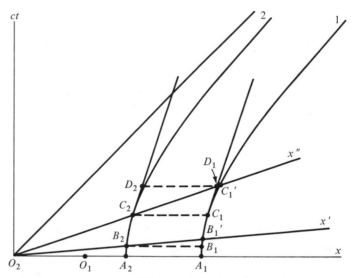

FIGURE 16.15   World lines of two identical rockets fired simultaneously from a leading and a trailing position in Σ.

The inertial frame in which the second rocket is at rest at B_2 is labeled Σ'. But since the x' axis drawn from the center of the trailer's hyperbola intersects the other world line at B_1', an observer riding the trailing rocket at B_2 would then locate the leading rocket at B_1'.

For the leader to be at rest in Σ', its hyperbola would also have to be centered at the origin as figure 16.14 shows. Since it is actually centered on O_1 to the right of the origin, rocket 1 is traveling away from B_2 in Σ'. Consequently, the observer in the trailing rocket would suppose that the leader was accelerating faster. A similar effect would be noted at each point on the string; the motion would put it under stress.

The observer in Σ would interpret this stress as arising because the relativistic contraction of (16-70) had been prevented. The stress would remain if both rockets cut off at the same time t, say at points C_2 and C_1. The rockets would then coast along the tangent lines $C_2 D_2$ and $C_1 D_1$.

EXAMPLE 16.8   Relate the coordinates of C_2 and D_1 to the velocity of the trailing rocket at C_2.

Introduce inertial frame Σ'' with its x'' axis running from origin O_2 through C_2 as figure 16.15 shows. Note that this axis cuts the world line of the leading rocket at D_1.

On this axis, t'' is zero and equation (16-60) yields

$$ct = \frac{v}{c} x$$

where v is the velocity of Σ'', and of the trailing rocket at C_2, with respect to Σ. If the x coordinates of C_2 and D_1 are p and q, the ct coordinates are $(v/c)p$ and $(v/c)q$.

Putting the coordinates of C_2 into the world line for the first rocket,

$$x^2 - (ct)^2 = a^2$$

then yields

$$\left(1 - \frac{v^2}{c^2}\right)p^2 = a^2 \quad \text{or} \quad p = \frac{a}{(1 - v^2/c^2)^{1/2}}.$$

The slope of the tangent through C_1 equals that through C_2, which is

$$\frac{d(ct)}{dx} = c\frac{dt}{dx} = c\frac{1}{v} = \frac{c}{v},$$

while the coordinates of C_1 are $p + L$, $(v/c)p$. Equating this slope to the ratio calculated from the coordinates obtained for D_1, C_1,

$$\frac{c}{v} = \frac{(v/c)q - (v/c)p}{q - (p + L)},$$

and rearranging, we get

$$q = p + \frac{L}{1 - v^2/c^2}.$$

EXAMPLE 16.9   What is the proper or rest length of the string in the coasting system?

Apply the Pythagorean theorem to the coordinates of C_2 and D_1,

$$(q - p)^2 + \left(\frac{v}{c}q - \frac{v}{c}p\right)^2 = \left(1 + \frac{v^2}{c^2}\right)(q - p)^2$$

$$= \frac{(1 + v^2/c^2)L^2}{(1 - v^2/c^2)^2},$$

to get the square of the displayed distance between the points. Then substitute the square root of this result into equation (16-115) and set $\tan \beta$ equal to v/c:

$$\Delta s = \frac{L(1 + v^2/c^2)^{1/2}}{(1 - v^2/c^2)}\frac{(1 - v^2/c^2)^{1/2}}{(1 + v^2/c^2)^{1/2}} = \frac{L}{(1 - v^2/c^2)^{1/2}} = \gamma L.$$

Note that L equals distance C_2C_1, the string's length as measured by the observer in Σ, who moves at speed v with respect to the object, while Δs is its rest length. The final result could have been obtained directly from equation (16-70).

16.12   Key ideas

Accurate measurements indicate that space and time are homogeneous and uniform, and that different inertial frames are equivalent. As a consequence, the Cartesian spatial and time coordinates in one inertial frame are linear functions of those in another.

It is also found that the Pythagorean theorem is valid under ordinary circumstances. Furthermore, the same units of space and time may be employed in different inertial systems. The primed inertial coordinates of a particle traveling at speed c from a beginning position at the origin satisfy the equation

$$x'^2 + y'^2 + z'^2 = c^2 t'^2. \tag{16-118}$$

If the particle should travel at the same speed in the unprimed system, we would also have

$$x^2 + y^2 + z^2 = c^2 t^2. \tag{16-119}$$

The linear transformation equations do allow for a single speed that is invariant in this way. Experiments show that the speed of light in free space is the speed c.

Now, the linear relationships that transform equation (16-118) to equation (16-119) also transform

$$x'^2 + y'^2 + z'^2 - c^2 t'^2 = r'^2 \tag{16-120}$$

to

$$x^2 + y^2 + z^2 - c^2 t^2 = r^2 \tag{16-121}$$

without altering r:

$$r' = r. \tag{16-122}$$

And they transform

$$dx'^2 + dy'^2 + dz'^2 - c^2 dt'^2 = ds'^2 \tag{16-123}$$

to

$$dx^2 + dy^2 + dz^2 - c^2 dt^2 = ds^2 \tag{16-124}$$

without altering ds:

$$ds' = ds. \tag{16-125}$$

Since the discussion in section 12.9 does not require the continuum to be Euclidean, the formulas apply to the systems of space and time that we are describing.

Indeed, equation (16-124) agrees with equation (12-54) if we take the *contravariant components* of ds to be

$$dx, \qquad dy, \qquad dz, \qquad c\, dt, \tag{16-126}$$

and the *covariant components* to be

$$dx, \qquad dy, \qquad dz, \qquad -c\, dt. \tag{16-127}$$

If we let ds be the displacement of a particle in the continuum, the derivatives

$$\frac{dx}{dt}, \quad \frac{dy}{dt}, \quad \frac{dz}{dt} \tag{16-128}$$

are the Σ components of its velocity v, and equation (16-124) can be rewritten as

$$ds^2 = dt^2 \left[\left(\frac{dx}{dt}\right)^2 + \left(\frac{dy}{dt}\right)^2 + \left(\frac{dz}{dt}\right)^2 - c^2 \right]$$

$$= dt^2(v^2 - c^2) = -c^2 dt^2 \left(1 - \frac{v^2}{c^2}\right). \tag{16-129}$$

If the inertial system Σ' is moving momentarily with the particle

$$dx' = dy' = dz' = 0, \tag{16-130}$$

equation (16-123) reduces to

$$ds'^2 = -c^2 \, dt'^2 = -c^2 \, d\tau^2 \tag{16-131}$$

where τ is the time recorded by a clock moving with the particle, the proper time for the particle.

Combining equations (16-131), (16-125), and (16-129) yields

$$-c^2 \, d\tau^2 = -c^2 \, dt^2 \left(1 - \frac{v^2}{c^2} \right) \tag{16-132}$$

or

$$d\tau = dt \left(1 - \frac{v^2}{c^2} \right)^{1/2}, \tag{16-133}$$

whence

$$\frac{dt}{d\tau} = \frac{1}{(1 - v^2/c^2)^{1/2}} = \gamma. \tag{16-134}$$

Since the proper time

$$\int d\tau = \tau \tag{16-135}$$

depends on the path followed by the particle, the differential

$$d\tau \qquad \text{is not exact.} \tag{16-136}$$

On the other hand, differential $d\tau$ is $i \, ds/c$ and ds is invariant during any transformation from one inertial frame to another. Therefore,

$$d\tau \qquad \text{is invariant.} \tag{16-137}$$

DISCUSSION QUESTIONS

16.1 May the homogeneity of space be an illusion?

16.2 Is space exactly Euclidean?

16.3 Can time be studied separate from space?

16.4 How is an inertial frame identified? Why can the time in one frame differ from that in another?

16.5 Cite (a) philosophic and (b) scientific evidence for the relativity principle. Recall how Newtonian mechanics is consistent with the principle.

16.6 Why can the coordinate systems be chosen as in figure 16.1 without loss of generality?

16.7 Explain why $\Delta x'$ and $\Delta t'$ are independent of x_1, y_1, z_1, and t_1. Why does doubling Δx and Δt double $\Delta x'$ and $\Delta t'$? Why are these connected by linear relationships?

16.8 Show how the linear relationships among x', t' and x, t let a force-free body travel along a rectilinear path at the same speed in both Σ and Σ'. How do we show that this fundamental speed is the speed of light in free space?

16.9 What displacement is invariant in going from one inertial frame to another? What interval is invariant?

16.10 Why are we concerned with infinitesimal quantities in physical applications?

16.11 Explain what inertial clock records the increase in age of an accelerating particle during each infinitesimal interval in its life.

16.12 How can we represent the conditions

$$a^2 - e^2 = 1,$$
$$f^2 - b^2 = 1,$$
$$ab = ef,$$

with hyperbolic functions? How do we relate these to observables?

16.13 What would absolute simultaneity imply?

16.14 Explain how a length projects from one inertial frame onto another.

16.15 How does a time interval project?

16.16 How does the image of a relativistically moving object appear?

16.17 How do parallel velocities combine? How is the speed of light in a dielectric affected by motion of the medium?

16.18 Why can't a Minkowski diagram be constructed when c is infinite?

16.19 What is the significance of the light cone erected on a given point?

16.20 Describe the loci of points (a) a constant proper distance and (b) a constant proper time from the vertex of a light cone.

16.21 How does a Lorentz transformation rotate the axes of a Minkowski coordinate system?

16.22 Derive a formula for the proper time represented by a straight line of length l in the 4-dimensional Euclidean space.

16.23 Explain what would cause the world line of a particle to be a hyperbola asymptotic to a light cone.

16.24 What transformations move events along a hyperbolic world line?

16.25 How do we diagram the longitudinal motion of a properly rigid measuring rod?

16.26 How can we keep a string or chain rigid with respect to a given inertial frame while it is accelerating in the frame?

16.27 Discuss the nature of $d\tau$. Show that

$$\frac{dt}{d\tau} = \gamma.$$

PROBLEMS

16.1 Solve the Lorentz transformation equations for x, y, z, and t.

16.2 Subject the wave equation

$$\frac{\partial^2 E}{\partial x^2} = \frac{1}{c^2} \frac{\partial^2 E}{\partial t^2}$$

to a Lorentz transformation.

16.3 A rod of proper length L rests in the xy plane inclined at angle θ from the x axis. What does an observer moving at speed u along the x axis find for its length and angle of inclination?

16.4 If a cosmic-ray muon travels at $0.980c$ and if its mean life against radioactive decay is 2.20×10^{-6} sec in its own rest system, what is its mean life as observed from the earth?

16.5 Suppose the distant moving cube in figure 16.6 is rotated about an axis passing through the middle of the top and bottom faces by angle α. How does its image now appear to the observer?

16.6 Transform the three Cartesian velocity components v_x, v_y, v_z in Σ to the components v_x', v_y', v_z' in Σ'.

16.7 How may the equality between the two expansions of r^2 in (16-25) be rearranged so that only positive terms appear? Among what components does the Pythagorean theorem hold? What kind of plot is suggested by these considerations?

16.8 Assuming r^2 to be negative, diagram a rotation and a subsequent Lorentz transformation which allow x', y', and z' to be 0.

16.9 Space travelers A and B have four plans for moving through space and time from a common origin: (a) From the starting point, both move in opposite directions with the same high speed. At some proper interval from the origin, they suddenly reverse their directions with no change in speed and return to their initial position. (b) Observer B travels at a constant high velocity for a given proper interval and then stops. At the same proper interval from the origin, A takes off at the previous velocity of B and finally joins him. (c) B takes off with constant speed in one direction until he reaches a distant spot. Then he immediately reverses his direction, keeping his speed constant, and returns to the starting position, where A remained. (d) B moves off at constant velocity. Sometime later A takes off in the same direction at a speed high enough to overtake B at a proper time equal to twice his starting time from the origin.

For each plan, sketch a Minkowski diagram and compare the times spent along the two paths.

16.10 If a properly rigid body moves at velocity v with respect to the laboratory frame, what is the observed size of an element whose proper volume is dV_0?

— — —

16.11 Subject the expression

$$d\phi = \frac{\partial \phi}{\partial x}\, dx + \frac{\partial \phi}{\partial y}\, dy + \frac{\partial \phi}{\partial z}\, dz + \frac{\partial \phi}{\partial t}\, dt$$

to a Lorentz transformation.

16.12 Show when

$$\text{(a)} \quad \gamma \simeq 1 + \frac{u^2}{2c^2} \qquad \text{(b)} \quad \gamma \simeq \sqrt{\frac{c}{2(c - u)}}\,.$$

16.13 In the coordinate system based on its star, a planet traces out a circular orbit with the period T. What is its period in an inertial frame in which the orbit is an ellipse with one principal axis twice as long as the other?

16.14 Find an expression for the velocity of light in a fluid moving in the same direction if the speed of light in free space were different from the fundamental speed c.

16.15 Relate the conventional acceleration of a particle moving along the x' axis of Σ' to its acceleration in Σ.

16.16 From the answer to problem 16.15, obtain the inertial system in which the acceleration of a particle is greatest.

16.17 Find the u/c for which the accelerations in Σ' and Σ are the same. Then put this into the velocity addition law to determine how the particle is moving with respect to each frame.

16.18 With the answer to problem 16.15, relate the accelerations in Σ' and Σ'' to the acceleration in the Σ for which the particle's instantaneous velocity v is zero. Express each result in terms of velocities v' and v''. Then obtain an expression for the joint invariant of acceleration and velocity.

16.19 A space man leaves the earth at constant velocity on January 1, 1990, spends 2 years speeding to his target, and immediately returns at the same speed to reach the earth on January 1, 2000. Calculate the speed of travel. Then construct a Minkowski diagram for the trip.

16.20 Write the displacement $d\mathbf{s}$ as

$$d\mathbf{s} = \mathbf{e}_1\, dx + \mathbf{e}_2\, dy + \mathbf{e}_3\, dz + \mathbf{e}_4 c\, dt.$$

Then form the scalar

$$ds^2 = d\mathbf{s} \cdot d\mathbf{s}$$

and determine what conditions the base vectors must satisfy for formula (16-33) to result.

16.21 Expand $dt/d\tau$ as a power series in v/c.

REFERENCES

BOOKS

Band, W. *Introduction to Mathematical Physics*, pp. 235–270. D. Van Nostrand Company, Inc., Princeton, N.J., 1959.

Einstein, A., Lorentz, H. A., Minkowski, H., and **Weyl, H.** *The Principle of Relativity*, pp. 1–96. Dover Publications, Inc., New York, 1958.

Schwartz, J. T. *Relativity in Illustrations*, pp. 1–117. New York University Press, New York, 1962.

Shadowitz, A. *Special Relativity*, pp. 1–83. W. B. Saunders Company, Philadelphia, 1968.

Yilmaz, H. *Introduction to the Theory of Relativity and the Principles of Modern Physics*, pp. 1–41. Blaisdell Publishing Company, New York, 1965.

ARTICLES

Albergotti, J. C., and **Larkin, J. H.** The Brehme Diagram in Two Spatial Dimensions. *Am. J. Phys.*, **39**, 193 (1971).

Anderson, J. L., and **Gautreau, R.** Operational Approach to Space and Time Measurements in Flat Space. *Am. J. Phys.*, **37**, 178 (1969).

Augustynek, A. Homogeneity of Time. *Am. J. Phys.*, **36**, 126 (1968).

Bergström, A. Geometrical Model of Relativistic *xtv* Space. *Am. J. Phys.*, **36**, 393 (1968).

Brehme, R. W. Geometrization of the Relativistic Velocity Addition Formula. *Am. J. Phys.*, **37**, 360 (1969).

Cushing, J. T. Vector Lorentz Transformations. *Am. J. Phys.*, **35**, 858 (1967).

Duguay, M. A. Light Photographed in Flight. *Am. Scientist*, **59**, 550 (1971).

Eisenberg, L. J. Necessity of the Linearity of Relativistic Transformations between Inertial Systems. *Am. J. Phys.*, **35**, 649 (1967).

Eisenlohr, H. Another Note on the Twin Paradox. *Am. J. Phys.*, **36,** 635 (1968).

Evett, A. A. An Aid for Clarifying Space-Time Concepts in Special Relativity. *Am. J. Phys.*, **39,** 44 (1971).

Frisch, D. H., and **Smith, J. H.** Measurement of the Relativistic Time Dilation Using μ-Mesons. *Am. J. Phys.*, **31,** 342 (1963).

Greenberger, D. M. The Reality of the Twin Paradox Effect. *Am. J. Phys.*, **40,** 750 (1972).

Guess, A. W. "Inverse" Terrell Effect. *Phys. Rev.*, **161,** 1295 (1967).

Holton, G. Einstein and the "Crucial" Experiment. *Am. J. Phys.*, **37,** 968 (1969).

Lang, D. W. The Meter Stick in the Match Box. *Am. J. Phys.*, **38,** 1181 (1970).

Lass, H. Accelerating Frames of Reference and the Clock Paradox. *Am. J. Phys.*, **31,** 274 (1963).

Muller, R. A. The Twin Paradox in Special Relativity. *Am. J. Phys.*, **40,** 966 (1972).

Newton, R. G. Causality Effects of Particles that Travel Faster Than Light. *Phys. Rev.*, **162,** 1274 (1967).

Parker, L., and **Schmieg, G. M.** A Useful Form of the Minkowski Diagram. *Am. J. Phys.*, **38,** 1298 (1970).

Quale, A. On the Dynamical Approach to the "Clock Paradox." *Am. J. Phys.*, **39,** 434 (1971).

Rekveld, J. New Aspects of the Teaching of Special Relativity. *Am. J. Phys.*, **37,** 716 (1969).

Rindler, W. Einstein's Priority in Recognizing Time Dilation Physically. *Am. J. Phys.*, **38,** 1111 (1970).

Romain, J. E. A Geometrical Approach to Relativistic Paradoxes. *Am. J. Phys.*, **31,** 576 (1963).

Romain, J. E. On Some Misconceptions about Relativistic Coordinate Transformations. *Nuovo Cimento*, **30,** 1254 (1963).

Sama, N. On the Ehrenfest Paradox. *Am. J. Phys.*, **40,** 415 (1972).

Schwartz, H. M. A New Method of Clock Synchronization without Light Signals. *Am. J. Phys.*, **39,** 1269 (1971).

Schwartz, H. M. Poincare's Rendiconti Paper on Relativity. Part I, *Am. J. Phys.*, **39,** 1287 (1971); Part II, *Am. J. Phys.*, **40,** 862 (1972).

Scott, G. D., and **van Driel, H. J.** Geometrical Appearances at Relativistic Speeds. *Am. J. Phys.*, **38,** 971 (1970).

Shankland, R. S. Conversations with Albert Einstein. *Am. J. Phys.*, **31,** 47 (1963).

Shankland, R. S. Michelson-Morley Experiment. *Am. J. Phys.*, **32,** 16 (1964).

Strakhovskii, G. M., and **Uspenskii, A. V.** Experimental Verification of the Theory of Relativity. *Soviet Phys. Uspekhi*, **8,** 505 (1966).

Suvorov, S. G. Einstein's Philosophical Views and Their Relation to His Physical Opinions. *Soviet Phys. Uspekhi*, **8,** 578 (1966).

Weinberger, H., and **Mossel, M.** Theory for a Unidirectional Interferometric Test of Special Relativity. *Am. J. Phys.*, **39,** 606 (1971).

17 / Relativistic Mechanics

17.1 Interactions in space and time

We have seen how the universe can be subdivided into very small apparently separate units called particles. When isolated from its surroundings, such a unit exhibits inertial motion, but when it is not isolated, it tends to interact with its neighbors.

A particle presumably influences a neighbor whenever it runs into the neighbor, in a collision. The participants may simply rebound from each other, or they may react. A particle may also interact with others at a distance by exchanging transient particles, by distorting and expanding the surrounding space, or by establishing a field that affects the neighbors.

While such processes may be consistent with Einstein's theory, direct action at a distance, as Newton envisaged in gravitation, is not, for the action does not remain instantaneous and direct under transformation. Rather in some inertial frames it becomes retarded, in others advanced.

Newtonian mechanics tells us that the changes produced by the interactions must satisfy conservation-of-momentum and conservation-of-energy laws. The momenta and energies are associated with the particles and with the fields. Each momentum is a 3-dimensional vector related to the velocity 3-vector and the energy scalar.

But since a 3-vector is not invariant during a Lorentz transformation, Newtonian laws distinguish between inertial frames. To correct the laws, we must substitute expressions that are suitably invariant.

17.2 4-Vector movements

We have also seen how the difference between the square of the Euclidean displacement of an effect and the square of the displacement of a particle traveling at the fundamental speed c over the same time interval is the same in all inertial frames. Let us identify this difference with the square of the magnitude of a 4-vector. The covariant components of the 4-vector are chosen so they combine with the contravariant components to yield the correct form.

507

If the origins of inertial systems Σ and Σ' are placed at the initial point and if the coordinates of the final point are x, y, z at time t and x', y', z' at time t', then equations (16-25) describe the relationship:

$$r^2 = x^2 + y^2 + z^2 - c^2t^2 = x'^2 + y'^2 + z'^2 - c^2t'^2. \tag{17-1}$$

This can be rewritten in the matrix form

$$r^2 = (x \quad y \quad z \quad -ct) \begin{pmatrix} x \\ y \\ z \\ ct \end{pmatrix} = (x' \quad y' \quad z' \quad -ct') \begin{pmatrix} x' \\ y' \\ z' \\ ct' \end{pmatrix}. \tag{17-2}$$

We can construct the 4-dimensional vector

$$\mathbf{r} = x\mathbf{e}_1 + y\mathbf{e}_2 + z\mathbf{e}_3 + ct\mathbf{e}_4$$
$$= x'\mathbf{e}_1' + y'\mathbf{e}_2' + z'\mathbf{e}_3' + ct'\mathbf{e}_4'. \tag{17-3}$$

Indeed if we assume

$$\mathbf{e}_1 \cdot \mathbf{e}_1 = \mathbf{e}_2 \cdot \mathbf{e}_2 = \mathbf{e}_3 \cdot \mathbf{e}_3 = -\mathbf{e}_4 \cdot \mathbf{e}_4 = 1, \tag{17-4}$$

$$\mathbf{e}_1' \cdot \mathbf{e}_1' = \mathbf{e}_2' \cdot \mathbf{e}_2' = \mathbf{e}_3' \cdot \mathbf{e}_3' = -\mathbf{e}_4' \cdot \mathbf{e}_4' = 1, \tag{17-5}$$

and

$$\mathbf{e}_1 \cdot \mathbf{e}_2 = \mathbf{e}_1 \cdot \mathbf{e}_3 = \mathbf{e}_1 \cdot \mathbf{e}_4 = \mathbf{e}_2 \cdot \mathbf{e}_3 = \mathbf{e}_2 \cdot \mathbf{e}_4 = \mathbf{e}_3 \cdot \mathbf{e}_4 = 0, \tag{17-6}$$

$$\mathbf{e}_1' \cdot \mathbf{e}_2' = \mathbf{e}_1' \cdot \mathbf{e}_3' = \mathbf{e}_1' \cdot \mathbf{e}_4' = \mathbf{e}_2' \cdot \mathbf{e}_3' = \mathbf{e}_2' \cdot \mathbf{e}_4' = \mathbf{e}_3' \cdot \mathbf{e}_4' = 0, \tag{17-7}$$

we find that

$$\mathbf{r} \cdot \mathbf{r} = x^2 + y^2 + z^2 - c^2t^2$$
$$= x'^2 + y'^2 + z'^2 - c^2t'^2. \tag{17-8}$$

Following the procedure in section 12.9, we call x, y, z, ct and x', y', z', ct' contravariant components of 4-vector $\mathbf{r}$. The covariant components must then be taken as x, y, z, $-ct$ or x', y', z', $-ct'$ to satisfy equation (17-1). The row matrices in (17-2) contain the covariant components of $\mathbf{r}$; the column matrices the contravariant components.

Similarly, the element $d\mathbf{s}$ between successive events in the life of a particle has the components listed in lines (16-126) and (16-127). We have

$$ds^2 = (dx \quad dy \quad dz \quad -c\,dt) \begin{pmatrix} dx \\ dy \\ dz \\ c\,dt \end{pmatrix} = (dx' \quad dy' \quad dz' \quad -c\,dt') \begin{pmatrix} dx' \\ dy' \\ dz' \\ c\,dt' \end{pmatrix}. \tag{17.9}$$

In a given inertial frame, the first three components of a 4-vector behave as the components of a 3-dimensional vector. Consequently we may describe the property by such a 3-vector and the remaining scalar. If we let

$$d\mathbf{r} = \mathbf{e}_1 \, dx + \mathbf{e}_2 \, dy + \mathbf{e}_3 \, dz \tag{17-10}$$

and

$$d\mathbf{r}' = \mathbf{e}_1' \, dx' + \mathbf{e}_2' \, dy' + \mathbf{e}_3' \, dz', \tag{17-11}$$

then $d\mathbf{s}$ is represented by the combination

$$(d\mathbf{r}, \, c \, dt) \tag{17-12}$$

in Σ and by

$$(d\mathbf{r}', \, c \, dt') \tag{17-13}$$

in Σ'.

EXAMPLE 17.1   Compute the square of the sum of two 4-vectors.

Let $\mathbf{e}_1$ and $\mathbf{e}_1'$ be unit vectors giving the directions of the spatial parts of 4-vectors $\mathbf{r}$ and $\mathbf{s}$ in the same inertial frame. Also let the spatial and temporal parts of $\mathbf{r}$ and $\mathbf{s}$ have the magnitudes r_1, r_4 and s_1, s_4. Then

$$\mathbf{r} = r_1\mathbf{e}_1 + r_4\mathbf{e}_4,$$
$$\mathbf{s} = s_1\mathbf{e}_1' + s_4\mathbf{e}_4;$$

so

$$\mathbf{r} + \mathbf{s} = r_1\mathbf{e}_1 + s_1\mathbf{e}_1' + (r_4 + s_4)\mathbf{e}_4$$

and

$$\begin{aligned}
(\mathbf{r} + \mathbf{s}) \cdot (\mathbf{r} + \mathbf{s}) &= r_1^2\mathbf{e}_1 \cdot \mathbf{e}_1 + s_1^2\mathbf{e}_1' \cdot \mathbf{e}_1' + (r_4 + s_4)^2\mathbf{e}_4 \cdot \mathbf{e}_4 + 2r_1s_1\mathbf{e}_1 \cdot \mathbf{e}_1' \\
&\quad + 2r_1(r_4 + s_4)\mathbf{e}_1 \cdot \mathbf{e}_4 + 2s_1(r_4 + s_4)\mathbf{e}_1' \cdot \mathbf{e}_4 \\
&= r_1^2 + s_1^2 - (r_4 + s_4)^2 + 2r_1s_1 \cos \theta.
\end{aligned}$$

Note that θ is the angle between $\mathbf{e}_1$ and $\mathbf{e}_1'$.

17.3   Allowed homogeneous linear transformations

In section 16.2, we found that every transformation from one inertial frame to another is linear. Furthermore, if the space-time origin of the second system is made coincident with that of the first, the transformation is homogeneous.

Such a transformation can always be represented by a matrix equation. Indeed if we take

$$\mathbf{r}' = \begin{pmatrix} x' \\ y' \\ z' \\ ct' \end{pmatrix}, \qquad \mathbf{r} = \begin{pmatrix} x \\ y \\ z \\ ct \end{pmatrix}, \tag{17-14}$$

we have

$$\mathbf{r}' = \mathbf{A}\mathbf{r} \tag{17-15}$$

where

$$\mathbf{A} = \begin{pmatrix} A_{11} & A_{12} & A_{13} & A_{14} \\ A_{21} & A_{22} & A_{23} & A_{24} \\ A_{31} & A_{32} & A_{33} & A_{34} \\ A_{41} & A_{42} & A_{43} & A_{44} \end{pmatrix}. \tag{17-16}$$

Let us define the *adjoints* of **r**, **r'** as the matrices obtained on transposing **r**, **r'** and changing the sign of the temporal component:

$$\mathbf{r}^{\times} = (x \quad y \quad z \quad -ct), \tag{17-17}$$

$$\mathbf{r'}^{\times} = (x' \quad y' \quad z' \quad -ct'). \tag{17-18}$$

We also define $\mathbf{A}^{\times}$ so the transformation

$$\mathbf{r'}^{\times} = \mathbf{r}^{\times}\mathbf{A}^{\times} \tag{17-19}$$

agrees with (17-15):

$$\mathbf{A}^{\times} = \begin{pmatrix} A_{11} & A_{21} & A_{31} & -A_{41} \\ A_{12} & A_{22} & A_{32} & -A_{42} \\ A_{12} & A_{23} & A_{33} & -A_{43} \\ -A_{14} & -A_{24} & -A_{34} & A_{44} \end{pmatrix}. \tag{17-20}$$

The desired invariance (17-1) is obtained when

$$r^2 = \mathbf{r}^{\times}\mathbf{r} = \mathbf{r'}^{\times}\mathbf{r'} = r'^2. \tag{17-21}$$

Equations (17-19) and (17-15) change the third expression as follows:

$$\mathbf{r'}^{\times}\mathbf{r'} = \mathbf{r}^{\times}\mathbf{A}^{\times}\mathbf{A}\mathbf{r}. \tag{17-22}$$

For the right side to equal $\mathbf{r}^{\times}\mathbf{r}$ in general, we must have

$$\mathbf{A}^{\times}\mathbf{A} = \mathbf{E}. \tag{17-23}$$

Matrix equation (17-23) yields the relationships

$$A_{lk}A_{l\mu} - A_{4k}A_{4\mu} = \delta_{k\mu}, \tag{17-24}$$

$$-A_{l4}A_{l\mu} + A_{44}A_{4\mu} = \delta_{4\mu}, \tag{17-25}$$

in which

$$k = 1, 2, 3, \quad l = 1, 2, 3, \quad \mu = 1, 2, 3, 4. \tag{17-26}$$

Note that only 10 of these relationships are different from each other.

EXAMPLE 17.2   Show how condition (17-23) yields the transformation matrix when Σ' is related to Σ as in figure 16.1.

As long as the primed axes stay parallel to the unprimed ones and the origin of Σ' moves along the x axis of Σ, coordinate y' is independent of x, z, ct and coordinate

z' is independent of x, y, ct. All off-diagonal elements in **A** are then zero except those in the corners, and equation (17-23) reduces to

$$\begin{pmatrix} A_{11} & 0 & 0 & -A_{41} \\ 0 & A_{22} & 0 & 0 \\ 0 & 0 & A_{33} & 0 \\ -A_{14} & 0 & 0 & A_{44} \end{pmatrix} \begin{pmatrix} A_{11} & 0 & 0 & A_{14} \\ 0 & A_{22} & 0 & 0 \\ 0 & 0 & A_{33} & 0 \\ A_{41} & 0 & 0 & A_{44} \end{pmatrix} = \begin{pmatrix} 1 & 0 & 0 & 0 \\ 0 & 1 & 0 & 0 \\ 0 & 0 & 1 & 0 \\ 0 & 0 & 0 & 1 \end{pmatrix}.$$

Multiplying out the left side and comparing with the right yields the equations

$$A_{11}{}^2 - A_{41}{}^2 = 1,$$
$$A_{22}{}^2 = 1,$$
$$A_{33}{}^2 = 1,$$
$$-A_{14}{}^2 + A_{44}{}^2 = 1,$$
$$A_{11}A_{14} - A_{41}A_{44} = 0.$$

These are satisfied when

$$A_{11} = \cosh \alpha, \qquad A_{41} = -\sinh \alpha,$$
$$A_{22} = 1,$$
$$A_{33} = 1,$$
$$A_{44} = \cosh \alpha, \qquad A_{14} = -\sinh \alpha.$$

So the conventional Lorentz transformation is represented by the matrix

$$\mathbf{A} = \begin{pmatrix} \cosh \alpha & 0 & 0 & -\sinh \alpha \\ 0 & 1 & 0 & 0 \\ 0 & 0 & 1 & 0 \\ -\sinh \alpha & 0 & 0 & \cosh \alpha \end{pmatrix}$$

with

$$\tanh \alpha = \frac{u}{c}$$

as before.

17.4 The 4-vector velocity

In Newtonian mechanics, the displacement of a particle, the corresponding velocity, and the corresponding momentum are all independent of the inertial frame being used. Furthermore, since the coordinate time is universal, it is the proper time for the particle, regardless of how fast the particle is moving.

But in relativistic theory, the spatial displacement, the conventional velocity, and the conventional momentum are not invariant. Differential $d\mathbf{r}$ of Newtonian theory and differential dt, being elements of distance and time, do vary from one inertial frame to another. (We will continue to use $d\mathbf{r}$ in the same sense as in section 2.11. Differential $d\mathbf{s}$ will be used for the vector displacement along a world line.) So we have to consider how a suitable displacement, velocity, and momentum can be constructed.

Let the Cartesian coordinates of the given particle be x, y, z at time t in inertial frame Σ and x', y', z' at time t' in inertial frame Σ'. Let these change by amounts dx, dy, dz, dt and dx', dy', dz', dt' in going between successive events.

The first set of spatial differentials combine in the familiar manner:

$$d\mathbf{r} = \mathbf{i}\, dx + \mathbf{j}\, dy + \mathbf{k}\, dz. \tag{17-27}$$

Equation (17-3) shows how this 3-dimensional vector (called a 3-vector) combines with the scalar $c\, dt$ to yield the 4-dimensional vector (called a 4-vector):

$$d\mathbf{s} = \mathbf{e}_1\, dx + \mathbf{e}_2\, dy + \mathbf{e}_3\, dz + \mathbf{e}_4 c\, dt$$
$$= d\mathbf{r} + \mathbf{e}_4 c\, dt. \tag{17-28}$$

Similarly, from the second set we get

$$d\mathbf{r}' = \mathbf{i}'\, dx' + \mathbf{j}'\, dy' + \mathbf{k}'\, dz', \tag{17-29}$$

whence

$$d\mathbf{s} = \mathbf{e}_1'\, dx' + \mathbf{e}_2'\, dy' + \mathbf{e}_3'\, dz' + \mathbf{e}_4' c\, dt'$$
$$= d\mathbf{r}' + \mathbf{e}_4' c\, dt'. \tag{17-30}$$

The contravariant components of $d\mathbf{s}$ are the coefficients of the unit vectors, being

$$dx, \qquad dy, \qquad dz, \qquad c\, dt, \tag{17-31}$$

in Σ and

$$dx', \qquad dy', \qquad dz', \qquad c\, dt', \tag{17-32}$$

in Σ'. From equations (16-131) and (16-125), a particle ages by the amount

$$d\tau = \frac{i\, ds}{c} \tag{17-33}$$

when it moves along element $d\mathbf{s}$ of its world line.

Formally dividing the contravariant components by the differential of proper time, an *invariant* scalar, we get the corresponding components for the relativistic velocity $d\mathbf{s}/d\tau$:

$$\frac{dx}{d\tau} = \frac{dt}{d\tau}\frac{dx}{dt} = \gamma v_x, \tag{17-34}$$

$$\frac{dy}{d\tau} = \frac{dt}{d\tau}\frac{dy}{dt} = \gamma v_y, \tag{17-35}$$

$$\frac{dz}{d\tau} = \frac{dt}{d\tau}\frac{dz}{dt} = \gamma v_z, \tag{17-36}$$

$$\frac{c\, dt}{d\tau} = \frac{dt}{d\tau}c = \gamma c. \tag{17-37}$$

Note that v_x, v_y, and v_z represent the conventional x, y, and z components of velocity in inertial frame Σ while γ is the expression for $dt/d\tau$ given by equation (16-134).

Similarly dividing the covariant components of $d\mathbf{s}$ by $d\tau$ yields

$$\gamma v_x, \qquad \gamma v_y, \qquad \gamma v_z, \qquad -\gamma c. \tag{17-38}$$

The square of the magnitude of the 4-vector velocity is obtained on adding the products of corresponding components:

$$(\gamma v_x \quad \gamma v_y \quad \gamma v_z \quad -\gamma c) \begin{pmatrix} \gamma v_x \\ \gamma v_y \\ \gamma v_z \\ \gamma c \end{pmatrix} = \gamma^2(v_x{}^2 + v_y{}^2 + v_z{}^2 - c^2)$$

$$= -c^2\gamma^2 \left(1 - \frac{v^2}{c^2}\right) = -c^2. \tag{17-39}$$

The magnitude itself is the invariant ic.

EXAMPLE 17.3 How are the distance moved by a particle in an infinitesimal time interval and the time interval related to the invariant ds?

First, solve equation (16-134) for dt and eliminate $d\tau$ with equation (17-33)

$$dt = \gamma \, d\tau = i \frac{\gamma}{c} \, ds.$$

Note that γ is related to the speed v of the particle by the equation

$$\gamma = \frac{1}{(1 - v^2/c^2)^{1/2}} \cdot$$

Then orient the axes so $dy = dz = 0$ and equation (16-124) reduces to

$$dx^2 - c^2 \, dt^2 = ds^2,$$

whence

$$dx^2 = ds^2 + c^2 \, dt^2.$$

Introduce the expression obtained for dt

$$dx^2 = ds^2 - \gamma^2 \, ds^2 = ds^2(1 - \gamma^2)$$

and solve for dx

$$dx = ds(1 - \gamma^2)^{1/2} = i \, ds(\gamma^2 - 1)^{1/2}.$$

17.5 The 4-vector momentum

In Newtonian mechanics, each particle exhibits a momentum equal to its mass times its velocity. The mass is independent of the particle's speed at the given instant of time.

In relativistic theory, this multiplier of velocity does not have to be invariant. We need only assume that the classical mass determined by an inertial observer momentarily at rest with respect to the particle is a scalar m_0. This mass is called the *proper mass* or *rest mass* of the particle.

As before, we let 4-vector $d\mathbf{s}$ be the displacement of the particle along its world line in proper time $d\tau$. Formally multiplying $d\mathbf{s}/d\tau$ by the scalar m_0 then yields the *relativistic momentum*

$$m_0 \frac{d\mathbf{s}}{d\tau} = m_0 \frac{dt}{d\tau} \frac{d\mathbf{s}}{dt} = m \frac{d\mathbf{s}}{dt}. \tag{17-40}$$

Quantity m is defined by the equation

$$m = m_0 \frac{dt}{d\tau}, \tag{17-41}$$

which is converted to

$$m = \frac{m_0}{(1 - v^2/c^2)^{1/2}} = \gamma m_0 \tag{17-42}$$

by formulas (16-134). Let us call m the *relativistic mass*.

The 4-velocity and 4-momentum are both 4-vectors because each is defined as a scalar times a 4-vector. Table 17.1 shows how the components are related to those of $d\mathbf{s}$. Each 4-vector can be broken down into a 3-vector and a so-called scalar part, as table 17.2 shows.

The contravariant components of a 4-vector are changed to covariant components by reversing the sign of the fourth component.

TABLE 17.1   Contravariant Components of the 4-Vectors Involved in Describing the Movement of a Particle

Infinitesimal Displacement along the World Line	4-Velocity	4-Momentum
$dx^1 = dx$	$v^1 = \dfrac{dx}{d\tau}$	$p^1 = m_0 \dfrac{dx}{d\tau} = mv_x$
$dx^2 = dy$	$v^2 = \dfrac{dy}{d\tau}$	$p^2 = m_0 \dfrac{dy}{d\tau} = mv_y$
$dx^3 = dz$	$v^3 = \dfrac{dz}{d\tau}$	$p^3 = m_0 \dfrac{dz}{d\tau} = mv_z$
$dx^4 = c\,dt$	$v^4 = c\dfrac{dt}{d\tau}$	$p^4 = m_0 c\dfrac{dt}{d\tau} = mc$

TABLE 17.2   Analysis of the 4-Vectors into 3-Vector and Scalar Parts

Infinitesimal Displacement	4-Velocity	4-Momentum
$d\mathbf{r}$	$\dfrac{d\mathbf{r}}{d\tau} = \mathbf{v}\dfrac{dt}{d\tau}$	$m_0\dfrac{d\mathbf{r}}{d\tau} = m\dfrac{d\mathbf{r}}{dt} = \mathbf{p}$
$c\,dt$	$c\dfrac{dt}{d\tau}$	$mc = \dfrac{E}{c}$

The square of the magnitude of the 4-vector momentum is obtained on adding the products of corresponding covariant and contravariant components:

$$(mv_x \quad mv_y \quad mv_z \quad -mc)\begin{pmatrix} mv_x \\ mv_y \\ mv_z \\ mc \end{pmatrix} = m^2(v_x{}^2 + v_y{}^2 + v_z{}^2 - c^2) = (mv)^2 - (mc)^2$$

$$= -m^2c^2\left(1 - \frac{v^2}{c^2}\right) = -m_0{}^2c^2. \qquad (17\text{-}43)$$

The last step involves eliminating m with equation (17-42).

Multiplying two of the equivalent forms in (17-43) by c^2,

$$(mv)^2c^2 - (mc^2)^2 = -m_0{}^2c^4, \qquad (17\text{-}44)$$

replacing mv with p, and mc^2 with E leads to

$$p^2c^2 - E^2 = -m_0{}^2c^4 \qquad (17\text{-}45)$$

or

$$p^2c^2 + m_0{}^2c^4 = E^2. \qquad (17\text{-}46)$$

Relationship (17-46) is represented on the right triangle of figure 17.1. Expression p represents the magnitude of the 3-momentum of the particle; but we still have to identify E as its energy.

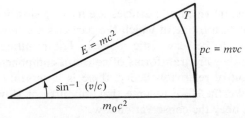

FIGURE 17.1   Geometric description of the relationship between the relativistic 3-momentum p and energy E.

17.6   Generalizing the conservation of momentum and mass laws

Newtonian equations do describe the behavior of a system of macroscopic particles moving slowly in a small region of space. But the equations are not suitably covariant; they do not allow for a finite fundamental speed c. Therefore, we pair 3-vectors from the theory with "scalars" and assume that the resulting 4-vectors transform as displacement $d\mathbf{s}$ transforms.

From equation (1-102) and the definitions of v_x, v_y, v_z, the conventional velocity $\mathbf{v}$ of a particle is the 3-vector

$$\mathbf{v} = v_x\mathbf{i} + v_y\mathbf{j} + v_z\mathbf{k}. \tag{17-47}$$

Multiplying $\mathbf{v}$ by the relativistic mass m of the particle yields the expressions

$$\mathbf{p} = m\mathbf{v} = mv_x\mathbf{i} + mv_y\mathbf{j} + mv_z\mathbf{k}$$
$$= p^1\mathbf{i} + p^2\mathbf{j} + p^3\mathbf{k} \tag{17-48}$$

that reduce to the classical momentum when v is very small with respect to c.

Let the particles in the given set be numbered $1, 2, \ldots, N$. The momentum of the jth particle is then represented by equation (17-48) in the form

$$\mathbf{p}_j = p_j{}^1\mathbf{i} + p_j{}^2\mathbf{j} + p_j{}^3\mathbf{k}. \tag{17-49}$$

When all N particles move slowly enough with respect to an inertial frame, Newtonian theory applies. If, in addition, the set is isolated, each component of the total momentum $\mathbf{P}$ is conserved:

$$p_1{}^a + p_2{}^a + \cdots + p_N{}^a = P^a = \text{constant}, \qquad a = 1, 2, 3. \tag{17-50}$$

Note that the summation is over all particles in the isolated set.

From table 17.2, the scalar to be paired with $\mathbf{p}_j$ is $m_j c$, where m_j is the relativistic mass of the jth particle. At small speeds, parameter m_j reduces to the Newtonian mass and the conservation-of-mass law yields

$$m_1 c + m_2 c + \cdots + m_N c = p_1{}^4 + p_2{}^4 + \cdots + p_N{}^4$$
$$= Mc = \text{constant} \tag{17-51}$$

where M is the total mass.

Equations (17-50) and (17-51) imply that the 4-vector

$$(\mathbf{P}, Mc) \tag{17-52}$$

is conserved in a frame in which all particles are moving slowly. However, a person can refer this 4-vector to a frame in which the particles are moving very rapidly, by considering the masses to be relativistic masses. The resulting components would stay constant because they are transforms of constant components.

From the standpoint of relativity theory, there is no essential difference between the two situations. We therefore assume that mixtures of slowly moving and fast moving particles also obey the conservation law

$$p_1{}^\alpha + p_2{}^\alpha + \cdots + p_N{}^\alpha = P^\alpha = \text{constant}, \qquad \alpha = 1, 2, 3, 4. \tag{17-53}$$

A large number of experiments support this generalization.

17.7 The significance of $m_j c^2$

An examination of consequences of annihilating a particle in a set of slowly moving particles now enables us to interpret c times the fourth contravariant component of 4-momentum.

We assume that the set of particles is isolated, so its behavior is governed by equation (17-53). Letting α be 4 then yields

$$p_1{}^4 + p_2{}^4 + \cdots + p_N{}^4 = \text{constant} \qquad (17\text{-}54)$$

or

$$m_1 c + m_2 c + \cdots + m_N c = \frac{E_1}{c} + \frac{E_2}{c} + \cdots + \frac{E_N}{c} = Mc = \frac{E}{c}, \qquad (17\text{-}55)$$

whence

$$m_1 c^2 + m_2 c^2 + \cdots + m_N c^2 = E. \qquad (17\text{-}56)$$

Here m_1 is the mass of the first particle, m_2 that of the second, and so on.

Variation-of-mass law (17-42) converts equation (17-56) to

$$\sum \frac{(m_j)_{\text{rest}} c^2}{(1 - v_j{}^2/c^2)^{1/2}} = E \qquad (17\text{-}57)$$

where $(m_j)_{\text{rest}}$ is the rest mass of the jth particle and v_j is its speed. Introducing the binomial expansion

$$\left(1 - \frac{v_j{}^2}{c^2}\right)^{-1/2} = 1 + \frac{1}{2}\frac{v_j{}^2}{c^2} + \cdots \qquad (17\text{-}58)$$

changes (17-57) to

$$\sum \left[(m_j)_{\text{rest}} c^2 + \tfrac{1}{2}(m_j)_{\text{rest}} v_j{}^2 + \cdots \right] = E. \qquad (17\text{-}59)$$

Since the higher terms contain negative powers of c, they are negligible when all particle speeds are small. Then

$$\sum (m_j)_{\text{rest}} c^2 + \sum \tfrac{1}{2}(m_j)_{\text{rest}} v_j{}^2 \simeq E. \qquad (17\text{-}60)$$

Note that each sum is over all particles in the set, and the second sum is the Newtonian kinetic energy.

Conservation law (17-53) states that E/c, and consequently E itself, are constant. If an interaction should cause the kth particle to disappear, amount $(m_k)_{\text{rest}} c^2$ by which the first sum decreases would have to go into the second sum. Annihilation of the kth particle thus causes this amount to be added to the kinetic energy of the system.

Thus, we interpret $(m_k)_{\text{rest}} c^2$ as the energy associated with the mass of the kth particle when it is at rest, and E is interpreted as the total energy. In equation (17-56), expression $m_j c^2$ is the mechanical and rest energy of the jth particle.

17.8 Conversion of kinetic energy to potential or rest energy

The simplest nontrivial mechanical system consists of two particles that interact appreciably only when they are very close together. Depending on their properties and the initial conditions, the two may rebound away from each other, or they may interchange matter and then separate, or they may remain united. In the second process, rest energy may either increase or decrease; in the third process, it must increase.

Indeed since relative movement of parts requires kinetic energy, maximum conversion to rest energy is achieved only when such movement has been eliminated. The over-all process is then

$$A + B \rightarrow M \tag{17-61}$$

where M designates A and B moving together in some form.

Since the components of momentum are conserved, the 4-momentum of M equals the sum of that of A and B initially. To get the rest energy of M, we need only transform the initial momenta to the system in which M is at rest, the center-of-mass system, and multiply the fourth component by c.

Suppose that the inertial frame in which the collision is observed is Σ, with the x axis pointing in the direction of movement of the center of mass. The Σ' of figure 16.1 may then be the system in which this center is at rest.

Since the 4-vector momentum

$$\left(\mathbf{p}, \frac{E}{c} \right) \tag{17-62}$$

behaves as $d\mathbf{s}$ does, it is transformed by the matrix in example 17.2,

$$\mathbf{L} = \begin{pmatrix} \cosh\alpha & 0 & 0 & -\sinh\alpha \\ 0 & 1 & 0 & 0 \\ 0 & 0 & 1 & 0 \\ -\sinh\alpha & 0 & 0 & \cosh\alpha \end{pmatrix} = \begin{pmatrix} \gamma & 0 & 0 & -\beta\gamma \\ 0 & 1 & 0 & 0 \\ 0 & 0 & 1 & 0 \\ -\beta\gamma & 0 & 0 & \gamma \end{pmatrix}, \tag{17-63}$$

where

$$\cosh\alpha = \frac{1}{(1 - u^2/c^2)^{1/2}} = \gamma, \tag{17-64}$$

$$\sinh\alpha = \frac{u/c}{(1 - u^2/c^2)^{1/2}} = \beta\gamma, \tag{17-65}$$

$$\beta = \frac{u}{c}. \tag{17-66}$$

If the components in Σ are p^1, p^2, p^3, p^4, the components in Σ' are the elements of the column matrix

$$\begin{pmatrix} \gamma & 0 & 0 & -\beta\gamma \\ 0 & 1 & 0 & 0 \\ 0 & 0 & 1 & 0 \\ -\beta\gamma & 0 & 0 & \gamma \end{pmatrix} \begin{pmatrix} p^1 \\ p^2 \\ p^3 \\ p^4 \end{pmatrix} = \begin{pmatrix} \gamma p^1 - \beta\gamma p^4 \\ p^2 \\ p^3 \\ -\beta\gamma p^1 + \gamma p^4 \end{pmatrix} \tag{17-67}$$

Applying equation (17-67) to the sum of the initial momenta

$$\left(\mathbf{p_A} + \mathbf{p_B}, \frac{E_A}{c} + \frac{E_B}{c} \right) \tag{17-68}$$

and letting Σ' be the frame in which M is at rest, we obtain

$$\gamma(p_A{}^1 + p_B{}^1) - \beta\gamma(p_A{}^4 + p_B{}^4) = 0, \tag{17-69}$$

$$-\beta\gamma(p_A{}^1 + p_B{}^1) + \gamma(p_A{}^4 + p_B{}^4) = \frac{E_0}{c}. \tag{17-70}$$

Equation (17-69) shows that

$$\frac{u}{c} = \beta = \frac{p_A{}^1 + p_B{}^1}{p_A{}^4 + p_B{}^4} = \frac{p_A{}^1 c + p_B{}^1 c}{E_A + E_B}. \tag{17-71}$$

Substituting this result into equation (17-70) then yields

$$(-\beta^2 + 1)\gamma(p_A{}^4 + p_B{}^4) = \frac{p_A{}^4 + p_B{}^4}{\gamma} = \frac{E_A/c + E_B/c}{\gamma}$$

$$= \frac{E_0}{c} \tag{17-72}$$

or

$$E_0 = \frac{1}{\gamma}(E_A + E_B). \tag{17-73}$$

The gamma for equation (17-73) is obtained by substituting the beta from equation (17-71) into

$$\gamma = \frac{1}{(1 - u^2/c^2)^{1/2}} = \frac{1}{(1 - \beta^2)^{1/2}}. \tag{17-74}$$

EXAMPLE 17.4   How is the square of the magnitude of the 4-momentum of a particle related to its rest energy?

The 4-momentum of a particle has been defined as

$$m_0 \frac{d\mathbf{s}}{d\tau}$$

where m_0 and τ are scalars, being the rest mass and proper time for the particle, while $d\mathbf{s}$ is the pertinent 4-vector displacement along the world line. Since

$$m = m_0 \frac{dt}{d\tau},$$

the 3-vector from the first three components may be written as

$$\mathbf{p} = \begin{pmatrix} mv_x \\ mv_y \\ mv_z \end{pmatrix} \quad \text{or} \quad \tilde{\mathbf{p}} = (mv_x \quad mv_y \quad mv_z).$$

Since parameter E has been introduced by the equation

$$\frac{E}{c} = mc = m_0 \frac{c\,dt}{d\tau},$$

the fourth contravariant component is E/c, the fourth covariant component $-E/c$.

Because the sum of the products of corresponding covariant and contravariant components for $d\mathbf{s}$ forms a scalar; the similar sum for the proportional 4-vector momentum is also invariant. This may be represented as

$$\left(\tilde{\mathbf{p}} \quad \frac{-E}{c} \right) \begin{pmatrix} \mathbf{p} \\ \frac{E}{c} \end{pmatrix} = \tilde{\mathbf{p}}\mathbf{p} - \frac{E^2}{c^2} = p^2 - \frac{E^2}{c^2}$$

in Σ, or as

$$\left(\tilde{\mathbf{p}}' \quad \frac{-E'}{c} \right) \begin{pmatrix} \mathbf{p}' \\ \frac{E'}{c} \end{pmatrix} = p'^2 - \frac{E'^2}{c^2}$$

in Σ'.

If we choose Σ' as the frame in which the particle is momentarily at rest, p' is zero, E' is rest energy E_0, and

$$p^2 - \frac{E^2}{c^2} = -\frac{E_0^2}{c^2}$$

or

$$p^2 c^2 = E^2 - E_0^2.$$

EXAMPLE 17.5   When a target containing protons is bombarded with a beam of negative pions, a resonance is observed at 600 MeV. What is the rest mass of the resonance particle?

Let us employ the conventional symbols, with index A indicating the property is that of a pion, B that of a proton.

Before the collision, the energy of the pion is 600 MeV plus its rest energy:

$$E_A = 139.58 + 600 = 739.58 \text{ MeV}.$$

The square of its momentum times c is

$$\begin{aligned} (p_A{}^1 c)^2 &= E_A{}^2 - E_{A0}{}^2 = (739.58)^2 - (139.58)^2 \\ &= 527{,}496 \text{ MeV}^2 \end{aligned}$$

according to the equation derived in example 17.4.

On the other hand, the reactant proton is effectively at rest, so its energy is its rest energy

$$E_B = E_{B0} = 938.26 \text{ MeV}$$

and its momentum is zero

$$p_B{}^1 c = 0.$$

Substituting into the square of equation (17-71),

$$\beta^2 = \frac{(p_A{}^1c + p_B{}^1c)^2}{(E_A + E_B)^2} = \frac{527,496}{(1677.84)^2} = 0.18738,$$

and putting the result into equation (17-73) yields

$$E_0 = \gamma^{-1}(E_A + E_B) = (1 - \beta^2)^{1/2}(E_A + E_B)$$
$$= (0.81262)^{1/2}(1677.84) = 1512 \text{ MeV}.$$

17.9   Binary collisions

As long as the rest mass of a particle is not changed, interaction with another particle leaves the square of its 4-momentum invariant. The transformation of the 4-vector is not only linear and homogeneous, but it has the same form as the general Lorentz transformation of section 17.3. Furthermore, we will find that in a binary collision both initial 4-momenta are subject to the same transformation.

Let us suppose that the particles come together, interact briefly, and then separate according to the equation

$$A + B \rightarrow C + D. \tag{17-75}$$

Let us also erect a nonrotating coordinate system on the point about which the moment of relativistic mass is zero.

If $\mathbf{r}_j$ is the radius vector drawn from the origin of this center-of-mass system to particle j and m_j is the mass of the particle, then

$$m_A\mathbf{r}_A + m_B\mathbf{r}_B = 0 \tag{17-76}$$

before the collision, while

$$m_C\mathbf{r}_C + m_D\mathbf{r}_D = 0 \tag{17-77}$$

afterwards.

Differentiating equations (17-76) and (17-77) with respect to the coordinate time and labeling the 3-momentum of particle j as $\mathbf{p}_j$, we obtain

$$\mathbf{p}_A + \mathbf{p}_B = 0 \tag{17-78}$$

before, and

$$\mathbf{p}_C + \mathbf{p}_D = 0 \tag{17-79}$$

after the collision. The 3-momenta are equal but oppositely directed, as figure 17.2 shows.

During the interaction, $\mathbf{p}_A$ and $\mathbf{p}_B$ must be deflected by the same angle ϕ. Furthermore, if the magnitudes change, they must change by the same amount. Thus, we have

$$\frac{p_C}{p_A} = \frac{p_D}{p_A} = \frac{p_C}{p_B} = \frac{p_D}{p_B} = \sigma \tag{17-80}$$

in the center-of-mass frame. Although the energies may change,

$$\frac{E_C}{E_A} = \rho_A, \tag{17-81}$$

$$\frac{E_D}{E_B} = \rho_B, \tag{17-82}$$

they are subject to the conservation law

$$E_A + E_B = E_C + E_D. \tag{17-83}$$

When the collision is elastic, σ, ρ_A, and ρ_B all equal one. The only influence on $\mathbf{p}_A$ and $\mathbf{p}_B$ is then rotation. With axes chosen as in figure 17.2, the rotation affects the first two components of each 4-momentum as operation C_n affects x and y in equation (4-29). The operation is implemented by the matrix

$$\mathbf{R'} = \begin{pmatrix} \cos\phi & -\sin\phi & 0 & 0 \\ \sin\phi & \cos\phi & 0 & 0 \\ 0 & 0 & 1 & 0 \\ 0 & 0 & 0 & 1 \end{pmatrix} \tag{17-84}$$

However, the coordinate system based on the center of mass is generally moving. In the inertial frame in which the laboratory is momentarily at rest, let us join initial positions of the two particles by the x axis, so that the center of mass moves along this axis in the positive direction. Determine the plane in which the deflections occur and place the y axis in this plane. The x' axis in the center-of-mass frame then

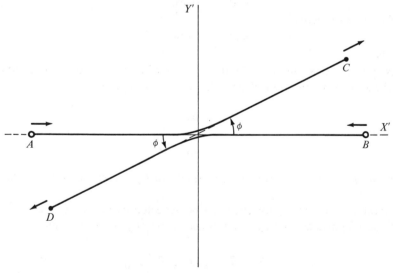

FIGURE 17.2   Collision between A and B as observed from the center-of-mass frame.

lies along the x axis, and the y' axis moves parallel to the y axis, the z' axis parallel to the z axis. The systems are related as the Σ and Σ' systems of figure 16.1.

Consequently, the transformation from the laboratory system to the center-of-mass system is implemented by matrix (17-63), while the inverse transformation is effected by

$$\mathbf{L}^{-1} = \begin{pmatrix} \gamma & 0 & 0 & \beta\gamma \\ 0 & 1 & 0 & 0 \\ 0 & 0 & 1 & 0 \\ \beta\gamma & 0 & 0 & \gamma \end{pmatrix}. \tag{17-85}$$

Multiplying these together yields the matrix

$$\mathbf{R} = \mathbf{L}^{-1}\mathbf{R}'\mathbf{L} = \begin{pmatrix} \gamma^2\cos\phi - \beta^2\gamma^2 & -\gamma\sin\phi & 0 & -\beta\gamma^2\cos\phi + \beta\gamma^2 \\ \gamma\sin\phi & \cos\phi & 0 & -\beta\gamma\sin\phi \\ 0 & 0 & 1 & 0 \\ \beta\gamma^2\cos\phi - \beta\gamma^2 & -\beta\gamma\sin\phi & 0 & -\beta^2\gamma^2\cos\phi + \gamma^2 \end{pmatrix} \tag{17-86}$$

for transforming the 4-momenta in the laboratory frame.

EXAMPLE 17.6   If particle B receives its kinetic energy from being struck by A and if the angle of deflection of each particle is ϕ in the center-of-mass system, what is the final energy of B?

Since the initial velocity of B in the inertial system in which the laboratory is momentarily at rest is zero, the initial 4-momentum of B is

$$\begin{pmatrix} 0 \\ 0 \\ 0 \\ m_{B0}c \end{pmatrix}.$$

Acting on this with matrix (17-86) gives us the 4-momentum of B after it is struck.

The final energy of B equals c times the fourth component in the resulting matrix

$$\begin{aligned} E_B'' &= c(-\beta^2\gamma^2\cos\phi + \gamma^2)m_{B0}c \\ &= m_{B0}c^2 + m_{B0}c^2(\gamma^2 - 1 - \gamma^2\beta^2\cos\phi) \\ &= m_{B0}c^2 + m_{B0}c^2\gamma^2\beta^2(1 - \cos\phi). \end{aligned}$$

The last step involves the substitution

$$\gamma^2 - 1 = \frac{1 - (1 - \beta^2)}{1 - \beta^2} = \frac{\beta^2}{1 - \beta^2} = \gamma^2\beta^2.$$

Since B is initially at rest, p_B is zero and the square of (17-71) reduces to

$$\beta^2 = \frac{p_A{}^2c^2}{(E_A + E_B)^2}.$$

This expression converts the square of (17-74) to

$$\gamma^2 = \frac{(E_A + E_B)^2}{(E_A + E_B)^2 - p_A^2 c^2} ;$$

so

$$\gamma^2 \beta^2 = \frac{p_A^2 c^2}{(E_A + E_B)^2 - p_A^2 c^2}$$

$$= \frac{E_A^2 - m_{A0}^2 c^4}{E_A^2 + 2E_A E_B + E_B^2 - E_A^2 + m_{A0}^2 c^4}$$

$$= \frac{E_A^2 - m_{A0}^2 c^4}{c^2(2E_A m_{B0} + m_{B0}^2 c^2 + m_{A0}^2 c^2)}$$

and

$$E_B'' = m_{B0} c^2 + \frac{m_{B0}(E_A^2 - m_{A0}^2 c^4)(1 - \cos \phi)}{2E_A m_{B0} + m_{B0}^2 c^2 + m_{A0}^2 c^2} .$$

17.10   Compton scattering

In 1905, Einstein revived the corpuscular theory of light, showing how it explained the observed distribution law for black-body radiation, the rule for photoluminescence, and the photoelectric effect. He assumed that the energy in electromagnetic radiation travels in small packets at the velocity of light. Since this rate of propagation equals the fundamental speed c, each of the packets behaves as a particle having no rest mass.

The energy E of a packet, or *photon*, was assumed to be proportional to the frequency v of the radiation

$$E = hv. \tag{17-87}$$

Note that h is Planck's constant. If the relativistic mass m of the photon is then related to its energy by equation (17-56) in the form

$$mc = \frac{E}{c}, \tag{17-88}$$

a 4-momentum can be defined with the 3-vector part

$$m\mathbf{c} = \mathbf{p}. \tag{17-89}$$

Arthur H. Compton found that this part enters conservation laws just as the 3-momentum of a particle with rest mass does.

Compton studied the deflection of high energy photons by matter. Those scattered by heavy particles, nuclei, were little changed. Those scattered by electrons underwent a predictable wave length shift.

Consider a single photon of initial momentum $m_1 c$ being deflected through angle θ in the laboratory frame of reference. The struck electron is given momentum $m_e v$ at angle ϕ in the same plane, while the final momentum of the photon is $m_2 c$ (see figure 17.3).

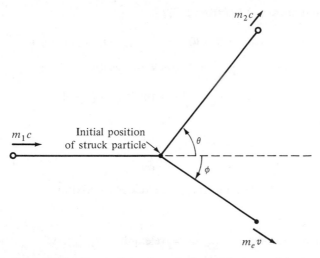

FIGURE 17.3   Deflection of a photon through angle θ on giving a particle the momentum $m_e v$.

Since the photon loses the same amount of horizontal momentum as the struck particle gains, we have

$$m_1 c - m_2 c \cos \theta = m_e v \cos \phi. \tag{17-90}$$

Since the vertical component that the photon gains is equal to, but oppositely directed from, that of the struck particle, we also have

$$m_2 c \sin \theta = m_e v \sin \phi. \tag{17-91}$$

Squaring each of these equations and adding to eliminate ϕ, we obtain

$$m_1^2 c^2 - 2 m_1 m_2 c^2 \cos \theta + m_2^2 c^2 = m_e^2 v^2. \tag{17-92}$$

The mass of the target particle obeys relativistic law (17-42)

$$m_e = \frac{(m_e)_{\text{rest}}}{(1 - v^2/c^2)^{1/2}} = \frac{(m_e)_{\text{rest}} c}{(c^2 - v^2)^{1/2}}. \tag{17-93}$$

Clear this of fractions, square,

$$m_e^2 c^2 - m_e^2 v^2 = (m_e)_{\text{rest}}^2 c^2, \tag{17-94}$$

and solve for $m_e^2 v^2$. Substitute the result into equation (17-92) and cancel c^2 to get

$$m_1^2 - 2 m_1 m_2 \cos \theta + m_2^2 = m_e^2 - (m_e)_{\text{rest}}^2. \tag{17-95}$$

In relativistic mechanics, the conservation-of-energy law is the conservation-of-mass law. Here the loss in mass of the photon equals the gain in mass of the struck particle:

$$m_1 - m_2 = m_e - (m_e)_{\text{rest}} \tag{17-96}$$

or

$$m_1^2 - 2 m_1 m_2 + m_2^2 = m_e^2 - 2 m_e (m_e)_{\text{rest}} + (m_e)_{\text{rest}}^2. \tag{17-97}$$

Subtracting equation (17-97) from (17-95),

$$2m_1m_2(1 - \cos \theta) = 2(m_e)_{\text{rest}} [m_e - (m_e)_{\text{rest}}], \qquad (17\text{-}98)$$

introducing (17-96) on the right, and canceling 2 yields the result

$$m_1m_2(1 - \cos \theta) = (m_e)_{\text{rest}} (m_1 - m_2) \qquad (17\text{-}99)$$

which rearranges to

$$\frac{m_1 - m_2}{m_1m_2} = \frac{1}{(m_e)_{\text{rest}}} (1 - \cos \theta). \qquad (17\text{-}100)$$

Equations (17-88), (17-87), and the phase velocity equation

$$c = \lambda v \qquad (17\text{-}101)$$

let us relate the photon mass m to the wavelength λ:

$$m = \frac{E}{c^2} = \frac{hv}{c(\lambda v)} = \frac{h}{c\lambda}. \qquad (17\text{-}102)$$

This result reduces equation (17-100) to

$$\lambda_2 - \lambda_1 = \frac{h}{c(m_e)_{\text{rest}}} (1 - \cos \theta) \qquad (17\text{-}103)$$

where λ_2 is the wavelength in the scattered beam while λ_1 is the wavelength in the incident beam. Compton found that equation (17-103) did describe the observed wavelength shift.

The behavior of a photon in a collision is governed by its momentum

$$p = mc; \qquad (17\text{-}104)$$

but combining (17-102) and (17-104) yields the formula

$$p = \frac{h}{c\lambda} c = \frac{h}{\lambda} = h\sigma \qquad (17\text{-}105)$$

relating this fundamental particle property to a fundamental wave property. Note that the wave number σ is defined as the reciprocal of the wavelength:

$$\sigma = \frac{1}{\lambda}. \qquad (17\text{-}106)$$

In 1923, Louis V. de Broglie suggested that a formula equivalent to (17-105) should also apply to particles that have a rest mass. This generalization was confirmed when a beam of monoenergetic electrons was found to be diffracted by crystalline matter as a beam of radiation with the wavelength λ.

17.11   Summary

In accessible regions where the gravitational field is not too strong, the Pythagorean theorem holds very accurately. Furthermore, no evidence for any nonuniformity in the passage of time through any inertial frame exists. There is no fundamental difference between inertial frames; a person cannot say that any one of them is absolutely at rest. As a consequence, space and time are linked together to form the continuum described in chapter 16. In particular, the difference between the square of the spatial displacement of an effect and the square of the displacement of an entity traveling at fundamental speed c over the same coordinate-time interval is the same in all inertial frames. Relating events infinitesimally close is the invariant

$$ds^2 = dx^2 + dy^2 + dz^2 - c^2 \, dt^2$$
$$= dx'^2 + dy'^2 + dz'^2 - c^2 \, dt'^2. \qquad (17\text{-}107)$$

Here x, y, z are coordinates of the effect in Σ at time t while x', y', z' are its coordinates in Σ' at time t'.

We speak of dx, dy, dz, $c\,dt$ and dx', dy', dz', $c\,dt'$ as the unprimed and the primed contravariant components of a 4-dimensional vector $d\mathbf{s}$. To get the invariant form (17-107), we then have to let the covariant components be dx, dy, dz, $-c\,dt$ and dx', dy', dz', $-c\,dt'$.

Since the proper time $d\tau$ and rest mass m_0 for a particle are invariant, the 4-velocity and 4-momentum

$$\frac{d\mathbf{s}}{d\tau}, \qquad m_0 \frac{d\mathbf{s}}{d\tau} \qquad (17\text{-}108)$$

are independent of inertial frame. They are characteristic of the physical system itself.

The Newtonian conservation-of-momentum and -mass laws apply when all particles are moving slowly. These transform into a conservation-of-4-momentum law that must hold when all particles are traveling near any given velocity with respect to an inertial frame. Since the law is valid whether particles are traveling slowly or fast, we assume it applies to mixtures of slowly and fast moving particles.

If we assume that one particle can be annihilated while interacting with slowly moving neighboring particles, we find that the kinetic energy of these increases by the amount

$$mc^2 \qquad (17\text{-}109)$$

where m is the relativistic mass of the disappearing particle. We interpret (17-109) as the energy of the particle.

The conservation laws govern the processes that occur in collisions between particles. Continuous interaction of a particle with a simple field will be studied in the next chapter.

DISCUSSION QUESTIONS

17.1 Use a Minkowski diagram to show how direct action at a distance transforms to an effect following the cause in some inertial frames and to an effect preceding the cause in others. What is wrong with Newton's law of gravitation?

17.2 What invariances characterize the space-time continuum? What is wrong with the classical conservation-of-momentum and conservation-of-energy laws?

17.3 How is the transformation from one inertial frame to another expressible in matrix form? What condition must the transformation matrix satisfy? Is this analogous to the unitary condition?

17.4 Why can the momentum $m\,d\mathbf{r}/dt$ be a physically significant expression in Newtonian mechanics?

17.5 Explain why the 4-dimensional $d\mathbf{s}/dt$ is not invariant. Why is $d\mathbf{s}/d\tau$ invariant?

17.6 What is rest mass? Why is $m_0\,d\mathbf{s}/d\tau$ invariant? When does $m_0\,d\mathbf{s}/d\tau$ change?

17.7 Explain how the conservation-of-momentum and -mass laws combine to form a single conservation law.

17.8 In what sense is $m_j c^2$ the energy of the jth particle? Does relativity theory imply that the jth particle can be annihilated?

17.9 In what frame can all kinetic energy be converted to rest energy? How is this conversion accomplished?

17.10 Why are the first three components of 4-momentum of the composite M zero in the center-of-mass frame? Why is this frame an inertial frame?

17.11 How does a binary collision appear in the center-of-mass frame? What transforms a 4-momentum from the laboratory to the center-of-mass frame?

17.12 How does a binary elastic collision affect the 4-momentum of a given particle?

17-13 How may the Einstein equation for the energy of a photon be justified?

17.14 Why is the ratio of this energy to the fundamental speed the fourth component of a 4-vector? What do the other three components form?

17.15 How did Compton study the behavior of the 3-momentum of a photon?

17.16 Show that photons obey de Broglie's equation. What was de Broglie's generalization?

PROBLEMS

17.1 If **s** is a 4 × 1 matrix transforming as the matrix **r** transforms, what function of its elements is invariant?

17.2 Derive an expression that shows how the 4-dimensional ds/dt varies from one inertial frame to another.

17.3 Show that for a single particle

$$\frac{v}{c} = \frac{pc}{E}.$$

17.4 By how much must the nonrelativistic kinetic energy of a particle be multiplied to get its true kinetic energy when v/c is (a) 0.100, (b) 0.700, (c) 0.9900?

17.5 In a scattering experiment, particles with 4-momenta $\mathbf{k}_1$ and $\mathbf{k}_2$ interact to produce particles with 4-momenta $-\mathbf{k}_3$ and $-\mathbf{k}_4$. Show that

$$(\mathbf{k}_1 + \mathbf{k}_2)^2 = (\mathbf{k}_3 + \mathbf{k}_4)^2,$$
$$(\mathbf{k}_1 + \mathbf{k}_4)^2 = (\mathbf{k}_2 + \mathbf{k}_3)^2,$$
$$(\mathbf{k}_1 + \mathbf{k}_3)^2 = (\mathbf{k}_2 + \mathbf{k}_4)^2.$$

17.6 A resonance is observed in the bombardment of protons with positive pions around 845 MeV. What is the rest mass of the resonance particle?

17.7 Antiprotons are to be produced by the reaction

$$p + p \rightarrow p + p + p + \bar{p}$$

in which a moving proton strikes another at rest. How much kinetic energy must the bombarding proton possess?

17.8 If a positive pion is to combine with a proton at rest to form a Δ^{++} with 1236 MeV rest mass, what must the initial kinetic energy of the pion be?

17.9 Show that a photon cannot materialize, that is, become an electron-positron pair, in free space.

— — —

17.10 Prove that if **s** is a 4-element column matrix and if $\mathbf{r}^{\times}\mathbf{r}$ and $\mathbf{r}^{\times}\mathbf{s}$ are invariant for any radius 4-vector **r**, then **s** is a 4-vector.

17.11 At what fraction of c does (a) a 200 MeV π^+, (b) a 200 MeV p^+, (c) a 200 MeV Ω^- move?

17.12 At what v/c is the true kinetic energy of a particle (a) 2.000 times, (b) 10.000 times the nonrelativistic value?

17.13 Particles with 4-momenta $\mathbf{k}_1$ and $\mathbf{k}_2$ collide to produce particles with 4-momenta $-\mathbf{k}_3$ and $-\mathbf{k}_4$. Show that the sum of the quantities in problem 17.5 equals the sum of the squares of all four rest masses multiplied by minus c^2

$$(\mathbf{k}_1 + \mathbf{k}_2)^2 + (\mathbf{k}_1 + \mathbf{k}_4)^2 + (\mathbf{k}_1 + \mathbf{k}_3)^2 = -c^2 m_{j0} m_{j0}.$$

17.14 Let q_1 and q_3 be the center-of-mass 3-momenta of the first and third particles of problem 17.13, while θ is the angle between these momenta, and show that

$$(\mathbf{k}_1 + \mathbf{k}_3)^2 = -m_{10}{}^2 c^2 - m_{30}{}^2 c^2$$
$$+ 2(q_1{}^2 + m_{10}{}^2 c^2)^{1/2}(q_3{}^2 + m_{30}{}^2 c^2)^{1/2} - 2q_1 q_3 \cos \theta.$$

17.15 When the kinetic energy of negative pions striking protons at rest reaches 610 MeV, a resonance is observed. What is the rest mass of the resonance particle?

17.16 Find the threshold energy for the reaction

$$\pi^+ + p^+ \rightarrow \Sigma^+ + K^+$$

in which the proton is initially at rest.

17.17 Find the threshold energy for the reaction

$$\gamma + e^- \rightarrow e^- + e^- + e^+$$

in which a photon strikes an electron at rest to produce an electron-positron pair.

17.18 Relate the recoil energy given to the electron in a Compton collision to the initial energy of the photon and the angle through which the photon is deflected.

REFERENCES

BOOKS

Bradbury, T. C. *Theoretical Mechanics*, pp. 562–623. John Wiley & Sons, Inc., New York, 1968.

Schwartz, H. M. *Introduction to Special Relativity*, pp. 1–130. McGraw-Hill Book Company, New York, 1968.

Synge, J. L. *Relativity: The Special Theory*, 2nd ed., pp. 162–206. North-Holland Publishing Company, Amsterdam, 1964.

Wangsness, R. K. *Introductory Topics in Theoretical Physics*, pp. 7–56. John Wiley & Sons, Inc., New York, 1963.

ARTICLES

Aranoff, S. Torques and Angular Momentum on a System at Equilibrium in Special Relativity. *Am. J. Phys.*, **37,** 453 (1969).

Brehme, R. W. The Advantage of Teaching Relativity with Four-Vectors. *Am. J. Phys.*, **36,** 896 (1968).

Callen, H., and **Horwitz, G.** Relativistic Thermodynamics. *Am. J. Phys.*, **39,** 938 (1971).

Chow, Y. The Kinematics of Two-Particle Collisions in Special Relativity. *Am. J. Phys.*, **37,** 397 (1969).

Danburg, J. S., and **Kalbfleisch, G. R.** Limits on the Rate of Emission of Negative-Energy Tachyons. *Phys. Rev. D*, **5,** 1575 (1972).

Ehlers, J., Rindler, W., and **Penrose, R.** Energy Conservation as the Basis of Relativistic Mechanics. II. *Am. J. Phys.*, **33,** 995 (1965).

Henry, J., and **Barrabes, C.** Covariant Equations for the Motion of a Point Body with Variable Rest Mass. *Am. J. Phys.*, **40,** 724 (1972).

Landau, B. V., and **Sampanthar, S.** A New Derivation of the Lorentz Transformation. *Am. J. Phys.*, **40,** 599 (1972).

Lapidus, I. R. Motion of a Relativistic Particle Acted Upon by a Constant Force and a Uniform Gravitational Field. *Am. J. Phys.*, **40,** 984 (1972).

Leibovitz, C. Rest Mass in Special Relativity. *Am. J. Phys.*, **37,** 834 (1969).

Mariwalla, K. H. On Tachyon Lorentz Transformation. *Am. J. Phys.*, **37,** 1281 (1969).

Pomeranz, K. B. The Equation of Motion for Relativistic Particles and Systems with a Variable Rest Mass. *Am. J. Phys.*, **32,** 955 (1964).

Proriol, J., and **Laverriere, R.** Relativistic Calculations of Angular Distributions and Correlations in a Three-Body Phase Space with Constraint Functions. *Am. J. Phys.*, **39,** 89 (1971).

Rosen, G. Galilean Invariance and the General Covariance of Nonrelativistic Laws. *Am. J. Phys.*, **40,** 683 (1972).

Schwartz, H. M. Axiomatic Deduction of the General Lorentz Transformations. *Am. J. Phys.*, **30,** 697 (1962).

Schwartz, H. M. Generalization of an Elementary Formula in Relativistic Kinematics Due to Pauli. *Am. J. Phys.*, **38,** 927 (1970).

Sears, F. W. Another Relativistic Paradox. *Am. J. Phys.*, **40,** 771 (1972).

Terrall, J. R. Elementary Treatment of Relativistic Cross Sections. *Am. J. Phys.*, **38,** 1460 (1970).

Yuen, C. K. Lorentz Transformation of Thermodynamic Quantities. *Am. J. Phys.*, **38,** 246 (1970).

18 / *Covariant Electrodynamics*

18.1   Relating magnetism to electricity

In Maxwellian theory, effects that depend merely on the position of a source charge are considered independent of effects produced by its motion. The fundamental equations governing electric fields do not, in any sense, determine those governing magnetism.

But in Einsteinian theory, each inertial frame is equivalent, and a charge can be momentarily at rest in only one of these. In the other frames, the charge does move and does produce a magnetic field.

We have seen how the magnitude of a 4-vector is invariant. Such a magnitude appears as a scalar with a sign in the frame in which the 3-vector part is zero. The scalar transforms into a vector-scalar combination on going to any other inertial frame. In a similar way, a classical vector property may transform into a 4×4 array that is a 4-dimensional tensor.

In applying these concepts to the field produced by a charge, we will start from scratch, imagining what the possibilities are at each stage and choosing what seems to be the simplest nontrivial alternative. Each choice that we will make is supported by extensive experiments.

18.2   Static charges

Let us begin with an inertial frame in an otherwise empty universe. At some point in this frame let us create a source for a field, a charge. Let us also assume that the resulting system is unique.

The field must then spread out equally in all directions from the source. If we let $\mathbf{r}$ be the vector distance of the field point from the source point, while $\mathbf{l}$ is $\mathbf{r}/r$, the field intensity becomes

$$\mathbf{E} = E(r)\mathbf{l}. \tag{18-1}$$

Let us next create a variable charge at the chosen field point. Let a scalar quantity q be defined that is proportional to the force $\mathbf{F}$ with which the field of the first charge acts

$$\mathbf{F} = q\mathbf{E}. \tag{18-2}$$

532

By symmetry, the force acting on the first charge is then

$$\mathbf{F}_1 = q_1 \mathbf{E}_1 \tag{18-3}$$

where $\mathbf{E}_1$ is the field at the first charge caused by the second and q_1 is the number of units of charge originally created.

Let us assume that fields from different sources act independently, the various contributions superposing on each other. Under such circumstances, reducing the charge at the origin from q_1 to 1 reduces the effect $\mathbf{E}$ at $\mathbf{r}$ by $1/q_1$, while increasing the source charge from 1 to q_1 multiplies the effect by q_1.

We also presume that the field spreads at a uniform rate from the charge. It can then be represented by a flux fanning out from a constant source. This is governed by an inverse square law

$$\mathbf{E} = \frac{q_1}{kr^2} \mathbf{l}. \tag{18-4}$$

In practice, we choose units so k is $4\pi\varepsilon$, and we let $\varepsilon\mathbf{E}$ be $\mathbf{D}$:

$$\mathbf{D} = \frac{q_1}{4\pi r^2} \mathbf{l}. \tag{18-5}$$

Adding source charges $q_2, \ldots, q_n$ at other points introduces fields that are described similarly. Superposing these, we obtain the displacement

$$\mathbf{D} = \frac{q_j \mathbf{l}_j}{4\pi r_j^2}. \tag{18-6}$$

Note that $\mathbf{r}_j$ is drawn from the jth source to the field point and $\mathbf{l}_j$ is a unit vector pointing in the same direction.

A continuous distribution of density ρ_0 can be treated as in section 14.3, with the same result as before

$$\nabla \cdot \mathbf{D} = \rho_0. \tag{18-7}$$

We can rewrite this equation in the form

$$\nabla \cdot \frac{\mathbf{E}}{c} = \frac{\rho_0}{c\varepsilon}. \tag{18-8}$$

Finally, equation (18-2) implies that the force per unit volume acting on a small element is

$$\mathbf{f} = \rho_0 \mathbf{E}. \tag{18-9}$$

18.3   Moving charges

A small element of charge is measured by a scalar quantity in the inertial frame in which it is momentarily at rest, just as a small element of mass is. The quantity $\rho_0/c\varepsilon$ appearing on the right of equation (18-8) is analogous to rest mass m_0 times fundamental speed c. Consequently, we expect it to transform in the same way.

According to equations (17-40), (16-134), and table 17.1, the 4-momentum of a particle is

$$(\mathbf{0}, m_0 c) \tag{18-10}$$

in the inertial frame in which it is momentarily at rest and

$$\left(\frac{\gamma m_0 c \mathbf{v}}{c}, \gamma m_0 c\right) \tag{18-11}$$

in the frame in which it is moving at conventional velocity **v**. Transforming the 4-vector

$$\left(\mathbf{0}, \frac{\rho_0}{c\varepsilon}\right) \tag{18-12}$$

similarly yields

$$\left(\frac{\gamma \rho_0 \mathbf{v}}{c^2 \varepsilon}, \frac{\gamma \rho_0}{c\varepsilon}\right). \tag{18-13}$$

From problem 16.10, we know that the volume of an element transforms according to the law

$$\gamma \, dV = dV_0 \tag{18-14}$$

if dV_0 is its proper volume. Multiplying both sides by the proper density ρ_0 then yields

$$\gamma \rho_0 \, dV = \rho_0 \, dV_0. \tag{18-15}$$

Product $\gamma \rho_0$ is the density of charge ρ in the new inertial frame. We may rewrite (18-13) in the form

$$\left(\frac{\rho \mathbf{v}}{c^2 \varepsilon}, \frac{\rho}{c\varepsilon}\right). \tag{18-16}$$

Identifying the charge density times its velocity as the current density

$$\rho \mathbf{v} = \mathbf{J} \tag{18-17}$$

and the coefficient of this as a new parameter

$$\frac{1}{c^2 \varepsilon} = \mu \tag{18-18}$$

converts the 4-vector to

$$\left(\mu \mathbf{J}, \frac{\rho}{c\varepsilon}\right). \tag{18-19}$$

Because the charge is conserved, equation (12-20) is satisfied:

$$\frac{\partial \rho}{\partial t} + \nabla \cdot \mathbf{J} = 0. \tag{18-20}$$

Dividing by $c^2\varepsilon$ converts this equation to

$$\frac{\partial}{\partial(ct)}\left(\frac{\rho}{c\varepsilon}\right) + \nabla \cdot \mu\mathbf{J} = 0. \tag{18-21}$$

If the αth contravariant component of 4-vector (18-19) is written as μJ^α and the αth coordinate as x^α, equation (18-21) becomes

$$\frac{\partial}{\partial x^\alpha}(\mu J^\alpha) = 0. \tag{18-22}$$

EXAMPLE 18.1   Project the product of charge density with fundamental speed onto inertial frames momentarily at rest and in motion with respect to an element of the charge. Construct the scalar product of the 4-vector current density with itself for each system and equate the results.

Choosing an element of charge, determining its proper density, and multiplying the result by the fundamental speed yields the quantity $\rho_0 c$. In the frame in which the element is at rest, this quantity appears as the scalar $m_0 c$; it is paired with a zero 3-vector

$$(\mathbf{0}, \rho_0 c).$$

Transforming this pair as (18-10) was transformed yields

$$(\gamma\rho_0\mathbf{v}, \gamma\rho_0 c).$$

But since

$$\gamma\rho_0 = \rho,$$

the general 4-vector is

$$(\rho\mathbf{v}, \rho c)$$

or

$$(\mathbf{J}, \rho c).$$

Multiplying (18-12) and (18-16) by $c^2\varepsilon$ yields the same forms.

The Cartesian components of $\mathbf{J}$ are J_x, J_y, J_z, while the fourth component of the 4-vector is ρc. Let us multiply each of these by the corresponding base vector and add to get the resolution

$$J_x\mathbf{e}_1 + J_y\mathbf{e}_2 + J_z\mathbf{e}_3 + \rho c\mathbf{e}_4.$$

In the inertial frame in which the element of charge is momentarily at rest, the components of $\mathbf{J}$ are 0, while the fourth component is $\rho_0 c$. Then the 4-vector becomes

$$\rho_0 c\mathbf{e}_4'.$$

The square of the magnitude of the 4-current density is the dot product of either of these expansions with itself. Since the square is the same in either inertial frame, we have

$$J_x^2 + J_y^2 + J_z^2 - \rho^2 c^2 = -\rho_0^2 c^2$$

if equations (17-4)–(17-7) are introduced after the scalar products are formed.

Writing the sum of the first three terms as the square of the magnitude of 3-vector **J**, converts this result to the equation

$$J^2 - \rho^2 c^2 = -\rho_0^2 c^2$$

analogous to (17-43).

18.4   The field tensor

The simplest equations into which the divergence equation transforms will now be constructed.

We assume that $\nabla \cdot \mathbf{D}$ is still the charge density when the charge about the pertinent point is moving. Equation (18-8) then becomes

$$\frac{\partial}{\partial x^1}\frac{E^1}{c} + \frac{\partial}{\partial x^2}\frac{E^2}{c} + \frac{\partial}{\partial x^3}\frac{E^3}{c} + \frac{\partial}{\partial x^4}0 = \frac{\rho}{c\varepsilon}. \tag{18-23}$$

The right side of equation (18-23) is the scalar part of the 4-vector that we have designated μJ^α. The left side must be the scalar part of a 4-vector also; and component E^β/c must be the scalar part of a 4-vector $F^{\alpha\beta}$

$$\frac{E^\beta}{c} = F^{4\beta}. \tag{18-24}$$

Equation (18-23) is thus the fourth component of the set

$$\frac{\partial F^{\alpha\beta}}{\partial x^\beta} = \mu J^\alpha. \tag{18-25}$$

The sixteen $F^{\alpha\beta}$'s are said to constitute the *field tensor* in the given inertial frame.
Because

$$F^{44} = 0 \tag{18-26}$$

in every inertial frame, we expect the other diagonal elements to be zero

$$F^{11} = F^{22} = F^{33} = 0 \tag{18-27}$$

and $F^{\alpha\beta}$ to be antisymmetric

$$F^{\alpha\beta} = -F^{\beta\alpha}. \tag{18-28}$$

Differentiating (18-28) with respect to both x^α, x^β, and summing over α and β, we get

$$\frac{\partial^2 F^{\alpha\beta}}{\partial x^\alpha\, \partial x^\beta} = -\frac{\partial^2 F^{\beta\alpha}}{\partial x^\alpha\, \partial x^\beta} = -\frac{\partial^2 F^{\alpha\beta}}{\partial x^\beta\, \partial x^\alpha} = -\frac{\partial^2 F^{\alpha\beta}}{\partial x^\alpha\, \partial x^\beta} \tag{18-29}$$

or

$$\frac{\partial^2 F^{\alpha\beta}}{\partial x^\alpha\, \partial x^\beta} = 0. \tag{18-30}$$

Differentiating equation (18-25) with respect to x^α and summing leads to

$$\frac{\partial^2 F^{\alpha\beta}}{\partial x^\alpha\, \partial x^\beta} = \frac{\partial}{\partial x^\alpha}(\mu J^\alpha). \tag{18-31}$$

If the field tensor is antisymmetric, as equation (18-26) indicates, then equation (18-30) is satisfied, and (18-31) reduces to

$$\frac{\partial}{\partial x^\alpha} (\mu J^\alpha) = 0, \tag{18-32}$$

the charge conservation law (18-22).

Choosing $F^{\alpha\beta}$ to be antisymmetric, so equation (18-20) is satisfied, and setting

$$F^{12} = B^3, \qquad F^{23} = B^1, \qquad F^{31} = B^2, \tag{18-33}$$

converts equations (18-25) to

$$\frac{\partial B^3}{\partial x^2} - \frac{\partial B^2}{\partial x^3} - \frac{1}{c} \frac{\partial E^1}{\partial (ct)} = \mu J^1, \tag{18-34}$$

$$-\frac{\partial B^3}{\partial x^1} + \frac{\partial B^1}{\partial x^3} - \frac{1}{c} \frac{\partial E^2}{\partial (ct)} = \mu J^2, \tag{18-35}$$

$$\frac{\partial B^2}{\partial x^1} - \frac{\partial B^1}{\partial x^2} - \frac{1}{c} \frac{\partial E^3}{\partial (ct)} = \mu J^3, \tag{18-36}$$

$$\frac{1}{c} \frac{\partial E^1}{\partial x^1} + \frac{1}{c} \frac{\partial E^2}{\partial x^2} + \frac{1}{c} \frac{\partial E^3}{\partial x^3} = \frac{\rho}{c\varepsilon}. \tag{18-37}$$

Equations (18-34) through (18-36) are the components of

$$\nabla \times \mathbf{B} = \mu \mathbf{J} + \mu\varepsilon \frac{\partial \mathbf{E}}{\partial t}, \tag{18-38}$$

a form of equation (14-78).

These differential equations describe the change in field caused by the charge that happens to be near a given event in the space-time continuum. The inertial frame in which this charge momentarily rests is generally not the same as the rest frame for other elements of charge at a given coordinate time. Nevertheless, the behavior of each of these elements governs the derivatives of $F^{\alpha\beta}$ about its space-time point. The field tensor itself varies smoothly from point to point.

EXAMPLE 18.2   What happens to the matrix of the field tensor when the observer goes from one inertial frame to another?

We have defined the field tensor in terms of its purely contravariant components. Multiplying these by corresponding covariant components of a vector and adding yields the contravariant components of a vector

$$F^{\alpha\beta} x_\beta = y^\alpha.$$

If we let $\mathbf{F}$ be the 4×4 matrix in which $F^{\alpha\beta}$ is a typical element, while $\mathbf{y}$ is the column matrix containing the contravariant components of $\mathbf{y}$, and $\mathbf{x}$ is the column matrix containing the covariant components of $\mathbf{x}$, then these equations appear as

$$\mathbf{F}\mathbf{x}_. = \mathbf{y}.$$

Each contravariant column matrix is transformed from one inertial frame to another by matrix $\mathbf{A}$ of section 17.3:

$$\mathbf{Ay} = \mathbf{y}',$$
$$\mathbf{Ax} = \mathbf{x}'.$$

Since $\mathbf{x}_.$ differs from $\mathbf{x}$ in the sign of the fourth component, an equation that produces the same result as the last one is

$$\widetilde{\mathbf{A}^\times}\mathbf{x}_. = \mathbf{x}_.'.$$

In the transformation, $\mathbf{F}$ is changed to $\mathbf{F}'$ and the second equation becomes

$$\mathbf{F}'\mathbf{x}_.' = \mathbf{y}'.$$

Substituting the preceding expressions for $\mathbf{x}_.'$ and $\mathbf{y}'$ into this equation yields

$$\mathbf{F}'\widetilde{\mathbf{A}^\times}\mathbf{x}_. = \mathbf{Ay}$$

whence

$$\mathbf{A}^{-1}\mathbf{F}'\widetilde{\mathbf{A}^\times}\mathbf{x}_. = \mathbf{y}.$$

For this equation to be the same as the second one, $\mathbf{Fx}_. = \mathbf{y}$, we must have

$$\mathbf{A}^{-1}\mathbf{F}'\widetilde{\mathbf{A}^\times} = \mathbf{F}$$

or

$$\mathbf{F}' = \mathbf{A}\mathbf{F}\tilde{\mathbf{A}},$$

since the transposed matrix $\tilde{\mathbf{A}}$ is the inverse of $\widetilde{\mathbf{A}^\times}$.

18.5   The dual of the field tensor

As the equation for the divergence of $\mathbf{D}$ implies that equations (18-34), (18-35), and (18-36) restricting $\mathbf{B}$ exist, a condition on the divergence of $\mathbf{B}$ leads to three additional equations restricting $\mathbf{D}$ and $\mathbf{E}$.

First, we recall how the inverse square law for $\mathbf{E} = \mathbf{D}/\varepsilon$ can be rewritten in the form

$$\nabla \cdot \mathbf{D} = \rho. \tag{18-39}$$

If a similar inverse square law existed for the magnetic intensity $\mathbf{H}$, we would also have

$$\nabla \cdot \mathbf{B} = \rho_m \tag{18-40}$$

where $\mathbf{B}$ is $\mu\mathbf{H}$ and ρ_m is the density of magnetic monopoles.

Since monopoles are not observed, we assume

$$\nabla \cdot \mathbf{B} = 0. \tag{18-41}$$

Elements of the field tensor are related to components of $\mathbf{B}$ by equations (18-33). Introducing these converts the divergence equation to

$$\frac{\partial F^{23}}{\partial x^1} + \frac{\partial F^{31}}{\partial x^2} + \frac{\partial F^{12}}{\partial x^3} = 0. \tag{18-42}$$

Since equation (18-42) is related to (18-41) as the fourth equation (18-25) is related to (18-8), we expect it to be the fourth of a set. The components F^{23}, F^{31}, F^{12} must then be $4b$ (or $b4$) elements of a tensor.

Suitable expressions can be constructed easily. We multiply the $\gamma\delta$ component of the field tensor by one-half the 4-dimensional permutation symbol

$$\varepsilon_{\alpha\beta\gamma\delta} = 1 \qquad \text{when } \alpha\beta\gamma\delta \text{ is an even permutation of 1234,} \qquad (18\text{-}43)$$

$$\varepsilon_{\alpha\beta\gamma\delta} = -1 \qquad \text{when } \alpha\beta\gamma\delta \text{ is an odd permutation of 1234,} \qquad (18\text{-}44)$$

$$\varepsilon_{\alpha\beta\gamma\delta} = 0 \qquad \text{otherwise,} \qquad (18\text{-}45)$$

and sum over the repeated indices

$$G_{\alpha\beta} = \tfrac{1}{2}\varepsilon_{\alpha\beta\gamma\delta}F^{\gamma\delta}. \qquad (18\text{-}46)$$

Tensor **G** is the *dual* of the field tensor. Calculating its 4β components,

$$G_{41} = \tfrac{1}{2}F^{32} - \tfrac{1}{2}F^{23} = -F^{23}, \qquad (18\text{-}47)$$

$$G_{42} = \tfrac{1}{2}F^{13} - \tfrac{1}{2}F^{31} = -F^{31}, \qquad (18\text{-}48)$$

$$G_{43} = \tfrac{1}{2}F^{21} - \tfrac{1}{2}F^{12} = -F^{12}, \qquad (18\text{-}49)$$

$$G_{44} = 0, \qquad (18\text{-}50)$$

substituting them into equation (18-42), reversing the signs throughout, and changing the contravariant coordinates into covariant ones leads to

$$\frac{\partial G_{41}}{\partial x_1} + \frac{\partial G_{42}}{\partial x_2} + \frac{\partial G_{43}}{\partial x_3} + \frac{\partial G_{44}}{\partial x_4} = 0. \qquad (18\text{-}51)$$

The right side of the resulting equation is the fourth component of a null 4-vector. Therefore, (18-51) is the fourth equation of the set

$$\frac{\partial G_{\alpha\beta}}{\partial x_\beta} = 0. \qquad (18\text{-}52)$$

The first equation is

$$\frac{\partial G_{11}}{\partial x_1} + \frac{\partial G_{12}}{\partial x_2} + \frac{\partial G_{13}}{\partial x_3} + \frac{\partial G_{14}}{\partial x_4} = 0. \qquad (18\text{-}53)$$

Equations (18-46), (18-24), and (18-33) show that

$$G_{11} = 0, \qquad (18\text{-}54)$$

$$G_{12} = \tfrac{1}{2}F^{34} - \tfrac{1}{2}F^{43} = -F^{43} = -\frac{E^3}{c}, \qquad (18\text{-}55)$$

$$G_{13} = \tfrac{1}{2}F^{42} - \tfrac{1}{2}F^{24} = F^{42} = \frac{E^2}{c}, \qquad (18\text{-}56)$$

$$G_{14} = \tfrac{1}{2}F^{23} - \tfrac{1}{2}F^{32} = F^{23} = B^1; \qquad (18\text{-}57)$$

so equation (18-53) involves

$$-\frac{\partial}{\partial x_2}\frac{E^3}{c} + \frac{\partial}{\partial x_3}\frac{E^2}{c} + \frac{\partial B^1}{\partial x_4} = 0 \tag{18-58}$$

or

$$\frac{\partial E_y}{\partial z} - \frac{\partial E_z}{\partial y} - \frac{\partial B_x}{\partial t} = 0. \tag{18-59}$$

When α is 2 and 3, we similarly get

$$\frac{\partial E_z}{\partial x} - \frac{\partial E_x}{\partial z} - \frac{\partial B_y}{\partial t} = 0, \tag{18-60}$$

$$\frac{\partial E_x}{\partial y} - \frac{\partial E_y}{\partial x} - \frac{\partial B_z}{\partial t} = 0. \tag{18-61}$$

Equations (18-59), (18-60), and (18-61) are the components of

$$-\nabla \times \mathbf{E} - \frac{\partial \mathbf{B}}{\partial t} = 0 \tag{18-62}$$

whence

$$\nabla \times \mathbf{E} = -\frac{\partial \mathbf{B}}{\partial t}. \tag{18-63}$$

Thus, equation (14-59) is related to equation (14-49).

EXAMPLE 18.3   Express the matrix of (a) the field tensor, and (b) the dual of the field tensor, in terms of components of the electromagnetic field.
    From equations (18-24), (18-28), and (18-33), we obtain

$$\mathbf{F} = \begin{pmatrix} 0 & B^3 & -B^2 & -\dfrac{E^1}{c} \\[2mm] -B^3 & 0 & B^1 & -\dfrac{E^2}{c} \\[2mm] B^2 & -B^1 & 0 & -\dfrac{E^3}{c} \\[2mm] \dfrac{E^1}{c} & \dfrac{E^2}{c} & \dfrac{E^3}{c} & 0 \end{pmatrix}.$$

Substituting the elements of **F** into equations (18-46) similarly yields the elements of **G**

$$
\mathbf{G} = \begin{pmatrix}
0 & -\dfrac{E^3}{c} & \dfrac{E^2}{c} & B^1 \\[2ex]
\dfrac{E^3}{c} & 0 & -\dfrac{E^1}{c} & B^2 \\[2ex]
-\dfrac{E^2}{c} & \dfrac{E^1}{c} & 0 & B^3 \\[2ex]
-B^1 & -B^2 & -B^3 & 0
\end{pmatrix}
$$

18.6 Equations governing the motion of charged particles

While Maxwell's equations are consistent with the special theory of relativity, Newton's equations are not. As a consequence, formula (14-89) is approximate and must be corrected.

Let us consider a particle carrying charge q, in the inertial frame in which it is momentarily at rest. Nonrelativistic laws then apply and the force acting on the particle is given by equation (18-2).

The components of **E** are related to elements of the field tensor by equation (18-24) in the form

$$E^a = cF^{4a}. \tag{18-64}$$

Substituting this into (18-2) leads to

$$F^a = qcF^{4a} = F^{a4}(-qc). \tag{18-65}$$

In the 4-dimensional Minkowski space, the final factor must be a 4-vector with the covariant parts

$$(0, -qc). \tag{18-66}$$

Transforming these to another frame as (18-10) was transformed yields

$$\left(q\mathbf{v}\gamma, -qc\gamma\right) = \left(q\mathbf{v}\,\frac{dt}{d\tau}, -qc\,\frac{dt}{d\tau}\right). \tag{18-67}$$

A comparison with table 17.2 shows that (18-67) contains the vector-scalar parts of q times the 4-velocity.

Transforming equations (18-65) to an inertial frame in which q is moving produces

$$F^\alpha = F^{\alpha\beta}qv_\beta. \tag{18-68}$$

The components of the field tensor are given by (18-24) and (18-33). Substituting these and (17-38) into equation (18-68), we get

$$F^1 = 0 + B^3 q\gamma \frac{dx^2}{dt} - B^2 q\gamma \frac{dx^3}{dt} - \frac{E^1}{c}(-q\gamma c)$$

$$= \gamma q \left(E^1 + \frac{dx^2}{dt} B^3 - \frac{dx^3}{dt} B^2 \right). \tag{18-69}$$

Similar expressions are obtained for F^2 and F^3. In vector notation, these become

$$\mathbf{F} = \gamma q(\mathbf{E} + \mathbf{v} \times \mathbf{B}). \tag{18-70}$$

For the scalar part, we get

$$F^4 = F^{4a} q\gamma \frac{dx^a}{dt} = \frac{\mathbf{E}}{c} \cdot q\gamma \mathbf{v} = \gamma \frac{q\mathbf{E} \cdot \mathbf{v}}{c}. \tag{18-71}$$

Since

$$\mathbf{v} \times \mathbf{B} \cdot \mathbf{v} = 0, \tag{18-72}$$

equation (18-71) can be rewritten in the form

$$F^4 = \frac{\gamma}{c} \left[q(\mathbf{E} + \mathbf{v} \times \mathbf{B}) \right] \cdot \mathbf{v}. \tag{18-73}$$

We interpret

$$q(\mathbf{E} + \mathbf{v} \times \mathbf{B}) \tag{18-74}$$

as the *Lorentz force* acting on the particle. Dotting this with $\mathbf{v}$ gives the time rate at which work is done on the particle. Assuming the energy goes to increase the mass, according to equation (17-56), we write

$$F^4 = \frac{\gamma}{c} \frac{d}{dt} (mc^2) = \frac{d}{d\tau} (mc). \tag{18-75}$$

But mc is the scalar part of the 4-momentum

$$(m\mathbf{v}, mc), \tag{18-76}$$

while F^4 is paired with $\mathbf{F}$; so existence of (18-75) implies existence of the equation

$$\mathbf{F} = \frac{d}{d\tau} (m\mathbf{v}) = \gamma \frac{d}{dt} (m\mathbf{v}). \tag{18-77}$$

This formula is the relativistic generalization of Newton's second law.

Since

$$m\mathbf{v} = \gamma m_0 \frac{d\mathbf{r}}{dt} = m_0 \frac{d\mathbf{r}}{d\tau}, \tag{18-78}$$

equation (18-77) can be rewritten as

$$F^a = \frac{d}{d\tau} (m_0 v^a) = m_0 \frac{dv^a}{d\tau}. \tag{18-79}$$

Equation (18-75) may also appear as

$$F^4 = \frac{d}{d\tau}(m_0\gamma c) = m_0 \frac{dv^4}{d\tau}.$$ (18-80)

Combining these results with formula (18-68) for the force, we obtain the equation of motion

$$m_0 \frac{dv^\alpha}{d\tau} = qF^{\alpha\beta}v_\beta.$$ (18-81)

18.7   A 4-potential for the field

Whenever a particle energy mc^2 increases because of interaction with an electromagnetic field, we may presume it to increase at the expense of a potential energy $q\phi$, if q is the charge on the particle. Alternatively, we may suppose mc itself to increase at the expense of $(q/c)\phi$. But mc is the fourth component of a 4-vector, so the postulated $(q/c)\phi$ must be the corresponding component of another 4-vector. The spatial part of this is designated $q\mathbf{A}$.

We can readily deduce the nature of $\mathbf{A}$ in any system small enough to allow practically instantaneous transmission of effects. First, consider a single source charge q_1 and a test charge q in the inertial frame in which q_1 is momentarily at rest. Suppose the test particle has mass m and is moving at velocity $\mathbf{w}$ in this frame (see figure 18.1).

From equation (18-4) and the argument in section 14.5, the corresponding potential is

$$\phi_0 = \frac{q_1}{4\pi\varepsilon|\mathbf{r} - \mathbf{r}_1|}$$ (18-82)

where radius vector $\mathbf{r}$ locates the test charge and $\mathbf{r}_1$ the source charge q_1.

If the test particle were uncharged, the components of its 4-momentum would be

$$\mathbf{p} = \frac{m_0\mathbf{w}}{(1 - w^2/c^2)^{1/2}},$$ (18-83)

$$\frac{E'}{c} = \frac{m_0 c}{(1 - w^2/c^2)^{1/2}},$$ (18-84)

where m_0 is the rest mass of the particle and E' its kinetic plus rest energy (see table 17.2).

FIGURE 18.1   How the charges are moving at a given moment in the first inertial frame.

Since motion of a test particle does not affect the electrostatic force acting on it at any given time t, the charge q does not contribute to the momentum $\mathbf{p}$ here. The charge does affect the total energy E, however, adding quantity $q\phi_0$ to the kinetic plus rest energy. Hence the components of the 4-momentum are

$$\mathbf{p} = \frac{m_0 \mathbf{w}}{(1 - w^2/c^2)^{1/2}}, \tag{18-85}$$

$$\frac{E}{c} = \frac{m_0 c}{(1 - w^2/c^2)^{1/2}} + \frac{q}{c}\phi_0, \tag{18-86}$$

in the frame momentarily fixed with respect to q_1.

Let us next shift to the inertial frame in which the source q_1 and its accompanying electric field move at 3-velocity $\mathbf{v}$ while the particle carrying the charge q moves at 3-velocity $\mathbf{u}$ as in figure 18.2. How the term $(q/c)\phi_0$ transforms is to be deduced from known 4-vector behavior.

We recall how a particle of rest mass m_0 has the 4-momentum components

$$\mathbf{p} = 0, \tag{18-87}$$

$$\frac{E}{c} = m_0 c, \tag{18-88}$$

in the reference frame in which it is at rest. But in the inertial frame in which it moves at 3-velocity $\mathbf{v}$, it has the components

$$\mathbf{p} = \frac{m_0 \mathbf{v}}{(1 - v^2/c^2)^{1/2}}, \tag{18-89}$$

$$\frac{E}{c} = \frac{m_0 c}{(1 - v^2/c^2)^{1/2}}. \tag{18-90}$$

The scalar part equals $m_0 c$ divided by $(1 - v^2/c^2)^{1/2}$ while the 3-vector part equals this scalar part multiplied by $\mathbf{v}/c$.

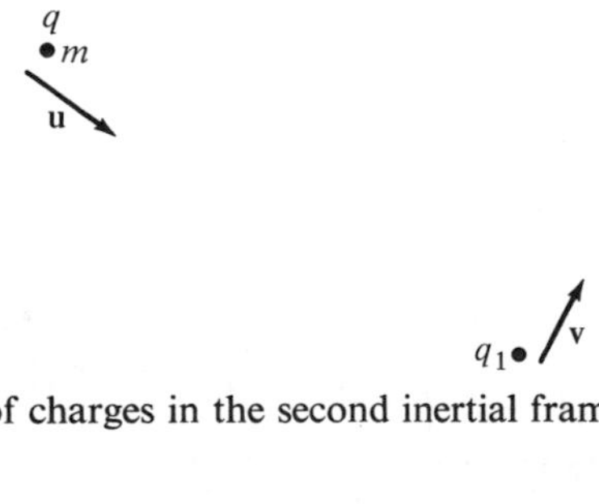

FIGURE 18.2   Movement of charges in the second inertial frame.

We treat the term $(q/c)\phi_0$ similarly in forms (18-85) and (18-86). We also replace **w** and w in the mechanical momentum-energy terms by **u** and u, the velocity and speed of m_0 in the final inertial frame, getting

$$\mathbf{p} = \frac{m_0\mathbf{u}}{(1 - u^2/c^2)^{1/2}} + \frac{q}{c^2}\frac{\phi_0\mathbf{v}}{(1 - v^2/c^2)^{1/2}} = m\mathbf{u} + q\mathbf{A}, \tag{18-91}$$

$$\frac{E}{c} = \frac{m_0 c}{(1 - u^2/c^2)^{1/2}} + \frac{q}{c}\frac{\phi_0}{(1 - v^2/c^2)^{1/2}} = mc + \frac{q}{c}\phi. \tag{18-92}$$

Note how the vector potential **A** and scalar potential ϕ have been introduced by definition in the last equality of each line.

Generally, there are many source charges moving in various directions with various speeds at a given instant in the reference inertial frame. If ϕ_{0j} is the electrostatic potential caused by the jth charge in the inertial frame in which this source is momentarily at rest, and if $\mathbf{v}_j$ is the instantaneous velocity of this source charge in the given reference frame, addition of the effects yields

$$\mathbf{p} = \frac{m_0\mathbf{u}}{(1 - u^2/c^2)^{1/2}} + \frac{q}{c^2}\frac{\phi_{0j}\mathbf{v}_j}{(1 - v_j^2/c^2)^{1/2}} = m\mathbf{u} + q\mathbf{A}, \tag{18-93}$$

$$\frac{E}{c} = \frac{m_0 c}{(1 - u^2/c^2)^{1/2}} + \frac{q}{c}\frac{\phi_{0j}}{(1 - v_j^2/c^2)^{1/2}} = mc + \frac{q}{c}\phi, \tag{18-94}$$

where **A** and ϕ are the indicated sums.

When the system is too large for the preceding discussion to be valid, existence of ϕ/c still implies existence of **A** and we have

$$\mathbf{p} = m\mathbf{u} + q\mathbf{A}, \tag{18-95}$$

$$\frac{E}{c} = mc + \frac{q}{c}\phi. \tag{18-96}$$

How the effects propagate can be determined from the differential equations (Maxwell equations) obtained in earlier sections.

18.8 Hamiltonian and Lagrangian functions for a moving charge

Since the equations governing electromagnetic fields come from transforming the equations governing purely electric fields, the general forms of the Hamiltonian and Lagrangian for a charged particle can be obtained from relationships that prevail in an electric field.

First, consider a particle carrying charge q at a small velocity in a purely electric field. Newtonian theory is then adequate and equation (10-25) is satisfied. The Hamiltonian H equals the total energy E.

But this appears in equation (18-96); so we have

$$\frac{H}{c} = mc + \frac{q}{c}\phi \qquad (18\text{-}97)$$

or

$$\frac{H}{c} - \frac{q}{c}\phi = mc. \qquad (18\text{-}98)$$

A Lorentz transformation converts the field into a mixed electric and magnetic system, but it does not alter the form of equation (18-96). As a result, we assume that equation (18-96) holds in general, paired with equation (18-95); and equation (18-98) is valid, paired with

$$\mathbf{p} - q\mathbf{A} = m\mathbf{u}. \qquad (18\text{-}99)$$

But $m\mathbf{u}$ and mc are the 3-vector and scalar parts of the 4-momentum. Multiplying corresponding covariant and contravariant components and adding, we get

$$m^2 u^2 - m^2 c^2 = -m_0{}^2 c^2 \qquad (18\text{-}100)$$

as in equation (17-43). Using equations (18-99) and (18-98) to replace $m\mathbf{u}$ and mc, we then obtain

$$(\mathbf{p} - q\mathbf{A})^2 - \left(\frac{H}{c} - \frac{q}{c}\phi\right)^2 = -m_0{}^2 c^2 \qquad (18\text{-}101)$$

or

$$(H - q\phi)^2 = c^2(\mathbf{p} - q\mathbf{A})^2 + m_0{}^2 c^4. \qquad (18\text{-}102)$$

In nonrelativistic mechanics, the Hamiltonian is related to the Lagrangian by equation (10-13). This may be rewritten in the form

$$L = \mathbf{p} \cdot \mathbf{u} - H. \qquad (18\text{-}103)$$

Introducing (18-95) for $\mathbf{p}$ and (18-97) for H yields

$$L = mu^2 + q\mathbf{A} \cdot \mathbf{u} - mc^2 - q\phi$$

$$= -m_0 c^2 \left(1 - \frac{u^2}{c^2}\right)^{1/2} + q\mathbf{A} \cdot \mathbf{u} - q\phi. \qquad (18\text{-}104)$$

Multiplying this Lagrangian by differential dt yields a quantity that is the same in all inertial frames. As a result, it is suitable for use in Hamilton's principle. Indeed, note that

$$\left(1 - \frac{u^2}{c^2}\right)^{1/2} = \frac{1}{\gamma} \qquad (18\text{-}105)$$

and

$$\mathbf{A} \cdot \mathbf{u} - \phi = \frac{1}{\gamma}\gamma\mathbf{u} \cdot \mathbf{A} - \frac{1}{\gamma}\gamma c \frac{\phi}{c} = \frac{1}{\gamma}u^\alpha A_\alpha; \qquad (18\text{-}106)$$

so

$$L \, dt = - \frac{1}{\gamma} (m_0 c^2 - u^\alpha q A_\alpha) \, dt$$

$$= (u^\alpha q A_\alpha - m_0 c^2) \, d\tau \qquad (18\text{-}107)$$

and

$$\int_{t_1}^{t_2} L \, dt = \int_{\tau_1}^{\tau_2} (u^\alpha q A_\alpha - m_0 c^2) \, d\tau. \qquad (18\text{-}108)$$

The expressions for γ came from equation (16-134), the components u^α from table 17.1, A_4 from (18-96) and the connection between the contravariant and covariant components.

According to Hamilton's principle, the charged particle moves so that

$$\int_{t_1}^{t_2} L \, dt \qquad (18\text{-}109)$$

is an extremum. The Euler equation that causes the variation of the integral to vanish is equation (11-42) with f replaced by L. Section 5.7 tells us that this equation can be written in the form

$$\frac{d\mathbf{p}}{dt} = \nabla L. \qquad (18\text{-}110)$$

EXAMPLE 18.4   Obtain the canonical $\mathbf{p}$ from the relativistic Lagrangian.

The generalized momentum p_k equals the partial derivative of L with respect to the kth component of the velocity. Differentiating (18-104),

$$p_k = \frac{\partial L}{\partial u_k} = -m_0 c^2 \frac{-2u_k/c^2}{2(1 - u^2/c^2)^{1/2}} + q A_k$$

$$= m u_k + q A_k,$$

and combining the components to form 3-vectors on each side, we obtain

$$\mathbf{p} = m\mathbf{u} + q\mathbf{A},$$

an expression that agrees with (18-95).

18.9   Dependence of components of the field on the 4-potential

Substitution of the Lagrangian for a charged particle into Lagrange's equation yields an equation of motion containing parts of the 4-potential. Rearranging the result allows us to identify the electric and magnetic force terms and to get expressions for $\mathbf{E}$ and $\mathbf{B}$.

We consider a particle of mass m and charge q traveling at velocity $\mathbf{u}$ in the field described by $\mathbf{A}$ and ϕ. The generalized momentum $\mathbf{p}$ is then given by equation (18-95), the Lagrangian by equation (18-104). Mechanical behavior of the particle is governed by Lagrange equation (18-110).

The differentiation rule from example 5.1 shows that

$$\frac{d}{dt} = \frac{\partial}{\partial t} + \frac{dx}{dt}\frac{\partial}{\partial x} + \frac{dy}{dt}\frac{\partial}{\partial y} + \frac{dz}{dt}\frac{\partial}{\partial z}$$

$$= \frac{\partial}{\partial t} + \mathbf{u} \cdot \nabla. \tag{18-111}$$

Consequently,

$$\frac{d\mathbf{p}}{dt} = \frac{d}{dt}(m\mathbf{u} + q\mathbf{A}) = \frac{d(m\mathbf{u})}{dt} + q\frac{d\mathbf{A}}{dt}$$

$$= \frac{d(m\mathbf{u})}{dt} + q\frac{\partial\mathbf{A}}{\partial t} + q(\mathbf{u}\cdot\nabla)\mathbf{A}. \tag{18-112}$$

In the partial differentiations imposed by ∇, the velocity components are kept constant. The first term in L is then constant and

$$\nabla L = q\nabla(\mathbf{A}\cdot\mathbf{u}) - q\nabla\phi. \tag{18-113}$$

Applying the second formula in problem 12.14 to the gradient of $\mathbf{A}\cdot\mathbf{u}$ and reducing, we obtain

$$q[\nabla(\mathbf{A}\cdot\mathbf{u})] = q[(\mathbf{A}\cdot\nabla)\mathbf{u} + (\mathbf{u}\cdot\nabla)\mathbf{A} + \mathbf{u}\times(\nabla\times\mathbf{A}) + \mathbf{A}\times(\nabla\times\mathbf{u})]$$

$$= q[(\mathbf{u}\cdot\nabla)\mathbf{A} + \mathbf{u}\times(\nabla\times\mathbf{A})]. \tag{18-114}$$

Note that the two terms drop out because of constancy of $\mathbf{u}$ in the differentiations. Substituting these forms into equation (18-110)

$$\frac{d(m\mathbf{u})}{dt} + q\frac{\partial\mathbf{A}}{\partial t} + q(\mathbf{u}\cdot\nabla)\mathbf{A} = q(\mathbf{u}\cdot\nabla)\mathbf{A} + q\mathbf{u}\times(\nabla\times\mathbf{A}) - q\nabla\phi \tag{18-115}$$

and reducing, we find

$$\frac{d(m\mathbf{u})}{dt} = -q\frac{\partial\mathbf{A}}{\partial t} - q\nabla\phi + q\mathbf{u}\times(\nabla\times\mathbf{A}). \tag{18-116}$$

We interpret $d(m\mathbf{u})/dt$ as the force acting on the particle. Electric intensity $\mathbf{E}$ is the force per unit charge if the test charge were at rest

$$\mathbf{E} = -\frac{\partial\mathbf{A}}{\partial t} - \nabla\phi, \tag{18-117}$$

while magnetic induction $\mathbf{B}$ is what $q\mathbf{u}$ is crossed with to get the magnetic force

$$\mathbf{B} = \nabla\times\mathbf{A}. \tag{18-118}$$

These substitutions put equation (18-116) in the form

$$\frac{d(m\mathbf{u})}{dt} = q(\mathbf{E} + \mathbf{u} \times \mathbf{B}) \tag{18-119}$$

which also comes from eliminating $\mathbf{F}/\gamma$ from equations (18-70) and (18-77).

18.10 Supplementary comments

The simplest charge we can picture consists of an interacting point characterized by a signed scalar quantity. Such a charge is spherically symmetric in free space.

If the field produced by the charge is unique, it must also be spherically symmetric. In group theoretical terms, the field must be a basis for the completely symmetric representation of the full rotation group. The field is made manifest by the force it exerts on a test charge. But such a force is represented by a 3-vector; the field intensity is a 3-vector.

Because forces add and the fields from different sources act independently, we must add the field intensities from the sources to get the over-all intensity. The superposition principle is thus satisfied.

Each constituent electrostatic field is represented by a signed flux fanning out symmetrically, so the divergence of $\mathbf{E}/c$ where the source charge density is ρ_0 is $\rho_0/c\varepsilon$. This relationship transforms into the equation for the field tensor on going to any other inertial frame. The matrix representing this tensor transforms according to the law

$$\mathbf{F}' = \mathbf{A}\mathbf{F}\tilde{\mathbf{A}}. \tag{18-120}$$

Tensor $\mathbf{F}$ is antisymmetric because its 44 component must always be 0, and $\mathbf{A}$ represents a general Lorentz transformation. This antisymmetric property causes the charge to be conserved in physical processes. The 23, 31, 12 components of $\mathbf{F}$ are identified with the 1, 2, 3 components of $\mathbf{B}$. The conditions on $\mathbf{F}$ then yield an equation for the curl of $\mathbf{B}$.

The definition of $\mathbf{B}$ is completed by taking its divergence equal to zero. Transforming this condition then leads to Faraday's law for the curl of $\mathbf{E}$.

We define the force acting on a particle as the 3-vector with correct transformation properties whose dot product with a displacement of the particle equals the energy added to the particle during the displacement. A form for the electromagnetic force is obtained on introducing the proper components of the field tensor into the electrostatic force law, transforming to a general inertial frame, and dividing by γ.

The energy carried by a particle equals its relativistic mass times the square of the fundamental speed. We may assume that energy is conserved, so the particle energy increases at the expense of a potential energy $q\phi$. But existence of a scalar potential implies existence of a vector potential $\mathbf{A}$ with the conventional properties.

In prerelativistic theory, the Hamiltonian of a charged particle equals its net energy, as long as the field is conservative. But in relativistic theory, a Lorentz transformation does not destroy this equality, although it does introduce the vector potential.

A Lagrangian can be set up, related to the Hamiltonian as before. It is found to yield an invariant integrand in Hamilton's principle. The corresponding Euler equation yields an equation of motion that contains the Lorentz force law.

DISCUSSION QUESTIONS

18.1 What kind of magnetic field can a Lorentz transformation eliminate?

18.2 Can an inertial frame exist in an empty universe?

18.3 Why is charge described by a scalar quantity?

18.4 Why is the field of an isolated charge spherically symmetric?

18.5 Does the field produced by one charge vary with the position or movement of another? What is the superposition principle?

18.6 Why does the field **D** vary inversely with the square of the distance from its source? How does this variation lead to the divergence equation?

18.7 Why does charge transform as mass? Explain how $\rho_0/c\varepsilon$ transforms.

18.8 Explain why the equation

$$\frac{\partial F^{\alpha\beta}}{\partial x^\beta} = \mu J^\alpha$$

applies when different elements of charge are moving differently.

18.9 How does the charge conservation law

$$\frac{\partial \rho}{\partial t} + \nabla \cdot \mathbf{J} = 0$$

follow from the antisymmetric nature of the field tensor?

18.10 In what way is the equation for the curl of **B** related to the equation for the divergence of **D**?

18.11 Can the condition

$$\nabla \cdot \mathbf{B} = 0$$

be based on the criterion of simplicity?

18.12 Explain how the equation for the curl of **E** is related to the equation for the divergence of **B**.

18.13 Why is the dual of the field tensor antisymmetric?

18.14 How can particles serve as probes for studying a given field?

18.15 Discuss the relativistic generalization of Newton's second law.

18.16 What is the relativistic equation of motion for a charged particle in an electromagnetic field?

18.17 How does existence of a scalar potential ϕ for a charge imply existence of a vector potential **A** for the same charge?

18.18 Why does our derivation of the form for **A** require the system to be small?

18.19 Explain how the transformation law for a purely electric potential ϕ_0 is obtained.

18.20 How is Hamiltonian H identified with the energy E of a charged particle?

18.21 Discuss how the Lagrangian L for a charged particle is obtained. Why should $L\,dt$ be invariant?

18.22 Why is

$$m\mathbf{u} + q\mathbf{A}$$

the canonical momentum?

18.23 How is the Hamiltonian for a charged particle related to its canonical momentum?

18.24 How is the Lorentz force law derived from Hamilton's principle?

PROBLEMS

18.1 Arrange appropriate covariant and contravariant components of the 4-current density in row and column matrices. Then construct a matrix equation describing the invariance derived in example 18.1.

18.2 Subject the field tensor

$$\begin{pmatrix} F^{11} & F^{14} \\ F^{41} & F^{44} \end{pmatrix}$$

to a Lorentz transformation. Show that if the 44 component is always zero, then F^{11} must be zero and F^{14} must equal $-F^{41}$.

18.3 Express the phase

$$\phi = \mathbf{k}\cdot\mathbf{r} - \omega t$$

of a monochromatic electromagnetic wave as the product of two 4-vectors. How does the propagation 4-vector transform? What do **k** and ω behave as?

18.4 Reduce the determinant of the field tensor to products of **B** and **E**.

18.5 Show that the determinant of the field tensor is invariant to transformation from one inertial frame to another.

18.6 Express $F^{\alpha\beta}G_{\alpha\beta}$ in terms of **B** and **E**. Show that the result is invariant.

18.7 Derive a differential equation of motion for a free particle from the appropriate relativistic Lagrangian.

18.8 If a particle moves along the x axis with

$$L = -\frac{1}{\gamma} m_0 c^2 + m_0 gx,$$

what differential equation governs its motion?

18.9 Let v_x/c be tanh θ in the differential equation of problem 18.8 and in the expressions for the components of the 4-velocity of the particle. Integrate the former, substitute the result into the latter, and again integrate to obtain parametric equations for the motion.

— — —

18.10 Determine how mass density transforms from one inertial frame to another.

18.11 How does the matrix of the covariant components of the field tensor transform? For convenience, this matrix may be designated $\mathbf{F}$...

18.12 From the form for ds^2, determine the metric-tensor components that are used to lower indices. Use these to determine elements $F_{\alpha\beta}$.

18.13 After solving problem 18.12, reduce the invariant sum $F_{\alpha\beta}F^{\alpha\beta}$ to a function of $\mathbf{B}$ and $\mathbf{E}$.

18.14 Calculate the determinant of the dual of the field tensor. Is this expression invariant?

18.15 How would equation (18-63) be changed if the continuum contained magnetic monopoles of density ρ_m determining the divergence of $\mathbf{B}$ according to equation (18-40)?

18.16 Use results from previous problems to show that (a) if the angle between $\mathbf{B}$ and $\mathbf{E}$ is acute (or obtuse) in one inertial frame, it is acute (or obtuse) in any other inertial frame, (b) if $|\mathbf{E}| > |c\mathbf{B}|$ (or $|\mathbf{E}| < |c\mathbf{B}|$) in one inertial frame, then $|\mathbf{E}| > |c\mathbf{B}|$ (or $|\mathbf{E}| < |c\mathbf{B}|$) in any other inertial frame, (c) if $\mathbf{E}$ is perpendicular to $\mathbf{B}$ and $|\mathbf{E}| \neq |c\mathbf{B}|$, there is an inertial frame in which either $\mathbf{E}$ or $\mathbf{B}$ vanishes.

18.17 From the transformation law for $\mathbf{F}$, determine how a Lorentz transformation along the x axis changes an electromagnetic field.

18.18 If a particle moves along the x axis with

$$L = -\frac{1}{\gamma} m_0 c^2 - \tfrac{1}{2}kx^2,$$

what differential equation governs its motion? Show that this equation integrates to

$$E = m_0 c^2 + \tfrac{1}{2}ka^2.$$

REFERENCES

BOOKS

Landau, L. D., and **Lifshitz, E. M.** (Translated by M. Hamermesh) *The Classical Theory of Fields*, 2nd ed., pp. 1–99. Addison-Wesley Publishing Company, Inc., Reading, Mass., 1962.

Panofsky, W. K. H., and **Phillips, M.** *Classical Electricity and Magnetism*, 2nd ed., pp. 324–340, 425–445. Addison-Wesley Publishing Company, Inc., Reading, Mass., 1962.

ARTICLES

Atwater, H. A. Radiation from a Uniformly Accelerated Charge. *Am. J. Phys.*, **38,** 1447 (1970).

Boyer, T. H. Energy and Momentum in Electromagnetic Field for Charged Particles Moving with Constant Velocities. *Am. J. Phys.*, **39,** 257 (1971).

Brehme, R. W. The Relativistic Lagrangian. *Am. J. Phys.*, **39,** 275 (1971).

Cohn, J. Electromagnetic Momentum of Charged Particles. *Am. J. Phys.*, **38,** 502 (1970).

Denman, H. H., and **Kiefer, H. M.** Invariance and Conservation Laws for a Classical Relativistic Particle. *Am. J. Phys.*, **38,** 300 (1970).

Dewan, E. M. The Magnetic Field as a Mechanism to Preserve Relativistic Momentum. *Am. J. Phys.*, **40,** 755 (1972).

Fisher, G. P. The Electric Dipole Moment of a Moving Magnetic Dipole. *Am. J. Phys.*, **39,** 1528 (1971).

Furry, W. H. Examples of Momentum Distributions in the Electromagnetic Field and in Matter. *Am. J. Phys.*, **37,** 621 (1969).

Hauser, W. On the Fundamental Equations of Electromagnetism. *Am. J. Phys.*, **38,** 80 (1970).

Kennedy, F. J. Approximately Relativistic Interactions. *Am. J. Phys.*, **40,** 63 (1972).

Krefetz, E. A "Derivation" of Maxwell's Equations. *Am. J. Phys.*, **38,** 513 (1970).

Lehmberg, R. H. Axiomatic Development of the Laws of Vacuum Electrodynamics. *Am. J. Phys.*, **29,** 584 (1961).

Parker, L., and **Schmieg, G. M.** Special Relativity and Diagonal Transformations. *Am. J. Phys.*, **38,** 218 (1970).

Pfleiderer, J. Lorentz Transformation Derived from First-Order Experiments. *Am. J. Phys.*, **37,** 1131 (1969).

Puri, S. P. Noninteraction Between Static Electric and Magnetic Fields. *Am. J. Phys.*, **34,** 162 (1966).

Rohrlich, F. Electromagnetic Momentum, Energy, and Mass. *Am. J. Phys.*, **38,** 1310 (1970).

Rosen, G. Conformal Invariance of Maxwell's Equations. *Am. J. Phys.*, **40,** 1023 (1972).

Schaffner, K. F. The Lorentz Electron Theory of Relativity. *Am. J. Phys.*, **37,** 498 (1969).

Zink, J. W. Relativity and the Classical Electron. *Am. J. Phys.*, **39,** 1403 (1971).

19 / *Spinors*

19.1 Possible forms for representing physical properties

In determining the character of time and space, an observer must interact with other physical bodies. Such interactions presumably have some influences on the properties studied. When the bodies are of appreciable (classical) size, however, the resulting perturbations can be made small enough so that their effects can be neglected.

A person then obtains the following results.

1. An object free of influences experiences a smooth uniform passage of time.

2. A nonrotating Euclidean space can be erected on the center of mass of this object and all other existing bodies located in the resulting space. Such a frame of reference is called an inertial system.

3. Physical laws are the same in all inertial systems. In particular, fundamental speed c is invariant.

Each coordinate system is a construct of the intellect. Physical reality, on the other hand, resides in the physical objects themselves. They possess properties that exist in all possible inertial frames. As a consequence, such properties must be represented by invariant operands: (a) scalars, (b) 4-vectors, (c) 4^n-component tensors.

Matrix **A** of section 17.3 links the coordinates of one inertial frame to those of another; it also effects the transformation of the column matrix representing a 4-vector. A 16-component tensor, on the other hand, transforms according to the law in example 18.2.

A similar transformation law effects the transformation of a 2-index tensor in a 2-dimensional complex-coordinate space. We will find that such a law is sufficient to describe all allowed physical transformations from one inertial frame to another.

A vector in the 2-dimensional complex space is called a *spinor* of the *first* rank. A combination of two spinors that transform together is called a *bispinor*. The 2-index tensor is called a spinor of the *second* rank.

Since the spinors are invariant operands in the same sense as the 4^n-component tensors are, they can also represent physical properties. Indeed, the state of a neutrino is described by a 2-element spinor; the state of an electron by a bispinor.

554

Furthermore, a spinor transformation is defined by 4 complex numbers that satisfy the unimodular condition. Matrix **A** of section 17.3, on the other hand, contains 16 parameters. Choosing these to fit conditions (17-24) and (17-25) is much more difficult than constructing the corresponding spinor transformation.

19.2   Spinors constructed from coordinates

Over extensive regions where the gravitational field is not too intense, the Pythagorean theorem is satisfied. Inertial systems Σ and Σ' can then be erected as Figure 16.1 shows and the rectangular coordinates of a particle traveling at speed c from the origin determined. These coordinates are found to obey the equations

$$x^2 + y^2 + z^2 - c^2 t^2 = 0, \tag{19-1}$$

$$x'^2 + y'^2 + z'^2 - c^2 t'^2 = 0. \tag{19-2}$$

The linear relationships that transform (19-1) to (19-2) also transform

$$x^2 + y^2 + z^2 - c^2 t^2 = r^2 \tag{19-3}$$

to

$$x'^2 + y'^2 + z'^2 - c^2 t'^2 = r'^2 \tag{19-4}$$

without altering r

$$r = r'. \tag{19-5}$$

If, by some trick, we obtain simpler laws for carrying out the former transformation, we will have a simpler procedure for effecting the latter. But first, we need to define the requisite expressions.

Equation (19-3) can be rewritten as the sum of two products

$$\begin{aligned} r^2 &= (x + iy)(x - iy) + (z + ct)(z - ct) \\ &= x_+ x_- + z_+ z_- \end{aligned} \tag{19-6}$$

if

$$x_+ = x + iy, \tag{19-7}$$

$$x_- = x - iy, \tag{19-8}$$

$$z_+ = z + ct, \tag{19-9}$$

$$z_- = z - ct. \tag{19-10}$$

When r is 0, that is, when the coordinates are somewhere on the light cone based on the origin, the sum is 0 and

$$x_+ x_- = -z_+ z_-. \tag{19-11}$$

Since the left side of equation (19-11) is positive, quantities z_+ and $-z_-$ are either both positive or both negative.

Any positive quantity can be represented as the product of a number and its complex conjugate. When z_+ and $-z_-$ are both positive, we can set

$$z_+ = a^2 = S^{1*}S^1, \tag{19-12}$$

$$-z_- = b^2 = S^2 S^{2*}; \tag{19-13}$$

so equation (19-11) becomes

$$x_+ x_- = S^{1*}S^1 S^2 S^{2*}. \tag{19-14}$$

The most general complex numbers x_+ and x_- satisfying this relationship are then obtained by taking

$$x_+ = S^{1*}S^2 = abe^{i\alpha}, \tag{19-15}$$

$$x_- = S^1 S^{2*} = abe^{-i\alpha}. \tag{19-16}$$

The phases of S^1 and S^2 need not be specified; only their difference α is significant.

Each of these components may be plotted in a 2-dimensional Euclidean space (see figure 19.1). We call the ordered number pair

$$\mathbf{S}^{\cdot} = \begin{pmatrix} S^1 \\ S^2 \end{pmatrix} \tag{19-17}$$

an *upper-index spinor*.

Events for which z_+ and $-z_-$ are both positive and r is zero lie on the future half of the light cone. Events on the past branch of the light cone yield negative z_+ and $-z_-$. Each of the former is located by an $\mathbf{S}^{\cdot}$, as we have just seen.

To locate any of the latter, we set

$$-z_+ = S_2 S_2^{*}, \tag{19-18}$$

$$z_- = S_1^{*}S_1; \tag{19-19}$$

so

$$x_+ x_- = S_1^{*}S_1 S_2 S_2^{*}. \tag{19-20}$$

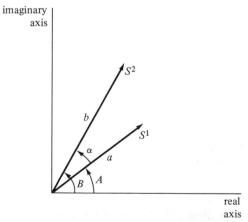

FIGURE 19.1   Components of spinor $\mathbf{S}^{\cdot}$ plotted in the complex plane.

Pertinent x_+ and x_- are then obtained by assuming

$$x_+ = S_1{}^*S_2, \tag{19-21}$$

$$x_- = S_1S_2{}^*. \tag{19-22}$$

The ordered number pair

$$\mathbf{S}. = \begin{pmatrix} S_1 \\ S_2 \end{pmatrix} \tag{19-23}$$

is called a *lower-index spinor*.

The Hermitian adjoint of matrix (19-17) is

$$\mathbf{S}^{\cdot\dagger} = (S^{1*} \quad S^{2*}), \tag{19-24}$$

while the Hermitian adjoint of (19-23) is

$$\mathbf{S}^{\cdot\dagger} = (S_1{}^* \quad S_2{}^*). \tag{19-25}$$

Premultiplying either of the adjoints by the original spinor yields a 2×2 matrix containing the combinations $x_+, x_-, \pm z_+, \mp z_-$.

The product

$$\mathbf{M}^{\cdot\cdot} = \mathbf{S}^{\cdot}\mathbf{S}^{\cdot\dagger} = \begin{pmatrix} S^1 \\ S^2 \end{pmatrix} (S^{1*} \quad S^{2*}) = \begin{pmatrix} S^1 S^{1*} & S^1 S^{2*} \\ S^2 S^{1*} & S^2 S^{2*} \end{pmatrix}$$

$$= \begin{pmatrix} z_+ & x_- \\ x_+ & -z_- \end{pmatrix} = \begin{pmatrix} z + ct & x - iy \\ x + iy & -z + ct \end{pmatrix} \tag{19-26}$$

is called a *second-rank* upper-index *spinor*.

The second expression

$$\mathbf{M}.. = \mathbf{S}.\mathbf{S}.^{\dagger} = \begin{pmatrix} S_1 \\ S_2 \end{pmatrix} (S_1{}^* \quad S_2{}^*) = \begin{pmatrix} S_1 S_1{}^* & S_1 S_2{}^* \\ S_2 S_1{}^* & S_2 S_2{}^* \end{pmatrix}$$

$$= \begin{pmatrix} z_- & x_- \\ x_+ & -z_+ \end{pmatrix} = \begin{pmatrix} z - ct & x - iy \\ x + iy & -z - ct \end{pmatrix} \tag{19-27}$$

is called a *second-rank* lower-index *spinor*.

Either of the column matrices may also multiply the adjoint of the other. When the first factor has upper indices and the second factor lower ones, we obtain

$$\mathbf{M}^{\cdot}. = \mathbf{S}^{\cdot}\mathbf{S}.^{\dagger} = \begin{pmatrix} S^1 \\ S^2 \end{pmatrix} (S_1{}^* \quad S_2{}^*) = \begin{pmatrix} S^1 S_1{}^* & S^1 S_2{}^* \\ S^2 S_1{}^* & S^2 S_2{}^* \end{pmatrix}. \tag{19-28}$$

Alternatively, we have

$$\mathbf{M}.^{\cdot} = \mathbf{S}.\mathbf{S}^{\cdot\dagger} = \begin{pmatrix} S_1 \\ S_2 \end{pmatrix} (S^{1*} \quad S^{2*}) = \begin{pmatrix} S_1 S^{1*} & S_1 S^{2*} \\ S_2 S^{1*} & S_2 S^{2*} \end{pmatrix}. \tag{19-29}$$

Expressions (19-28) and (19-29) are called *mixed* second-rank spinors.

19.3   Expressing linear transformations

Every homogeneous linear transformation of the elements in a column matrix is effected by a conformable square matrix. Such a transformation of a first-rank spinor implies a tensor transformation of the corresponding second-rank spinor.

When the first-rank spinor has an upper index, the transformed spinor has the form

$$(\mathbf{S}^{\cdot})' = \mathbf{AS}^{\cdot} \tag{19-30}$$

while the Hermitian adjoint is

$$(\mathbf{S}^{\cdot\dagger})' = \mathbf{S}^{\cdot\dagger}\mathbf{A}^{\dagger}. \tag{19-31}$$

The corresponding second-rank upper-index spinor is obtained on multiplying these:

$$(\mathbf{S}^{\cdot})'(\mathbf{S}^{\cdot\dagger})' = \mathbf{AS}^{\cdot}\mathbf{S}^{\cdot\dagger}\mathbf{A}^{\dagger}. \tag{19-32}$$

The first equality in (19-26) applied to the final and initial forms then yields

$$(\mathbf{M}^{\cdot\cdot})' = \mathbf{AM}^{\cdot\cdot}\mathbf{A}^{\dagger}. \tag{19-33}$$

Likewise, every homogeneous linear transformation of elements in a column lower-index spinor is implemented by a conformable square matrix:

$$(\mathbf{S}_{.})' = \mathbf{BS}_{..} \tag{19-34}$$

Then

$$(\mathbf{S}_{.}^{\dagger})' = \mathbf{S}_{.}^{\dagger}\mathbf{B}^{\dagger} \tag{19-35}$$

and

$$(\mathbf{S}_{.})'(\mathbf{S}_{.}^{\dagger})' = \mathbf{BS}_{.}\mathbf{S}_{.}^{\dagger}\mathbf{B}^{\dagger}. \tag{19-36}$$

The first equality in (19-27) applied to the final and the initial forms leads to

$$(\mathbf{M}_{..})' = \mathbf{BM}_{..}\mathbf{B}^{\dagger}. \tag{19-37}$$

Any 2-element column matrix $\mathbf{T}^{\cdot}$ (or $\mathbf{T}_{.}$) that transforms as $\mathbf{S}^{\cdot}$ (or $\mathbf{S}_{.}$) does

$$(\mathbf{T}^{\cdot})' = \mathbf{AT}^{\cdot}, \tag{19-38}$$

$$(\mathbf{T}_{.})' = \mathbf{BT}_{.}, \tag{19-39}$$

is called a first-rank spinor, while any 2×2 matrix $\mathbf{N}^{\cdot\cdot}$ (or $\mathbf{N}_{..}$) transforming as $\mathbf{M}^{\cdot\cdot}$ (or $\mathbf{M}_{..}$) does

$$(\mathbf{N}^{\cdot\cdot})' = \mathbf{AN}^{\cdot\cdot}\mathbf{A}^{\dagger}, \tag{19-40}$$

$$(\mathbf{N}_{..})' = \mathbf{BN}_{..}\mathbf{B}^{\dagger}, \tag{19-41}$$

is called a second-rank spinor.

19.4   Transformations that preserve *r*

Operators $\mathbf{A}$ and $\mathbf{B}$ must meet certain conditions if they are to transform the spinors from one inertial system to another. Elements x_{+}, x_{-}, z_{+}, z_{-} are then transformed in a relatively simple manner.

Let us first write matrix **A** as

$$\begin{pmatrix} \alpha & \beta \\ \gamma & \delta \end{pmatrix} \tag{19-42}$$

so the transformation of **S** ̇, equation (19-30), appears as

$$\begin{pmatrix} (S^1)' \\ (S^2)' \end{pmatrix} = \begin{pmatrix} \alpha & \beta \\ \gamma & \delta \end{pmatrix} \begin{pmatrix} S^1 \\ S^2 \end{pmatrix} = \begin{pmatrix} \alpha S^1 + \beta S^2 \\ \gamma S^1 + \delta S^2 \end{pmatrix}. \tag{19-43}$$

Substituting the resulting primed spinor elements into the primed form of (19-15) yields

$$\begin{aligned} x_+' &= (\alpha S^1 + \beta S^2)^*(\gamma S^1 + \delta S^2) \\ &= \alpha^*\delta S^{1*}S^2 + \beta^*\gamma S^{2*}S^1 + \alpha^*\gamma S^{1*}S^1 + \beta^*\delta S^{2*}S^2 \\ &= \alpha^*\delta x_+ + \beta^*\gamma x_- + \alpha^*\gamma z_+ - \beta^*\delta z_-. \end{aligned} \tag{19-44}$$

Similarly, the primed forms of (19-16), (19-12), and (19-13) lead to

$$\begin{aligned} x_-' &= (\alpha S^1 + \beta S^2)(\gamma S^1 + \delta S^2)^* \\ &= \beta\gamma^* x_+ + \alpha\delta^* x_- + \alpha\gamma^* z_+ - \beta\delta^* z_-, \end{aligned} \tag{19-45}$$

$$\begin{aligned} z_+' &= (\alpha S^1 + \beta S^2)^*(\alpha S^1 + \beta S^2) \\ &= \alpha^*\beta x_+ + \alpha\beta^* x_- + \alpha^*\alpha z_+ - \beta\beta^* z_-, \end{aligned} \tag{19-46}$$

$$\begin{aligned} -z_-' &= (\gamma S^1 + \delta S^2)(\gamma S^1 + \delta S^2)^* \\ &= \gamma^*\delta x_+ + \gamma\delta^* x_- + \gamma^*\gamma z_+ - \delta\delta^* z_-. \end{aligned} \tag{19-47}$$

When these expressions are substituted into the second-rank upper-index coordinate spinor, the result factors into three matrices:

$$\begin{pmatrix} z_+' & x_-' \\ x_+' & -z_-' \end{pmatrix} = \begin{pmatrix} \alpha & \beta \\ \gamma & \delta \end{pmatrix} \begin{pmatrix} z_+ & x_- \\ x_+ & -z_- \end{pmatrix} \begin{pmatrix} \alpha^* & \gamma^* \\ \beta^* & \delta^* \end{pmatrix}. \tag{19-48}$$

Taking the determinant of each side of (19-48) then yields

$$-z_+'z_-' - x_+'x_-' = (\alpha\delta - \beta\gamma)(-z_+z_- - x_+x_-)(\alpha^*\delta^* - \beta^*\gamma^*). \tag{19-49}$$

If both Σ and Σ' are inertial systems, quantity r^2 in equation (19-6) is invariant and the product of the outside factors on the right of (19-49) must equal 1. For simplicity, we assign zero phase to each factor; so

$$\alpha\delta - \beta\gamma = 1. \tag{19-50}$$

Note that the square of the radius vector need not be zero. Equations (19-44) through (19-47) transform *any* event, whether or not it lies on the light cone. Thus, the second-rank spinor (19-33) is not limited to the cone.

In the equations relating the first-order spinor components to the set x_+, x_-, z_+, z_-,

$$S^1 \qquad \text{plays the same role as} \qquad S_2{}^*,$$
$$S^{1*} \qquad \text{plays the same role as} \qquad -S_2,$$
$$S^2 \qquad \text{plays the same role as} \qquad -S_1{}^*,$$

and

$$S^{2*} \qquad \text{plays the same role as} \qquad S_1.$$

Consequently, equation (19-43) implies that

$$\begin{pmatrix} (S_2{}^*)' \\ -(S_1{}^*)' \end{pmatrix} = \begin{pmatrix} \alpha & \beta \\ \gamma & \delta \end{pmatrix} \begin{pmatrix} S_2{}^* \\ -S_1{}^* \end{pmatrix} \tag{19-51}$$

or

$$\begin{pmatrix} (S_2{}^*)' \\ (S_1{}^*)' \end{pmatrix} = \begin{pmatrix} \alpha & -\beta \\ -\gamma & \delta \end{pmatrix} \begin{pmatrix} S_2{}^* \\ S_1{}^* \end{pmatrix}. \tag{19-52}$$

Rearranging the elements

$$\begin{pmatrix} (S_1{}^*)' \\ (S_2{}^*)' \end{pmatrix} = \begin{pmatrix} \delta & -\gamma \\ -\beta & \alpha \end{pmatrix} \begin{pmatrix} S_1{}^* \\ S_2{}^* \end{pmatrix} \tag{19-53}$$

and taking the complex conjugate, we get

$$\begin{pmatrix} (S_1)' \\ (S_2)' \end{pmatrix} = \begin{pmatrix} \delta^* & -\gamma^* \\ -\beta^* & \alpha^* \end{pmatrix} \begin{pmatrix} S_1 \\ S_2 \end{pmatrix} \tag{19-54}$$

whence

$$\mathbf{B} = \begin{pmatrix} \delta^* & -\gamma^* \\ -\beta^* & \alpha^* \end{pmatrix}. \tag{19-55}$$

Equation (19-34) effects the same transformation as equation (19-43) when $\mathbf{B}$ is related to the parameters α, β, γ, δ as equation (19-55) states. The operation transforms x_+, x_-, z_+, z_- from one inertial frame to another when the parameters satisfy the unimodular condition (19-50).

Since

$$\begin{pmatrix} \delta & -\beta \\ -\gamma & \alpha \end{pmatrix} \begin{pmatrix} \alpha & \beta \\ \gamma & \delta \end{pmatrix} = \begin{pmatrix} \alpha\delta - \beta\gamma & 0 \\ 0 & -\beta\gamma + \alpha\delta \end{pmatrix}, \tag{19-56}$$

the relationship between $\mathbf{B}$ and $\mathbf{A}$ reduces to the form

$$\mathbf{B}^\dagger \mathbf{A} = \mathbf{E}. \tag{19-57}$$

19.5 Parameterizing the transformations

The 4×4 real matrix in equation (17-15) and the 2×2 complex matrix in equation (19-43) both cause rectangular coordinates to be transformed linearly. For the radius 4-vector $\mathbf{r}$ to be preserved, however, elements of these matrices have to meet certain conditions.

The 4×4 matrix contains 16 real numbers that must satisfy equations (17-24) and (17-25). As we noted before, 10 of these equations are independent; so the transformation involves 6 independent parameters. The 2 × 2 matrix contains 8 real numbers. One pair of these must be chosen to satisfy unimodular condition (19-50). Again, we have 6 independent parameters.

It is much simpler to choose 8 numbers consistent with condition (19-50) than to choose 16 numbers consistent with conditions (17-24) and (17-25), for we may pick 3 pairs arbitrarily and merely adjust the last pair to fit equation (19-50).

In our examples, we will introduce an additional simplification. We will only consider transformations that do not mix the first-rank spinor components. Thus, we will take β and γ to be zero. Parameter δ is then the reciprocal of α.

EXAMPLE 19.1   What happens to x, y, z, ct when the second component of the coordinate spinor $\mathbf{S}^{\cdot}$ turns by angle $\tfrac{1}{2}\phi$ while the first component turns by minus this angle?

For unprimed coordinates on the future branch of the light cone, we have

$$S^1 = ae^{iA},$$
$$S^2 = be^{iB},$$

as figure 19.1 shows. The given rotation of S^1 then decreases angle A by $\tfrac{1}{2}\phi$

$$(S^1)' = ae^{i(A-\phi/2)}$$

while the rotation of S^2 increases angle B by $\tfrac{1}{2}\phi$

$$(S^2)' = be^{i(B+\phi/2)}.$$

These equations can be combined in the form

$$\begin{pmatrix} (S^1)' \\ (S^2)' \end{pmatrix} = \begin{pmatrix} e^{-i\phi/2} & 0 \\ 0 & e^{i\phi/2} \end{pmatrix} \begin{pmatrix} S^1 \\ S^2 \end{pmatrix}$$

that appears in (19-43). The elements in the operator transforming $\mathbf{S}^{\cdot}$ are then

$$\alpha = e^{-i\phi/2}, \qquad \beta = 0, \qquad \gamma = 0, \qquad \delta = e^{i\phi/2}.$$

Since

$$\alpha\delta - \beta\gamma = e^{-i\phi/2}e^{i\phi/2} - 0 = 1,$$

the 2 × 2 matrix is unimodular. Substituting it and its Hermitian adjoint into (19-48) transforms the coordinates in the second-rank spinor $\mathbf{M}^{\cdot\cdot}$ from one inertial frame to another. These coordinates need not be on the light cone; they may locate *any* event.

The parameters we have obtained reduce equations (19-44) through (19-47) to

$$x_+' = e^{i\phi}x_+ = (\cos\phi + i\sin\phi)x_+,$$
$$x_-' = e^{-i\phi}x_- = (\cos\phi - i\sin\phi)x_-,$$

$$z_+' = z_+,$$
$$z_-' = z_-.$$

The definitions of x_+ and x_- convert the first two equations to

$$x' + iy' = (\cos \phi + i \sin \phi)x - (\sin \phi - i \cos \phi)y,$$
$$x' - iy' = (\cos \phi - i \sin \phi)x - (\sin \phi + i \cos \phi)y.$$

Solving for x' and for y' yields

$$x' = x \cos \phi - y \sin \phi,$$
$$y' = x \sin \phi + y \cos \phi,$$

while solving the other two equations for z' and ct' yields

$$z' = z,$$
$$ct' = ct.$$

These results can be combined in the form

$$\begin{pmatrix} x' \\ y' \\ z' \\ ct' \end{pmatrix} = \begin{pmatrix} \cos \phi & -\sin \phi & 0 & 0 \\ \sin \phi & \cos \phi & 0 & 0 \\ 0 & 0 & 1 & 0 \\ 0 & 0 & 0 & 1 \end{pmatrix} \begin{pmatrix} x \\ y \\ z \\ ct \end{pmatrix}.$$

The temporal coordinate is not affected, but the spatial coordinates change as in equation (4-29); the transformation is an ordinary rotation by angle ϕ in the xy plane.

EXAMPLE 19.2   What happens to x, y, z, ct when the second component of the coordinate spinor $S\dot{}$ is multiplied by $e^{\kappa/2}$ while the first component is multiplied by the reciprocal of this amount?

Spinor $S\dot{}$ changes in the given manner if

$$\begin{pmatrix} (S^1)' \\ (S^2)' \end{pmatrix} = \begin{pmatrix} e^{-\kappa/2} & 0 \\ 0 & e^{\kappa/2} \end{pmatrix} \begin{pmatrix} S^1 \\ S^2 \end{pmatrix}.$$

Transformation matrix (19-42) then has the components

$$\alpha = e^{-\kappa/2}, \qquad \beta = 0, \qquad \gamma = 0, \qquad \delta = e^{\kappa/2},$$

and

$$\alpha\delta - \beta\gamma = e^{-\kappa/2}e^{\kappa/2} - 0 = 1;$$

the matrix is unimodular.

The resulting parameters reduce equations (19-44) through (19-47) to

$$x_+' = x_+,$$
$$x_-' = x_-,$$
$$z_+' = e^{-\kappa}z_+ = (\cosh \kappa - \sinh \kappa)z_+,$$
$$z_-' = e^{\kappa}z_- = (\cosh \kappa + \sinh \kappa)z_-,$$

whence

$$x' = x,$$

$$y' = y,$$

$$z' = (\cosh \kappa)z - (\sinh \kappa)ct,$$

$$ct' = -(\sinh \kappa)z + (\cosh \kappa)ct,$$

or

$$\begin{pmatrix} x' \\ y' \\ z' \\ ct' \end{pmatrix} = \begin{pmatrix} 1 & 0 & 0 & 0 \\ 0 & 1 & 0 & 0 \\ 0 & 0 & \cosh \kappa & -\sinh \kappa \\ 0 & 0 & -\sinh \kappa & \cosh \kappa \end{pmatrix} \begin{pmatrix} x \\ y \\ z \\ ct \end{pmatrix}.$$

On comparing this equation with equation (16-50), we see that the process is a Lorentz transformation for which the origin of Σ' moves along the z axis of Σ.

19.6   Representing 4-vectors

Any four numbers that transform from one inertial system to another, as components x, y, z, ct of radius vector $\mathbf{r}$ do, form a 4-vector. Such dimensional numbers may replace the components of $\mathbf{r}$ in a matrix without changing the properties of the matrix.

Indeed, let us start with the upper-index spinor (19-26)

$$\mathbf{M}^{\cdot\cdot} = \begin{pmatrix} z + ct & x - iy \\ x + iy & -z + ct \end{pmatrix}, \tag{19-58}$$

the lower-index spinor (19-27)

$$\mathbf{M}_{\cdot\cdot} = \begin{pmatrix} z - ct & x - iy \\ x + iy & -z - ct \end{pmatrix}, \tag{19-59}$$

and a 4-vector $\mathbf{V}$ with the rectangular components

$$V_x, \qquad V_y, \qquad V_z, \qquad V_{ct}, \tag{19-60}$$

in inertial frame Σ.

Replacing the components of $\mathbf{r}$ with the corresponding components of $\mathbf{V}$ yields

$$\mathbf{N}^{\cdot\cdot} = \begin{pmatrix} V_z + V_{ct} & V_x - iV_y \\ V_x + iV_y & -V_z + V_{ct} \end{pmatrix} \tag{19-61}$$

and

$$\mathbf{N}_{\cdot\cdot} = \begin{pmatrix} V_z - V_{ct} & V_x - iV_y \\ V_x + iV_y & -V_z - V_{ct} \end{pmatrix}. \tag{19-62}$$

Because V_x, V_y, V_z, V_{ct} behave as x, y, z, ct do, matrix $\mathbf{N}^{\cdot\cdot}$ is an upper-index second-rank spinor while matrix $\mathbf{N}_{\cdot\cdot}$ is a lower-index second-rank spinor.

If we let

$$\sigma_x = \begin{pmatrix} 0 & 1 \\ 1 & 0 \end{pmatrix}, \tag{19-63}$$

$$\sigma_y = \begin{pmatrix} 0 & -i \\ i & 0 \end{pmatrix}, \tag{19-64}$$

$$\sigma_z = \begin{pmatrix} 1 & 0 \\ 0 & -1 \end{pmatrix}, \tag{19-65}$$

$$\mathbf{1} = \begin{pmatrix} 1 & 0 \\ 0 & 1 \end{pmatrix}, \tag{19-66}$$

then formulas (19-61) and (19-62) can be rewritten as

$$\mathbf{N}^{\cdot\cdot} = V_x\sigma_x + V_y\sigma_y + V_z\sigma_z + V_{ct}\mathbf{1} \tag{19-67}$$

and

$$\mathbf{N}_{\cdot\cdot} = V_x\sigma_x + V_y\sigma_y + V_z\sigma_z - V_{ct}\mathbf{1}. \tag{19-68}$$

Since the bases σ_x, σ_y, and σ_z were introduced by Wolfgang Pauli in his work on electron spin, they are called the *Pauli spin matrices*.

EXAMPLE 19.3   Construct a contravariant spinor representing the 4-momentum of a particle.

From table 17.2, the 4-momentum consists of

$$\left(\mathbf{p}, \frac{E}{c} \right)$$

with

$$\mathbf{p} = p_x\mathbf{i} + p_y\mathbf{j} + p_z\mathbf{k}.$$

Substituting the components into (19-61) yields

$$\begin{pmatrix} p_z + \dfrac{E}{c} & p_x - ip_y \\[2mm] p_x + ip_y & -p_z + \dfrac{E}{c} \end{pmatrix}$$

or

$$p_x\sigma_x + p_y\sigma_y + p_z\sigma_z + \frac{E}{c}\mathbf{1}.$$

19.7   4-Vector and spinor differential operators

A physical property may be described by a scalar function varying from event to event in the continuum. The infinitesimal change associated with each infinitesimal displacement factors into the gradient of the function dotted with the displacement.

The displacement may be defined by either its contravariant or its covariant rectangular components. The other factor in each term is the corresponding component of the gradient. But this equals the analogous part of the del operator acting on the function. From the parts, we can construct 4-vector and spinor differential operators.

Indeed, differentiating a general scalar function of the rectangular coordinates of inertial frame Σ

$$\phi(x, y, z, ct) \tag{19-69}$$

yields

$$d\phi = \frac{\partial\phi}{\partial x}\, dx + \frac{\partial\phi}{\partial y}\, dy + \frac{\partial\phi}{\partial z}\, dz + \frac{\partial\phi}{\partial(ct)}\, d(ct), \tag{19-70}$$

whence

$$d\phi = \left(\frac{\partial\phi}{\partial x} \quad \frac{\partial\phi}{\partial y} \quad \frac{\partial\phi}{\partial z} \quad \frac{\partial\phi}{\partial(ct)}\right) \begin{pmatrix} dx \\ dy \\ dz \\ d(ct) \end{pmatrix} \tag{19-71}$$

or

$$d\phi = \left(\frac{\partial\phi}{\partial x} \quad \frac{\partial\phi}{\partial y} \quad \frac{\partial\phi}{\partial z} \quad -\frac{\partial\phi}{\partial(ct)}\right) \begin{pmatrix} dx \\ dy \\ dz \\ -d(ct) \end{pmatrix}. \tag{19-72}$$

In the second expression in line (12-54), we can make the first $d\mathbf{s}$ differ from the second. The final expression then equals the sum of covariant components of one vector times the corresponding contravariant components of the other. But the result is a number; whenever covariant components of one vector are multiplied by contravariant components of another and the terms added, we obtain a scalar.

Since $d\phi$ is a scalar and the components in the second factor of expression (19-71) are contravariant, those in the first factor must be covariant, and the row *covariant* differential operator is

$$\left(\frac{\partial}{\partial x} \quad \frac{\partial}{\partial y} \quad \frac{\partial}{\partial z} \quad \frac{\partial}{\partial(ct)}\right). \tag{19-73}$$

Since the components in the second factor of expression (19-72) are covariant, the row *contravariant* differential operator is

$$\left(\frac{\partial}{\partial x} \quad \frac{\partial}{\partial y} \quad \frac{\partial}{\partial z} \quad -\frac{\partial}{\partial(ct)}\right). \tag{19-74}$$

The gradient itself is a 4-vector with the contravariant components

$$\left(\frac{\partial\phi}{\partial x} \quad \frac{\partial\phi}{\partial y} \quad \frac{\partial\phi}{\partial z} \quad -\frac{\partial\phi}{\partial(ct)}\right). \tag{19-75}$$

Substituting these into formula (19-61) yields the upper-index second-rank spinor

$$
\mathbf{D}^{\cdot\cdot}\phi =
\begin{pmatrix}
\dfrac{\partial\phi}{\partial z} - \dfrac{\partial\phi}{\partial(ct)} & \dfrac{\partial\phi}{\partial x} - i\dfrac{\partial\phi}{\partial y} \\[2ex]
\dfrac{\partial\phi}{\partial x} + i\dfrac{\partial\phi}{\partial y} & -\dfrac{\partial\phi}{\partial z} - \dfrac{\partial\phi}{\partial(ct)}
\end{pmatrix}
\tag{19-76}
$$

whence

$$
\mathbf{D}^{\cdot\cdot} =
\begin{pmatrix}
\dfrac{\partial}{\partial z} - \dfrac{\partial}{\partial(ct)} & \dfrac{\partial}{\partial x} - i\dfrac{\partial}{\partial y} \\[2ex]
\dfrac{\partial}{\partial x} + i\dfrac{\partial}{\partial y} & -\dfrac{\partial}{\partial z} - \dfrac{\partial}{\partial(ct)}
\end{pmatrix}.
\tag{19-77}
$$

From (19-62), we similarly obtain the lower-index differential operator

$$
\mathbf{D}_{\cdot\cdot} =
\begin{pmatrix}
\dfrac{\partial}{\partial z} + \dfrac{\partial}{\partial(ct)} & \dfrac{\partial}{\partial x} - i\dfrac{\partial}{\partial y} \\[2ex]
\dfrac{\partial}{\partial x} + i\dfrac{\partial}{\partial y} & -\dfrac{\partial}{\partial z} + \dfrac{\partial}{\partial(ct)}
\end{pmatrix}.
\tag{19-78}
$$

Introducing the Pauli spin matrices lets us rewrite these spinor differential operators in the simple forms

$$
\mathbf{D}^{\cdot\cdot} = \sigma_x\frac{\partial}{\partial x} + \sigma_y\frac{\partial}{\partial y} + \sigma_z\frac{\partial}{\partial z} - \mathbf{1}\frac{\partial}{\partial(ct)}
\tag{19-79}
$$

and

$$
\mathbf{D}_{\cdot\cdot} = \sigma_x\frac{\partial}{\partial x} + \sigma_y\frac{\partial}{\partial y} + \sigma_z\frac{\partial}{\partial z} + \mathbf{1}\frac{\partial}{\partial(ct)}.
\tag{19-80}
$$

EXAMPLE 19.4   Show that the product of a second-rank upper-index spinor with a first-rank lower-index spinor is a first-rank upper-index spinor.

Let $\mathbf{N}^{\cdot\cdot}$ and $\mathbf{T}_{\cdot}$ be the matrices representing the given spinors in inertial frame Σ while $(\mathbf{N}^{\cdot\cdot})'$ and $(\mathbf{T}_{\cdot})'$ are the matrices representing the same spinors in inertial frame Σ'. The relationships among these are then given by equations (19-40) and (19-39)

$$
(\mathbf{N}^{\cdot\cdot})' = \mathbf{A}\mathbf{N}^{\cdot\cdot}\mathbf{A}^\dagger
$$

$$
(\mathbf{T}_{\cdot})' = \mathbf{B}\mathbf{T}_{\cdot\cdot}
$$

The Hermitian adjoint of equation (19-57) is

$$
\mathbf{A}^\dagger\mathbf{B} = \mathbf{E};
$$

so the product of the transformed spinors reduces to the same form

$$(\mathbf{N}^{\cdot\cdot})'(\mathbf{T}.)' = \mathbf{A}\mathbf{N}^{\cdot\cdot}\mathbf{A}^{\dagger}\mathbf{B}\mathbf{T}. = \mathbf{A}\mathbf{N}^{\cdot\cdot}\mathbf{T}.$$

that appears in equation (19-38). The combination $\mathbf{N}^{\cdot\cdot}\mathbf{T}.$ behaves as a first-rank upper-index spinor.

19.8 Attributes of 4-vectors and spinors

A typical 4-vector is $\mathbf{r}$, the geodesic line drawn from the origin to an event (x, y, z, ct) in the Minkowski continuum.

If the event lies within the light cone based on the origin, quantity

$$x^2 + y^2 + z^2 \tag{19-81}$$

is less than quantity

$$c^2 t^2 \tag{19-82}$$

and

$$\mathbf{r} \cdot \mathbf{r} = x^2 + y^2 + z^2 - c^2 t^2 \tag{19-83}$$

is negative. If the event lies on the cone itself, (19-81) equals (19-82) and r^2 is zero. If the event lies elsewhere, (19-81) is greater than (19-82) and r^2 is positive.

Vanishing of r^2 does not imply that x, y, z, and ct are zero, so vanishing of the dot product of any 4-vector with itself does not imply that its components are zero in a given inertial frame.

Such vanishing does let the 4-vector be represented by a first-rank spinor. Matrix $\mathbf{S}^{\cdot}$ (or $\mathbf{S}.$) describes an $\mathbf{r}$ drawn to the future (or past) half of the light cone. Matrix $\mathbf{T}^{\cdot}$ (or $\mathbf{T}.$) similarly describes a 4-vector $\mathbf{V}$ for which $\mathbf{V} \cdot \mathbf{V}$ is zero.

If one of the elements of $\mathbf{S}^{\cdot}$ (or $\mathbf{S}.$) is zero, x_+ and x_- are zero and the spatial part of the radius vector lies along the z axis. If one element of $\mathbf{T}^{\cdot}$ (or $\mathbf{T}.$) is zero, the 4-vector $\mathbf{V}$ has zero x and y components.

Thus, if one element of a first-rank spinorial derivative is zero, the function acted on cannot change in the x and y directions; it must propagate in the $\pm z$ directions. If there need be no difference between these two directions, the other element must also be zero.

19.9 Functions whose spinorial derivative vanishes

Let us now determine possible forms for a first-rank spinor function with a null derivative and discuss physical disturbances that such forms represent.

Let us seek elements ϕ^1 and ϕ^2 that satisfy

$$\mathbf{D}..\phi^{\cdot} = 0 \tag{19-84}$$

or

$$\left(\sigma_x \frac{\partial}{\partial x} + \sigma_y \frac{\partial}{\partial y} + \sigma_z \frac{\partial}{\partial z} + \mathbf{1} \frac{\partial}{\partial(ct)}\right)\begin{pmatrix} \phi^1 \\ \phi^2 \end{pmatrix} = 0. \tag{19-85}$$

Since differential equation (19-85) is linear, homogeneous, and first-order, the elements may be expressed as a linear combination of exponentials. A typical term appears as

$$\phi^j = u^j e^{i(\mathbf{k}\cdot\mathbf{r}-\omega t)} \tag{19-86}$$

where

$$\mathbf{k} = k_x \mathbf{e}_1 + k_y \mathbf{e}_2 + k_z \mathbf{e}_3, \tag{19-87}$$

$$\mathbf{r} = x\mathbf{e}_1 + y\mathbf{e}_2 + z\mathbf{e}_3. \tag{19-88}$$

Substituting (19-86) into (19-85), carrying out the indicated differentiations,

$$\left(\sigma_x k_x + \sigma_y k_y + \sigma_z k_z - 1\frac{\omega}{c}\right)\begin{pmatrix}u^1\\u^2\end{pmatrix} e^{i(\mathbf{k}\cdot\mathbf{r}-\omega t)} = 0, \tag{19-89}$$

and introducing the forms of the Pauli spin matrices yield

$$\begin{pmatrix}k_z - \dfrac{\omega}{c} & k_x - ik_y \\ k_x + ik_y & -k_z - \dfrac{\omega}{c}\end{pmatrix}\begin{pmatrix}u^1\\u^2\end{pmatrix} e^{i(\mathbf{k}\cdot\mathbf{r}-\omega t)} = 0. \tag{19-90}$$

In a term in which u^1 is not zero and u^2 is zero, we must have

$$k_z - \frac{\omega}{c} = 0 \quad \text{or} \quad k_z = \frac{\omega}{c}, \tag{19-91}$$

$$k_x + ik_y = 0 \quad \text{or} \quad k_x = 0, \quad k_y = 0. \tag{19-92}$$

In a term in which u^2 is not zero but u^1 is zero, we must have

$$k_z + \frac{\omega}{c} = 0 \quad \text{or} \quad k_z = -\frac{\omega}{c}, \tag{19-93}$$

$$k_x - ik_y = 0 \quad \text{or} \quad k_x = 0, \quad k_y = 0. \tag{19-94}$$

Since ϕ^j is a function only of the angle in the exponent, a given phase propagates according to the law

$$\mathbf{k}\cdot\mathbf{r} - \omega t = \text{constant} \tag{19-95}$$

whence

$$k_x x + k_y y + k_z z = \omega t + \text{constant}. \tag{19-96}$$

Substituting relationships (19-91) and (19-92) into equation (19-96) yields

$$z = ct + \text{constant}' \tag{19-97}$$

while (19-93), (19-94) convert (19-96) to

$$z = -ct + \text{constant}''. \tag{19-98}$$

Consequently, a disturbance described by (19-84) and (19-86) would move in the same manner in both the $+z$ and $-z$ directions. This isotropy between opposite directions remains after transforming to another inertial frame because such a transformation does not alter the form of equation (19-84). A physical disturbance that behaves in this way is a stream of (a) electron neutrinos or (b) muon neutrinos.

Experiments show that a uniform stream of particles is diffracted as a beam of light with the wavelength

$$\lambda = \frac{h}{p} \tag{19-99}$$

where h is Planck's constant and p the particle momentum. If ϕ^j is interpreted as a wave in which k is $2\pi/\lambda$, we have

$$\mathbf{k} = 2\pi \frac{\mathbf{p}}{h}. \tag{19-100}$$

Then equations (19-92) and (19-94) imply that

$$p_x = p_y = 0. \tag{19-101}$$

If we also assume that the energy of a particle is related to the frequency by the relationship

$$E = h\nu = \frac{h}{2\pi} \omega \tag{19-102}$$

that Einstein assumed for photons, then equation (19-91) implies that

$$2\pi \frac{p_z}{h} = \frac{2\pi}{h} \frac{E}{c} \tag{19-103}$$

or

$$p_z = \frac{E}{c} \tag{19-104}$$

while equation (19-93) implies that

$$p_z = -\frac{E}{c}. \tag{19-105}$$

Since the relativistic energy is mc^2, we have

$$p = mc. \tag{19-106}$$

The particles making up the stream travel at speed c.

19.10 The simplest coupled spinor differential equations

A disturbance that travels more slowly does change on the light cone based on the origin. Instead of setting the spinorial derivative of the corresponding function equal to zero, we may then make it proportional to a second spinor whose derivative behaves similarly.

Indeed, let us consider the coupled equations

$$D..\phi^{\cdot} = -A\chi., \tag{19-107}$$

$$D^{\cdot\cdot}\chi. = A\phi^{\cdot} \tag{19-108}$$

in which A is a number to be determined.

On dividing (19-107) by the negative of this number, substituting into (19-108), and rearranging the result, we get

$$D^{\cdot\cdot}D..\phi^{\cdot} = -A^2\phi^{\cdot}. \tag{19-109}$$

Similarly eliminating ϕ' yields

$$D..D^{\cdot\cdot}\chi. = -A^2\chi.. \tag{19-110}$$

Since these equations are linear, homogeneous, and second order, the elements are linear combinations of exponentials. Typically

$$\phi^j = u^j e^{i(\mathbf{k} \cdot \mathbf{r} - \omega t)} \tag{19-111}$$

with $\mathbf{k}$ and $\mathbf{r}$ given by (19-87) and (19-88).

Substituting (19-111) into (19-109) leads to

$$\begin{pmatrix} k_z + \dfrac{\omega}{c} & k_x - ik_y \\[2mm] k_x + ik_y & -k_z + \dfrac{\omega}{c} \end{pmatrix} \begin{pmatrix} k_z - \dfrac{\omega}{c} & k_x - ik_y \\[2mm] k_x + ik_y & -k_z - \dfrac{\omega}{c} \end{pmatrix} \begin{pmatrix} u^1 \\ u^2 \end{pmatrix} e^{i(\mathbf{k} \cdot \mathbf{r} - \omega t)}$$

$$= A^2 \begin{pmatrix} u^1 \\ u^2 \end{pmatrix} e^{i(\mathbf{k} \cdot \mathbf{r} - \omega t)}. \tag{19-112}$$

For arbitrary u^1 and u^2, the product of the two 2×2 matrices on the left must equal A^2 times the unit 2×2 matrix. This condition is satisfied if

$$k_z^2 - \frac{\omega^2}{c^2} + k_x^2 + k_y^2 = A^2. \tag{19-113}$$

No restriction is imposed on k_x, k_y, or k_z.

But a uniform stream of particles is diffracted as a beam of light with the propagation vector

$$\mathbf{k} = 2\pi \frac{\mathbf{p}}{h} = \frac{\mathbf{p}}{\hbar} \tag{19-114}$$

if $\mathbf{p}$ is the momentum of a typical particle and $\hbar$ is Planck's constant divided by 2π. Furthermore, the energy of a particle is related to the frequency by the Einstein relationship

$$E = h\nu = \hbar\omega. \tag{19-115}$$

In the inertial frame in which the momentum of each particle in the stream is zero, k is zero and equation (19-113) reduces to

$$-\frac{\omega^2}{c^2} = A^2. \tag{19-116}$$

From equation (19-115) and the relativistic energy equation, we also have

$$\omega = \frac{2\pi}{h} E = \frac{2\pi}{h} m_0 c^2 = \frac{m_0 c^2}{\hbar}. \tag{19-117}$$

Therefore,

$$A^2 = -\frac{m_0^2 c^2}{\hbar^2} \tag{19-118}$$

and

$$A = \frac{im_0 c}{\hbar}. \tag{19-119}$$

If forms (19-111) and

$$\chi_j = v_j e^{i(\mathbf{k}\cdot\mathbf{r}-\omega t)} \tag{19-120}$$

are substituted into equations (19-107) and (19-108), the differentiations performed, and relationships (19-114) and (19-115) introduced, we obtain

$$\begin{pmatrix} p_z - \dfrac{E}{c} & p_x - ip_y \\ p_x + ip_y & -p_z - \dfrac{E}{c} \end{pmatrix} \phi^{\cdot} = -m_0 c\chi, \tag{19-121}$$

$$\begin{pmatrix} p_z + \dfrac{E}{c} & p_x - ip_y \\ p_x + ip_y & -p_z + \dfrac{E}{c} \end{pmatrix} \chi_{\cdot} = m_0 c\phi^{\cdot}. \tag{19-122}$$

Following example 19.3, the 2 × 2 matrices reduce to

$$\boldsymbol{\sigma}\cdot\mathbf{p} \mp 1\frac{E}{c}. \tag{19-123}$$

and equations (19-121) and (19-122) can be rewritten as

$$-\left(\boldsymbol{\sigma}\cdot\mathbf{p} - \frac{E}{c}\right)\phi^{\cdot} - m_0 c\chi_{\cdot} = 0, \tag{19-124}$$

$$\left(\boldsymbol{\sigma}\cdot\mathbf{p} + \frac{E}{c}\right)\chi_{\cdot} - m_0 c\phi^{\cdot} = 0, \tag{19-125}$$

or

$$\left[\begin{pmatrix} \dfrac{E}{c} - \boldsymbol{\sigma}\cdot\mathbf{p} & 0 \\ 0 & \dfrac{E}{c} + \boldsymbol{\sigma}\cdot\mathbf{p} \end{pmatrix} - \begin{pmatrix} 0 & m_0 c \\ m_0 c & 0 \end{pmatrix}\right]\begin{pmatrix} \phi^{\cdot} \\ \chi_{\cdot} \end{pmatrix} = 0. \tag{19-126}$$

Particles with a rest mass travel at less than speed c. The function describing their motion may consequently be of the kind described here. This consists of two spinors $\phi^{\cdot}$ and $\chi_{\cdot}$. that transform together—a combination called a bispinor.

19.11   The Dirac equation

Related to (19-126) is the equation formulated by Paul A. M. Dirac for describing the relativistic behavior of an electron.

In deriving it, we first multiply the sum of equations (19-124) and (19-125) by c:

$$(E - m_0c^2)(\phi^{\cdot} + \chi_{\cdot}) - c\boldsymbol{\sigma} \cdot \mathbf{p}(\phi^{\cdot} - \chi_{\cdot}) = 0. \qquad (19\text{-}127)$$

Let us also multiply the difference of equations (19-124) and (19-125) by c to obtain

$$(E + m_0c^2)(\phi^{\cdot} - \chi_{\cdot}) - c\boldsymbol{\sigma} \cdot \mathbf{p}(\phi^{\cdot} + \chi_{\cdot}) = 0. \qquad (19\text{-}128)$$

If we then let

$$\boldsymbol{\alpha} = \begin{pmatrix} 0 & \boldsymbol{\sigma} \\ \boldsymbol{\sigma} & 0 \end{pmatrix}, \qquad (19\text{-}129)$$

$$\beta = \begin{pmatrix} 1 & 0 \\ 0 & -1 \end{pmatrix}, \qquad (19\text{-}130)$$

$$\Psi = \begin{pmatrix} \phi^{\cdot} + \chi_{\cdot} \\ \phi^{\cdot} - \chi_{\cdot} \end{pmatrix}, \qquad (19\text{-}131)$$

the equations combine in the form

$$(E - c\boldsymbol{\alpha} \cdot \mathbf{p} - \beta m_0c^2)\Psi = 0 \qquad (19\text{-}132)$$

whence

$$(c\boldsymbol{\alpha} \cdot \mathbf{p} + \beta m_0c^2)\Psi = E\Psi. \qquad (19\text{-}133)$$

Equation (19-133) is used to describe the behavior of a free particle with rest mass m_0. When the charge on the particle is q and the 4-potential is not zero, $-q\phi$ is associated with E, and $-q\mathbf{A}$ with $\mathbf{p}$, as in equation (18-102). Then we have

$$[c\boldsymbol{\alpha} \cdot (\mathbf{p} - q\mathbf{A}) + \beta m_0c^2]\Psi = (E - q\phi)\Psi. \qquad (19\text{-}134)$$

Equation (19-134) is called Dirac's equation.

Dirac did not employ spinorial theory in setting up (19-134). Instead, he worked from energy equations (17-46) and (18-102), determining a linear square root of each side and allowing the resulting operators to act on the state function Ψ.

EXAMPLE 19.5   Determine the properties that the hypercomplex numbers α_x, α_y, α_z, β must have so the linear expression

$$c\boldsymbol{\alpha} \cdot \mathbf{p} + \beta m_0c^2 = E$$

is the square root of

$$c^2p^2 + m_0{}^2c^4 = E^2.$$

Being ordinary numbers, the components of **p** and c, m_0 can be placed anywhere in a given product. The relative positions of α_x, α_y, α_z, and β must be preserved, however. The square of the linear expression is

$$E^2 = (c\boldsymbol{\alpha} \cdot \mathbf{p} + \beta m_0 c^2)(c\boldsymbol{\alpha} \cdot \mathbf{p} + \beta m_0 c^2)$$
$$= c^2[\alpha_x^2 p_x^2 + \alpha_y^2 p_y^2 + \alpha_z^2 p_z^2 + (\alpha_x \alpha_y + \alpha_y \alpha_x)p_x p_y$$
$$+ (\alpha_y \alpha_z + \alpha_z \alpha_y)p_y p_z + (\alpha_z \alpha_x + \alpha_x \alpha_z)p_z p_x]$$
$$+ \beta^2 m_0^2 c^4 + m_0 c^3[(\alpha_x \beta + \beta \alpha_x)p_x + (\alpha_y \beta + \beta \alpha_y)p_y + (\alpha_z \beta + \beta \alpha_z)p_z].$$

This agrees with the formula for E^2 when

$$\alpha_x^2 = \alpha_y^2 = \alpha_z^2 = \beta^2 = 1,$$
$$\alpha_x \alpha_y + \alpha_y \alpha_x = \alpha_y \alpha_z + \alpha_z \alpha_y = \alpha_z \alpha_x + \alpha_x \alpha_z = 0,$$
$$\alpha_x \beta + \beta \alpha_x = \alpha_y \beta + \beta \alpha_y = \alpha_z \beta + \beta \alpha_z = 0.$$

Since

$$\sigma_x^2 = \sigma_y^2 = \sigma_z^2 = 1$$

and

$$\sigma_x \sigma_y + \sigma_y \sigma_x = \sigma_y \sigma_z + \sigma_z \sigma_y = \sigma_z \sigma_x + \sigma_x \sigma_z = 0,$$

the components of (19-129) and matrix (19-130) meet these requirements. Other matrices that also meet these requirements could also be employed.

EXAMPLE 19.6   How is equation (19-133) related to equation (17-46)?

From example 19.5, we can write the square root of equation (17-46) as

$$c\boldsymbol{\alpha} \cdot \mathbf{p} + \beta m_0 c^2 = E.$$

Letting each side act on a function Ψ that governs the state of the particle of mass m_0, we obtain

$$(c\boldsymbol{\alpha} \cdot \mathbf{p} + \beta m_0 c^2)\Psi = E\Psi.$$

19.12   Synopsis

In any compact region away from large concentrations of matter, a person can find an inertial frame in which a Cartesian coordinate system can be erected. It is possible to measure time with respect to the frame. Furthermore, each event occurring at a given time t in the region can be located by the coordinates x, y, z of the point where it happens. These four real quantities needed to distinguish the event are representable by two complex quantities, a number pair. The two ordered numbers define a point in a 2-dimensional complex space.

If points in this space are to correspond to events in the given region, invariance of the form

$$x^2 + y^2 + z^2 - c^2 t^2 \tag{19-135}$$

must not be violated. A suitable, yet simple, procedure involves mapping only the points of the light cone based on the origin of the space-time system. The linear relationships that transform these also transform all other events in the region properly.

Specifically, we let

$$x + iy = S^{1*}S^2, \tag{19-136}$$

$$x - iy = S^1 S^{2*}, \tag{19-137}$$

and

$$z + ct = S^{1*}S^1, \tag{19-138}$$

$$z - ct = -S^2 S^{2*}. \tag{19-139}$$

Consequently,

$$\begin{aligned}
x^2 + y^2 + z^2 - c^2 t^2 &= (x + iy)(x - iy) + (z + ct)(z - ct) \\
&= S^{1*}S^2 S^1 S^{2*} - S^{1*}S^1 S^2 S^{2*} \\
&= 0. \tag{19-140}
\end{aligned}$$

The ordered number pair

$$\mathbf{S}^{\cdot} = \begin{pmatrix} S^1 \\ S^2 \end{pmatrix} \tag{19-141}$$

is called a first-rank upper-index spinor.

We also let

$$x + iy = S_1 {}^* S_2, \tag{19-142}$$

$$x - iy = S_1 S_2 {}^*, \tag{19-143}$$

and

$$z + ct = -S_2 S_2 {}^*, \tag{19-144}$$

$$z - ct = S_1 {}^* S_1. \tag{19-145}$$

Then

$$\begin{aligned}
x^2 + y^2 + z^2 - c^2 t^2 &= (x + iy)(x - iy) + (z + ct)(z - ct) \\
&= S_1 {}^* S_2 S_1 S_2 {}^* - S_2 S_2 {}^* S_1 {}^* S_1 \\
&= 0. \tag{19-146}
\end{aligned}$$

The ordered number pair

$$\mathbf{S}_{\cdot} = \begin{pmatrix} S_1 \\ S_2 \end{pmatrix} \tag{19-147}$$

is called a first-rank lower-index spinor.

We can subject either spinor to a homogeneous linear transformation. The process leaves form (19-135) invariant when the coefficients satisfy the uni-modular condition (19-50). Physical rotations and Lorentz transformations are represented; reflections and inversions are not. A spinor transformation represents an operation that can actually be carried out on a physical set of axes, or on an extended body.

Turning the vector representing S^2 through angle $\frac{1}{2}\phi$ and the vector representing S^1 through minus this angle rotates the radius vector locating a point in physical space through angle ϕ. Since the effect of this rotation is indistinguishable from the effect of rotation by $\phi + 2\pi$, turning S^2 through $\frac{1}{2}\phi + \pi$ and S^1 through minus this angle represents the same physical operation. Two spinor transformations correspond to one physical operation.

The transformation properties of a spinor expression are not changed when x, y, z, ct in the defining relationships are replaced by the corresponding components V_x, V_y, V_z, V_{ct} of a 4-vector. The second-rank spinors can be expanded in terms of the Pauli spin matrices:

$$\mathbf{N}^{..} = V_x\sigma_x + V_y\sigma_y + V_z\sigma_z + V_{ct}\mathbf{1}, \tag{19-148}$$

$$\mathbf{N}_{..} = V_x\sigma_x + V_y\sigma_y + V_z\sigma_z - V_{ct}\mathbf{1}. \tag{19-149}$$

The transformation properties are also not changed when x, y, z, ct are replaced by $\partial/\partial x$, $\partial/\partial y$, $\partial/\partial z$, $-\partial/\partial(ct)$. The resulting spinorial operators combine with spinors to give spinorial derivatives.

Letting the spinorial derivative of a first-rank spinor function equal zero yields the condition governing the propagation of a neutrino. Making it proportional to a second spinor whose spinorial derivative is proportional to the first spinor yields the condition governing the propagation of an electron.

DISCUSSION QUESTIONS

19.1 Describe the mathematical expressions that can represent physical properties.

19.2 How can the transformation law for points on the light cone describe how any given event transforms, regardless of its location?

19.3 To what extent do the equations

$$z_+ = S^{1*}S^1, \qquad -z_- = S^2S^{2*}$$

limit z and t? To what extent are the spatial coordinates limited when the equations

$$x_+ = S^{1*}S^2, \qquad x_- = S^1S^{2*}$$

are added to the set? Explain.

19.4 What is (a) $\mathbf{S}^{\cdot\dagger}\mathbf{S}^{\cdot}$, (b) $\mathbf{S}^{\cdot}\mathbf{S}^{\cdot\dagger}$, (c) $\mathbf{S}_{.}^{\dagger}\mathbf{S}_{.}$, (d) $\mathbf{S}_{.}\mathbf{S}_{.}^{\dagger}$? Does the Hermitian adjoint of a spinor contain any information that the original spinor does not contain? Why?

19.5 How are the transformation matrices $\mathbf{A}$ and $\mathbf{B}$, for $\mathbf{S}^{\cdot}$ and $\mathbf{S}_{.}$, related?

19.6 What do the matrix formulas for transforming $\mathbf{S}^{\cdot}$ and $\mathbf{S}_{.}$ imply about the corresponding transformations of second-rank spinors? Can equation (19-48) be written from the general result?

19.7 Show that each allowed homogeneous transformation, combination of physical rotation and Lorentz transformation, depends on only six parameters.

19.8 Explain how the first-rank spinors for an event are changed when the corresponding point moves (a) through angle ϕ around the z axis, (b) from one inertial frame to a parallel frame traveling at speed u along the z axis.

19.9 Show that

$$\begin{pmatrix} V_z \pm V_{ct} & V_x - iV_y \\ V_x + iV_y & -V_z \pm V_{ct} \end{pmatrix}$$

is an upper-index (or lower-index) second-rank spinor when V_x, V_y, V_z, V_{ct} are rectangular components of a 4-vector.

19.10 Show that $\mathbf{M}^{\cdot\cdot}\mathbf{M}_{\cdot\cdot}$ is a scalar.

19.11 Obtain rectangular components of the gradient of a function in the Minkowski continuum. Formulate the corresponding spinors.

19.12 Discuss the spinorial differential equations that govern the propagation of (a) neutrinos, (b) electrons.

19.13 How do equations (19-44) through (19-47) show that two spinor transformations correspond to one physical operation?

PROBLEMS

19.1 Determine why we cannot consider x and y, z and ct, to be real and imaginary parts, respectively, of rectangular coordinates in a 2-dimensional space.

19.2 Show that a single reflection in a plane cannot be represented by a spinor transformation.

19.3 What coordinate transformation is produced by a change in sign of both S^1 and S^2?

19.4 What transformation is produced by an interchange of S^1 and S^2, with a change in sign of either quantity? Why is the sign change necessary?

19.5 What kind of spinor is

$$\begin{pmatrix} S^1(z - ct) + S^2(x - iy) \\ S^1(x + iy) - S^2(z + ct) \end{pmatrix} ?$$

19.6 Express the operator $\mathbf{A}$ that determines how spinor $\mathbf{S}^{\cdot}$ changes when the radius vector locating a point rotates $-\delta\phi$ about the z axis in terms of the appropriate unit matrix and the Pauli spin matrices.

19.7 If operator $\mathbf{A}$ acting on $\mathbf{S}^{\cdot}$ has the form

$$\mathbf{1} \cos \frac{\theta}{2} + i\sigma_x \sin \frac{\theta}{2}$$

what does it represent?

19.8 Determine what the spinor transformation of example 19.1 does to the complex vector $\mathbf{F}$ when

$$\begin{pmatrix} F_z & F_x - iF_y \\ F_x + iF_y & -F_z \end{pmatrix} = \mathbf{N}^{\cdot}_{\cdot\cdot}.$$

19.9 Find the linear square root of the form

$$E^2 = c^2(p_x^2 + p_y^2 + p_z^2)$$

employing scalars and 2×2 matrices only. To what particle is the result applicable?

— — —

19.10 What simple nontrivial function of the coordinates is a scalar during homogeneous Galilean transformations? How would spinors be constructed to preserve invariance of this form?

19.11 Show that an inversion cannot be represented by a spinor transformation.

19.12 If a physical transformation merely multiplies each S^1 by k, what does it do to an S^2?

19.13 If operator **A** acting on **S**˙ has the form

$$\mathbf{1} \cosh \frac{\kappa}{2} + \sigma_x \sinh \frac{\kappa}{2},$$

what process does it represent?

19.14 What changes do the operators

(a) $$\mathbf{A} = \mathbf{1} + \frac{i}{2}\, \sigma_x\, \delta\theta$$

(b) $$\mathbf{A} = \mathbf{1} + \frac{i}{2}\, \sigma_y\, \delta\phi$$

effect?

19.15 Show that

$$\mathbf{D}^{\cdot\cdot} = \begin{pmatrix} 2 \dfrac{\partial}{\partial(S_1 S_1^{\,*})} & 2 \dfrac{\partial}{\partial(S_2 S_1^{\,*})} \\[4mm] 2 \dfrac{\partial}{\partial(S_1 S_2^{\,*})} & 2 \dfrac{\partial}{\partial(S_2 S_2^{\,*})} \end{pmatrix}.$$

19.16 Combine Maxwell's equations to obtain components of

$$\left(\mu \mathbf{J}, \frac{1}{c\varepsilon}\, \rho \right)$$

as operations on

$$\mathbf{F} = \mathbf{B} + i\, \frac{\mathbf{E}}{c}.$$

Assume that μ and ε are constant.

19.17 Show that the spinor equation

$$\mathbf{D}^{\cdot\cdot}\mathbf{N}^{\cdot}. = i\mathbf{I}^{\cdot\cdot}$$

where $\mathbf{N}^{\cdot}.$ is defined as in problem 19.8 while $\mathbf{I}^{\cdot\cdot}$ is the second-rank spinor built from components of

$$\left(\mu\mathbf{J}, \frac{1}{c\varepsilon}\rho\right),$$

represents the two equations obtained in problem 19.16.

19.18 Prove that no 2×2 matrix anticommutes with σ_x, σ_y, and σ_z. As a result, we must employ larger matrices for α_x, α_y, α_z, β.

REFERENCES

BOOKS

Synge, J. L. *Relativity: The Special Theory*, 2nd ed., pp. 102–110. North-Holland Publishing Company, Amsterdam, 1964.

ARTICLES

Payne, W. T. Spinor Theory and Relativity. I. *Am. J. Phys.*, **23,** 526 (1955).
Payne, W. T. Spinor Theory and Relativity. II. *Am. J. Phys.*, **27,** 318 (1959).
Rindler, W. What are Spinors? *Am. J. Phys.*, **34,** 937 (1966).
Walker, M. J. Representation of Rotations in Three-Space by Complex 2×2 Matrices. *Am. J. Phys.*, **24,** 205 (1956).

20 / *The Schwarzschild Gravitational Field*

20.1   The equivalence principle and its significance

Coulomb's law for electricity and Newton's law for gravitation are both inverse-square laws. Each by itself implies direct action at a distance, and so is inexact. Adding magnetism as in chapters 14 and 18 makes Coulomb's law consistent with relativity. Our problem in this chapter is to correct Newton's law.

We first recall how each mass used in equation (2-18) is not essentially different from the mass we would employ in equations (2-1) and (2-2). The quantity analogous to electric charge is equivalent to the number determining the inertia of a body. We call the former the *gravitational mass*, the latter the *inertial mass*.

Because of this *equivalence*, the acceleration of a freely falling body does not depend on its mass, but only on the strength of the gravitational field where it is moving. Thus, the path traversed by the body depends only on its initial position, velocity, and on the field at each point along its path—things that may be represented geometrically in space-time.

The unique path between two events is the one that is stationary with respect to arbitrary small variations, the *geodesic* curve. In Minkowski space-time each geodesic is a straight line. Furthermore the path followed by a force-free body is such a line.

In general relativity, the freely falling body is also assumed to be force-free. We consider it to move along the unique geometric path, the geodesic, in a space-time to be determined. Since the structure of the continuum depends on the form for path element ds, a generalization of equation (16-124) will be sought.

20.2   Elements of time and space between neighboring events

In our generalization, we will suppose that time is bound up with space as another dimension. Furthermore, events may be located throughout this space-time so that a measure of the separation between any two, infinitesimally close, is determinable and invariant.

Indeed, it is assumed that an observer can survey any pertinent region of the continuum with measuring rods and clocks in the conventional manner. We particularly consider points fixed with respect to a body that is uninfluenced by other bodies or by fields (except as they might determine its inertial properties). Coordinate time t is taken as the time shown by a clock at one of the points in the weak-field region.

Close to any path between neighboring points, we may construct a rectangular lattice such that all angles between adjacent edges proceeding from any given point are the same. Distances may then be measured with respect to edges drawn from an event on the path. The independent components dx, dy, dz between it and another event time dt later can be determined.

We suppose that the nth power of the corresponding displacement depends on the initial event P and the differentials:

$$ds^n = f(P, dx, dy, dz, dt). \tag{20-1}$$

Equation (16-124) is a special form of this relationship.

Now, the units of space and time are arbitrary. Altering them both by the same ratio while leaving the initial and final events unchanged alters the separation ds by the same ratio. So if

$$dx' = \lambda\, dx, \qquad dy' = \lambda\, dy, \qquad dz' = \lambda\, dz, \tag{20-2}$$

and

$$dt' = \lambda\, dt, \tag{20-3}$$

then

$$ds' = \lambda\, ds. \tag{20-4}$$

In the primed system, equation (20-1) has the form

$$ds'^n = f(P, dx', dy', dz', dt'). \tag{20-5}$$

Combining this with equations (20-2) and (20-3), then with (20-4) and the original (20-1), yields

$$
\begin{aligned}
ds'^n &= f(P, \lambda\, dx, \lambda\, dy, \lambda\, dz, \lambda\, dt) \\
&= \lambda^n f(P, dx, dy, dz, dt).
\end{aligned}
\tag{20-6}
$$

As a consequence, quantity ds^n is a homogeneous function of degree n in the differentials.

Representing the function as a power series, we obtain

$$
\begin{aligned}
ds^n = {}& g_{1\ldots 1}\, dx^n + g_{2\ldots 2}\, dy^n + \cdots \\
& + g_{1\ldots 12}\, dx^{n-1}\, dy + g_{1\ldots 21}\, dx^{n-2}\, dy\, dx + \cdots
\end{aligned}
\tag{20-7}
$$

where each g may vary with point P. Since the differentials commute, we make the different g's having permutations of the same numbers as subscripts equal

$$
\begin{matrix}
g_{1\ldots 12} = g_{1\ldots 21} = \cdots \\
\vdots \qquad\qquad \vdots
\end{matrix}
\tag{20-8}
$$

without loss of generality.

When all differentials dx, dy, dz, dt are changed in sign, ds must remain unaltered, if the separation is not to have directional properties. Therefore, degree n is even $(2, 4, 6, \ldots)$. The simplest possibility is

$$n = 2. \tag{20-9}$$

Then

$$ds^2 = g_{11}\,dx^2 + g_{22}\,dy^2 + g_{33}\,dz^2 + g_{44}\,dt^2$$
$$+ 2g_{12}\,dx\,dy + 2g_{13}\,dx\,dz + 2g_{14}\,dx\,dt$$
$$+ 2g_{23}\,dy\,dz + 2g_{24}\,dy\,dt + 2g_{34}\,dz\,dt. \tag{20-10}$$

Away from singularities, the g's presumably vary slowly. In the immediate neighborhood of an event, we may then assume them to be constant. The resulting quadratic form for ds^2 can be put into diagonal form and units chosen so that

$$ds^2 \simeq dx'^2 + dy'^2 + dz'^2 - c^2\,dt'^2 \tag{20-11}$$

near the event.

If n had to be greater than 2, this reduction would not be possible; but accurate measurements show that in small regions, the Pythagorean theorem does hold— equation (20-11) is valid.

This tangential property enables us to define proper time as in special relativity. We merely transform equation (20-11) to a double-primed coordinate system in which a particle traveling between the given events is momentarily at rest:

$$ds^2 \simeq 0 + 0 + 0 - c^2\,dt''^2. \tag{20-12}$$

Since a coordinate clock moves with the particle over the given interval, we identify the time it records with the *proper time* τ of the particle; we have

$$-ds^2 = c^2\,d\tau^2 \tag{20-13}$$

as in equation (16-131).

EXAMPLE 20.1 Set up an expression for the square of the displacement in a continuum possessing spherical symmetry.

Let us start with a 3-dimensional Euclidean space at a given coordinate time t and introduce those generalizations that do not destroy spherical symmetry about the origin.

The square of a displacement in the Minkowski continuum satisfies equation (16-124). In conventional spherical coordinates, this expression becomes

$$ds^2 = dr^2 + r^2\,d\theta^2 + r^2 \sin^2\theta\,d\phi^2 - c^2\,dt^2$$
$$= dr^2 + r^2\,d\omega^2 - c^2\,dt^2$$

where $d\omega$ is the angle subtended in space by radius vectors to the initial and final positions.

Any change that preserves the assumed symmetry does not alter the surface that was initially distance r from the center; so it does not alter the element of length $r\,d\omega$ on the surface (see figure 20.1). But since it may alter each radial element and each element of time, the first term has to be replaced by $g_{11}\,dr^2$ and the last term by $g_{44}\,dt^2$ where g_{11} and g_{44} are functions of r and t. We obtain

$$ds^2 = g_{11}\,dr^2 + r^2\,d\omega^2 + g_{44}\,dt^2.$$

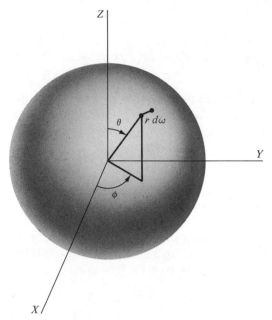

FIGURE 20.1   Element of length $r\, d\omega$ on a Euclidean sphere in a non-Euclidean space. Radial elements may be smaller or larger than the corresponding Euclidean ones, so their integral from the center is not distance r.

20.3   A spherically symmetric region

In our development, we also assume that space by itself is isotropic—a vacuum cannot act to reduce the symmetry of a region. The field of an isolated spherically symmetric star is spherically symmetric.

About the star at any given proper distance, we can therefore construct a sphere whose surface is like the surface of a sphere in ordinary 3-dimensional Euclidean space. Coordinate r is defined by setting the area of such a sphere equal to $4\pi r^2$; so the square of any element of length thereon has the form

$$r^2\, d\omega^2 = r^2\, d\theta^2 + r^2 \sin^2 \theta\, d\phi^2. \tag{20-14}$$

We find that r is not the distance of the surface from the origin in the continuum.

Time t is chosen as the time recorded by a clock at rest with respect to the star and out far enough so that the field has negligible effect on it. In this region the continuum is presumably Minkowskian and we have

$$ds^2 = dr^2 + r^2\, d\omega^2 - c^2\, dt^2 \tag{20-15}$$

where r is large.

Wherever the field is appreciable, the elements perpendicular to the corresponding Euclidean sphere may differ from dr and $c\, dt$. As discussed in example 20.1, we then have

$$ds^2 = g_{11}\, dr^2 + r^2\, d\omega^2 + g_{44}\, dt^2. \tag{20-16}$$

Since the field is assumed to be spherically symmetric, coefficients g_{11} and g_{44} do not vary with θ or ϕ; they depend only on r and t.

From equations (20-15) and (20-16), we see that

$$g_{11} = 1 \quad \text{and} \quad g_{44} = -c^2 \tag{20-17}$$

where r is large, as long as our presumption about space-time at large r's is correct.

EXAMPLE 20.2   What coordinate changes leave the line element for a spherically symmetric region invariant?

When spherical coordinates based on the center of symmetry are employed, the square of the line element is given by equation (20-16) combined with equation (20-14):

$$ds^2 = g_{11}\, dr^2 + r^2\, d\theta^2 + r^2 \sin^2 \theta\, d\phi^2 + g_{44}\, dt^2.$$

The gravitational coefficients g_{11} and g_{44} vary only with r and t, so coordinate ϕ appears only in the differential in the next-to-last term. Increasing ϕ by any constant does not alter $d\phi$. Consequently, such a change has no effect on ds.

In applications, we may make the xy plane pass through two events on the trajectory as well as through the center of symmetry. Then, all the geodesic lies in this plane; θ can be set equal to 90° and ϕ identified with ω. As in Newtonian physics, we expect the invariance with respect to change in this angle to lead to a conservation law.

In a static field, the g's do not vary with time, and coordinate t appears only in the last differential. Replacing t with $t +$ constant then leaves ds unchanged. This symmetry yields a second conservation law.

20.4   Differential equations of motion for a particle in the region

Because of the equivalence between gravitational and inertial mass, the acceleration of a body caused by gravity depends only on the local field—not on properties of the body. Other bodies in the immediate locality are accelerated in the same manner. So within a freely falling small spacecraft, each particle subject only to gravitation appears to be free of force; it behaves as a free particle in the Minkowski continuum. Such a body moves along a geodesic between given events in its life.

Einstein's theory of gravitation involves a generalization of this geometric result. It is assumed that each particle moves along the geodesic path between events no matter how far apart the points are separated. (Certain regions are, of course, inaccessible to a particle already at a certain point.) The field manifests itself as distortions in the geometry.

About a spherically symmetric star, the field is spherically symmetric. An element ds from the path of a test particle, small enough not to disturb the symmetry, then satisfies equations (20-16) and (20-14).

If we label events along the path by a parameter λ and let

$$\frac{dr}{d\lambda} = \dot{r}, \quad \frac{d\theta}{d\lambda} = \dot{\theta}, \quad \frac{d\phi}{d\lambda} = \dot{\phi}, \quad \frac{dt}{d\lambda} = \dot{t}, \tag{20-18}$$

the quadratic form becomes

$$ds^2 = (g_{11}\dot{r}^2 + r^2\dot{\theta}^2 + r^2 \sin^2 \theta \, \dot{\phi}^2 + g_{44}\dot{t}^2) \, d\lambda^2. \tag{20-19}$$

The condition that ds lie along a geodesic means that functions r, θ, ϕ, and t have to be chosen so

$$\delta \int [g_{11}\dot{r}^2 + r^2\dot{\theta}^2 + r^2 \sin^2 \theta \, \dot{\phi}^2 + g_{44}\dot{t}^2]^{1/2} \, d\lambda = \delta \int f \, d\lambda = \delta \int ds = 0, \tag{20-20}$$

or by equations (11-42) so

$$\frac{d}{d\lambda} \frac{\partial f}{\partial \dot{r}} - \frac{\partial f}{\partial r} = 0, \tag{20-21}$$

$$\frac{d}{d\lambda} \frac{\partial f}{\partial \dot{\theta}} - \frac{\partial f}{\partial \theta} = 0, \tag{20-22}$$

$$\frac{d}{d\lambda} \frac{\partial f}{\partial \dot{\phi}} - \frac{\partial f}{\partial \phi} = 0, \tag{20-23}$$

$$\frac{d}{d\lambda} \frac{\partial f}{\partial \dot{t}} - \frac{\partial f}{\partial t} = 0. \tag{20-24}$$

Differentiating f with respect to θ and $\dot{\theta}$ yields

$$\frac{\partial f}{\partial \theta} = \frac{r^2\dot{\phi}^2 \sin \theta \cos \theta}{f}, \tag{20-25}$$

$$\frac{\partial f}{\partial \dot{\theta}} = \frac{r^2\dot{\theta}}{f} = r^2 \frac{d\theta}{f \, d\lambda} = r^2 \frac{d\theta}{ds}, \tag{20-26}$$

while differentiating it with respect to ϕ and $\dot{\phi}$ gives us

$$\frac{\partial f}{\partial \phi} = 0, \tag{20-27}$$

$$\frac{\partial f}{\partial \dot{\phi}} = \frac{r^2 \sin^2 \theta \, \dot{\phi}}{f} = r^2 \sin^2 \theta \frac{d\phi}{ds}. \tag{20-28}$$

When ϕ is constant, $\dot{\phi}$ is zero and the partial derivatives in (20-25) and (20-28) are both zero. Then equation (20-23) reduces to

$$\frac{d}{d\lambda} 0 - 0 = 0 \tag{20-29}$$

while equation (20-22) becomes

$$\frac{d}{d\lambda} \left(r^2 \frac{d\theta}{ds} \right) = 0. \tag{20-30}$$

Integrating (20-30), we obtain

$$r^2 \frac{d\theta}{ds} = k_\omega. \tag{20-31}$$

Another first integral, which is not essentially different, results when θ stays at 90°. For then, equation (20-22) becomes

$$\frac{d}{d\lambda} 0 - 0 = 0 \tag{20-32}$$

while equation (20-23) reduces to

$$\frac{d}{d\lambda} \left(r^2 \frac{d\phi}{ds} \right) = 0, \tag{20-33}$$

whence

$$r^2 \frac{d\phi}{ds} = k_\omega. \tag{20-34}$$

Since the orientation of axes about the center of symmetry is arbitrary, we have an equation like (20-31) or (20-34) for motion in *any* plane passing through the center. Thus

$$r^2 \frac{d\omega}{ds} = k_\omega \tag{20-35}$$

where ω is the angle between initial and present events in the plane.

This conservation law follows from the spherical symmetry surrounding the source of the field. It is analogous to Kepler's second law, equation (2-5), because ds is $ic\,d\tau$. Thus, equation (20-35) is a generalized form of the conservation-of-angular-momentum law.

If we take the field to be *static*, gravitational coefficients g_{11} and g_{44} do not depend on t and

$$\frac{\partial f}{\partial t} = 0. \tag{20-36}$$

Then equation (20-24) integrates to give

$$\frac{\partial f}{\partial \dot{t}} = k_t. \tag{20-37}$$

Differentiating function f with respect to $\dot{t}$,

$$\frac{\partial f}{\partial \dot{t}} = \frac{g_{44}\dot{t}}{f} = g_{44} \frac{dt}{f\,d\lambda} = g_{44} \frac{dt}{ds}, \tag{20-38}$$

and using the result to eliminate $\partial f/\partial \dot{t}$ in equation (20-37) leads to

$$g_{44} \frac{dt}{ds} = k_t. \tag{20-39}$$

Equation (20-39) follows from the assumed constancy of the field with respect to coordinate time t. In section 20.5, we will interpret it as an energy conservation law.

EXAMPLE 20.3   Describe how equation (20-34) reduces to Kepler's second law.

The line element ds is related to the element of proper time $d\tau$ by equation (20-13), whence

$$ds = ic\, d\tau.$$

But in the Newtonian approximation, time is universal; proper time does not differ appreciably from coordinate time t. So

$$ds \simeq ic\, dt.$$

Also in the Newtonian approximation, the radius of the sphere of area $4\pi r^2$ is r, while ϕ is the conventional azimuthal angle. Equation (20-34) reduces to

$$r^2 \frac{d\phi}{dt} \simeq ick_\omega = \text{constant},$$

a form of equation (2-5).

20.5   Energy of the particle

Away from singularities, the immediate neighborhood of any given event in the life of a particle fits a Minkowski continuum. In the small region, components of the 4-momentum are readily recognizable. We can find how p^4 varies, relate it to integration constant k_t, and, where the field is weak, to energy E. We thus get a relationship between k_t and E.

Let us consider a particle moving between two events infinitesimally close in the spherically symmetric continuum. Along the short line traced out, we presume expressions g_{11}, r^2, and g_{44} are approximately constant. Then the mutually perpendicular elements $g_{11}^{1/2}\, dr$, $r\, d\omega$, $-ig_{44}^{1/2}\, dt$ correspond to Cartesian components dx', dy', $c\, dt'$ in the quadratic form

$$ds^2 = dx'^2 + dy'^2 - c^2\, dt'^2. \tag{20-40}$$

We may treat these elements as the components in the first column of table 17.1 were treated. Formally dividing each by the corresponding element of proper time $d\tau$ yields the appropriate component of 4-velocity. Also multiplying each by the rest mass of the test particle yields a component of 4-momentum. See table 20.1.

Into the expression for the fourth component of **p**, let us introduce formula (20-13) for $d\tau$. Let us also assume that the field is static and introduce formula (20-39) for $g_{44}(dt/ds)$:

$$p^4 = -im_0 g_{44}^{1/2} \frac{dt}{d\tau} = m_0 g_{44}^{1/2} c \frac{dt}{ds} = m_0 c \frac{k_t}{g_{44}^{1/2}}. \tag{20-41}$$

Out where the field has negligible effect on a clock, equations (20-17) hold. Furthermore, p^4 reduces to the form in table 17.2. Then

$$p^4 = m_0 c \frac{k_t}{ic} = -im_0 k_t = \frac{E}{c}, \tag{20-42}$$

TABLE 20.1 Rectangular Components of Pertinent 4-Vectors in the Spherically Symmetric Gravitational Field

Infinitesimal Displacement along the World Line	4-Velocity	4-Momentum
$g_{11}^{1/2} \, dr$	$g_{11}^{1/2} \dfrac{dr}{d\tau}$	$m_0 g_{11}^{1/2} \dfrac{dr}{d\tau}$
$r \, d\omega$	$r \dfrac{d\omega}{d\tau}$	$m_0 r \dfrac{d\omega}{d\tau}$
$-ig_{44}^{1/2} \, dt$	$-ig_{44}^{1/2} \dfrac{dt}{d\tau}$	$-im_0 g_{44}^{1/2} \dfrac{dt}{d\tau}$

whence

$$k_t = \frac{iE}{m_0 c}. \tag{20-43}$$

The derivation of equation (20-37) shows that k_t is constant along the path of a particle. If we also assume that particle energy E is conserved when the field is static, equation (20-43) remains valid as the particle penetrates the strong field.

20.6 A suitable gravitational law

Our next step is to assume that the gravitational coefficients are governed by a local law that is indistinguishable from the Newtonian form out where the field has negligible effect on a clock and on each local radial element of distance.

From equation (2-92), the acceleration of a particle falling radially in the Newtonian field of a spherically symmetric mass M is

$$\frac{d^2 r}{dt^2} = -\frac{GM}{r^2} \tag{20-44}$$

where G is the gravitational constant, t the coordinate time, and r the Euclidean distance from the center of symmetry.

In the corresponding relativistic theory, we must reinterpret the variables t and r. We assume that both possess a local, rather than an extensive, significance, since the influence of mass M on the particle is not exerted instantaneously and directly, but is propagated through the field.

Therefore, we replace t by the proper time τ for the particle. Furthermore, we let r be the radius of curvature of the Euclidean spherical surface that intersects the given event in the life of the particle. This radius can be determined from measurements made in the local region.

The law for free fall directed toward the center of symmetry thus becomes

$$\frac{d^2r}{d\tau^2} = -\frac{GM}{r^2} \tag{20-45}$$

where r is the curvature coordinate introduced in example 20.1 and section 20.3 while τ is the proper time. Since

$$-c^2\,d\tau^2 = ds^2, \tag{20-46}$$

the gravitational law can be rewritten in the form

$$\frac{d^2r}{ds^2} = \frac{GM}{c^2r^2}. \tag{20-47}$$

20.7   Gravitational coefficients for the spherically symmetric field

The relationships that we have set up can be manipulated to yield g_{11} and g_{44}.

Indeed, the square of the displacement of a test particle moving in a gravitational field spherically symmetric about the origin is given by equation (20-16). Formally dividing this by ds^2 yields

$$1 = g_{11}\left(\frac{dr}{ds}\right)^2 + r^2\left(\frac{d\omega}{ds}\right)^2 + g_{44}\left(\frac{dt}{ds}\right)^2. \tag{20-48}$$

Assuming that the particle travels along a geodesic and that the field is static gave us conservation laws (20-35) and (20-39). Introducing these converts (20-48) to

$$1 = g_{11}\left(\frac{dr}{ds}\right)^2 + \frac{k_\omega^2}{r^2} + \frac{k_t^2}{g_{44}} \tag{20-49}$$

or

$$\left(\frac{dr}{ds}\right)^2 = \frac{1}{g_{11}} - \frac{k_\omega^2}{g_{11}r^2} - \frac{k_t^2}{g_{11}g_{44}}. \tag{20-50}$$

Differentiating equation (20-50) and canceling dr/ds from each term then leads to

$$2\frac{d^2r}{ds^2} = \frac{d}{dr}\left(\frac{1}{g_{11}}\right) - k_\omega^2\frac{d}{dr}\left(\frac{1}{g_{11}r^2}\right) - k_t^2\frac{d}{dr}\left(\frac{1}{g_{11}g_{44}}\right). \tag{20-51}$$

According to equation (20-43), parameter k_t is proportional to the particle energy E. Increasing the velocity of a particle at a given position in the field increases its kinetic energy and so increases k_t along the imaginary axis.

Equation (20-47) implies that acceleration d^2r/ds^2 depends only on the position of the particle along the chosen radius. It does not depend on the particle's velocity. If this independence actually exists, as experiments indicate, the last term in equation

(20-51) must be zero. Since $-k_t^2$ is not zero, we must have

$$\frac{d}{dr}\left(\frac{1}{g_{11}g_{44}}\right) = 0 \qquad (20\text{-}52)$$

or

$$g_{11}g_{44} = \text{constant.} \qquad (20\text{-}53)$$

Out where the field has negligible effect on elements of radial distance and time, equation (20-15) is valid and

$$g_{11}g_{44} = -c^2. \qquad (20\text{-}54)$$

For this product to be fixed everywhere, the constant in (20-53) must be $-c^2$; equation (20-54) must hold generally.

Furthermore, when the test particle falls vertically, parameter k_ω is 0 and equation (20-51) reduces to

$$2\frac{d^2r}{ds^2} = \frac{d}{dr}\left(\frac{1}{g_{11}}\right). \qquad (20\text{-}55)$$

Substituting the acceleration from (20-47) into this equation

$$\frac{d}{dr}\left(\frac{1}{g_{11}}\right) = \frac{2GM}{c^2r^2} \qquad (20\text{-}56)$$

and integrating yields

$$\frac{1}{g_{11}} = A - \frac{2GM}{c^2r}. \qquad (20\text{-}57)$$

When $r = \infty$, the field vanishes and $g_{11} = 1$. Therefore

$$A = 1 \qquad (20\text{-}58)$$

and

$$g_{11} = \frac{1}{1 - (2GM)/(c^2r)} = \frac{1}{1 - r_g/r}. \qquad (20\text{-}59)$$

The quantity

$$r_g = \frac{2GM}{c^2} \qquad (20\text{-}60)$$

is called the *gravitational radius* for mass M.

Substituting (20-59) into (20-54) and solving for the remaining gravitational coefficient, we obtain

$$g_{44} = -\frac{c^2}{g_{11}} = -c^2\left(1 - \frac{r_g}{r}\right). \qquad (20\text{-}61)$$

The resulting form for the square of the displacement

$$ds^2 = \left(1 - \frac{r_g}{r}\right)^{-1} dr^2 + r^2\,d\omega^2 - c^2\left(1 - \frac{r_g}{r}\right)dt^2 \qquad (20\text{-}62)$$

was first found by Karl *Schwarzschild*.

EXAMPLE 20.4   Describe how the classical potential and kinetic energy terms come out of the Schwarzschild line element.

We assume that positions are measured relative to the source mass and that time is measured with respect to a neighboring clock at rest in the field. We also assume that the field is spherically symmetric, so the given particle or small body and the reference clock move along a world line whose element squared is

$$ds^2 = g_{11}\, dr^2 + r^2\, d\omega^2 + g_{44}\, dt^2.$$

Since differentials dr and $d\omega$ for the fixed clock are zero, the proper time τ_0 that it records obeys the formula

$$-c^2\, d\tau_0{}^2 = ds^2 = 0 + 0 + g_{44}\, dt^2,$$

whence

$$c\, d\tau_0 = -ig_{44}^{1/2}\, dt.$$

Since velocity v of the particle multiplied by the pertinent differential time equals the distance traversed, and this is given by the square root of the sum of the first two terms in the appropriate ds^2, we have

$$v^2\, d\tau_0{}^2 = g_{11}\, dr^2 + r^2\, d\omega^2.$$

Substituting the expression for square of distance traversed and for $g_{44}\, dt^2$ into ds^2 for the particle

$$ds^2 = v^2\, d\tau_0{}^2 - c^2\, d\tau_0{}^2 = -d\tau_0{}^2(c^2 - v^2)$$

and solving for the pertinent derivative, we obtain

$$\frac{d\tau_0}{ds} = \frac{1}{i(c^2 - v^2)^{1/2}} = \frac{1}{ic(1 - v^2/c^2)^{1/2}}.$$

This relationship converts the energy equation from (20-43) and (20-39)

$$E = -im_0 c k_t = m_0 c (g_{44}^{1/2}) \frac{-ig_{44}^{1/2}\, dt}{ds} = m_0 c^2 (g_{44}^{1/2}) \frac{d\tau_0}{ds}$$

to

$$E = m_0 c (-ig_{44}^{1/2}) \frac{1}{(1 - v^2/c^2)^{1/2}}$$

$$= m_0 c (-ig_{44}^{1/2}) \left(1 + \frac{1}{2}\frac{v^2}{c^2} + \cdots \right).$$

Taking the square root of equation (20-61), rearranging, expanding the binomial

$$-ig_{44}^{1/2} = c \left(1 - \frac{2GM}{c^2 r} \right)^{1/2} = c - \frac{GM}{cr} + \cdots$$

and substituting the result into the expansion of E above gives us

$$E = m_0 c^2 - \frac{GM m_0}{r} + \tfrac{1}{2} m_0 v^2 + \cdots.$$

Note that the first term is the relativistic rest energy, while the second term corresponds to the classical potential, the third term the classical kinetic energy. The higher terms involving negative powers of c are small except when v is very large or r very small.

20.8 Path of a planet in a Schwarzschild field

A body bound in a spherically symmetric Newtonian field traces out an ellipse. Correcting the field for relativistic effects, as we have done, alters the orbit so that it no longer closes exactly. Instead, the point of closest approach to the center of the field progresses slowly with time, as figure 20.2 shows.

In deriving an expression for this advance, let us begin with equation (20-16) in the derivative form

$$1 = g_{11} \left(\frac{dr}{ds} \right)^2 + r^2 \left(\frac{d\omega}{ds} \right)^2 + g_{44} \left(\frac{dt}{ds} \right)^2 \qquad (20\text{-}63)$$

with the variables chosen as in figure 20.1. We presume that the movement is all in one plane, so we can orient the axes to make $d\omega$ equal $d\phi$ and choose ϕ to be the independent variable. Introducing the expansion

$$\frac{dr}{ds} = \frac{dr}{d\phi} \frac{d\phi}{ds} \qquad (20\text{-}64)$$

and eliminating all derivatives with respect to s by means of conservation laws (20-34), (20-35), and (20-39) then leads to

$$1 = g_{11} \left(\frac{dr}{d\phi} \right)^2 \frac{k_\omega^2}{r^4} + \frac{k_\omega^2}{r^2} + \frac{k_t^2}{g_{44}}. \qquad (20\text{-}65)$$

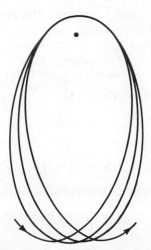

FIGURE 20.2   Orbit whose perihelion and aphelion slowly advance.

But in the Schwarzschild field

$$g_{44} = -c^2 \left(1 - \frac{r_g}{r} \right) \tag{20-66}$$

and

$$g_{11}g_{44} = -c^2. \tag{20-67}$$

These gravitational coefficients convert g_{44} times equation (20-65) to

$$-c^2 \left(1 - \frac{r_g}{r} \right) = -c^2 \left(\frac{dr}{d\phi} \right)^2 \frac{k_\omega^2}{r^4} - c^2 \left(1 - \frac{r_g}{r} \right) \frac{k_\omega^2}{r^2} + k_t^2. \tag{20-68}$$

If we divide by $-c^2$ and let

$$u = \frac{1}{r} \quad \text{and} \quad \frac{du}{d\phi} = -\frac{1}{r^2} \frac{dr}{d\phi}, \tag{20-69}$$

the differential equation becomes

$$1 - r_g u = k_\omega^2 \left(\frac{du}{d\phi} \right)^2 + k_\omega^2 u^2 - r_g k_\omega^2 u^3 - \frac{k_t^2}{c^2}. \tag{20-70}$$

To eliminate k_t, we differentiate (20-70) with respect to ϕ. Let us also cancel $2k_\omega^2\, du/d\phi$ and rearrange to get

$$\frac{d^2u}{d\phi^2} + u = -\frac{r_g}{2k_\omega^2} + \tfrac{3}{2}r_g u^2$$

$$= A + \tfrac{3}{2}r_g u^2 \tag{20-71}$$

where

$$A = -\frac{r_g}{2k_\omega^2}. \tag{20-72}$$

As long as coordinate r is much larger than the gravitational radius r_g, the last term in equation (20-71) is relatively small and we have

$$\frac{d^2u}{d\phi^2} + u \simeq A. \tag{20-73}$$

The particular integral of equation (20-73) is

$$u_1 = A \tag{20-74}$$

while the complementary function is

$$u_2 = A(B \cos \phi + C \sin \phi). \tag{20-75}$$

Choosing the axes so the maximum in the complementary function occurs at $\phi = 0$ yields

$$C = 0 \tag{20-76}$$

and

$$u = u_1 + u_2 = A(1 + B \cos \phi). \tag{20-77}$$

Since u is $1/r$, this equation has the same form as (2-4) and the second-from-last equation in example 2.1, that is,

$$\frac{1}{r} = \frac{1}{a(1 - e^2)} (1 + e \cos \phi). \tag{20-78}$$

Our approximation has led to the Kepler solution.

To determine how the orbit does not close, consider the part of the neglected term which alters the $AB \cos \phi$ contribution to the solution. Let

$$u = A + u_3 \tag{20-79}$$

and form

$$\frac{d^2 u}{d\phi^2} = \frac{d^2 u_3}{d\phi^2}, \tag{20-80}$$

$$u^2 = A^2 + 2Au_3 + u_3{}^2. \tag{20-81}$$

Then equation (20-71) becomes

$$\frac{d^2 u_3}{d\phi^2} + (1 - 3Ar_g)u_3 = \tfrac{3}{2}r_g A^2 + \tfrac{3}{2}r_g u_3{}^2. \tag{20-82}$$

As before, the last term is small. Neglecting it, we obtain the particular integral

$$u_4 = \frac{3r_g A^2}{2(1 - 3Ar_g)} \tag{20-83}$$

and the complementary function

$$u_5 = AB \cos (1 - 3Ar_g)^{1/2}\phi \simeq AB \cos (1 - \tfrac{3}{2}Ar_g)\phi. \tag{20-84}$$

For the orbits being considered, Ar_g is relatively small and the truncated expansion of the radical is valid.

Note how the correction to the coefficient of ϕ depends on the particular integral from the preceding iteration. Repeating the process would introduce a correction of order $A^2 r_g{}^2$. But terms of this order are being neglected, so we have

$$u = A[1 + B \cos (1 - \tfrac{3}{2}Ar_g)\phi] + \text{small terms}. \tag{20-85}$$

When the orbit is nearly elliptical, we may approximate A and B by comparing equations (20-77) and (20-78)

$$A \simeq \frac{1}{a(1 - \varepsilon^2)}, \tag{20-86}$$

$$B \simeq \varepsilon. \tag{20-87}$$

The change in coordinate ϕ as the planet moves from one point of closest approach to the center on to the next is designated $2\pi + \Delta\phi$. But in this cycle, the argument of the cosine in (20-85) increases by 2π; so

$$(1 - \tfrac{3}{2}Ar_g)(\phi_0 + 2\pi + \Delta\phi) = (1 - \tfrac{3}{2}Ar_g)\phi_0 + 2\pi, \qquad (20\text{-}88)$$

whence

$$-\tfrac{3}{2}Ar_g(2\pi) + (1 - \tfrac{3}{2}Ar_g)\,\Delta\phi = 0. \qquad (20\text{-}89)$$

Since $\tfrac{3}{2}Ar_g$ is small with respect to 1 (under the assumed conditions), equation (20-89) reduces to

$$\Delta\phi = 3\pi Ar_g. \qquad (20\text{-}90)$$

EXAMPLE 20.5   Calculate the gravitational radius of the sun.

Substitute the gravitational constant, mass of the sun, and the speed of light in free space

$$G = 6.673 \times 10^{-20} \text{ km}^3 \text{ kg}^{-1} \text{ sec}^{-2}$$

$$M = 1.98732 \times 10^{30} \text{ kg}$$

$$c = 2.99792 \times 10^5 \text{ km sec}^{-1}$$

into equation (20-60):

$$r_g = \frac{2(6.673 \times 10^{-20} \text{ km}^3 \text{ kg}^{-1} \text{ sec}^{-2})(1.98732 \times 10^{30} \text{ kg})}{(2.99792 \times 10^5 \text{ km sec}^{-1})^2}$$

$$= 2.951 \text{ km.}$$

At this radius, g_{44} vanishes while g_{11} increases without limit. But on approaching the center of the solar system, the surface of the sun is reached long before the radius of curvature r of the corresponding Euclidean layer falls to this value.

EXAMPLE 20.6   Calculate the relativistic contribution to motion of the perihelion of Mercury.

Substituting expression (20-86) for A into (20-90), we obtain

$$\Delta\phi = \frac{3\pi r_g}{a(1 - \varepsilon^2)}.$$

But the orbit of Mercury has

$$a = 5.788 \times 10^7 \text{ km,}$$

$$\varepsilon = 0.205627.$$

Consequently,

$$\Delta\phi = \frac{(9.42478)(2.951 \text{ km})}{(5.788 \times 10^7 \text{ km})(0.957718)} = 5.01735 \times 10^{-7} \text{ radians}$$

or

$$\Delta\phi = 0.10349 \text{ sec.}$$

In one century, the advance amounts to

$$\Delta\phi = 0.10349 \frac{365.256}{87.969} 10^2 = 42.97 \text{ sec.}$$

The observed advance in the perihelion is 5599.74 sec century^{-1}, while the amounts attributable to interaction with other planets, solar oblateness, and the general precession add up to 5557.18 sec century^{-1}. The difference, 42.56 sec century^{-1}, checks with the relativistic result 42.97 sec century^{-1}.

20.9 Deflection of a light ray by a massive body

Even photons, traveling at local speed c, are deflected by the gravitational field of a massive body. From outside the region of strong gravitational interaction, a ray appears as figure 20.3 indicates.

Because of the equivalence principle, all particles in a given small neighborhood are subject to the same gravitational acceleration. If the particles are being observed in a freely falling space ship, they appear to be independent of the gravitational field. If there are no other significant interactions present, the particles behave as free particles in a Minkowski frame moving with the ship.

With respect to this moving frame, each photon travels at speed c, and its path is made up of elements for which

$$ds' = 0. \tag{20-91}$$

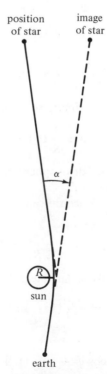

FIGURE 20.3 Shift in the apparent position of a star caused by the observed photons having passed near the sun.

A local Lorentz transformation converts the moving frame at a given instant of time t to a frame momentarily at rest in the field. The resulting continuum is tangential to the continuum that describes the field, at the event about which the photon is being studied. Since this transformation does not alter the element of displacement, we have

$$ds = 0 \qquad (20\text{-}92)$$

along a light ray in the gravitational field.

Let us consider the field to be spherically symmetric and static. The element of displacement squared is then given by equation (20-16), with g_{11} and g_{44} independent of the coordinate time t. Let us orient the axes so $d\omega$ is $d\phi$. Let us introduce the conservation laws that follow from the geodesic hypothesis and employ the gravitational coefficients for the Schwarzschild field. Parameter k_t is then eliminated by differentiation and equation (20-71) obtained, as before.

Conservation equation (20-34) can be rewritten as

$$r^2 \, d\phi = k_\omega (ic \, d\tau). \qquad (20\text{-}93)$$

Depressing the change $d\tau$ for a given $d\phi$ causes k_ω to move down the negative i axis. In the limit when ds is zero and $d\tau$ is zero, we have

$$k_\omega = -\infty i. \qquad (20\text{-}94)$$

The first term on the right of equation (20-71) then drops out and the equation reduces to

$$\frac{d^2 u}{d\phi^2} + u = \tfrac{3}{2} r_g u^2. \qquad (20\text{-}95)$$

Furthermore, the first term in approximate solution (20-77) vanishes and we have

$$u \simeq \frac{1}{R} \cos \phi \qquad (20\text{-}96)$$

where R is the curvature of the Euclidean sphere passing through the point of closest approach to the center of the field.

Proceeding by iteration, we introduce this approximate solution into the right side of equation (20-95),

$$\frac{d^2 u}{d\phi^2} + u = \frac{3}{2} \frac{r_g}{R^2} \cos^2 \phi, \qquad (20\text{-}97)$$

and solve. Thus, we obtain the particular integral

$$u = \frac{r_g}{2R^2} (1 + \sin^2 \phi) \qquad (20\text{-}98)$$

and the improved solution

$$u = \frac{1}{R} \cos \phi + \frac{r_g}{2R^2} (1 + \sin^2 \phi) \qquad (20\text{-}99)$$

which is sufficiently accurate for our purposes.

Close to the source of the ray, u is nearly zero and

$$0 \simeq \frac{1}{R} \cos \phi_1 + \frac{r_g}{2R^2} (1 + \sin^2 \phi_1); \qquad (20\text{-}100)$$

while near the earth, u is also nearly zero and

$$0 \simeq \frac{1}{R} \cos \phi_2 + \frac{r_g}{2R^2} (1 + \sin^2 \phi_2). \qquad (20\text{-}101)$$

Since the trajectory is symmetric about the point of closest approach where $\phi = 0$, we have

$$\phi_1 = \frac{\pi}{2} + \frac{\alpha}{2}. \qquad (20\text{-}102)$$

Note that α is the total angular deflection shown in figure 20.3. Because α is small,

$$\cos \phi_1 \simeq -\frac{\alpha}{2}, \qquad \sin \phi_1 \simeq 1, \qquad (20\text{-}103)$$

equation (20-100) reduces to

$$0 = -\frac{1}{R}\frac{\alpha}{2} + \frac{r_g}{2R^2}(2). \qquad (20\text{-}104)$$

Hence the total deflection is

$$\alpha = \frac{2r_g}{R}. \qquad (20\text{-}105)$$

Note that figure 20.3 is a map of the path on a plane in a Euclidean space projected by the observer on earth. In it the curvature R appears as the distance of closest approach to the center of the massive body.

EXAMPLE 20.7   How much is the ray from a star displaced when it grazes the surface of the sun?
Substitute the gravitational radius

$$r_g = 2.951 \text{ km}$$

and the conventional radius

$$R = 6.9565 \times 10^5 \text{ km}$$

of the sun into equation (20-105) and convert from radian to angular measure.

$$\alpha = \frac{2(2.951 \text{ km})(2.06265 \times 10^5 \text{ sec radian}^{-1})}{6.9565 \times 10^5 \text{ km}}$$

$$= 1.750 \text{ sec}.$$

For rays in general, the displacement is inversely proportional to the curvature R at the point of closest approach. The effect is appreciable only for stars that appear close to the sun. The displacement is then measured by comparing an eclipse plate with another taken, say, six months later when the light does not pass through appreciably curved space about the sun. Fair agreement with theory has been obtained.

20.10   Dilation of time by a gravitational field

The effect of a massive body on the flow of time at any neighboring point can be determined from the rate of a known periodic process, located at the desired point and observed from the outside. The results agree with those deduced from the form for the displacement ds.

Let us consider a mechanism that undergoes cyclic changes in empty space at frequency v. In proper time $d\tau$, it then goes through $v\, d\tau$ cycle. Bringing the mechanism into the gravitational field in a space ship and then letting the ship fall freely leaves the mechanism in a state that appears to a person on board like that in free space, because of the equivalence principle. The system still goes through $v\, d\tau$ cycle in proper time $d\tau$. But this procedure can be arranged so the mechanism is left momentarily at rest at the desired position in the field. We presume that it then behaves as the same mechanism would if fixed there, over the small interval $d\tau$.

When the cyclic mechanism is at rest in the gravitational field, we have

$$dx = dy = dz = 0 \qquad (20\text{-}106)$$

in (20-10) and the equation reduces to

$$ds^2 = 0 + g_{44}\, dt^2. \qquad (20\text{-}107)$$

The proper time τ is related to the displacement by equation (20-13)

$$ds^2 = -c^2\, d\tau^2. \qquad (20\text{-}108)$$

Eliminating ds from (20-107) and (20-108) and taking the square root yields

$$c\, d\tau = -ig_{44}^{1/2}\, dt. \qquad (20\text{-}109)$$

Let us suppose that the mechanism emits a signal at its frequency v. This would be received outside the field over period Δt with the altered frequency v_A. Since the number of oscillations does not change in the transmission, we have

$$v\, \Delta\tau = v_A\, \Delta t, \qquad (20\text{-}110)$$

whence

$$\frac{v_A}{v} = \frac{\Delta\tau}{\Delta t} = \frac{d\tau}{dt}. \qquad (20\text{-}111)$$

Formula (20-109) relating $d\tau$ to dt reduces this frequency ratio to

$$\frac{v_A}{v} = \frac{-ig_{44}^{1/2}}{c}, \tag{20-112}$$

whence

$$\frac{\Delta v}{v} = \frac{v - v_A}{v} = \frac{c + ig_{44}^{1/2}}{c}. \tag{20-113}$$

With the Schwarzschild metric, we obtain

$$\frac{\Delta v}{v} = \frac{c - c(1 - r_g/r)^{1/2}}{c} = 1 - 1 + \frac{r_g}{2r} + \cdots$$

$$\simeq \frac{r_g}{2r}. \tag{20-114}$$

EXAMPLE 20.8   How much is the frequency of a photon altered when the particle rises distance h above the earth's surface?

Let v_0 be the frequency of the radiation at the surface, v_h the frequency received at height h, and v_A the frequency that would be observed outside the field. Then from equation (20-112), construct

$$\frac{v_0}{v_A} - \frac{v_h}{v_A} = \frac{c}{(-ig_{44}^{1/2})_0} - \frac{c}{(-ig_{44}^{1/2})_h}$$

with 0 and h labeling the positions in the field.

Introduce the Schwarzschild metric,

$$\frac{v_0 - v_h}{v_A} = \left(1 - \frac{2GM}{c^2 r_0}\right)^{-1/2} - \left(1 - \frac{2GM}{c^2 r_h}\right)^{-1/2},$$

expand the binomials,

$$\frac{\Delta v}{v} \simeq 1 + \frac{GM}{c^2 r_0} - 1 - \frac{GM}{c^2 r_h} = \frac{GM}{c^2 r_0 r_h}(r_h - r_0),$$

and identify the magnitude of g from formula (2-94):

$$\Delta v \simeq \frac{vg}{c^2} h.$$

For emission at the earth's surface, the coefficient of vh is given by

$$\frac{g}{c^2} = \frac{980 \text{ cm sec}^{-2}}{(3.00 \times 10^{10} \text{ cm sec}^{-1})^2} = 1.09 \times 10^{-18} \text{ sec cm}^{-1}.$$

With the technique developed by Rudolf L. Mössbauer, the shift in frequency of gamma radiation has been measured and the theoretical equation confirmed.

20.11   Rate of free fall

The compression of space and the dilation of time about a massive body reduce the velocity at which a particle or small body appears to fall. Indeed, if a star were to collapse within its gravitational radius $2GM/c^2$, a falling test particle would appear to come to rest at this radius.

In considering the effect, let us assume that the gravitational field is spherically symmetric and that the particle falls vertically. An element ds along its world line is then given by

$$ds^2 = g_{11} \, dr^2 + g_{44} \, dt^2, \tag{20-115}$$

whence

$$\frac{dr}{dt} = \frac{\left[-g_{44} + \left(\frac{ds}{dt} \right)^2 \right]^{1/2}}{g_{11}^{1/2}} = \frac{\left[-\frac{g_{44}}{c^2} + \frac{1}{c^2} \left(\frac{ds}{dt} \right)^2 \right]^{1/2}}{g_{11}^{1/2}} \, c. \tag{20-116}$$

From energy conservation law (20-39), we have

$$\frac{1}{c} \frac{ds}{dt} = \frac{g_{44}}{ck_t}, \tag{20-117}$$

so

$$\frac{dr}{dt} = \frac{\left[-\frac{g_{44}}{c^2} \left(1 - \frac{g_{44}}{k_t^2} \right) \right]^{1/2}}{g_{11}^{1/2}} \, c. \tag{20-118}$$

Since dr/dt is zero at the radius of curvature r_0 of the Euclidean layer from which the fall began, we set

$$k_t^2 = g_{44} \qquad \text{at} \qquad r = r_0. \tag{20-119}$$

Now the Schwarzschild line element squared is

$$ds^2 = \left(1 - \frac{r_g}{r} \right)^{-1} dr^2 + r^2 \, d\omega^2 - c^2 \left(1 - \frac{r_g}{r} \right) dt^2. \tag{20-120}$$

Consequently, we set

$$g_{44} = -c^2 \left(1 - \frac{r_g}{r_0} \right) \tag{20-121}$$

at $r = r_0$, while otherwise

$$g_{44} = -c^2 \left(1 - \frac{r_g}{r} \right) \tag{20-122}$$

and

$$\frac{1}{g_{11}^{1/2}} = \left(1 - \frac{r_g}{r} \right)^{1/2}. \tag{20-123}$$

These expressions convert equation (20-118) to

$$\frac{dr}{dt} = \left(1 - \frac{r_g}{r}\right)^{1/2} \left[\left(1 - \frac{r_g}{r}\right)\left(1 - \frac{1 - r_g/r}{1 - r_g/r_0}\right)\right]^{1/2} c$$

$$= \left(1 - \frac{r_g}{r}\right)\left(1 - \frac{1 - r_g/r}{1 - r_g/r_0}\right)^{1/2} c. \tag{20-124}$$

Note that as r approaches r_g, rate dr/dt decreases to zero.

If a star is large enough, it tends to collapse after it uses up its nuclear fuel. If the system contains little rotational energy, particles at its surface then move approximately as freely falling particles. A distant observer will see the apparent radius approach r_g very quickly. By equation (20-124), dr/dt will then tend to zero and the process will seem to stop.

EXAMPLE 20.9   What rates exhibit singular behavior in the Schwarzschild field at $r = r_g$?

According to table 20.1 and example 20.4, an observer in a spherically symmetric field records the radial displacement

$$dl = g_{11}^{1/2} \, dr$$

and the transverse displacement

$$dw = r \, d\omega$$

during the time

$$d\tau_0 = \frac{-ig_{44}^{1/2}}{c} \, dt.$$

If the field also satisfies equations (20-59) and (20-61), approaching r_g from above causes g_{11} to increase without limit and causes g_{44} to go to zero.

Then dl/dr, the radial displacement per unit change in coordinate r, and $dt/d\tau_0$, the rate of increase of coordinate time with local time τ_0, both grow without limit as a particle approaches the surface on which r equals the gravitational radius.

20.12   Completing the Schwarzschild field

When the spherically symmetric source is small enough, the gravitational field extends within the sphere at r_g. To link the inside region to the field without, coordinates that do not become singular at the gravitational radius have to be introduced. A suitable choice will be considered here.

For convenience, we first let

$$ct = T \tag{20-125}$$

and rewrite equation (20-62) as

$$ds^2 = \left(1 - \frac{r_g}{r}\right)^{-1} dr^2 + r^2 \, d\omega^2 - \left(1 - \frac{r_g}{r}\right) dT^2. \tag{20-126}$$

Wherever $r > r_g$, we then set

$$u = \pm \left(\frac{r}{r_g} - 1\right)^{1/2} \exp \frac{r}{2r_g} \cosh \frac{T}{2r_g}, \qquad (20\text{-}127)$$

$$v = \pm \left(\frac{r}{r_g} - 1\right)^{1/2} \exp \frac{r}{2r_g} \sinh \frac{T}{2r_g}. \qquad (20\text{-}128)$$

Squaring each of these and subtracting v^2 from u^2 yields

$$u^2 - v^2 = \left(\frac{r}{r_g} - 1\right) \exp \frac{r}{r_g}, \qquad (20\text{-}129)$$

whence

$$2u\,du - 2v\,dv = \frac{r}{r_g} \exp \frac{r}{r_g} \frac{dr}{r_g} \qquad (20\text{-}130)$$

or

$$\frac{2r_g^{\,2}}{r} \exp\left(-\frac{r}{r_g}\right)(u\,du - v\,dv) = dr. \qquad (20\text{-}131)$$

Dividing equation (20-128) by (20-127) yields

$$\frac{v}{u} = \tanh \frac{T}{2r_g} \qquad (20\text{-}132)$$

or

$$2r_g \tanh^{-1} \frac{v}{u} = T, \qquad (20\text{-}133)$$

whence

$$2r_g \frac{u\,dv - v\,du}{u^2 - v^2} = dT. \qquad (20\text{-}134)$$

Introducing equation (20-129) and reducing now gives us

$$\frac{2r_g^{\,2}/r}{1 - r_g/r} \exp\left(-\frac{r}{r_g}\right)(u\,dv - v\,du) = dT. \qquad (20\text{-}135)$$

Substituting these expressions for dr and dT into equation (20-126) and reducing with equation (20-129),

$$\begin{aligned}
ds^2 &= \left(1 - \frac{r_g}{r}\right)^{-1} \frac{4r_g^{\,4}}{r^2} \exp\left(-\frac{2r}{r_g}\right)(u^2\,du^2 - 2uv\,du\,dv + v^2\,dv^2) \\
&\quad + r^2\,d\omega^2 - \left(1 - \frac{r_g}{r}\right)\left(1 - \frac{r_g}{r}\right)^{-2} \frac{4r_g^{\,4}}{r^2} \exp\left(-\frac{2r}{r_g}\right) \\
&\quad \times (u^2\,dv^2 - 2uv\,du\,dv + v^2\,du^2) \\
&= \frac{4r_g^{\,3}}{r} \exp\left(-\frac{r}{r_g}\right)\left[\left(1 - \frac{r_g}{r}\right)^{-1} \exp\left(-\frac{r}{r_g}\right)\frac{r_g}{r}(u^2 - v^2)\right] \\
&\quad \times (du^2 - dv^2) + r^2\,d\omega^2 \\
&= \frac{4r_g^{\,3}}{r} \exp\left(-\frac{r}{r_g}\right)(du^2 - dv^2) + r^2\,d\omega^2, \qquad (20\text{-}136)
\end{aligned}$$

lead to a line element squared having the form

$$ds^2 = A(du^2 - dv^2) + r^2\,d\omega^2 \tag{20-137}$$

where

$$A = \frac{4r_g^{\,3}}{r}\exp\left(-\frac{r}{r_g}\right). \tag{20-138}$$

Wherever only u changes, ds^2 is positive; whenever only v changes, ds^2 is negative. Consequently, the coordinate u is spacelike while v is timelike.

To extend this result within the singular sphere, we take

$$u = \pm\left(1 - \frac{r}{r_g}\right)^{1/2}\exp\frac{r}{2r_g}\sinh\frac{T}{2r_g}, \tag{20-139}$$

$$v = \pm\left(1 - \frac{r}{r_g}\right)^{1/2}\exp\frac{r}{2r_g}\cosh\frac{T}{2r_g}, \tag{20-140}$$

wherever $r < r_g$. Thus defined, coordinates u and v remain real. Furthermore, equations (20-129) and (20-134) still hold and the Schwarzschild line element squared (20-126) is converted to the same form

$$ds^2 = A(du^2 - dv^2) + r^2\,d\omega^2 \tag{20-141}$$

obtained before.

Equations (20-127), (20-128), (20-139), and (20-140) map the (r, T) half plane of figure 20.4 into the (u, v) full plane of figure 20.5. In the latter, the two representations of the outside region are connected by a smooth throat. When the source is a point mass of fixed magnitude, this throat is constant.

By equation (20-129), events at a given r lie on a hyperbola in the u, v space (see figure 20.5). By equation (20-132) and the equation

$$\frac{u}{v} = \tanh\frac{T}{2r_g} \tag{20-142}$$

derived from (20-139) and (20-140), events at a given time t lie on either of two straight lines passing through the origin (see figure 20.6).

Along a light ray, we have

$$ds = 0. \tag{20-143}$$

If the ray travels along a radius, that is, if

$$d\omega = 0, \tag{20-144}$$

then equation (20-137) yields

$$du = \pm dv. \tag{20-145}$$

In the (u, v) plane these geodesics are inclined at 45° everywhere, as in figure 20.7. Timelike geodesics have a greater slope, while spacelike geodesics have a smaller slope at any given point (see figures 20.8 and 20.9).

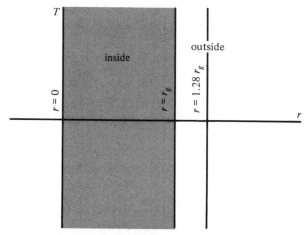

FIGURE 20.4   Regions inside and outside the Schwarzschild singularity plotted in the (r, T) plane.

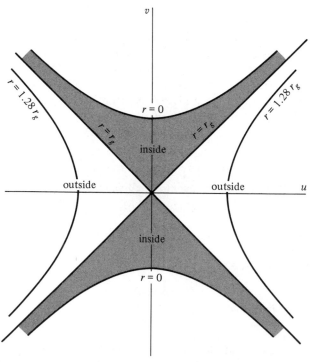

FIGURE 20.5   Corresponding regions plotted in the (u, v) plane.

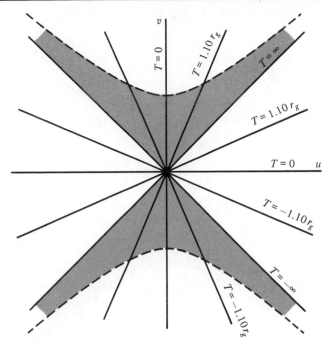

FIGURE 20.6   Equitime lines in the (u, v) plane.

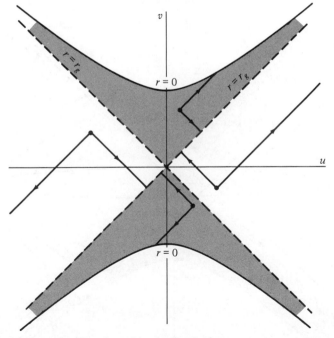

FIGURE 20.7   Radial light rays plotted in the (u, v) plane.

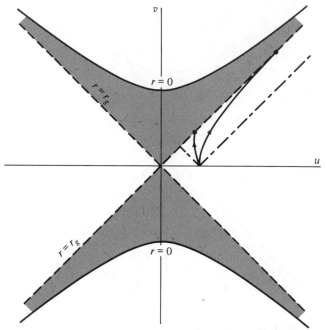

Figure 20.8   Sketch of timelike geodesics starting from a point on the *u* axis. Note how these lie within a light cone based on the initial point.

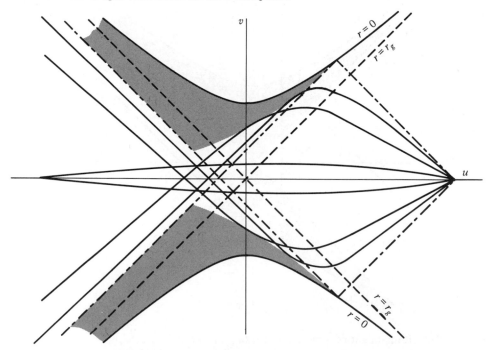

Figure 20.9   Spacelike geodesics extending through the Schwarzschild throat.

Because $T = \infty$ on the line $r = r_g$, light beginning within the Schwarzschild singularity never escapes. Furthermore, light beginning in the right outside region (right quadrant) cannot reach the left outside region (left quadrant).

Any particle traveling radially toward the sphere at r_g reaches it at $T = \infty$, after an infinite amount of coordinate time. Therefore, no particle can pass from the right outside region to the left one. No particle, initially inside, can get out.

At the surface of a nonrotating star, particles fall as shown in figure 20.8. Coordinate time T increases, following equation (20-132). At the end of the tracks shown, T is infinite. There is no time left for the particle to go farther. In the strong gravitational field, much space has been compressed and an enormous dilation of time has been introduced.

20.13   Electric charge

John A. Wheeler and his coworkers have also represented electric charges in a geometric manner.

Their fundamental idea is that lines of force can get trapped in the topology of space. To see how this may happen, suppose that everywhere throughout a given Minkowski space, virtual electric lines of force exist, canceling one another as in figure 20.10. Now if sufficient energy is concentrated over a certain region, space is compressed there as about a point mass.

But in the infinite region at the surface corresponding to $r = r_g$, neighboring lines can become separated. A part of the energy may then act to increase the separation with the result pictured in the sequence, figures 20.11, 20.12.

From the outside, only the region to the singular surface can be observed. Thus, the handle linking the spot where the lines of force enter to that where they emerge is not seen. We say that negative charge exists where the lines enter and that positive charge exists where they emerge.

This theory does not explain the quantization of charge—the fact that charge occurs in multiples of a unit charge. But quantization is what distinguishes modern from classical theoretical physics. Further discussion of it would take us beyond the scope of this book.

FIGURE 20.10 Virtual electric lines of force which cancel each other in free space.

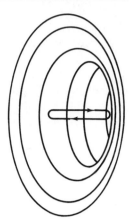

Figure 20.11a

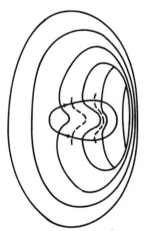

Figure 20.11b

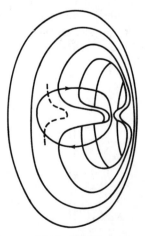

Figure 20.11c

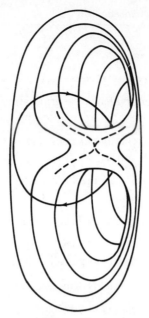

Figure 20.11d

FIGURE 20.11   Schematic description of the formation of electric charge by separation of adjacent compensating lines of force near a varying singular surface.

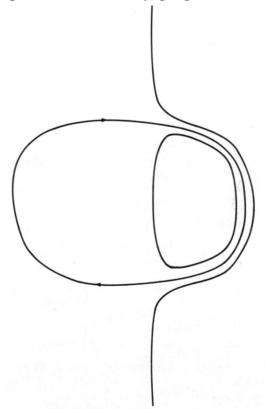

FIGURE 20.12   Lines of electric force trapped in the handle formed as in figure 20.11.

DISCUSSION QUESTIONS

20.1 What is a space-time continuum?

20.2 Explain how the equivalence of gravitational and inertial mass implies that a gravitational field can be represented by the geometry of space-time.

20.3 How can we set up rectangular coordinates for describing the motion of a particle in a general continuum?

20.4 What is ds? Why do we expect ds^n to be a function of position and co-ordinate differentials?

20.5 Why is ds^n a homogeneous function of the rectangular differentials? Why do we take n to be 2?

20.6 How is the proper time for a particle obtained?

20.7 Explain how we formulate ds^2 for a spherically symmetric continuum.

20.8 How can we determine radius of curvature r of the Euclidean spherical surface from local measurements?

20.9 Compare a geodesic in the Minkowski continuum with its variations.

20.10 Justify the hypothesis that a particle subject to gravitational attractions moves along a geodesic.

20.11 Derive the conservation equation that is a generalization of Kepler's second law. Why is the movement in a plane containing the source of the field?

20.12 Discuss the derivation of the condition

$$g_{44}\,\frac{dt}{ds} = k_t.$$

20.13 Explain how k_t is related to the energy of the particle.

20.14 Why is

$$g_{11}g_{44} = -c^2$$

throughout the spherically symmetric continuum?

20.15 Consider how the Newtonian law for acceleration of a small planet can be recast in suitable relativistic form.

20.16 From one of these forms derive the formula for the square of the Schwarzschild line element.

20.17 How is the classical orbit for a particle moving about a massive spherically symmetric body altered by relativistic considerations?

20.18 Explain how light rays are bent by a spherically symmetric gravitational field.

20.19 In what way can we compare the rate of a process within and without a field? How much does a Schwarzschild field slow down a process?

20.20 Derive an expression for the velocity at which a body falls radially towards a point mass.

20.21 In what sense is there an infinite amount of space available about the spherical singularity at r_g? What happens to the proper time of a particle in this region?

20.22 Explain how the continuum within r_g is linked to that without.

20.23 What is electric charge?

20.24 Why are the lines of force unobservable where they pass through the handle (or wormhole)?

PROBLEMS

20.1 Use the geodesic hypothesis to obtain a conservation-of-linear-momentum law for a particle moving in the continuum described by

$$ds^2 = dx^2 + dy^2 + dz^2 - c^2 \, dt^2.$$

20.2 Transform the spherically symmetric line element squared to rectangular coordinates.

20.3 Relate the angular velocity of a small planet traversing a circular orbit in a Schwarzschild field to the radius of curvature r of the circle.

20.4 From the Schwarzschild line element, show that the local velocity of light, as it grazes the sun, is c.

20.5 According to quantum theory, the energy of a photon is proportional to its frequency, while the relativistic equation for its mass is $E = mc^2$. Using these relationships and the classical formula for gravitational potential energy, calculate how much the frequency of a gamma particle shifts as it rises height l in a uniform gravitational field.

20.6 Show that equation (20-124) reduces to a Newtonian formula when $r \gg r_g$.

20.7 Find the gravitational potential energy of a particle at rest on the singular sphere at r_g.

20.8 Calculate the time t for a photon to travel radially inward from r_0 to r.

20.9 In a weak gravitational field, the Pythagorean theorem holds and

$$g_{jk} = \delta_{jk}, \quad \begin{matrix} j = 1, 2, 3 \\ k = 1, 2, 3 \end{matrix}$$

in equation (20-10). Furthermore, we have

$$g_{44} = -c^2.$$

Show that, for the rate of a clock to be unaffected when it is transported slowly from $(x_1, 0, 0)$ at t_1 to $(x_2, 0, 0)$ at t_2, one must also have

$$g_{14} = 0.$$

20.10 Set up the square of the line element for a continuum possessing cylindrical symmetry.

20.11 Obtain the conservation laws for a particle moving in the continuum for which

$$ds^2 = g_{11} (dr^2 + r^2 \, d\theta^2 + r^2 \sin^2 \theta \, d\phi^2) + g_{44} \, dt^2$$

if g_{11} and g_{44} are functions of r alone.

20.12 Transform the Schwarzschild line element to the form in problem 20.11 (the isotropic form) by the substitution

$$r = \left(1 + \frac{r_g}{4r'}\right)^2 r'.$$

20.13 Show that the local velocity of light, anywhere in a gravitational field, is c.

20.14 From the semimajor axis of Mercury's orbit 5.788×10^7 km, its period $(87.969)(8.61641 \times 10^4)$ sec, the eccentricity 0.205627, and the speed of light, calculate the relativistic advance of its perihelion.

20.15 Show that the local speed of a freely falling body approaches c as it nears the gravitational radius r_g.

REFERENCES

BOOKS

Adler, R., Bazin, M., and **Schiffer, M.** *Introduction to General Relativity*, pp. 164–209. McGraw-Hill Book Company, New York, 1965.

DeWitt, C., and **DeWitt, B. S.** (eds.). *Relativity, Groups, and Topology*, pp. 165–520. Gordon and Breach, Science Publishers, Inc., New York, 1964.

Graves, J. C. *The Conceptual Foundations of Contemporary Relativity Theory*, pp. 119–319. The MIT Press, Cambridge, Mass., 1971.

Wheeler, J. A. *Geometrodynamics*, pp. 9–87. Academic Press, Inc., New York, 1962.

Zeldovich, Ya. B., and **Novikov, I. D.** (Translated by E. Arlock, edited by K. S. Thorne and W. D. Arnett) *Stars and Relativity*, vol. 1, pp. 64–128. University of Chicago Press, Chicago, 1971.

ARTICLES

Boyer, R. H. Some Uses of Hyperbolic Velocity Space. *Am. J. Phys.*, **33,** 910 (1965).

Braginskii, V. B., and **Rudenko, V. N.** Relativistic Gravitational Experiments. *Soviet Phys. Uspekhi,* **13,** 165 (1970).

Broucke, R. Linearization of Schwarzschild's Line Element: Application to the Clock Paradox. *Am. J. Phys.,* **39,** 1461 (1971).

Broucke, R. Lorentz Transformations in the Presence of a Uniform Gravitational Field. *Am. J. Phys.,* **39,** 1465 (1971).

Chandrasekhar, S. On the "Derivation" of Einstein's Field Equations. *Am. J. Phys.,* **40,** 224 (1972).

Fronsdal, C. Completion and Embedding of the Schwarzschild Solution. *Phys. Rev.,* **116,** 778 (1959).

Fuller, R. W., and **Wheeler, J. A.** Causality and Multiply Connected Space-Time. *Phys. Rev.,* **128,** 919 (1962).

Hafele, J. C. Relativistic Time for Terrestrial Circumnavigations. *Am. J. Phys.,* **40,** 81 (1972).

Harvey, A. L. Brief Review of Lorentz-Covariant Scalar Theories of Gravitation. *Am. J. Phys.,* **33,** 449 (1965).

Hernandez, W. C., Jr. Static, Axially Symmetric, Interior Solution in General Relativity. *Phys. Rev.,* **153,** 1359 (1967).

Holstein, B. R., and **Swift, A. R.** The Relativity Twins in Free Fall. *Am. J. Phys.,* **40,** 746 (1972).

Israel, W. Gravitational Collapse and Causality. *Phys. Rev.,* **153,** 1388 (1967).

Kruskal, M. D. Maximal Extension of Schwarzschild Metric. *Phys. Rev.,* **119,** 1743 (1960).

Marsh, L. M. Relativistic Accelerated Systems. *Am. J. Phys.,* **33,** 934 (1965).

Parker, L. Motion in a Schwarzschild Field. I. Precession of a Moving Gyroscope. *Am. J. Phys.,* **37,** 309 (1969).

Rindler, W. Counterexample to the Tangherlini Argument. *Am. J. Phys.,* **37,** 72 (1969).

Tangherlini, F. R. An Introduction to the General Theory of Relativity. *Suppl. Nuovo Cimento,* **20,** 1 (1961).

Tangherlini, F. R. Postulational Approach to Schwarzschild's Exterior Solution with Application to a Class of Interior Solutions. *Nuovo Cimento,* **25,** 1081 (1962).

Wheeler, J. A. Problems on the Frontiers between General Relativity and Differential Geometry. *Revs. Modern Phys.,* **34,** 873 (1962).

Zel'dovich, Ya. B., and **Novikov, I. D.** Relativistic Astrophysics. I. *Soviet Phys. Uspekhi,* **7,** 763 (1965); II. *Soviet Phys. Uspekhi,* **8,** 522 (1966).

Appendix / Characters and Bases for Irreducible Representations of Symmetry Groups

A.1 Tabulated expressions

The boldface symbol in the upper left corner of each table identifies the symmetry group, while each letter under it identifies an irreducible representation of the group. The numbers in a row are the characters; the vectors and dyads are bases for the representation. The heading for each column of numbers labels the class (by a typical operation) and states the number of covering operations therein.

The symbols for the operations are explained in sections 4.3 and 4.4. How a group may be identified is discussed in section 7.4. The concept of class is introduced in section 4.8, that of reducibility in section 8.5.

Vectors $\mathbf{i}$, $\mathbf{j}$, and $\mathbf{k}$ are the conventional Cartesian base vectors drawn from the center of the given symmetric system. The pertinent orientations of axes may be determined from the characters listed. The sums containing ω and its powers are not generally normalized to one.

In the first column, letters A, B, C, D are used for one-dimensional representations, E for two-dimensional irreducible representations, F for three-dimensional irreducible representations, G for four-dimensional irreducible representations, H for five-dimensional irreducible representations.

One-dimensional representations that are neither antisymmetric nor symmetric with respect to the smallest rotation, or to a smaller rotation-reflection when present, are labeled C and D; the remaining one-dimensional representations that are antisymmetric to the smallest rotation, or to a smaller rotation-reflection, when present, when the smallest rotation is even, are labeled B (except in $\mathbf{O}$ and $\mathbf{O}_h$); and the others are designated A. The subscript a on a C or D representation indicates that the operand is multiplied by $\exp(\pm 2\pi i a/n)$ when the system is subject to C_n. In the $\mathbf{D}_2$ and $\mathbf{D}_{2h}$ groups, subscript 1 indicates the B representation that is symmetric with respect to rotation about the z axis, 2 indicates the one that is symmetric about the y axis, 3 represents the one that is symmetric about the x axis. Otherwise, subscripts 1 and 2 on A or B designate the representations that are symmetric or antisymmetric with respect to a C_2 perpendicular to the principal axis, or when such an axis is missing, to a vertical plane of symmetry.

Primes and double primes indicate symmetry or antisymmetry with respect to σ_h. The subscripts g and u indicate symmetry or antisymmetry to inversion. The tables indicate how the subscripts are used on the other letters.

C_1	E			
A	1	$\mathbf{i, j, k}$	$\mathbf{j \times k, k \times i, i \times j}$	$\mathbf{ii, jj, kk, jk, kj, ki, ik, ij, ji}$

$C_s \equiv S_1$	E	σ_h			
A'	1	1	$\mathbf{i, j}$	$\mathbf{i \times j}$	$\mathbf{ii, jj, kk, ij, ji}$
A''	1	-1	$\mathbf{k}$	$\mathbf{j \times k, k \times i}$	$\mathbf{jk, kj, ki, ik}$

$C_i \equiv S_2$	E	i			
A_g	1	1		$\mathbf{j \times k, k \times i, i \times j}$	$\mathbf{ii, jj, kk, jk, kj, ki, ik, ij, ji}$
A_u	1	-1	$\mathbf{i, j, k}$		

S_4	E	S_4	C_2	$S_4^{\,3}$			
A	1	1	1	1		$\mathbf{i \times j}$	$\dfrac{1}{\sqrt{2}}(\mathbf{ii+jj}),\ \mathbf{kk},\ \dfrac{1}{\sqrt{2}}(\mathbf{ij-ji})$
B	1	-1	1	-1	$\mathbf{k}$		$\dfrac{1}{\sqrt{2}}(\mathbf{ii-jj}),\ \dfrac{1}{\sqrt{2}}(\mathbf{ij+ji})$
C	1	i	-1	$-i$	$\dfrac{1}{\sqrt{2}}(\mathbf{i}-i\mathbf{j})$	$\dfrac{1}{\sqrt{2}}(\mathbf{j \times k} + i\mathbf{k \times i})$	$\dfrac{1}{\sqrt{2}}\mathbf{k}(\mathbf{i}+i\mathbf{j}),\ \dfrac{1}{\sqrt{2}}(\mathbf{i}+i\mathbf{j})\mathbf{k}$
D	1	$-i$	-1	i	$\dfrac{1}{\sqrt{2}}(\mathbf{i}+i\mathbf{j})$	$\dfrac{1}{\sqrt{2}}(\mathbf{j \times k} - i\mathbf{k \times i})$	$\dfrac{1}{\sqrt{2}}\mathbf{k}(\mathbf{i}-i\mathbf{j}),\ \dfrac{1}{\sqrt{2}}(\mathbf{i}-i\mathbf{j})\mathbf{k}$

S_6	E	C_3	C_3^2	i	S_6^5	S_6		
A_g	1	1	1	1	1	1	$\mathbf{i \times j}$	$\frac{1}{\sqrt{2}}(\mathbf{ii}+\mathbf{jj}),\ \mathbf{kk},\ \frac{1}{\sqrt{2}}(\mathbf{ij}-\mathbf{ji})$
C_g	1	ω	ω^2	1	ω	ω^2	$\frac{1}{\sqrt{3}}(\mathbf{e_1}'+\omega^2\mathbf{e_2}'+\omega\mathbf{e_3}')$	$\frac{1}{\sqrt{3}}(\mathbf{e_1 e_1}+\omega^2\mathbf{e_2 e_2}+\omega\mathbf{e_3 e_3})$, $\frac{1}{\sqrt{3}}\mathbf{k}(\mathbf{e_1}+\omega^2\mathbf{e_2}+\omega\mathbf{e_3})$, $\frac{1}{\sqrt{3}}(\mathbf{e_1}+\omega^2\mathbf{e_2}+\omega\mathbf{e_3})\mathbf{k}$
D_g	1	ω^2	ω	1	ω^2	ω	$\frac{1}{\sqrt{3}}(\mathbf{e_1}'+\omega\mathbf{e_2}'+\omega^2\mathbf{e_3}')$	$\frac{1}{\sqrt{3}}(\mathbf{e_1 e_1}+\omega\mathbf{e_2 e_2}+\omega^2\mathbf{e_3 e_3})$, $\frac{1}{\sqrt{3}}\mathbf{k}(\mathbf{e_1}+\omega\mathbf{e_2}+\omega^2\mathbf{e_3})$, $\frac{1}{\sqrt{3}}(\mathbf{e_1}+\omega\mathbf{e_2}+\omega^2\mathbf{e_3})\mathbf{k}$
A_u	1	1	1	-1	-1	-1	$\mathbf{k}$	
C_u	1	ω	ω^2	-1	$-\omega$	$-\omega^2$	$\frac{1}{\sqrt{3}}(\mathbf{e_1}+\omega^2\mathbf{e_2}+\omega\mathbf{e_3})$	
D_u	1	ω^2	ω	-1	$-\omega^2$	$-\omega$	$\frac{1}{\sqrt{3}}(\mathbf{e_1}+\omega\mathbf{e_2}+\omega^2\mathbf{e_3})$	

$\omega = \exp(2\pi i/3)$

$\mathbf{e_1 = i},\quad \mathbf{e_2} = C_3\mathbf{i},\quad \mathbf{e_3} = C_3^2\mathbf{i},\quad \mathbf{e_1}' = \mathbf{j \times k},\quad \mathbf{e_2}' = C_3\mathbf{e_1}',\quad \mathbf{e_3}' = C_3^2\mathbf{e_1}'$

C_2	E	C_2			
A	1	1	$\mathbf{k}$	$\mathbf{i}\times\mathbf{j}$	$\mathbf{ii},\mathbf{jj},\mathbf{kk},\mathbf{ij},\mathbf{ji}$
B	1	-1	$\mathbf{i},\mathbf{j}$	$\mathbf{j}\times\mathbf{k},\mathbf{k}\times\mathbf{i}$	$\mathbf{ki},\mathbf{ik},\mathbf{jk},\mathbf{kj}$

C_3	E	C_3	C_3^2			
A	1	1	1	$\mathbf{k}$	$\mathbf{i}\times\mathbf{j}$	$\dfrac{1}{\sqrt{2}}(\mathbf{ii}+\mathbf{jj}),\ \mathbf{kk},\ \dfrac{1}{\sqrt{2}}(\mathbf{ij}-\mathbf{ji})$
C	1	ω	ω^2	$\dfrac{1}{\sqrt{3}}(\mathbf{e}_1+\omega^2\mathbf{e}_2+\omega\mathbf{e}_3)$	$\dfrac{1}{\sqrt{3}}(\mathbf{e}_1'+\omega^2\mathbf{e}_2'+\omega\mathbf{e}_3')$	$\dfrac{1}{\sqrt{3}}(\mathbf{e}_1\mathbf{e}_1+\omega^2\mathbf{e}_2\mathbf{e}_2+\omega\mathbf{e}_3\mathbf{e}_3)$
						$\dfrac{1}{\sqrt{3}}\mathbf{k}(\mathbf{e}_1+\omega^2\mathbf{e}_2+\omega\mathbf{e}_3)$
						$\dfrac{1}{\sqrt{3}}(\mathbf{e}_1+\omega^2\mathbf{e}_2+\omega\mathbf{e}_3)\mathbf{k}$
D	1	ω^2	ω	$\dfrac{1}{\sqrt{3}}(\mathbf{e}_1+\omega\mathbf{e}_2+\omega^2\mathbf{e}_3)$	$\dfrac{1}{\sqrt{3}}(\mathbf{e}_1'+\omega\mathbf{e}_2'+\omega^2\mathbf{e}_3')$	$\dfrac{1}{\sqrt{3}}(\mathbf{e}_1\mathbf{e}_1+\omega\mathbf{e}_2\mathbf{e}_2+\omega^2\mathbf{e}_3\mathbf{e}_3)$
						$\dfrac{1}{\sqrt{3}}\mathbf{k}(\mathbf{e}_1+\omega\mathbf{e}_2+\omega^2\mathbf{e}_3)$
						$\dfrac{1}{\sqrt{3}}(\mathbf{e}_1+\omega\mathbf{e}_2+\omega^2\mathbf{e}_3)\mathbf{k}$

$\omega = \exp(2\pi i/3)$

$\mathbf{e}_1=\mathbf{i},\quad \mathbf{e}_2=C_3\mathbf{i},\quad \mathbf{e}_3=C_3^2\mathbf{i},\quad \mathbf{e}_1'=\mathbf{j}\times\mathbf{k},\quad \mathbf{e}_2'=C_3\mathbf{e}_1',\quad \mathbf{e}_3'=C_3^2\mathbf{e}_1'$

C_4	E	C_4	C_2	$C_4{}^3$
A	1	1	1	1
B	1	-1	1	-1
C	1	i	-1	$-i$
D	1	$-i$	-1	i

C_5	E	C_5	$C_5{}^2$	$C_5{}^3$	$C_5{}^4$
A	1	1	1	1	1
C_1	1	ω	ω^2	ω^3	ω^4
D_1	1	ω^4	ω^3	ω^2	ω
C_2	1	ω^2	ω^4	ω	ω^3
D_2	1	ω^3	ω	ω^4	ω^2

$\omega = \exp(2\pi i/5)$

C_4			
A	$\mathbf{k}$	$\mathbf{i} \times \mathbf{j}$	$\frac{1}{\sqrt{2}}(\mathbf{ii} + \mathbf{jj})$, $\mathbf{kk}$, $\frac{1}{\sqrt{2}}(\mathbf{ij} - \mathbf{ji})$
B			$\frac{1}{\sqrt{2}}(\mathbf{ii} - \mathbf{jj})$, $\frac{1}{\sqrt{2}}(\mathbf{ij} + \mathbf{ji})$
C	$\frac{1}{\sqrt{2}}(\mathbf{i} - i\mathbf{j})$	$\frac{1}{\sqrt{2}}(\mathbf{j} \times \mathbf{k} - i\mathbf{k} \times \mathbf{i})$	$\frac{1}{\sqrt{2}}\mathbf{k}(\mathbf{i} - i\mathbf{j})$, $\frac{1}{\sqrt{2}}(\mathbf{i} - i\mathbf{j})\mathbf{k}$
D	$\frac{1}{\sqrt{2}}(\mathbf{i} + i\mathbf{j})$	$\frac{1}{\sqrt{2}}(\mathbf{j} \times \mathbf{k} + i\mathbf{k} \times \mathbf{i})$	$\frac{1}{\sqrt{2}}\mathbf{k}(\mathbf{i} + i\mathbf{j})$, $\frac{1}{\sqrt{2}}(\mathbf{i} + i\mathbf{j})\mathbf{k}$

C_5		
A	**k**	$\frac{1}{\sqrt{2}}(\mathbf{ii}+\mathbf{jj})$, **kk**, $\frac{1}{\sqrt{2}}(\mathbf{ij}-\mathbf{ji})$
	$\mathbf{i}\times\mathbf{j}$	
C_1	$\frac{1}{\sqrt{5}}(\mathbf{e}_1+\omega^4\mathbf{e}_2+\omega^3\mathbf{e}_3+\omega^2\mathbf{e}_4+\omega\mathbf{e}_5)=\mathbf{f}_1$	$\mathbf{kf}_1, \mathbf{f}_1\mathbf{k}$
D_1	$\frac{1}{\sqrt{5}}(\mathbf{e}_1+\omega\mathbf{e}_2+\omega^2\mathbf{e}_3+\omega^3\mathbf{e}_4+\omega^4\mathbf{e}_5)=\mathbf{f}_2$	$\mathbf{kf}_2, \mathbf{f}_2\mathbf{k}$
C_2		$\frac{1}{\sqrt{5}}(\mathbf{e}_1\mathbf{e}_1+\omega^3\mathbf{e}_2\mathbf{e}_2+\omega\mathbf{e}_3\mathbf{e}_3+\omega^4\mathbf{e}_4\mathbf{e}_4+\omega^2\mathbf{e}_5\mathbf{e}_5)$
D_2		$\frac{1}{\sqrt{5}}(\mathbf{e}_1\mathbf{e}_1+\omega^2\mathbf{e}_2\mathbf{e}_2+\omega^4\mathbf{e}_3\mathbf{e}_3+\omega\mathbf{e}_4\mathbf{e}_4+\omega^3\mathbf{e}_5\mathbf{e}_5)$

$$\mathbf{e}_1 = \mathbf{i} \text{ or } \mathbf{e}_1 = \mathbf{j}\times\mathbf{k}, \quad \mathbf{e}_2 = C_5\mathbf{e}_1, \quad \mathbf{e}_3 = C_5{}^2\mathbf{e}_1, \quad \mathbf{e}_4 = C_5{}^3\mathbf{e}_1, \quad \mathbf{e}_5 = C_5{}^4\mathbf{e}_1$$

C_6	E	C_6	C_3	C_2	C_3^2	C_6^5	$\mathbf{k}$	$\mathbf{i} \times \mathbf{j}$	
A	1	1	1	1	1	1	$\mathbf{k}$	$\mathbf{i} \times \mathbf{j}$	$\dfrac{1}{\sqrt{2}}(\mathbf{ii}+\mathbf{jj}), \mathbf{kk}, \dfrac{1}{\sqrt{2}}(\mathbf{ij}-\mathbf{ji})$
B	1	-1	1	-1	1	-1			
C_1	1	$-\omega^2$	ω	-1	ω^2	$-\omega$	$\dfrac{1}{\sqrt{3}}(\mathbf{e}_1 + \omega^2\mathbf{e}_2 + \omega\mathbf{e}_3)$	$\dfrac{1}{\sqrt{3}}(\mathbf{e}_1' + \omega^2\mathbf{e}_2' + \omega\mathbf{e}_3')$	$\dfrac{1}{\sqrt{3}}\mathbf{k}(\mathbf{e}_1 + \omega^2\mathbf{e}_2 + \omega\mathbf{e}_3)$
									$\dfrac{1}{\sqrt{3}}(\mathbf{e}_1 + \omega^2\mathbf{e}_2 + \omega\mathbf{e}_3)\mathbf{k}$
D_1	1	$-\omega$	ω^2	-1	ω	$-\omega^2$	$\dfrac{1}{\sqrt{3}}(\mathbf{e}_1 + \omega\mathbf{e}_2 + \omega^2\mathbf{e}_3)$	$\dfrac{1}{\sqrt{3}}(\mathbf{e}_1' + \omega\mathbf{e}_2' + \omega^2\mathbf{e}_3')$	$\dfrac{1}{\sqrt{3}}\mathbf{k}(\mathbf{e}_1 + \omega\mathbf{e}_2 + \omega^2\mathbf{e}_3)$
									$\dfrac{1}{\sqrt{3}}(\mathbf{e}_1 + \omega\mathbf{e}_2 + \omega^2\mathbf{e}_3)\mathbf{k}$
C_2	1	ω	ω^2	1	ω	ω^2			$\dfrac{1}{\sqrt{3}}(\mathbf{e}_1\mathbf{e}_1 + \omega\mathbf{e}_2\mathbf{e}_2 + \omega^2\mathbf{e}_3\mathbf{e}_3)$
D_2	1	ω^2	ω	1	ω^2	ω			$\dfrac{1}{\sqrt{3}}(\mathbf{e}_1\mathbf{e}_1 + \omega^2\mathbf{e}_2\mathbf{e}_2 + \omega\mathbf{e}_3\mathbf{e}_3)$

$$\mathbf{e}_1 = \mathbf{i}, \quad \mathbf{e}_2 = C_3\mathbf{i}, \quad \mathbf{e}_3 = C_3^2\mathbf{i}, \quad \mathbf{e}_1' = \mathbf{j} \times \mathbf{k}, \quad \mathbf{e}_2' = C_3\mathbf{e}_1', \quad \mathbf{e}_3' = C_3^2\mathbf{e}_1'$$

$$\omega = \exp(2\pi i/3)$$

C_{2v}	E	C_2	$\sigma_v(xz)$	$\sigma_v'(yz)$			
A_1	1	1	1	1	**k**		**ii, jj, kk**
A_2	1	1	-1	-1		**i × j**	**ij, ji**
B_1	1	-1	1	-1	**i**	**k × i**	**ki, ik**
B_2	1	-1	-1	1	**j**	**j × k**	**jk, kj**

C_{3v}	E	$2C_3$	$3\sigma_v$			
A_1	1	1	1	**k**		$\dfrac{1}{\sqrt{2}}(\mathbf{ii + jj})$, **kk**
A_2	1	1	-1		**i × j**	$\dfrac{1}{\sqrt{2}}(\mathbf{ij - ji})$
E	2	-1	0	**(i, j)**	**(j × k, k × i)**	$\left[\dfrac{1}{\sqrt{2}}(\mathbf{ii - jj}), \dfrac{1}{\sqrt{2}}(\mathbf{ij + ji})\right]$, **(ik, jk), (ki, kj)**

C_{4v}	E	$2C_4$	C_2	$2\sigma_v$	$2\sigma_d$			
A_1	1	1	1	1	1	**k**		$\frac{1}{\sqrt{2}}(\mathbf{ii}+\mathbf{jj})$, **kk**
A_2	1	1	1	−1	−1		$\mathbf{i}\times\mathbf{j}$	$\frac{1}{\sqrt{2}}(\mathbf{ij}-\mathbf{ji})$
B_1	1	−1	1	1	−1			$\frac{1}{\sqrt{2}}(\mathbf{ii}-\mathbf{jj})$
B_2	1	−1	1	−1	1			$\frac{1}{\sqrt{2}}(\mathbf{ij}+\mathbf{ji})$
E	2	0	−2	0	0	$(\mathbf{i},\mathbf{j})$	$(\mathbf{j}\times\mathbf{k}, \mathbf{k}\times\mathbf{i})$	$(\mathbf{ik},\mathbf{jk}), (\mathbf{ki},\mathbf{kj})$

C_{5v}	E	$2C_5$	$2C_5{}^2$	$5\sigma_v$			
A_1	1	1	1	1	**k**		$\frac{1}{\sqrt{2}}(\mathbf{ii}+\mathbf{jj})$, **kk**
A_2	1	1	1	−1		$\mathbf{i}\times\mathbf{j}$	$\frac{1}{\sqrt{2}}(\mathbf{ij}-\mathbf{ji})$
E_1	2	$2\cos 72°$	$2\cos 144°$	0	$(\mathbf{i},\mathbf{j})$	$(\mathbf{j}\times\mathbf{k}, \mathbf{k}\times\mathbf{i})$	$(\mathbf{ik},\mathbf{jk}), (\mathbf{ki},\mathbf{kj})$
E_2	2	$2\cos 144°$	$2\cos 72°$	0			$\left[\frac{1}{\sqrt{2}}(\mathbf{ii}-\mathbf{jj}), \frac{1}{\sqrt{2}}(\mathbf{ij}+\mathbf{ji})\right]$

C_{6v}	E	$2C_6$	$2C_3$	C_2	$3\sigma_v$	$3\sigma_d$		
A_1	1	1	1	1	1	1	$\mathbf{k}$	$\frac{1}{\sqrt{2}}(\mathbf{ii}+\mathbf{jj})$, $\mathbf{kk}$
A_2	1	1	1	1	-1	-1	$\mathbf{i}\times\mathbf{j}$	$\frac{1}{\sqrt{2}}(\mathbf{ij}-\mathbf{ji})$
B_1	1	-1	1	-1	1	-1		
B_2	1	-1	1	-1	-1	1		
E_1	2	1	-1	-2	0	0	$(\mathbf{i},\mathbf{j})$ $(\mathbf{j}\times\mathbf{k},\mathbf{k}\times\mathbf{i})$	$(\mathbf{ik},\mathbf{jk})$, $(\mathbf{ki},\mathbf{kj})$
E_2	2	-1	-1	2	0	0		$\left[\frac{1}{\sqrt{2}}(\mathbf{ii}-\mathbf{jj}), \frac{1}{\sqrt{2}}(\mathbf{ij}+\mathbf{ji})\right]$

C_{2h}	E	C_2	i	σ_h		
A_g	1	1	1	1	$\mathbf{i}\times\mathbf{j}$	$\mathbf{ii},\mathbf{jj},\mathbf{kk},\mathbf{ij},\mathbf{ji}$
B_g	1	-1	1	-1	$\mathbf{j}\times\mathbf{k},\mathbf{k}\times\mathbf{i}$	$\mathbf{ki},\mathbf{ik},\mathbf{jk},\mathbf{kj}$
A_u	1	1	-1	-1	$\mathbf{k}$	
B_u	1	-1	-1	1	$\mathbf{i},\mathbf{j}$	

C_{3h}	E	C_3	$C_3{}^2$	σ_h	S_3	$S_3{}^5$			
A'	1	1	1	1	1	1		$\mathbf{i \times j}$	$\dfrac{1}{\sqrt{2}}(\mathbf{ii + jj})$, $\mathbf{kk}$
C'	1	ω	ω^2	1	ω	ω^2	$\dfrac{1}{\sqrt{3}}(\mathbf{e}_1 + \omega^2\mathbf{e}_2 + \omega\mathbf{e}_3)$		$\dfrac{1}{\sqrt{3}}(\mathbf{e}_1\mathbf{e}_1 + \omega^2\mathbf{e}_2\mathbf{e}_2 + \omega\mathbf{e}_3\mathbf{e}_3)$
D'	1	ω^2	ω	1	ω^2	ω	$\dfrac{1}{\sqrt{3}}(\mathbf{e}_1 + \omega\mathbf{e}_2 + \omega^2\mathbf{e}_3)$		$\dfrac{1}{\sqrt{3}}(\mathbf{e}_1\mathbf{e}_1 + \omega\mathbf{e}_2\mathbf{e}_2 + \omega^2\mathbf{e}_3\mathbf{e}_3)$
A''	1	1	1	-1	-1	-1	$\mathbf{k}$		
C''	1	ω	ω^2	-1	$-\omega$	$-\omega^2$		$\dfrac{1}{\sqrt{3}}(\mathbf{e}_1' + \omega^2\mathbf{e}_2' + \omega\mathbf{e}_3')$,	$\dfrac{1}{\sqrt{3}}\mathbf{k}(\mathbf{e}_1 + \omega^2\mathbf{e}_2 + \omega\mathbf{e}_3)$, $\dfrac{1}{\sqrt{3}}(\mathbf{e}_1 + \omega^2\mathbf{e}_2 + \omega\mathbf{e}_3)\mathbf{k}$
D''	1	ω^2	ω	-1	$-\omega^2$	$-\omega$		$\dfrac{1}{\sqrt{3}}(\mathbf{e}_1' + \omega\mathbf{e}_2' + \omega^2\mathbf{e}_3')$,	$\dfrac{1}{\sqrt{3}}\mathbf{k}(\mathbf{e}_1 + \omega\mathbf{e}_2 + \omega^2\mathbf{e}_3)$, $\dfrac{1}{\sqrt{3}}(\mathbf{e}_1 + \omega\mathbf{e}_2 + \omega^2\mathbf{e}_3)\mathbf{k}$

$\omega = \exp(2\pi i/3)$

$\mathbf{e}_1 = \mathbf{i}, \quad \mathbf{e}_2 = C_3\mathbf{i}, \quad \mathbf{e}_3 = C_3{}^2\mathbf{i}, \quad \mathbf{e}_1' = \mathbf{j \times k}, \quad \mathbf{e}_2' = C_3\mathbf{e}_1', \quad \mathbf{e}_3' = C_3{}^2\mathbf{e}_1'$

C_{4h}	E	C_4	C_2	$C_4^{\,3}$	i	$S_4^{\,3}$	σ_h	S_4		
A_g	1	1	1	1	1	1	1	1	$\mathbf{i}\times\mathbf{j}$	$\frac{1}{\sqrt{2}}(\mathbf{ii}+\mathbf{jj})$, $\mathbf{kk}$, $\frac{1}{\sqrt{2}}(\mathbf{ij}-\mathbf{ji})$
B_g	1	-1	1	-1	1	-1	1	-1		$\frac{1}{\sqrt{2}}(\mathbf{ii}-\mathbf{jj})$, $\frac{1}{\sqrt{2}}(\mathbf{ij}+\mathbf{ji})$
C_g	1	i	-1	$-i$	1	i	-1	$-i$	$\frac{1}{\sqrt{2}}(\mathbf{j}\times\mathbf{k}-i\mathbf{k}\times\mathbf{i})$	$\frac{1}{\sqrt{2}}(\mathbf{ik}-i\mathbf{jk})$, $\frac{1}{\sqrt{2}}(\mathbf{ki}-i\mathbf{kj})$
D_g	1	$-i$	-1	i	1	$-i$	-1	i	$\frac{1}{\sqrt{2}}(\mathbf{j}\times\mathbf{k}+i\mathbf{k}\times\mathbf{i})$	$\frac{1}{\sqrt{2}}(\mathbf{ik}+i\mathbf{jk})$, $\frac{1}{\sqrt{2}}(\mathbf{ki}+i\mathbf{kj})$
A_u	1	1	1	1	-1	-1	-1	-1	$\mathbf{k}$	
B_u	1	-1	1	-1	-1	1	-1	1		
C_u	1	i	-1	$-i$	-1	$-i$	1	i	$\frac{1}{\sqrt{2}}(\mathbf{i}-i\mathbf{j})$	
D_u	1	$-i$	-1	i	-1	i	1	$-i$	$\frac{1}{\sqrt{2}}(\mathbf{i}+i\mathbf{j})$	

C_{5h}	E	C_5	$C_5{}^2$	$C_5{}^3$	$C_5{}^4$	σ_h	S_5	$S_5{}^7$	$S_5{}^3$	$S_5{}^9$
A'	1	1	1	1	1	1	1	1	1	1
$C_1{}'$	1	ω	ω^2	ω^3	ω^4	1	ω	ω^2	ω^3	ω^4
$D_1{}'$	1	ω^4	ω^3	ω^2	ω	1	ω^4	ω^3	ω^2	ω
$C_2{}'$	1	ω^2	ω^4	ω	ω^3	1	ω^2	ω^4	ω	ω^3
$D_2{}'$	1	ω^3	ω	ω^4	ω^2	1	ω^3	ω	ω^4	ω^2
A''	1	1	1	1	1	-1	-1	-1	-1	-1
$C_1{}''$	1	ω	ω^2	ω^3	ω^4	-1	$-\omega$	$-\omega^2$	$-\omega^3$	$-\omega^4$
$D_1{}''$	1	ω^4	ω^3	ω^2	ω	-1	$-\omega^4$	$-\omega^3$	$-\omega^2$	$-\omega$
$C_2{}''$	1	ω^2	ω^4	ω	ω^3	-1	$-\omega^2$	$-\omega^4$	$-\omega$	$-\omega^3$
$D_2{}''$	1	ω^3	ω	ω^4	ω^2	-1	$-\omega^3$	$-\omega$	$-\omega^4$	$-\omega^2$

$$\omega = \exp(2\pi i/5)$$

C_{5h}		$\mathbf{i} \times \mathbf{j}$	
A'			$\dfrac{1}{\sqrt{2}}(\mathbf{ii}+\mathbf{jj}),\ \mathbf{kk},\ \dfrac{1}{\sqrt{2}}(\mathbf{ij}-\mathbf{ji})$
$C_1{}'$		$\dfrac{1}{\sqrt{5}}(\mathbf{e}_1 + \omega^4\mathbf{e}_2 + \omega^3\mathbf{e}_3 + \omega^2\mathbf{e}_4 + \omega\mathbf{e}_5) = \mathbf{f}_1$	
$D_1{}'$		$\dfrac{1}{\sqrt{5}}(\mathbf{e}_1 + \omega\mathbf{e}_2 + \omega^2\mathbf{e}_3 + \omega^3\mathbf{e}_4 + \omega^4\mathbf{e}_5) = \mathbf{f}_2$	
$C_2{}'$			$\dfrac{1}{\sqrt{5}}(\mathbf{e}_1\mathbf{e}_1 + \omega^3\mathbf{e}_2\mathbf{e}_2 + \omega\mathbf{e}_3\mathbf{e}_3 + \omega^4\mathbf{e}_4\mathbf{e}_4 + \omega^2\mathbf{e}_5\mathbf{e}_5)$
$D_2{}'$			$\dfrac{1}{\sqrt{5}}(\mathbf{e}_1\mathbf{e}_1 + \omega^2\mathbf{e}_2\mathbf{e}_2 + \omega^4\mathbf{e}_3\mathbf{e}_3 + \omega\mathbf{e}_4\mathbf{e}_4 + \omega^3\mathbf{e}_5\mathbf{e}_5)$
A''	$\mathbf{k}$		
$C_1{}''$		$\dfrac{1}{\sqrt{5}}(\mathbf{e}_1' + \omega^4\mathbf{e}_2' + \omega^3\mathbf{e}_3' + \omega^2\mathbf{e}_4' + \omega\mathbf{e}_5')$	$\mathbf{kf}_1,\ \mathbf{f}_1\mathbf{k}$
$D_1{}''$		$\dfrac{1}{\sqrt{5}}(\mathbf{e}_1' + \omega\mathbf{e}_2' + \omega^2\mathbf{e}_3' + \omega^3\mathbf{e}_4' + \omega^4\mathbf{e}_5')$	$\mathbf{kf}_2,\ \mathbf{f}_2\mathbf{k}$
$C_2{}''$			
$D_2{}''$			

$$\mathbf{e}_1 = \mathbf{i}, \quad \mathbf{e}_2 = C_5\mathbf{e}_1, \quad \mathbf{e}_3 = C_5{}^2\mathbf{e}_1, \quad \mathbf{e}_4 = C_5{}^3\mathbf{e}_1, \quad \mathbf{e}_5 = C_5{}^4\mathbf{e}_1$$

$$\mathbf{e}_1' = \mathbf{j} \times \mathbf{k}, \quad \mathbf{e}_2' = C_5\mathbf{e}_1', \quad \mathbf{e}_3' = C_5{}^2\mathbf{e}_1', \quad \mathbf{e}_4' = C_5{}^3\mathbf{e}_1', \quad \mathbf{e}_5' = C_5{}^4\mathbf{e}_1'$$

C_{6h}	E	C_6	C_3	C_2	C_3^2	C_6^5	i	S_3^5	S_6^5	σ_h	S_6	S_3
A_g	1	1	1	1	1	1	1	1	1	1	1	1
B_g	1	-1	1	-1	1	-1	1	-1	1	-1	1	-1
C_{1g}	1	$-\omega^2$	ω	-1	ω^2	$-\omega$	1	$-\omega^2$	ω	-1	ω^2	$-\omega$
D_{1g}	1	$-\omega$	ω^2	-1	ω	$-\omega^2$	1	$-\omega$	ω^2	-1	ω	$-\omega^2$
C_{2g}	1	ω	ω^2	1	ω	ω^2	1	ω	ω^2	1	ω	ω^2
D_{2g}	1	ω^2	ω	1	ω^2	ω	1	ω^2	ω	1	ω^2	ω
A_u	1	1	1	1	1	1	-1	-1	-1	-1	-1	-1
B_u	1	-1	1	-1	1	-1	-1	1	-1	1	-1	1
C_{1u}	1	$-\omega^2$	ω	-1	ω^2	$-\omega$	-1	ω^2	$-\omega$	1	$-\omega^2$	ω
D_{1u}	1	$-\omega$	ω^2	-1	ω	$-\omega^2$	-1	ω	$-\omega^2$	1	$-\omega$	ω^2
C_{2u}	1	ω	ω^2	1	ω	ω^2	-1	$-\omega$	$-\omega^2$	-1	$-\omega$	$-\omega^2$
D_{2u}	1	ω^2	ω	1	ω^2	ω	-1	$-\omega^2$	$-\omega$	-1	$-\omega^2$	$-\omega$

$$\omega = \exp(2\pi i/3)$$

C_{6h}			
A_g		$\mathbf{i}\times\mathbf{j}$	$\frac{1}{\sqrt{2}}(\mathbf{ii}+\mathbf{jj}),\ \mathbf{kk},\ \frac{1}{\sqrt{2}}(\mathbf{ij}-\mathbf{ji})$
B_g			
C_{1g}		$\frac{1}{\sqrt{3}}(\mathbf{e}_1'+\omega^2\mathbf{e}_2'+\omega\mathbf{e}_3')$	$\frac{1}{\sqrt{3}}\mathbf{k}(\mathbf{e}_1+\omega^2\mathbf{e}_2+\omega\mathbf{e}_3)$ $\frac{1}{\sqrt{3}}(\mathbf{e}_1+\omega^2\mathbf{e}_2+\omega\mathbf{e}_3)\mathbf{k}$
D_{1g}		$\frac{1}{\sqrt{3}}(\mathbf{e}_1'+\omega\mathbf{e}_2'+\omega^2\mathbf{e}_3')$	$\frac{1}{\sqrt{3}}\mathbf{k}(\mathbf{e}_1+\omega\mathbf{e}_2+\omega^2\mathbf{e}_3)\mathbf{k}$ $\frac{1}{\sqrt{3}}(\mathbf{e}_1+\omega\mathbf{e}_2+\omega^2\mathbf{e}_3)\mathbf{k}$
C_{2g}			$\frac{1}{\sqrt{3}}(\mathbf{e}_1\mathbf{e}_1+\omega\mathbf{e}_2\mathbf{e}_2+\omega^2\mathbf{e}_3\mathbf{e}_3)$
D_{2g}			$\frac{1}{\sqrt{3}}(\mathbf{e}_1\mathbf{e}_1+\omega^2\mathbf{e}_2\mathbf{e}_2+\omega\mathbf{e}_3\mathbf{e}_3)$
A_u	$\mathbf{k}$		
B_u			
C_{1u}	$\frac{1}{\sqrt{3}}(\mathbf{e}_1+\omega^2\mathbf{e}_2+\omega\mathbf{e}_3)$		
D_{1u}	$\frac{1}{\sqrt{3}}(\mathbf{e}_1+\omega\mathbf{e}_2+\omega^2\mathbf{e}_3)$		
C_{2u}			
D_{2u}			

$\mathbf{e}_1=\mathbf{i},\quad \mathbf{e}_2=C_3\mathbf{i},\quad \mathbf{e}_3=C_3^2\mathbf{i},\quad \mathbf{e}_1'=\mathbf{j}\times\mathbf{k},\quad \mathbf{e}_2'=C_3\mathbf{e}_1',\quad \mathbf{e}_3'=C_3^2\mathbf{e}_1'$

$\mathbf{D}_2$	E	$C_2(z)$	$C_2(y)$	$C_2(x)$			
A	1	1	1	1			**ii, jj, kk**
B_1	1	1	-1	-1	**k**	$\mathbf{i \times j}$	**ij, ji**
B_2	1	-1	1	-1	**j**	$\mathbf{k \times i}$	**ki, ik**
B_3	1	-1	-1	1	**i**	$\mathbf{j \times k}$	**jk, kj**

$\mathbf{D}_3$	E	$2C_3$	$3C_2$			
A_1	1	1	1			$\dfrac{1}{\sqrt{2}}\,(\mathbf{ii + jj}),\ \mathbf{kk}$
A_2	1	1	-1	**k**	$\mathbf{i \times j}$	$\dfrac{1}{\sqrt{2}}\,(\mathbf{ij - ji})$
E	2	-1	0	$(\mathbf{i, j})$	$(\mathbf{j \times k, k \times i})$	$\left[\dfrac{1}{\sqrt{2}}\,(\mathbf{ii - jj}),\ \dfrac{1}{\sqrt{2}}\,(\mathbf{ij + ji})\right]$, $(\mathbf{ik, jk}),\ (\mathbf{ki, kj})$

$\mathbf{D_4}$	E	$2C_4$	C_2	$2C_2{}'$	$2C_2{}''$			
A_1	1	1	1	1	1			$\frac{1}{\sqrt{2}}(\mathbf{ii} + \mathbf{jj})$, $\mathbf{kk}$
A_2	1	1	1	-1	-1	$\mathbf{k}$	$\mathbf{i} \times \mathbf{j}$	$\frac{1}{\sqrt{2}}(\mathbf{ij} - \mathbf{ji})$
B_1	1	-1	1	1	-1			$\frac{1}{\sqrt{2}}(\mathbf{ii} - \mathbf{jj})$
B_2	1	-1	1	-1	1			$\frac{1}{\sqrt{2}}(\mathbf{ij} + \mathbf{ji})$
E	2	0	-2	0	0	$(\mathbf{i}, \mathbf{j})$	$(\mathbf{j} \times \mathbf{k}, \mathbf{k} \times \mathbf{i})$	$(\mathbf{ik}, \mathbf{jk})$, $(\mathbf{ki}, \mathbf{kj})$

$\mathbf{D_5}$	E	$2C_5$	$2C_5{}^2$	$5C_2$			
A_1	1	1	1	1			$\frac{1}{\sqrt{2}}(\mathbf{ii} + \mathbf{jj})$, $\mathbf{kk}$
A_2	1	1	1	-1	$\mathbf{k}$	$\mathbf{i} \times \mathbf{j}$	$\frac{1}{\sqrt{2}}(\mathbf{ij} - \mathbf{ji})$
E_1	2	$2 \cos 72°$	$2 \cos 144°$	0	$(\mathbf{i}, \mathbf{j})$	$(\mathbf{j} \times \mathbf{k}, \mathbf{k} \times \mathbf{i})$	$(\mathbf{ik}, \mathbf{jk})$, $(\mathbf{ki}, \mathbf{kj})$
E_2	2	$2 \cos 144°$	$2 \cos 72°$	0			$\left[\frac{1}{\sqrt{2}}(\mathbf{ii} - \mathbf{jj}), \frac{1}{\sqrt{2}}(\mathbf{ij} + \mathbf{ji})\right]$

$\mathbf{D}_6$	E	$2C_6$	$2C_3$	C_2	$3C_2'$	$3C_2''$			
A_1	1	1	1	1	1	1			$\frac{1}{\sqrt{2}}(ii+jj),\ kk$
A_2	1	1	1	1	-1	-1	k	$i \times j$	$\frac{1}{\sqrt{2}}(ij-ji)$
B_1	1	-1	1	-1	1	-1			
B_2	1	-1	1	-1	-1	1			
E_1	2	1	-1	-2	0	0	(i, j)	$(j \times k,\ k \times i)$	$(ik, jk),\ (ki, kj)$
E_2	2	-1	-1	2	0	0			$\left[\frac{1}{\sqrt{2}}(ii-jj),\ \frac{1}{\sqrt{2}}(ij+ji)\right]$

$\mathbf{D}_{2d}$	E	$2S_4$	C_2	$2C_2'$	$2\sigma_d$			
A_1	1	1	1	1	1			$\frac{1}{\sqrt{2}}(ii+jj),\ kk$
A_2	1	1	1	-1	-1		$i \times j$	$\frac{1}{\sqrt{2}}(ij-ji)$
B_1	1	-1	1	1	-1			$\frac{1}{\sqrt{2}}(ii-jj)$
B_2	1	-1	1	-1	1	k		$\frac{1}{\sqrt{2}}(ij+ji)$
E	2	0	-2	0	0	(i, j)	$(j \times k,\ k \times i)$	$(ik, jk),\ (ki, kj)$

$\mathbf{D}_{3d}$	E	$2C_3$	$3C_2$	i	$2S_6$	$3\sigma_d$		
A_{1g}	1	1	1	1	1	1		$\frac{1}{\sqrt{2}}(\mathbf{ii}+\mathbf{jj})$, $\mathbf{kk}$
A_{2g}	1	1	−1	1	1	−1	$\mathbf{i}\times\mathbf{j}$	$\frac{1}{\sqrt{2}}(\mathbf{ij}-\mathbf{ji})$
E_g	2	−1	0	2	−1	0	$(\mathbf{j}\times\mathbf{k}, \mathbf{k}\times\mathbf{i})$	$\left[\frac{1}{\sqrt{2}}(\mathbf{ii}-\mathbf{jj}), \frac{1}{\sqrt{2}}(\mathbf{ij}+\mathbf{ji})\right]$, $(\mathbf{ik}, \mathbf{jk}), (\mathbf{ki}, \mathbf{kj})$
A_{1u}	1	1	1	−1	−1	−1		
A_{2u}	1	1	−1	−1	−1	1	$\mathbf{k}$	
E_u	2	−1	0	−2	1	0	$(\mathbf{i}, \mathbf{j})$	

$\mathbf{D}_{4d}$	E	$2S_8$	$2C_4$	$2S_8{}^3$	C_2	$4C_2'$	$4\sigma_d$		
A_1	1	1	1	1	1	1	1		$\frac{1}{\sqrt{2}}(\mathbf{ii}+\mathbf{jj})$, $\mathbf{kk}$
A_2	1	1	1	1	1	−1	−1	$\mathbf{i}\times\mathbf{j}$	$\frac{1}{\sqrt{2}}(\mathbf{ij}-\mathbf{ji})$
B_1	1	−1	1	−1	1	1	−1		
B_2	1	−1	1	−1	1	−1	1	$\mathbf{k}$	
E_1	2	$\sqrt{2}$	0	$-\sqrt{2}$	−2	0	0	$(\mathbf{i}, \mathbf{j})$	
E_2	2	0	−2	0	2	0	0		
E_3	2	$-\sqrt{2}$	0	$\sqrt{2}$	−2	0	0	$(\mathbf{j}\times\mathbf{k}, \mathbf{k}\times\mathbf{i})$	$\left[\frac{1}{\sqrt{2}}(\mathbf{ii}-\mathbf{jj}), \frac{1}{\sqrt{2}}(\mathbf{ij}+\mathbf{ji})\right]$, $(\mathbf{ik}, \mathbf{jk}), (\mathbf{ki}, \mathbf{kj})$

$\mathbf{D}_{5d}$	E	$2C_5$	$2C_5{}^2$	$5C_2$	i	$2S_{10}{}^3$	$2S_{10}$	$5\sigma_d$	
A_{1g}	1	1	1	1	1	1	1	1	$\frac{1}{\sqrt{2}}(\mathbf{ii}+\mathbf{jj})$, $\mathbf{kk}$
A_{2g}	1	1	1	−1	1	1	1	−1	$\frac{1}{\sqrt{2}}(\mathbf{ij}-\mathbf{ji})$; $\mathbf{i}\times\mathbf{j}$
E_{1g}	2	$2\cos72°$	$2\cos144°$	0	2	$2\cos144°$	$2\cos72°$	0	$(\mathbf{j}\times\mathbf{k},\ \mathbf{k}\times\mathbf{i})$; $(\mathbf{ik},\mathbf{jk}),(\mathbf{ki},\mathbf{kj})$
E_{2g}	2	$2\cos144°$	$2\cos72°$	0	2	$2\cos72°$	$2\cos144°$	0	$\left[\frac{1}{\sqrt{2}}(\mathbf{ii}-\mathbf{jj}),\frac{1}{\sqrt{2}}(\mathbf{ij}+\mathbf{ji})\right]$
A_{1u}	1	1	1	1	−1	−1	−1	−1	
A_{2u}	1	1	1	−1	−1	−1	−1	1	$\mathbf{k}$
E_{1u}	2	$2\cos72°$	$2\cos144°$	0	−2	$-2\cos72°$	$-2\cos144°$	0	$(\mathbf{i},\mathbf{j})$
E_{2u}	2	$2\cos144°$	$2\cos72°$	0	−2	$-2\cos144°$	$-2\cos72°$	0	

D_{6d}	E	$2S_{12}$	$2C_6$	$2S_4$	$2C_3$	$2S_{12}{}^5$	C_2	$6C_2{}'$	$6\sigma_d$		
A_1	1	1	1	1	1	1	1	1	1		$\frac{1}{\sqrt{2}}(\mathbf{ii}+\mathbf{jj})$, $\mathbf{kk}$
A_2	1	1	1	1	1	1	1	-1	-1	$\mathbf{i}\times\mathbf{j}$	$\frac{1}{\sqrt{2}}(\mathbf{ij}-\mathbf{ji})$
B_1	1	-1	1	-1	1	-1	1	1	-1		
B_2	1	-1	1	-1	1	-1	1	-1	1		
E_1	2	$\sqrt{3}$	1	0	-1	$-\sqrt{3}$	-2	0	0	$\mathbf{k}$	
E_2	2	1	-1	-2	-1	1	2	0	0	$(\mathbf{i},\mathbf{j})$	
E_3	2	0	-2	0	2	0	-2	0	0		
E_4	2	-1	-1	2	-1	-1	2	0	0		
E_5	2	$-\sqrt{3}$	1	0	-1	$\sqrt{3}$	-2	0	0	$(\mathbf{j}\times\mathbf{k},\ \mathbf{k}\times\mathbf{i})$	$\left[\frac{1}{\sqrt{2}}(\mathbf{ii}-\mathbf{jj}),\ \frac{1}{\sqrt{2}}(\mathbf{ij}+\mathbf{ji})\right]$, $(\mathbf{ik},\mathbf{jk}),(\mathbf{ki},\mathbf{kj})$

D_{2h}	E	$C_2(z)$	$C_2(y)$	$C_2(x)$	i	$\sigma(xy)$	$\sigma(zx)$	$\sigma(yz)$			
A_g	1	1	1	1	1	1	1	1			ii, jj, kk
B_{1g}	1	1	−1	−1	1	1	−1	−1	i × j		ij, ji
B_{2g}	1	−1	1	−1	1	−1	1	−1	k × i		ki, ik
B_{3g}	1	−1	−1	1	1	−1	−1	1	j × k		jk, kj
A_u	1	1	1	1	−1	−1	−1	−1			
B_{1u}	1	1	−1	−1	−1	−1	1	1		k	
B_{2u}	1	−1	1	−1	−1	1	−1	1		j	
B_{3u}	1	−1	−1	1	−1	1	1	−1		i	

D_{3h}	E	$2C_3$	$3C_2$	σ_h	$2S_3$	$3\sigma_v$			
A_1'	1	1	1	1	1	1			$\dfrac{1}{\sqrt{2}}(\mathrm{ii}+\mathrm{jj})$, kk
A_2'	1	1	−1	1	1	−1		i × j	$\dfrac{1}{\sqrt{2}}(\mathrm{ij}-\mathrm{ji})$
E'	2	−1	0	2	−1	0	(i, j)		$\left[\dfrac{1}{\sqrt{2}}(\mathrm{ii}-\mathrm{jj}),\ \dfrac{1}{\sqrt{2}}(\mathrm{ij}+\mathrm{ji})\right]$
A_1''	1	1	1	−1	−1	−1			
A_2''	1	1	−1	−1	−1	1	k		
E''	2	−1	0	−2	1	0		(j × k, k × i)	(ik, jk), (ki, kj)

$\mathbf{D}_{4h}$	E	$2C_4$	C_2	$2C_2'$	$2C_2''$	i	$2S_4$	σ_h	$2\sigma_v$	$2\sigma_d$			
A_{1g}	1	1	1	1	1	1	1	1	1	1			$\frac{1}{\sqrt{2}}(\mathbf{ii}+\mathbf{jj})$, $\mathbf{kk}$
A_{2g}	1	1	1	-1	-1	1	1	1	-1	-1		$\mathbf{i}\times\mathbf{j}$	$\frac{1}{\sqrt{2}}(\mathbf{ij}-\mathbf{ji})$
B_{1g}	1	-1	1	1	-1	1	-1	1	1	-1			$\frac{1}{\sqrt{2}}(\mathbf{ii}-\mathbf{jj})$
B_{2g}	1	-1	1	-1	1	1	-1	1	-1	1			$\frac{1}{\sqrt{2}}(\mathbf{ij}+\mathbf{ji})$
E_g	2	0	-2	0	0	2	0	-2	0	0		$(\mathbf{j}\times\mathbf{k}, \mathbf{k}\times\mathbf{i})$	$(\mathbf{ik}, \mathbf{jk})$, $(\mathbf{ki}, \mathbf{kj})$
A_{1u}	1	1	1	1	1	-1	-1	-1	-1	-1			
A_{2u}	1	1	1	-1	-1	-1	-1	-1	1	1	$\mathbf{k}$		
B_{1u}	1	-1	1	1	-1	-1	1	-1	-1	1			
B_{2u}	1	-1	1	-1	1	-1	1	-1	1	-1			
E_u	2	0	-2	0	0	-2	0	2	0	0	$(\mathbf{i}, \mathbf{j})$		

$\mathbf{D}_{5h}$	E	$2C_5$	$2C_5{}^2$	$5C_2$	σ_h	$2S_5$	$2S_5{}^3$	$5\sigma_v$		
$A_1{}'$	1	1	1	1	1	1	1	1		$\frac{1}{\sqrt{2}}(\mathbf{ii}+\mathbf{jj})$, $\mathbf{kk}$
$A_2{}'$	1	1	1	-1	1	1	1	-1	$\mathbf{i}\times\mathbf{j}$	$\frac{1}{\sqrt{2}}(\mathbf{ij}-\mathbf{ji})$
$E_1{}'$	2	$2\cos 72°$	$2\cos 144°$	0	2	$2\cos 72°$	$2\cos 144°$	0	$(\mathbf{i},\mathbf{j})$	
$E_2{}'$	2	$2\cos 144°$	$2\cos 72°$	0	2	$2\cos 144°$	$2\cos 72°$	0		$\left[\frac{1}{\sqrt{2}}(\mathbf{ii}-\mathbf{jj}),\ \frac{1}{\sqrt{2}}(\mathbf{ij}+\mathbf{ji})\right]$
$A_1{}''$	1	1	1	1	-1	-1	-1	-1		
$A_2{}''$	1	1	1	-1	-1	-1	-1	1	$\mathbf{k}$	
$E_1{}''$	2	$2\cos 72°$	$2\cos 144°$	0	-2	$-2\cos 72°$	$-2\cos 144°$	0	$(\mathbf{j}\times\mathbf{k},\ \mathbf{k}\times\mathbf{i})$	
$E_2{}''$	2	$2\cos 144°$	$2\cos 72°$	0	-2	$-2\cos 144°$	$-2\cos 72°$	0	$(\mathbf{ik},\mathbf{jk}),\ (\mathbf{ki},\mathbf{kj})$	

D_{6h}	E	$2C_6$	$2C_3$	C_2	$3C_2'$	$3C_2''$	i	$2S_3$	$2S_6$	σ_h	$3\sigma_d$	$3\sigma_v$		
A_{1g}	1	1	1	1	1	1	1	1	1	1	1	1		$\frac{1}{\sqrt{2}}(\mathbf{ii}+\mathbf{jj})$, $\mathbf{kk}$
A_{2g}	1	1	1	1	-1	-1	1	1	1	1	-1	-1	$\mathbf{i}\times\mathbf{j}$	$\frac{1}{\sqrt{2}}(\mathbf{ij}-\mathbf{ji})$
B_{1g}	1	-1	1	-1	1	-1	1	-1	1	-1	1	-1		
B_{2g}	1	-1	1	-1	-1	1	1	-1	1	-1	-1	1		
E_{1g}	2	1	-1	-2	0	0	2	1	-1	-2	0	0	$(\mathbf{j}\times\mathbf{k},\ \mathbf{k}\times\mathbf{i})$	$(\mathbf{ik},\ \mathbf{jk}),\ (\mathbf{ki},\ \mathbf{kj})$
E_{2g}	2	-1	-1	2	0	0	2	-1	-1	2	0	0		$\left[\frac{1}{\sqrt{2}}(\mathbf{ii}-\mathbf{jj}),\ \frac{1}{\sqrt{2}}(\mathbf{ij}+\mathbf{ji})\right]$
A_{1u}	1	1	1	1	1	1	-1	-1	-1	-1	-1	-1		
A_{2u}	1	1	1	1	-1	-1	-1	-1	-1	-1	1	1	$\mathbf{k}$	
B_{1u}	1	-1	1	-1	1	-1	-1	1	-1	1	-1	1		
B_{2u}	1	-1	1	-1	-1	1	-1	1	-1	1	1	-1		
E_{1u}	2	1	-1	-2	0	0	-2	-1	1	2	0	0	$(\mathbf{i},\ \mathbf{j})$	
E_{2u}	2	-1	-1	2	0	0	-2	1	1	-2	0	0		

$\mathbf{T}$	E	$4C_3$	$4C_3{}^2$	$3C_2$		
A	1	1	1	1		$\frac{1}{\sqrt{3}}(\mathbf{ii} + \mathbf{jj} + \mathbf{kk})$
C	1	ω	ω^2	1		$\frac{1}{\sqrt{3}}(\mathbf{ii} + \omega^2\mathbf{jj} + \omega\mathbf{kk})$
D	1	ω^2	ω	1		$\frac{1}{\sqrt{3}}(\mathbf{ii} + \omega\mathbf{jj} + \omega^2\mathbf{kk})$
F	3	0	0	-1	$(\mathbf{i}, \mathbf{j}, \mathbf{k})$ $(\mathbf{j}\times\mathbf{k}, \mathbf{k}\times\mathbf{i}, \mathbf{i}\times\mathbf{j})$	$(\mathbf{jk}, \mathbf{ki}, \mathbf{ij}),\ (\mathbf{kj}, \mathbf{ik}, \mathbf{ji})$

$\omega = \exp{(2\pi i/3)}$

$\mathbf{T}_d$	E	$8C_3$	$3C_2$	$6S_4$	$6\sigma_d$		
A_1	1	1	1	1	1		$\frac{1}{\sqrt{3}}(\mathbf{ii} + \mathbf{jj} + \mathbf{kk})$
A_2	1	1	1	-1	-1		
E	2	-1	2	0	0		$\left[\frac{1}{\sqrt{6}}(2\mathbf{kk} - \mathbf{ii} - \mathbf{jj}), \frac{1}{\sqrt{2}}(\mathbf{ii} - \mathbf{jj})\right]$
F_1	3	0	-1	1	-1	$(\mathbf{j}\times\mathbf{k}, \mathbf{k}\times\mathbf{i}, \mathbf{i}\times\mathbf{j})$	$\left[\frac{1}{\sqrt{2}}(\mathbf{jk} - \mathbf{kj}), \frac{1}{\sqrt{2}}(\mathbf{ki} - \mathbf{ik}), \frac{1}{\sqrt{2}}(\mathbf{ij} - \mathbf{ji})\right]$
F_2	3	0	-1	-1	1	$(\mathbf{i}, \mathbf{j}, \mathbf{k})$	$\left[\frac{1}{\sqrt{2}}(\mathbf{jk} + \mathbf{kj}), \frac{1}{\sqrt{2}}(\mathbf{ki} + \mathbf{ik}), \frac{1}{\sqrt{2}}(\mathbf{ij} + \mathbf{ji})\right]$

T_h	E	$4C_3$	$4C_3^2$	$3C_2$	i	$4S_6$	$4S_6^2$	$3\sigma_d$		
A_g	1	1	1	1	1	1	1	1		$\frac{1}{\sqrt{3}}(\mathbf{ii} + \mathbf{jj} + \mathbf{kk})$
C_g	1	ω	ω^2	1	1	ω	ω^2	1		$\frac{1}{\sqrt{3}}(\mathbf{ii} + \omega^2\mathbf{jj} + \omega\mathbf{kk})$
D_g	1	ω^2	ω	1	1	ω^2	ω	1		$\frac{1}{\sqrt{3}}(\mathbf{ii} + \omega\mathbf{jj} + \omega^2\mathbf{kk})$
F_g	3	0	0	−1	3	0	0	−1	$(\mathbf{j}\times\mathbf{k}, \mathbf{k}\times\mathbf{i}, \mathbf{i}\times\mathbf{j})$	$(\mathbf{jk}, \mathbf{ki}, \mathbf{ij}), (\mathbf{kj}, \mathbf{ik}, \mathbf{ji})$
A_u	1	1	1	1	−1	−1	−1	−1		
C_u	1	ω	ω^2	1	−1	−ω	−ω^2	−1		
D_u	1	ω^2	ω	1	−1	−ω^2	−ω	−1		
F_u	3	0	0	−1	−3	0	0	1	$(\mathbf{i}, \mathbf{j}, \mathbf{k})$	

$$\omega = \exp(2\pi i/3)$$

O	E	$8C_3$	$6C_2$	$6C_4$	$3C_4^2$		
A_1	1	1	1	1	1		$\frac{1}{\sqrt{3}}(\mathbf{ii} + \mathbf{jj} + \mathbf{kk})$
A_2	1	1	−1	−1	1		
E	2	−1	0	0	2		$\left[\frac{1}{\sqrt{6}}(2\mathbf{kk} - \mathbf{ii} - \mathbf{jj}), \frac{1}{\sqrt{2}}(\mathbf{ii} - \mathbf{jj})\right]$
F_1	3	0	−1	1	−1	$(\mathbf{i}, \mathbf{j}, \mathbf{k})$ $(\mathbf{j}\times\mathbf{k}, \mathbf{k}\times\mathbf{i}, \mathbf{i}\times\mathbf{j})$	$\left[\frac{1}{\sqrt{2}}(\mathbf{jk} - \mathbf{kj}), \frac{1}{\sqrt{2}}(\mathbf{ki} - \mathbf{ik}), \frac{1}{\sqrt{2}}(\mathbf{ij} - \mathbf{ji})\right]$
F_2	3	0	1	−1	−1		$\left[\frac{1}{\sqrt{2}}(\mathbf{jk} + \mathbf{kj}), \frac{1}{\sqrt{2}}(\mathbf{ki} + \mathbf{ik}), \frac{1}{\sqrt{2}}(\mathbf{ij} + \mathbf{ji})\right]$

O_h	E	$8C_3$	$6C_2$	$6C_4$	$3C_4^2$	i	$6S_4$	$8S_6$	$3\sigma_h$	$6\sigma_d$
A_{1g}	1	1	1	1	1	1	1	1	1	1
A_{2g}	1	1	-1	-1	1	1	-1	1	1	-1
E_g	2	-1	0	0	2	2	0	-1	2	0
F_{1g}	3	0	-1	1	-1	3	1	0	-1	-1
F_{2g}	3	0	1	-1	-1	3	-1	0	-1	1
A_{1u}	1	1	1	1	1	-1	-1	-1	-1	-1
A_{2u}	1	1	-1	-1	1	-1	1	-1	-1	1
E_u	2	-1	0	0	2	-2	0	1	-2	0
F_{1u}	3	0	-1	1	-1	-3	-1	0	1	1
F_{2u}	3	0	1	-1	-1	-3	1	0	1	-1

O_h	
A_{1g}	$\dfrac{1}{\sqrt{3}}(\mathbf{ii} + \mathbf{jj} + \mathbf{kk})$
A_{2g}	
E_g	$\left[\dfrac{1}{\sqrt{6}}(2\mathbf{kk} - \mathbf{ii} - \mathbf{jj}), \dfrac{1}{\sqrt{2}}(\mathbf{ii} - \mathbf{jj})\right]$
F_{1g}	$\left[\dfrac{1}{\sqrt{2}}(\mathbf{jk} - \mathbf{kj}), \dfrac{1}{\sqrt{2}}(\mathbf{ki} - \mathbf{ik}), \dfrac{1}{\sqrt{2}}(\mathbf{ij} - \mathbf{ji})\right]$ $(\mathbf{j} \times \mathbf{k}, \mathbf{k} \times \mathbf{i}, \mathbf{i} \times \mathbf{j})$
F_{2g}	$\left[\dfrac{1}{\sqrt{2}}(\mathbf{jk} + \mathbf{kj}), \dfrac{1}{\sqrt{2}}(\mathbf{ki} + \mathbf{ik}), \dfrac{1}{\sqrt{2}}(\mathbf{ij} + \mathbf{ji})\right]$
A_{1u}	
A_{2u}	
E_u	
F_{1u}	$(\mathbf{i}, \mathbf{j}, \mathbf{k})$
F_{2u}	

I	E	$12C_5$	$12C_5{}^2$	$20C_3$	$15C_2$	
A	1	1	1	1	1	$\dfrac{1}{\sqrt{3}}(\mathbf{ii}+\mathbf{jj}+\mathbf{kk})$
F_1	3	$-2\cos144°$	$-2\cos72°$	0	-1	$(\mathbf{i},\mathbf{j},\mathbf{k})\ (\mathbf{j}\times\mathbf{k},\ \mathbf{k}\times\mathbf{i},\ \mathbf{i}\times\mathbf{j})\quad \left[\dfrac{1}{\sqrt{2}}(\mathbf{jk}-\mathbf{kj}),\ \dfrac{1}{\sqrt{2}}(\mathbf{ki}-\mathbf{ik}),\ \dfrac{1}{\sqrt{2}}(\mathbf{ij}-\mathbf{ji})\right]$
F_2	3	$-2\cos72°$	$-2\cos144°$	0	-1	
G	4	-1	-1	1	0	
H	5	0	0	-1	1	$\left[\dfrac{1}{\sqrt{6}}(2\mathbf{kk}-\mathbf{ii}-\mathbf{jj}),\ \dfrac{1}{\sqrt{2}}(\mathbf{ii}-\mathbf{jj}),\ \dfrac{1}{\sqrt{2}}(\mathbf{jk}+\mathbf{kj}),\ \dfrac{1}{\sqrt{2}}(\mathbf{ki}+\mathbf{ik}),\ \dfrac{1}{\sqrt{2}}(\mathbf{ij}+\mathbf{ji})\right]$

I$_h$	E	$12C_5$	$12C_5{}^2$	$20C_3$	$15C_2$	i	$12S_{10}$	$12S_{10}{}^3$	$20S_6$	15σ
A_g	1	1	1	1	1	1	1	1	1	1
F_{1g}	3	$-2\cos144°$	$-2\cos72°$	0	-1	3	$-2\cos72°$	$-2\cos144°$	0	-1
F_{2g}	3	$-2\cos72°$	$-2\cos144°$	0	-1	3	$-2\cos144°$	$-2\cos72°$	0	-1
G_g	4	-1	-1	1	0	4	-1	-1	1	0
H_g	5	0	0	-1	1	5	0	0	-1	1
A_u	1	1	1	1	1	-1	-1	-1	-1	-1
F_{1u}	3	$-2\cos144°$	$-2\cos72°$	0	-1	-3	$2\cos72°$	$2\cos144°$	0	1
F_{2u}	3	$-2\cos72°$	$-2\cos144°$	0	-1	-3	$2\cos144°$	$2\cos72°$	0	1
G_u	4	-1	-1	1	0	-4	1	1	-1	0
H_u	5	0	0	-1	1	-5	0	0	1	-1

I_h		
A_g		$\dfrac{1}{\sqrt{3}}(\mathbf{ii} + \mathbf{jj} + \mathbf{kk})$
F_{1g}	$(\mathbf{j} \times \mathbf{k}, \; \mathbf{k} \times \mathbf{i}, \; \mathbf{i} \times \mathbf{j})$	$\left[\dfrac{1}{\sqrt{2}}(\mathbf{jk} - \mathbf{kj}), \; \dfrac{1}{\sqrt{2}}(\mathbf{ki} - \mathbf{ik}), \; \dfrac{1}{\sqrt{2}}(\mathbf{ij} - \mathbf{ji})\right]$
F_{2g}		
G_g		
H_g		$\left[\dfrac{1}{\sqrt{6}}(2\mathbf{kk} - \mathbf{ii} - \mathbf{jj}), \; \dfrac{1}{\sqrt{2}}(\mathbf{ii} - \mathbf{jj}),\right.$
		$\left.\dfrac{1}{\sqrt{2}}(\mathbf{jk} + \mathbf{kj}), \; \dfrac{1}{\sqrt{2}}(\mathbf{ki} + \mathbf{ik}), \; \dfrac{1}{\sqrt{2}}(\mathbf{ij} + \mathbf{ji})\right]$
A_u		
F_{1u}	$(\mathbf{i}, \mathbf{j}, \mathbf{k})$	
F_{2u}		
G_u		
H_u		

$C_{\infty v}$	E	$2C_{2\pi/\phi}$	$\cdots$	$\infty\sigma_v$			
$A_1 \equiv \Sigma^+$	1	1	$\cdots$	1	$\mathbf{k}$		$\frac{1}{\sqrt{2}}(\mathbf{ii+jj}),\ \mathbf{kk}$
$A_2 \equiv \Sigma^-$	1	1	$\cdots$	-1		$\mathbf{i\times j}$	$\frac{1}{\sqrt{2}}(\mathbf{ij-ji})$
$E_1 \equiv \Pi$	2	$2\cos\phi$	$\cdots$	0	$(\mathbf{i,j})$	$(\mathbf{j\times k,\ k\times i})$	$(\mathbf{ik,\ jk}),\ (\mathbf{ki,\ kj})$
$E_2 \equiv \Delta$	2	$2\cos 2\phi$	$\cdots$	0			$\left[\frac{1}{\sqrt{2}}(\mathbf{ii-jj}),\ \frac{1}{\sqrt{2}}(\mathbf{ij+ji})\right]$
$E_3 \equiv \Phi$	2	$2\cos 3\phi$	$\cdots$	0			
$\cdots$	$\cdots$	$\cdots$					

$D_{\infty h}$	E	$2C_{2\pi/\phi}$	$\cdots$	$\infty C_2'$	i	$\cdots$	$2S_{2\pi/\phi}$	$\cdots$	σ_h	$\infty\sigma_v$
$A_{1g} \equiv \Sigma_g^+$	1	1	$\cdots$	1	1	$\cdots$	1	$\cdots$	1	1
$A_{2g} \equiv \Sigma_g^-$	1	1	$\cdots$	-1	1	$\cdots$	1	$\cdots$	1	-1
$E_{1g} \equiv \Pi_g$	2	$2\cos\phi$	$\cdots$	0	2	$\cdots$	$-2\cos\phi$	$\cdots$	-2	0
$E_{2g} \equiv \Delta_g$	2	$2\cos 2\phi$	$\cdots$	0	2	$\cdots$	$2\cos 2\phi$	$\cdots$	2	0
$E_{3g} \equiv \Phi_g$	2	$2\cos 3\phi$	$\cdots$	0	2	$\cdots$	$-2\cos 3\phi$	$\cdots$	-2	0
$\cdots$	$\cdots$	$\cdots$			$\cdots$		$\cdots$		$\cdots$	$\cdots$
$A_{1u} \equiv \Sigma_u^+$	1	1	$\cdots$	-1	-1	$\cdots$	-1	$\cdots$	-1	1
$A_{2u} \equiv \Sigma_u^-$	1	1	$\cdots$	1	-1	$\cdots$	-1	$\cdots$	-1	-1
$E_{1u} \equiv \Pi_u$	2	$2\cos\phi$	$\cdots$	0	-2	$\cdots$	$2\cos\phi$	$\cdots$	2	0
$E_{2u} \equiv \Delta_u$	2	$2\cos 2\phi$	$\cdots$	0	-2	$\cdots$	$-2\cos 2\phi$	$\cdots$	-2	0
$E_{3u} \equiv \Phi_u$	2	$2\cos 3\phi$	$\cdots$	0	-2	$\cdots$	$-2\cos 3\phi$	$\cdots$	2	0
$\cdots$	$\cdots$	$\cdots$			$\cdots$		$\cdots$		$\cdots$	$\cdots$

$\mathbf{D}_{\infty h}$		
$A_{1g} \equiv \Sigma_g^+$		$\frac{1}{\sqrt{2}}(\mathbf{ii} + \mathbf{jj})$, $\mathbf{kk}$
$A_{2g} \equiv \Sigma_g^-$	$\mathbf{i} \times \mathbf{j}$	$\frac{1}{\sqrt{2}}(\mathbf{ij} - \mathbf{ji})$
$E_{1g} \equiv \Pi_g$	$(\mathbf{j} \times \mathbf{k}, \mathbf{k} \times \mathbf{i})$	$(\mathbf{ik}, \mathbf{jk})$, $(\mathbf{ki}, \mathbf{kj})$
$E_{2g} \equiv \Delta_g$		$\left[\frac{1}{\sqrt{2}}(\mathbf{ii} - \mathbf{jj}), \frac{1}{\sqrt{2}}(\mathbf{ij} + \mathbf{ji})\right]$
$E_{3g} \equiv \Phi_g \cdots$		
$A_{1u} \equiv \Sigma_u^+$		
$A_{2u} \equiv \Sigma_u^-$	$\mathbf{k}$	
$E_{1u} \equiv \Pi_u$	$(\mathbf{i}, \mathbf{j})$	
$E_{2u} \equiv \Delta_u$		
$E_{3u} \equiv \Phi_u \cdots$		

$\mathbf{N}_{3p}$	E	$\cdots$	$\infty C_{2\pi/\phi}$	$\cdots$
$\cdots$	$\cdots$	$\cdots$	$\cdots$	$\cdots$
Γ^j	$2j + 1$	$\cdots$	$\dfrac{\sin(j + \frac{1}{2})\phi}{\sin \frac{1}{2}\phi}$	$\cdots$
$\cdots$	$\cdots$	$\cdots$	$\cdots$	$\cdots$

N_{3p}		
$A \equiv \Gamma^0$	$(\mathbf{i}, \mathbf{j}, \mathbf{k})$	$\dfrac{1}{\sqrt{3}}(\mathbf{ii} + \mathbf{jj} + \mathbf{kk})$
$F \equiv \Gamma^1$	$(\mathbf{j} \times \mathbf{k}, \mathbf{k} \times \mathbf{i}, \mathbf{i} \times \mathbf{j})$	$\left[\dfrac{1}{\sqrt{2}}(\mathbf{jk} - \mathbf{kj}), \dfrac{1}{\sqrt{2}}(\mathbf{ki} - \mathbf{ik}), \dfrac{1}{\sqrt{2}}(\mathbf{ij} - \mathbf{ji}) \right]$
$H \equiv \Gamma^2$		$\left[\dfrac{1}{\sqrt{6}}(2\mathbf{kk} - \mathbf{ii} - \mathbf{jj}), \dfrac{1}{\sqrt{2}}(\mathbf{ii} - \mathbf{jj}), \right.$
		$\left. \dfrac{1}{\sqrt{2}}(\mathbf{jk} + \mathbf{kj}), \dfrac{1}{\sqrt{2}}(\mathbf{ki} + \mathbf{ik}), \dfrac{1}{\sqrt{2}}(\mathbf{ij} + \mathbf{ji}) \right]$
$\dots$		

Answers to Problems

Chapter 1

1.4 $\cos^{-1}(-\frac{1}{3}) \simeq 109° \, 28'$.   1.8 $\dfrac{d^2\mathbf{r}}{d\phi^2} + \mathbf{r} = \mathbf{C}$.

1.9 $\dot{\mathbf{A}} = (\dot{A}_r - A_\theta \dot{\theta} - A_\phi \sin\theta \dot{\phi})\mathbf{l} + (\dot{A}_\theta + A_r \dot{\theta} - A_\phi \cos\phi\dot{\phi})\mathbf{m} +$
$(\dot{A}_\phi + A_r \sin\theta\dot{\phi} + A_\theta \cos\theta\dot{\phi})\mathbf{n}$.   1.10   (a) $-k\{\sqrt{5} - 1 - \ln[(1 + \sqrt{5})/2]\}$,
(b) $-k\pi/\sqrt{3}$.   1.11 $v = cr, \mathbf{a} = -c^2\mathbf{r}$.   1.15   60°.

1.19 $\mathbf{n} = \dfrac{x}{a^2}\mathbf{i} + \dfrac{y}{b^2}\mathbf{j} + \dfrac{z}{c^2}\mathbf{k}$.

1.20 $\dot{\mathbf{A}} = \dot{A}_z\mathbf{k} + (\dot{A}_r - A_\phi\dot{\phi})\mathbf{l} + (\dot{A}_\phi + A_r\dot{\phi})\mathbf{n}$.   1.21   (a) $k \ln \frac{1}{2}(1 + \sqrt{5})$,
(b) $\pi k/3\sqrt{3}$.   1.22 $v = c(2kr)^{1/2}, a = kc^2(5 + 4\cos\phi)^{1/2}$.

Chapter 2

2.1   686.97 days.   2.2 1.046×10^3.

2.3 1.499×10^7 force units dyne^{-1}.   2.4   1 earth radius.

2.5 $r = \dfrac{gm}{k}\left\{t - \dfrac{m}{k}\left[1 - \exp\left(-\dfrac{kt}{m}\right)\right]\right\}$.   2.6 $r = A\cos\left[\left(\dfrac{g}{R}\right)^{1/2}t + \alpha\right]$.

2.7 5.96×10^{24} kg.   2.8 $-c/r^3$.   2.9 $V = -kz^2 \ln(r/a)$,
$F_z = 2kz \ln(r/a)$.   2.10   (b) and (c).   2.11   0.7233.

2.12 $y^2 = -2cx + c^2$.   2.13 $-c/r^3$.   2.14 $-c/r^5$.

2.15 $s = \dfrac{m\dot{s}_0}{k}\left[1 - \exp\left(-\dfrac{kt}{m}\right)\right]$.   2.16 $s = \dfrac{m}{k}\ln\cosh\left[\left(\dfrac{kg}{m}\right)^{1/2}t\right]$.

2.17 $-kmr$.   2.18   0.872.   2.19 $\sqrt{2v_0}$.

2.20 $-ar\cos\theta\cos\phi, ar\sin\phi$.

Chapter 3

3.1 1.602×10^{-13} joule.   3.2   0.09 MeV, 3.25 MeV.

3.3 $m_A/(m_A + m_B)$.   3.4 $2v$, 3.3 keV.

3.5 $\dfrac{m_1}{m_2} = 1 + \dfrac{2 \sin \theta_2 \cos (\theta_1 + \theta_2)}{\sin \theta_1}.$ 3.6 $mv^2.$

3.7 $m\left(\dfrac{a^2}{12} + \dfrac{b^2}{12}\right) + ml^2.$ 3.8 $2\pi \left(\dfrac{k^2 + l^2}{gl}\right)^{1/2}.$ 3.10   300 days.

3.11   0.0233.   3.12   0.117 MeV, 6.207 MeV.   3.13   1.64 MeV.

3.14   4.6 MeV.   3.15   30°.   3.16 $\cos^{-1}\left[1 - 0.293 \left(\dfrac{m_1}{m_1 + m_2}\right)^2\right].$

3.17 $\tfrac{2}{5}Ma^2.$ 3.18   14.3 rev min^{-1}.   3.19 $2\pi \left(\dfrac{2k}{g}\right)^{1/2}$

where $k = \left(\dfrac{I_b}{M}\right)^{1/2}.$

Chapter 4

4.2 $\sigma_j\sigma_k = i\sigma_l, \ \sigma_1\sigma_1 = E, \ \sigma_2\sigma_2 = E, \ \sigma_3\sigma_3 = E.$

4.4 $\begin{pmatrix} 1 & 0 & 0 \\ 0 & \cos 72° & -\sin 72° \\ 0 & \sin 72° & \cos 72° \end{pmatrix}.$ 4.5   (a) Rotation by $\tfrac{4}{5}(2\pi)$,

(b) rotation by $\tfrac{3}{5}(2\pi)$.   4.6   (a) $\begin{pmatrix} \dfrac{\sqrt{3}-1}{2} \\ \dfrac{\sqrt{3}+1}{2} \\ 1 \end{pmatrix}$, (b) $\begin{pmatrix} -\dfrac{\sqrt{3}+1}{2} \\ -\dfrac{\sqrt{3}-1}{2} \\ 1 \end{pmatrix}.$

4.7 $I_{xy} = m\dfrac{b^2 - a^2}{24}$, when $a = b$.   4.9 $-1, (3 \pm\sqrt{7}\,i)/4.$

4.10 $\begin{pmatrix} 0 \\ 0 \\ 1 \end{pmatrix}, \dfrac{1}{\sqrt{2}}\begin{pmatrix} 1 \\ \pm i \\ 0 \end{pmatrix}.$ 4.11 $\tfrac{1}{2}.$ 4.13 $a\begin{pmatrix} 1 & 0 \\ 0 & 1 \end{pmatrix} + b\begin{pmatrix} 0 & 1 \\ -1 & 0 \end{pmatrix}.$

4.14   They may differ only in the first column.   4.15 $\dfrac{d\mathbf{A}}{dt}\mathbf{A} + \mathbf{A}\dfrac{d\mathbf{A}}{dt}.$

4.17 $\begin{pmatrix} -\tfrac{1}{2} & \mp\dfrac{\sqrt{3}}{2} & 0 \\ \mp\dfrac{\sqrt{3}}{2} & \tfrac{1}{2} & 0 \\ 0 & 0 & 1 \end{pmatrix}.$ 4.18 $\begin{pmatrix} -\tfrac{1}{4} & \tfrac{1}{4} & \tfrac{1}{4} \\ \tfrac{5}{4} & -\tfrac{1}{4} & -\tfrac{5}{4} \\ -\tfrac{5}{4} & \tfrac{1}{4} & \tfrac{7}{4} \end{pmatrix}.$

4.19   When $s = b$.   4.20 E, C_4 and C_4^{-1}, C_2, σ_v and σ_v', σ_d and $\sigma_d'.$

4.21 $1, -0.4154 \pm 0.9096i.$ 4.22 $\dfrac{1}{\sqrt{3}}\begin{pmatrix} 1 \\ 1 \\ 1 \end{pmatrix}, \dfrac{1}{\sqrt{3}}\begin{pmatrix} 1 \\ \exp(\pm 2\pi i/3) \\ \exp(\mp 2\pi i/3) \end{pmatrix}.$

4.23 $\begin{pmatrix} 1 & A_{12} & A_{13} + \frac{1}{2}A_{12}A_{23} \\ 0 & 1 & A_{23} \\ 0 & 0 & 0 \end{pmatrix}$.

Chapter 5

5.1 (a) $\frac{1}{2}ma^2(\dot{\theta}^2 + \sin^2\theta\dot{\phi}^2)$, (b) $\frac{1}{2}m(\dot{r}^2 + r^2\dot{\phi}^2)$. 5.2 $-c/r^3$.

5.3 (a) $-(g/r)\sin\theta$, (b) $mg\cos\theta + mr\dot{\theta}^2$. 5.4 $-(\frac{2}{27})g$.

5.5 $g/2$. 5.6 28.6 ft $\sec^{-1}$. 5.7 $2\ddot{r} + g - \dfrac{A^2}{r^3} = 0$.

5.8 $\ddot{\theta} + \dfrac{b\omega^2}{a}\sin\theta = 0$. 5.9 $\ddot{x}_1 = \dfrac{m_1 - m_2}{m_1 + m_2}g$.

5.10 $dx - a\sin\theta\,d\phi = 0$, $dy + a\cos\theta\,d\phi = 0$. 5.11 $(g/2)\sin\phi$.

5.12 $\dfrac{mg}{6\pi\eta r}$. 5.13 $\frac{1}{2}\mathscr{L}_1\dot{q}_1{}^2 - \dfrac{q_2{}^2}{2\mathscr{C}} + \mathscr{E}(q_1 + q_2)$.

5.14 $\frac{1}{2}\mathscr{L}_1\dot{q}_1{}^2 + \mathscr{M}_{12}\dot{q}_1\dot{q}_2 + \frac{1}{2}\mathscr{L}_2\dot{q}_2{}^2 - \dfrac{q_1{}^2}{2\mathscr{C}} + \mathscr{E}_2 q_2$. 5.15 $\ddot{x} + \omega_0{}^2 x = 0$.

5.16 $\frac{1}{2}m(h_1{}^2\dot{q}_1{}^2 + h_2{}^2\dot{q}_2{}^2 + h_3{}^2\dot{q}_3{}^2)$. 5.17 $\frac{1}{2}ma^2(\sinh^2 u + \sin^2 v)(\dot{u}^2 + \dot{v}^2)$.

5.18 $-mb^2/r^3$. 5.19 $-2A^2b^2m/r^5$. 5.20 kA^3.

5.21 $0.244g$. 5.22 $\dfrac{a\omega^2}{g - l\omega^2}\cos\omega t$.

5.23 $r = Ae^{(\sin\alpha)\omega t} + Be^{-(\sin\alpha)\omega t} + \dfrac{g\cos\alpha}{\omega^2\sin^2\alpha}$.

5.25 $2(R - r)\ddot{\theta}_2 - g\sin\theta = 0$. 5.26 $a > h$.

5.27 $\mathscr{L}_1\ddot{q}_1 = \mathscr{E}$, $\mathscr{R}_2\dot{q}_2 = \mathscr{E}$.

Chapter 6

6.1 $\omega_1 = 0$, $\omega_2 = (k/\mu)^{1/2}$ where $\mu = mM/(m + M)$.

6.2 $q_1 = m\eta_1 + M\eta_2$, $q_2 = \eta_1 - \eta_2$: 6.3 $\omega_1 = 0$,

$\omega_2 = \left(\dfrac{k}{m}\right)^{1/2}$, $\omega_3 = \left[\dfrac{k}{m}\left(1 + \dfrac{2m}{M}\right)\right]^{1/2}$. 6.4 $q_1 = m\eta_1 + M\eta_2 + m\eta_3$,

$q_2 = \eta_1 - \eta_3$, $q_3 = \eta_1 - 2\eta_2 + \eta_3$. 6.5 $(2 \pm \sqrt{2})^{1/2}\left(\dfrac{g}{l}\right)^{1/2}$.

6.6 $\theta_2 = \sqrt{2}\,\theta_1$, $\theta_2 = -\sqrt{2}\,\theta_1$. 6.7 $X_1 = 0.414X_2$.

6.8 $0.953\omega_0$, $0.903\omega_0$. 6.9 $\omega = \left(\dfrac{g}{l}\right)^{1/2}(1 - \frac{1}{8}\theta_{\max}^3)^{1/2}$.

6.10 $\omega = \left[\dfrac{1}{m}\left(a \pm \sqrt{\dfrac{m}{M}}\,c\right)\right]^{1/2}$. 6.11 $\sqrt{m}\,x \pm \sqrt{M}\,y$.

6.12 $\omega_1 = 0$, $\omega_2 = 0$, $\omega_3 = \left(2k\,\dfrac{M + 2m}{mM}\right)^{1/2}$.

6.13 (a) $m\eta_1 + M\eta_2 + m\eta_3$, (b) $\eta_1 - \eta_3$, (c) $\eta_1 - 2\eta_2 + \eta_3$.

6.14 $\omega_1 = 0$, $\omega_2 = 0$, $\omega_3 = 0$, $\omega_4 = (2a/m)^{1/2}$,

$\omega_{5,6} = \left\{ \dfrac{1}{m} \left[a + 2b \pm (a^2 - 2ab + 4b^2)^{1/2} \right] \right\}^{1/2}$.   6.15 $q_1 = \xi_1 + \xi_2 + \xi_3$,

$q_2 = \eta_1 + \eta_2 + \eta_3$, $q_3 = \xi_1 - \eta_3$, $q_4 = \xi_3 - \xi_2 - \eta_1 + \eta_2$.

6.16 $X_2 = \pm (M/m)^{1/2} X_1$.   6.17 $Bte^{-ft/2}$.

6.18 $m\ddot{x} + \dfrac{2ak}{l + a} x + \dfrac{kl}{(l + a)^3} x^3 = 0$.

Chapter 7

7.1 C_n.   7.2 T_d, O_h, O_h, I_h, I_h.   7.3   (a) $C_2, C_4{}^{-1}, E$, (b) $C_3{}^2, \sigma_h$,
$C_3, S_3{}^{-1}, E$.   7.4   (a) $C_{\infty v}$, (b) C_{2v}, (c) S_4.   7.6   Both are hexagonal.
7.7 Λ^0.   7.9   (2, 1), 15.   7.10   27.   7.11 $S_3, \sigma_h, S_4{}^{-1}$,
$S_3{}^{-1}$; T_d.   7.12   (a) C_{3v}, (b) D_{6h}, (c) D_{3h}, (d) Eclipsed: D_{3h},
staggered: D_{3d}.   7.13 C_3.   7.14 D_{3h}.   7.15 $l\mathbf{a} + mC_6\mathbf{a}$.
7.16   Monoclinic.   7.17 $n + n, n + p, \pi^0 + n, \pi^- + p$.

7.18   , .   7.19   15 particles.   7.20   42 particles.

Chapter 8

8.3 $\dfrac{1}{\sqrt{2}} (\mathbf{e}_1 - i\mathbf{e}_2)$, reducible.   8.4   Representation is reducible.

8.5   (1), (1), (−1).   8.7 $A_{1g} + B_{2g} + E_u$.

8.8 $\frac{1}{2}(\mathbf{r}_1 + \mathbf{r}_2 + \mathbf{r}_3 + \mathbf{r}_4)$, $\frac{1}{2}(\mathbf{r}_1 - \mathbf{r}_2 + \mathbf{r}_3 - \mathbf{r}_4)$, $\dfrac{1}{\sqrt{2}} (\mathbf{r}_1 - \mathbf{r}_3)$, $\dfrac{1}{\sqrt{2}} (\mathbf{r}_2 - \mathbf{r}_4)$.

8.10   *B.*   8.14   (1), (1), (−1), (1), (−1).   8.17 $A + C + D + 3F$.

8.18 $\dfrac{1}{\sqrt{12}} (\mathbf{u}_1 + \mathbf{u}_4 + \mathbf{u}_7 + \mathbf{u}_{10} + \mathbf{u}_2 + \mathbf{u}_5 + \mathbf{u}_8 + \mathbf{u}_{11} + \mathbf{u}_3 + \mathbf{u}_6 + \mathbf{u}_9 + \mathbf{u}_{12})$,

$\dfrac{1}{\sqrt{12}} [\mathbf{u}_1 + \mathbf{u}_4 + \mathbf{u}_7 + \mathbf{u}_{10} + \omega(\mathbf{u}_2 + \mathbf{u}_5 + \mathbf{u}_8 + \mathbf{u}_{11}) + \omega^2(\mathbf{u}_3 + \mathbf{u}_6 + \mathbf{u}_9 + \mathbf{u}_{12})]$,

$\dfrac{1}{\sqrt{12}} [\mathbf{u}_1 + \mathbf{u}_4 + \mathbf{u}_7 + \mathbf{u}_{10} + \omega^2(\mathbf{u}_2 + \mathbf{u}_5 + \mathbf{u}_8 + \mathbf{u}_{11}) + \omega(\mathbf{u}_3 + \mathbf{u}_6 + \mathbf{u}_9 + \mathbf{u}_{12})]$,

$\dfrac{1}{\sqrt{8}} (\mathbf{u}_1 + \mathbf{u}_2 + \mathbf{u}_4 + \mathbf{u}_5 - \mathbf{u}_7 - \mathbf{u}_8 - \mathbf{u}_{10} - \mathbf{u}_{11}), \dots,$

$\dfrac{1}{\sqrt{8}} (\mathbf{u}_1 - \mathbf{u}_2 + \mathbf{u}_4 - \mathbf{u}_5 - \mathbf{u}_7 + \mathbf{u}_8 - \mathbf{u}_{10} + \mathbf{u}_{11}), \dots, \frac{1}{2}(\mathbf{u}_1 - \mathbf{u}_4 - \mathbf{u}_7 + \mathbf{u}_{10}), \dots .$

8.19 $A + C + D$.   8.20   The completely symmetric representation.

Chapter 9

9.1 Motion of linked bars: A_1, motion of center against countermotion of linked bars: B_1, B_2 (in group $\mathbf{C}_{2v}$).

9.2 $A_1: \dfrac{1}{2\pi}\left(\dfrac{k_1 + k_2}{2m_1}\right)^{1/2}$, $B_1: \dfrac{1}{2\pi}\left(\dfrac{2k_1}{\mu}\right)^{1/2}$,

$B_2: \dfrac{1}{2\pi}\left(\dfrac{2k_2}{\mu}\right)^{1/2}$, $\mu = \dfrac{4m_1 m_2}{4m_1 + m_2}$.

9.3 $k_3 = 2k_2$.

9.4 $A_1: \mathbf{u}_2 + \mathbf{u}_3$, $B_1: \mathbf{u}_2 - \mathbf{u}_3$, $\mathbf{u}_1$.

9.5 $A_1: \dfrac{1}{2\pi}\left(\dfrac{k}{m_2}\right)^{1/2}$,

$B_1: 0, \dfrac{1}{2\pi}\left(\dfrac{m_1 + 6m_2}{m_1 m_2} k\right)^{1/2}$.

9.6 Motion of linked bars: B_1, E_2, motion of center against countermotion of linked bars: E_1.

9.7 $\dfrac{1}{2\pi}\left(\dfrac{3k}{\mu}\right)^{1/2}$ where

$\mu = \dfrac{6m_1 m_2}{6m_1 + m_2}$.

9.8 $B_1: \dfrac{1}{2\pi}\left(\dfrac{4k}{3m_1}\right)^{1/2}$, $E_2: \dfrac{1}{2\pi}\left(\dfrac{27}{26}\dfrac{k}{m_1}\right)^{1/2}$.

9.9 $m\ddot{q} + 2k_2 q + \dfrac{2k_1}{l^2} q^3 = 0$.

9.10 A_2, B_2, E_1, E_2.

9.12 $0, \dfrac{1}{2\pi}3\left(\dfrac{k}{m}\right)^{1/2}, \dfrac{1}{2\pi}\dfrac{3}{2}\left(\dfrac{k}{m}\right)^{1/2}, \dfrac{1}{2\pi}\dfrac{3\sqrt{3}}{2}\left(\dfrac{k}{m}\right)^{1/2}$.

9.13 $\tfrac{1}{2}k'[(\theta_2 - \theta_1)^2 + (\theta_3 - \theta_2)^2 + (\theta_4 - \theta_3)^2 + (\theta_5 - \theta_4)^2 + (\theta_6 - \theta_5)^2 + (\theta_1 - \theta_6)^2]$.

9.18 $A_1': \dfrac{1}{2\pi}\left(\dfrac{3k}{m}\right)^{1/2}$, $E': \dfrac{1}{2\pi}\left(\dfrac{3k}{2m}\right)^{1/2}$.

9.19 $\ddot{q} + \dfrac{3k}{m}q = 0$.

Chapter 10

10.1 $\phi - \phi_0 = \dfrac{gk}{2(r^2 + k^2)}(t - t_0)^2$.

10.2 $\cos\theta = -1$,

$\dfrac{1}{2\pi}\left(\dfrac{g}{r} - \omega^2\right)^{1/2}$; $\cos\theta = -\dfrac{g}{r\omega^2}, \dfrac{1}{2\pi}\left(\omega^2 - \dfrac{g^2}{r^2\omega^2}\right)^{1/2}$.

10.3 $\dfrac{1}{2\pi}\left(\dfrac{15}{26}\dfrac{g}{r}\right)^{1/2}$.

10.4 $\pi\left(\dfrac{R}{g}\right)^{1/2}$.

10.6 -2 cm.

10.8 $a_{jl}a_{kl} = \delta_{jk}$.

10.11 $\theta = \sin^{-1}\dfrac{1}{\sqrt{3}}$.

10.12 $\dfrac{1}{2\pi}\left[\dfrac{(m + M)g \sin^2\theta}{(m + 2M \sin^2\theta)l \cos\theta}\right]^{1/2}$.

10.13 $\dfrac{1}{2\pi}\left(\dfrac{g}{l}\right)^{1/2}$.

10.14 $\dfrac{1}{2\pi}\left[\dfrac{12g(R - a/2)}{3a^2 + l^2}\right]^{1/2}$.

10.16 $10, -8, 6,$ $-4, 2, 0$ cm.

10.17 Equal.

Chapter 11

11.1 $y = mx + b$.

11.3 When the intermediate point lies on $\mathbf{c}$.

11.4 Between lines $x = a$ and $x = b$, $y = c$.

11.5 Circle.

11.6 $y = c \cosh \dfrac{x - b}{c}$.   11.7 $z \sin \dfrac{\phi + a}{b} = c$.

11.8 $\dfrac{d^2}{dt^2} \dfrac{\partial f}{\partial \ddot{q}} - \dfrac{d}{dt} \dfrac{\partial f}{\partial \dot{q}} + \dfrac{\partial f}{\partial q} = 0$.   11.9 $m\ddot{x} = -kx$ and $\tfrac{1}{2}m\dot{x}^2 + \tfrac{1}{2}kx^2 = E$.

11.10 $(1 + z_y{}^2)z_{xx} + (1 + z_x{}^2)z_{yy} - 2z_x z_y z_{xy} = 0$.

11.11 $(x^2 + y^2)^{3/2} + \lambda x = 0$.   11.12   Arc of circle.

11.15 $q = at + b$.   11.16   When $\dfrac{d^2 f}{d\dot{q}^2} \neq 0$.   11.17 $\theta = a$.

11.18 $z = b\theta + c$.   11.19 $y = c \cosh \dfrac{x - b}{c}$.

11.20 $r^3 = c \sec 3(\phi - a)$.   11.21 $\dfrac{\partial f}{\partial q} - \dfrac{d^3}{dt^3} \dfrac{\partial f}{\partial \dddot{q}} = 0$.

11.23 $q_k = a_k t + b_k$, $k = 1, 2, 3$.   11.25 $\nabla^2 u = k^2 u$.

11.26 $r = $ constant.

Chapter 12

12.1   (a) $\mathbf{i} + \mathbf{j}$, (b) $2\mathbf{r}$, (c) $\mathbf{r}/r^2$.   12.2 $\dfrac{1}{\sqrt{5}}(\mathbf{j} + 2\mathbf{k})$.   12.3   (a) $2x$, 0;

(b) 0, 2ω.   12.5 $-\tfrac{1}{2}c^3$.   12.6 $\nabla \cdot \mathbf{v} = \dfrac{1}{r} \dfrac{\partial(rv_r)}{\partial r} + \dfrac{1}{r} \dfrac{\partial v_\phi}{\partial \phi} + \dfrac{\partial v_z}{\partial z}$,

$\nabla^2 \varphi = \dfrac{1}{r} \dfrac{\partial}{\partial r} \left(r \dfrac{\partial \varphi}{\partial r} \right) + \dfrac{1}{r^2} \dfrac{\partial^2 \varphi}{\partial \phi^2} + \dfrac{\partial^2 \varphi}{\partial z^2}$,

$\nabla \times \mathbf{v} = \left(\dfrac{1}{r} \dfrac{\partial v_z}{\partial \phi} - \dfrac{\partial v_\phi}{\partial z} \right) \mathbf{l} + \left(\dfrac{\partial v_r}{\partial z} - \dfrac{\partial v_z}{\partial r} \right) \mathbf{n} + \dfrac{1}{r} \left[\dfrac{\partial}{\partial r} (rv_\phi) - \dfrac{\partial v_r}{\partial \phi} \right] \mathbf{k}$.

12.7 $a^1 = \ddot{q}^1$, $a^2 = \ddot{q}^2$, $a^3 = \ddot{q}^3$, $a_1 = \ddot{q}^1 + \cos\theta\ddot{q}^2$, $a_2 = \cos\theta\ddot{q}^1 + \ddot{q}^2$, $a_3 = \ddot{q}^3$.

12.8 $\cos^{-1} \dfrac{g_{12}}{\sqrt{g_{11}g_{22}}}$, 0, 0.   12.9 $\bar{\mathbf{g}}_l = \dfrac{\partial q^k}{\partial \bar{q}^l} \mathbf{g}_k$, $\bar{g}_{jk} = \dfrac{\partial q^l}{\partial \bar{q}^j} \dfrac{\partial q^m}{\partial \bar{q}^k} g_{lm}$.

12.11   (a) $\mathbf{k}$, (b) $-\mathbf{r}/r^3$, (c) $nr^{n-2}\mathbf{r}$.   12.12   (a) Impossible,

(b) $-(x^2 + y^2)^{-1/2} + A$.   12.13   (a) 0, $-nx^{n-1}\mathbf{j}$; (b) $(3 - n)/r^n$, 0.

12.16 $\nabla^2 \varphi = \dfrac{1}{a^2(\sinh^2 u + \sin^2 v)} \left(\dfrac{\partial^2 \varphi}{\partial u^2} + \dfrac{\partial^2 \varphi}{\partial v^2} \right) + \dfrac{\partial^2 \varphi}{\partial z^2}$.

12.17 $\nabla^2 \varphi = \dfrac{1}{\xi(\xi^2 + \eta^2)} \dfrac{\partial}{\partial \xi} \left(\xi \dfrac{\partial \varphi}{\partial \xi} \right) + \dfrac{1}{\eta(\xi^2 + \eta^2)} \dfrac{\partial}{\partial \eta} \left(\eta \dfrac{\partial \varphi}{\partial \eta} \right) + \dfrac{1}{\xi\eta} \dfrac{\partial}{\partial \phi} \left(\dfrac{1}{\xi\eta} \dfrac{\partial \varphi}{\partial \phi} \right)$.

12.18 $\nabla^2 \varphi = \dfrac{1}{a^2(\mu^2 - v^2)}$

$\times \left[\dfrac{\partial}{\partial \mu} (\mu^2 - 1) \dfrac{\partial \varphi}{\partial \mu} + \dfrac{\partial}{\partial v} (1 - v^2) \dfrac{\partial \varphi}{\partial v} + \dfrac{\mu^2 - v^2}{(\mu^2 - 1)(1 - v^2)} \dfrac{\partial^2 \varphi}{\partial \phi^2} \right]$.

12.20 $\bar{\Gamma}_{jk}{}^n = \dfrac{\partial u^n}{\partial q^r} \dfrac{\partial q^t}{\partial u^j} \dfrac{\partial q^v}{\partial u^k} \Gamma_{tv}{}^r + \dfrac{\partial u^n}{\partial q^t} \dfrac{\partial^2 q^t}{\partial u^j \partial u^k}$.

Chapter 13

13.1 $\dfrac{1}{r^2}\dfrac{\partial}{\partial r}(\rho r^2 v_r) + \dfrac{1}{r\sin\theta}\dfrac{\partial}{\partial\theta}(\rho\sin\theta\, v_\theta) + \dfrac{1}{r\sin\theta}\dfrac{\partial}{\partial\phi}(\rho v_\phi) = -\dfrac{\partial\rho}{\partial t}.$

13.2 $v_r = \dfrac{A}{r^2}$, source supplying $\dfrac{dV}{dt} = 4\pi A.$ **13.3**   Equation of continuity is

satisfied; $\omega = \dfrac{k}{2r}.$ **13.5**   All systems in which the unique principal axis is a

coordinate axis.   **13.6** $-p - \dfrac{4}{3}\dfrac{\mu}{\rho}\dfrac{\partial\rho}{\partial t}.$ **13.7** $\sigma = a\mathbf{E} = \sigma'.$

13.8 $v_x = \dfrac{1}{2\mu}\dfrac{d}{dx}(p + \rho g\zeta)(z^2 - hz) + \dfrac{u}{h}z.$

13.9 $\dfrac{1}{r}\dfrac{\partial}{\partial r}(\rho r v_r) + \dfrac{1}{r}\dfrac{\partial}{\partial\phi}(\rho v_\phi) + \dfrac{\partial}{\partial z}(\rho v_z) = -\dfrac{\partial\rho}{\partial t}.$ **13.10** $v_r = \dfrac{A}{r},$

$\dfrac{dV}{dt} = 2\pi A.$ **13.11**   Yes, $\frac{1}{2}k(n + 1)r^{n-1}.$ **13.12**   When $d = -a,$

$cx^2 - 2axy - by^2 = A$ (a conic section).   **13.14** $A\colon \sigma_{11} + \sigma_{22}, \sigma_{33};$
$B\colon \sigma_{11} - \sigma_{22}.$ **13.15** $A_{1g} + E_g.$

13.16 $v_z = \dfrac{1}{2\mu}\dfrac{dp}{dz}\dfrac{(x/a)^2 + (y/b)^2 - 1}{(1/a)^2 + (1/b)^2}.$

Chapter 14

14.1 $9 \times 10^9 q^2 \cos\theta = 4l^2 mg \sin^3\theta.$ **14.2**   At any finite $x, y,$ and $z,$
$\rho = 0.$ **14.3** 3.33×10^{-6} coul.   **14.4** $60°.$

14.5 $\dfrac{A}{2\varepsilon}(x^2 + y^2 - 2z^2) + C.$ **14.6**   Through 3rd order.

14.7 (a) $\dfrac{Ir}{2\pi a^2}$, (b) $\dfrac{I}{2\pi r}.$ **14.8** $\dfrac{\mu I a^2}{2(a^2 + x^2)^{3/2}}.$ **14.9** $\dfrac{3\mu I a^2 xy}{4(a^2 + x^2)^{5/2}}.$

14.10 $\dfrac{\mu N^2 a^2}{4r}I^2.$ **14.11** $-\dfrac{q^2}{\pi\varepsilon b^3}x.$ **14.12** 5.31×10^{-3} coul.

14.13 $\dfrac{q}{2\pi\varepsilon r}.$ **14.14** $\rho = 0,$ charges on infinitely separated planes,
$\varphi = -Ax - By - Cz + D.$ **14.15** q at origin surrounded by
$\rho = -\dfrac{q}{4\pi a^2 r}e^{-r/a}.$ **14.16**   Through 2nd order.

14.17 $\dfrac{\mu I k}{2\pi}\ln\dfrac{r_2}{r_1}.$ **14.18** $\dfrac{\mu I}{2\pi r}.$ **14.19** $\dfrac{\pi}{4}\mu n I.$

14.20 $B_x = \dfrac{\mu I a^2}{2(a^2 + x^2)^{3/2}} + \dfrac{3\mu I a^2(y^2 + z^2)(a^2 - 4x^2)}{8(a^2 + x^2)^{7/2}}.$

Chapter 15

15.2   0.787 volt sec m^{-2}.   15.3   0.184 m.   15.4 $69.0(T/n_e)^{1/2}$.

15.5 $B_0(1 + a^2)/a^2$.   15.6 $(v/B_0) \oint (B_0 - B)^{1/2} \, dz$.

15.7   Ionized H, 0.00027.   15.8 9.72×10^6 cm^{-3}.   15.9 V_e^2.

15.11 1.866×10^{10} sec^{-1}.   15.12 4.3×10^{-5} cm.

15.13 $2\pi m M/q^2$.   15.14   0.127 cm.   15.15 5.64×10^{10} sec^{-1}.

15.16   2090 Å.   15.17 $k^2 V_i^2 - \omega^2 + \Omega_p^2 = 0$.   15.18 V_i^2, V_s^2.

Chapter 16

16.1 $x = \gamma(x' + ut'), y = y', z = z', t = \gamma\left(t' + \dfrac{u}{c^2} x'\right)$.

16.2 $\dfrac{\partial^2 E}{\partial x'^2} = \dfrac{1}{c^2} \dfrac{\partial^2 E}{\partial t'^2}$.   16.3 $L(\gamma^{-2} \cos^2 \theta + \sin^2 \theta)^{1/2}, \tan^{-1}(\gamma \tan \theta)$.

16.4 1.11×10^{-5} sec.   16.5   Rotated by angle $\alpha + \theta$.

16.6 $\dfrac{v_x - u}{1 - uv_x/c^2}, \dfrac{v_y}{\gamma(1 - uv_x/c^2)}, \dfrac{v_z}{\gamma(1 - uv_x/c^2)}$.   16.7   In one Cartesian

system: x, y, z, ct'; in a rotated Cartesian system: x', y', z', ct.

16.10 $\left(1 - \dfrac{v^2}{c^2}\right)^{1/2} dV_0$.   16.11   Form is unchanged.

16.12 $u \ll c, u \simeq c$.   16.13 $2T$.   16.14 $\dfrac{c'}{n} + u\left(1 - \dfrac{1}{n^2} \dfrac{c'^2}{c^2}\right)$.

16.15 $\dfrac{a'}{a} = \dfrac{(1 - u^2/c^2)^{3/2}}{(1 - uv/c^2)^3}$.   16.16 $u = v$.   16.17 $v' = -v$.

16.18 $\dfrac{a'}{(1 - v'^2/c^2)^{3/2}} = \dfrac{a''}{(1 - v''^2/c^2)^{3/2}}$.   16.19 $0.92c$.

16.20 $\mathbf{e}_j \cdot \mathbf{e}_\kappa = \mathbf{e}_\kappa \cdot \mathbf{e}_j = \delta_{j\kappa}, \mathbf{e}_4 \cdot \mathbf{e}_4 = -1; j = 1, 2, 3; \kappa = 1, 2, 3, 4$.

16.21 $1 + \dfrac{1}{2}\left(\dfrac{v}{c}\right)^2 + \dfrac{3}{8}\left(\dfrac{v}{c}\right)^4 + \dfrac{5}{16}\left(\dfrac{v}{c}\right)^6 + \dfrac{35}{128}\left(\dfrac{v}{c}\right)^8 + \cdots$.

Chapter 17

17.1 $s_1^2 + s_2^2 + s_3^2 - s_4^2$.   17.2 $\dfrac{ds}{dt} = \left(1 - \dfrac{v^2}{c^2}\right)^{1/2} \dfrac{ds}{d\tau}$.

17.4   (a) 1.00756, (b) 1.6338, (c) 12.42.   17.6   1657 MeV.

17.7 $6m_pc^2$.   17.8   195 MeV.   17.11   (a) 0.9116, (b) 0.5662,

(c) 0.4497.   17.12   (a) 0.786, (b) 0.9853.   17.15   1519 MeV.

17.16   891 MeV.   17.17 $4m_ec^2$.   17.18 $\dfrac{E_\gamma^2}{E_{e \text{ rest}}}(1 - \cos \theta)$.

Chapter 18

18.1 $(J_x \quad J_y \quad J_z \quad -\rho c) \begin{pmatrix} J_x \\ J_y \\ J_z \\ \rho c \end{pmatrix} = -\rho_0{}^2 c^2.$

18.3 $\left(k_x \quad k_y \quad k_z \quad -\dfrac{\omega}{c}\right) \begin{pmatrix} x \\ y \\ z \\ ct \end{pmatrix}.$ 18.4 $\left(\dfrac{\mathbf{B} \cdot \mathbf{E}}{c}\right)^2.$

18.6 $-4\dfrac{\mathbf{B} \cdot \mathbf{E}}{c}.$ 18.7 $\dfrac{d}{dt}(mu) = 0.$ 18.8 $\dfrac{d}{d\tau}(\gamma u) = \gamma g.$

18.9 $x = \dfrac{c^2}{g}\cosh\dfrac{g}{c}\tau,\ t = \dfrac{c}{g}\sinh\dfrac{g}{c}\tau.$ 18.10 $\gamma^{-2}\rho = \rho_0.$

18.11 $(\mathbf{F}_{\cdot\cdot})' = \tilde{\mathbf{A}}^\times \mathbf{F}_{\cdot\cdot} \mathbf{A}^\times.$ 18.13 $2\left(B^2 - \dfrac{E^2}{c^2}\right).$

18.14 $\left(\dfrac{\mathbf{B} \cdot \mathbf{E}}{c}\right)^2,$ yes.   18.15 $\nabla \times \mathbf{E} = -\dfrac{\partial \mathbf{B}}{\partial t} - \mathbf{J}_m.$

18.17 $E_x' = E_x,\ E_y' = \dfrac{E_y - vB_z}{(1 - v^2/c^2)^{1/2}},\ E_z' = \dfrac{E_z + vB_y}{(1 - v^2/c^2)^{1/2}},$

$B_x' = B_x,\ B_y' = \dfrac{B_y - (v/c^2)E_x}{(1 - v^2/c^2)^{1/2}},\ B_z' = \dfrac{B_z - (v/c^2)E_y}{(1 - v^2/c^2)^{1/2}}.$

Chapter 19

19.3   None.   19.4   Rotation by 180° about y axis.   19.5   First-rank lower-index.   19.6 $1 + \dfrac{i}{2}(\delta\phi)\sigma_z.$ 19.7   Rotation by $-\theta$ about x axis.

19.8   Rotates $\mathbf{F}$ by ϕ about z axis.   19.9 $c(\sigma_x p_x + \sigma_y p_y + \sigma_z p_z).$

19.10   Construct the second-rank spinor $\begin{pmatrix} z & x - iy \\ x + iy & -z \end{pmatrix}$ and have $\alpha\delta - \beta\gamma = 1.$

19.12   Multiplies it by $k^{-1}.$ 19.13   Lorentz transformation along x axis.
19.14   (a) Rotation by $-\delta\theta$ about x axis, (b) rotation by $-\delta\phi$ about y axis.

Chapter 20

20.2 $\dfrac{g_{11} - 1}{r^2}(x^2\,dx^2 + y^2\,dy^2 + z^2\,dz^2 + 2xy\,dx\,dy + 2yz\,dy\,dz + 2zx\,dz\,dx) +$

$dx^2 + dy^2 + dz^2 + g_{44}\,dt^2.$ 20.3 $\dot{\omega}^2 = \dfrac{GM}{r^3}.$ 20.5 $\dfrac{g}{c^2}vl.$

20.6 $\frac{1}{2}m\dot{r}^2 = \frac{GMm}{r} - \frac{GMm}{r_0}$.   20.7 $-m_0c^2$.

20.8 $\frac{r_0 - r}{c} + \frac{r_g}{c} \ln \frac{r_0 - r_g}{r - r_g}$.

20.10 $ds^2 = g_{11}\,dr^2 + g_{22}\,dz^2 + r^2\,d\phi^2 + g_{44}\,dt^2$.   20.11 $g_{11}r^2 \dfrac{d\omega}{ds} = k_\omega$,

$g_{44}\dfrac{dt}{ds} = k_t$.   20.12 $\left(1 + \dfrac{r_g}{4r'}\right)^4 (dr'^2 + r'^2\,d\omega^2) - c^2 \dfrac{(1 - r_g/4r')^2}{(1 + r_g/4r')^2}\,dt^2$.

20.14 $\dfrac{24\pi^3 a^2}{c^2 T^2(1 - e^2)} = 5.041 \times 10^{-7}$ radians rev^{-1}.

Name Index

Subject Index

Contraction of moving length, 482–483
 reality of, 498
Contravariant components, 364
 of displacement along world line, 500
 of 4-momentum, 514–515
 of 4-velocity, 514
Conversion as symmetry operation,
  204–205
Coordinates
 artificiality of, 30, 554
 classical
  center of mass, 77
  generalized, 146
  inertial, 36–38
  symmetry-adapted, 267
 relativistic
  general, 579–581
   as cause of singularity, 601–603
  special, 474–475
Coriolis acceleration, 30
Coriolis energy, 198
Correlating
 modes of motion in related structures,
  282–283
 representations of related groups,
  282–283
Corresponding points
 in variation of property, 320
Coulomb
 as unit, 414
Coulomb's law, 44–45, 414
Coupled pendulums, 180–182
 excitation of, 186
 normal coordinates for, 182–183
Covariance, 38, 364
Covariant algebraic form
 for Euclidean space, 133
 for Minkowski space-time, 478
Covariant components
 of n-dimensional vectors, 364–365
 of 4-vectors, 508, 514–515
Covariant derivative, 367
Covariant differential form
 for Minkowski space-time, 478–479
Covering operation, 205
 effect on basis expressions of, 237
Cramer's formula, 120

zero numerator in, 126
Crankshaft model, 88
Critical damping, 188
Cross product, 6
 expansion of, 13
  with permutation symbol, 18–20
 as product of matrices, 108
Cross section, 55–57
Crystal, 206
 symmetry of, 206–208
C_3 group
 vector bases for, 241–244
C_{3v} group
 1-dimensional representations of,
  250–251
 2-dimensional representation of,
  246–247
Curl, 360, 377
 divergence of, 380
 of electric intensity, 429
 of fluid velocity
  density of power dissipation in,
   403–404
 in generalized coordinates, 376
 integral of, 361–362
 of magnetic intensity, 433
 of vector potential, 425
Current density, 432–433
 generalized Ohm's law for, 436
 in plasma, 452
 relativistic 4-vector, 535–536
  product with permeability, 534–537
Curvature of Euclidean sphere
 coordinate r as, 582
Cutoff frequency, 465
Cyclic indices, 115
Cyclic mechanism as clock, 37
Cyclotron frequency, 444, 459
Cylinder, circular
 covering operations of
  subgroups of, 206
Cylindrical coordinates
 base vectors for, 13–14
 transformation to
  using principal function, 307
Cylindrically symmetric flow
 slow, steady, 406–408

E

Eccentricity, 40
  of orbit in inverse-square force field,
    51–53
Eigenvalues
  of matrix, 126
    condition for reality of, 131
    use in diagonalization of, 127–130
Eigenvectors
  of matrix, 126
    orthogonality of, 131–132
    possible reality of, 132
Einstein equation
  for electron, 570
  for neutrino, 569
  for photon, 524
Electric charge, 413–415, 532–533
  bound, 413
    polarization of, 418–420
  free, 413–415
    current elements of, 422
      interactions among, 422–423
      magnetic field due to, 423–425
    current per unit area of, 432–433,
      436
    4-vector, 534–536
  geometric interpretation of, 607–609
  in motion, 423, 435
    field of, 543–544
    Hamiltonian for, 545–546
    Lagrangian for, 546–547
    4-momentum of, 545
Electric circuit, 166–167
Electric energy density, 435
Electric field, 414–415
  curl of, 429
    from divergence of magnetic induc-
      tion, 540
  dependence on 4-potential, 548
  divergence of, 416–417
  fundamental aspects of, 532–533
Electrodynamics of charged particle,
  435–436, 442–443
Electromagnetic wave, 436–437
  in plasma, 465

Electromotive force
  induced, 429
Electron, 211
  relativistic equation for, 570–572
Electron volt, 101–102, 445
Electron wave, 463
  dispersion equation for, 466–467
Electrostatic force
  law governing, 414, 532–533
Element
  of fluid, 387–388
  of group, 237–238
  of length, 26, 364, 580–581
  of matrix, 106
  of symmetry, 203–205
Ellipse, 40
  as orbit of planet, 39
  polar equation for, 41
Ellipsoid
  for angular momentum, 92–95
  for rotational energy, 92–95
End point
  effect of variation at, 328–329
Energy
  classical, 64, 76, 85
    associated with
      movement of center of mass, 64,
        86
      rotation, 87
      vibration, 177–178
    conservation of, 85, 298
      in collision, 76
    dissipation of, 163–165
      in fluid, 403–406
    electromagnetic, 433–435
    Hamiltonian, 296–297
    Lagrangian, 163
  relativistic, 517
    conversion in collision, 520–524
    dependence on momentum,
      519–520
    Hamiltonian, 545–546
    Lagrangian, 546–548
    of particle in gravitational field,
      586–587
Energy-momentum 4-vector, 514–515
Ensemble of systems, 300–303

Potential energy (*continued*)
in central field, 47
generalized, 163
as power series, 178
with quadratic terms only, 178
in matrix form, 178
from Schwarzschild line element, 590–591
Potential forces, 162
Power
of matrix, 127–128
Power dissipation
description by Rayleigh function of, 165–166
in fluid, 403–404
Power flow
in electromagnetic field, 433–435
Poynting vector, 435
Precession, 98–99
in crankshaft motion, 88
Pressure, partial
exerted by constituent in gas, 452
Pressure gradient, partial
in plasma, 453–454
Primitive cell, 207
Principal axis
of symmetric dyadic, 92–93, 132–133
for symmetry operation, 111
Principal function
Hamilton's, 306–307
Principal moments of inertia, 92–93, 132–133
Principal rates of strain, 390
Principal stresses, 394–395
Product
dyadic
of two vectors, 9
matrix
from scalar and matrix, 106
from two matrices, 106–107
in block form, 125
scalar
of two scalars, 1, 19–20
of two vectors, 5–6, 17–18
vector
from scalar and vector, 2

from two vectors, 6, 18–20
from vector and dyadic, 16
Product of inertia, 89
Projection
of length onto moving inertial frame, 482–483
of operand onto bases for representation, 255–256
of period onto moving inertial frame, 483–485
of vector in given direction, 5
Propagation vector
for disturbance
in crystal, 208
in plasma, 455
complex nature of, 456
Proper acceleration, 496
Proper density, 533–534
Proper length, 482
Proper mass, 513–514
Proper rigidity, 496–497
Proper time, 479–480
in general relativity, 581
inexactness of differential of, 501
Proper volume, 534
Pseudo scalar, 24
Pseudo vector, 24
Pulley
second pulley and weights supported by, 155–156
weights supported by, 151–153
Push or negative pull
as force, 38
Pythagorean theorem, 10
effect on transformations of, 476
local consistency with, 473, 581

Q

Quadratic form
for kinetic energy, 177–178
as matrix product, 105
for potential energy, 178
for small displacements, 580–581
Quark, 218–219
behavior of, 220–221

Vector (*continued*)
    as part of 4 × 4 tensor, 536
    spinor representation of, 563
Vector form
    for Lagrange equations, 164
Vector function
    of position, 354
        curl of, 359–360
        divergence of, 355–356
        fluid representing, 354–356
    of time
        differentiation of, 25–26
Vector potential
    for fluid velocity, 404
    for magnetic induction, 425
        and electric scalar potential, 545
Vectors
    addition of, 2–4, 12
    multiplication of
        to form dyadic, 9
        to form scalar, 5–6, 17–18
        to form vector, 6, 18–20
    permutation of, 114–116
Velocity
    classical, 26
        components of
            Cartesian, 26
            cylindrical, 28–30
            spherical, 30
        derivative with respect to time of,
            26–30
        fluid, 383
            as curl of vector function, 404
            del of, transposed, 388–389
            del of, transposed, rate-of-strain
                part of, 389–390
            del of, transposed, vorticity part
                of, 389–391
    relativistic
        addition of, 488–489
        radial
            in spherically symmetric field,
                600–601
        4-vector, 511–513
Vibration
    independence from other motions of,
        197–199

linear, 176
    for multibody system, 179–184
        with damping, 186–188
        with damping, and forcing,
            188–189
    nonlinear
        fundamental of, 189–193, 196–197
        harmonics for, 190–191, 196–197
        subharmonics for, 193–195
    separating modes of
        with secular equation, 179–180
        by symmetry considerations, 267,
            280
Virtual small quantity, 150–151
Viscosity
    bulk coefficient of, 397
    shear coefficient of, 399
Viscous damping, 62, 165
Volume
    Lorentz transformation of, 534
    of solid of revolution
        extremizing, 340–342
Vortex line
    circulation around, 358
    curl of velocity at, 359–360
Vorticity
    matrix for, 389
        dependence on angular velocity of,
            390–391
    power loss caused by, 403–404
    around sphere, 404–406

W

Waves
    in plasma
        algebraic equations governing,
            457–462
        dispersion equations
            for longitudinal, 463–467
            for transverse, 463
        interactions among, 467–468
Weak interaction, 211
Weak variation, 346
Work
    done on system, 27–28

during arbitrary small change, 150–151

    for row of representation, 266

done on test charge, 420–421

World line, 480

    curvature and slope of, 490

    hyperbolic, 494–496

Y

Young diagram, 223, 230

Young table, 223, 230

    for (λ, μ) multiplet, 225–226

    of baryons, 226–227

    of mesons, 230–231

Z

Zero frequency

    for mode of oscillation, 177, 179

Zero numerator

    in Cramer's formula, 126, 180